International
Energy Agency

WORLD ENERGY OUTLOOK 2011

INTERNATIONAL ENERGY AGENCY

The International Energy Agency (IEA), an autonomous agency, was established in November 1974. Its primary mandate was – and is – two-fold: to promote energy security amongst its member countries through collective response to physical disruptions in oil supply, and provide authoritative research and analysis on ways to ensure reliable, affordable and clean energy for its 28 member countries and beyond. The IEA carries out a comprehensive programme of energy co-operation among its member countries, each of which is obliged to hold oil stocks equivalent to 90 days of its net imports. The Agency's aims include the following objectives:

■ Secure member countries' access to reliable and ample supplies of all forms of energy; in particular, through maintaining effective emergency response capabilities in case of oil supply disruptions.

■ Promote sustainable energy policies that spur economic growth and environmental protection in a global context – particularly in terms of reducing greenhouse-gas emissions that contribute to climate change.

■ Improve transparency of international markets through collection and analysis of energy data.

■ Support global collaboration on energy technology to secure future energy supplies and mitigate their environmental impact, including through improved energy efficiency and development and deployment of low-carbon technologies.

■ Find solutions to global energy challenges through engagement and dialogue with non-member countries, industry, international organisations and other stakeholders.

IEA member countries:

Australia
Austria
Belgium
Canada
Czech Republic
Denmark
Finland
France
Germany
Greece
Hungary
Ireland
Italy
Japan
Korea (Republic of)
Luxembourg
Netherlands
New Zealand
Norway
Poland
Portugal
Slovak Republic
Spain
Sweden
Switzerland
Turkey
United Kingdom
United States

The European Commission also participates in the work of the IEA.

International Energy Agency

It is the job of governments to take the decisions that will deliver a secure and sustainable energy future. They can make wise decisions only if they are well-informed and advised. That is the job of bodies such as the International Energy Agency (IEA), which it has been my privilege to lead since September 2011.

The *World Energy Outlook* (*WEO*) team, under the Agency's Chief Economist, Fatih Birol, has again done an outstanding job in *WEO-2011* to underpin sound energy decision making. With the invaluable help of many global experts inside and outside the Agency, the team has given us a wealth of current energy information, analysis and perspectives on the future.

For example, we find:

- what it will cost to bring modern energy to every citizen of the world by 2030 and how to finance it;

- that, provided governments honour their existing intentions, renewable energy is set to provide half the new power generating capacity required between now and 2035;

- that, by virtue of their size and distribution, natural gas resources contribute encouragingly to future energy security, casting a golden glow over the outlook for natural gas;

- how oil markets will be shaped by increasing demand for mobility and rising upstream costs – and the consequences of any shortfall of investment in the Middle East and North Africa;

- what would have to be done (and whether that is realistic) if the nuclear component of future energy supply were to be halved, or if the availability of carbon capture and storage technology slipped by ten years;

- the extent of the overwhelming dominance of China in global energy in 25 years time and the global significance of the choices China makes to meet its needs; and

- how much energy Russia can save simply by matching the energy efficiency standards of OECD countries and how that would serve both Russia's national objectives and the interests of global energy supply.

As a former minister, my background is that of a politician. My mission at the IEA is to bridge the divide between analysts and politicians so that the right energy policy decisions are made by governments across the world, both members and non-members of the IEA alike. The Agency will not aspire to determine those decisions which lie outside its area of executive responsibility for its members' energy security. But I will push the remit to identify the decisions which others need, or which would be wise, to adopt. The *WEO* is an invaluable tool to this end.

The starkest decisions are those which must be taken without delay. I end by highlighting one area squarely in this category: the energy decisions necessary to contain the rise in the average global temperature to 2° Celsius. We read here of the way carbon emissions are already "locked-in" because of the nature of the plant and equipment which we continue to build. If we do not change course, by 2015 over 90% of the permissible energy sector emissions to 2035 will already be locked in. By 2017, 100%. We can still act in time to preserve a plausible path to a sustainable energy future; but each year the necessary measures get progressively tougher and viciously more expensive. So, let's not wait any longer!

Maria van der Hoeven
Executive Director

This publication has been produced under the authority of the Executive Director of the International Energy Agency. The views expressed do not necessarily reflect the views or policies of individual IEA member countries

ACKNOWLEDGEMENTS

This study was prepared by the Office of the Chief Economist (OCE) of the International Energy Agency in co-operation with other offices of the Agency. It was designed and directed by **Fatih Birol**, Chief Economist of the IEA. **Laura Cozzi** co-ordinated the analysis of climate policy and modelling; **Amos Bromhead** co-ordinated the analysis of subsidies and nuclear; **John Corben** co-ordinated the analysis of oil and natural gas; **Marco Baroni** co-ordinated the power-generation analysis; **Tim Gould** co-ordinated the Russia analysis; **Paweł Olejarnik** co-ordinated the coal analysis; **Dan Dorner** contributed to the analysis of global trends and co-ordinated the focus on energy access. Other colleagues in the Office of the Chief Economist contributed to multiple aspects of the analysis and were instrumental in delivering the study: **Prasoon Agarwal** (transport and buildings), **Maria Argiri** (natural gas), **Christian Besson** (oil and Russia), **Alessandro Blasi** (Russia), **Raffaella Centurelli** (energy access and buildings), **Michel D'Ausilio** (power and renewables), **Dafydd Elis** (power and climate analysis), **Matthew Frank** (natural gas and subsidies), **Timur Gül** (transport and climate analysis), **Shinichi Kihara** (nuclear), **Kate Kumaria** (climate analysis), **Jung Woo Lee** (subsidies and nuclear), **Katrin Schaber** (power and renewables), **Tatsuya Tomie** (coal and power), **Timur Topalgoekceli** (oil and natural gas), **Brent Wanner** (power and subsidies), **David Wilkinson** (power, climate analysis and nuclear), **Peter Wood** (oil and natural gas), **Akira Yanagisawa** (subsidies, industry and natural gas) and **Tatiana Zhitenko** (Russia). **Sandra Mooney** provided essential support.

Robert Priddle carried editorial responsibility.

The study benefitted from input provided by numerous IEA experts in the Energy Statistics Division, the Energy Markets and Security Directorate, the Directorate of Global Energy Dialogue, and the Energy Policy and Technology Directorate. The Communication and Information Office was instrumental in bringing the book to completion. In particular, Christina Hood, Carlos Fernandez Alvarez, Isabel Murray, Moritz Paulus, Johannes Truby, Nathalie Trudeau and Dennis Volk provided valuable contributions. Experts from several directorates of the OECD and the Nuclear Energy Agency also contributed to the report, particularly Ron Cameron, Rob Dellink, Bertrand Magné, Helen Mountford, Ron Steenblik and Robert Vance. Thanks also go to Debra Justus for proofreading the text.

Ian Cronshaw and Trevor Morgan provided valuable input to the analysis. The following experts also contributed: Igor Bashmakov, Jan-Hein Jesse, Tatiana Mitrova and Liu Qiang.

The work could not have been achieved without the substantial support and co-operation provided by many government bodies, international organisations and energy companies worldwide, notably:

The Department of Energy and Climate Change, United Kingdom; Department of State, United States; Edison; Electric Power Development Co. Ltd, Japan; Enel; Energy Forecasting Agency (APBE), Russia; Energy Studies Institute, Singapore; Eni; Foreign and Commonwealth Office, United Kingdom; Fortum Corporation; IEA Coal Industry Advisory Board (CIAB);

Institute of Energy Economics at the University of Cologne, Germany; International Institute for Applied Systems Analysis, Austria; Intergovernmental Panel on Climate Change (IPCC); Ministère de l'écologie, du développement durable, des transports et du logement, France; Ministry of Economic Affairs, The Netherlands; Ministry of Energy, Russian Federation; Ministry of Petroleum and Energy, Norway; Norwegian Agency for Development Cooperation (NORAD); Royal Ministry of Foreign Affairs, Norway; Peabody Energy; Ministry of Knowledge Economy, Korea; Japan Gas Association, Japan; Schlumberger; Shell; Shenhua Group; Statoil; Toyota Motor Corporation; United Nations Development Programme (UNDP) and the United Nations Industrial Development Organization (UNIDO).

Many international experts provided input, commented on the underlying analytical work and reviewed early drafts of each chapter. Their comments and suggestions were of great value. They include:

Saleh Abdurrahman	National Energy Council of Indonesia
Ali Aissaoui	APICORP
Marco Arcelli	Enel
Gabriela Elizondo Azuela	World Bank
Pepukaye Bardouille	International Finance Corporation, United States
Andrew Barfour	Institution of Engineers, Ghana
Andrew Barnett	The Policy Practice, United Kingdom
Yuriy Baron	Ministry of Energy, Russian Federation
Paul Baruya	IEA Clean Coal Centre, United Kingdom
Igor Bashmakov	Centre for Energy Efficiency, Russian Federation
Nicolas Bauer	Potsdam Institute for Climate Impact Research, Germany
Georg Bäuml	Volkswagen
Johannes Baur	European Commission
Nazim Bayraktar	Energy Market Regulatory Agency, EPDK, Turkey
Chris Beaton	International Institute for Sustainable Development, Switzerland
Carmen Becerril	Acciona Energia
Rachid Bencherif	OPEC Fund for International Development
Kamel Bennaceur	Schlumberger
Bruno Bensasson	GDF Suez
Mikul Bhatia	World Bank
Sankar Bhattacharya	Monash University, Australia

Alexey Biteryakov	GazpromExport
Roberto Bocca	World Economic Forum
Jean-Paul Bouttes	EdF
Albert Bressand	Columbia School of International and Public Affairs, United States
Nigel Bruce	World Health Organization, Switzerland
Peter Brun	Vestas
Michael Buffier	Xstrata Coal
David Cachot	Trafigura Beheer B.V.
Guy Caruso	Center for Strategic and International Studies, United States
Milton Catelin	World Coal Association
Sharat Chand	The Energy and Resources Institute, India
Chris Charles	International Institute for Sustainable Development, Switzerland
Hela Cheikhrouhou	African Development Bank
Elisabeth Clemens	Norwegian Agency for Development Cooperation
Ben Clements	International Monetary Fund
Janusz Cofala	International Institute for Applied Systems Analysis, Austria
Dean Cooper	United Nations Environment Programme
Alan Copeland	Bureau of Resources and Energy Economics, Australia
Hans Daniels	Alpha Natural Resources, United States
Christian De Gromard	Agence Française de Developpement
Michel Deelen	Ministry of Foreign Affairs, The Netherlands
Jos Delbeke	European Commission
Carmine Difiglio	Department of Energy, United States
Mark Dominik	
Gina Downes	Eskom
Józef Dubinski	Central Mining Institute, Poland
Mohamed El-Ashry	UN Foundation
Kari Espegren	Institute for Energy Technology, Norway

Donald Ewart	Marston
Vladimir Feygin	Institute of Energy and Finance, Russian Federation
Christiana Figueres	UN Framework Convention on Climate Change
Peter Fraser	Ontario Energy Board, Canada
Irene Freudenschuss-Reichl	Ambassador, Austria
Hari Kumar Gadde	World Bank
Dario Garofalo	Enel
Francesco Gattei	Eni
Dolf Gielen	International Renewable Energy Agency
Duleep Gopalakrishnan	ICF International
Rainer Görgen	Federal Ministry of Economics and Technology, Germany
Michael Grubb	University of Cambridge, United Kingdom
Howard Gruenspecht	Energy Information Administration, United States
Antoine Halff	Department of Energy, United States
Ian Hall	Anglo American
Wenke Han	Energy Research Institute, China
Brian Heath	Coal Industry Advisory Board, International Energy Agency
Sigurd Heiberg	Statoil
James Henderson	Oxford Institute for Energy Studies, United Kingdom
Antonio Hernández Garcia	Ministry of Industry, Tourism and Trade, Spain
James Hewlett	Department of Energy, United States
Masazumi Hirono	The Japan Gas Association
Neil Hirst	Grantham Institute on Climate Change, United Kingdom
Takashi Hongo	Japan Bank for International Cooperation
Trevor Houser	Peterson Institute for International Economics
Tom Howes	European Commission
Steve Hulton	Wood Mackenzie
Esa Hyvärinen	Fortum Corporation
Catherine Inglehearn	Foreign and Commonwealth Office, United Kingdom
Fumiaki Ishida	New Energy and Industrial Technology Development Organization, Japan

James Jensen	Jensen Associates
Jan-Hein Jesse	JOSCO Energy Finance & Strategy Advisors
Marianne Kah	ConocoPhillips
Bob Kamandanu	Indonesian Coal Mining Association
Shaanti Kapila	Asian Development Bank
Ryan Katofsky	Navigant Consulting
Marlin Kees	GIZ, Germany
Hisham Khatib	Honorary Vice Chairman, World Energy Council; and former Minister of Energy, Jordan
Lucy Kitson	International Institute for Sustainable Development, Switzerland
Mikhail Klubnichkin	PricewaterhouseCoopers Russia B.V.
David Knapp	Energy Intelligence
Oliver Knight	World Bank
Kenji Kobayashi	Asia Pacific Energy Research Centre, Japan
Masami Kojima	World Bank
Hans-Jorgen Koch	Ministry of Transportation and Energy, Denmark
Doug Koplow	Earth Track, Inc.
Ken Koyama	The Institute of Energy Economics, Japan
Igor Kozhukovsky	Energy Forecasting Agency, Russian Federation
Natalia Kulichenko-Lotz	World Bank
Rakesh Kumar	PTC India
Takayuki Kusajima	Toyota Motor Corporation
Sarah Ladislaw	Center for Strategic and International Studies, United States
Georgette Lalis	European Commission
Kerryn Lang	International Institute for Sustainable Development, Switzerland
Richard Lavergne	Ministry of Economy, Finance and Industry, France
Rima Le Gocuic	Agence Française de Developpement
Man-ki Lee	Korea Atomic Energy Research Institute
Christian Lelong	BHP Billiton Energy Coal

Steve Lennon	Eskom
Michael Liebreich	Bloomberg New Energy Finance
Qiang Liu	Energy Research Institute, China
Massimo Lombardini	European Commission
Philip Lowe	European Commission
Matthew Lynch	World Business Council for Sustainable Development, Switzerland
Joan MacNaughton	Alstom Power Systems
Teresa Malyshev	UNDP
Claude Mandil	Former IEA Executive Director
Samantha McCulloch	Australian Coal Association
Michael Mellish	Energy Information Administration, United States
Lawrence Metzroth	Arch Coal
Ryo Minami	Ministry of Economy, Trade and Industry, Japan
Tatiana Mitrova	Skolkovo Energy Centre, Russian Federation
Klaus Mohn	Statoil
Lucio Monari	World Bank
Koji Morita	The Institute of Energy Economics, Japan
Yuji Morita	The Institute of Energy Economics, Japan
Ed Morse	CitiGroup
Richard Morse	Stanford University, United States
Dong-Woo Noh	Korea Energy Economics Institute
Petter Nore	Norwegian Agency for Development Cooperation
Martin Oettinger	Global CCS Institute
Patrick Oliva	Michelin
Simon-Erik Ollus	Fortum Corporation
Ayse Yasemin Örücü	Ministry of Energy and Natural Resources, Turkey
Shonali Pachauri	International Institute for Applied Systems Analysis, Austria
Jay Paidipati	Navigant Consulting
Binu Parthan	Renewable Energy & Energy Efficiency Partnership, Austria
Christian Pichat	AREVA

Jeff Piper	European Commission
Oleg Pluzhnikov	Ministry of Economic Development, Russian Federation
Roberto Potì	Edison
Ireneusz Pyc	Siemens
Maggi Rademacher	E.ON
Gustav Resch	Vienna University of Technology, Austria
Brian Ricketts	Euracoal
Hans-Holger Rogner	International Atomic Energy Agency
David Rolfe	Department of Energy and Climate Change, United Kingdom
Bert Roukens	Ministry of Foreign Affairs, The Netherlands
Assaad Saab	EdF
Bernard Saincy	GDF Suez
Alain Sanglerat	GDF Suez
Steve Sawyer	Global Wind Energy Council, Belgium
Wendy Schallom	Arch Coal
Hans-Wilhelm Schiffer	RWE
Sandro Schmidt	Federal Institute for Geosciences and Natural Resources, Germany
Philippe Schulz	Renault
Adnan Shihab-Eldin	Arabdar Consultants
P.R. Shukla	Indian Institute of Management
Maria Sicilia Salvatores	Iberdrola
Adam Sieminski	Deutsche Bank
Laura Solanko	Bank of Finland
Benjamin Sporton	World Coal Association
Robert Stavins	Harvard University, United States
James Steele	Department of State, United States
Jonathan Stern	Oxford Institute for Energy Studies, United Kingdom
Michael Stoppard	IHS CERA
Ulrik Stridbaek	Dong Energy
Greg Stringham	Canadian Association of Petroleum Producers

Supriatna Suhala	Indonesian Coal Mining Association
Cartan Sumner	Peabody Energy
Philip Swanson	Energy consultant
Minoru Takada	United Nations
Kuniharu Takemata	Electric Power Development Co., Ltd. (J-POWER)
Nobuo Tanaka	The Institute of Energy Economics, Japan
Bernard Terlinden	GDF Suez
Sven Teske	Greenpeace International
Wim Thomas	Shell
Elspeth Thomson	Energy Studies Institute, Singapore
Simon Trace	Practical Action, United Kingdom
Samuel Tumiwa	Asian Development Bank
Jo Tyndall	Ministry of Foreign Affairs and Trade, New Zealand
Oras Tynkkynen	Member of Parliament, Finland
Maria Vagliasindi	World Bank
Coby Van der Linde	Clingendael Institute, The Netherlands
Noe van Hulst	International Energy Forum, Saudi Arabia
Wim J. Van Nes	SNV Netherland Development Organisation
Adnan Vatansever	Carnegie Endowment for International Peace, United States
Umberto Vergine	Eni
Stefan Vergote	European Commission
Frank Verrastro	Center for Strategic and International Studies, United States
Heike Volkmer	GIZ, Germany
Graham Weale	RWE
Peter Wells	Cardiff Business School, United Kingdom
Liu Wenge	China Coal Information Institute
Jacob Williams	Peabody Energy
Steven Winberg	CONSOL Energy
Peter Wooders	International Institute for Sustainable Development, Switzerland

Liu Xiaoli	Energy Research Institute of NDRC
Vitaly Yermakov	IHS CERA
Shigehiro Yoshino	Nippon Export and Investment Insurance, Japan
Alex Zapantis	Rio Tinto

The individuals and organisations that contributed to this study are not responsible for any opinions or judgements contained in this study. All errors and omissions are solely the responsibility of the IEA.

WORKSHOPS

A number of workshops and meetings were held to gather essential input to this study. The workshop participants have contributed valuable new insights, feedback and data for this analysis.

Outlook for Coal Industry and Markets, Beijing: 14 April 2011

Russia Energy Outlook, Moscow: 20 April 2011

Energy for All: Financing Access for the Poor, Paris: 13 May 2011

TABLE OF CONTENTS

List of figures

Part A: GLOBAL ENERGY TRENDS

Chapter 1: Context and analytical framework

Chapter 2: Energy projections to 2035

Chapter 5: Power and renewables outlook

Chapter 6: Climate change and the 450 Scenario

Part B: OUTLOOK FOR RUSSIAN ENERGY

Chapter 7: Russian domestic energy prospects

Chapter 8: Russian resources and supply potential

Part D: SPECIAL TOPICS

Chapter 12: The implications of less nuclear power

Chapter 13: Energy for all

List of tables

List of boxes

Part A: GLOBAL ENERGY TRENDS

Part B: OUTLOOK FOR RUSSIAN ENERGY

List of spotlights

Comments and questions are welcome and should be
addressed to:

Dr. Fatih Birol
Chief Economist
Director, Office of the Chief Economist
International Energy Agency
9, rue de la Fédération
75739 Paris Cedex 15
France

Telephone: (33-1) 4057 6670
Fax: (33-1) 4057 6509
Email: weo@iea.org

More information about the *World Energy Outlook* is available at
www.worldenergyoutlook.org.

"If we don't change direction soon, we'll end up where we're heading"

There are few signs that the urgently needed change in direction in global energy trends is underway. Although the recovery in the world economy since 2009 has been uneven, and future economic prospects remain uncertain, global primary energy demand rebounded by a remarkable 5% in 2010, pushing CO_2 emissions to a new high. Subsidies that encourage wasteful consumption of fossil fuels jumped to over $400 billion. The number of people without access to electricity remained unacceptably high at 1.3 billion, around 20% of the world's population. Despite the priority in many countries to increase energy efficiency, global energy intensity worsened for the second straight year. Against this unpromising background, events such as those at the Fukushima Daiichi nuclear power plant and the turmoil in parts of the Middle East and North Africa (MENA) have cast doubts on the reliability of energy supply, while concerns about sovereign financial integrity have shifted the focus of government attention away from energy policy and limited their means of policy intervention, boding ill for agreed global climate change objectives.

This *Outlook* assesses the threats and opportunities facing the global energy system based on a rigorous quantitative analysis of energy and climate trends. The analysis includes three global scenarios and multiple case studies. The central scenario for this *Outlook* is the New Policies Scenario, in which recent government policy commitments are assumed to be implemented in a cautious manner – even if they are not yet backed up by firm measures. Comparison with the results of the Current Policies Scenario, which assumes no new policies are added to those in place as of mid-2011, illustrates the value of these commitments and plans. From another angle, comparison is also instructive with the 450 Scenario, which works back from the international goal of limiting the long-term increase in the global mean temperature to two degrees Celsius (2°C) above pre-industrial levels, in order to trace a plausible pathway to that goal. **The wide difference in outcomes between these three scenarios underlines the critical role of governments to define the objectives and implement the policies necessary to shape our energy future.**

Short-term uncertainty does little to alter the longer-term picture

Despite uncertainty over the prospects for short-term economic growth, demand for energy in the New Policies Scenario grows strongly, increasing by one-third from 2010 to 2035. The assumptions of a global population that increases by 1.7 billion people and 3.5% annual average growth in the global economy generate ever-higher demand for energy services and mobility. A lower rate of global GDP growth in the short-term than assumed in this *Outlook* would make only a marginal difference to longer-term trends.

The dynamics of energy markets are increasingly determined by countries outside the OECD. Non-OECD countries account for 90% of population growth, 70% of the increase in economic output and 90% of energy demand growth over the period from 2010 to 2035.

China consolidates its position as the world's largest energy consumer: in 2035 it consumes nearly 70% more energy than the United States, the second-largest consumer, even though, by then, per-capita energy consumption in China is still less than half the level in the United States. The rates of growth in energy consumption in India, Indonesia, Brazil and the Middle East are even faster than in China.

Global investment in energy supply infrastructure of $38 trillion (in year-2010 dollars) is required over the period 2011 to 2035. Almost two-thirds of the total investment is in countries outside of the OECD. Oil and gas collectively account for almost $20 trillion, as both the need for upstream investment and the associated cost rise in the medium and long term. The power sector claims most of the remainder, with over 40% of this being for transmission and distribution networks.

The age of fossil fuels is far from over, but their dominance declines. Demand for all fuels rises, but the share of fossil fuels in global primary energy consumption falls slightly from 81% in 2010 to 75% in 2035; natural gas is the only fossil fuel to increase its share in the global mix over the period to 2035. In the power sector, renewable energy technologies, led by hydropower and wind, account for half of the new capacity installed to meet growing demand.

Steps in the right direction, but the door to 2°C is closing

We cannot afford to delay further action to tackle climate change if the long-term target of limiting the global average temperature increase to 2°C, as analysed in the 450 Scenario, is to be achieved at reasonable cost. In the New Policies Scenario, the world is on a trajectory that results in a level of emissions consistent with a long-term average temperature increase of more than 3.5°C. Without these new policies, we are on an even more dangerous track, for a temperature increase of 6°C or more.

Four-fifths of the total energy-related CO_2 emissions permissible by 2035 in the 450 Scenario are already "locked-in" by our existing capital stock (power plants, buildings, factories, etc.). If stringent new action is not forthcoming by 2017, the energy-related infrastructure then in place will generate all the CO_2 emissions allowed in the 450 Scenario up to 2035, leaving no room for additional power plants, factories and other infrastructure unless they are zero-carbon, which would be extremely costly. Delaying action is a false economy: for every $1 of investment avoided in the power sector before 2020 an additional $4.3 would need to be spent after 2020 to compensate for the increased emissions.

New energy efficiency measures make a difference, but much more is required. Energy efficiency improves in the New Policies Scenario at a rate twice as high as that seen over the last two-and-a-half decades, stimulated by tighter standards across all sectors and a partial phase-out of subsidies to fossil fuels. In the 450 Scenario, we need to achieve an even higher pace of change, with efficiency improvements accounting for half of the additional reduction in emissions. The most important contribution to reaching energy security and climate goals comes from the energy that we do not consume.

Rising transport demand and upstream costs reconfirm the end of cheap oil

Short-term pressures on oil markets may be eased by slower economic growth and by the expected return of Libyan oil to the market, but trends on both the oil demand and supply sides maintain pressure on prices. We assume that the average IEA crude oil import price remains high, approaching $120/barrel (in year-2010 dollars) in 2035 (over $210/barrel in nominal terms) in the New Policies Scenario although, in practice, price volatility is likely to remain.

All of the net increase in oil demand comes from the transport sector in emerging economies, as economic growth pushes up demand for personal mobility and freight. Oil demand (excluding biofuels) rises from 87 million barrels per day (mb/d) in 2010 to 99 mb/d in 2035. The total number of passenger cars doubles to almost 1.7 billion in 2035. Sales in non-OECD markets exceed those in the OECD by 2020, with the centre of gravity of car manufacturing shifting to non-OECD countries before 2015. The rise in oil use comes despite some impressive gains in fuel economy in many regions, notably for passenger vehicles in Europe and for heavy freight in the United States. Alternative vehicle technologies emerge that use oil much more efficiently or not at all, such as electric vehicles, but it takes time for them to become commercially viable and penetrate markets. With limited potential for substitution for oil as a transportation fuel, the concentration of oil demand in the transport sector makes demand less responsive to changes in the oil price (especially where oil products are subsidised).

The cost of bringing oil to market rises as oil companies are forced to turn to more difficult and costly sources to replace lost capacity and meet rising demand. Production of conventional crude oil – the largest single component of oil supply – remains at current levels before declining slightly to around 68 mb/d by 2035. To compensate for declining crude oil production at existing fields, 47 mb/d of gross capacity additions are required, twice the current total oil production of all OPEC countries in the Middle East. A growing share of output comes from natural gas liquids (over 18 mb/d in 2035) and unconventional sources (10 mb/d). The largest increase in oil production comes from Iraq, followed by Saudi Arabia, Brazil, Kazakhstan and Canada. Biofuels supply triples to the equivalent of more than 4 mb/d, bolstered by $1.4 trillion in subsidies over the projection period.

Oil imports to the United States, currently the world's biggest importer, drop as efficiency gains reduce demand and new supplies such as light tight oil are developed, but increasing reliance on oil imports elsewhere heightens concerns about the cost of imports and supply security. Four-fifths of oil consumed in non-OECD Asia comes from imports in 2035, compared with just over half in 2010. Globally, reliance grows on a relatively small number of producers, mainly in the MENA region, with oil shipped along vulnerable supply routes. In aggregate, the increase in production from this region is over 90% of the required growth in world oil output, pushing the share of OPEC in global production above 50% in 2035.

A shortfall in upstream investment in the MENA region could have far-reaching consequences for global energy markets. Such a shortfall could result from a variety of factors, including higher perceived investment risks, deliberate government policies to develop production capacity more slowly or constraints on upstream domestic capital flows because priority is given to spending on other public programmes. If, between 2011 and 2015, investment in the MENA region runs one-third lower than the $100 billion per year required in the New Policies Scenario, consumers could face a substantial near-term rise in the oil price to $150/barrel (in year-2010 dollars).

Golden prospects for natural gas

There is much less uncertainty over the outlook for natural gas: factors both on the supply and demand sides point to a bright future, even a golden age, for natural gas. Our *Outlook* reinforces the main conclusions of a *WEO* special report released in June 2011: gas consumption rises in all three scenarios, underlining how gas does well under a wide range of future policy directions. In the New Policies Scenario, demand for gas all but reaches that for coal, with 80% of the additional demand coming from non-OECD countries. Policies promoting fuel diversification support a major expansion of gas use in China; this is met through higher domestic production and through an increasing share of LNG trade and Eurasian pipeline imports. Global trade doubles and more than one-third of the increase goes to China. Russia remains the largest gas producer in 2035 and makes the largest contribution to global supply growth, followed by China, Qatar, the United States and Australia.

Unconventional gas now accounts for half of the estimated natural gas resource base and it is more widely dispersed than conventional resources, a fact that has positive implications for gas security. The share of unconventional gas rises to one-fifth of total gas production by 2035, although the pace of this development varies considerably by region. The growth in output will also depend on the gas industry dealing successfully with the environmental challenges: a golden age of gas will require golden standards for production. Natural gas is the cleanest of the fossil fuels, but increased use of gas in itself (without carbon capture and storage) will not be enough to put us on a carbon emissions path consistent with limiting the rise in average global temperatures to 2°C.

Renewables are pushed towards centre stage

The share of non-hydro renewables in power generation increases from 3% in 2009 to 15% in 2035, underpinned by annual subsidies to renewables that rise almost five-times to $180 billion. China and the European Union drive this expansion, providing nearly half of the growth. Even though the subsidy cost per unit of output is expected to decline, most renewable-energy sources need continued support throughout the projection period in order to compete in electricity markets. While this will be costly, it is expected to bring lasting benefits in terms of energy security and environmental protection. Accommodating more electricity from renewable sources, sometimes in remote locations, will require additional investment in transmission networks amounting to 10% of total transmission investment: in the European Union, 25% of the investment in transmission networks is needed for this purpose. The contribution of hydropower to global power generation remains at around 15%, with China, India and Brazil accounting for almost half of the 680 gigawatts of new capacity.

Treading water or full steam ahead for coal?

Coal has met almost half of the increase in global energy demand over the last decade. Whether this trend alters and how quickly is among the most important questions for the future of the global energy economy. Maintaining current policies would see coal use rise by a further 65% by 2035, overtaking oil as the largest fuel in the global energy mix. In the New Policies Scenario, global coal use rises for the next ten years, but then levels off to finish 25% above the levels of 2009. Realisation of the 450 Scenario requires coal consumption to peak well before 2020 and then decline. The range of projections for coal demand in 2035 across the three scenarios is nearly as large as total world coal demand in 2009. The implications of policy and technology choices for the global climate are huge.

China's consumption of coal is almost half of global demand and its Five-Year Plan for 2011 to 2015, which aims to reduce the energy and carbon intensity of the economy, will be a determining factor for world coal markets. China's emergence as a net coal importer in 2009 led to rising prices and new investment in exporting countries, including Australia, Indonesia, Russia and Mongolia. In the New Policies Scenario, the main market for traded coal continues to shift from the Atlantic to the Pacific, but the scale and direction of international trade flows are highly uncertain, particularly after 2020. It would take only a relatively small shift in domestic demand or supply for China to become a net-exporter again, competing for markets against the countries that are now investing to supply its needs. India's coal use doubles in the New Policies Scenario, so that India displaces the United States as the world's second-largest coal consumer and becomes the largest coal importer in the 2020s.

Widespread deployment of more efficient coal-fired power plants and carbon capture and storage (CCS) technology could boost the long-term prospects for coal, but there are still considerable hurdles. If the average efficiency of all coal-fired power plants were to be five percentage points higher than in the New Policies Scenario in 2035, such an accelerated move away from the least efficient combustion technologies would lower CO_2 emissions from the power sector by 8% and reduce local air pollution. Opting for more efficient technology for new coal power plants would require relatively small additional investments, but improving efficiency levels at existing plants would come at a much higher cost. In the New Policies Scenario, CCS plays a role only towards the end of the projection period. Nonetheless, CCS is a key abatement option in the 450 Scenario, accounting for almost one-fifth of the additional reductions in emissions that are required. If CCS is not widely deployed in the 2020s, an extraordinary burden would rest on other low-carbon technologies to deliver lower emissions in line with global climate objectives.

Second thoughts on nuclear would have far-reaching consequences

Events at Fukushima Daiichi have raised questions about the future role of nuclear power, although it has not changed policies in countries such as China, India, Russia and Korea that are driving its expansion. In the New Policies Scenario, nuclear output rises by more than 70% over the period to 2035, only slightly less than projected last year. However, we also examine the possible implications of a more substantial shift away from nuclear power in a Low Nuclear Case, which assumes that no new OECD reactors are built, that non-OECD countries build only half of the additions projected in our New Policies Scenario and that the operating lifespan of existing nuclear plants is shortened. While creating opportunities for renewables, such a low-nuclear future would also boost demand for fossil fuels: the increase in global coal demand is equal to twice the level of Australia's current steam coal exports and the rise in gas demand is equivalent to two-thirds of Russia's current natural gas exports. The net result would be to put additional upward pressure on energy prices, raise additional concerns about energy security and make it harder and more expensive to combat climate change. The consequences would be particularly severe for those countries with limited indigenous energy resources which have been planning to rely relatively heavily on nuclear power. It would also make it considerably more challenging for emerging economies to satisfy their rapidly growing demand for electricity.

The world needs Russian energy, while Russia needs to use less

Russia's large energy resources underpin its continuing role as a cornerstone of the global energy economy over the coming decades. High prospective demand and international prices for fossil fuels might appear to guarantee a positive outlook for Russia, but the challenges facing Russia are, in many ways, no less impressive than the size of its resources. Russia's core oil and gas fields in Western Siberia will decline and a new generation of higher-cost fields need to be developed, both in the traditional production areas of Western Siberia and in the new frontiers of Eastern Siberia and the Arctic. A responsive Russian fiscal regime will be needed to provide sufficient incentives for investment. Oil production plateaus around 10.5 mb/d before starting a slight decline to 9.7 mb/d in 2035; gas production increases by 35% to 860 billion cubic metres (bcm) in 2035, with the Yamal peninsula becoming the new anchor of Russian supply.

As the geography of Russian oil and gas production changes, so does the geography of export. The majority of Russia's exports continue to go westwards to traditional markets in Europe, but a shift towards Asian markets gathers momentum. Russia gains greater diversity of export revenues as a result: the share of China in Russia's total fossil-fuel export earnings rises from 2% in 2010 to 20% in 2035, while the share of the European Union falls from 61% to 48%.

Russia aims to create a more efficient economy, less dependent on oil and gas, but needs to pick up the pace of change. If Russia increased its energy efficiency in each sector to the levels of comparable OECD countries, it could save almost one-third of its annual primary energy use, an amount similar to the energy used in one year by the United Kingdom. Potential savings of natural gas alone, at 180 bcm, are close to Russia's net exports in 2010. New energy efficiency policies and continued price reforms for gas and electricity bring some improvements but, in our analysis, do not unlock more than a part of Russia's efficiency potential. Faster implementation of efficiency improvements and energy market reforms would accelerate the modernisation of the Russian economy and thereby loosen its dependency on movements in international commodity prices.

Achieving energy for all will not cost the earth

We estimate that, in 2009, around $9 billion was invested globally to provide first access to modern energy, but more than five-times this amount, $48 billion, needs to be invested each year if universal access is to be achieved by 2030. Providing energy access for all by 2030 is a key goal announced by the UN Secretary-General. Today, 1.3 billion people do not have electricity and 2.7 billion people still rely on the traditional use of biomass for cooking. The investment required is equivalent to around 3% of total energy investment to 2030. Without this increase, the global picture in 2030 is projected to change little from today and in sub-Saharan Africa it gets worse. Some existing policies designed to help the poorest miss their mark. Only 8% of the subsidies to fossil-fuel consumption in 2010 reached the poorest 20% of the population.

International concern about the issue of energy access is growing. The United Nations has declared 2012 to be the "International Year of Sustainable Energy for All" and the Rio+20 Summit represents an important opportunity for action. More finance, from many sources and in many forms, is needed to provide modern energy for all, with solutions matched to the particular challenges, risks and returns of each category of project. Private sector investment needs to grow the most, but this will not happen unless national governments adopt strong governance and regulatory frameworks and invest in capacity building. The public sector, including donors, needs to use its tools to leverage greater private sector investment where the commercial case would otherwise be marginal. Universal access by 2030 would increase global demand for fossil fuels and related CO_2 emissions by less than 1%, a trivial amount in relation to the contribution made to human development and welfare.

PREFACE

Part A of this *WEO* (Chapters 1 to 6) presents a comprehensive overview of our energy projections to 2035, in a fuel-by-fuel format which will be familiar to regular readers.

As last year, we present the results of modelling three scenarios: the *Current Policies Scenario*, the *New Policies Scenario* and the *450 Scenario* (a title derived from the objective of limiting the concentration of greenhouse gases in the atmosphere to no more than 450 parts per million).

We also examine a number of plausible variations on these central scenarios: a Low GDP Case, a Deferred Investment Case (deferred investment in oil in the Middle East and North Africa), a Golden Age of Gas (a more optimistic scenario for natural gas, summarised and updated from a separate publication in the *WEO* series in June 2011), a Delayed Carbon Capture and Storage Case, and a Low Nuclear Case (see Part D, Chapter 12).

Chapter 1 describes the methodological framework and the assumptions that underpin the scenarios, commenting on changes since *WEO-2010*. Chapter 2 surveys global trends in the demand and supply of all fuels in all scenarios, but with a special emphasis on the New Policies Scenario. The results are given fuel-by-fuel, sector-by-sector and region-by-region. Production prospects, investment in the energy supply infrastructure, changes in inter-regional trade and the outlook for energy-related emissions are all covered.

Chapters 3 to 6 deal in detail with oil, natural gas, electricity and renewables and climate change (with special emphasis in Chapter 6 on the 450 Scenario). This year coal has been given special attention in its own right (see Part C).

CHAPTER 1

CONTEXT AND ANALYTICAL FRAMEWORK
What's driving energy markets?

H I G H L I G H T S

- *WEO-2011* assesses the implications for global energy markets to 2035 of alternative assumptions about energy and climate policy. The New Policies Scenario – the central scenario in this *Outlook* – takes into account recently announced commitments and plans, even if they are yet to be formally adopted and implemented. The Current Policies Scenario takes account only of those policies that had been enacted by mid-2011. The 450 Scenario, sets out an illustrative energy pathway that is consistent with a 50% chance of meeting the goal of limiting the increase in average global temperature to 2°C.

- The rate of growth in world GDP – a fundamental driver of energy demand – is assumed to average 3.6% per year over the period 2009 to 2035 in all scenarios. Non-OECD countries account for over 70% of the increase in global economic output, pushing their share of global GDP from almost 45% today to over 60% in 2035. China alone makes up 31% of the increase in global GDP to 2035 and India a further 15%.

- Population growth will continue to underpin rising energy demand. The world's population is assumed to increase by 26%, from 6.8 billion in 2009 to 8.6 billion in 2035, with over 90% of the increase in non-OECD regions. The annual increase in the world's population slows progressively, from 78 million in 2010 to 56 million in 2035.

- Energy prices will continue to have a major impact on future demand and supply patterns. In the New Policies Scenario, the IEA crude oil import price is assumed to approach $120/barrel (in year-2010 dollars) in 2035. The price rises more rapidly in the Current Policies Scenario and more slowly in the 450 Scenario. Natural gas prices broadly follow the trend in oil prices, but on an energy-equivalent basis their ratio remains lower than the historical average. Coal prices rise much less than oil and gas prices. In the 450 Scenario, it is assumed that retail prices for oil-based transport fuels, despite lower global demand, are held by government action at levels similar to the Current Policies Scenario.

- Some countries have already imposed a price on CO_2 emissions and certain others are assumed to follow suit (through taxation, cap-and-trade schemes or some equivalent). The CO_2 prices in 2035 are assumed to range from $30 to $45/tonne (in year-2010 dollars) in the New Policies Scenario, and from $95 to $120/tonne in the 450 Scenario.

- Rates of technological development and deployment, and their impact on energy efficiency, vary across the three scenarios. For example, carbon capture and storage technologies and electric vehicles are deployed on a very limited scale in the New Policies Scenario, but both play more significant roles in the 450 Scenario. No completely new technologies are assumed to be deployed at substantial levels.

Introduction

Although the years 2008 to 2010 were considered to be a turbulent period in global energy markets, new uncertainties have arisen from events in 2011. The "Arab Spring" – an outbreak of civil unrest and protests that started in Tunisia in December 2010 and quickly spread across parts of the Middle East and North Africa – had major implications throughout the region and internationally, including by leading to an almost complete halt to the supply of oil from Libya, prompting International Energy Agency (IEA) member countries to release emergency stocks for only the third time in the organisation's history. In addition to their tragic human consequences, the devastating earthquake and tsunami in Japan in March accentuated the tightening of global energy markets that was already underway and raised new questions about the longer-term prospects for nuclear power in Japan and in other parts of the world. Energy prices remained persistently high throughout the first half of 2011, before falling in late August at the prospect of Libyan crude oil returning to the market and growing doubts about the state of the global economic recovery. And, in the midst of global irresolution about the scale and urgency of the challenge we face in managing the risks from climate change, data became available that indicate that carbon-dioxide (CO_2) emissions are rising at the quickest rate in history.

Incorporating these and other developments, *WEO-2011* provides a full update of energy demand and supply projections to 2035. The first part (Part A) analyses the possible evolution of energy markets to 2035 under three different scenarios. Consistent with past editions of the *Outlook*, the core scenarios rest on common assumptions about macroeconomic conditions and population growth, while their assumptions about government policy differ – which in turn affects energy prices and technology deployment. For each of the scenarios, detailed projections are presented of trends in energy demand and supply, as well as energy infrastructure investment, by fuel, region and sector.[1] We also investigate the implications of these trends for CO_2 emissions and local pollution, and analyse issues surrounding high-carbon infrastructure "lock-in", including the cost that would be incurred if it became necessary to retire early, or retrofit, energy-related capital with long lifetimes in order to meet climate imperatives. The results of this analysis are intended to provide a sound quantitative framework for assessing and comparing possible future trends in energy markets and the cost-effectiveness of new policies to tackle energy security and environmental concerns.

The uncertainty facing the world today makes it wise to consider how unexpected events might change the energy landscape. Therefore, in addition to the three full scenarios, Part A (Chapters 1 to 6), also includes numerous sensitivity cases analysing the effects of possible high-impact events that could dramatically change the future course of energy markets. For example, in Chapter 2, we analyse the impact of slower economic growth in the period to 2015, to illustrate the possible consequences for the energy sector of another economic downturn. In Chapter 3, we present a Deferred Investment Case, which, in the wake of a number of developments including the Arab Spring, looks at the implications of a possible shortfall over the next few years in upstream oil and gas investment in the Middle

---

1. Since the last *Outlook*, regional disaggregation has been updated to enable account to be taken of the accession of Chile, Estonia, Israel and Slovenia to the OECD; see Annex C for full details of the new groupings.

East and North Africa, a region that is expected to deliver a growing share of the world's hydrocarbons in the coming decades. In Chapter 4 we consider the implications of a "Golden Age of Gas" alongside more conservative projections. And in Chapters 6 and 12 we analyse the implications for achieving ambitious climate goals if certain key energy technologies, including carbon capture and storage (CCS) and nuclear power, are developed and deployed more slowly.

Consistent with past practice, we offer in Part B of this year's *Outlook* an in-depth analysis of the prospects for energy supply and use in a single country, this time Russia, which is the one of the world's largest energy producers. This includes an assessment of the country's domestic energy needs and the outlook for the production and export of oil and gas, investment needs and constraints and the implications of energy developments in the country for global energy security and environmental sustainability. In a manner comparable to the analysis of renewables in 2010, natural gas in 2009 and oil in 2008, Part C provides an expanded assessment of the prospects for a particular fuel, this time coal. We look at global demand and supply issues, the evolution of traded coal markets and the adequacy of investment through the production and delivery chain. Many would argue that this focus on coal is long overdue: in 2000 to 2010, the increase in global coal demand was almost equal to that of all other forms of energy combined.

Part D of the report takes a detailed look at three special topics which are of high current relevance. First, in Chapter 12, we present a Low Nuclear Case, which investigates the implications for global energy balances of a possible collapse in the expansion of nuclear power capacity worldwide, following the accident at the Fukushima Daiichi nuclear plant in Japan. Second, in Chapter 13, we intensify the established *WEO* practice of highlighting the key strategic challenge of energy poverty, this time by identifying possible means of raising and administering the finance required to deliver energy to those who would otherwise have no access to it. This analysis was first released as a special input to a high-level meeting hosted by the government of Norway, in Oslo on 10 October 2011, that brought together Heads of State, international institutions and other key stakeholders to consider how to accelerate progress towards universal energy access as part of a broader push to achieve health and development goals. In Chapter 14, we provide an update of the IEA's continuing work on subsidies to both fossil fuels and renewables. This chapter is aimed at encouraging action to reform fossil-fuel subsidies globally, while highlighting the important role that well-designed incentives can play in developing and deploying cleaner and more efficient technologies in order to reduce greenhouse-gas emissions and pollution, and to diversify the energy mix.

Defining the scenarios

Three scenarios are presented in this year's *Outlook*: the New Policies Scenario, the Current Policies Scenario and the 450 Scenario. In each case, what is offered is a set of internally consistent projections: none should be considered as a forecast. The projection period runs to 2035. The starting year is 2010, as historical market data for all countries were available at the time of writing only up to 2009 (although preliminary data for 2010 were available

Table 1.1 ● Selected key policy assumptions by scenario and region*

	Current Policies Scenario	New Policies Scenario	450 Scenario
OECD			Staggered introduction of CO_2 prices in all countries; $100 billion annual financing provided to non-OECD countries by 2020; on-road PLDV emissions average 65 g/km in 2035.
United States	New appliance standards; state-level support schemes for renewables; enhanced CAFE standards; tax credits for renewable energy sources.	Shadow price of carbon adopted from 2015, affecting investment decisions in power generation; new HDV standards for each model year from 2014 to 2018; EPA regulations on mercury and other pollutants in the power sector.	A 17% reduction in CO_2 emissions compared with 2005 by 2020; CO_2 pricing implemented from 2020.
Japan	Long-Term Outlook on Energy Supply and Demand (2009), including reforms in steel manufacturing, support for renewables generation and improved fuel efficiency for vehicles.	Implementation of Strategic Energy Plan**; shadow price of carbon implemented from 2015, affecting new investment in power generation.	A 25% reduction in emissions compared with 1990 by 2020; CO_2 pricing implemented from 2020.
European Union	ETS, covering power, industry and (from 2012) aviation; Energy Performance of Buildings directive; emissions standards for PLDVs. A 20% reduction in emissions compared with 1990 by 2020; renewables to reach 20% share in energy demand in 2020.	ETS, covering power, industry and (from 2012) aviation; new LCV standards; more stringent PLDV standards.	A 30% reduction in emissions compared with 1990 by 2020; ETS strengthened in line with the 2050 roadmap.
Australia, New Zealand	New Zealand: Domestic ETS from 2010.	Australia: A 5% reduction in emissions compared with 2000 by 2020; carbon tax from mid-2012 and domestic emissions trading from 2015. New Zealand: A 10% cut in emissions compared with 1990 by 2020.	Australia: A 25% reduction in emissions compared with 2000 by 2020; New Zealand: A 20% reduction in emissions compared with 1990 by 2020.
Korea		A 30% reduction in emissions compared with business-as-usual by 2020; CO_2 pricing from 2015.	A 30% reduction in emissions compared with business-as-usual by 2020; higher CO_2 pricing.

*See Annex B for more details of the policy assumptions across the three scenarios. **Following the Great East Japan Earthquake, Japan is undertaking a full review of its Strategic Energy Plan with results expected in 2012.

Table 1.1 ● Selected key policy assumptions by scenario and region (continued)

	Current Policies Scenario	New Policies Scenario	450 Scenario
Non-OECD	Fossil-fuel subsidies are phased out in countries that already have policies in place to do so.	Fossil-fuel subsidies are phased out in all net-importing regions by 2020 (at the latest) and in net-exporting regions where specific policies have already been announced.	Receipt of finance to support domestic mitigation action; on-road PLDV emissions average 100 g/km in 2035. International sectoral agreements for iron and steel, and cement. Fossil-fuel subsidies are phased out in net-importing regions by 2020 and in net-exporting regions by 2035. ***
China	Implementation of measures in 12th Five-Year Plan, including 17% cut in CO_2 intensity by 2015; solar additions of 5 GW by 2015; wind additions of 70 GW by 2015 and start construction of 120 GW of hydropower by 2015.	A 40% reduction in carbon intensity compared with 2005 by 2020; CO_2 pricing from 2020; a 15% share of non-fossil energy in total energy supply by 2020; 70 to 80 GW of nuclear power by 2020; 12th Five-Year Plan renewables targets exceeded. PLDV fuel economy targets by 2015.	A 45% reduction in carbon intensity compared with 2005 by 2020; higher CO_2 pricing; enhanced support for renewables.
India	Trading of renewable energy certificates; measures under national solar mission and national mission on enhanced energy efficiency (Perform Achieve and Trade [PAT] scheme for industry).	A 20% reduction in CO_2 intensity compared with 2005 by 2020; proposed auto fuel-efficiency standards; 20 GW of solar energy production capacity by 2022.	A 25% reduction in CO_2 intensity compared with 2005 by 2020; expanded feed-in tariffs for renewables.
Brazil	Solar incentives; ethanol targets in road transport (20% to 25%).	A 36% reduction in emissions compared with business-as-usual by 2020; increase of electricity generation from renewable sources.	A 39% reduction in emissions compared with business-as-usual by 2020; increased generation from renewable sources; CO_2 pricing from 2020.
Russia (see also Chapter 7)	Gradual real increases in residential gas and electricity prices (1% per year) and average gas price paid by industry (1.5%); implementation of 2009 energy efficiency legislation.	A 2% per year real increase in residential gas and electricity prices; industrial gas prices reach the equivalent of export prices (minus taxes and transportation) in 2020; implementation of measures included in the 2010 energy efficiency state programme.	Faster liberalisation of residential gas and electricity prices; CO_2 pricing from 2020; full implementation of measures included in the 2010 energy efficiency state programme; stronger support for nuclear power and renewables.

***Except the Middle East where subsidisation rates are assumed to decline to a maximum of 20% by 2035.

Notes: Pricing of CO_2 emissions is either by an emissions trading scheme (ETS) or carbon taxes. PLDV = Passenger light-duty vehicle; LCV= Light commercial vehicle; HDV = Heavy-duty vehicle; and CAFE = Corporate Average Fuel Economy.

1

in many cases and have been incorporated).[2] Assumptions about government policies are critical to the three scenarios and over 3 000 policies and measures in OECD and non-OECD countries have been considered during their preparation.[3] A summary of some of the key policy targets and measures, by scenario, is set out in Table 1.1; more detailed assumptions can be found in Annex B.

The **New Policies Scenario** – the central scenario of this *Outlook* – incorporates the broad policy commitments and plans that have been announced by countries around the world to tackle energy insecurity, climate change and local pollution, and other pressing energy-related challenges, even where the specific measures to implement these commitments have yet to be announced. Those commitments include renewable energy and energy-efficiency targets and support, programmes relating to nuclear phase-out or additions, national pledges to reduce greenhouse-gas emissions communicated officially under the Cancun Agreements and the initiatives taken by G-20 and APEC economies to phase out inefficient fossil-fuel subsidies that encourage wasteful consumption. This scenario provides a benchmark to assess the achievements and limitations of recent developments in climate and energy policy. As many of the formal commitments that have been modelled in the New Policies Scenario relate to the period to 2020, we have assumed that additional unspecified measures are introduced that maintain through to 2035 a similar trajectory of global decline in carbon intensity – measured as emissions per dollar of gross domestic product. International sectoral agreements are assumed to be implemented across several industries, including cement and light-duty vehicles.

As only limited details are available for many of the initiatives considered in the New Policies Scenario, the extent to which they will actually be implemented is uncertain. Some targets are conditional on financial transfers from Annex I to non-Annex I countries or commitment to comparable emissions reductions by a set of countries, while other commitments involve a range. Some pledges relate to energy or carbon intensity, rather than absolute reductions in emissions. As a result, it is far from certain what these commitments will mean for emissions, even if they are met fully. Similarly, following the G-20 and APEC commitments on fossil-fuel subsidies, many countries have started to implement, or have proposed, reforms to bring their domestic energy prices into line with the levels that would prevail in a less distorted market, but the success of these reforms is very uncertain, in the face of steep economic, political and social hurdles. Furthermore, it is uncertain what new action governments may decide to take in the coming years (particularly post-2020) as perceptions of risk and threat change, and what implications these policies might have for greenhouse-gas emissions. To allow for all these uncertainties, the New Policies Scenario adopts a pragmatic approach by assuming cautious implementation of recently announced commitments and plans. In countries where uncertainty over climate policy is very high, it is assumed that the policies adopted are insufficient to reach their target.

---

2. As a result of the 2008/2009 economic crisis, world primary energy consumption fell by 1.1% in 2009, the first decline of any significance since 1981. Although the economic recovery is still rather sluggish, preliminary data suggest that global energy demand rebounded by a remarkable 5% in 2010.

3. The WEO Policy Database is available at www.iea.org/textbase/pm/?mode=weo .

The New Policies Scenario should not be read as a forecast. Even though it is likely that many governments around the world will take firm policy action to tackle increasing energy insecurity, local pollution, climate change and other energy-related problems, the policies that are actually put in place in the coming years will certainly differ from those assumed in this scenario. On the one hand, governments may decide to take stronger action to implement their current commitments than assumed in this scenario and/or may adopt even more stringent targets. On the other hand, it is quite possible that some governments will fail to implement the policies required to meet even their current pledges.

WEO-2011 also presents updated projections for the **Current Policies Scenario** (called the Reference Scenario prior to *WEO-2010*) to show how the future might look on the basis of the perpetuation, without change, of the government policies and measures that had been enacted or adopted by mid-2011. A number of the policy commitments and plans that were included in the New Policies Scenario in *WEO-2010* (IEA, 2010a) have since been enacted so are now included in the Current Policies Scenario in this *Outlook*. These include, for example, China's 12th Five-Year Plan for the period 2011 to 2015; a new scheme in India that enables trading of renewable energy certificates and a new programme of support for alternative fuel vehicles; new EU directives covering the energy performance of buildings and emissions standards for light-commercial vehicles; and new appliance standards in the United States.

The *Outlook* presents, in Chapter 6, updated projections for the **450 Scenario**, which sets out an energy pathway that is consistent with a 50% chance of meeting the goal of limiting the increase in average global temperature to two degrees Celsius (2°C), compared with pre-industrial levels. According to climate experts, to meet this goal it will be necessary to limit the long-term concentration of greenhouse gases in the atmosphere to around 450 parts per million of carbon-dioxide equivalent (ppm CO_2-eq). For the period to 2020, the 450 Scenario assumes more vigorous policy action to implement fully the Cancun Agreements than is assumed in the New Policies Scenario (which assumes cautious implementation). After 2020, OECD countries and other major economies are assumed to set economy-wide emissions targets for 2035 and beyond that collectively ensure an emissions trajectory consistent with stabilisation of the greenhouse-gas concentration at 450 ppm.

The most significant change in the 450 Scenario compared with *WEO-2010* relates to the starting point for emissions: after a dip in 2009, caused by the global financial crisis, emissions climbed by a record 5.3% in 2010, reaching 30.4 gigatonnes (Gt). This higher starting point means that even bigger emissions reductions will be needed in the future in order to limit the global increase in temperature to 2°C. Furthermore, with another year passing, the difficulty and cost of meeting ambitious climate goals has been increased as ongoing investments have "locked in" high-carbon infrastructure. The 450 Scenario is now more demanding than it was just twelve months ago.

Main non-policy assumptions

Economic growth

Today, four years on from the start of the global financial crisis, there remain persistent doubts about the sustainability of the global economic recovery. Projections from the

International Monetary Fund (IMF), at the time of writing, show global gross domestic product (GDP) growing by around 4.0% in 2011 and 2012 from over 5.0% in 2010 (IMF, 2011). These aggregate numbers mask considerable divergence in expected performance across regions: the advanced economies (essentially those of the OECD) are projected to grow by just 1.5% in 2011, compared with 6.0% in the emerging economies. Moreover, it appears that the risks are to the downside, and it is quite possible that the IMF and other forecasting bodies will revise down their projections for GDP growth for 2011 and 2012 (see Box 2.1 in Chapter 2).

The economic outlook has been clouded primarily by worries in OECD countries about sovereign and private-sector indebtedness, the effect on the banking sector of possible sovereign debt defaults and how plans to reduce deficits and debts will affect future economic growth. In many advanced economies, fiscal adjustment in the wake of the global financial crisis is underway, reducing budget deficits and stemming the rise of government debt to GDP ratios; but some countries are continuing to struggle to put their finances in order. Recent economic concerns have focused particularly on weaker than expected economic activity in the United States and related uncertainty over plans for fiscal consolidation, and the fiscal challenges facing a number of countries in Europe. Economic risks are also evident in some emerging and developing economies, with signs of potential overheating.

The rise in oil prices since September 2010 has fed concerns about near-term economic prospects. Oil prices dropped somewhat from early August 2011, following weaker economic data from United States and Europe and on promising signs of a resumption of oil exports from Libya, but they remain high by historical standards. Were the average IEA crude oil import prices of $100 per barrel over the first half of the year to persist through to the end of the year, spending on imports by the OECD – which is set to import almost 60% of its oil needs in 2011 – would amount to 2.2% of its GDP. The share of GDP spent on oil imports is generally even higher in oil-importing developing countries, because their economies are typically more oil intensive. Higher oil prices have weighed on growth in oil-importing countries by consuming a greater proportion of household and business expenditure. They have also put upward pressure on inflation, both directly, through increases in fuel prices, and indirectly, as prices of other goods have risen to reflect the higher input costs. The recent growth in production of biofuels that compete with food use, like corn-based ethanol, has arguably strengthened these links. Inflationary impacts have been most pronounced in the emerging economies, particularly in Asia, energy weighs relatively heavily in domestic consumer price indices. Globally, the net economic loss engendered by higher prices has been only partly offset by the oil-producing countries "recycling" some of their surplus export revenues back into the global economy in the form of increased imports of goods and services.

The civil unrest that has swept through some parts of the Middle East and North Africa (MENA) region, which is responsible for around 35% and 20% of the world's oil and natural gas output respectively, since December 2010 has contributed to higher energy prices and any worsening of the unrest or its spread to major exporting countries in the region could lead to a surge in prices sufficient to tip the global economy back into recession. The unrest has led a number of Gulf States to boost significantly public spending, including spending on welfare programmes. While this is likely to lead to greater demand for imported goods (which will reduce the global economic impact of high prices) it also means that oil producers may need yet higher oil prices in the future in order to keep their budgets in balance. In many OPEC countries, the current budget breakeven oil price is typically between $70/barrel and $90/barrel (see Chapter 3).

In this *Outlook*, world GDP (expressed in real purchasing power parity, or PPP, terms[4]) is assumed to grow by an average of 3.6% per year over the period 2009 to 2035, compared with 3.1% from 1990 to 2009 (Table 1.2). Assumed growth is somewhat higher than in last year's *Outlook*, in part due to the lower 2009 base. The medium-term GDP growth assumptions are based primarily on IMF projections, with some adjustments to reflect more recent information available for the OECD (OECD, 2011) and other countries from national and other sources. Longer-term GDP assumptions are derived from an assessment of historical growth rates and expectations for the growth in labour supply and the speed at which labour productivity improves. The risk of slower GDP growth in the near term is examined in a sensitivity analysis, the results of which are set out in Chapter 2.

Consistent with the pattern of recent decades, the non-OECD countries are assumed to remain the main engine of global economic growth, lifting their share of world GDP from 44% in 2010 to 61% in 2035 (compared with 33% in 1985). China and India are expected to continue to grow faster than countries in other regions, followed by the Middle East countries. China's growth rate slows from 8.1% per year in the period 2009 to 2020 to 4.3% per year in 2020 to 2035, less than half the rate at which it has been growing in recent years. India displaces China in the early 2020s as the fastest-growing country, the result of demographic factors and its earlier stage of economic development. India's growth nonetheless slows from 7.7% per year in 2009 to 2020 to 5.8% per year in 2020 to 2035. Among the OECD regions, OECD Americas continues to grow fastest, at 2.4% per year on average over the projection period, buoyed by more rapid growth in its population and labour force. Europe and Asia Oceania are expected to see the lowest GDP growth of any of the major regions.

The energy projections in the *Outlook* are highly sensitive to these underlying assumptions about GDP growth. Historically, energy demand has tended to rise broadly in line with GDP and economic downturns have been linked with a flattening or reduction of growth in energy usage. The so-called income elasticity of energy – the increase in energy demand relative to GDP – has gradually declined over time and has actually reversed in some developed countries. In most cases, it tends to be higher for countries at an early stage of economic development than for more mature economies. This can be explained by differences in the structure of economic output and the curtailing of income-driven increases in demand through efficiency improvements and saturation effects.

---

4. Purchasing power parities (PPPs) measure the amount of a given currency needed to buy the same basket of goods and services, traded and non-traded, as one unit of the reference currency, in this report, the US dollar. By adjusting for differences in price levels, PPPs, in principle, can provide a more reliable indicator than market exchange rates of the true level of economic activity globally or regionally and, thus, help in analysing the main drivers of energy demand and comparing energy intensities across countries and regions.

Table 1.2 ● Real GDP assumptions by region ($2010 trillion)

	2009	2015	2020	2035	2009-2020*	2009-2035*
OECD	**39.9**	**46.5**	**51.8**	**69.4**	**2.4%**	**2.2%**
Americas	17.3	20.5	23.2	32.4	2.7%	2.4%
United States	*14.3*	*16.8*	*18.9*	*26.2*	*2.6%*	*2.4%*
Europe	16.1	18.3	20.3	26.7	2.1%	2.0%
Asia Oceania	6.5	7.6	8.4	10.3	2.3%	1.8%
Japan	*4.1*	*4.7*	*5.0*	*5.9*	*1.7%*	*1.4%*
Non-OECD	**30.9**	**45.5**	**59.4**	**106.8**	**6.1%**	**4.9%**
E. Europe/Eurasia	3.7	4.8	5.8	9.3	4.1%	3.6%
Russia	*2.1*	*2.8*	*3.3*	*5.4*	*4.1%*	*3.6%*
Asia	17.6	28.2	38.5	73.6	7.4%	5.7%
China	*9.4*	*15.9*	*22.3*	*41.6*	*8.1%*	*5.9%*
India	*3.7*	*6.0*	*8.3*	*19.4*	*7.7%*	*6.6%*
Middle East	2.4	3.1	3.8	6.6	4.3%	4.0%
Africa	2.9	3.9	4.7	7.3	4.6%	3.7%
Latin America	4.3	5.6	6.6	10.0	4.0%	3.3%
Brazil	*2.0*	*2.7*	*3.2*	*5.1*	*4.3%*	*3.6%*
World	**70.8**	**92.0**	**111.2**	**176.2**	**4.2%**	**3.6%**
European Union	*14.9*	*16.8*	*18.5*	*24.3*	*2.0%*	*1.9%*

*Compound average annual growth rate.

Notes: Calculated based on GDP expressed in year-2010 dollars at constant purchasing power parity. The assumed rates of economic growth for 2009 to 2035 are the same for all three scenarios presented in this *Outlook*.

Sources: IMF, OECD and World Bank databases; IEA databases and analysis.

Assumptions about GDP growth are the same in each of the three scenarios modelled in *WEO-2011*. In reality, the fundamental energy system transformation that takes place in the 450 Scenario or the higher energy prices in the Current Policies Scenario could have at least temporary adverse impacts on GDP growth. In the case of the 450 Scenario, the higher economic cost associated with the shift in investment to low-carbon technologies could depress GDP growth, though the economic benefits that would accrue from reduced environmental damage could offset or perhaps even outweigh these direct GDP losses. For reasons such as this and to facilitate comparison between scenarios, the simplifying assumption has been adopted that GDP growth rates remain unchanged.

Population

Population growth is a key driver of future energy trends as the level of population has a direct effect on the size and composition of energy demand and an indirect effect by influencing economic growth and development. Based on the latest United Nations projections, in each of the three scenarios world population grows from an estimated

6.8 billion in 2009 to around 8.6 billion in 2035, an average rate of increase of 0.9% per year (Table 1.3) (UNPD, 2011). In line with the long-term historical trend, the population growth rate slows progressively over the projection period, from 1.1% per year in 2009 to 2020 to 0.8% in 2020 to 2035. However, this means that by 2035, the annual increase in population is still 56 million, compared with 78 million in 2010.

Table 1.3 ● **Population and urbanisation assumptions by region**

	Population growth*			Population (million)		Urbanisation rate (%)	
	2009-2020	2020-2035	2009-2035	2009	2035	2009	2035
OECD	**0.5%**	**0.3%**	**0.4%**	**1 229**	**1 373**	**77%**	**84%**
Americas	0.9%	0.7%	0.8%	470	571	81%	87%
United States	*0.8%*	*0.7%*	*0.7%*	*312*	*377*	*82%*	*88%*
Europe	0.4%	0.2%	0.3%	557	599	74%	82%
Asia Oceania	0.2%	-0.1%	0.0%	203	203	73%	81%
Japan	*-0.1%*	*-0.4%*	*-0.3%*	*127*	*118*	*67%*	*75%*
Non-OECD	**1.2%**	**0.9%**	**1.0%**	**5 536**	**7 183**	**44%**	**57%**
E. Europe/Eurasia	0.1%	-0.1%	0.0%	335	331	63%	70%
Russia	*-0.1%*	*-0.4%*	*-0.3%*	*142*	*133*	*73%*	*78%*
Asia	0.9%	0.6%	0.7%	3 546	4 271	38%	53%
China	*0.4%*	*0.0%*	*0.1%*	*1 338*	*1 387*	*46%*	*65%*
India	*1.3%*	*0.9%*	*1.0%*	*1 155*	*1 511*	*30%*	*43%*
Middle East	1.9%	1.4%	1.6%	195	293	66%	74%
Africa	2.3%	2.0%	2.1%	1 009	1 730	39%	53%
Latin America	1.0%	0.7%	0.8%	451	558	79%	86%
Brazil	*0.8%*	*0.4%*	*0.6%*	*194*	*224*	*86%*	*92%*
World	**1.1%**	**0.8%**	**0.9%**	**6 765**	**8 556**	**50%**	**61%**
European Union	*0.3%*	*0.1%*	*0.2%*	*501*	*521*	*74%*	*81%*

*Compound average annual growth rates.

Note: The assumed rates of population growth for 2009 to 2035 are the same for all three scenarios presented in this *Outlook*.

Sources: UNPD and World Bank databases; IEA analysis.

The increase in global population is expected to occur overwhelmingly in non-OECD countries, mainly in Asia and Africa, pushing the OECD's share of global population down from 18% today to just 16% in 2035. Both India (1.5 billion people) and China (1.4 billion people) will have larger populations than the entire OECD by 2035. The population of Russia falls marginally through the *Outlook* period. The population of the OECD increases, but by only 0.4% per year on average over 2009 to 2035. Most of the increase in the OECD occurs in the OECD Americas. Slower growth in population contributes to slower projected rates of increase in energy demand in the OECD than in the rest of the world.

The number of people living in urban areas worldwide is projected to grow by 1.9 billion, from 3.4 billion in 2009 to 5.3 billion in 2035. Rates of urbanisation remain an important determinant of energy demand, as energy use is related to income, and city and town dwellers in the developing world tend to have higher incomes and better access to energy services. The income effect usually outweighs any efficiency gains that come from higher density settlements in urban areas. Because of the rapid growth of cities and towns in the developing world, the pattern of energy use in towns will increasingly shape global energy use, though the rural/urban differential is expected to diminish somewhat. By contrast, city and rural residents in developed countries tend to enjoy similar levels of energy service.

Box 1.1 ● How does the IEA model long-term energy trends?

The projections for all three scenarios presented in *WEO-2011* are derived from the IEA's World Energy Model (WEM) – a partial equilibrium model designed to replicate how energy markets function over the medium and longer term. Developed over many years, the WEM consists of six main modules: (i) final energy demand (with sub-models covering residential, services, agriculture, industry, transport and non-energy use); (ii) power generation and heat; (iii) refining/petrochemicals and other transformation; (iv) oil, natural gas, coal, and biofuels supply; (v) CO_2 emissions; and (vi) investment.

The WEM is designed to analyse:

● Global energy prospects: These include trends in demand, supply availability and constraints, international trade and energy balances by sector and by fuel (currently through to 2035).

● Environmental effects of energy use: CO_2 emissions from fuel combustion are derived from the detailed projections of energy consumption.

● Effects of policy actions and technological changes: Scenarios and cases are used to analyse the impact of policy actions and technological developments on energy demand, supply, trade, investment and emissions.

● Investment in the energy sector: The model evaluates the investment requirements in the fuel-supply chain needed to satisfy projected energy demand. It also evaluates demand-side investment requirements.

Since the last *Outlook*, regional disaggregation of the WEM has been enhanced to take account of the accession of Chile, Estonia, Israel and Slovenia to the OECD. The model now produces energy-demand projections for eleven specific countries and a further fourteen regional groupings. On the supply side, projections for oil, gas, coal and biofuels are derived for all major producers.

Assumptions based on analysis of the latest developments in energy markets, the broader economy and energy and climate policy, are used as inputs to the WEM, together with huge quantities of historical data on economic and energy variables. Much of the data is obtained from the IEA's own databases of energy and economic statistics (www.iea.org/statistics). Additional data from a wide range of external sources is also used. These sources are indicated in the relevant sections of this report.

Energy prices

The evolution of energy prices is a key determinant of energy trends, as the actual prices paid by energy consumers affect how much of each fuel they wish to consume and how much it is worth investing in improving the efficiency of a particular technology used to provide an energy service. International energy prices are an exogenous (*i.e.* external) determinant of energy demand and supply in the World Energy Model, the model used to derive the *WEO-2011* projections (Box 1.1). However, the model is run in an iterative manner, adjusting prices so as to ensure demand and supply are in balance in each year of the projection period.

The assumed energy price paths should not be interpreted as forecasts. Rather, they reflect our judgement of the prices that would be needed to encourage sufficient investment in supply to meet projected demand over the *Outlook* period. Although the price paths follow smooth trends, this should not be taken as a prediction of stable energy markets. Prices will, in reality, deviate from these assumed trends, widely at times, in response to short-term fluctuations in demand and supply and to geopolitical events. The price assumptions differ across the three scenarios, reflecting the impact policy-driven reductions in demand would be likely to have.

The *WEO-2011* projections are based on the average retail prices of each fuel used in end-use, power generation and other transformation sectors. These prices are derived from assumptions about the international prices of fossil fuels and take into account domestic supply and demand conditions (as these continue to set prices in many cases, such as, for example, natural gas in the United States and coal in parts of China). End-user prices take account of any taxes, excise duties and carbon-dioxide emissions pricing, as well as any subsidies. In all scenarios, the rates of value-added taxes and excise duties on fuels are assumed to remain unchanged throughout the *Outlook* period. End-user electricity prices are based on the costs of generation, the costs of transmission and distribution and the costs associated with operating the electricity system and supplying customers (such as billing and metering). In addition, electricity prices in each region are increased or reduced to reflect taxes and subsidies for end-users and, in some cases, the inclusion in prices of a contribution towards renewables subsidies.

Oil prices

Oil prices have remained extremely volatile over the past year. By late September 2011, prices for benchmark Brent and West Texas Intermediate (WTI) futures were trading at around $110/barrel and $86/barrel respectively. Prices rose strongly over the second half of 2010 and through to April 2011, driven by tightening supply and demand fundamentals (in part, due to the loss of most production in Libya) and fears that the turmoil in parts of the Middle East and North Africa could spread to other major producers.

In this *Outlook*, oil prices are assumed to rise steadily to 2035 in all but the 450 Scenario, as rising global demand requires the development of increasingly expensive sources of oil (see Chapter 3). The level of prices needed to match oil supply and demand varies with the degree of policy effort made to curb demand growth (Figure 1.1). In the New Policies Scenario, the average IEA crude oil import price – a proxy for international oil prices – reaches $109/barrel (in real 2010 dollars) in 2020 and $120/barrel in 2035 (Table 1.4).[5, 6] In nominal terms, the price essentially doubles compared with current levels, to $212/barrel in 2035. In our Deferred Investment Case presented in Chapter 3, in which we assume that MENA upstream investment is one-third below the level called for in the New Policies Scenario over the period 2011 to 2020, real prices jump to $150/barrel, before falling back as production recovers. In the Current Policies Scenario, substantially higher oil prices than those in the New Policies Scenario are needed to balance supply with the higher level of demand. The crude oil price rises briskly, especially after 2020, reaching $118/barrel in 2020 and $140/barrel in 2035. In the 450 Scenario, by contrast, lower oil demand means there is less need to produce oil from costly fields higher up the supply curve in non-OPEC countries (see Chapter 6 for details of the drivers of oil demand in this scenario). As a result, the oil price is assumed to level off at about $97/barrel by 2015 and to maintain that level through to 2035. Importantly, in the 450 Scenario, administrative arrangements are assumed to be put in place to keep end-user prices for oil-based transport fuels at a level similar to the Current Policies Scenario. This assumption ensures that the lower international prices that result from policy action over and above that assumed in the Current Policies Scenario do not lead to a rebound in transport demand through lower end-user prices.

Figure 1.1 ● Average IEA crude oil import price

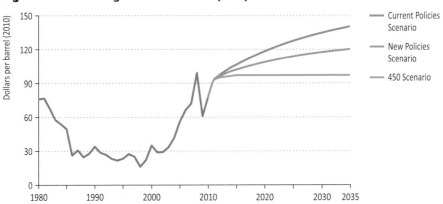

---

5. In 2010, the average IEA crude oil import price was $1.50/barrel lower than first-month WTI and $2.20/barrel lower than Brent.

6. The price assumptions utilised in *WEO-2011* differ from those in the IEA's *Medium-Term Oil and Gas Markets 2011 (MTOGM)* (IEA, 2011). The main reason is that the *MTOGM* assumes prices in line with those prevailing on the futures curve out to 2016 at the time the projections are made, in order to identify what this price trajectory (together with other assumptions, notably about GDP) would mean for market balances. So the projections are not intended to reflect a market equilibrium. By contrast, markets are assumed to balance in the *WEO-2011* projections. For this reason, the WEO and MTOMR projections are not strictly comparable.

Natural gas prices

Historically, natural gas prices in OECD countries have been closely correlated to oil prices through indexation clauses in long-term supply contracts or indirectly through competition between gas and oil products in power generation and end-use markets. In Europe, some two-thirds of the continent's gas is supplied under long-term contracts, whereby the gas price is indexed to oil prices, with a lag of several months (although a degree of indexation to spot gas prices has been introduced in some contracts). Prices in Asia are also predominately set under long-term contracts that are indexed to crude oil. However, in a growing number of markets, gas prices are set freely in a competitive gas market, an approach known as gas-to-gas competition. Prices are set in this way in North America, the United Kingdom and Australia and, increasingly, in continental Europe, accounting for some three-quarters of total OECD gas use.

Differences in pricing mechanisms inevitably lead to differences in the actual level of prices. When oil prices are high, as they are today, oil-indexed gas prices tend to be high. The level of gas prices under gas-to-gas competition depends on the supply/demand balance in each regional market, including the prices of all competing fuels. In the past two to three years, gas prices set this way have been significantly lower than oil-indexed prices both in the United States and in continental Europe. Nonetheless, differentials have narrowed in Europe as spot prices have risen, with the rebound in demand, and the price of alternative fuels (especially coal) has risen. Over the past few years, US natural gas prices have fallen relative to oil prices, because of a glut of gas caused by the boom in unconventional gas. We assume that North American gas prices recover slightly, relative to oil prices, over the projection period, largely reflecting our expectation that gas production costs there will tend to rise as production shifts to more costly basins. The growing share of liquefied natural gas (LNG) in global gas supply and increasing opportunities for short-term trading of LNG are expected to contribute to a degree of convergence in regional prices over the projection period, but significant price differentials between the United States, Europe and Japan are expected to remain, reflecting the relative isolation of these markets from one another and the cost of transport between regions.

In this year's *Outlook*, we have revised down our natural gas price assumptions in the three main scenarios compared with *WEO-2010*, because of improved prospects for the commercial production of unconventional gas. Although gas prices broadly follow the trend in oil prices, the ratio of gas to oil prices on an energy-equivalent basis remains below historical averages in all regions, particularly in North America (Figure 1.2). Gas prices are assumed to vary across the three scenarios in line with the degree of policy effort to curb growth in energy demand. In the New Policies Scenario, prices reach $12 per million British thermal units (MBtu) in Europe, $14/MBtu in the Pacific and $9/MBtu in North America by 2035 (in real 2010 dollars). In the Golden Age of Gas Scenario (GAS Scenario) presented in Chapter 4, prices are assumed to be up to $1.50/MBtu lower than in the New Policies Scenario, largely because of the scenario's more optimistic assumptions about future gas supply.

Table 1.4 • Fossil-fuel import price assumptions by scenario (dollars per unit)

	Unit	2010	New Policies Scenario					Current Policies Scenario					450 Scenario				
			2015	2020	2025	2030	2035	2015	2020	2025	2030	2035	2015	2020	2025	2030	2035
Real terms (2010 prices)																	
IEA crude oil imports	barrel	78.1	102.0	108.6	113.6	117.3	120.0	106.3	118.1	127.3	134.5	140.0	97.0	97.0	97.0	97.0	97.0
Natural gas imports																	
United States	MBtu	4.4	6.0	6.7	7.3	7.9	8.6	6.1	7.0	7.7	8.4	9.0	5.9	6.5	8.0	8.4	7.8
Europe	MBtu	7.5	9.6	10.4	11.1	11.7	12.1	9.8	11.0	11.9	12.6	13.0	9.4	9.8	9.8	9.7	9.4
Japan	MBtu	11.0	12.2	12.9	13.4	13.9	14.3	12.7	13.5	14.2	14.8	15.2	11.9	12.0	12.0	12.1	12.1
OECD steam coal imports	tonne	99.2	103.7	106.3	108.1	109.3	110.0	104.6	109.0	112.8	115.9	118.4	100.3	93.3	83.2	73.7	67.7
Nominal terms																	
IEA crude oil imports	barrel	78.1	114.3	136.4	159.8	184.9	211.9	119.1	148.2	179.1	211.9	247.2	108.7	121.8	136.4	152.9	171.3
Natural gas imports																	
United States	MBtu	4.4	6.8	8.4	10.3	12.5	15.1	6.9	8.7	10.9	13.2	16.0	6.6	8.2	11.2	13.3	13.8
Europe	MBtu	7.5	10.8	13.0	15.6	18.4	21.3	10.9	13.8	16.8	19.9	23.0	10.5	12.3	13.8	15.3	16.6
Japan	MBtu	11.0	13.7	16.2	18.9	21.9	25.2	14.2	17.0	20.0	23.4	26.8	13.4	15.1	16.9	19.1	21.4
OECD steam coal imports	tonne	99.2	116.2	133.5	152.0	172.2	194.2	117.2	136.8	158.6	182.6	209.0	112.4	117.2	117.0	116.1	119.5

Notes: Gas prices are weighted averages expressed on a gross calorific-value basis. All prices are for bulk supplies exclusive of tax. The US natural gas import price is used as a proxy for prices prevailing on the domestic market. Nominal prices assume inflation of 2.3% per year from 2010.

Figure 1.2 ● Ratio of average natural gas and coal import prices to crude oil prices in the New Policies Scenario

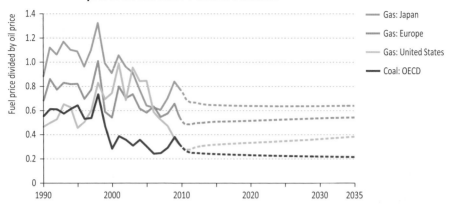

Note: Calculated on an energy-equivalent basis.

Steam coal prices

Historically, seaborne traded steam coal prices have broadly tended to follow oil and gas prices, reflecting the dynamics of inter-fuel competition and the importance of oil in the cost of transporting and mining coal, particularly in opencast mines. By contrast, prices for land-based traded steam coal do not always have a close link to oil and gas prices. Coal prices weakened relative to both oil and gas prices in the decade to 2010, partly as a result of differences in market conditions among the different fuels and growing environmental constraints on coal use in OECD countries. However, coal prices have recently rebounded, with surging demand in China and other emerging economies.

Our international steam-coal price assumptions (which drive our assumptions about coking coal and other coal qualities) vary markedly across the three scenarios presented in *WEO-2011*. Prices averaged $99 per tonne in 2010. In the New Policies Scenario, they are assumed to rise gradually throughout the projection period, reaching $110/tonne (in year-2010 dollars) by 2035. The increase is much less in percentage terms than that for oil or gas partly because coal production costs are expected to remain low and because coal demand flattens out by 2020. Prices are higher in the Current Policies Scenario, reaching $118/tonne by 2035. By contrast, they are significantly lower in the 450 Scenario, dropping to $93/tonne in 2020 and $68/tonne in 2035, as a result of a widespread and large-scale shift away from coal to cleaner fuels.

CO_2 prices

The pricing of CO_2 emissions (either through cap-and-trade schemes or carbon taxes) affects investment decisions in the energy sector by altering the relative costs of competing fuels. A number of countries have implemented emissions trading schemes to set prices for CO_2, while many others have schemes under development, with some being in an advanced stage of design. Other countries have introduced carbon taxes (taxes on fuels linked to their emissions) or are considering doing so.

In the Current Policies Scenario, carbon pricing through cap-and-trade is limited to the existing EU Emission Trading System, which covers the 27 member states of the European Union, and the New Zealand Emissions Trading Scheme. The price of CO_2 under the EU Emission Trading System is assumed to reach $30/tonne in 2020 (in year-2010 dollars) and $45/tonne in 2035 (Table 1.5). In the New Policies Scenario, it is assumed that measures to put a price on CO_2 are also established in Australia from 2012, Korea from 2015 and China from 2020, with the breadth of sectoral coverage varying from country to country.[7] Although neither the United States nor Canada introduces carbon pricing at the federal level in the New Policies Scenario, we have assumed that from 2015 onwards all investment decisions in the power sector in these countries factor in an implicit or "shadow" price for carbon; a shadow price is also assumed in Japan's power sector from 2015. In the countries that are assumed to adopt shadow pricing for carbon emissions, power projects are approved only if they remain profitable under the assumption that a carbon price is introduced. Given the uncertainty that surrounds future climate policy, many companies around the world already use such an approach as a means of ensuring they are well prepared against the contingency of the introduction of a carbon tax or a cap-and-trade programme.

Table 1.5 ● CO$_2$ price assumptions in selected regions by scenario
($2010 per tonne)

	Region	Sectors	2020	2030	2035
Current Policies Scenario	European Union	Power, industry and aviation	30	40	45
New Policies Scenario	European Union	Power, industry and aviation	30	40	45
	Korea	Power and industry	18	36	45
	Australia, New Zealand	All	30	40	45
	China	All	10	23	30
450 Scenario	United States, Canada	Power and industry	20	87	120
	European Union	Power, industry and aviation	45	95	120
	Japan, Korea, Australia, New Zealand	Power and industry *	35	90	120
	China, Russia, Brazil, South Africa	Power and industry**	10	65	95

*All sectors in Australia and New Zealand. **All sectors in China.

Note: In the New Policies Scenario, the United States, Canada and Japan are assumed to adopt a shadow price for CO_2 in the power sector as of 2015; it starts at $15/tonne, rising to $35/tonne in 2035.

---

7. China has announced plans to introduce city and provincial level pilot carbon emission trading schemes in the near future and to develop an economy-wide scheme within the current decade.

In the 450 Scenario, we assume that pricing of CO_2 emissions (either through cap-and-trade schemes or carbon taxes) is eventually established in all OECD countries and that CO_2 prices in these markets begin to converge from 2025, reaching $120/tonne in 2035. A growing number of key non-OECD countries are also assumed to put a price on CO_2 emissions. Although we assume no direct link between these systems before the end of the projection period, all systems have access to offsets, which is likely to lead to convergence of carbon prices to some degree. The CO_2 price in China is assumed to rise from $10/tonne in 2020 to $95/tonne in 2035. Similar CO_2 price levels are reached in 2035 in Russia, South Africa and Brazil.

Technology

The *WEO* projections of energy demand are sensitive to assumptions about improvements in the efficiency of current energy technologies and the adoption of new ones. While no completely new energy technologies, beyond those known today, are assumed to be deployed before the end of the projection period, it is assumed that existing end-use technologies become steadily more energy efficient. The pace of efficiency gains varies for each fuel and each sector, depending on our assessment of the potential for improvements and the stage reached in technology development and commercialisation. Similarly, technological advances are also assumed to improve the efficiency of producing and supplying energy. In most cases they are expected to lower the cost of energy supply and lead to new and cleaner ways of producing and delivering energy services.

The energy sector has a relatively slow rate of capital replacement in general, due to the long lifetime of many of the capital assets used for producing, delivering and using energy (Figure 1.3). As a result, it can take many years for more efficient technologies to become widely used, which limits the speed with which technological progress can lower the amount of energy needed to provide a particular energy service. Most cars and trucks, heating and cooling systems and industrial boilers will be replaced well before 2035. On the other hand, most existing buildings, roads, railways and airports, as well as many power stations and refineries, will still be in use then. This can have significant implications for efforts to combat climate change, as emissions that will come from much of the infrastructure that is currently in place or under construction can be thought of as "locked-in". In other words, it would be inordinately expensive to retire early or to retrofit that infrastructure, or allow it to stand idle. This does not mean that such emissions are unavoidable, but rather that a very strong policy intervention would be required – and require justification – before it would make economic sense to replace or modify that capacity (see Chapter 6).

Our assumptions for technology development and deployment vary by scenario, as both are heavily influenced by government policies and energy prices (IEA, 2010b). Technological change is fastest in the 450 Scenario, thanks to the effect of various types of government support, including economic instruments (such as carbon pricing, energy taxes and subsidies), regulatory measures (such as standards and mandates) and direct public-sector investment. Technological change is slowest in the Current Policies Scenario, because no new public policy actions are assumed. Yet, even in this scenario, significant technological improvements do occur, aided by higher energy prices. The extent to which energy assets are retired before the end of their normal lives (or the stock is modernised to reduce energy needs) is limited in the

Current Policies Scenario; it is greatest in the 450 Scenario. The reason for this, as indicated, is that retiring these assets before the end of their normal lives is very costly and would only occur as a result of very strong government policy incentives or regulations.

Figure 1.3 ● Typical lifetime of energy-related capital stock

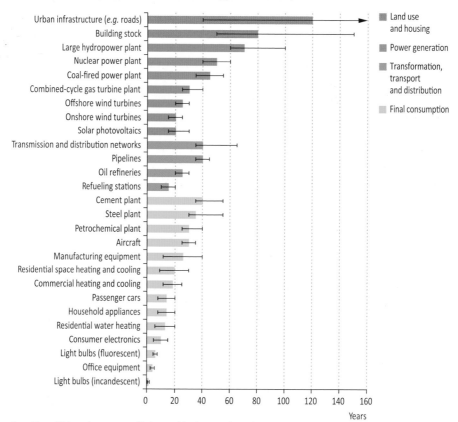

Note: The solid bars show average lifetimes while the range lines show typical variations.

Although no completely new technologies are assumed to be deployed in any of the three scenarios, some that are currently approaching the commercialisation phase are assumed to become available and to be deployed to some degree before the end of the projection period. For example, carbon capture and storage technology is expected to be deployed, on a very limited scale in the New Policies Scenario, but much more widely in the 450 Scenario (stimulated by stronger CO_2 price signals). Advanced biofuels, including those from ligno-cellulosic feedstock, are assumed to reach commercialisation, by around 2015 in the 450 Scenario and by 2025 in the Current Policies Scenario. Hydrogen fuel cells based on natural gas are expected to start to become economically attractive after 2020 in some small-scale power generation applications and, to a much lesser extent, in the transport sector. Exploration and production techniques for oil and gas are also expected to improve, lowering the unit production costs and opening up new opportunities for development.

CHAPTER 2

ENERGY PROJECTIONS TO 2035

Re-energising the global economy?

H I G H L I G H T S

- In the New Policies Scenario, our central scenario, global energy demand increases by 40% between 2009 and 2035. Demand grows for all energy sources. Oil demand increases by 18% and is driven by transport. Coal demand, dictated largely by non-OECD countries, increases for around the next ten years but then stabilises, ending around 25% higher than 2009. Absolute growth in natural gas demand is nearly equal to that of oil and coal combined. Nuclear power generation grows by more than 70%, led by China, Korea and India. Modern renewables grow faster than any other energy form in relative terms, but in absolute terms total demand is still not close to the level of any single fossil fuel in 2035.

- The focus of growth in both energy demand and supply switches away from the OECD. Nearly 90% of global energy demand growth is in non-OECD countries. OPEC oil production reaches more than half of the world total in 2035. Non-OECD countries account for more than 70% of global gas production in 2035, focused in the largest existing gas producers, such as Russia, the Caspian and Qatar. China consumes nearly 70% more energy than the United States in 2035, is the largest oil consumer and oil importer, and continues to consume nearly half of world coal production. Despite this, China's per-capita energy consumption is less than half the level of the United States in 2035.

- Inter-regional trade in oil increases by around 30% and is equivalent to more than half of world oil consumption in 2035. Trade in natural gas nearly doubles, with gas from Russia and the Caspian region going increasingly to Asia. India becomes the largest coal importer by around 2020, but China remains the determining factor in global coal markets. The OECD share of inter-regional fossil fuel trade declines from 42% in 2009 to 29% in 2035 and becomes more focused on natural gas and oil.

- Global investment in the energy supply infrastructure of $38 trillion is required over the period 2011 to 2035. Two-thirds of this is needed in non-OECD countries. The power sector claims nearly $17 trillion of the total investment. Oil and gas combined require nearly $20 trillion, increasing to reflect higher costs and a need for upstream investment to rise in the medium and long term. Coal and biofuels account for the remaining investment.

- In the New Policies Scenario, global energy-related CO_2 emissions increase by 20%, following a trajectory consistent with a long-term rise in the average global temperature in excess of 3.5°C. Around 45% of the emissions in 2035 are already locked-in, coming from capital stock which either exists now or is under construction and will still be operating in 2035.

Overview of energy trends by scenario

Nothing is as certain as change, and the energy landscape has changed significantly over the last year: turmoil in the Middle East and North Africa, the tsunami in Japan and consequent damage to the Fukushima Daiichi nuclear power plant, the release of China's 12th Five-Year Plan (2011 to 2015), new moratoria on shale-gas drilling and, despite some progress, the failure to achieve a legally binding agreement to limit global greenhouse-gas emissions. There is also the critical issue of the fragile health of the global economy and its uncertain future prognosis. But what impact has all of this change had on the outlook for our energy future?

WEO-2011 considers three scenarios, based on differing assumptions (see Chapter 1), with the results varying significantly, but all contributing their own messages to policy makers and analysts. The New Policies Scenario is our central scenario; it takes account of both existing government policies and declared policy intentions. The Current Policies Scenario looks at a future in which the government policies and measures enacted or adopted by mid-2011 remain unchanged. In contrast to the other scenarios, the 450 Scenario is an outcome-driven scenario, illustrating a global energy pathway with a 50% chance of limiting the increase in the average global temperature to 2°C. This would require the long-term concentration of greenhouse gases in the atmosphere to be limited to around 450 parts per million of carbon-dioxide equivalent (ppm CO_2-eq).

In the New Policies Scenario, world primary energy demand is projected to increase from 12 150 million tonnes of oil equivalent (Mtoe) in 2009 to 16 950 Mtoe in 2035, an increase of 40%, or 1.3% growth per year (Figure 2.1).[1] Global energy demand increases more quickly in the Current Policies Scenario, reaching 18 300 Mtoe in 2035, 51% higher than 2009, and representing average growth of 1.6% per year.

Figure 2.1 ● **World primary energy demand by scenario**

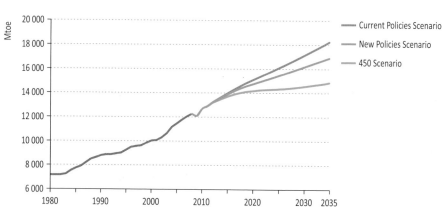

---

1. Compound average annual growth rate.

In the 450 Scenario, global energy demand still increases between 2009 and 2035, reaching 14 850 Mtoe in 2035, an increase of 23% or 0.8% per year. In 2035, energy demand in the 450 Scenario is 19% lower than in the Current Policies Scenario and more than 12% lower than in the New Policies Scenario, due mainly to the differing extent to which policies are implemented to improve energy efficiency.

In all scenarios, fossil fuels (oil, coal and natural gas) remain the dominant sources of energy in 2035 (Table 2.1), but their share of the energy mix varies: the share of fossil fuels decreases from 81% of world primary energy supply in 2009 to 80% in 2035 in the Current Policies Scenario, 75% in the New Policies Scenario and 62% in the 450 Scenario.

Table 2.1 ● **World primary energy demand by fuel and scenario** (Mtoe)

	1980	2009	New Policies Scenario		Current Policies Scenario		450 Scenario	
			2020	2035	2020	2035	2020	2035
Coal	1 792	3 294	4 083	4 101	4 416	5 419	3 716	2 316
Oil	3 097	3 987	4 384	4 645	4 482	4 992	4 182	3 671
Gas	1 234	2 539	3 214	3 928	3 247	4 206	3 030	3 208
Nuclear	186	703	929	1 212	908	1 054	973	1 664
Hydro	148	280	377	475	366	442	391	520
Biomass and waste*	749	1 230	1 495	1 911	1 449	1 707	1 554	2 329
Other renewables	12	99	287	690	256	481	339	1 161
Total	7 219	12 132	14 769	16 961	15 124	18 302	14 185	14 870

* Includes traditional and modern uses.

The outlook for each fuel differs across the scenarios, in some cases markedly (Figure 2.2). Compared with 2009, demand in 2035 for all forms of energy increases in the New Policies Scenario and the Current Policies Scenario, but the extent differs. In the New Policies Scenario, absolute growth in demand for natural gas is the strongest of all fuels. Global gas demand nearly reaches the level of coal demand by the end of the *Outlook* period. In the Current Policies Scenario, coal demand grows the most in absolute terms and overtakes oil to capture the largest single share of the energy mix before 2035. In the 450 Scenario, demand for coal and oil declines, compared with 2009, while the outlook is generally more positive for natural gas, nuclear power and renewables.

The make-up of global energy demand in 2035 across the different scenarios is illustrated in Figure 2.3. The strength with which energy and environmental policies are implemented in the different scenarios has a particularly strong impact on the outlook for coal and renewables, but in opposite directions. In the Current Policies Scenario, coal represents nearly 30% of the energy mix and renewables 14% in 2035. In the 450 Scenario, the share of coal in total energy demand declines to less than 16% in 2035, while that of renewables increases to 27%.

Figure 2.2 ● World primary energy demand by fuel and scenario, 2009 and 2035 (Mtoe)

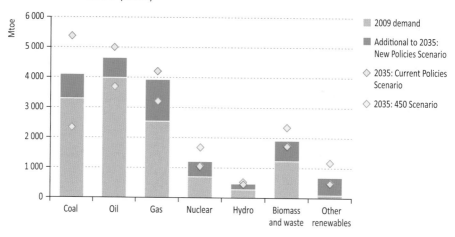

In 2010, global energy-related emissions of carbon dioxide (CO_2), the main component of the energy-related greenhouse gases contributing to climate change, were 30.4 gigatonnes (Gt), more than 5% higher than in 2009. The three scenarios have a dramatically different impact on the future level of world energy-related CO_2 emissions (Figure 2.4). By 2035, global energy-related CO_2 emissions are projected to increase to 36.4 Gt in the New Policies Scenario and 43.3 Gt in the Current Policies Scenario, but decrease to 21.6 Gt in the 450 Scenario (see Chapter 6 for more on climate change and the 450 Scenario). In terms of CO_2 emissions, cautious implementation of announced policies, as in the New Policies Scenario, achieves only one-third of the cumulative change required over the *Outlook* period to achieve a trajectory consistent with limiting the average global temperature rise to 2°C.

Figure 2.3 ● Shares of energy sources in world primary demand by scenario, 2035

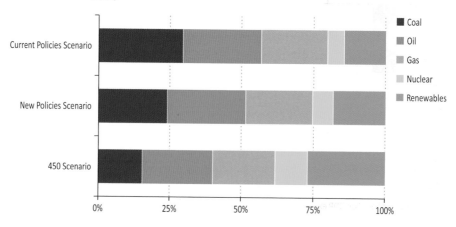

Figure 2.4 ● World energy-related CO$_2$ emissions by scenario

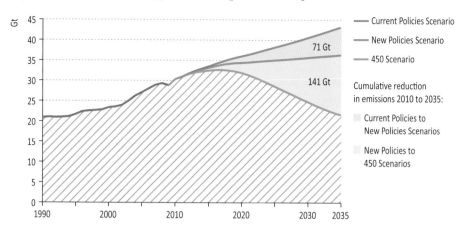

Global energy intensity – the amount of fuel needed to generate each unit of gross domestic product (GDP) – is projected to decline across all *WEO-2011* scenarios over the *Outlook* period, and in all cases by more than the average level observed from 1985 to 2009 (Figure 2.5). In 2035, global energy intensity is projected to have decreased by 36% in the New Policies Scenario, 31% in the Current Policies Scenario and 44% in the 450 Scenario. The main change across scenarios is the extent to which energy intensity reduces in non-OECD countries, particularly China and, to a lesser extent, India and Russia. Despite being interrupted three times in the last decade, the decline in energy intensity is a longstanding trend, driven primarily by improved energy efficiency and structural changes in the global economy.

Figure 2.5 ● Average annual percentage change in global primary energy intensity by scenario (2009-2035) and region

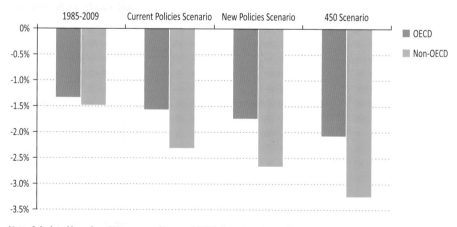

Note: Calculated based on GDP expressed in year-2010 dollars at market exchange rates.

Energy trends in the New Policies Scenario

This section concentrates on the results of the New Policies Scenario, our central scenario.[2] In doing so, it details the impact of existing and planned policies, when implemented cautiously, on the key energy trends in demand, supply, trade, investment and emissions in the period up to 2035.

Primary energy mix

Major events of the last year have had an impact on short and medium term energy trends, but have done little to quench the world's increasing thirst for energy in the long-term. Based on preliminary data, total primary energy demand is estimated to have increased by 4.7% in 2010, easily recouping the 1.1% decline of 2009 and representing a very large rebound in demand by historical standards. Short-term economic uncertainty plays a prominent role in the near-term outlook for energy demand. However, while slower economic growth than assumed (see Chapter 1) in the next couple of years would dampen demand growth temporarily, it would have little impact on longer-term trends, especially if the downturn were to be followed by a period of accelerated growth (Box 2.1).

In the New Policies Scenario, total primary energy demand increases by 40% over the *Outlook* period to reach more than 16 950 Mtoe in 2035 (Table 2.2). While demand increases consistently throughout the *Outlook* period, the growth rate slows from an average of 1.8% per year in the period to 2020 to 0.9% per year from 2020 to 2035. This shift is attributed largely to a tempering in global economic and population growth, and improved levels of energy efficiency. Many large OECD economies, such as the United States and Japan, see very modest energy demand growth, averaging 0.2% and 0.05% per year respectively. This contrasts with many large non-OECD economies: China and India experience energy demand growth of 2% and 3.1% per year respectively.

Table 2.2 ● **World primary energy demand by fuel in the New Policies Scenario** (Mtoe)

	1980	2009	2015	2020	2030	2035	2009-2035*
Coal	1 792	3 294	3 944	4 083	4 099	4 101	0.8%
Oil	3 097	3 987	4 322	4 384	4 546	4 645	0.6%
Gas	1 234	2 539	2 945	3 214	3 698	3 928	1.7%
Nuclear	186	703	796	929	1 128	1 212	2.1%
Hydro	148	280	334	377	450	475	2.1%
Biomass and waste	749	1 230	1 375	1 495	1 761	1 911	1.7%
Other renewables	12	99	197	287	524	690	7.8%
Total	7 219	12 132	13 913	14 769	16 206	16 961	1.3%

* Compound average annual growth rate.

---

2. Annex A provides detailed projections of energy demand by fuel, sector and region for all three scenarios.

Box 2.1 ● The impact of lower near-term economic growth on energy demand

At the time of writing, the latest data and forecasts reveal an increasingly pessimistic assessment of global economic activity. They point towards a slowdown in global growth, attributed largely to those advanced economies that are facing fiscal and financial sector balance-sheet problems. This worsening outlook means that there is a risk of weaker growth in GDP than assumed in the New Policies Scenario – over the next couple of years at least. This would inevitably affect the outlook for global energy demand and, in the light of this uncertainty, we have prepared a Low GDP Case to test the sensitivity of global energy demand to weaker economic growth in the period to 2015. This deliberately takes a more negative view of near-term economic growth than the New Policies Scenario, but is in line with the latest revision of economic forecasts by international organisations.

In the Low GDP Case, global GDP is assumed to grow by 3.8% in 2011 (against 4.3% in the New Policies Scenario), 4.1% in 2012 (4.4%) and then gradually to return to the same level as the New Policies Scenario after 2015 – following the path for that scenario for the remainder of the *Outlook* period. This results in global GDP being 0.8% lower than in the New Policies Scenario in 2015, and 0.7% lower in 2035. Most of the slowdown occurs in OECD countries. All other assumptions made in the New Policies Scenario, including energy prices, are unchanged in the Low GDP Case, in order to isolate the effects.

Unsurprisingly, in the Low GDP Case world primary energy demand grows more slowly than in the New Policies Scenario, reaching 13 860 Mtoe in 2015 – 50 Mtoe, or 0.4%, lower. By 2035, primary energy demand in the Low GDP Case is 60 Mtoe, also 0.4%, lower. The proportional impact on energy demand is greatest in OECD economies, with the United States experiencing the largest reduction. However, the relatively less energy-intensive nature of the OECD economies mitigates the overall impact of this Low GDP Case on global energy demand. Furthermore, the fact that projected energy demand growth in the New Policies Scenario is expected to be driven strongly by non-OECD economies means that a slowdown in the near-term economic prospects of the OECD economies does not have a large impact on the global picture, particularly when looking out to 2035.

In reality, the longer term impact on global energy demand of such a slowdown could be expected to be even lower than estimated here. This is because, if the constraint that other assumptions do not change is relaxed, the weaker economic growth in the Low GDP Case would be expected to put downward pressure on energy prices and so mitigate, in part, the reduction in energy demand projected here. In addition, an economic slowdown might well be followed by a period of more rapid economic growth. This is not allowed for in the Low GDP Case, but has often been observed in the past.

The global recession and high prices have had only a relatively small impact on demand for oil and it is likely to be temporary (Spotlight considers the economic impacts of high oil prices). In the New Policies Scenario, demand increases for all energy sources over the *Outlook* period, but the pace and trend varies. Fossil fuels account for 59% of the increase in global energy demand from 2009 to 2035, or an additional 2 850 Mtoe. This growth in demand comes in spite of increasing fuel prices and additional policy measures, such as energy efficiency measures in Brazil, China, India and Russia. In the New Policies Scenario, global demand for oil increases from 84 million barrels per day (mb/d) in 2009 (almost 87 mb/d in 2010) to 99 mb/d in 2035 (Figure 2.6). Notwithstanding, this growth, its share in the primary energy mix actually decreases from 33% in 2009 to 27% in 2035. Significantly higher average oil prices and switching away from oil in the power generation and industrial sectors do not offset increasing demand in the transport sector, where demand is relatively price inelastic and substitution possibilities are limited (see Chapter 3). All of the net growth in global oil demand in the New Policies Scenario comes from the transport sector in the non-OECD countries, growth being particularly strong in India, China and the Middle East. In short, despite energy security and climate concerns, oil demand continues to grow and the global economy relies on oil more than on any other fuel (see Chapter 3 for the oil market outlook).

Figure 2.6 ● **World primary energy demand by fuel in the New Policies Scenario**

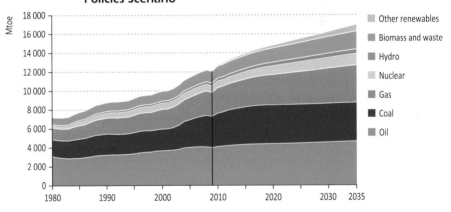

Global coal demand increased, on average, by slightly more than 1% per year from 1980 to 2000, but then grew by over 4% per year from 2000 to 2009. This acceleration in demand growth over the last decade was driven overwhelmingly by China and, to a lesser extent, India and other emerging economies. In the New Policies Scenario, demand for coal increases by 25%, to reach 5 860 million tonnes of coal equivalent (Mtce)[3] in 2035, but the pace of this growth differs markedly over time. In the period to 2020, global coal demand experiences strong growth but it then slows rapidly, with the level of global demand remaining broadly flat for much of the rest of the *Outlook* period, before then flirting tentatively with decline as 2035 approaches. The share of coal in the global energy mix peaks at 28% early in the *Outlook* period, but then declines to 24% by 2035.

---

3. 1 Mtce is equal to 0.7 Mtoe.

What are the economic impacts of high oil prices?

The economic impacts of high oil prices differ depending on the nature of the price increase, the oil intensity and import dependence of an economy, economic conditions at the time, and the government and consumer response. In the past, high oil prices have often been associated with economic recession, as at the time of the oil shocks of the 1970s, but this association has weakened in more recent times. While there are many reasons for this, probably the most important is that the oil intensity of the global economy has fallen significantly.

Higher oil prices increase production costs for many goods and services, and thereby put pressure on price levels generally. Price inflation is manifest both directly through increased fuel prices (including natural gas, if gas contracts are linked to oil prices) and indirectly as the prices of other goods rise to reflect higher input costs. The inflationary impact is more pronounced in relatively energy-intensive economies, typically developing countries. Higher costs and inflation, and lower profit margins, can damage demand and employment and put upward pressure on wages. Such changes in economic activity in turn influence financial markets, interest rates and exchange rates. Depending on the speed, scale and expected duration of the oil price increases, consumer and producer confidence can be impaired materially.

While oil-import bills of key consuming countries are at high levels, net oil exporters are benefitting from a redistribution of income. The majority of oil exporting nations run balance of payments current account surpluses that grow as oil revenues rise, unless they are recycled through increased imports. For every country experiencing a rising current account surplus, there will be others with widening current account deficits or declining surpluses. The propensity to spend has historically been much lower in net oil-exporting countries compared with oil-importing countries. This means that overall global demand is likely to decline as a result of this redistribution, even though many oil-exporting countries have increased public spending in the light of higher oil revenues. The sharp increase in public expenditure means that the oil price required to balance their budgets has risen and a sudden drop significantly below this level could result in abrupt action to cut spending (see Chapter 3).

Sound economic policies may not eliminate the adverse impacts of high oil prices on net oil-importing economies, but they can moderate them. Inflationary pressures derived from high oil prices emphasise the need to adopt a coherent monetary policy that anchors price expectations effectively. But consumers and businesses can also be tempted to try to ride the effects of higher oil prices, by reducing short-term saving or increasing the use of credit, and governments may do so through an offsetting tax cut or by subsidising the price of oil and other energy products. However, the ability of governments to act in this way depends critically on their fiscal starting point and such actions quickly become costly and difficult to undo. The G-20, APEC and other countries recognised the shortcomings of such action by committing to phase out inefficient fossil fuel subsidies that promote wasteful consumption. The most obvious action that oil-importing countries can take to reduce the impact of high prices on their economies is to reduce their reliance on oil.

China will be pivotal in determining the evolution of global coal markets (the Spotlight later in this chapter considers China's role in traded coal markets). While its demand for coal increases apace during the period covered by its 12th Five-Year Plan (Box 2.2), the reorientation of the energy mix set in train by the Five-Year Plan provides the groundwork for the slowdown in coal demand growth around 2020. Later in the *Outlook* period, interventions by governments, such as carbon pricing, have an impact on coal demand, as does the increasing preference in many countries for natural gas in industrial uses and the power sector (see Part C for the coal market outlook).

Box 2.2 ● China's 12th Five-Year Plan (2011-2015)

China's 12th Five-Year Plan (the Plan) covers the period 2011 to 2015 and, while focused on China's domestic energy landscape, it has profound implications for the global energy picture. It concentrates on energy efficiency and the use of cleaner energy sources to mitigate the effects of rapidly rising energy demand, which would otherwise increase China's dependence on imports and exacerbate local pollution. Reducing energy and carbon intensity are two key goals of the Plan. Targets are set to cut energy consumption per unit of GDP by 16% by 2015 and to cut CO_2 emissions per unit of GDP by 17%. The CO_2 intensity target, included for the first time, is in line with China's Copenhagen pledge to achieve 40% to 45% reductions below 2005 levels by 2020. The Plan also establishes new targets intended to diversify the primary energy mix. The proportion of non-fossil fuels in primary energy consumption is targeted to reach 11.4% by 2015, up from 8.3% in 2010. Natural gas, nuclear and renewables are expected to be aggressively promoted. The Plan also includes targets to reduce emissions of major pollutants by between 8% and 10%. Many of the policies reflected in the 12th Five-Year Plan are expected to be taken forward through other detailed plans and targets. In some cases, *WEO-2011* anticipates such targets, based on credible reports at the time of writing, to provide further depth to the analysis.

Compared with the previous year, global gas demand decreased by 2% in 2009, but it is estimated to have rebounded strongly in 2010, increasing by nearly 7%. Absolute growth in natural gas demand continues to exceed that of all other fuels, and is nearly equal to that of oil and coal combined over the *Outlook* period, increasing by 54% to reach 4 750 billion cubic metres (bcm) in 2035. The flexibility of natural gas as a fuel, together with its environmental and energy security attributes, makes it an attractive fuel in a number of countries and sectors. For example, in the United States, where unconventional gas supply is increasingly abundant, demand grows in power generation. In Asia, demand increases across a range of sectors, including industry and residential, and in the Middle East power generation and gas-to-liquids (GTL) see notable demand growth. The strong growth in demand for natural gas is supported by relatively low prices, policy support in China and substitution for nuclear capacity in some cases. The share of natural gas in the global energy mix increases from 21% in 2009 to 23% in 2035. Demand for natural gas rises close to the level of coal demand over the *Outlook* period and is near to overtaking it, to become the second most important fuel in the primary energy mix, by 2035 (Figure 2.7) (see Chapter 4 for the natural gas market outlook).

Figure 2.7 ● Shares of energy sources in world primary energy demand in the New Policies Scenario

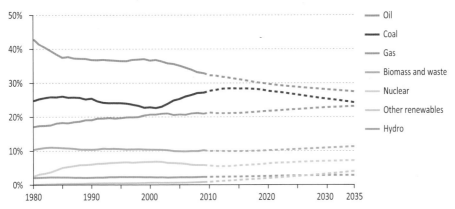

Attitudes toward nuclear power have evolved over the last year, though it is as yet far from certain exactly how energy policies and plans will change. While many countries are reassessing the future role of nuclear power, expectations over the *Outlook* period have diminished only slightly. In the New Policies Scenario, nuclear power generation increases by 73% over the *Outlook* period, growing from 2 700 terawatt-hours (TWh) in 2009 to 4 660 TWh in 2035, only 5% lower than projected in *WEO-2010*.

Non-OECD countries are responsible for nearly 80% of the global increase in installed nuclear power generating capacity over the *Outlook* period. Building over 110 gigawatts (GW), China alone accounts for nearly half of the global increase in nuclear capacity to 2035, a projection that has increased in response to the 12th Five-Year Plan. Nuclear capacity in the OECD is expected to increase by 53 GW, reaching 380 GW by the end of the *Outlook* period. This projection is 23 GW lower than last year, for a number of reasons, including reduced nuclear competitiveness, fewer than expected capacity additions and some plant delays. Nuclear plant retirements in Germany also now occur significantly earlier in the *Outlook* period. The share of nuclear power in total primary energy demand increases slightly over the projection period, from 6% in 2009 to 7% in 2035, lower than the 8% projected in *WEO-2010* (see Chapter 5 for the power sector outlook and Chapter 12 for the implications of less nuclear power).

Demand for modern renewable energy – including wind, solar, geothermal, marine, modern biomass and hydro – grows from 860 Mtoe in 2009 to 2 365 Mtoe in 2035. The share of modern renewable energy in the primary energy mix increases from 7% in 2009 to 14% in 2035. Despite this strong growth, global demand for all sources of renewable energy collectively will still not be close to that for any single fossil fuel in 2035. Demand for renewable energy increases substantially in all regions, but different types achieve greater or lesser penetration in different locations. For example, hydropower accounts for more than 60% of electricity generation in Latin America in 2035, while wind power accounts for nearly 20% of generation in the European Union. In 2035, the share of renewables (including traditional biomass use) in total primary energy demand is 23% in the European Union, 20% in India, 16% in the United States and 13% in China.

Regional trends

Primary energy demand in non-OECD countries grows from 6 600 Mtoe in 2009 to 10 800 Mtoe in 2035, representing nearly 90% of all energy demand growth over the *Outlook* period (Figure 2.8). Non-OECD countries represent 54% of global energy demand in 2009 and this grows to 64% in 2035. While average annual growth in energy demand in non-OECD countries is 1.9% over the *Outlook* period, growth slows from 2.7% per year to 2020, to 1.4% per year from 2020 to 2035. Higher energy demand growth in non-OECD countries relative to the OECD is consistent with the faster rates of growth of population, economic activity and urbanisation expected over the *Outlook* period. For example, non-OECD countries are expected to represent 84% of the global population in 2035 (up from 82% in 2009), and the proportion of people residing in urban areas in these countries increases from 44% to 57% (see Chapter 1 for population assumptions). While non-OECD countries account for all of the global growth in coal and oil demand over the *Outlook* period, they also account for 73% of the global increase in nuclear power, 55% of the increase in non-hydro renewable energy and 88% of the increase in hydropower generation.

Figure 2.8 ● **World primary energy demand by region in the New Policies Scenario**

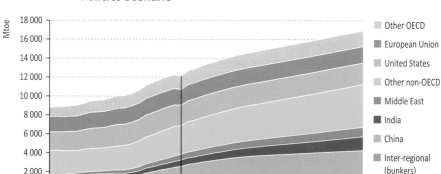

China has consolidated its position as the world's largest energy consumer (Table 2.3). From consuming less than half as much energy as the United States in 2000, it now consumes slightly more and is projected to consume nearly 70% more than the United States in 2035. In the New Policies Scenario, China accounts for more than 30% of global growth in energy demand from 2009 to 2035. Its share of global energy demand, having increased from 11% in 2000 to 19% in 2009, is projected to be 23% in 2035. China's per-capita energy consumption overtakes the world average early in the *Outlook* period, but is still at less than half the level of the United States in 2035.

China overtakes the United States in terms of oil imports shortly after 2020 and becomes the largest oil consumer in the world around 2030, consuming 15 mb/d by 2035, nearly double the level of 2009. While the share of coal in China's primary energy mix declines to one-half of the total in 2035, the absolute amount of coal consumed, at 2 800 Mtce, is still nearly two-and-a-half times more than that of the OECD. Even with the signalled policy shift away from coal, China accounts for 48% of cumulative world coal consumption over the *Outlook* period, or around 74 300 Mtce. China also leads the world by some margin in the supply of renewable energy in 2035.

Table 2.3 ● **World primary energy demand by region in the New Policies Scenario** (Mtoe)

	1980	2000	2009	2015	2020	2030	2035	2009-2035*
OECD	**4 067**	**5 292**	**5 236**	**5 549**	**5 575**	**5 640**	**5 681**	**0.3%**
Americas	2 102	2 695	2 620	2 780	2 787	2 835	2 864	0.3%
United States	*1 802*	*2 270*	*2 160*	*2 285*	*2 264*	*2 262*	*2 265*	*0.2%*
Europe	1 501	1 765	1 766	1 863	1 876	1 890	1 904	0.3%
Asia Oceania	464	832	850	906	912	914	912	0.3%
Japan	*345*	*519*	*472*	*498*	*490*	*481*	*478*	*0.0%*
Non-OECD	**2 981**	**4 475**	**6 567**	**8 013**	**8 818**	**10 141**	**10 826**	**1.9%**
E. Europe/Eurasia	1 242	1 001	1 051	1 163	1 211	1 314	1 371	1.0%
Russia	*n.a.*	*620*	*648*	*719*	*744*	*799*	*833*	*1.0%*
Asia	1 066	2 172	3 724	4 761	5 341	6 226	6 711	2.3%
China	*603*	*1 108*	*2 271*	*3 002*	*3 345*	*3 687*	*3 835*	*2.0%*
India	*208*	*460*	*669*	*810*	*945*	*1 256*	*1 464*	*3.1%*
Middle East	114	364	589	705	775	936	1 000	2.1%
Africa	274	505	665	739	790	878	915	1.2%
Latin America	284	432	538	644	700	787	829	1.7%
Brazil	*114*	*185*	*237*	*300*	*336*	*393*	*421*	*2.2%*
World**	**7 219**	**10 034**	**12 132**	**13 913**	**14 769**	**16 206**	**16 961**	**1.3%**
European Union	*n.a.*	*1 683*	*1 654*	*1 731*	*1 734*	*1 724*	*1 731*	*0.2%*

*Compound average annual growth rate.
**World includes international marine and aviation bunkers (not included in regional totals).

Energy demand in India was higher than that of Russia in 2009, making it the third-largest energy consumer in the world. Despite this, many millions of people in the country still have no access to modern energy services (see Chapter 13). India's energy demand more than doubles between 2009 to 2035, rising from less than 700 Mtoe to nearly 1 500 Mtoe, accounting for 16% of the increase in global energy demand over the *Outlook* period. At an average of 3.1% per year, India's rate of energy demand growth is very high over the *Outlook* period. The increase in energy demand is primarily driven by rapid economic growth and population growth. Coal continues to dominate the energy picture in India, representing around 42% of total energy demand in 2035, practically the same share as 2009. Demand for natural gas trebles, but it still plays a relatively modest role in the overall energy mix, at around 10% in 2035. The total number of passenger light-duty vehicles (PLDVs) per 1 000 people increases from 10 in 2009 to over 100 in 2035, driving a four-times increase in demand for oil in the transport sector, which reaches more than 190 Mtoe (3.9 mb/d) – slightly more than half of total primary oil demand in 2035 (7.4 mb/d). Electricity demand per capita nearly trebles over the *Outlook* period, driving increased energy demand for power generation.

Primary energy demand in Russia decreased by 6% in 2009, but preliminary data suggest it has regained most of this ground in 2010. In the New Policies Scenario, energy demand in Russia increases by 28% over the *Outlook* period (estimated increase of 21% from 2010 to 2035), to reach around 830 Mtoe in 2035. The rise in energy demand is constrained by a slight decline in population, and by energy efficiency and pricing policies that begin to tap into Russia's large potential for energy saving. While Russia's energy intensity is projected to decline by 49% over the *Outlook* period (based on GDP in year-2010 dollars at market exchange rates), it is still nearly three-times the average level in the OECD in 2035 (Figure 2.9). Natural gas continues to dominate Russia's energy mix, representing 52% of total primary energy demand in 2035. While there is a large relative increase in the use of renewables in Russia, they still only represent 7% of its primary energy mix in 2035, much lower than the 18% share in global primary energy demand (see Part B for the energy outlook for Russia).

Figure 2.9 ● *Energy intensity in selected countries and regions in the New Policies Scenario, 1990-2035*

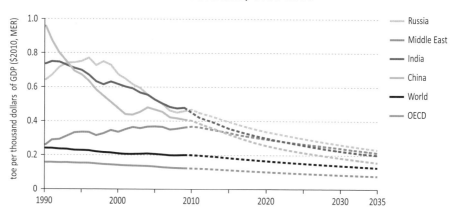

Primary energy demand in the Middle East increases by 70%, to reach 1 000 Mtoe in 2035. Oil demand in the Middle East grows by an average of 1.3% per year to reach 9.2 mb/d in 2035, driven largely by the road-transport sector. The Middle East accounts for 17% of world oil demand growth from 2009 to 2035. Despite this demand growth, the share of oil in the energy mix diminishes from 51% in 2009 to 43% by 2035, relinquishing share primarily to natural gas. Demand for natural gas overtakes that of oil before 2025. Increases in demand for natural gas are highest in power generation, desalinisation and in use as a feedstock for petrochemicals. Nuclear power also emerges gradually over the *Outlook* period.

In Latin America, primary energy demand is projected to grow on average by 1.7% per year and to reach 830 Mtoe by 2035. Demand for renewables in the power generation sector in Latin America doubles over the *Outlook* period, to reach nearly 140 Mtoe in 2035, and represents over 60% of total demand in this sector. Biofuels use in road transport increases by more than 4% per year, to reach 1 mb/d by 2035, representing around 25% of fuel consumption in the sector. Primary energy demand in Brazil grows to 420 Mtoe in 2035, representing more than half of the regional total. Consumption of natural gas in Brazil increases by nearly 6% per year, to reach around 90 bcm in 2035, stimulated by greater levels of domestic gas supply. Over the *Outlook* period, the share of natural gas in the power sector in Brazil grows from 5% to 20% and its share in industry grows from 10% to 19%.

In Africa, the population is projected to grow by 71% from 2009 to 2035 but there is only a 38% increase in energy demand over the same period. This means that, while Africa's share of the global population increases from 15% in 2009 to 20% in 2035, its share of global energy demand declines fractionally to 5.4% over the same timeframe. Furthermore, one-third of total energy demand in Africa in 2035 takes the form of traditional biomass use in the residential sector. A large share of the population in countries such as Nigeria, Ethiopia and Tanzania continue to live without access to electricity and clean cooking facilities (see Chapter 13 on energy access).

In the New Policies Scenario, primary energy demand in OECD countries grows from around 5 200 Mtoe in 2009 to nearly 5 700 Mtoe in 2035, an increase of 8%. Oil demand declines by 0.6% per year on average, going from 42 mb/d in 2009 to 36 mb/d in 2035, mainly as a result of fuel economy policies and saturating vehicle markets. Many OECD countries see coal demand decline significantly as policies to reduce carbon emissions take effect, particularly after 2020. By 2035, OECD countries consume 22% less coal than in 2009. Natural gas is an important fuel for power generation in the OECD in 2035 and also plays a prominent role in the industrial, service and residential sectors. Nuclear power use in the OECD increases by 23% over the *Outlook* period.

While the United States remains the second-largest energy consumer in the world, its total energy demand, at around 2 270 Mtoe in 2035, is only slightly higher than in 2009. Oil demand declines, mainly as a result of improved fuel economy in the road-transport sector, but is still projected to be 14.5 mb/d in 2035 – slightly less than 15% of global oil demand. Coal demand in the United States also declines over the *Outlook* period, but by less than expected in *WEO-2010*. The absence of a carbon price means that there is less incentive to move away from coal-powered plants in power generation, but the prospect of such a policy being introduced in the future means that a shadow price of carbon (see Chapter 1) does affect new investment decisions over the *Outlook* period. Thanks to improving supply prospects and lower expected prices, natural gas demand increases over the *Outlook* period, albeit at a modest pace. The outlook for nuclear power, which has declined slightly from *WEO-2010*, continues to be influenced significantly by the availability of policy support. Consumption of renewable energy increases by 4% per year over the *Outlook* period and represents 16% of the energy mix in 2035.

Total primary energy demand in the European Union increases by less than 5% from 2009 levels by 2035, with all of the growth happening before 2020. Demand for coal decreases by nearly 50% over the *Outlook* period and its share in the energy mix declines from 16% in 2009 to 8% in 2035 (Figure 2.10). In contrast, natural gas demand increases by 24% over the *Outlook* period, largely in the power sector but also in industry and heating in buildings. The share of natural gas in the primary energy mix increases from 25% in 2009 to 30% in 2035. Energy efficiency policies play an important role in the decline in demand for oil in the transport sector. However, oil still dominates energy consumption in the transport sector in 2035, accounting for 83% of the total. In the European Union, attitudes towards nuclear power have seen a negative shift over the past year. This shift results in several nuclear plant retirements occurring earlier in the *Outlook* period than projected in *WEO-2010*, principally in Germany in the period 2020 to 2025. Installed nuclear capacity is estimated to be 129 GW in 2035, around 8 GW (6%) lower than projected in *WEO-2010*, with the gap being filled largely by natural gas and renewables. Consumption of renewable energy sees strong growth of around 3.5% per year over the *Outlook* period. The share of renewables in the energy mix increases from 10% in 2009 to 23% in 2035, with wind power and, to a lesser extent, solar photovoltaic (PV) becoming prominent.

Figure 2.10 ● **Energy mix in selected countries and regions in the New Policies Scenario, 2035**

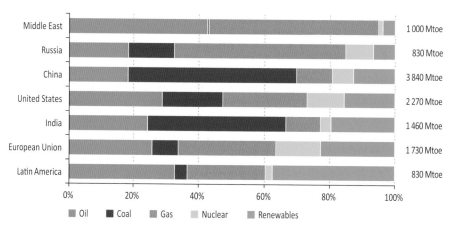

The energy outlook for Japan has been impacted by a number of factors over the last year, many derived from the earthquakes and tsunami that struck the country and the resultant damage to the Fukushima Daiichi nuclear power plant. Total energy demand is expected to continue to recover in the short term, before resuming a longer-term declining trend, and reaches almost 480 Mtoe in 2035. An upward revision to population estimates and a less certain outlook for climate-related policies contribute to total energy demand being 2% higher in 2035 than in *WEO-2010*. The share of fossil fuels in Japan's energy mix declines from 81% in 2009 to 70% in 2035 but, within this, demand for coal and oil drops significantly, while demand for natural gas increases. Demand for oil declines by more than one-quarter, underpinned by a 31% decline in energy demand in the transport sector. This results from market saturation in road transport, improved fuel efficiency and increased adoption of hybrids (32% of passenger light-duty vehicle sales in 2035) and electric vehicles and plug-in hybrids (14% of sales in the same year). Nuclear power accounts for 33% of electricity generated in 2035, with natural gas accounting for 30% and coal 15%, though these projections must be regarded as particularly tentative, pending the outcome of public and policy debate in Japan.

Sectoral trends

Energy demand in the power sector was affected by the global recession, but only slightly, and continues to grow more strongly than in any other sector in the New Policies Scenario. Over the *Outlook* period, energy demand in this sector increases by 57%, from about 4 600 Mtoe in 2009 to just below 7 200 Mtoe in 2035, and accounts for over half of all growth in primary energy demand (see Chapter 5 for the power sector outlook). Of the energy demand growth for power generation, 87% is in non-OECD countries. A slightly improved economic outlook and lower gas prices are important drivers of increased energy demand in this sector, relative to *WEO-2010*. A negative shift in stance toward nuclear power in some countries and the prospects for carbon pricing in many countries are also important influences. The power sector accounts for 38% of global primary energy demand in 2009 and this share increases to 42% in 2035.

There is a slightly stronger outlook for fossil fuels in the power sector compared with last year. Projected demand for coal has increased, as attitudes towards nuclear power and carbon pricing shift, and so has demand for natural gas, mainly as a result of China's 12th Five-Year Plan and the presumption of lower gas prices. While coal demand continues to be heavily focused on power generation, demand from the power sector accounts for less than half of total natural gas demand in 2035 (Figure 2.11). Over the *Outlook* period, the increase in demand for coal in power generation in non-OECD countries is more than three-times as great as the corresponding decline in the OECD.

Figure 2.11 ● **World primary energy demand by fuel and sector in the New Policies Scenario, 2035**

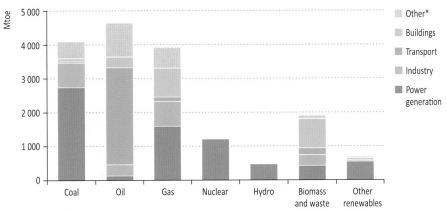

*Other includes other energy sector, agriculture and non-energy use.

The outlook for nuclear power has been affected by the Fukushima Daiichi accident and the subsequent policy response to it in a number of countries. Global use of nuclear power is 5% lower than projected in *WEO-2010*, though it still increases, to 4 660 TWh, in 2035. The reduction in the OECD is driven by changes in Japan, Europe and the United States. However, the outlook in non-OECD countries has been impacted much less, with existing plans delayed rather than discarded. The increase in the use of nuclear power in non-OECD countries is more than two-and-a-half times that of the increase in the OECD over the *Outlook* period. In 2035, total installed nuclear capacity in non-OECD countries reaches more than 250 GW, compared with 380 GW in the OECD.

Total electricity generation from renewable sources increases from 3 900 TWh in 2009 to 11 100 TWh in 2035, its share of total generation growing from 19% to 31%. Throughout the *Outlook* period, hydropower is the largest renewable source of electricity generation. Electricity generated from wind power increases by nearly ten-times over the *Outlook* period, reaching 2 700 TWh in 2035. China's consumption of electricity generated from wind power surpasses that of the European Union soon after 2030. Solar generated electricity (solar PV and concentrating solar power [CSP]) sees strong growth globally, particularly in the second half of the *Outlook* period, and reaches 1 050 TWh in 2035. Global electricity generating capacity using renewable sources increases from 1 250 GW in 2009 to 3 600 GW in 2035.

Total final consumption[4] by end-users is projected to grow by 1.3% per year, reaching about 11 600 Mtoe by 2035. Global demand in the industry sector grows by 49% over the *Outlook* period (Figure 2.12), increasing its share of total final consumption marginally, from 27% in 2009 to 29% in 2035. Non-OECD countries dominate demand growth in this sector, with China alone accounting for 36% of global growth and India a further 17%. OECD industrial energy demand increases through to 2020, before dropping back slightly to just under 890 Mtoe in 2035. Natural gas, electricity and renewables experience the strongest growth in this sector, all averaging growth per year of 2% or more, whereas coal sees more modest demand growth of 0.7% per year and oil is 0.1% per year. Coal continues to dominate energy input to iron and steel production. In the chemicals sector, use of electricity increases significantly from 2009 to 2035.

Energy demand in the transport sector increases by 43% to reach 3 260 Mtoe in 2035. Demand in non-OECD countries increases by 2.9% per year on average and, together with inter-regional transport, accounts for all of the net energy demand growth in the transport sector. China alone accounts for more than one-third of global demand growth in the transport sector. India sees demand increase progressively over the *Outlook* period, with the average annual rate of growth going from around 3.5% up to 2020 to more than 7% from 2020 to 2035. In contrast, the OECD sees a slight decline in demand over the *Outlook* period, with a notable reduction in Japan.

---

4. Total final consumption (TFC) is the sum of consumption by the various end-use sectors. TFC is broken down into energy demand in the following sectors: industry, transport, buildings (including residential and services) and other (including agriculture and non-energy use). It excludes international marine and aviation bunkers, except at world level where it is included in the transport sector.

Figure 2.12 ● **Incremental energy demand by sector and region in the New Policies Scenario, 2009-2035**

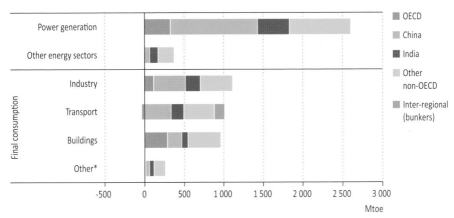

* Includes agriculture and non-energy use.

Road transport demand continues to dominate in the transport sector, representing 75% of demand in 2035. Oil demand in the road transport sector increases by 32% over the *Outlook* period, from 35 mb/d in 2009 to 45 mb/d in 2035. While many countries have adopted fuel efficiency standards, the growth in demand in non-OECD countries more than offsets the effect of these improvements. The global stock of road transport vehicles nearly doubles between 2009 and 2035, but the adoption of stronger efficiency standards, a shift in market focus toward non-OECD countries with lower average vehicle usage levels and, to some extent, the increased use of alternative vehicle technologies, means that the increase in energy demand is only around 42%. While the picture improves, the penetration of electric vehicles remains relatively small globally in 2035.

In 2009, an average of around 40 in every 1 000 people in non-OECD countries own a passenger light-duty vehicle (PLDV). In the OECD, the average approached 500 PLDVs per 1 000 people. In the New Policies Scenario, annual PLDV sales in non-OECD countries are projected to overtake those in OECD countries by around 2020 and reach more than 100 million by 2035. Both China and India see a huge increase in the average level of vehicle ownership from 2009 to 2035. The PLDV stock in non-OECD countries overtakes that of the OECD in the early 2030s. However, vehicle ownership levels in non-OECD countries, at about 125 PLDVs per 1 000 people in 2035, remain well below the OECD levels of almost 550 per 1 000 people in 2035 (Figure 2.13).

In the buildings sector, global energy demand increases by 34%, driven largely by new-build in non-OECD countries, but the sector's share of total final consumption declines marginally. Electricity consumption grows on average by 2.2% a year in this sector, benefitting from increased competitiveness and greater appliance penetration in non-OECD countries. Policy intervention facilitates the growth of renewables in this sector (largely solar thermal) both in the OECD and non-OECD countries.

Global demand for electricity is projected to increase by nearly 85% over the *Outlook* period, reaching over 31 700 TWh in 2035 (see Chapter 5). More than four-fifths of the growth in global electricity demand arises in non-OECD countries, as a result of greater use of household appliances and of electrical equipment in the industry and services sectors, as access to electricity increases and prosperity rises (see Chapter 13). Despite this growth, average electricity demand per capita in non-OECD countries in 2035 is still only 31% that of the OECD. Developing Asia sees its share of global electricity consumption increase from 28% in 2009 to 44% in 2035. China alone accounts for almost 30% of the world's total electricity consumption by the end of the *Outlook* period. In the OECD, the slower growth in electricity demand stems from lower rates of economic and population expansion, greater energy efficiency improvements and the adoption of measures such as carbon pricing.

Figure 2.13 ● **Number of PLDVs per thousand people by region, 2009 and 2035, and the change in oil demand in road transport**

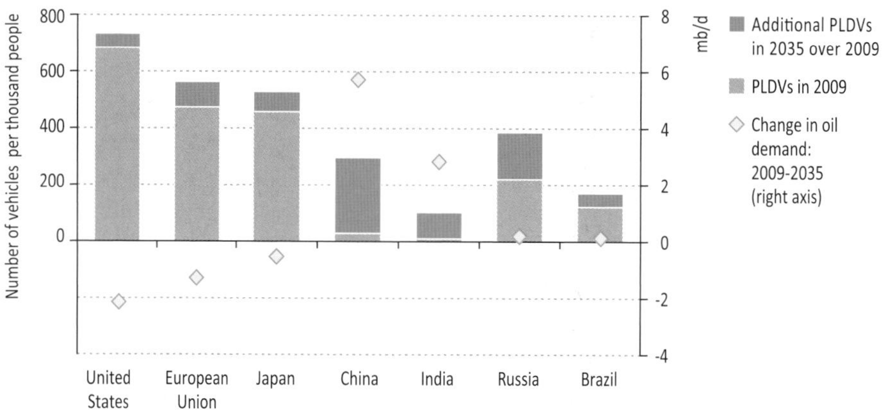

Energy production and trade

The events of the past year, while casting shadows over future energy supply prospects at least as menacing as those obscuring the pattern of future demand, do not call into question one thing: the world's endowment of economically exploitable energy resources is more than sufficient to satisfy the projected level of global consumption over the *Outlook* period and well beyond. The global pattern of production will change, with non-OECD countries accounting for all of the net global increase in oil and coal production between 2009 and 2035 in the New Policies Scenario and most of the increase in natural gas production (Figure 2.14). The details are examined, fuel-by-fuel and sector-by-sector, in subsequent chapters. This section briefly surveys the production prospects, setting the scene for the subsequent focus, in the concluding sections of the chapter, on four issues closely related to the future pattern of production: the changing profile of inter-regional energy trade; spending on imports; the scale and adequacy of investment in the energy-supply infrastructure; and energy-related emissions, notably of CO_2.

Figure 2.14 ● Incremental world energy supply by fuel in the New Policies Scenario, 2009-2035

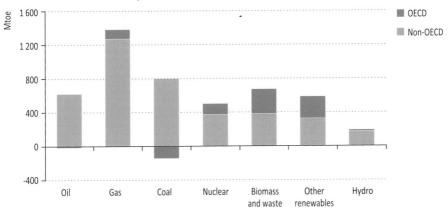

Resources and production prospects

Total world oil production, excluding processing gains and biofuels, was nearly 84 mb/d in 2010 and is slightly more than 96 mb/d by 2035 in the New Policies Scenario, an increase of 15%. A growing share of this production comes from unconventional sources, including oil sands and extra-heavy oil (see Chapter 3). Proven reserves of oil increased to 1.47 trillion barrels at the end of 2010 according to the *Oil and Gas Journal* (O&GJ, 2010) or 48 years of production at existing levels. Remaining recoverable resources are estimated to be much larger and could reach nearly 5.5 trillion barrels. Total non-OPEC oil production is projected to peak at 51 mb/d shortly after 2015 and then fall to less than 48 mb/d in 2035. Production declines in most non-OPEC countries, with Brazil, Canada and Kazakhstan being notable exceptions. In contrast, OPEC oil production continues to grow over the *Outlook* period, reaching nearly 49 mb/d in 2035, pushing its share of world production from 42% in 2010 to 51% in 2035.[5] The Middle East region is expected to increase oil production by around 12 mb/d, reaching a total of 37 mb/d by the end of the *Outlook* period (Figure 2.15). While Saudi Arabia continues to be the largest oil producer in the region, Iraq experiences the largest growth in production over the *Outlook* period. Though its potential to increase oil production from existing levels is large, in the short term, production growth in Iraq will be constrained by the need for infrastructure investment. Heightened geopolitical tension in the Middle East and North Africa continues to provide a more challenging backdrop against which to make long-term investment decisions. The fundamental concern is being able to judge the implications of these developments on short-term oil supply, and whether they might serve to defer investment essential to provide supply in the medium to long term (Chapter 3 analyses the implications of a Deferred Investment Case).

Global coal production is projected to rise by 19% between 2009 and 2035, with most of the growth occurring before 2020, reaching 5 860 Mtce by the end of the *Outlook* period.

---

5. OPEC is the Organization of the Petroleum Exporting Countries. As of mid-2011, its member countries included Algeria, Angola, Ecuador, Iran, Iraq, Kuwait, Libya, Nigeria, Qatar, Saudi Arabia, United Arab Emirates and Venezuela. For purposes of comparison, *WEO-2011* assumes that this membership remains unchanged throughout the *Outlook* period.

Nearly all of the production growth comes from non-OECD countries and, at more than 1 000 Mtce, the total increase in non-OECD countries is more than five-times the size of the offsetting net decrease in the OECD. The share of global coal production in non-OECD countries increases from 72% in 2009 to almost 80% in 2035. China sees the biggest increase in absolute terms with output growing by around 540 Mtce, although the rate of increase in production is higher in India and Indonesia. Until 2020, production of steam coal grows at around the same rate as that of coking (or metallurgical)[6] coal, reflecting similar trends in the growth in coal-fired power generation capacity and steel demand. Coal is estimated to be the world's most abundant fossil fuel. Proven reserves of coal are estimated to be around 1 trillion tonnes, equivalent to 150 years of world production in 2009 (BGR, 2010). Remaining resources are estimated to be over 21 trillion tonnes.

Figure 2.15 ● **Oil production in selected regions in the New Policies Scenario, 2010 and 2035**

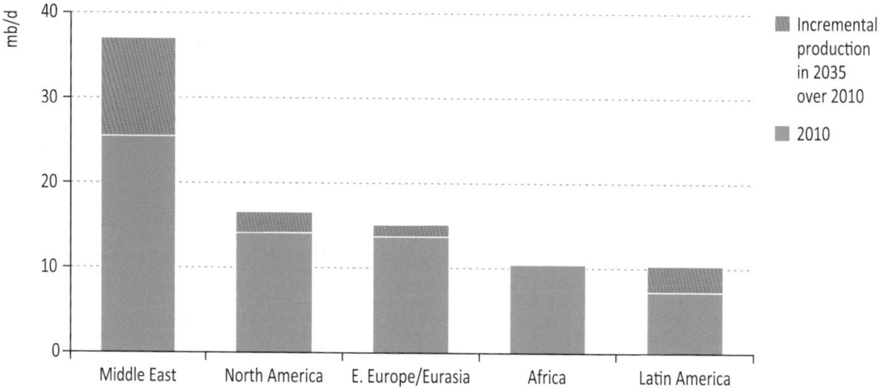

In the New Policies Scenario, global natural gas production increases by 56% from 2009, to reach 4 750 bcm in 2035. Non-OECD countries account for 91% of the projected increase, most of it coming from the largest existing gas producers. The Middle East, with the largest reserves and lowest production costs, sees the biggest increase in absolute terms, though Russia and the Caspian also see a significant increase. China more than triples its production of natural gas over the *Outlook* period. In the OECD, the United States and Australia are the main areas of gas production growth, mainly in the form of unconventional gas. The global production of unconventional gas increases by about two-and-a-half times, to exceed 1 000 bcm before 2035. Unconventional gas represents 39% of the growth in gas production over the *Outlook* period – its share of total gas production increases from 13% in 2009 to 22% in 2035.

Proven natural gas reserves are estimated to stand at 190 trillion cubic metres (tcm) (Cedigaz, 2010), around twice the amount of gas produced to date (see Chapter 4). Total recoverable global natural gas resources are estimated to be equivalent to around 265 years of the consumption level in 2009. Further exploration, development and

---

6. Strictly speaking, metallurgical coal includes all types of coal used in the metals sectors. Although coking coal accounts for most of this, it also includes small quantities of high-quality steam coal.

production continue to clarify the scale of unconventional gas resources worldwide and the prospects for their exploitation. Unconventional gas resources are now estimated to be of comparable size to conventional resources. However, the use of hydraulic fracturing in unconventional gas production has raised environmental concerns and is challenging the adequacy of existing regulatory regimes.

The estimated resource base of uranium, the raw material for nuclear fuel, remains sufficient to fuel the world's nuclear reactors at current consumption rates well beyond the *Outlook* period (NEA and IAEA, 2010).

In resource terms, significant potential remains for expanding energy production from renewables, but the impact of the global recession has tested the readiness and ability of some governments to maintain levels of support. In the New Policies Scenario, total electricity generation from renewables increases by more than 180% over the *Outlook* period, with wind power, hydropower and biomass being the largest sources of growth. While the amount of electricity generated from solar PV, CSP, geothermal and marine also grows significantly over the *Outlook* period, their starting point is low and their collective contribution to satisfying global electricity demand remains relatively small in 2035. In the OECD, the share of total electricity generation coming from renewables increases from 18% in 2009 to 33% in 2035. This compares to an increase from 21% to 29% across non-OECD countries, on average. Globally, China is the largest producer of electricity from renewables in 2035, with its output from hydropower alone surpassing the electricity generated from all forms of renewables collectively in any other single country (Figure 2.16).

Figure 2.16 ● Largest producers of electricity from renewables in the New Policies Scenario, 2035

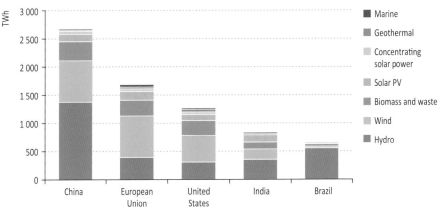

Inter-regional trade

Inter-regional energy trade plays an understated, but vital, role in the broader energy system and depends on the provision of adequate energy transport and storage infrastructure. Increased inter-connectivity provides the energy system with flexibility and redundancy, but it also means that sudden changes in one part may have a broad impact. While very different in their nature, sudden, unexpected events, such as those at Fukushima and in

Libya, demonstrated both the capacity of the global energy system to react to changing global supply-demand patterns and also some of its limitations.

Three-quarters of all energy is consumed within the country or region in which it is produced. However, the proportion of energy traded across regions, the patterns of trade and the structure of the market differ significantly by fuel. In the New Policies Scenario, the OECD share of total inter-regional energy trade declines from 42% in 2009 to 29% in 2035 and becomes increasingly concentrated on natural gas and oil, to the detriment of coal. Outside of the OECD, Asia accounts for an increasing proportion of global energy trade in all fossil fuels.

The amount of oil traded between major regions increases by about 30% to 53 mb/d in 2035, the share in total global consumption growing to 54%. Net imports of oil into the OECD decline consistently across the *Outlook* period, from 24 mb/d in 2010 to 17 mb/d in 2035. Net imports into the United States decrease significantly over the *Outlook* period, due both to decreasing demand and a slight increase in domestic supply (Figure 2.17). Oil imports into the European Union remain steady at 9.8 mb/d until around 2020, before declining gradually, to reach 8.8 mb/d in 2035. China's oil imports increase by more than two-and-a-half times over the *Outlook* period to reach 12.6 mb/d in 2035, nearly twice the level of Russia's oil exports, or around one-third of OPEC exports, in the same year. China overtakes the United States to become the world's largest oil importer shortly after 2020. Its import dependence increases from 54% in 2010 to 84% in 2035. India's oil imports increase by more than 4% per year over the *Outlook* period to reach 6.8 mb/d, its reliance on imports increasing from 73% in 2010 to 92% in 2035. Such increased reliance on oil imports in many countries (particularly non-OECD countries) is likely to lead to attention being given to the issue of security of supply. Over the *Outlook* period, OPEC oil exports increase by 42% to 38.4 mb/d. Russia's net oil exports decline from 7.5 mb/d in 2010 to 6.4 mb/d in 2035, driven primarily by decreasing levels of production.

Figure 2.17 ● Oil demand and the share of imports by region in the New Policies Scenario, 2010 and 2035

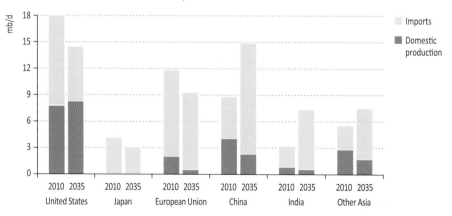

In the New Policies Scenario, most coal is consumed domestically, meaning that global coal trade remains small relative to overall demand. Nonetheless, trade in hard coal between major regions is projected to increase from 750 Mtce in 2009, growing rapidly to 2020 but stabilising thereafter, to settle around 1 000 Mtce in 2035 – around 18% of world hard coal production. The pattern of trade will continue to shift towards Asia and away from Atlantic Basin markets. The OECD as a whole ceases to be an importer of hard coal, becoming a net exporter around 2030. Japan, the largest coal importer in 2009, sees its import requirement peak early in the *Outlook* period and then decline gradually, to reach 115 Mtce in 2035. A coal exporter until recently, China sees its import requirement exceed that of Japan around 2015, peak at nearly 200 Mtce shortly after 2015 and then decline to around 80 Mtce in 2035. However, the scale of China's coal appetite is so huge, relative to others, that even quite a small shift in its domestic demand-supply balance can have major implications for the global picture (Spotlight on the role of China in traded coal markets). India's hard coal imports increase by more than 6% per year over the *Outlook* period, becoming the world's largest importer soon after 2020 and importing nearly 300 Mtce in 2035, nearly five-times the level of 2009. India is expected to look first to Indonesia, Australia and South Africa to satisfy its import needs. Australia sees its hard coal exports peak before 2020 and then gradually decline to around 300 Mtce in 2035, still 18% higher than 2009. Indonesia sees its hard coal exports increase from 190 Mtce in 2009 to around 280 Mtce in 2035, but are on a declining path later in the *Outlook* period.

Inter-regional trade in natural gas nearly doubles over the *Outlook* period, increasing from 590 bcm in 2009 to around 1 150 bcm in 2035. The expansion occurs in both pipeline gas and liquefied natural gas (LNG). The proportion of gas that is traded across regions increases from 19% in 2009 to 25% in 2035. The market for natural gas becomes more globalised over the *Outlook* period, but only gradually. The need for natural gas imports into the European Union grows from 310 bcm in 2009 to 540 bcm in 2035 and its dependence on imports increases from 61% to 86% (Figure 2.18). Reflecting the growing availability of domestic unconventional gas, natural gas imports into the United States decline from early in the *Outlook* period and remain relatively small throughout. Developing Asia moves from being a marginal exporter of natural gas in 2009 to importing nearly 300 bcm in 2035. China accounts for around 210 bcm of these imports in 2035 and its share of imports increases from 8% to 42%.

Figure 2.18 ● **Natural gas demand and the share of imports by region in the New Policies Scenario, 2009 and 2035**

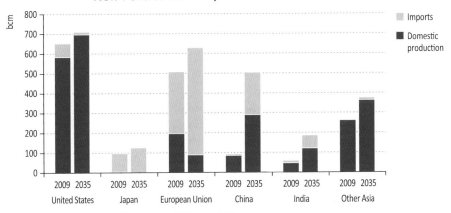

Note: Other Asia had net natural gas exports of 56 bcm in 2009.

Natural gas exports from Russia more than double over the *Outlook* period, reaching around 330 bcm by 2035. Exports from the Caspian region more than treble, to reach around 135 bcm in 2035. Gas supplies from both Russia and the Caspian are increasingly pulled eastwards, mainly to satisfy the rapidly growing demand of China. Natural gas exports from Africa increase by more than 4% per year, reaching 280 bcm in 2035, and go both by pipeline to Europe and further afield as LNG. With the benefit of the giant Gorgon project and others, Australia becomes an increasingly important gas exporter. LNG exports from Australia increase from 14 bcm in 2009 to 85 bcm in 2020 and 115 bcm in 2035. Led by Qatar, natural gas exports from the Middle East increase significantly. In 2035, the Middle East is exporting around 150 bcm of natural gas, through a mix of pipelines and LNG, to Europe and Asia.

S P O T L I G H T

China's role in traded coal markets – the ultimate uncertainty?

China has long been a net exporter of coal, but became a net importer in 2009 – a development foreseen in *WEO-2007* (IEA, 2007). China's net coal imports surged by around 45% in 2010 to an estimated 126 Mtce, making it the world's second largest importer after Japan. Based on preliminary data, net imports may fall back in 2011, but they are still likely to stay above 100 Mtce. As China is by far the world's biggest coal producer and has massive resources, China as a coal importer could be seen in some ways as analogous to Saudi Arabia as an oil importer. The speed and magnitude of this abrupt turnaround has had a major impact on traded coal markets.

China's coal imports have been driven primarily by price. In late 2008, world coal prices and freight rates fell sharply, as the financial crisis took hold. Delivered prices for China's domestic coal also fell, but to a lesser extent as a result of transport bottlenecks and supply constraints. This made imported coal cheaper than domestic grades in some locations. For example, by October 2009, steam coal delivered to the east coast of China from Indonesia was up to 40% cheaper than domestic coal on a quality-adjusted basis. Imports were also necessary to meet the need for high-quality hard coking coal for steel making. The surge in imports has created opportunities for coal exporters within reasonable reach of China, including Indonesia, Australia, Mongolia, Russia, the United States and Canada. Many exporters are now increasing production in the expectation of future Chinese demand growth, but this strategy is not without risk. Despite the massive scale of China's imports – estimated to be 15% of globally traded coal in 2010 – they met just 5% of its demand. Therefore, very slight changes in China's domestic coal demand or supply could push it back to becoming a net exporter, competing against the countries that are now investing to supply its needs. For example, net imports would fall to zero by 2015 (compared with our projection of around 185 Mtce) if China's output were just 1.2% per year above the level in the New Policies Scenario. Many people are asking: is China's status as a net importer temporary or a new structural feature of the market?

The answer will depend on China's ability to meet rising demand from domestic production and how that affects the spread between domestic and international prices. At present, the competitiveness of China's coal production is hindered by insufficient rail infrastructure between producing regions in the north and west of the country and demand centres in the east and south. In some locations, coal has to be transported by road at a higher cost. But some of these pressures look set to ease. Transport capacity is set to expand under China's 12th Five-Year Plan and large "coal power bases" are to be built to convert coal to electricity close to where it is mined for transmission to demand centres. There are also plans for further coal-to-liquids projects. The consolidation of China's coal sector is also expected to improve its productivity and competitiveness. On the other hand, mining in China is moving further west, as Inner Mongolia overtakes Shanxi as the major coal area, and is going deeper and exploiting poorer quality seams.

Most analysts agree that China will remain a net importer over the medium term, at least, as it is unlikely to expand its mining and transport infrastructure quickly enough. But the exact level of imports is uncertain, even more so post-2020 and particularly steam coal. What is certain is that developments in China's coal market will remain one of the most important factors in determining global coal prices and trade patterns.

Spending on imports

A combination of increasing reliance on energy imports and relatively high energy prices has seen the energy import bill weigh more heavily on many national accounts in recent years. Global annual expenditure on oil and gas imports is expected to move sharply higher in 2011 than 2010, driven primarily by oil prices (Chapter 1 details oil and natural gas price assumptions). If oil prices (IEA import price) average $100 per barrel over 2011, it is likely to be the first year in which collective OPEC net oil revenues exceed $1 trillion.

Annual global expenditure on oil and gas imports more than doubles over the *Outlook* period, reaching nearly $2.9 trillion in 2035 under our assumptions. The share of natural gas within total spending also increases over time. Even with the measures that are assumed to be introduced to restrain growth in energy demand, the New Policies Scenario implies a persistently high level of spending on oil and gas imports by many countries. As a proportion of GDP (at market exchange rates), India's projected spending on energy is highest and it remains above 4% in 2035 (Figure 2.19). In absolute terms, India sees its oil and gas- import bill overtake that of Japan before 2030 and overtake that of the United States by the end of the *Outlook* period, standing at around $330 billion in 2035 (Figure 6.11 in Chapter 6 shows oil-import bills in absolute terms in selected countries by scenario). China sees its oil and gas-import bill increase from under $100 billion in 2009 to $660 billion in 2035 (in 2010 dollars), overtaking the European Union soon after 2030. China's import bill remains around 3% of GDP. In the United States, the increasing prominence of domestic supplies in meeting demand means that expenditure on oil and gas imports declines in absolute terms, to around $275 billion in 2035, and also as a share of GDP.

Figure 2.19 ● Expenditure on net imports of oil and gas as a share of real GDP by region in the New Policies Scenario, 1980-2035

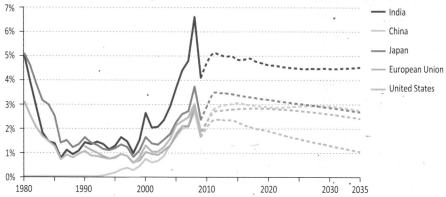

Note: Calculated as the value of net imports at the prevailing average international prices. The split between crude/refined products and LNG/piped gas is not taken into account. GDP is measured at market exchange rates in year-2010 dollars.

Investment in energy-supply infrastructure

Much of the investment required in the energy sector is large-scale and the financial returns come over a long time period. Even in what seen relatively tranquil times, investment decisions have to be taken in conditions of great uncertainty. Such uncertainty can, for example, relate to the economic outlook, to developments in climate and other environmental policies, to depletion policies in key producing governments and changes to legal, fiscal and regulatory regimes.

In the New Policies Scenario, cumulative investment of nearly $38 trillion (year-2010 dollars) is required in energy-supply infrastructure over the *Outlook* period, a significant increase from *WEO-2010* (Figure 2.20). This investment provides for the replacement of the reserves and production facilities that are exhausted or retired, and for the expansion of production and transport capacity to meet the increase in energy demand over the *Outlook* period. Global investment in the power sector from 2011 to 2035 totals $16.9 trillion, 58% of which goes to new power plants and 42% to transmission and distribution. New power generating capacity of 5 900 GW is added worldwide. Globally, renewables account for 60% of the investment in the power sector over the *Outlook* period. Investment in new power plants in non-OECD countries, particularly China, increasingly outstrips that in the OECD over time. China adds more new generating capacity powered by coal, gas, nuclear, hydropower, biomass, wind and solar than any other country.

Global upstream investment for oil is projected to rise by around 9% in 2011, reaching a new all-time high level of over $550 billion. In the New Policies Scenario, cumulative investment in oil-supply infrastructure is $10 trillion over the *Outlook* period, with a need for heavy upstream oil investment in the medium and longer term. The largest investment is required in North America, Latin America and Africa (Table 2.4). An important factor driving the increase in the investment required is the observed rise in the related costs, which is projected to continue. There is also a slightly greater emphasis than in *WEO-2010* on non-OPEC supply in the early part of the *Outlook* period, which is relatively more expensive to produce. Of particular importance is the Middle East and North Africa region, where investment needs are not as high, but required production increases are substantial: it is crucial that these investments are made in a timely manner (see the Deferred Investment Case in Chapter 3). Upstream investment in conventional oil is more than ten-times greater than investment in unconventional oil over the *Outlook* period. The largest investments in refining are expected to take place in China, India, the Middle East and the United States.

Figure 2.20 ● **Cumulative investment in energy-supply infrastructure by fuel in the New Policies Scenario, 2011-2035** (in year-2010 dollars)

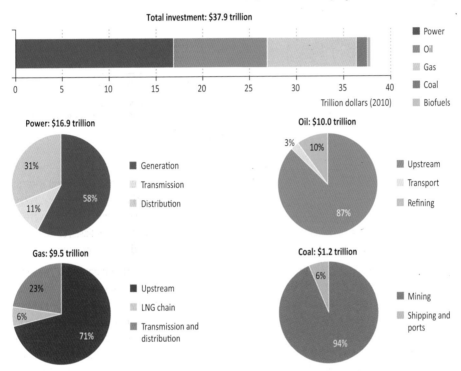

Total cumulative investment requirements in the coal industry are nearly $1.2 trillion over the *Outlook* period, or $47 billion per year. Around 94% of this sum is for mining, which includes both opening new mines and developing existing mines. China's investment needs are much greater than those of any other single country and this investment must be sustained throughout the *Outlook* period, even after domestic demand has peaked. In the OECD, Australia requires the largest investment in coal, followed by the United States.

Cumulative investment in the natural gas supply chain is projected to be $9.5 trillion. Exploration and development of gas fields, including bringing new fields on stream and sustaining output at existing fields, will absorb 71% of total gas investment, while transmission and distribution absorbs the rest. One-quarter of total upstream investment goes to unconventional gas, the remainder funding development and production of conventional gas.

The OECD accounts for around 36% of the cumulative investment required in energy supply over the *Outlook* period. The relatively high level of investment required in the OECD countries, compared with the OECD share of energy demand growth, is attributable to the need to retire and replace more ageing energy infrastructure, the relatively more capital-intensive energy mix and the higher average cost of its capacity additions in each category. The United States accounts for 14% of global cumulative energy supply investment over the *Outlook* period. China accounts for around 15% of global cumulative investment, amounting to $5.8 trillion and is heavily focused on the power sector (Figure 2.21). Latin America, Africa the Middle East and Russia all require significant levels of investment, particularly in oil and gas, over the *Outlook* period.

Table 2.4 ● Cumulative investment in energy-supply infrastructure by fuel and region in the New Policies Scenario, 2011-2035 (billion in year-2010 dollars)

	Coal	Oil	Gas	Power	Biofuels	Total
OECD	175	2 703	3 756	6 897	216	13 746
Americas	78	2 100	2 172	3 009	142	7 501
Europe	7	511	1 019	2 892	72	4 501
Asia Oceania	90	91	565	996	2	1 745
Non-OECD	934	7 027	5 661	9 986	136	23 744
E. Europe/Eurasia	38	1 398	1 562	1 029	6	4 033
Russia	*24*	*787*	*1 077*	*614*	*0*	*2 502*
Asia	812	963	1 664	7 018	60	10 518
China	*647*	*510*	*638*	*3 968*	*31*	*5 794*
India	*87*	*203*	*266*	*1 631*	*16*	*2 203*
Middle East	0	1 137	510	583	0	2 230
Africa	52	1 557	1 316	638	3	3 564
Latin America	32	1 971	609	718	68	3 399
Inter-regional transport	55	268	80	-	4	407
World	1 164	9 997	9 497	16 883	356	37 897

Figure 2.21 ● Cumulative investment in energy-supply infrastructure by region in the New Policies Scenario, 2011-2035

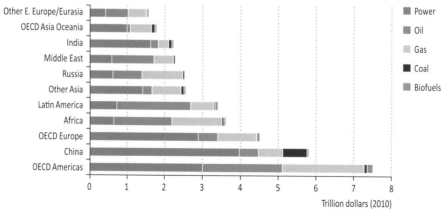

Energy-related emissions

Greenhouse gases

The decline in energy-related CO_2 emissions caused by the global economic recession has been short-lived, erased by an increase in 2010 of more than 5%, to 30.4 Gt. In the New Policies Scenario, energy-related CO_2 emissions continue to increase, rising by nearly 20%, to 36.4 Gt in 2035 (Figure 2.22). This leads to an emissions trajectory consistent with a long-term global temperature increase of more than 3.5°C. The energy sector was responsible for 65% of all greenhouse-gas emissions globally in 2010 and is expected to see its share increase to 72% in 2035.

Energy consumed in power plants, factories, buildings and other energy-using equipment existing or under construction in 2010 account for 70%, or 23.9 Gt, of global CO_2 emissions in 2020. This share declines to 44% in 2035. These figures illustrate the long lifetimes associated with energy-using equipment and infrastructure, the power generation and industry sectors exhibiting the largest "lock-in" of emissions. Unless there is widespread retrofitting with carbon capture and storage, in these sectors in particular, or early retirement of the capital stock, these emissions represent large-scale immoveable inertia that is felt to 2035 and beyond (see Chapter 6 for analysis on the lock-in of CO_2 emissions in the energy sector).

Figure 2.22 ● **Energy-related CO_2 emissions in the New Policies Scenario by fuel, 1980-2035**

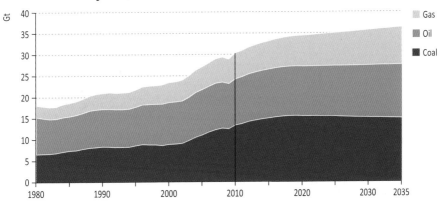

Energy-related CO_2 emissions from coal increase from 13.3 Gt in 2010 to 14.9 Gt in 2035, having actually peaked at around 15.5 Gt before 2020 and then declined gradually thereafter. This is due to the induced switching to natural gas in many applications and to sustained support for renewables in the power sector. CO_2 emissions from oil increase throughout the *Outlook* period, going from 10.9 Gt in 2010 to 12.6 Gt in 2035, but the rate of growth is slower after 2020. While natural gas emits the least CO_2 of all fossil fuels when combusted, relatively strong demand growth throughout the *Outlook* period results in CO_2 emissions from natural gas increasing by an average of 1.5% per year, rising from 6.1 Gt in 2010 to 8.9 Gt in 2035.

In the New Policies Scenario, cumulative energy-related global emissions over the next 25 years are projected to be three-quarters of the total amount emitted over the past 110 years. While the majority of historical energy-related emissions came from OECD countries (Figure 2.23), non-OECD countries account for all the projected growth in emissions over the *Outlook* period. OECD energy-related emissions are expected to peak early in the *Outlook* period at nearly 12.5 Gt and then decline steadily to 10.5 Gt in 2035. This trend is higher than that projected in *WEO-2010*, mainly reflecting a loss in momentum in the implementation of climate-related policies, the rebound in 2010 emissions and lower reliance on nuclear in countries such as Germany and Japan.

Figure 2.23 ● **Cumulative energy-related CO_2 emissions in selected countries and regions, 1900-2035**

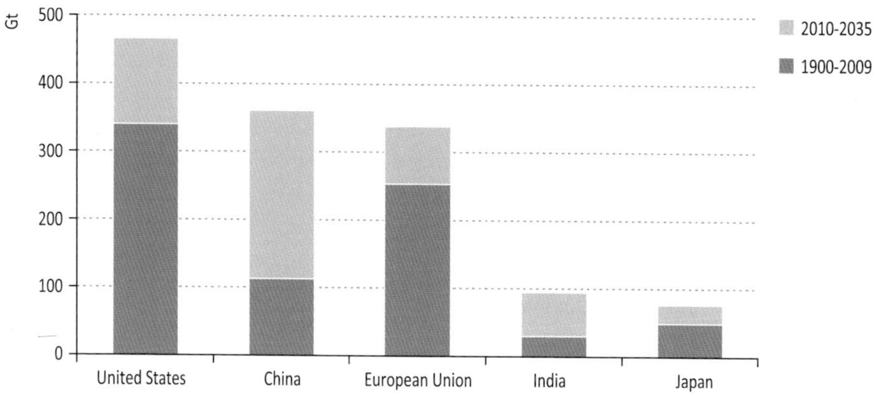

In non-OECD countries, China alone will see its CO_2 emissions increase from 7.5 Gt in 2010 to 10.3 Gt in 2035 (Figure 2.24). However, the growth in China's emissions is expected to slow significantly over time. On a per-capita basis, China's rising emissions are expected to converge with average OECD per-capita emissions, which follow a declining trend. Both approach around 7.5 tonnes per capita in 2035. Despite convergence in some cases, large discrepancies will continue to persist in others. For example, the average citizen in the United States is expected to be responsible for around 12 tonnes of CO_2 emissions in 2035, while a citizen in India is expected to be responsible for less than one-fifth of this level.

Energy-related CO_2 emissions are expected to rise in all sectors over the *Outlook* period, but the largest increase is expected in power generation, where emissions increase by 2.3 Gt, and transport, where emissions increase by 2.1 Gt. These two sectors combined account for almost 75% of the increase in energy-related emissions.

Local pollutants

Emissions of other pollutants follow heterogeneous regional paths. Sulphur dioxide (SO_2) emissions, the main cause of acid rain, are expected to fall in almost all major countries, as emissions standards become stricter and coal use either peaks or declines. India is an exception, as regulation is expected to be delayed and coal use is rising steadily. Nitrogen oxides (NO_x)

Figure 2.24 ● Energy-related CO_2 emissions by region in 2035 in the New Policies Scenario and the change from 2010

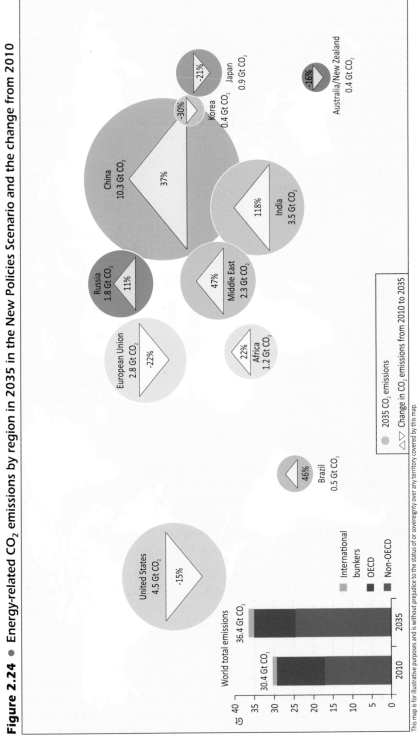

This map is for illustrative purposes and is without prejudice to the status of or sovereignty over any territory covered by this map.

Notes: The circles reflect the relative volume of energy-related CO_2 emissions from selected countries and regions in 2035. The arrows indicate the change in these emissions from 2010 to 2035. The bar chart shows world energy-related CO_2 emissions and the split between the OECD, non-OECD countries and international bunkers.

are expected to follow a declining trend in OECD countries, but rise in non-OECD countries. Transport is the main cause of such emissions, and the scale of increasing mobility in developing countries outpaces the effect of the emission standards that many of these countries are implementing, though such regulation is nonetheless very effective in decoupling the growth in vehicle ownership from an equivalent increase in NO_x. Trends for particulate matter vary by region, depending on the extent of reliance on biomass as a household fuel and regulation of emissions from cars. The trends in the New Policies Scenario for particulate matter imply a worsening of the health impact – in India, for example, a reduction of life expectancy of more than six months per person by 2035, compared with current levels (IIASA, 2011). In 2009, the global cost of air pollution controls was around $280 billion and this cost increases to more than $550 billion by 2035, due to higher activity levels and the greater stringency of the controls.

CHAPTER 3

OIL MARKET OUTLOOK

Will investment come fast enough?

H I G H L I G H T S

- Policy action to curb demand and a continuing ability to develop new supplies will be critical to the mid- and long-term outlook for international oil markets. Global oil demand in the New Policies Scenario increases slowly over the *Outlook* period, from 87 mb/d in 2010 to 99 mb/d in 2035. The crude oil price rises to $120/barrel (in year-2010 dollars) in 2035. All the net growth in demand comes from non-OECD countries, mostly in Asia; OECD oil use falls.

- All of the net growth in global oil demand in the New Policies Scenario comes from the transport sector in emerging economies. Non-OECD car markets expand substantially – vehicle production there is projected to overtake the OECD before 2015 and car sales to exceed those in the OECD by 2020 – forcing up oil consumption despite impressive gains in vehicle fuel economy. Alternative vehicle technologies are emerging that use oil much more efficiently or not at all, such as electric vehicles, but it will take time for them to become commercially viable and penetrate markets.

- Oil production (net of processing gains) reaches 96 mb/d in 2035, a rise of 13 mb/d on 2010 levels in the New Policies Scenario. Crude oil supply increases marginally to a plateau of around 69 mb/d (just below the historic high of 70 mb/d in 2008) and then declines slightly to around 68 mb/d by 2035. Nonetheless, gross capacity additions of 47 mb/d – twice the current OPEC Middle East production – are needed just to compensate for declining production at existing fields. A growing share of global output comes from natural gas liquids, unconventional sources and light tight oil.

- Non-OPEC production falls marginally, while OPEC's market share expands from 42% in 2010 to 51% in 2035. Output grows in most Middle East OPEC countries (led by Iraq and Saudi Arabia), but falls in most non-OPEC countries (the main exceptions being Brazil, Canada and Kazakhstan). Increasing reliance on imports in the importing non-OECD regions, notably Asia, will inevitably heighten concerns about the cost of imports and supply security. To meet the projected oil supply requirements in the New Policies Scenario calls for growing levels of capital spending – cumulative upstream investment of $8.7 trillion (in year-2010 dollars) is needed in 2011 to 2035.

- In a Deferred Investment Case, we examine the implications of MENA upstream investment running one-third below the level in the New Policies Scenario in 2011 to 2015. MENA production is, as a result, more than 6 mb/d lower by 2020 and prices jump to $150/barrel, before falling back as production rises. MENA exporters earn more in the near term, thanks to higher prices, but less in the longer term, as they lose market share.

Demand

Primary oil demand trends

The outlook for oil demand differs markedly between the three scenarios presented in this *Outlook*, primarily as a result of the different assumptions about government policies, such as fuel efficiency standards, removal of end-user subsidies and support for alternative fuels, and the extent to which they succeed in curbing oil demand. After two consecutive years of decline during the economic crisis, global primary demand for oil (excluding biofuels) rebounded in 2010 with economic recovery, reaching 86.7 million barrels per day (mb/d), a 3.1% increase over 2009. Demand continues to grow over the *Outlook* period in both the Current Policies and New Policies Scenarios, driven mainly by developing countries, but falls in the 450 Scenario (Figure 3.1 and Table 3.1).

Figure 3.1 ● World primary oil demand and oil price* by scenario

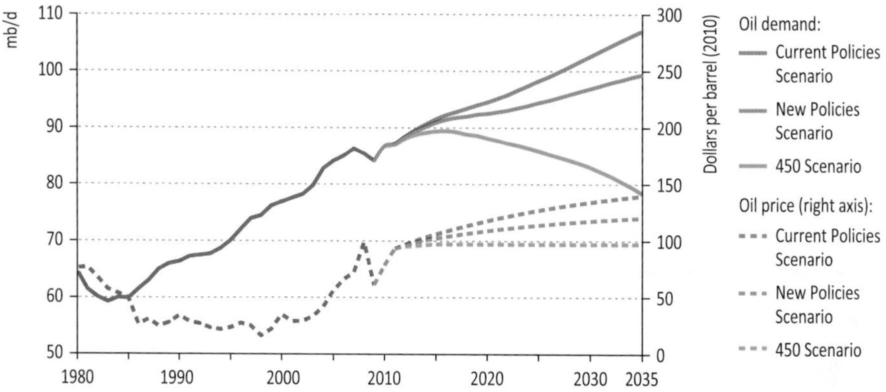

* Average IEA crude oil import price.

The Current Policies Scenario, in which only existing policies are assumed to be in effect, sees oil demand reaching 107 mb/d by 2035, a 24% increase over 2010 levels, or an average annual increase of 0.8%. In the New Policies Scenario, which takes account of policy commitments and cautious implementation of published targets, oil demand reaches 99 mb/d by 2035, a 15% increase over year 2010 levels (0.5% per year). Lower oil demand in the latter scenario is largely a result of policy action that promotes more efficient oil use, switching to other fuels and higher end-user prices as a result of reduced subsidies in some emerging major consuming countries. In the 450 Scenario, oil demand falls between 2010 and 2035 as a result of strong policy action to limit carbon-dioxide (CO_2) emissions; oil demand peaks before 2020 at just below 90 mb/d and declines to 78 mb/d by the end of the projection period, over 8 mb/d, or almost 10%, below 2010 levels. While higher-than-expected demand in 2010 and revised assumptions for economic and population growth have altered the short- and medium-term trajectories, oil demand trends to 2035 are broadly similar to those projected in last year's *Outlook* in all three scenarios. Taking account of the sharp increases in oil prices since late 2010 and the assumed higher GDP growth, however, the oil prices needed to balance demand with supply are somewhat higher than assumed in *WEO-2010*. This is despite several new policies to reduce oil demand announced over the last year, which are taken into account in the 2011 New Policies Scenario, such as the new fuel economy standards under discussion by the government of India for passenger vehicles (see Chapter 1 and Annex B for details).

The average IEA crude oil import price to balance demand and supply, which is derived after numerous iterations of the World Energy Model, reaches $118 per barrel (in year-2010 dollars) in 2020 and $140/barrel in 2035 in the Current Policies Scenario. In the New Policies Scenario, the price rises more slowly, to $109/barrel in 2020 and $120/barrel in 2035, as demand grows less rapidly as a result of stronger policy action. There are both supply- and demand-side reasons for these assumed oil-price increases. On the demand side, the increasing dominance of the transport sector in global oil demand tends to reduce the sensitivity of demand to oil prices in the medium term, as the economic competitiveness of conventional technologies limits the potential for substitution of oil as a transportation fuel. In addition, the cost of supply rises progressively through the projection period, as existing sources are depleted and oil companies are forced to turn to more difficult and costly sources to replace lost capacity in order to meet rising demand. Moreover, constraints on how quickly investment can be stepped up also limit the rate of growth of production, contributing to upward pressure on prices. In the 450 Scenario, prices are considerably lower than in the other scenarios, levelling off at just below $100/barrel in 2015, as a result of the persistent decline in demand.

Table 3.1 ● Primary oil demand by region and scenario (mb/d)

	1980	2010	New Policies Scenario		Current Policies Scenario		450 Scenario	
			2020	2035	2020	2035	2020	2035
OECD	40.9	42.5	40.0	35.8	40.8	38.5	38.2	26.5
Non-OECD	19.8	37.6	45.0	54.5	46.3	59.3	42.7	44.2
International bunkers*	3.5	6.6	7.5	9.1	7.5	9.4	7.2	7.7
World oil demand	**64.2**	**86.7**	**92.4**	**99.4**	**94.6**	**107.1**	**88.1**	**78.3**
Share of non-OECD	*33%*	*47%*	*53%*	*60%*	*53%*	*61%*	*53%*	*63%*
Biofuels demand**	0.0	1.3	2.3	4.4	2.1	3.4	2.7	7.8
World liquids demand	*64.2*	*88.0*	*94.7*	*103.7*	*96.7*	*110.6*	*90.8*	*86.1*

*Includes international marine and aviation fuel. **Expressed in energy equivalent volumes of gasoline and diesel.

Historically, economic activity has been the principal driver of oil demand, and it remains an important factor in all scenarios and regions. However, oil intensity – the volume of oil consumed for each dollar of gross domestic product (GDP) – has fallen considerably in recent decades and is expected to continue to fall even faster over the projection period: global oil intensity declined at an annual rate of 1.5% over the past 25 years and is projected to continue to fall by an average 2.8% per year until 2020 in the New Policies Scenario, amplified by considerable government efforts to curb oil demand (Figure 3.2). Oil intensity declines further thereafter, at a slower rate of 2.1% per year until 2035. The principal reasons for the projected decline in oil intensity at levels above the historical average are the assumed high oil prices required to balance supply and demand in the New Policies Scenario – on average, prices are nearly three times above the historical average of the last 25 years in real terms, promoting increased efficiency and conservation measures – and the continuation of numerous policy efforts to curb oil demand, both in OECD countries and in emerging economies such as China and India, prompted by oil and energy security concerns.

Oil intensity differs considerably across countries and regions as a result of a combination of structural, climatic and cultural factors and different pricing. The United States has historically had a higher intensity than the worldwide average, but it fell below the average in 2008 and the gap with the rest of the OECD is projected to narrow over the *Outlook* period. China's intensity has been, on average, 35% higher than the world average, but well below the figure for non-OECD countries as a whole. It is projected to fall below the world average by around 2020 and continue to decline gradually thereafter as a result of policy efforts to decrease the economy's emissions intensity and increasing efficiency in the transportation sector (Annex B). While the absolute differences in oil intensity between the major countries and regions are projected to narrow over the *Outlook* period, the relative rankings remain stable.

Figure 3.2 ● **Primary oil intensity* by region in the New Policies Scenario**

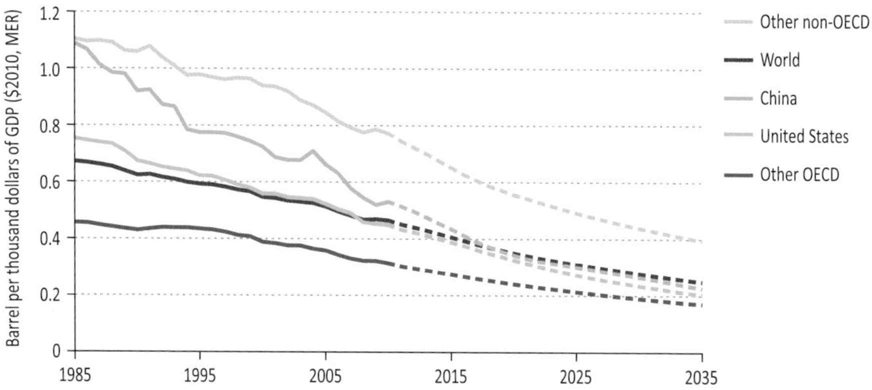

*Oil demand per unit of GDP at market exchange rates (MER). World excludes international marine and aviation bunkers.

Regional trends

The pattern of oil demand diverges over the projection period among the major regions, according to their level of economic development. Demand drops in all three OECD regions, but continues to expand in the non-OECD countries in aggregate (Table 3.2). In the New

Policies Scenario, OECD primary demand declines steadily through to 2035, from over 42 mb/d in 2010 to 40 mb/d in 2020 and 36 mb/d in 2035 – an average rate of decline of 0.7% per year. This results from high oil prices, saturation effects and a range of additional government measures to curb oil demand, particularly in the transportation sector. The largest decline in percentage terms occurs in Japan, averaging 1.2% per year, driven mainly by energy efficiency and alternative vehicle technology policy in the transport sector. The United States experiences the largest decline in absolute terms, with demand dropping 3.5 mb/d over the projection period. In 2010, only 15% of the 2.6 mb/d year-on-year rebound in global oil demand was attributable to OECD countries (89% of which was in the United States alone) and OECD demand remains below its historic peak in 2005. OECD demand declined by 2.4% per year on average between 2007 and 2010.

Table 3.2 ● Primary oil demand by region in the New Policies Scenario (mb/d)

	1980	2010	2015	2020	2025	2030	2035	2010-2035*
OECD	**40.9**	**42.5**	**41.7**	**40.0**	**38.2**	**36.9**	**35.8**	**-0.7%**
Americas	20.6	22.6	22.3	21.4	20.4	19.8	19.3	-0.6%
United States	*17.3*	*18.0*	*17.8*	*16.8*	*15.8*	*15.1*	*14.5*	*-0.9%*
Europe	14.2	12.7	12.5	12.0	11.5	11.0	10.6	-0.7%
Asia Oceania	6.2	7.2	7.0	6.6	6.4	6.1	5.9	-0.8%
Japan	*4.9*	*4.2*	*4.0*	*3.7*	*3.5*	*3.3*	*3.1*	*-1.2%*
Non-OECD	**19.8**	**37.6**	**42.1**	**45.0**	**48.2**	**51.5**	**54.5**	**1.5%**
E. Europe/Eurasia	9.4	4.6	4.9	5.0	5.2	5.3	5.4	0.7%
Russia	*n.a.*	*2.9*	*3.0*	*3.0*	*3.1*	*3.1*	*3.2*	*0.4%*
Asia	4.3	17.7	20.5	22.5	24.9	27.5	29.9	2.1%
China	*1.9*	*8.9*	*11.1*	*12.2*	*13.4*	*14.5*	*14.9*	*2.1%*
India	*0.7*	*3.3*	*3.6*	*4.2*	*4.9*	*6.0*	*7.4*	*3.4%*
Middle East	1.7	6.9	7.6	8.1	8.7	9.0	9.2	1.2%
Africa	1.2	3.1	3.5	3.6	3.7	3.9	4.0	1.0%
Latin America	3.1	5.2	5.7	5.7	5.8	5.8	5.9	0.5%
Brazil	*1.3*	*2.3*	*2.5*	*2.5*	*2.5*	*2.6*	*2.6*	*0.5%*
Bunkers**	3.5	6.6	7.0	7.5	7.9	8.5	9.1	1.3%
World oil demand	**64.2**	**86.7**	**90.8**	**92.4**	**94.4**	**96.9**	**99.4**	**0.5%**
European Union	*n.a.*	*11.9*	*11.5*	*11.0*	*10.4*	*9.8*	*9.3*	*-1.0%*
Biofuels demand***	0.0	1.3	1.8	2.3	2.9	3.6	4.4	5.0%
World liquids demand	*64.2*	*88.0*	*92.6*	*94.7*	*97.3*	*100.5*	*103.7*	*0.7%*

*Compound average annual growth rate. **Includes international marine and aviation fuel. *** Expressed in energy equivalent volumes of gasoline and diesel.

In contrast, non-OECD oil demand is set to continue to expand. It has been less affected by the economic recession, continuing to grow by an average of 3.5% per year in the three years to 2010. In the New Policies Scenario, non-OECD primary demand rises on average by 1.5% per year, from about 38 mb/d in 2010 to 45 mb/d in 2020 and almost 55 mb/d in 2035. Oil use in non-OECD countries overtakes that in the OECD by 2015. In absolute terms, the bulk of oil demand growth in non-OECD countries occurs in Asia, where it grows by 12 mb/d, or 70%, between 2010 and 2035. Half of the increase in Asian oil demand (and about 48% of the global increase) comes from China, where demand expands by 6 mb/d. Starting from a low base, India is the second-largest contributor to oil demand growth, accounting for 33% of the global increase. Demand in the Middle East grows by 2.4 mb/d, on the back of strong economic and population growth and widespread subsidies.

Sectoral trends

The transport sector will remain the main driver of global oil demand as economic growth increases demand for personal mobility and freight. Transport oil demand reaches almost 60 mb/d in 2035, a growth of about 14 mb/d over 2010 levels, outweighing a drop in demand in other sectors. Booming transport oil needs in non-OECD countries more than offset a decline in transport demand in the OECD (Figure 3.3). Assumed increases in oil prices provide a strong economic signal to reduce oil use in many sectors, either through increased efficiency or fuel substitution. This depresses oil demand, particularly in industry, where oil is substituted by natural gas (and other fuels) for heat and steam generation, as is the case in Europe and the United States today. With increasing divergence of oil and gas prices over the projection period, such inter-fuel substitution becomes more common in Asia as well, where to date relatively higher gas prices have limited its use.

Figure 3.3 ● Change in primary oil demand by sector and region in the New Policies Scenario, 2010-2035

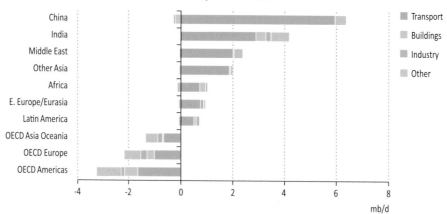

Changes in oil demand in the buildings sector reflect, to some degree, the stage of economic development. In the OECD countries and China, oil demand is reduced due to the shift away from kerosene and liquefied petroleum gas (LPG) towards electricity; but in Africa and India, where access to clean energy for cooking is currently limited, demand increases (see

Chapter 13 for a discussion of energy access). There is much more limited scope at present to substitute for oil-based fuels in the transport sector and the petrochemical industry. The petrochemical sector, especially in Asia and to a lesser extent in Europe, relies heavily on naphtha and LPG as feedstock, though natural gas use is increasing due to the expansion of ethane-based production in the Middle East. The transport sector depends almost entirely on oil products, with 93% of all the fuel used in the sector being oil-based in 2010. Alternative vehicle technologies that use oil much more efficiently (such as hybrid vehicles or plug-in hybrids) or not at all (such as electric vehicles) are emerging, but it will take time for them to become commercially viable on a large scale and for car makers and fuel suppliers to invest in new production facilities and distribution infrastructure, where needed, on the scale required to make a significant difference.

Focus on the transport sector

Road transport is expected to continue to dominate total oil demand in the transportation sector. In the New Policies Scenario, road transport is responsible for about 75% of global transport oil demand by 2035, down only slightly from 77% in 2010 (Figure 3.4). Oil demand for road freight grows fastest, by 1.7% per year on average, despite significant fuel-efficiency gains, especially in the United States where recent government proposals for heavy-duty vehicles aim at improving fuel efficiency between 10% and 17% through to 2018.

Figure 3.4 ● **World transportation oil demand by mode in the New Policies Scenario**

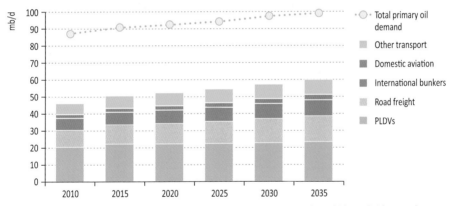

Note: PLDVs are passenger light-duty vehicles comprising passenger cars, sports utility vehicles and pick-up trucks.

Passenger light-duty vehicles (PLDVs) remain the single largest component of transport oil consumption, though their share shrinks from about 45% today to 39% by 2035. This trend is driven by major improvements in fuel economy in many countries, especially in the largest car markets in the United States, China, Europe and Japan – a result both of high international oil prices and of government policies, including fuel efficiency standards, labelling and research and development. These policies make a substantial contribution to curbing overall oil demand growth over the *Outlook* period. However, the July 2011 proposal by the United States government to further increase vehicle fuel-efficiency standards (from the current 35.5 miles-per-gallon [mpg] by 2016) to an average 54.5 mpg by 2025, which is equivalent to 4.3 litres per 100 km (l/100 km), is not taken into account in this year's projections; if confirmed it would have an important impact on longer-term US oil demand (though the long-term target will still be subject to an evaluation well before 2025).

Of the projected increase in transport oil demand between 2010 and 2035, 37% comes from road freight traffic, 21% from PLDVs, 18% from international bunkers, 7% from aviation and the remaining 17% from other modes. International aviation and marine bunkers, as well as domestic aviation and navigation, grow with increasing GDP, but the growth is moderated by fuel economy targets recently announced by the International Maritime Organization for shipping and by energy efficiency measures both in aircraft technology and flight logistics. The increase in demand from road freight traffic comes entirely from non-OECD countries, offsetting a decline in OECD countries resulting from efficiency gains and fuel switching.

Road-freight traffic is strongly correlated with economic growth, as increased levels of consumption lead to greater movement of goods. In non-OECD countries, road-freight tonne-kilometres increase by 3.7% annually, slightly more than the resulting oil demand, due to efficiency improvements (Figure 3.5). Although the tonne-kilometres operated by trucks and lorries grow by 0.5% per year on average until 2035 in the increasingly service-oriented OECD countries, the oil needed to fuel this growth drops in the New Policies Scenario as a result of efficiency improvements.

Figure 3.5 ● **Average annual change in transport oil demand and GDP by region and transport mode in the New Policies Scenario, 2010-2035**

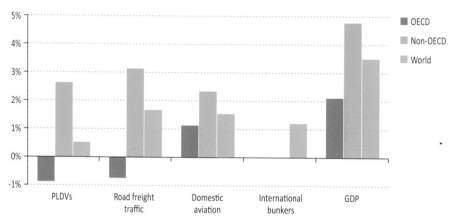

Increased use of alternatives to oil-based transport fuels (gasoline, diesel and LPG) also help to temper oil-demand growth, though to a much smaller degree than efficiency gains in vehicles with internal combustion engines. Biofuels make the biggest such contribution, as use grows from 1.3 million barrels of oil equivalent per day (Mboe/d) today to 4.4 Mboe/d in 2035, an annual rate of increase of 5%. The share of biofuels in total transport fuel demand rises from less than 3% today to just above 6% by 2035. Although biofuels are mainly used in the road transport sector, the aviation industry has recently done several tests on aviation biofuels and, if large-scale projects were successfully implemented, aviation demand could increase strongly. Natural gas also plays a growing role in the transport sector, its share rising from 3% to 4%. The use of natural gas grows most in road transport, where its share rises from 1% to 3%. Currently the dominant use of gas in the transport sector is in gas compression for pipeline transport and distribution.[1] While the economic case for natural gas vehicles is often promising, for example in the United States, there is often a lack of the policy support needed for a more significant uptake (IEA, 2011a). Electricity use is mainly confined to the railway sector in the New Policies Scenario. Electricity makes only minor contributions to the road transport energy mix, but the share of electricity in total transport fuel demand grows from 1% today to about 2% in 2035 (Figure 3.6).

Figure 3.6 ● World non-oil-based fuel use by transport mode in the New Policies Scenario

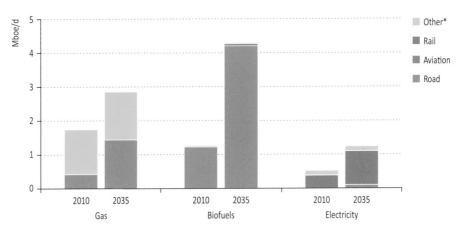

*Other includes navigation, pipeline and non-specified transport.

While theoretically many options exist for replacing oil-based fuels in road transport, for various reasons none of the potential candidates and technologies has so far grown out of niche markets. There are barriers to the uptake of each alternative fuel and vehicle technology, including their applicability to different road transport modes, the need to develop vehicle drive-trains to accommodate the specific properties of the fuel, their cost-competitiveness and their environmental performance relative to oil (Table 3.3). Where the alternative fuel cannot be used directly in existing oil distribution networks and applications, it requires the build-up of a dedicated infrastructure. To compete today, the majority of alternative fuels need government support of one form or another. Where such support is provided, it is often justified by the energy-security or environmental benefits that those fuels can bring (see Chapter 14 for a discussion of biofuels subsidies). For alternative fuels to grow faster than projected in the New Policies Scenario, stronger and more concerted policy action, improved international co-operation and long-term planning would be needed.

---

1. As per IEA definition of transport sector energy use.

Table 3.3 ● Factors affecting fuel choices for future road transport technologies

Fuel	Application	Cost competitiveness	Technology status 2011	Technology potential 2035	Environmental performance	Infrastructure availability	Government policies
Gasoline & diesel (including LPG, methanol and other additives)	All vehicle types.	Low cost of vehicle; running costs rise with oil price; LPG and methanol can lower fuel costs.	Mature fuels and vehicles; potential to use hybrid vehicles.	Vehicle fuel efficiency could increase by 40% (ICE) or 50% (hybrids).	Current emissions high, but potential to decrease with vehicle improvements.	Existing.	Mandates to increase fuel efficiency, labelling.
Biofuels	All vehicle types.	Low cost of vehicle; fuel costs depend on cost and supply of biomass.	Vehicles and conventional biofuels proven; scaling supply difficult.	Advanced fuels could be made; vehicle efficiency similar to gasoline and diesel.	Highly variable by fuel and region, CO_2 not always reduced vs. gasoline and diesel.	Fuels blended into gasoline or diesel using existing infrastructure.	Blending mandates, tax credits, R&D funding.
Natural gas	All vehicle types. LNG has more potential than CNG for HDVs.	Higher vehicle costs; fuel costs depend on oil-to-gas price difference.	CNG vehicles available and proven, LNG less deployed, but technology available.	Vehicle efficiency similar to gasoline and diesel.	Use of gas is less CO_2 emission and pollution intensive than gasoline and diesel.	Exists in some countries, but requires extensive roll-out in new markets.	Sustained policy guidance to grow market share is limited to a few countries.
Electricity	All vehicle types but battery size and weight limit applications.	Vehicle costs high (battery cost at least $500/kWh); low running cost.	Very efficient fuel use, some vehicles available, but no mass production.	Roll-out depends on battery cost reduction and consumer response.	No exhaust from pure EVs; WTW emissions depend on power generation mix.	Required for recharging, but grids available for long-distance transmission.	Subsidies for new cars in several countries.
Hydrogen	Feasible for all vehicles, but most practical for LDVs.	Vehicle costs high (fuel cell cost at least $500/kW).	Vehicles currently at R&D stage, only prototypes available.	Depends on pace of fuel cell and hydrogen storage developments.	Use of hydrogen is clean. WTW emissions depend on hydrogen source.	Required for production, refuelling and transmission.	Limited to R&D support.

Key	Very strong	Strong	Neutral	Weak

Note: Colours (see key) illustrate the relative strengths and weaknesses of fuels in each category.

CNG = compressed natural gas. EV = electric vehicle. HDV = heavy-duty vehicle. ICE = Internal combustion engine. LDV = light duty vehicle. LNG = liquefied natural gas. WTW = well-to-wheels.

Passenger light-duty vehicles

Passenger light-duty vehicles (PLDVs) currently use about 20 million barrels of oil each day, or about 60% of total road oil consumption, and remain the largest oil-consuming sub-sector over the *Outlook* period. The extent of the need for oil to fuel the PLDV fleet will hinge on four main factors: the rate of expansion of the fleet; average fuel economy; average vehicle usage; and the extent to which oil-based fuels are displaced by alternative fuels. In the New Policies Scenario, the projected expansion of the fleet would double PLDV oil consumption between 2010 and 2035 if there were no change in the fuel mix, vehicle fuel efficiencies or average vehicle-kilometres travelled; but the projected increase is limited to about 15%, as a result of switching to alternative fuels and, to a much larger extent, efficiency improvements and the decrease in average vehicle use as non-OECD vehicle markets (where average vehicle use today tends to be lower than in the OECD) become increasingly dominant (Figure 3.7).

Figure 3.7 • World PLDV oil demand in the New Policies Scenario

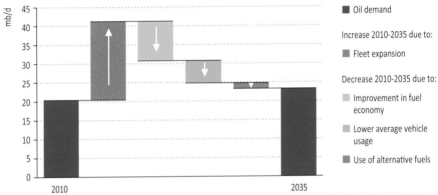

Note: The theoretical increase in oil use caused by fleet expansion assumes no change in the fuel mix, the vehicle fuel efficiency and the average vehicle-distance travelled.

Demand for mobility is strongly correlated with incomes and fuel prices. So as incomes rise – especially in the emerging economies – the size of the global car fleet will inevitably rise in the long term. However, vehicle usage patterns are also affected by incomes and prices. A rise in fuel prices (whether caused by higher prices on international markets or a rise in domestic prices) or a drop in incomes (such as during the global financial crisis) can lead to short-term changes in behaviour. But vehicle-miles travelled usually tend to rebound as consumers become accustomed to the new level of price or as the economy recovers. The United States, one of the largest car markets in the world, is a good example of this phenomenon, partly because public transport infrastructure is limited and most people rely on cars for commuting (Figure 3.8). Government policies to promote modal shifts, like the extension of rail and urban transport networks, can change the long-term picture. The growth in oil demand from expanding vehicle fleets in countries with large inter-city travel distances, such as China, will be critically influenced by the availability of non-road travel options. However, we assume little change in vehicle usage patterns over the *Outlook* period. Efficiency improvements, therefore, remain the main lever to reduce oil demand.

Figure 3.8 ● Change in road vehicle travel in relation to changes in GDP per capita and oil price in the United States, 1985-2010

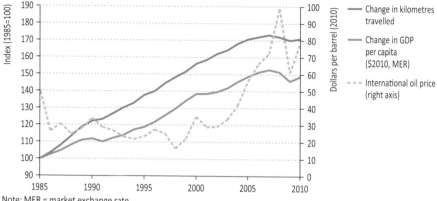

Note: MER = market exchange rate.
Sources: US Federal Highway Association database; IEA data and analysis.

The largest car markets in the world were historically to be found in OECD countries, with the United States the leader, followed by Japan and several European countries, headed by Germany and France (Box 3.1 looks at the regional shift in car making). However, the growth in vehicle sales in China over the past ten years has been spectacular, from less than 1 million in 2000 to almost 14 million in 2010, when China overtook the United States to become the world's largest single car market (Figure 3.9). Vehicle sales in the United States and some European countries plunged during the recession and, despite financial support by governments, have struggled to recover since.

Figure 3.9 ● PLDV sales in selected markets, 2000-2010

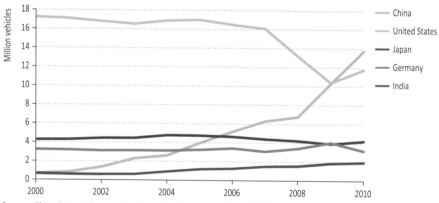

Sources: China Automotive Review statistics database; WardsAuto; IEA databases.

While the rate of growth in PLDV sales in China has been impressive over the last decade, they started in 2000 from a very low base. Even today, vehicle ownership in China is only just above 30 vehicles per 1 000 people, compared with close to 500 in the European Union and 700 in the United States (Figure 3.10). On average, only 40 out of 1 000 people in non-OECD countries own a car today, while average car ownership is approaching 500 per 1 000 people in OECD countries.

Figure 3.10 ● **PLDV ownership in selected markets in the New Policies Scenario**

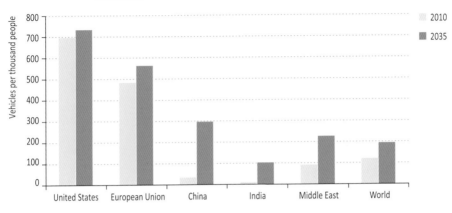

Despite a slowdown in sales in the first half of 2011, China is projected to reinforce its leading position, with PLDV sales reaching more than 50 million cars in 2035 in the New Policies Scenario. Yet the market will still be far from saturated then, with projected vehicle ownership levels of only about 300 vehicles per 1 000 inhabitants. China's total vehicle stock grows more than three-fold from 2010 to 2035 in the New Policies Scenario. The pace of growth of China's vehicle stock will be critical for world oil demand: if the entire stock (including all road-transport modes) grew 1% faster than at the projected level of 5% per year, oil demand in China would be almost 2 mb/d higher by 2035. How likely this is to happen is extremely uncertain. One reason is the heterogeneity of both vehicle ownership levels and the ratio of ownership to average incomes, even among the more developed coastal regions. For example, ownership rates in Beijing in 2009 were almost three times higher than in Shanghai, despite similar levels of GDP per capita.[2] This is partly the result of differences in urban planning and indicates that future vehicle ownership could well be lower than projected.

Non-OECD countries account for 80% of the projected doubling of the global PLDV fleet, their stock growing from around 200 million vehicles in 2010 to more than 900 million in 2035. In the New Policies Scenario, vehicle sales in non-OECD countries as a whole are projected to overtake those of OECD countries around 2020, reaching more than 100 million vehicles per year in 2035. The vehicle stock in non-OECD countries overtakes that in the OECD in the early 2030s (Figure 3.11). Nonetheless, at about 125 vehicles per 1 000 inhabitants in 2035, vehicle-ownership levels in non-OECD countries remain well below OECD levels, where they climb to almost 550 per 1 000 inhabitants. By then, around 55% of all the PLDVs on the road worldwide are in non-OECD countries, compared with less than 30% today.

---

2. According to data from the National Statistics Database of the National Bureau of Statistics of China (www.stats.gov.cn).

PLDV sales in the New Policies Scenario remain dominated by conventional internal combustion engine vehicles, though hybrid vehicles are projected to make significant inroads into global car markets. By 2035, 28% of all PLDV sales in OECD countries are projected to be hybrids, while in non-OECD countries the figure is 18%. The share is less than 2% worldwide today. As in previous *Outlooks*, the inroads made by other alternative vehicles remain marginal: the share of natural gas vehicles reaches about 2% and that of electric vehicles and plug-in hybrids combined is just below 4%.

Figure 3.11 ● PLDV sales and stocks in the New Policies Scenario

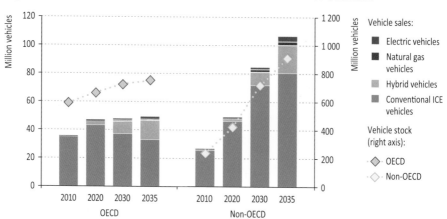

Notes: ICE = internal combustion engine. Electric vehicles include full electric vehicles and plug-in hybrids.

Vehicle fuel efficiency is the other key factor in the outlook for fuel use by PLDVs, and, more generally, transport sector oil demand as a whole. Today, the average tested fuel economy of a new PLDV worldwide is around 8 l/100km. With existing conventional internal combustion engine technology, modest hybridisation and non-engine improvements, such as tyres, efficiency improvements of up to 40% are possible, but not all of this potential is expected to be exploited. Advanced technologies, including hybrids and plug-in hybrids, are also set to contribute to overall improvements in fuel economy. In the New Policies Scenario, with recently adopted government policies and measures and high oil prices by historical standards, we project fuel requirements for all new PLDV sales combined to fall to 6 l/100 km by 2020 and just above 5 l/100 km by 2035. Many countries with large vehicle markets have taken action to improve the fuel economy of vehicles substantially, though in many cases the targets relate to new vehicle sales between 2015 and 2020. The effect on the average efficiency of the entire vehicle stock will not be fully felt until the existing stock is fully replaced. For post-2020 vehicle sales, there is currently little policy guidance, which leaves further efficiency improvements to the market and guides our cautious projections of efficiency gains. However, if policies, such as the currently discussed extension of the US Corporate Average Fuel Economy (CAFE) standards to 2025 were to materialise in a number of key vehicle markets, oil demand could be reduced further and the economic case for alternative fuels and technologies would improve.

Box 3.1 ● The future of car making

The recent growth of emerging markets such as China could generate a shift in global car manufacturing. While car manufacturing has historically been concentrated in the United States, Japan, the European Union (mostly Germany and France) and Korea, non-OECD car manufacturers, especially in China, are on the rise thanks to the boom in domestic sales and lower labour costs. The share of Chinese manufacturers in global car production more than doubled in the three years to 2009, from 6% to about 13% of all cars produced worldwide (OICA, 2011). Western companies are also profiting from this boom by building cars increasingly in non-OECD countries. The share of non-OECD countries in global car production grew from 23% in 2005 to 44% in 2010, with total output rising from 10.5 million to 25.6 million vehicles. OECD vehicle production, which was around 36 million vehicles per year in the mid-2000s, plunged to 28 million cars in 2009, but rebounded to close to 33 million in 2010. Although developments in car markets over the last five years have been influenced by the economic crisis and government support programmes, it is likely that the level of car production in non-OECD countries could overtake that of OECD countries before 2015.

Wherever these cars are built, the growing importance of non-OECD car markets means that non-OECD policy will become an increasingly important determinant of global fuel economy standards and therefore of global oil demand patterns. China has recently adopted a fuel economy target of 6.7 l/100 km for 2015, comparable to the 2015 target in the United States, and the government is considering further increasing this standard to 4.5 l/100 km (equivalent to the fuel consumption of a current hybrid) by 2025.

Government policies will be important in encouraging or mandating improvements in fuel economy. Generally, standards have an important influence on average fuel economies, given their binding nature. The impact of non-binding policies, such as voluntary agreements or labelling, can be hard to predict, as running costs are just one of many factors that private motorists take into account when deciding which car to buy and how to use it. The quickest and cheapest way to reduce oil use by PLDVs (other than by behavioural changes such as fuel-efficient driving or modal shifts towards increased use of public transport) is to utilise smaller cars. As an illustration of this effect, if US consumers today bought cars of the same size and weight on average as in Europe, new vehicle fuel economy would improve by about 30% per kilometre driven and total US oil demand would fall by 2.4 mb/d within about fifteen years. But vehicle size, comfort and status are important considerations in a consumer's decision to buy a car and so it is not certain that consumers will change purchasing or driving behaviour quickly in the absence of incentives that go beyond those already announced. Recent research has shown that US motorists tend to undervalue the potential benefits of fuel economy (Greene, 2010). Nonetheless, in countries where households devote a large share of their budget to spending on fuel for their cars, fuel economy has been shown to be an increasingly important consideration as fuel prices rise. In the United States, the share of household spending on vehicle fuel grew from about 3% in the early 1990s and 2000s to more than 5% in 2008, though it has varied since with changing oil prices.

The rate of penetration of alternative fuels is another factor affecting oil use in PLDVs. From a technology perspective, the share of biofuels is largely a supply-side issue, as in most cases biofuels are blended into conventional gasoline or diesel, requiring no change to today's vehicles if mixtures are kept within certain limits. Flex-fuel vehicles, *i.e.* vehicles that can use either gasoline or ethanol, are on sale in several markets, but some other fuels, notably electricity, require a fundamental change in vehicle technology and refuelling infrastructure. In the New Policies Scenario, electric vehicles and plug-in hybrid sales expand only slowly, amounting to 0.4 million vehicles in 2020 and just below 6 million in 2035. This slow increase is mainly the result of high vehicle purchase costs which, in the absence of sufficient support or incentives, make them less economically attractive than internal combustion engine vehicles. Even as new business models improving the economic case for electric vehicles emerge, consumer acceptance that is required for the widespread adoption of electric vehicles remains uncertain and guides our cautious outlook. In the projections, oil savings from electric vehicles worldwide amount to less than 20 thousand barrels per day (kb/d) in 2020 (Figure 3.12) and about 170 kb/d in 2035.

Figure 3.12 ● **Oil savings through electric vehicle sales: country targets,[3] industry plans and in the New Policies Scenario**

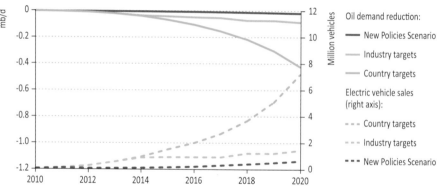

The projected levels of sales of electric vehicles and plug-in hybrids in the New Policies Scenario are well below targets for 2020 that have been adopted by several countries; in aggregate, these targets amount to a stock of almost 25 million electric vehicles and plug-in hybrids, implying vehicle sales to reach about seven million per year by 2020. If the targets were achieved, sales of electric vehicles would make up almost 8% of global PLDV sales and just over 2% of the global stock by 2020, saving up to 420 kb/d of oil in that year. However, the car industry does not appear to be prepared to produce that many electric vehicles yet (IEA, 2011b). Plans announced up to mid-2011 suggest that car manufacturers are likely to expand capacity in total to only 1.4 million electric vehicles per year by 2020, suggesting a

---

3. Countries included: Austria, Canada, China, Denmark, Finland, France, Germany, Ireland, Israel, Japan, Korea, Netherlands, Portugal, Spain, Sweden, Switzerland, the United Kingdom and the United States. Note that the country target of China that is used here is currently under revision. Some of these countries have launched the Electric Vehicles Initiative together with the IEA.

wait-and-see strategy in a still nascent industry, which in turn could mean that countries would collectively achieve only 20% of their targets, with resulting oil savings of less than 80 kb/d.

We estimate that investment in manufacturing capacity able to deliver the targeted number of electric vehicles (even if that target is unlikely to be met from today's perspective) would be of the order of $85 billion in the period to 2020.[4] Additional investment would be required to provide the recharging infrastructure (each electric vehicle typically requires 1.3 recharging points), amounting to roughly $50 billion.[5] The consumer would also be required to pay more for these vehicles, which currently cost at least $15 000 more than a conventional vehicle of equivalent size. Assuming that this cost increment could be reduced by 50% by 2020, then the additional spending on electric vehicles until 2020 would be about $230 billion. These costs would need to be carried by the consumers or, to the extent that they are subsidised, by governments. It is still unclear how much more the consumer is willing to pay for an electric vehicle or what is the desired payback period, *i.e.* the time it takes for the fuel savings to offset the higher upfront purchase price of the vehicle. Some recent tests of public acceptability by individual car manufacturers have given promising results; but for electric vehicle adoption to become widespread, it is estimated that payback times will need to be reduced by a factor of about three to four, or mitigated by innovative manufacturer-consumer business models.

Supply

Resources and reserves

Proven reserves of oil increased in 2010, by about 0.5% to 1 526 billion barrels at year-end according to the *BP Statistical Review of World Energy* (BP, 2011) and by 8.5% to 1 470 billion barrels according to the *Oil & Gas Journal* (O&GJ, 2010). The bigger increase recorded by the O&GJ reflects an increased estimate for Venezuela, based on recent results from the Magna Carta project, led by Petróleos de Venezuela (PDVSA), which reassessed reserves in the Orinoco extra-heavy oil belt (Box 3.2 defines different types of liquid fuels). Neither set of numbers includes the upward revisions of reserves announced in late 2010 by Iraq and Iran.

An estimated 16 billion barrels of crude oil reserves were discovered in 2010. This is less than the 25 billion barrels of crude oil produced in 2010, but well above the 8 billion barrels per year of discoveries required to satisfy production from fields "yet to be found" called for over the projection period in the New Policies Scenario (Figure 3.16).[6] The amount of crude oil discovered annually over the last decade averaged 14 billion barrels. The trends continue to be towards both smaller discoveries in mature basins and larger discoveries in more remote places, with the significant exception of light tight oil in North America (Spotlight). Although light tight oil has been known for many years, extraction of it has only recently become economically viable, through the application of new technology and the advent of higher oil prices.

---

4. This is derived from an assumed cost base of $12 200 per vehicle per year, based on recent Nissan/Renault announcements.
5. This calculation assumes costs of $750 for home recharging points and $5 000 for public recharging points. It further assumes that the bulk of recharging is done at home as evidenced in recent pilot projects, so that the required density of public recharging points is low.
6. These numbers do not include volumes of NGLs discovered and produced.

Box 3.2 ● **Definitions of different types of liquid fuels**

For the purposes of this chapter, the following definitions are used (Figure 3.13):

● Oil comprises crude, natural gas liquids, condensates and unconventional oil, but does not include biofuels (for the sake of completeness and to facilitate comparisons, relevant biofuel quantities are separately mentioned in some sections and tables).

● Crude oil makes up the bulk of the oil produced today; it is a mixture of hydrocarbons that exist in liquid phase under normal surface conditions. It includes light tight oil. It also includes condensates that are mixed-in with commercial crude oil streams.

● Natural gas liquids (NGLs) are light hydrocarbons that are contained in associated or non-associated natural gas in a hydrocarbon reservoir and are produced within a gas stream. They comprise ethane, propane, butanes, pentanes-plus and condensates.

● Condensates are light liquid hydrocarbons recovered from associated or non-associated gas reservoirs. They are composed mainly of pentanes and higher carbon number hydrocarbons. They normally have an American Petroleum Institute (API) gravity of between 50° and 85°.

● Conventional oil includes crude oil and NGLs.

● Unconventional oil includes extra-heavy oil, natural bitumen (oil sands), kerogen oil, gas-to-liquids (GTL), coal-to-liquids (CTL) and additives.

● Biofuels are liquid fuels derived from biomass, including ethanol and biodiesel.

Figure 3.13 ● **Liquid fuel schematic**

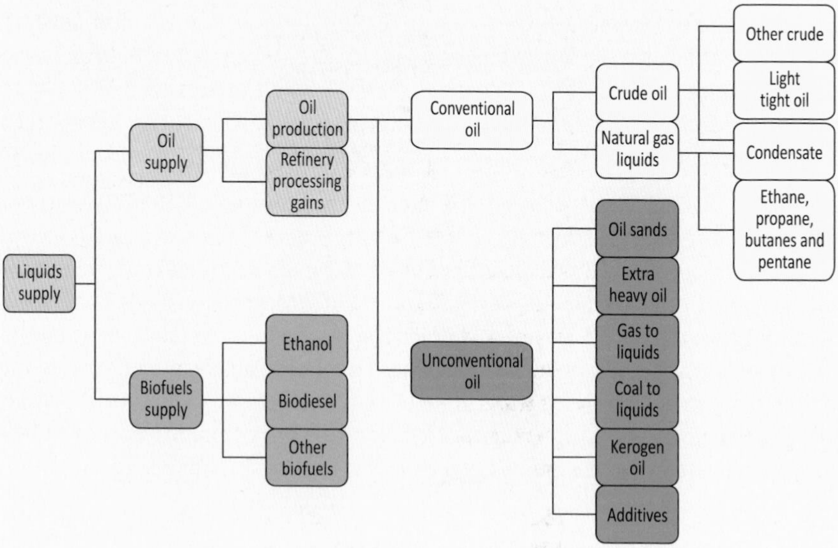

Upward revisions to reserves in fields already discovered, by both field extensions and enhanced oil recovery techniques, amounted to some 13 billion barrels in 2010. Thus, total reserves additions amounted to 29 billion barrels outstripping production by 4 billion barrels (as has been the case now for several decades). Overall, the geographic distribution of proven reserves remains largely unchanged from previous years, with the bulk of them located in the Middle East.

We now put remaining recoverable resources worldwide at nearly 5 500 billion barrels, with proven reserves[7] amounting to about one-quarter of the total (Figure 3.14). Other recoverable conventional oil resources amount to 1 300 billion barrels, while recoverable unconventional resources total over 2 700 billion barrels. Unconventional oil resources, including extra-heavy oil and kerogen oil, have a large potential; however, many technical, commercial and political obstacles need to be overcome before they can be fully developed. Recoverable resources, a more important factor in long-term production projections than proven reserve volumes, are estimated on the basis of data from various sources, notably the US Geological Survey (USGS) and the German Federal Institute for Geosciences and Natural Resources (BGR) (USGS, 2000, 2008a, 2009a and 2010; BGR, 2010).[8] For unconventional resources, these studies have tended to focus on large known accumulations or specific regions, particularly those where conventional resources have been heavily exploited. It is likely that regions that have received less attention, such as the Middle East or Africa, also have important endowments of unconventional resources.

Figure 3.14 ● Recoverable oil resources and production by region and type in the New Policies Scenario

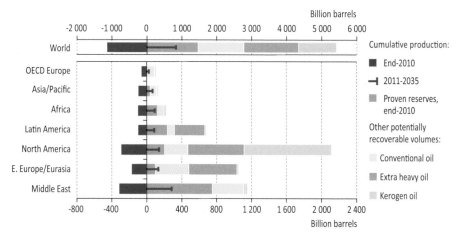

Note: Cumulative production is shown as a negative number, such that the total of the bars to the right indicate remaining recoverable resources.

Sources: BGR (2010); USGS (2000, 2008a, 2009a, 2009b and 2010); O&GJ (2010); IEA databases and analysis.

---

7. Proven reserves are usually defined as discovered volumes for which there is a 90% probability of profitable extraction (see *WEO-2010* for a detailed description) (IEA, 2010).

8. See *WEO-2010* for a detailed discussion of the methodology and definitions of resources, a detailed analysis of the prospects for production of unconventional oil, and a review of the sensitivity of sources of supply to resource availability and accessibility (IEA, 2010).

Production prospects

Global oil production, together with processing gains (the volume increase in supply that occurs during crude oil refining) matches demand in all three scenarios in this *Outlook*, and is of course always less than the installed production capacity.[9] Production trends vary markedly across the scenarios, rising to 104 mb/d in 2035 in the Current Policies Scenario and 96.4 mb/d in the New Policies Scenario, but peaking before 2020 and then falling to 76 mb/d in the 450 Scenario (Table 3.4). In the New Policies Scenario, production grows by 13 mb/d, or 15%, between 2010 and 2035 (Figure 3.15). The rate of growth falls gradually through the projection period, from 1.1% per year from 2010 to 2015 to 0.5% per year in the period 2030 to 2035, averaging 0.6% per year over the full projection period. Crude oil production – the largest single component of oil production – increases marginally to a plateau of around 69 mb/d (slightly below the historic high of 70 mb/d in 2008)[10] and then declines slowly to around 68 mb/d by 2035.

Table 3.4 ● Oil production and supply by source and scenario (mb/d)

	1980	2010	New Policies Scenario 2020	New Policies Scenario 2035	Current Policies Scenario 2020	Current Policies Scenario 2035	450 Scenario 2020	450 Scenario 2035
OPEC	25.5	34.8	39.6	48.7	41.1	53.4	36.7	36.5
Crude oil	24.7	29.0	29.8	34.7	31.0	37.6	27.8	26.7
Natural gas liquids	0.9	5.2	8.0	11.0	8.2	12.5	7.2	7.6
Unconventional	0.0	0.6	1.8	3.0	1.8	3.3	1.7	2.2
Non-OPEC	37.1	48.8	50.4	47.7	51.0	50.5	49.0	39.4
Crude oil	34.1	40.3	39.0	33.1	39.4	34.7	38.2	27.6
Natural gas liquids	2.8	6.4	7.3	7.5	7.4	7.8	6.9	6.3
Unconventional	0.2	2.1	4.1	7.0	4.2	8.0	3.9	5.5
World oil production	62.6	83.6	90.0	96.4	92.0	103.9	85.7	75.9
Crude oil	58.8	69.3	68.8	67.9	70.5	72.3	66.0	54.3
Natural gas liquids	3.7	11.7	15.3	18.5	15.6	20.3	14.1	13.9
Unconventional	0.2	2.6	5.9	10.0	6.0	11.3	5.7	7.8
Processing gains	*1.2*	*2.1*	*2.5*	*3.0*	*2.5*	*3.2*	*2.4*	*2.4*
World oil supply*	63.8	85.7	92.4	99.4	94.6	107.1	88.1	78.3
Biofuels	*0.0*	*1.3*	*2.3*	*4.4*	*2.1*	*3.4*	*2.7*	*7.8*
World liquids supply**	*63.9*	*87.0*	*94.7*	*103.7*	*96.7*	*110.6*	*90.8*	*86.1*

*Differences between historical supply and demand volumes are due to changes in stocks. **Includes biofuels, expressed in energy equivalent volumes of gasoline and diesel.

---

9. Production in this *Outlook* refers to volumes produced, not to installed production capacity.

10. Revisions made to production data since the publication of *WEO-2010* show worldwide production of crude oil in 2008 averaged 70.4 mb/d, slightly above the 70.2 mb/d in 2006.

Figure 3.15 ● World oil production in the New Policies Scenario, 2010 and 2035

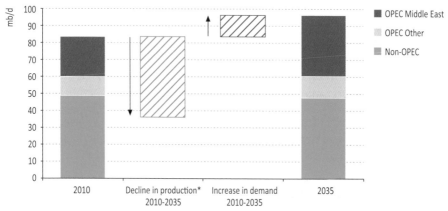

*Of oil fields producing in 2010.

The need to bring new production capacity on stream in the New Policies Scenario is much greater than the projected increase in production. This is because of the need to compensate for the decline in production from currently producing fields as they come off plateau, following the natural course of depletion.[11] We project that crude oil production from fields that were producing in 2010 will drop from 69 mb/d to 22 mb/d by 2035 – a fall of over two-thirds (Figure 3.16). This decline is twice the current oil production of all OPEC countries in the Middle East. As a result, the gross additional capacity needed to maintain the current production level is 17 mb/d by 2020 and 47 mb/d by 2035. The necessary capacity additions will come largely from fields already discovered but yet to be developed, mainly in OPEC countries. This projection is derived from detailed analysis of the production profiles of different types of fields in each region. Natural gas liquids (NGLs) production from currently producing fields also declines, though this is more than offset by the significant increase in new NGLs production associated with increased gas production (see Chapter 4).

Figure 3.16 ● World liquids supply by type in the New Policies Scenario

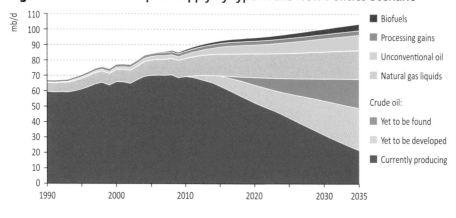

11. See *WEO-2008* for detailed analysis of field decline rates (IEA, 2008).

NGLs, produced together with natural gas, provide a growing contribution to global oil supply, driven by the projected growth in gas production and reduced gas flaring (associated gas is often rich in NGLs) and wetter gas production in some countries.[12] Unconventional oil, such as Canadian oil sands, Venezuelan extra-heavy oil, coal-to-liquids (CTL), gas-to-liquids (GTL), kerogen shales and additives, also play an increasingly important role. Current low gas prices in North America have triggered renewed interest in GTL there. For example, Sasol (a South Africa-based leader in GTL technology) is pursuing feasibility studies into the economic viability of GTL plants fed by shale gas located in Louisiana and separately in British Columbia with Talisman, an independent Canadian exploration and production company. Other companies are thought to be considering GTL projects too. Therefore, our GTL projections have been updated to include increased production in North America. Interest in CTL (as well as in methanol to be used as a gasoline blending agent) remains strong in China. High oil prices and growth in the sale of M15 gasoline (a blend of gasoline with 15% methanol by volume) and M85 gasoline (with 85% methanol) have triggered a boom in methanol production worldwide, with previously mothballed plants being reactivated in several countries. Methanol is often cheaper to produce than ethanol, but for technical reasons cannot be blended into gasoline in large concentrations without modifications to the vehicle engine and fuel system. It also requires additional safeguards for handling due to its high toxicity. For the purposes of our projections, it is classified as an additive, within the category of unconventional oil.

Figure 3.17 ● Major changes in liquids supply in the New Policies Scenario, 2010-2035

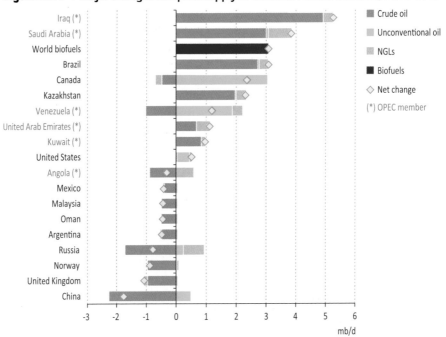

12. NGLs are added to crude oil when reporting production in volume terms. However, the balance between demand and supply is made on an energy equivalent basis, taking into account the lower energy per unit volume of NGLs compared with crude oil. Additional investment in polymerisation units in refineries is included in the investment projections to enable NGLs to replace crude oil, particularly in producing transport fuels.

By country, Iraq provides the biggest projected increase in production between 2010 and 2035. Most OPEC countries increase production over the *Outlook* period, while production falls in most non-OPEC countries. The main exceptions are Brazil, Canada and Kazakhstan, all of which see an increase in output of over 2 mb/d or more (Figure 3.17). The biggest projected declines occur in China, the United Kingdom, Norway and Russia.

Non-OPEC production

In the New Policies Scenario, non-OPEC oil production in aggregate declines slightly over the projection period, from 48.8 mb/d in 2010 to 47.7 mb/d in 2035. However, this trend masks significant underlying changes: the 5 mb/d increase of crude oil production from Brazil and Kazakhstan is more than offset by declines in most other non-OPEC producers, resulting in an overall decline of 7 mb/d of crude oil output. The balance is partially redressed by an increase of 5 mb/d in the output of unconventional oil, mainly from oil sands in Canada, CTL projects (principally in China, South Africa and to a lesser degree the United States) and an increase in NGL production of 1 mb/d (Table 3.5).

Latin American production decreases as mature declining basins are only partly offset by increased production from the deepwater pre-salt play off Brazil. In late 2010, the state oil company, Petrobras, launched the world's largest-ever share offering, raising around $67 billion, to support its plans to invest $128 billion over the five years to 2016 in developing pre-salt discoveries. This will involve building 11 floating production, storage and offloading vessels, up to 28 drilling rigs (in Brazilian yards) and 146 supply boats to support the drilling of up to 500 wells. This effort, along with Petrobras' other projects, is projected to drive up Brazil's production from 2.1 mb/d in 2010 to about 3 mb/d in 2015 and 4.4 mb/d by 2020. We project production to rise further to 5.2 mb/d by 2035.

The Caspian region strengthens its role as a key oil producer and exporter throughout the projection period.[13] Kazakhstan will be an important area of new oil production capacity, with production projected to increase from 1.6 mb/d in 2010 to nearly 4 mb/d in 2035. Kazakhstan's projected increase in production has been pushed back in this year's *Outlook*, due to continued wrangling over the Kashagan oilfield development. This super-giant field was discovered in 2000 and initial production is anticipated in late 2012 at the earliest. Production from the first development phase should ramp up to 300 kb/d and subsequent phases could see total production reach 1.3 mb/d by 2030. The development has been plagued by cost over-runs and schedule delays due to technical complexities. The Kazakh government did not approve the Phase 2 development plan in early 2011, citing concerns about the cost and schedule.

Canada sees rapid growth in oil production throughout the projection period, mainly from oil sands (output of which grows from 1.5 mb/d in 2010 to 4.5 mb/d in 2035). This could be hindered by continuing debate about the environmental effects of oil sands exploitation (IEA, 2010). However, the Canadian oil industry and provincial and federal governments continue to seek ways to reduce the environmental footprint and less intrusive in-situ extraction techniques are becoming more common. For example, the government of Alberta has committed to spend $2 billion on developing carbon capture and storage projects.

---

13. See *WEO-2010* for a detailed analysis of energy prospects for the Caspian region (IEA, 2010).

Table 3.5 ● Non-OPEC oil* production in the New Policies Scenario (mb/d)

	1980	2010	2015	2020	2025	2030	2035	2010-2035 Delta	2010-2035 %**
OECD	17.3	18.9	19.4	18.8	18.6	18.8	18.9	0.0	0.0%
Americas	14.2	14.1	15.0	15.2	15.6	16.1	16.6	2.4	0.6%
Canada	1.7	3.4	4.3	4.5	5.0	5.4	5.7	2.3	2.1%
Mexico	2.1	3.0	2.7	2.5	2.4	2.5	2.5	-0.4	-0.6%
United States	10.3	7.8	8.0	8.2	8.1	8.2	8.3	0.5	0.3%
Europe	2.6	4.2	3.7	2.9	2.4	2.1	1.8	-2.3	-3.2%
Asia Oceania	0.5	0.6	0.8	0.7	0.6	0.5	0.5	-0.1	-0.5%
Non-OECD	19.8	29.9	31.4	31.6	31.4	30.4	28.8	-1.1	-0.2%
E. Europe/Eurasia	12.5	13.7	14.2	14.0	14.6	15.2	15.1	1.4	0.4%
Russia	11.1	10.5	10.4	9.9	9.7	9.7	9.7	-0.8	-0.3%
Kazakhstan	0.5	1.6	1.9	2.3	3.3	3.9	3.9	2.3	3.6%
Asia	4.5	7.8	7.9	7.5	6.8	5.7	4.7	-3.1	-2.0%
China	2.1	4.1	4.2	4.2	3.8	3.0	2.3	-1.8	-2.2%
India	0.2	0.9	0.8	0.7	0.7	0.7	0.6	-0.3	-1.4%
Middle East	0.5	1.7	1.5	1.2	1.2	1.1	1.0	-0.7	-2.2%
Africa	1.0	2.6	2.6	2.4	2.2	2.0	1.8	-0.7	-1.3%
Latin America	1.3	4.1	5.3	6.4	6.6	6.4	6.2	2.1	1.7%
Brazil	0.2	2.1	3.0	4.4	5.1	5.2	5.2	3.1	3.6%
Total non-OPEC	37.1	48.8	50.8	50.4	50.0	49.2	47.7	-1.1	-0.1%
Non-OPEC market share	59%	58%	57%	56%	55%	52%	49%	n.a.	n.a.
Conventional	37.0	46.8	47.5	46.3	44.9	43.1	40.6	-6.1	-0.6%
Crude oil	34.1	40.3	40.5	39.0	37.5	35.6	33.1	-7.2	-0.8%
Natural gas liquids	2.8	6.4	7.0	7.3	7.4	7.5	7.5	1.1	0.6%
Unconventional	0.2	2.1	3.3	4.1	5.1	6.1	7.0	5.0	5.1%
Share of total non-OPEC	0%	4%	6%	8%	10%	12%	15%	n.a.	n.a.
Canada oil sands	0.1	1.5	2.5	2.9	3.6	4.0	4.5	3.0	4.4%
Gas-to-liquids	-	0.0	0.0	0.1	0.2	0.3	0.4	0.4	9.1%
Coal-to-liquids	0.0	0.2	0.2	0.4	0.7	0.9	1.2	1.1	8.5%

*Includes crude oil, NGLs and unconventional oil. **Compound average annual growth rate.

Russia, discussed in detail in Chapter 8, maintains its position as the largest oil producer in the world for the next few years, before it is overtaken again by Saudi Arabia. Initially relatively flat at above 10 mb/d, Russian oil production then slowly declines to 9.7 mb/d in 2035, as new greenfield developments struggle to keep pace with the increasing decline in production from older fields. Russian fiscal policy is a key determinant of when and how quickly Russian production will decline. Current terms limit the incentive to invest when prices rise; our projections assume sympathetic evolution of taxation.

In the United States, declines in production from mature conventional basins in Alaska and in the lower-48 states are outweighed by strong growth in supplies of light tight oil (Spotlight), rising NGL production and deepwater production in the Gulf of Mexico, which resumes its growth following the Deepwater Horizon disaster (Box 3.3). There is further potential for upside in production from all of these sources: if shale gas producers target plays with higher liquids content, NGLs would increase, additional light tight oil plays could be developed and more of the recent discoveries in the Gulf of Mexico (or future Arctic discoveries) could be developed. CTL and GTL also contribute to the growth in total supply, particularly in the later part of the projection period, when kerogen oil also adds to the production increase. Biofuels are an important contributor to total liquids production, with the United States currently producing almost half of the world's biofuels.

S P O T L I G H T

The new American revolution: light tight oil

Light tight oil provides another good example of how the industry continues to innovate, developing new techniques and technologies to tap previously uneconomic resources. The term refers to oil produced from shale, or other very low permeability rocks, with technologies similar to those used to produce shale gas, *i.e.* horizontal wells and multi-stage hydraulic fracturing (see Box 4.1). Geologically, light tight oil is an analogue of shale gas; oil or gas have either not been expelled from the (shale) source rock or have migrated only short distances into other, usually low permeability, rock formations, adjacent to or within the shale itself. This is why light tight oil is often referred to as shale oil. Unfortunately, the term shale oil is also often used to refer to oil produced by industrial heat treatment of shale, which is rich in certain types of kerogen – a mixture of solid organic material. To avoid confusion, we refer to this as kerogen oil, or kerogen shale. In line with industry practice, we classify light tight oil as conventional oil and kerogen oil as unconventional.

Interest in light tight oil started with the Bakken shale, a large formation underlying North Dakota and extending into Saskatchewan, Manitoba and Montana. Production from the Bakken in North Dakota began on a small scale in the early 1950s, reaching about 90 kb/d in the early 2000s, after operators began drilling and fracturing horizontal wells. After 2005, production increased more rapidly, to average some 310 kb/d in 2010 and reach a high of 423 kb/d in July 2011 (NDSG, 2011). The combination of success in the Bakken and the widening of the differential between oil and gas prices prompted a flurry of interest in developing light tight oil throughout North America (IEA, 2011c). Shale gas producers trying to improve investment returns began targeting acreage with wetter gas, *i.e.* higher liquid content, increasing their NGL production, and some progressed further to shales containing predominantly oil, like the Bakken. Drilling activity is also shifting from gas to oil in the Eagle Ford play in Texas, where oil production was on average only 10 kb/d in 2010 (RRC, 2011), and is also rising in the Niobrara play in Colorado, Utah and Wyoming, in various plays in California (including the Monterrey play) and in the Cardium and Exshaw plays in Canada. Other recently identified plays are also likely to be developed.

The size of light tight oil resources and how much can be technically and economically extracted are still poorly known. The United States Geological Survey estimated in the past that the Bakken held about 4 billion barrels of recoverable oil, but this is now widely considered to be an underestimate, and it is being re-evaluated (USGS, 2008b). Recoverable light tight oil resources in the lower 48 states were recently estimated to be at least 24 billion barrels (US DOE/EIA, 2011a), larger than the country's 22.3 billion barrels of proven reserves (US DOE/EIA, 2010).

Like shale gas, production from light tight oil wells declines rapidly. However, initial production rates vary widely, depending on geology, well lengths and the number of fractures carried out. Averaged over the first month, Bakken wells produce from 300 barrels per day (b/d) to more than 1 000 b/d, but within five years the great majority of wells are producing less than 100 b/d. Over their life-time they typically recover between 300 thousand barrels (kb) and 700 kb. Given the steep declines in production, new wells are constantly needed to maintain output; hence, constant drilling activity is essential to production growth. During 2010, the Bakken rig fleet doubled to 160 units, which drilled 700 new wells, increasing monthly average oil production between January and November by about 120 kb/d. However, in December winter weather disrupted activity and no new wells came on stream, resulting in monthly average oil production dropping by nearly 13 kb/d.

High oil prices have been a key driver of this growth, given that the breakeven oil price for a typical light tight oil development is around $50/barrel (including royalty payments). The surge in production has outpaced infrastructure developments. Oil pipelines are becoming congested, adding to the surplus of oil at the Cushing hub in Oklahoma, which serves as the delivery point for West Texas Intermediate (WTI) crude oil futures contracts. This is partly why WTI currently trades at an exceptionally large discount to other benchmark crudes, such as Brent. This discount is likely to ease in the longer term as new transport capacity is built, particularly to the refineries on the Gulf Coast. In North Dakota, the lack of gas treatment and transport facilities has led to an increase in flaring, with a record 29% of produced gas being flared in the first half of 2011.

To provide an indication of prospective light tight oil production in the United States, we have modelled the output potential of three plays which are currently active; the Bakken, Niobrara and Eagle Ford. Representative well production profiles were used for each of the plays and the assumption made that by 2013 the drilling activity in each play rises to the current level in the Bakken, is then maintained until 2020, after which it falls steadily, to zero by 2035. In total some 30 000 wells are drilled. Production reaches about 1.4 mb/d by 2020, peaking shortly thereafter, as drilling rates drop. Cumulative oil recovery by 2035 is about 11 billion barrels, or slightly less than half of the latest estimated recoverable light tight oil resources in the United

States (US DOE/EIA, 2011a). Of course, any upward reassessment of these resources could lead to higher production rates or maintenance of plateau production for longer, while new plays are likely to be developed if the current levels of profitability are sustained.

Figure 3.18 ● Light tight oil production potential from selected plays

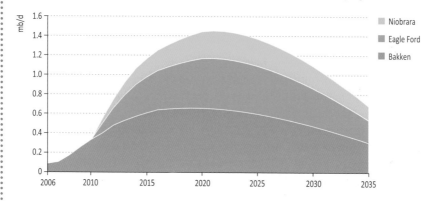

So does light tight oil represent a new energy revolution? It is certainly having an impact in the United States, where we estimate production could exceed 1.4 mb/d by 2020, somewhat reducing US imports; but this alone is unlikely to affect the dynamics of global oil supply significantly. For this new source of oil to have a wider impact, production would need to take off in other plays or in other parts of the world. The resources outside North America have not been quantified but, as with shale gas, it is likely that light tight oil is present in many locations worldwide. For example, the Paris basin in France could hold several billion barrels of recoverable resources, and tests are being conducted in the Neuquen basin in Argentina. As is the case for shale gas, environmental concerns or regulatory constraints could hinder developments, but the potential economic and energy security benefits could result in significantly higher growth of light tight oil production.

Mexico's production is projected to continue to decline in the first half of the projection period, due to the slow pace of new developments, and then sees slow growth in line with the country's resource potential, particularly in its deepwater sector of the Gulf of Mexico. For many years Mexico's production relied heavily on the super-giant offshore Cantarell field, discovered in 1976. Cantarell's production has declined by about 1.5 mb/d over the last decade, though this has been partially offset by production from the adjacent Ku Maloob Zaap complex, overall the country's output is still 0.9 mb/d lower than its previous peak. Legislation was changed in 2008 to allow Petróleos Mexicanos (PEMEX), the state oil company, to sign incentivised development contracts with other oil companies. The first three contracts for secondary development of small, mature onshore fields were awarded in August 2011. Larger contracts, which could have a more substantial impact on the country's production, are expected to be tendered in future.

Box 3.3 ● The long-term implications of the Deepwater Horizon disaster

Following the Macondo blowout, the US government imposed a moratorium in May 2010 on all deepwater drilling. Although the moratorium was lifted after five months, no further drilling permits were issued until February 2011, almost ten months after the blowout. As a result we estimate that deepwater Gulf of Mexico production in 2011 will be 300 kb/d lower than estimated prior to the disaster and that it could take until 2015 for production to catch-up with pre-Macondo projections.

Although activity in the Gulf of Mexico was at a standstill for much of 2010, both operators and contractors have maintained a strong interest and presence in the region. Chevron took a final investment decision on the $7.5 billion Jack/Saint Malo project in late 2010 and Shell did likewise with the Cardamom field, with 140 Mboe of recoverable volumes, in mid-2011. In June 2011, ExxonMobil announced a new discovery, with an estimated recoverable volume of 700 Mboe. Of the approximately 30 deepwater rigs active in the Gulf of Mexico in early 2010, only seven had moved to other regions by mid-2011, leaving sufficient capacity to deliver these new projects and continue exploration.

The Macondo disaster has led to extensive reappraisal of all aspects of deepwater development. Regulators and operators alike have been closely examining the findings of numerous inquiries, in an effort to avoid a repeat incident. While not all of the inquiries are complete, regulations are likely to tighten and operators will become more cautious. In the Gulf of Mexico, two consortia have formed organisations to provide emergency response capabilities for well-control incidents: ten operators, led by the super-majors, have created the Marine Well Containment Company; and 24 independents have clubbed together in the Helix Well Containment Group, which will utilise equipment used to cap and kill the Macondo well. While these developments inevitably lead to increased costs, this is not anticipated to greatly inflate deepwater costs, though some smaller companies, lacking depth of experience or the skill levels needed for deepwater developments, may choose to concentrate on other opportunities.

Outside the United States, there has been little discernable reduction in activity. Angola held a deepwater licensing round in 2010, awarding acreage in January 2011, and Trinidad and Tobago has announced its intention to hold a deepwater bid round later in 2011. Deepwater exploration continues in East Africa and Asia, with several discoveries reported. There has been a revival of interest in the South China Sea. Petrobras has continued exploration in Brazilian deepwater basins, finding the Libra and Franco fields in 2010 and pursuing a swathe of developments, with ambitious production targets – the largest deepwater development programme ever undertaken. Deepwater production increases in importance throughout the *Outlook* period, as exploration for and discoveries of new fields, particularly in non-OPEC countries, increasingly move to deeper, more difficult areas.

Europe's production decline is well established, with North Sea output dropping from just under 6 mb/d in 2000 to just over 3.5 mb/d in 2010. It is projected to fall further, to 1.6 mb/d by 2035, though the decline may be offset to some extent by increasing NGL production and possible development of new discoveries in the exploration frontiers of Greenland, the Barents Sea or exploitation of light tight oil on land. While recent discoveries offshore Norway could yield over a billion barrels of recoverable resources, these are insufficient to reduce the aggregate rate of decline of the North Sea significantly.

In the Asia-Pacific region, Australia's conventional oil production stays above 0.5 mb/d thanks to rising NGL output from the large liquefied natural gas (LNG) projects that are currently under construction, offsetting declines in crude production. China maintains production at its current level of around 4 mb/d until about 2025, before conventional resource limitations lead to declining output. This decline is offset to some extent by growing unconventional production, including CTL. There are some promising developments in non-OPEC Africa countries, notably offshore West Africa (Ghana, Cameroon and the more speculative Liberia, Sierra Leone and Sao Tome and Principe) and in East Africa (particularly in Uganda and Tanzania). Collectively, their contribution is projected to slow the decline in production of the region as a whole, from 2.6 mb/d in 2010 to 1.8 mb/d in 2035. In non-OPEC Middle East countries, there is renewed interest in developing the vast kerogen oil resources in Jordan, with at least three possible projects being studied (another kerogen oil project is being considered in Morocco, where pilot projects have been undertaken in the past). Omani and Syrian oil production continues to decline, with few new discoveries. Overall, production from these countries in aggregate is projected to decline from 1.7 mb/d in 2010 to 1 mb/d in 2035.

OPEC production

In the New Policies Scenario, OPEC countries in aggregate account for all of the increase in global oil production between 2010 and 2035. Consequently, OPEC's share of oil production grows steadily through the projection period, from 42% in 2010 to 44% in 2020 and 51% in 2035 – a level close to its historical peak just before the first oil crisis in 1973. Growth in OPEC output comes mainly from the Middle East, where production grows over 12 mb/d, compared with an increase outside the region of just under 1.7 mb/d (Table 3.6). This reflects the pattern of resource endowments and, in part, the onset of decline in Angolan production in the second half of the projection period, as the country's prolific deepwater discoveries, that were brought on stream in the 2000s, go into decline, along with declining crude production in Algeria and resource limitations in Ecuador. In the New Policies Scenario, the recent unrest in the Middle East and North Africa is assumed to have no significant long-term impact on investment and production in the region. The possibility of investments in that region being deferred is discussed later in this chapter.

Iraq's oil production potential is immense (Box 3.4), but exploiting it depends on consolidating the progress made in peace and stability in the country, including resolution of the debate about the legal status of contracts awarded by the Kurdistan Regional Government (KRG). Iraq remains one of the least explored major resource holders. Although about 80 fields have been discovered, only 30 are currently in production and large swathes of the Western Desert remain unexplored. Proven reserves total 115 billion barrels (O&GJ, 2010), but the resource potential is much larger. Oil production in mid-2011 averaged 2.8 mb/d (IMOO, 2011 and IEA, 2011d), including some 0.14 mb/d of production from the KRG area. According to Deputy Prime Minister al-Shahristani, it could reach 3 mb/d by the beginning of 2012. We project production to reach 5.4 mb/d in 2020 and 7.7 mb/d in 2035. Even so, this level is well below that targeted originally by the government.

Saudi Arabia is expected to contribute the second-largest increase in oil production worldwide between now and 2035, raising production by almost 40% from 10 mb/d in 2010 to nearly 14 mb/d by 2035. The national oil company, Saudi Aramco, will be responsible for most of the investment required to build up production in the coming decades, as international oil companies are involved in production only in the Partitioned Neutral Zone shared with Kuwait, and in gas exploration. Roughly half of Saudi production still comes from the Ghawar field – the largest oilfield in the world. Smaller fields, some of them super-giants in their own right, are now being developed. In some cases, these were discovered several decades ago, but not brought on stream due to technical or cost considerations and the global supply-demand balance. In 2010, Saudi Aramco completed the Khurais and Khursaniya projects, which together added 1.7 mb/d of new capacity. Saudi Aramco is also implementing a pilot carbon-dioxide enhanced oil recovery scheme on a small sector of Ghawar to demonstrate its technical viability – which could further extend the productive life of the field, by increasing recovery.

Box 3.4 ● Prospects for increased oil production in Iraq

The Iraqi government has awarded twelve Technical Service Contracts (TSCs) for the development of major oil fields. So far, operators have focused on "quick-win" activities, like refurbishing facilities, repairing wells and drilling a limited number of new wells, so as to raise production quickly by 10%, the threshold at which the government is contractually obliged to start making payments. The profit margin in the TSCs is fixed, providing a strong incentive to keep cost and investment down, but there is little opportunity to increase returns should oil prices rise. Continued investment will hinge on the success of these projects and opportunities elsewhere.

BP and the China National Petroleum Corporation (CNPC) have increased output at the Rumaila oil field to around 1.2 mb/d, triggering payment obligations by the government. CNPC has signed an oil export agreement with the government enabling it to take delivery of about two million barrels in May 2011 as payment in kind. ExxonMobil and ENI, operators of West Qurna and Zubair oilfields respectively, are also thought to have reached the trigger point for payments.

Iraqi production has grown from an average of 2.4 mb/d in 2010 to around 2.8 mb/d in mid-2011. Were all the TSCs to proceed according to their original schedule, Iraqi oil production capacity would reach more than 12 mb/d by 2017. However, given Iraq's infrastructure, logistical and security challenges, government officials have acknowledged that the target level of production capacity originally implied will not be achievable and a more realistic aspiration of reaching 6 mb/d to 7 mb/d by 2017 is being discussed. The KRG area contributes around 0.14 mb/d of total production under contracts awarded in the last decade, but the validity of these contracts is not yet recognised by the central government. This area has great exploration potential which has attracted a number of companies, but large investments are unlikely until the legal differences with Baghdad are resolved.

Slow progress is being made to develop infrastructure. The government is working on doubling the Gulf export terminal capacity to 3.6 mb/d by early 2012, but many believe this project will be delayed. Work is also progressing on repairing and upgrading a pipeline to the Turkish Mediterranean port of Ceyhan, to double capacity to around 1 mb/d in 2013. Export bottlenecks mean that production is likely to be constrained to less than 3 mb/d until 2012. Thereafter, it could start increasing, by around 300 kb/d annually, as repairs, upgrades and new facilities are completed. In the New Policies Scenario, production reaches 5.4 mb/d in 2020 and nearly 8 mb/d in 2035. Other supporting infrastructure is also required to facilitate the increased production. Shell is building a quay on the Shatt al-Arab waterway, to allow shipping of equipment to the Majnoon field. ExxonMobil is leading a consortium planning a "common seawater supply facility", a $10 billion project to deliver up to 12 mb/d of treated seawater for injection into the southern oilfields.

Roughly half of Iraq's associated gas production is currently flared. With rising oil production, flaring increased from 8.1 bcm in 2009 to 9.1 bcm in 2010 (GGFRP, 2010). In mid-2011 the government signed an agreement with Shell and Mitsubishi to form the Basra Gas Company which will gather and treat gas produced in the Basra region (such as that currently being flared) and market it. The government is keen to use the gas for power generation, so as to reduce shortages, which have been a significant economic and political issue for some time.

The United Arab Emirates' (UAE) oil production has increased over the last decade, from 2.6 mb/d in 2000 to 2.9 mb/d in 2010. Abu Dhabi, the UAE's principal oil producer, has stated its intention to increase sustainable oil production capacity to 3.5 mb/d by 2018. It recognises that new developments are becoming more complex and will require the continued support of international oil companies. These companies already participate in the UAE's upsteam sector through partial ownership of both Abu Dhabi Company for Onshore Oil Operations (ADCO) and Abu Dhabi Marine Operating Company (ADMA). ADCO's 75-year concession will expire in 2014 and ADMA's 65-year concession in 2018. It remains uncertain whether these contracts will be renegotiated, re-tendered or opened up and restructured. Major new investments are unlikely to be sanctioned until this matter is resolved. We project production to continue increasing steadily throughout the projection period, reaching 3.3 mb/d in 2020 and 4 mb/d by 2035.

Table 3.6 ● OPEC oil* production in the New Policies Scenario (mb/d)

	1980	2010	2015	2020	2025	2030	2035	2010-2035 Delta	2010-2035 %**
Middle East	**18.0**	**23.8**	**26.7**	**28.6**	**30.4**	**32.9**	**36.0**	**12.2**	**1.7%**
Iran	1.5	4.2	4.2	4.3	4.3	4.5	4.8	0.5	0.5%
Iraq	2.6	2.4	4.2	5.4	6.0	6.8	7.7	5.2	4.7%
Kuwait	1.4	2.5	2.6	2.7	2.9	3.1	3.5	1.0	1.3%
Qatar	0.5	1.7	2.0	2.1	2.1	2.2	2.3	0.5	1.1%
Saudi Arabia	10.0	10.0	10.5	10.9	11.6	12.6	13.9	3.8	1.3%
United Arab Emirates	2.0	2.9	3.3	3.3	3.4	3.7	4.0	1.1	1.3%
Non-Middle East	**7.6**	**11.0**	**11.0**	**10.9**	**11.4**	**11.9**	**12.7**	**1.7**	**0.6%**
Algeria	1.1	1.9	2.0	2.0	2.1	2.1	2.1	0.3	0.5%
Angola	0.2	1.8	1.8	1.8	1.8	1.7	1.5	-0.3	-0.7%
Ecuador	0.2	0.5	0.4	0.3	0.3	0.3	0.2	-0.2	-3.0%
Libya	1.9	1.7	1.6	1.6	1.6	1.8	2.0	0.4	0.8%
Nigeria	2.1	2.5	2.4	2.4	2.5	2.7	2.9	0.4	0.6%
Venezuela	2.2	2.7	2.8	2.8	3.1	3.4	3.9	1.2	1.4%
Total OPEC	**25.5**	**34.8**	**37.7**	**39.6**	**41.7**	**44.9**	**48.7**	**13.9**	**1.4%**
OPEC market share	41%	42%	43%	44%	45%	48%	51%	n.a.	n.a.
Conventional oil	**25.5**	**34.2**	**36.3**	**37.8**	**39.6**	**42.4**	**45.7**	**11.5**	**1.2%**
Crude oil	24.7	29.0	29.3	29.8	30.7	32.4	34.7	5.7	0.7%
Natural gas liquids	0.9	5.2	7.0	8.0	8.8	9.9	11.0	5.8	3.0%
Unconventional oil	**0.0**	**0.6**	**1.4**	**1.8**	**2.2**	**2.5**	**3.0**	**2.4**	**6.8%**
Venezuela extra-heavy oil	0.0	0.5	1.1	1.4	1.7	1.9	2.3	1.8	6.6%
Gas-to-liquids	-	0.0	0.2	0.2	0.3	0.4	0.5	0.4	11.7%

*Includes crude oil, NGLs and unconventional oil. **Compound average annual growth rate.

Kuwait's oil production has trended upwards in the last decade, from 2.2 mb/d in 2000 to 2.8 mb/d in 2008, before falling back in 2009 with OPEC production constraints in the face of weak demand. In the late 1990s, Kuwait offered contracts to international oil companies to re-develop and operate several heavy oil fields on its northern border with Iraq in a scheme originally known as "Project Kuwait". While this project had the potential to expand production significantly, it has stalled in the face of political opposition. Emphasis has now shifted away from heavy oil to developing the country's lighter oil reserves. In early 2011, the national company, Kuwait Oil Company, reaffirmed its target to boost production capacity by 1 mb/d to 4 mb/d by 2020 and to sustain that capacity until 2030. Achieving this will be contingent on securing the technical assistance of international companies and creating a climate conducive to large investments. We project production to increase gradually over the projection period, reaching 3.5 mb/d by 2035.

Iran is currently the Middle East's second-largest oil producer and has significant upside production potential, both for crude oil and NGLs. However, the current political isolation of the country and fiscal terms which discourage foreign investment, make it unlikely that this potential will be realised quickly. Currently, investment in the Iranian oil sector is barely sufficient to maintain production capacity as older producing fields decline. Most revenues from oil exports are consigned to finance other state programmes, starving the national oil company of funds to invest in developing new fields and infrastructure and in combating declines at existing fields. Thus, we project only a slow increase in overall oil output during the projection period, in large part driven by NGLs.

Qatari crude oil production has grown only modestly in the last decade, mainly through development of the offshore Al Shaheen field. However, NGL production has risen sharply in the last two years as a swathe of LNG projects have come on-stream. Qatar's North Field – the largest gas field in the world – has large condensate reserves, which are likely to fuel further growth in NGL production. However, there is currently a moratorium on new gas developments, which is not expected to be lifted before 2015. GTL production is also being boosted, with the Pearl GTL plant's production rising to 140 kb/d (together with 120 kb/d of NGLs) in 2012. In the longer term, more LNG export capacity is expected to be added and there is scope for new GTL projects beyond the current Oryx and Pearl plants, as a hedge against any decoupling of gas and oil prices. As a result of increased gas production, production of NGLs exceeded crude oil production for the first time in 2010 and is set to make up the greater part of higher oil production throughout the projection period. Total oil output is projected to reach 2.3 mb/d in 2035.

Venezuela's oil production has declined through much of the last decade, due to lack of investment. This has been due, in part, to a nationalistic policy, which has compelled international oil companies either to cede a significant amount of equity to the national company, PDVSA, or to leave the country altogether, as in the case of ExxonMobil and ConocoPhillips. We project a modest decline in conventional oil production over the projection period. However, this decline is more than offset by rapid growth in unconventional, extra-heavy oil from the Orinoco belt, which has attracted investment from a number of foreign companies, in partnership with PDVSA.[14] In aggregate, Venezuelan production rises from 2.7 mb/d in 2010 to 3.9 mb/d in 2035.

Nigeria is the largest oil producer in Africa, producing some 2.5 mb/d in 2010, from the Niger Delta and offshore in shallow and, to a lesser extent, deep water. Production has been sporadically disrupted by civil conflicts, which have led to a substantial amount of capacity being shut-in in recent years. Investment has also been slowed by the inability of the Nigerian National Petroleum Company, to covers its share of investment in joint ventures with international oil companies. Uncertainty about a draft Petroleum Industry Bill, first issued in 2008 and still under review, which could have a significant impact on upstream terms and conditions, is also damaging the investment climate. With these uncertainties in mind, we project a slow rate of investment in the near term and, therefore, a slight drop in production in the early part of the *Outlook* period. However, we project a rebound in output later in the projection period, on the assumption that the investment climate improves. An increase in NGL production also contributes to higher oil production in the longer term, as efforts to reduce gas flaring slowly bear fruit. Production drops to 2.4 mb/d in 2015, but recovers to 2.9 mb/d in 2035.

---

14. See last year's *Outlook* for a detailed discussion of the prospects for Venezuelan extra-heavy oil production (IEA, 2010).

Libyan oil production almost entirely stopped in early 2011 following the eruption of conflict. Over the decade to 2010, Libyan production fluctuated between 1.4 mb/d and 1.8 mb/d, and was averaging around 1.7 mb/d at the start of 2011. Libyan crude oil is generally of a "light sweet" quality (meaning it has relatively low density and low sulphur content), making it ideal for refining into low-sulphur transport fuels. Although (as of September 2011) the extent of damage to infrastructure is not fully known, some production has resumed and, if security improves, could rise towards pre-conflict levels over the coming two years. Our projections assume renewed investment before 2015, allowing production to increase gradually to 2 mb/d by 2035.

Other OPEC countries are expected to maintain more or less steady levels of production for a large part of the projection period, the variations reflecting their individual resource endowments. Production potential in both Angola and Ecuador is constrained by the currently limited extent of ultimately recoverable resource estimates, though new discoveries could alter this picture.

Trade

International trade in oil (including crude oil, NGLs, unconventional oil, additives and refined products) is poised to continue to grow. In the New Policies Scenario, trade between the major regions increases from 37 mb/d in 2010 to more than 48 mb/d in 2035, growing 11 mb/d over the projection period (Figure 3.19). This compares with a projected increase in global demand of nearly 13 mb/d. Net imports to China exceed 12.5 mb/d by 2035, up by almost 8 mb/d from current levels. India's net imports grow by over 4 mb/d to nearly 7 mb/d – the second-largest increase. The reliance on imports in developing Asia, as a whole increases from 56% of total oil needs in 2010 to 84% in 2035. European imports remain relatively flat at 9 mb/d, as demand and regional production decline almost in parallel.

Net imports to North America fall by nearly 6 mb/d to under 3 mb/d by 2035, on the back of rising indigenous production and falling demand. The United States accounts for most of this decline, with imports falling by 4 mb/d below today's level in 2035. Among the exporting regions, the Middle East sees the biggest increase in net exports, which expand by more than 9 mb/d to almost 28 mb/d by the end of the projection period. Exports from Latin American also rise strongly, by more than 2 mb/d between 2010 and 2035, thanks to the projected surge in production in Brazil.

The rising dependence on imports in some non-OECD regions, particularly in Asia, will inevitably heighten concerns about supply security, as reliance grows on supplies from a small number of producers, especially in the Middle East and North Africa (MENA)[15] region, which are shipped along vulnerable supply routes, particularly the straits of Hormuz in the Persian Gulf, the Malacca Straits in Southeast Asia and the Gulf of Aden. That message has been brought home by recent civil unrest in the region and the conflict in Libya. Net imports into all importing countries are much lower in the 450 Scenario, rising just over 2 mb/d by 2020 and then falling back to close to their current levels by 2035, demonstrating the energy-security benefits from concerted policy action to avert climate change.

---

15. The MENA region is defined as the following 17 countries: Algeria, Bahrain, Egypt, Iran, Iraq, Jordan, Kuwait, Lebanon, Libya, Morocco, Oman, Qatar, Saudi Arabia, Syria, Tunisia, United Arab Emirates and Yemen.

Figure 3.19 ● Regional oil demand and net trade in the New Policies Scenario (mb/d)

This map is for illustrative purposes and is without prejudice to the status of or sovereignty over any territory covered by this map.

Notes: Imports and exports show net volumes traded. International marine and aviation fuel use is not included.

■ Net imports (mb/d)
■ Net exports (mb/d)
■ Demand (mb/d)

Trends in oil and gas production costs

The costs of oil and gas production, both to operate current capacity and develop new supply, have been rising strongly in recent years and are assumed to continue increasing over the projection period, contributing to upward pressure on prices. Trends in costs depend on multiple factors. Over the past ten years, worldwide costs of developing production capacity have doubled, largely due to increases in the cost of materials, personnel, equipment and services. Such costs correlate closely to both oil prices and levels of exploration and development activity (Figure 3.20).

Figure 3.20 ● **IEA upstream investment cost index, oil price index and drilling activity**

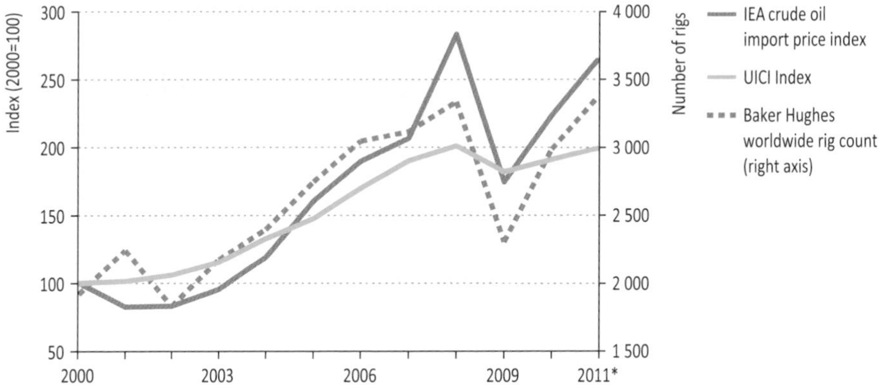

*Preliminary estimates based on trends in the first half of the year.

Notes: The IEA Upstream Investment Cost Index (UICI), set at 100 in 2000, measures the change in underlying capital costs for exploration and production. It uses weighted averages to remove the effects of spending on different types and locations of upstream projects. The IEA crude oil import price index is set at 100 in 2000.

Sources: IEA databases and analysis based on industry sources; Baker Hughes databases (BH, 2011).

The cost of incremental production tends to rise as more easily accessed resources are depleted, and development moves to more difficult resources with less favourable geology or in more remote locations. Developing such resources often requires more complex and energy-intensive processes, and sometimes new infrastructure, either to reach the resource or to extract and export oil to market, while the volumes of oil extracted per well tend to be lower.

Innovation, progress along learning curves and greater scale can act against these cost pressures. New technologies and processes help to lower the unit costs of like-for-like activities: this applies to both incremental reductions from improvements in existing technology and fundamental changes brought about by innovative solutions, which may also facilitate exploitation of new resource categories. An example of the latter is the application of horizontal wells and hydraulic fracturing to the exploitation of light tight oil.

The level of investment and activity undertaken in any year to maintain or expand production varies widely across the world. The United States has a long production history and most of the easily accessible onshore oil has already been produced. Consequently operators now target more difficult resources, like light tight oil, which require many more wells per unit of production than conventional resources. As a result, the United States now uses over half of the world's drilling rig fleet to produce just 9% of the world's oil (and 19% of the world's gas). This is largely because the average well in the United States produces about 20 barrels per day (b/d), compared with 3 400 b/d per well in Saudi Arabia (Table 3.7), which still has giant, discovered fields that have yet to be developed. Because of its high activity levels, the United States attracted nearly one-quarter of worldwide oil and gas investment in 2010. Russia, which produces nearly 40% more oil than the United States, also has a large oil and gas industry and employs the second-largest number of drilling rigs globally, albeit only about half the number of the United States.

Table 3.7 ● **Oil production, indicative development activity and investment in the United States, Russia and Saudi Arabia, 2010**

	United States	Russia	Saudi Arabia
Proven oil reserves* (billion barrels)	22	77	265
Crude oil and NGL production (mb/d)	7.6	10.5	9.9
Reserves to production ratio (years)	8	20	73
Number of producing oil wells	370 000	127 000	2 900
Average production per well (b/d)	20	80	3 400
Number of active oil and gas drilling rigs**	1 700	850	100
Oil wells drilled: Exploration	778	300	27
Production	18 138	3 700	148
Total upstream capital expenditure** ($ billion)	120	30	10
Employment in the oil and gas industry	2 200 000	1 100 000	90 000

*Definitions of proven reserves vary by country. **Drilling activity and expenditure cover both oil and gas. Capital expenditure for Saudi Arabia and Russia are estimates.

Sources: IEA databases and analysis based on industry sources, including: OPEC (2011), US DOE/EIA (2011b), US DOE/EIA (2011c), Otkritie (2011) and SA (2011) for number of producing wells and number of wells drilled; BP (2011) for oil reserves; BH (2011) for active number of drilling rigs; API (2011), ILO (2010) and Rosstat (2010) for employment.

Middle East oil resources remain the cheapest in the world to exploit, although cost pressures are increasing even there, as operators tackle more difficult resources. Information about costs in specific countries is patchy. BP announced in 2009 that it planned to invest, with its partner China National Petroleum Corporation, around $15 billion over the next 20 years in order to increase production from Iraq's Rumaila oilfield from about 1 mb/d to 2.85 mb/d, implying a capital cost of around $8 000 per barrel of daily capacity (/b/d) added. This would be close to the lowest cost anywhere in the world for a development project. The capital cost of increasing production from the Upper Zakum oilfield in offshore Abu Dhabi from about 500 kb/d to 750 kb/d, which is being undertaken by ExxonMobil, is expected to be about the same per barrel of daily capacity. In Saudi Arabia, the development of the heavy and sour Manifa field, the capacity of which is planned to reach a plateau of around 900 kb/d in 2014, is expected to cost about $15 000/b/d. The cost per daily barrel of capacity of developing the giant heavy Wafra oilfield in the Partitioned Neutral Zone, with capacity due to reach 600 kb/d, is expected to be similar; the field, which is being developed by Chevron, requires thermal stimulation with steam injection. Capital costs for development of deepwater oil are much higher, ranging from $40 000/b/d to $80 000/b/d.

The total cost of producing oil, including the amortisation of development costs but excluding taxes and profit margins, is well below the current market price of oil, generating significant economic rent that is captured by governments in taxes and royalties and by oil companies in profits (Figure 3.21). The OPEC Middle East countries have by far the lowest costs, followed by the main North African producers. However, to generate sufficient revenue to balance government budgets in OPEC countries (the budget breakeven) requires a much higher oil price and this figure has been rising in recent years. This is particularly the case in the Middle East, where large, youthful populations are putting pressure on education systems, housing and job creation schemes. In many of these countries, which rely heavily on oil export revenues to fund government budgets, the budget breakeven oil price is now above $80/barrel. This will become an increasingly important consideration in the formation of future oil prices.

Figure 3.21 • **Breakeven costs, budget breakeven and commercially attractive prices for current oil production for selected producers, mid-2011**

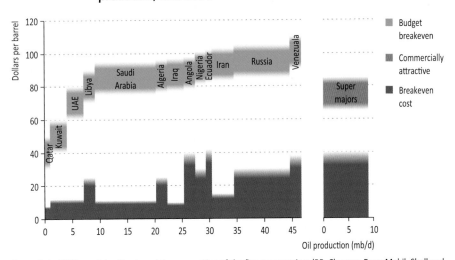

Notes: Only OPEC countries, Russia and the aggregation of the five super-majors (BP, Chevron, ExxonMobil, Shell and Total) are included. The breakeven cost is the realised oil price at which all operating expenses (excluding taxes) and capital costs (including a 10% capital discount rate), are fully recovered.

Sources: IEA databases and analysis based on industry sources: APICORP (2011), Deutsche Bank (2011), Credit Suisse (2011), IMF (2011), PFC (2011) and CGES (2011).

For countries and companies where development and production is driven primarily by commercial rather than fiscal motives, the key criterion for sustainable long-term investment is for income from production to cover capital cost recovery, operating costs and fiscal payments, together with a competitive commercial return. For developments in regions where international oil companies and other privately owned companies are the primary project developers and operators, oil prices in a range from $70/barrel to $90/barrel are currently needed to meet this criterion in most resource categories. This takes into account taxes, royalties and the risk-adjusted required returns on investment. Breakeven costs, which exclude taxes, vary by resource category and average around $40/barrel.

As development and production costs vary both regionally and by resource category, the IEA oil and gas supply models include cost inflation factors (correlated to oil price), regional cost inflation factors (to simulate cost increases as resources are depleted) and technology learning curves to compute costs on an annual basis for all regions and resource types over the projection period. Production costs then become input parameters of the supply-modelling. The worldwide increase in average production cost (as measured by the breakeven cost) for new oil and gas production is projected to exceed 16% (in real terms) between 2011 and 2035 (Figure 3.22). The MENA region has the largest production increase over this period and average breakeven costs there increase from just over $12 per barrel of oil equivalent ($/boe) in 2011 to more than $15/boe in 2035, as production increases by more than 17 Mboe/d. But costs increase by even more in most other regions. They increase most in Latin America, to over $50/boe by 2035, because of the growing importance of high-cost deepwater developments.

Figure 3.22 • **Oil and gas production and breakeven costs in the New Policies Scenario**

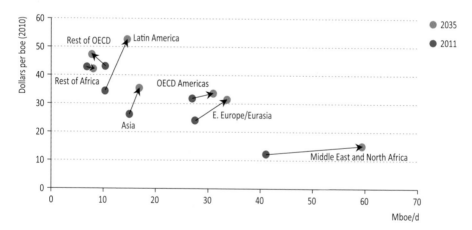

Oil and gas investment

Global upstream oil and gas investment is set to continue to grow strongly in 2011, hitting a new record of over $550 billion – 9% up on capital spending in 2010 (Table 3.8) and almost 10% higher than the previous peak in 2008.[16] These estimates are based on the announced spending plans of 70 leading oil and gas companies. Total upstream spending is calculated by adjusting upwards their announced spending, according to their estimated share of

---

16. Upstream investment is not always reported separately for oil and gas.

world oil and gas production in each year. The increase in spending reflects both rising costs (which are driving up the costs of current and newly launched projects) and higher prices (which have enhanced potential investment returns before tax). The spending plans of 25 leading companies suggest that global spending in the downstream sector (which for these companies represents over one-quarter of total oil and gas investment) may increase even faster than in the upstream sector in 2011.

Table 3.8 • Oil and gas industry investment by company (nominal dollars)

	Upstream			Total		
	2010 ($ billion)	2011 ($ billion)	Change 2010/2011	2010 ($ billion)	2011 ($ billion)	Change 2010/2011
Petrochina	23.6	27.9	18%	42.7	53.3	25%
Petrobras	23.9	24.0	0%	43.4	52.3	21%
ExxonMobil	27.3	28.8	6%	32.2	34.0	6%
Gazprom	26.9	24.8	-8%	29.2	27.0	-8%
Royal Dutch Shell	21.2	19.4	-9%	23.7	26.0	10%
Chevron	18.8	22.6	20%	19.6	26.0	33%
Pemex	17.4	18.9	9%	20.8	22.2	7%
BP	17.8	19.3	9%	18.4	20.0	9%
Total	14.8	16.0	8%	18.0	20.0	11%
Sinopec	8.2	8.3	1%	16.7	19.1	14%
Eni	12.9	12.9	0%	18.4	18.4	0%
Statoil	12.6	14.4	14%	14.0	16.0	14%
ConocoPhillips	8.5	12.0	41%	9.8	13.5	38%
Rosneft	6.1	8.0	31%	8.9	11.0	23%
Lukoil	4.9	6.9	41%	6.8	9.0	32%
CNOOC	5.1	8.8	73%	5.1	8.8	73%
BG Group	5.9	6.0	2%	7.7	8.4	9%
Apache	4.2	6.4	51%	5.4	8.1	51%
Repsol YPF	4.1	4.5	11%	6.8	7.5	11%
Suncor Energy Inc.	4.8	5.4	13%	5.8	6.8	17%
Occidental	3.1	4.9	56%	3.9	6.1	56%
Devon Energy Corp	5.9	5.5	-7%	6.5	6.0	-7%
Anadarko	4.7	5.2	12%	5.2	5.8	12%
Chesapeake	4.9	5.8	17%	4.7	5.8	22%
EnCana	4.5	4.4	-2%	4.8	4.7	-2%
Sub-total 25	292.0	320.9	10%	378.3	435.6	15%
Total 70 companies	408.3	446.7	9%	n.a.	n.a.	n.a.
World	**505.1**	**552.6**	**9%**	**n.a.**	**n.a.**	**n.a.**

Notes: Only publically available data has been included, but estimates of upstream spending, as a share of the total, have been made in cases where detailed breakdowns were unavailable (IEA databases include both public and non-public estimates for all major oil and gas producing companies). The world total for upstream investment is derived by prorating upwards the spending of the 70 leading companies, according to their estimated share of oil and gas production in each year. Pipeline investment by Gazprom is classified as upstream as it is required for the viability of projects.

Sources: Company reports and announcements; IEA analysis.

Trends in upstream spending differ somewhat according to the size and type of operating company. The 25 largest oil and gas companies plan to spend an estimated $320 billion in 2011, 10% more than in 2010. Their spending in 2010 turned out to be 8%, or $22 billion, higher than had been budgeted in the middle of the year. The five leading international oil companies (or super-majors) – BP, Chevron, ExxonMobil, Shell, and Total – spent a total of $100 billion in 2010 and have reported a 6% increase in budgeted spending for 2011. The super-majors' share of world upstream spending will fall to 19% of the total, while national oil company spending in 2011 is projected to rise by 8%, to over $220 billion, taking their share of world upstream spending to 40% (Figure 3.23). Other independent companies plan to increase capital spending by more than 12%, to more than $220 billion in 2011. There is a marked difference in capital efficiency – defined as the annual capital spending required to maintain output per barrel of oil equivalent produced – between the national companies, particularly those in the Middle East and North Africa with preferential access to low-cost reserves, and other companies.

Figure 3.23 ● **Worldwide upstream oil and gas investment and capital efficiency by company type**

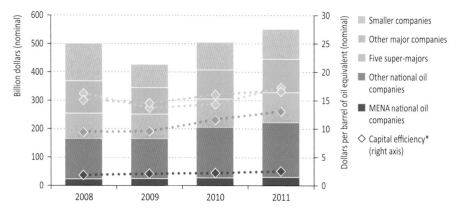

Notes: Capital efficiency is defined as the capital spending to maintain output per barrel of oil equivalent produced, it is not included for the group of smaller companies. 2011 data is based on company plans. The five super-majors are defined as BP, Chevron, ExxonMobil, Shell and Total.

Source: IEA databases and analysis.

Annual upstream investment in nominal terms more than quadrupled between 2000 and 2011. In real terms, *i.e.* adjusted for cost inflation,[17] it increased by 120% over this period, as investment shifted to more complex projects with higher costs per barrel per day of capacity added. As indicated above, the unit cost for additional capacity added has risen sharply. For example, over the last decade the aggregate production of the five super-majors has fluctuated between 16 Mboe/d and 18 Mboe/d with little discernable trend, despite a quadrupling of upstream investment in nominal terms.

---

17. Taking account of increases in input costs, such as drilling, equipment and material, and adjusted to remove the effects of changes in spending on different types and locations of upstream projects.

Based on the New Policies Scenario projections, upstream spending will need to continue to rise in the medium and longer term, for three reasons: rising demand; rising costs, as a result of both cost inflation and a need to develop more difficult resources; and a need to combat rising decline rates at existing fields. However, there is at present no evidence that this will occur, at least in the next five years. We have compiled data on the five-year upstream capital spending plans of 16 of the largest companies, which represent about 30% of 2011 planned spending and average over $10 billion annually. The average annual spending planned by these companies to 2015 is slightly lower than the average spending planned for 2011 by the same companies. The plans of only four of these companies show an increase in annual average spending to 2015 compared with 2011, with seven remaining flat and five showing a fall.

In the longer term, the required increase in the rate of investment accelerates, although the need to develop more costly resources and the high cost of unconventional oil, GTL and CTL plants in non-OPEC regions will be offset to some degree by the shift in production towards the Middle East and other regions, where development and production costs are much lower. Cumulative investment over the period 2011 to 2035 worldwide in the oil and gas sector in total is estimated to amount to $19.5 trillion in the New Policies Scenario (Table 3.9). Of this, upstream investment accounts for about 80%, or nearly $620 billion per year. This is considerably more than the average of $440 billion over 2010 to 2035 projected in last year's *WEO*. The increase is explained by faster growth in production in the near term and higher supply from more costly unconventional sources in non-OPEC countries in the long term, as well as higher cost inflation.

Table 3.9 ● Cumulative investment in oil and gas supply infrastructure by region in the New Policies Scenario, 2011-2035 ($2010 billion)

	Oil			Gas			Annual average upstream O&G
	Upstream	Refining	Total	Upstream	T&D*	Total	
OECD	2 438	265	2 703	2 632	933	3 565	203
Americas	1 975	126	2 100	1 679	492	2 172	146
United States	*1 330*	*96*	*1 426*	*1 287*	*n.a.*	*n.a.*	*105*
Europe	418	93	511	648	354	1 002	43
Asia Oceania	45	46	91	305	87	391	14
Non-OECD	6 282	745	7 027	4 126	1 139	5 265	416
E. Europe/Eurasia	1 310	88	1 398	1 084	380	1 464	96
Russia	*738*	*49*	*787*	*733*	*246*	*979*	*59*
Asia	526	438	963	1 180	385	1 564	68
China	*300*	*210*	*510*	*408*	*182*	*590*	*28*
India	*63*	*140*	*203*	*175*	*62*	*238*	*10*
Middle East	1 010	127	1 137	257	226	483	51
Africa	1 522	35	1 557	1 121	58	1 179	106
Latin America	1 914	57	1 971	483	91	574	96
Brazil	*1 350*	*29*	*1 379*	*165*	*n.a.*	*n.a.*	*61*
World**	8 720	1 010	9 997	6 758	2 072	9 497	619

*T&D is transmission and distribution and is calculated on a regional basis. **World total oil includes an additional $268 billion of investment in inter-regional transport infrastructure. World total gas includes an additional $587 billion of investment in LNG infrastructure and $80 billion of investment in LNG carriers.

Impact of deferred upstream investment in the Middle East and North Africa

A time of great uncertainty

The outlook for global energy markets is clouded by considerable uncertainty, notably the pace of economic recovery, the future of climate-change policies, the role of gas in global energy supply following the unconventional gas "revolution", the role of nuclear energy in the aftermath of Fukushima Daiichi and the impact of political changes in the Middle East and North Africa (MENA) region. The first four sources of uncertainty are examined elsewhere in this *Outlook*: economic growth in Chapter 2, climate polices in Chapter 6, the prospects for faster growth in gas use in Chapter 4 and the future of nuclear power in Chapter 12. This section addresses the implications should there be a shortfall in investment in the near term in upstream oil and gas investment in MENA.

In the New Policies Scenario – our central scenario – we project that MENA will contribute more than 90% of the required growth in oil production to 2035 (Table 3.10). To achieve this growth, upstream investment in MENA needs to average $100 billion per year from 2011 to 2020, and $115 billion per year from 2021 to 2035 (in year-2010 dollars). But it is far from certain that all of this investment will be forthcoming, for many different reasons affecting some or all of the countries in the region. The consequences for global energy markets could be far-reaching. Potential causes of lower investment in one or more countries include:

- Deliberate government policies to develop production capacity more slowly in order to hold back resources for future generations or to support the oil price in the near term.

- Constraints on capital flows to upstream development because priority is given to spending on other public programmes.

- Restricted, or higher-cost, access to loans or other forms of capital.

- Delays due to legal changes or renegotiation of existing agreements.

- Increased political instability and conflicts.

- Economic sanctions imposed by the international community.

- Higher perceived investment risks, whether political or stemming from uncertainties in demand.

- Constraints on inward investment as a result of stronger resource nationalism, particularly in regimes seeking to pre-empt popular uprisings.

- Delays due to physical damage to infrastructure during conflicts.

Similar kinds of risk exist in other major oil- and gas-producing countries and regions. Our analysis of the MENA region does not focus on any particular country, but considers the consequences of lower upstream investment across the region.

Table 3.10 ● The role of MENA in global oil and gas production in the New Policies Scenario

	MENA	World	Share of world
Conventional remaining recoverable oil resources (billion barrels)	1 178	2 401	49%
Remaining recoverable gas resources (tcm)	154	810	19%
Oil production 2010 (mb/d)	29.9	83.6	36%
Gas production 2010 (bcm)	632	3 275	19%
Oil production growth to 2035 (mb/d)*	12	12.8	93%
Gas production growth to 2035 (bcm)	424	1 474	29%
Upstream investment needed 2011-2035 ($2010 billion)	2 706	15 478	17%

*Oil production in this table and section refers to total oil production (crude oil, NGLs and unconventional oil, excluding biofuels).

Note: tcm = trillion cubic metres.

The Deferred Investment Case

The *Deferred Investment Case* analyses how global oil markets might evolve if investment in the upstream industry of MENA countries were to fall short of that required in the New Policies Scenario over the next few years. The key assumption for this case is that upstream oil (and gas) investment is reduced by one-third in all MENA countries, compared with that in the New Policies Scenario over the period 2011 to 2015. The assumed shortfall in investment is then made good gradually after 2015 so that by 2020 the level of investment is back to that provided for in the New Policies Scenario. In this way, the case illustrates what could happen if there were an "orderly" shortage of investment, rather than a sudden interruption resulting, for example, from a serious conflict in one or more of the major producing countries of the region.[18] The shortfall in upstream investment includes a shortfall in gas investment, so would affect oil and gas markets directly and have knock-on effects for other energy markets; but the prime focus of this analysis is on the oil market.

As a result of the assumed deferral in investment, MENA oil and gas production capacity falls progressively short of that projected in the New Policies Scenario, reflecting the delay in bringing new capacity on stream. We have assumed that the amount of capacity held unused (spare capacity) remains constant as a percentage of total available capacity, so the loss of capacity results directly in lower production. The shortfall in oil production in MENA, compared with the New Policies Scenario, reaches over 6 mb/d in 2020 – large enough to have a significant impact on global oil balances, so increasing oil and gas prices. That prompts higher investment in developing resources in other countries, though higher development costs there mean that the shortfall in MENA production is only partially made

---

18. The effects of the conflict in Libya are already included in all the scenarios.

good by other regions. Higher oil and gas prices also lead to some loss of demand, such that the reduced demand and increased non-MENA production compensate for the lower MENA production to keep the market in balance.

Box 3.5 ● *Assumptions and methodology of the Deferred Investment Case*

The assumptions on reduced investment in the upstream oil and gas industry in MENA were fed into the oil supply module of the World Energy Model (WEM), generating a lower level of output from already producing fields and fields to be developed in each country in the region. The impact on international oil prices was derived through several iterations of the WEM supply and demand modules to find the oil-price trajectory that brings non-MENA supply and global demand into equilibrium in each year of lower MENA supply.

With the exception of international fuel prices, assumptions about all the other factors driving energy demand and supply in the MENA region and elsewhere were kept the same as in the New Policies Scenario. End-user fuel subsidy policies in MENA and other countries were assumed to remain as they are in the New Policies Scenario. The terms of trade between consuming and producing nations would be likely to change over the projection period, as oil-producing countries would benefit in the short term from higher revenues from imports. However, our analysis shows that, over the projection period as a whole, the cumulative net revenues of MENA do not change significantly compared to the New Policies Scenario. The same holds true for import bills in importing countries. For these reasons, and to avoid unduly complicating comparisons with the New Policies Scenario, the GDP assumptions of the New Policies Scenario have not been changed. In practice, reduced investment in MENA and higher oil prices would be likely to affect the path of GDP growth to a different extent in different regions.

Results of the Deferred Investment Case

Oil prices

The international crude oil price, for which the average IEA crude oil import price serves as a proxy, increases rapidly in the short to medium term, as the investment shortfall becomes apparent to the market. The price peaks at around $150/barrel in 2016/2017 (in year-2010 dollars, equivalent to $176/barrel in nominal terms), when the investment shortfall relative to the New Policies Scenario is starting to decline. The price then starts to decrease gradually and converges with that of the New Policies Scenario by the end of the projection period (Figure 3.24). As the oil prices are annual averages, they have smooth trajectories; however the higher prices of the Deferred Investment Case would be likely to be accompanied by significantly increased price volatility.

Figure 3.24 ● Average IEA crude oil import price in the New Policies Scenario and Deferred Investment Case

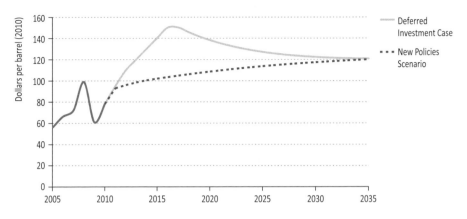

The upward part of the price trajectory in the Deferred Investment Case is similar to that which occurred in the last decade, resulting in a spike in 2008 – also the result, at least in part, of insufficient investment. The downward slope is comparable to the decline after the second oil shock in 1979/1980, which saw the same kind of market response, involving demand reduction through increased efficiency, fuel switching and increased investment in exploration and production in other regions, as that now assumed in the Deferred Investment Case (Figure 3.25).

Figure 3.25 ● Change in average IEA crude oil import price after initial fall in investment in the Deferred Investment Case, compared to past price shocks

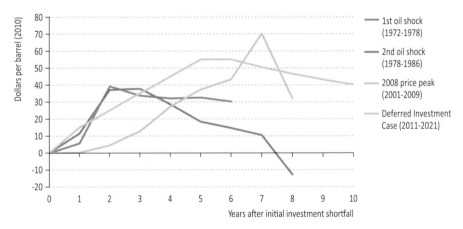

Gas and coal prices react to the increased oil price and the reduction in investment in gas supply capacity in MENA, though to a different extent across regions. In the period of elevated oil prices, gas and coal prices decouple to some degree from oil prices, so that the

ratios of gas and coal prices to oil prices are lower than in the New Policies Scenario. This is because the loss of gas supply, as a proportion of global supply, is significantly less than that of oil supply and because oil prices have only a moderate influence over coal prices.

Impact on oil demand

With sharply higher prices, global primary oil demand in the Deferred Investment Case barely increases in the medium term, reaching only about 88 mb/d in 2015 – fractionally higher than in 2010 and about 3.2 mb/d below the New Policies Scenario (Figure 3.26). Thereafter, demand picks up steadily, reaching 89.3 mb/d in 2020 (3.1 mb/d lower than in the New Policies Scenario) and 97.8 mb/d in 2035 (1.6 mb/d lower). The peak of the demand reduction occurs in 2017, when demand is 3.9 mb/d lower than in the New Policies Scenario. The change in demand is more pronounced in the near term, because oil prices rise quickly as the market reacts to the shortfall in investment. The higher price encourages energy conservation, notably via reduced driving distances. In the long term, higher oil prices encourage switching to alternative fuels – notably conventional and advanced biofuels and electric vehicles – and accelerate the commercialisation and sale of more efficient technologies, including more fuel-efficient vehicles. Coal use also sees a modest increase in demand, driven by increase in CTL production.

Figure 3.26 ● **World primary oil demand in the New Policies Scenario and the Deferred Investment Case**

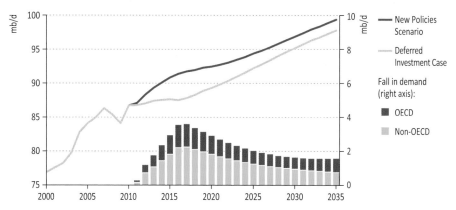

In the medium term, non-OECD countries account for most of the reduction in oil consumption in the Deferred Investment Case – nearly 60% (or 1.8 mb/d) in 2020 – but their share declines to less than 50% (or 0.8 mb/d) by 2035. Consumers in developing countries respond to higher prices by delaying the acquisition of their first car or by driving less in the cars they already own. The poorest consumers switch less rapidly to modern oil-based fuels. Demand for oil products in OECD countries is less affected initially by the increase in international prices than demand in non-OECD countries, mainly because end-user prices are a smaller share of total OECD prices due to higher taxes and because the cost of oil represents a smaller share of total disposable household income. However, sustained high prices at the level expected in the Deferred Investment Case make the production of many types of biofuels competitive (IEA, 2010). This stimulates investment in additional biofuels capacity – mostly in OECD countries – to replace a portion of oil demand throughout the projection period. The two largest oil-consuming and importing countries – the United States and China – play a key role in long-term demand reduction, their combined contribution growing from 30% of global reductions in 2020 to 50% in 2035. In the United States, the impact of higher prices on demand is marked, because taxes there are low compared with most of the rest of the world; in China, higher prices spur vehicle manufacturers to accelerate larger scale production of more efficient conventional and electric vehicles.

The transport sector is responsible for most of the reduction in oil demand, accounting for nearly 70% of the fall in primary demand in 2020 and 95% in 2035. Initially, demand falls as motorists drive less in response to higher prices. This effect disappears by 2035, as prices converge with those in the New Policies Scenario (Figure 3.27). The increase in fuel prices prompts consumers to purchase more efficient vehicles, but the effect on overall fleet efficiency takes time to materialize. Similarly, switching to alternative fuels grows in the long term as supplies increase and consumers modify and upgrade their vehicles. Biofuels use grows from 1.3 mb/d in 2010 to 5.5 mb/d in 2035 – 1.2 m/d, or 26%, above the New Policies Scenario.[19] In the longer term, persistently higher prices also accelerate the deployment of light-duty electric vehicles (including plug-ins): sales reach 7 million vehicles, or 4.5% of total passenger light-duty vehicles sales, in 2035 – one percentage point more than in the New Policies Scenario. Sales of natural gas vehicles do not change significantly, as gas prices also increase, even if not to the same extent as oil prices. Nonetheless, transport energy demand is only slightly more diversified than in the New Policies Scenario, with oil still accounting for 86% of total transport fuel use in 2035 – a reduction of just two percentage points.

Figure 3.27 ● Reduction in global oil demand in the transport sector by source in the Deferred Investment Case relative to the New Policies Scenario

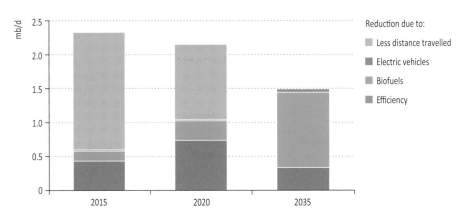

19. Much of the increase in biofuels use results from an increase in the volumes blended into conventional gasoline and diesel. Higher oil prices increase the incentive to invest in biofuels production facilities, which boosts supply and displaces demand for oil-based fuels.

Oil demand falls much less in volume terms, compared with the New Policies Scenario, in non-transport sectors. The share of buildings in oil use remains relatively constant at around 8%. Demand in industry and in non-energy use reacts more in the near term as a result of energy savings, but recovers briskly in the longer term, as prices converge with those in the New Policies Scenario, because the potential for energy-efficiency gains is more limited than in other sectors.

Impact on oil production

In the Deferred Investment Case, lower investment in MENA countries reduces global oil production by 3.8 mb/d in 2017 (at its peak) and 1.5 mb/d in 2035, compared with the New Policies Scenario. MENA production falls by 3.4 mb/d in 2015 (Figure 3.28). Between 2015 and 2020, the shortfall in MENA production increases further, to peak at around 6.2 mb/d in 2020, but this is partially offset by an increase in production of 3.2 mb/d in other regions. MENA investment is assumed to recover gradually after 2015 but, due to the decline in production from existing fields, total MENA production continues to fall until 2020. After 2020, MENA production begins to increase, attaining its 2010 level by 2023. Moreover, a combination of some permanent demand destruction and the long lifetime of the production capacity brought online in non-MENA countries (particularly unconventional oil projects) means that MENA production never returns to the production levels projected in the New Policies Scenario; in 2035, MENA production is still about 1.2 mb/d lower.

Figure 3.28 ● *Changes in global oil production and demand in the Deferred Investment Case relative to the New Policies Scenario*

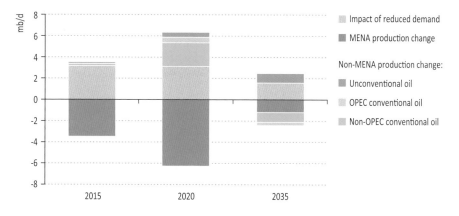

The recovery of production in the MENA region after 2015 is comparable to that observed after past supply disruptions, though in the Deferred Investment Case the loss of production is initially smaller, but then becomes larger in the longer term (Figure 3.29). Although the circumstances of the reduction in production are markedly different from those of an abrupt supply disruption, the comparison suggests that the projected rate of such recovery is plausible.

Initially, production in other regions struggles to increase because of the time involved in bringing new production online. Compared with the New Policies Scenario, the increase in non-MENA production is equivalent to about half of the shortfall in MENA production in 2020; lower demand makes good the difference. In the near to medium term, increased non-MENA production comes mainly from outside OPEC. The main contributors are Russia, Canada (with increased unconventional production) and Brazil, but smaller production increases come from many other countries. Nonetheless, towards the end of the projection period, non-MENA production of conventional oil is actually lower than in the New Policies Scenario, due to a combination of faster resource depletion, competition from unconventional capacity built-up earlier and competition from the resurging, lower cost, MENA production. Unconventional oil production – oil sands, extra-heavy oil, CTL, GTL and kerogen oil – accelerates over the projection period, relative to the New Policies Scenario. By 2035, it is 0.9 mb/d higher, compensating for three-quarters of the fall in MENA production.

Figure 3.29 ● **Profile of oil production recovery after disruption in Deferred Investment Case and past events**

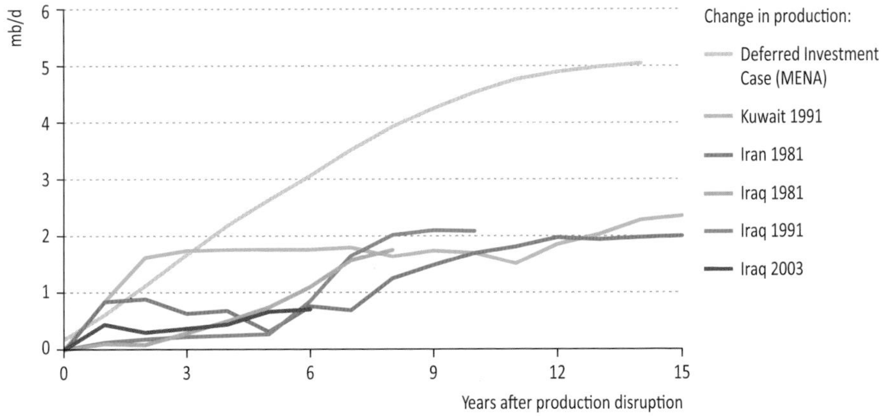

Overall upstream investment for the period 2011 to 2035 in the Deferred Investment Case, at around $15.3 trillion (in year-2010 dollars), is almost the same as in the New Policies Scenario – despite lower production over the entire period. This is because the capital costs per unit of production capacity in other regions are generally significantly higher than in MENA.

Impact on trade

In the Deferred Investment Case, MENA oil and gas exports are reduced by a volume similar to the decrease in production, in comparison with the New Policies Scenario (the fall is slightly less, as MENA oil and gas demand is also reduced marginally because of higher prices). In the first eight years of the projection period, when oil prices are highest, MENA cash flows from oil and gas exports (the value of exports, net of lifting costs and investment costs) are higher than in New Policies Scenario, because lower export volumes are more than offset by higher oil prices (and the call on the cash flow for investment is also lower). The difference in annual cash-flow gradually grows, reaching 23% in 2015, before dropping back to near zero in 2019. However, because MENA countries lose market share throughout the whole projection period, their total cumulative cash flow between 2011 and 2035 is lower in the Deferred Investment Case than in the New Policies Scenario: cash flow is $16.7 trillion, versus $17 trillion (in year-2010 dollars) (Figure 3.30). Gains in cash flow in the period to 2018 are more than offset by the loss in the period from 2019 to 2035, due to lower MENA production and declining prices; [20] this loss in annual cash-flow reaches 15% in 2023, before falling back to 5% in the 2030s.

Figure 3.30 ● **Oil and gas export cash flows and import costs by region in the New Policies Scenario and Deferred Investment Case, 2011-2035**

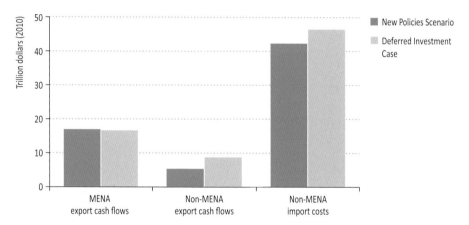

Note: Cash flows and cumulative import bills are shown undiscounted.

Other exporting countries obviously enjoy higher revenues in the Deferred Investment Case, as they replace some of the lost MENA production and benefit from higher oil prices. Non-MENA export cash flow is 60% higher than in the New Policies Scenario, at $8.8 trillion. Oil and gas import bills for importing countries increase in both the medium and long term, as the reduction in demand and increase in domestic production do not fully offset the effect of price increases, relative to the New Policies Scenario. Total import bills increase 10% to more than $46 trillion in the Deferred Investment Case; but oil supplies are more diversified, as MENA's share of inter-regional exports falls.

---

20. Discounting the value of these cash flows at a rate equal to the rate of growth of GDP per capita – as a proxy – reduces the loss of cash flow to about $70 billion.

CHAPTER 4

NATURAL GAS MARKET OUTLOOK

Nothing but blue skies?

H I G H L I G H T S

- Natural gas is projected to play an increasingly important role in the global energy economy. It is the only fossil fuel for which demand rises in all three *Outlook* scenarios. In the New Policies Scenario, world demand increases to 4.75 tcm in 2035 at an average rate of 1.7% per year. Global gas consumption catches up with coal consumption. Gas demand growth in the Current Policies Scenario is 2% per year, pulled up by higher total energy demand, but is only 0.9% in the 450 Scenario as demand peaks around 2030, before falling in favour of zero-carbon energy sources.

- Economic growth and energy policies in non-OECD countries will be the key determinant of future gas consumption: non-OECD countries account for 81% of demand growth in the New Policies Scenario. A major expansion of gas use in China pushes domestic demand above 500 bcm by 2035, from 110 bcm in 2010. Power generation takes the largest share of global consumption, increasing at 1.8% per year in the New Policies Scenario; but there is a broad-based rise in gas use across industry, buildings and (from a much lower base) also the transportation sectors.

- The world's remaining resources of natural gas can comfortably meet the projections of global demand to 2035 and well beyond. Unconventional gas accounts for roughly half the estimated resource base of over 800 tcm; its share in output rises in the New Policies Scenario from 13% in 2009 to above 20% in 2035, although the pace of this development varies considerably by region. Growth in unconventional output will depend on the industry dealing successfully with the environmental challenges.

- Russia is the largest gas producer in 2035 (reaching nearly 860 bcm) and makes the largest contribution to supply growth over the projection period. China emerges as a major producer in Asia (290 bcm in 2035), while output from the Middle East and Africa also increases rapidly. Europe remains the largest import market – EU imports reach 540 bcm in 2035 – but imports into China increase fastest. Supply and demand in North America remain broadly in balance throughout the *Outlook* period.

- Although based on somewhat different assumptions, the projections in this *Outlook* reinforce the main conclusions of a special report released earlier in 2011 which examined whether the world is entering a "Golden Age of Gas". Fundamental factors on both the supply and demand side point to an increased share of gas in the global energy mix. Although gas is the cleanest of the fossil fuels, increased use of gas in itself (without CCS) will not be enough to put us on a carbon emissions path consistent with limiting the rise in average global temperatures to 2°C.

Demand

Primary gas demand trends

In last year's *Outlook*, we asked whether the prospects for natural gas are now so bright that we are entering a "golden age" of natural gas. A special *WEO* report, released in June 2011, examined the prospects for, and implications of, higher gas use by developing a "Golden Age of Gas Scenario" (IEA, 2011); the key findings from this special *WEO* report are summarised at the end of this chapter. The modelling results presented below incorporate the insights from this special report, but not all its assumptions (which intentionally constituted a particularly favourable set of conditions for natural gas to 2035). We revert here to the assumptions of the three *WEO* scenarios as described in Chapter 1.

This year's *Outlook* nonetheless highlights the increasingly important role that natural gas is expected to play in the global energy mix. It is the only fossil fuel for which demand rises in all three scenarios, underlining one of the chief attractions of gas: it is a fuel that does well under a wide range of future policy directions. Moreover, gas demand in all scenarios in 2035 is higher than the *WEO-2010* projections. This reflects the impact on the years to 2015 of the 12th Five-Year Plan, announced by China in 2011, which envisages a major expansion of domestic use of natural gas. The global consequences of the damage to the Fukushima nuclear plant in Japan push up projections of future gas consumption, as natural gas is the fuel which benefits most readily from any switch away from nuclear power (see Chapters 5 and 12). Higher projected output of unconventional gas also acts to keep increases in the price of natural gas below the level envisaged in *WEO-2010*, increasing its competitiveness against other fuels. There is, nonetheless, still a large variation in the trajectories of gas demand between the three scenarios (Table 4.1).

Table 4.1 ● **Primary natural gas demand by region and scenario** (bcm)

	1980	2009	New Policies Scenario		Current Policies Scenario		450 Scenario	
			2020	2035	2020	2035	2020	2035
OECD	959	1 518	1 705	1 841	1 714	1 927	1 597	1 476
Non-OECD	557	1 558	2 183	2 909	2 215	3 160	2 068	2 400
World	**1 516**	**3 076**	**3 888**	**4 750**	**3 929**	**5 087**	**3 665**	**3 876**
Share of non-OECD	*37%*	*51%*	*56%*	*61%*	*56%*	*62%*	*56%*	*62%*

In the New Policies Scenario, demand for natural gas grows from 3.1 trillion cubic metres (tcm) in 2009[1] to 4.75 tcm in 2035, an increase of 55%. The pace of annual demand growth averages 1.7% per year over the entire period 2009 to 2035, but the rate of growth

---

1. Estimated global gas demand in 2010 rebounded to 3.3 tcm.

decreases through the projection period. The share of gas in the global energy mix rises from 21% in 2009 to 23% in 2035, catching up with the share of coal. Demand grows more quickly in the Current Policies Scenario, at an average of 2% per year, and reaches almost 5.1 tcm in 2035. In the 450 Scenario, however, demand growth is slower, at 0.9% per year, and tails off after 2030 as the market penetration of renewables increases. In this scenario, the share of gas in the global energy mix increases only slightly from 21% to 22%.

The projection of gas consumption in the Current Policies Scenario reflects higher overall energy demand in this scenario, which drives up demand for all fuels (Figure 4.1). Gas demand in the New Policies Scenario, which is within 7% of that in the Current Policies Scenario in 2035, is pushed higher by new measures that favour gas use relative to other fossil fuels, such as stricter regulation of emissions and pollutants (see Chapter 1 and Annex B). Demand in the 450 Scenario is 24% below the Current Policies Scenario in 2035 and 18% below the New Policies Scenario, due in part to lower demand for electricity in this scenario and to strong additional policy action to reach the goal of limiting the rise in greenhouse-gas emissions.

Natural gas demand has bounced back strongly from the decline seen in 2009; according to preliminary data, global gas demand rose by an estimated 6.6% in 2010, more than compensating for the earlier fall. An unusually cold winter in Europe and a hot summer in the Pacific region accounted for half of the 6% rise in OECD demand; data adjusted for average temperatures would put gas demand in Europe close to the levels seen in late 2007. However, non-OECD demand has jumped ahead: gas use in China, for example, rose by 19% in 2010, making it the fourth-largest gas consumer in the world (after the United States, Russia and Iran). Looking further ahead, our projections show demand for gas increasing in both OECD and non-OECD regions in all periods and all scenarios, with the exception of OECD gas demand after 2020 in the 450 Scenario, which contracts by 0.5% per year. Overall, economic growth and energy policies in non-OECD markets, China in particular, will be the key determinants of the overall increase in global gas demand.

Figure 4.1 ● World primary natural gas demand by scenario

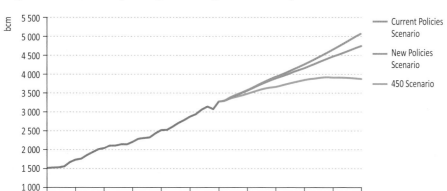

Regional and sectoral trends

Total demand from non-OECD countries overtook OECD demand in 2008 and is projected to grow at 2.4% per year over the period to 2035 in the New Policies Scenario, compared with 0.7% in the OECD. A disaggregated analysis of the projections confirms that the fastest growing individual gas markets across the world are all outside the OECD (where the markets are sizeable, but more mature) (Figure 4.2). The largest increments in demand in 2035, compared to 2009, are in China and the Middle East. Among the major non-OECD markets, only in Russia – where natural gas already accounts for more than 50% of primary energy use – is the growth in consumption more modest (see Chapter 7).

Figure 4.2 ● **Natural gas demand by selected region in the New Policies Scenario, 2009 and 2035**

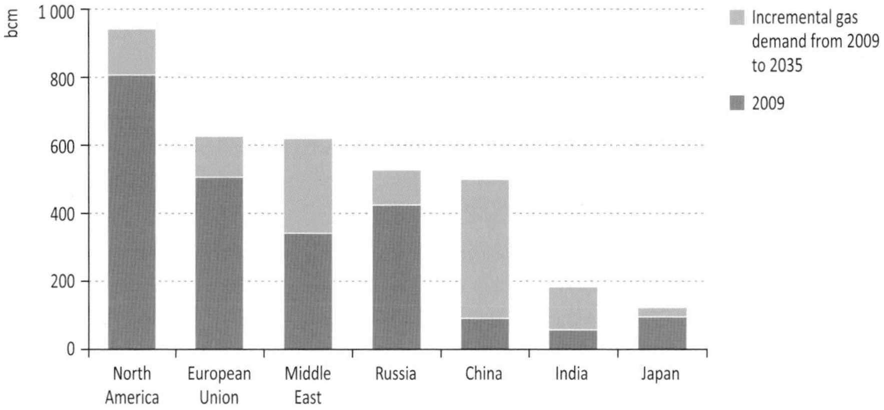

Notes: 2009 is the base year for gas projections. Rates of growth would be lower if 2010 figures were used as base year due to the impact of the economic crisis on gas demand in 2009.

Among the major countries and regions, China is the fastest growing, with annual average growth of 6.7%. Over the projection period as a whole, China accounts for a quarter of global gas demand growth (Figure 4.2 and Table 4.2). Only around 10% of residential households in China presently have access to natural gas, well below the global average of 40%, and national policies are increasingly supportive of an expanded role for gas in China's energy consumption as a way to diversify the energy mix and reduce local pollution. Our projection for China's demand for gas in 2015 is nearly 200 bcm, rising to 500 bcm in 2035, 11% of China's energy mix. Increases are spread across all of the main consuming sectors, with the largest increment in power generation.

Gas demand also expands quickly in other parts of Asia and in the Middle East: gas is a particularly attractive fuel for countries that are seeking to satisfy rapid growth in fast-growing cities. As in China, the Indian government plans to increase the share of gas in its energy mix. The domestic gas market is projected to grow from 59 bcm in 2009 to 190 bcm in 2035, 11% of total primary energy demand. In the Middle East, gas has become the preferred fuel for power generation, substituting for oil in many cases and freeing up the more valuable product for export. Consumption in gas to-liquids (GTL) plants in the

Middle East (included in "other energy sector" in Figure 4.3) also adds to the increase. Overall, gas use in the Middle East rises from 340 bcm in 2009 to 620 bcm in 2035, a growth rate of 2.3% per year. Demand in Russia grows more slowly, at an average of 0.8% per year, mainly because of continuing improvements in energy efficiency as the capital stock is renewed and end-user gas prices increase.

Table 4.2 ● **Primary natural gas demand by region in the New Policies Scenario** (bcm)

	1980	2009	2015	2020	2025	2030	2035	2009-2035*
OECD	959	1 518	1 654	1 705	1 746	1 804	1 841	0.7%
Americas	660	811	852	877	900	928	951	0.6%
United States	*581*	*652*	*680*	*685*	*692*	*703*	*710*	*0.3%*
Europe	264	537	604	627	644	666	671	0.9%
Asia Oceania	35	170	198	201	202	210	219	1.0%
Japan	*25*	*97*	*118*	*122*	*122*	*125*	*126*	*1.0%*
Non-OECD	557	1 558	1 911	2 183	2 417	2 668	2 909	2.4%
E. Europe/Eurasia	438	627	698	723	763	797	830	1.1%
Caspian	*n.a.*	*107*	*124*	*131*	*143*	*151*	*161*	*1.6%*
Russia	*n.a.*	*426*	*467*	*478*	*495*	*513*	*530*	*0.8%*
Asia	36	357	531	686	796	921	1063	4.3%
China	*14*	*93*	*197*	*301*	*366*	*435*	*502*	*6.7%*
India	*2*	*59*	*76*	*99*	*120*	*150*	*186*	*4.5%*
Middle East	35	343	402	450	509	578	622	2.3%
Africa	13	99	112	129	142	153	161	1.9%
Latin America	35	133	168	196	208	220	233	2.2%
Brazil	*1*	*20*	*41*	*60*	*70*	*80*	*91*	*5.9%*
World	**1 516**	**3 076**	**3 565**	**3 888**	**4 164**	**4 473**	**4 750**	**1.7%**
European Union	*n.a.*	*508*	*572*	*593*	*608*	*626*	*629*	*0.8%*

*Compound average annual growth rate.

In OECD countries, gas for the power sector accounts for the largest share of demand growth over the projection period; the share of gas in the OECD electricity mix rises by one percentage point to 24% in 2035. This continues a trend to gas as the predominant choice for new generation over the last ten years, albeit at a slower pace. Economic uncertainty, policy choices (including carbon pricing in Europe) and the addition of more variable renewables have all contributed to the attraction of gas-fired power, which has lower capital costs and shorter construction times than the main alternatives. Compared with *WEO-2010*, gas demand for power generation is higher because of decisions in some OECD countries in 2011, notably in Germany, to reduce or rule out the role of nuclear power in their energy mix. Consumption of natural gas in the European Union increases by an average of 0.8% per year from

2009 to 2035 in the New Policies Scenario and reaches 630 bcm in 2035, 30 bcm higher than the corresponding figure for 2035 in *WEO-2010*.

Figure 4.3 ● **Incremental primary natural gas demand by region and sector in the New Policies Scenario, 2009-2035**

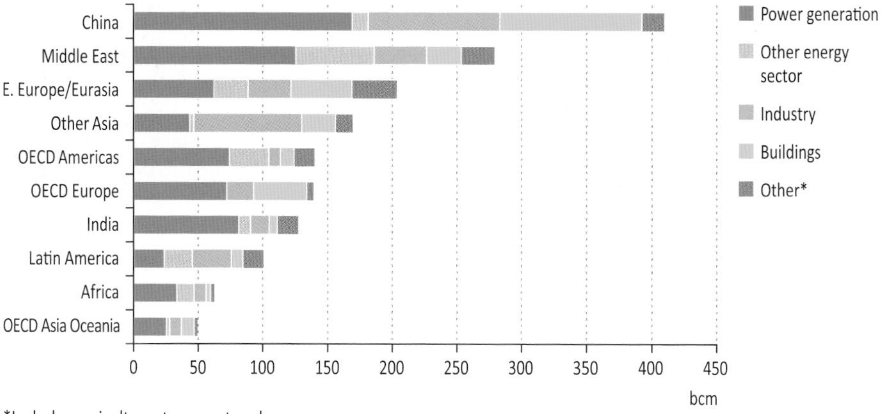

*Includes agriculture, transport and non-energy use.

The largest share of global gas demand comes from the power sector, where gas use for electricity generation rises to more than 1.9 tcm by 2035, at an annual rate of 1.8% (Figure 4.4). The share of gas in global electricity generation increases slightly from 21% today to 22% in 2035, while the share of coal declines from 40% in 2009 to 33% in 2035 and the share of oil falls from 5% to 1% over the same period. After power generation, the next largest consumption of gas in 2035 is in buildings, primarily for space and water heating; 70% of gas demand in the buildings sector currently comes from OECD countries. Global consumption of gas in buildings grows more slowly over the projection period than in other major end-use sectors. Demand growth is quicker in industry, where gas is used mainly for the production of steam for mechanical energy and supplying the heat needed to produce materials and commodities. If available, gas can be an attractive choice for industrial processes, since it is easy to handle, more efficient and has fewer adverse environmental impacts than other fossil fuels. Gas use in industry grows from 535 bcm in 2009 to 890 bcm in 2035, with petrochemicals, iron and steel, and non-metallic minerals sub-sectors such as cement, taking the largest shares.

At a global level, gas does not currently compete strongly in all markets or in all sectors, but it is making inroads. From a low base, gas use in road transportation is projected to increase more quickly than in any other sector, at an average rate of 5.3% per year in the New Policies Scenario. This is at present a relatively under-developed market for natural gas: more than 70% of the world stock of natural gas vehicles is in only five countries: Pakistan, Argentina, Iran, Brazil and India. Although natural gas in transportation typically brings considerable fuel-cost savings and emissions reductions, greater use has usually been held back by the limited availability of refuelling infrastructure. We project natural gas consumption in road transportation will quadruple to more than 80 bcm by 2035; even so, this is still only 3% of the total energy used for road transportation.

Figure 4.4 ● Primary natural gas demand by sector in the New Policies
Scenario, 2009 and 2035

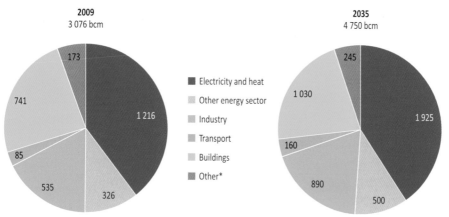

*Includes agriculture and non-energy use.

Supply

Resources and reserves

The world's remaining resources of natural gas can comfortably meet the projections of
global demand growth in this *Outlook* to 2035 and well beyond. This conclusion has been
reinforced over the last few years as our understanding of the recoverable resource base
has expanded, particularly of the size of unconventional gas resources – from coal beds
(coalbed methane), low-permeability reservoirs (tight gas) and shale formations (shale gas).[2]
Conventional recoverable resources of just over 400 tcm are equal to around 120 years
of production at 2010 levels; adding unconventional recoverable resources (which are of
similar size) brings this figure to nearly 250 years (Figure 4.5).

The fact that unconventional resources are more widely dispersed than conventional
resources has implications for gas security since all major regions now have total
recoverable gas resources equal to at least 75 years of current consumption. This does
not mean that gas is readily available in each region; resources require substantial
investment – sometimes over decades – before they can be produced and marketed.
But it does imply that countries and regions have the option, if they so wish, to develop
alternative, more diversified sources of gas supply; our analysis (IEA, 2011) suggests that
plentiful volumes of unconventional gas can be produced at costs similar to those of
North America (between $3 per million British thermal units [MBtu] and $7/MBtu). It is
worth comparing the outlook for gas with that for oil where, even with growing output of
unconventional oil, the trend is towards greater reliance on a small number of producers,
with oil delivered to markets along a limited number of potentially vulnerable supply routes
(see Chapter 3).

---

2. For example, the recent assessment of worldwide shale gas resources from the US Energy Information Administration
(US DOE/EIA, 2011).

Proven gas reserves of 190 tcm are estimated based on operators' public filings or government records, and are only a relatively small proportion of the total resources (Cedigaz, 2010). Most of the proven reserves are conventional. Unconventional proven reserves account for a significant share of the total only in the United States and Canada. Overall, 130 tcm out of the total proven reserves, a 70% share, are in Eastern Europe/Eurasia (mainly Russia) and the Middle East (Iran and Qatar); however, these regions account for a much smaller share (46%) of the total estimated recoverable resources.

Figure 4.5 ● **Recoverable gas resources and production by region and type, end-2010**

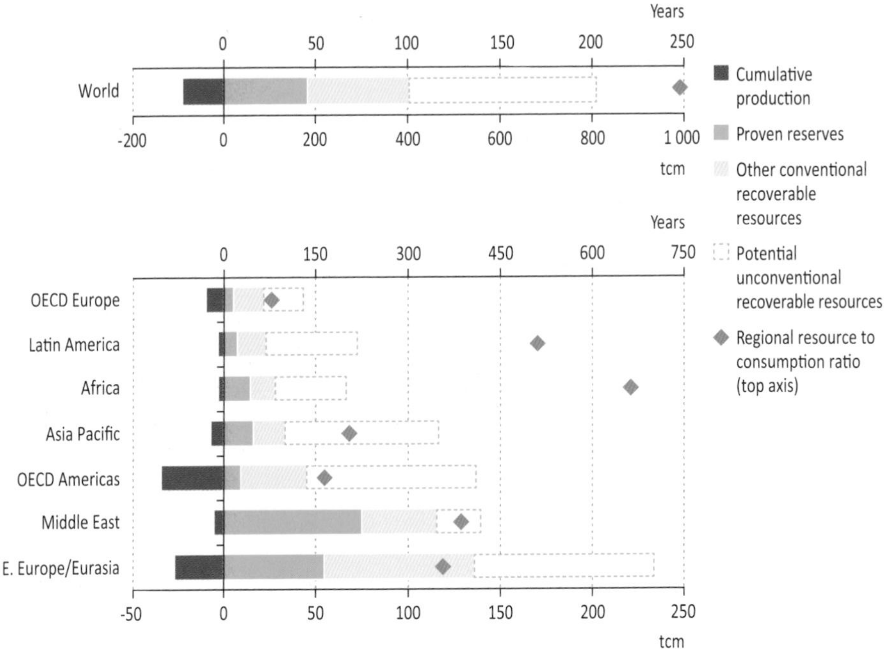

Notes: Cumulative production to date is shown as a negative number, so that the total of the bars to the right indicates remaining recoverable resources. Russian reserves are discussed in detail in Chapter 8.

Sources: Cedigaz (2010); USGS (2000 and 2008); BGR (2010); US DOE/EIA (2011); IEA estimates and analysis.

Production prospects

Gas production of between 3.9 tcm and 5.1 tcm is required to meet projected levels of consumption in 2035, depending on the scenario (Table 4.3). Gas prices are the main mechanism bringing demand and supply into balance. Higher prices in the Current Policies Scenario stimulate the production necessary to meet demand; on the other hand, lower prices in the 450 Scenario, brought about by far-reaching policy measures that weaken demand, result in lower investment and, consequently, lower output (see discussion of retail and end-user fuel prices in Chapter 1). In all scenarios, most of the increase in output comes from countries outside the OECD.

Table 4.3 ● Primary natural gas production by region, type and scenario (bcm)

	1980	2009	New Policies Scenario		Current Policies Scenario		450 Scenario	
			2020	2035	2020	2035	2020	2035
OECD	890	1 145	1 223	1 296	1 221	1 308	1 161	1 157
Non-OECD	639	1 906	2 666	3 454	2 708	3 780	2 504	2 719
World	1 528	3 051	3 888	4 750	3 929	5 087	3 665	3 876
% unconventional	0%	13%	15%	22%	n.a.	n.a.	n.a.	n.a.

Notes: Definitions and reporting of tight gas vary across countries and regions, so the split between conventional and unconventional gas production is approximate (and reported only for the New Policies Scenario). Differences between historical supply and demand volumes are due to changes in stocks.

Conventional gas will still account for the bulk of global gas production in 2035, but the share of unconventional gas rises from 13% in 2009 to 22% in 2035 and it provides 39% of incremental production over this period in the New Policies Scenario. Most of the increase in unconventional output in this scenario comes from shale gas and from coalbed methane, both of which reach a 9% share of global production in 2035. Tight gas has a lower share (4%).

Mergers, acquisitions and partnerships, reflecting confidence in the potential for unconventional output growth outside North America, are spreading expertise in its production. However, production projections are subject to a large degree of uncertainty, particularly in regions where little or no such production has been undertaken to date. Even where production is already taking place, environmental concerns could hold back or reduce output (Box 4.1). A range of factors needs to be positively aligned before unconventional production can make headway, including suitable geology, public acceptance, well-adapted regulatory and fiscal regimes and widespread access to experience and technology. For a relatively new industry, this points to a future in which the pace of unconventional development will vary considerably by country and by region.

Sources of incremental gas production in OECD Europe are increased conventional gas output from Norway and, later in the *Outlook* period, some unconventional gas production, most likely led by Poland. However, production from these sources is not enough to offset declines elsewhere so the overall output trend is down; production falls from nearly 300 bcm in 2009 to just above 200 bcm in 2035. The figures for the European Union (without Norway) are even more striking, with gas production declining from the 2009 level of nearly 200 bcm to less than 90 bcm in 2035, a 55% drop.

The share of unconventional gas production in North America is projected to increase steadily. It was already 56% in 2009 and rises to 64% in 2035, meaning that, in this market at least, it should no longer be strictly considered as "unconventional". After briefly overtaking Russia as the world's largest gas producer, the United States returns to being the second-largest global gas producer, a position that it retains throughout the projection period. Among other OECD producers, Australia increases gas output substantially, from 47 bcm in 2009 to almost 160 bcm in 2035, production being both conventional and unconventional (mainly coalbed methane).

Box 4.1 ● Environmental impact of unconventional gas

Although used as a production enhancement technique for many years, the rapid expansion of hydraulic fracturing to produce unconventional gas in the United States has put a spotlight on the environmental impact of shale gas production. Hydraulic fracturing involves pumping large volumes of water, mixed with sand and chemicals, into shale formations. Fracturing the rock is essential to stimulate the flow of gas into shale gas wells. The total volume of water injected ranges from 7 500 to 20 000 cubic metres per well. Given that each shale gas development requires drilling a large number of wells, concerns have been raised about the availability of water for hydraulic fracturing and the possible contamination of freshwater aquifers. Questions have also been raised about the level of greenhouse-gas emissions from shale gas production, compared with conventional production.

Best practice for shale gas production, backed up by effective regulation and monitoring, can mitigate the environmental risks, reducing the potential effects to a level similar to that for conventional gas production. The most important measures ensure that the well and the shale formation itself remain hydraulically isolated from other geological formations, especially freshwater aquifers. There is also the need to minimise water use, including by recycling it, and to ensure appropriate treatment and disposal of the water that is used.

Combustion is the main source of greenhouse-gas emissions from gas and combustion of shale gas is no different from combustion of natural gas from any other source. Nonetheless, shale gas produced to proper environmental standards has slightly higher well-to-burner emissions than conventional gas. The main incremental source of emissions is from the gas released during the process of completing wells, when some gas returns to the surface, together with the fracturing fluids. Depending on how these additional volumes are treated, whether they are vented or flared, the result overall is well-to-burner emissions for shale gas that are 3.5% to 12% higher than the equivalent for conventional gas (IEA, 2011). In most cases it is possible, using specialised equipment, to capture, treat and market the gas produced during the completion phase. Doing so brings overall emissions even closer to the levels of conventional gas production; while entailing a slight increase in costs, it also makes additional gas available for sale.

Gas production in Eastern Europe/Eurasia, from Russia and the Caspian countries, is greater than in any other *WEO* region over the period to 2035 (Table 4.4). The prospects for Russian gas, with output increasing from 570 bcm in 2009 to almost 860 bcm in 2035, are discussed in detail in Chapter 8. The development of Caspian resources is held back by the long distance to market and an uncertain investment climate in some countries. Turkmenistan is expected to continue its recovery from the collapse in its gas production in 2009, the rise in Turkmenistan output being driven by the super-giant South Yolotan field, which becomes the main source for export to China along the expanded Central Asia Gas Pipeline. Azerbaijan's gas production is projected to rise to over 55 bcm in 2035, more than three times current output, as incentives for upstream development improve with the anticipated opening of a southern gas corridor to European markets.

Table 4.4 ● Primary natural gas production by region in the New Policies Scenario (bcm)

	1980	2009	2015	2020	2025	2030	2035	2009-2035*
OECD	**890**	**1 148**	**1 181**	**1 227**	**1 242**	**1 275**	**1 297**	**0.5%**
Americas	658	796	814	840	865	905	932	0.6%
Canada	78	164	161	176	178	172	172	0.2%
Mexico	25	48	45	46	51	59	60	0.9%
United States	554	583	606	616	633	669	696	0.7%
Europe	219	294	279	259	240	222	204	-1.4%
Netherlands	96	79	83	67	54	41	28	-3.8%
Norway	26	106	109	117	122	124	120	0.5%
United Kingdom	37	62	37	27	17	12	10	-6.9%
Asia Oceania	12	55	84	124	134	146	159	4.2%
Australia	9	47	78	120	131	144	158	4.8%
Non-OECD	**639**	**1 903**	**2 384**	**2 661**	**2 921**	**3 197**	**3 452**	**2.3%**
E. Europe/Eurasia	485	753	909	957	1 069	1 138	1 197	1.8%
Russia	n.a.	572	679	692	779	822	858	1.6%
Turkmenistan	n.a.	38	71	89	98	109	120	4.5%
Azerbaijan	n.a.	16	22	39	48	55	56	4.8%
Asia	59	393	502	581	642	708	773	2.6%
China	14	85	135	176	212	252	290	4.8%
India	2	46	63	78	91	105	120	3.7%
Indonesia	17	77	95	102	106	112	119	1.7%
Malaysia	2	60	69	71	72	73	74	0.8%
Middle East	38	412	527	580	614	701	773	2.5%
Qatar	3	89	160	174	180	205	219	3.5%
Iran	4	137	137	151	165	195	225	1.9%
Iraq	1	1	9	28	41	57	70	17.1%
Saudi Arabia	11	75	89	95	97	108	116	1.7%
UAE	8	49	50	52	52	56	60	0.8%
Africa	22	196	260	320	361	399	442	3.2%
Algeria	13	78	107	134	147	160	171	3.1%
Nigeria	2	23	40	56	75	91	110	6.2%
Libya	5	16	15	20	25	35	49	4.4%
Latin America	35	152	190	228	238	253	269	2.2%
Brazil	1	12	24	55	73	88	99	8.5%
Venezuela	15	22	25	31	40	56	73	4.8%
Argentina	10	44	48	54	52	45	41	-0.3%
World	**1 528**	**3 051**	**3 565**	**3 888**	**4 164**	**4 473**	**4 750**	**1.7%**
European Union	n.a.	196	174	145	122	103	89	-3.0%

*Compound average annual growth rate.

In Asia, the most significant development is the emergence of China as a major gas producer (even though it continues to rely on imports to meet domestic demand) as national companies step up their efforts to increase output (Figure 4.6). China's gas production rises from 85 bcm in 2009 to 290 bcm in 2035. The bulk of this increase is expected to come from tight gas deposits, coalbed methane and shale gas (China's first licensing round for shale gas acreage took place in mid-2011.) Domestic production in India is estimated at more than 50 bcm in 2010 and output is projected to rise to 120 bcm in 2035. As in China, there is increased interest in unconventional gas: a fifth coalbed methane licensing round and a shale gas licensing round are scheduled for 2012.

Figure 4.6 ● *Change in annual natural gas production in selected countries in the New Policies Scenario*

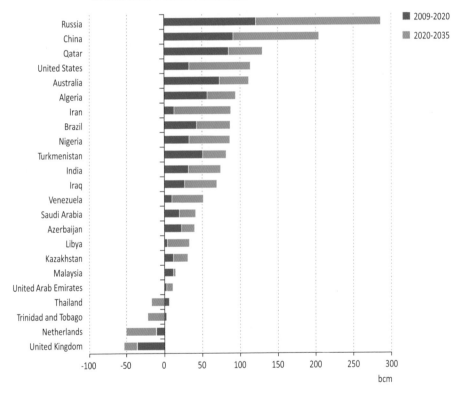

Gas production in the Middle East expands rapidly, rising at an average rate of 2.5 % per year throughout the projection period, to more than 770 bcm in 2035. The initial boost comes from Qatar, as newly built liquefied natural gas (LNG) plants increase their throughput and the Pearl GTL project ramps up. Production jumps from about 90 bcm in 2009 to more than 160 bcm in 2015, but then rises more slowly, under the influence of the moratorium on new development projects which has been put in place pending the outcome of a study of the effects of current projects on the world's largest gas field, the North Field. Sizeable production growth in Iran is likely only after 2020, as short-term prospects are held back by international sanctions that limit investment and technology transfer. The resource base in Iraq could support a substantial rise in gas output, but average annual increases of 17% come from a very low base. Rising production costs and generally low domestic price present a challenge to many countries across the region, particularly as gas starts to be produced from gas condensate fields rather than produced as low-cost gas together with oil production (associated gas).

The outlook for North African producers has been affected by the political upheavals in the region in 2011 but, even before this, there were signs of difficulty in maintaining production in some countries, notably Egypt, at the pace necessary to satisfy domestic demand and export commitments. We project that output from Egypt will peak around 2020, before falling back to slightly less than 60 bcm in 2035. Projections for Libya in the period to 2015 have been cut back substantially, compared to *WEO-2010,* as a result of the conflict. Elsewhere, Algeria remains the continent's largest producer, with output expected to rise to more than 170 bcm in 2035. But, overall, we expect the traditional predominance of North African producers to be challenged by a number of sub-Saharan developments, with Nigeria and Angola leading the increases in output. Production for the African continent as a whole rises to over 440 bcm in 2035, a 9% share of global supply. In Latin America, production of conventional gas continues to predominate. The main source of supply growth is Brazil, where development of the pre-salt fields is expected to boost production from 12 bcm in 2009 to 100 bcm in 2035. Brazil overtakes Argentina as the region's largest producer and becomes a net gas exporter before 2025.

Inter-regional trade

The volume of gas traded internationally (between *WEO* regions) is set to increase, with both pipeline gas and LNG playing important roles. The main growth in pipeline supply occurs in Eurasia, with the expansion of Russian and Caspian capacity for export, both to Europe and to China. LNG trade has expanded at an unprecedented rate in recent years, with global liquefaction capacity in mid-2011 estimated at 370 bcm, compared with 250 bcm in 2007. The pace of capacity additions is slowing down, but LNG still accounts for 42% of the projected growth in inter-regional gas trade during the period to 2035.

Rising exports from Russia, supplemented by increasing trade out of the Caspian region, mean that Eastern Europe/Eurasia strengthens its position as the largest net exporting region (Figure 4.7). An important shift over the projection period – and a major driver for production increases in this region – is that Russian and Caspian resources start to meet a much larger share of China's growing import needs, as well as supplying traditional markets in Europe (see Chapter 9). Overall, China accounts for 35% of the total growth in inter-regional trade, as its import requirement grows from less than 10 bcm in 2009 to 125 bcm in 2020 and over 210 bcm in 2035. China becomes the second-largest import market in the world after Europe. China's imports come from a variety of sources, by pipeline from Central Asia, Russia and also from Myanmar, and as LNG from a suite of global suppliers.

Figure 4.7 ● Net gas trade by major region in the New Policies Scenario (bcm)

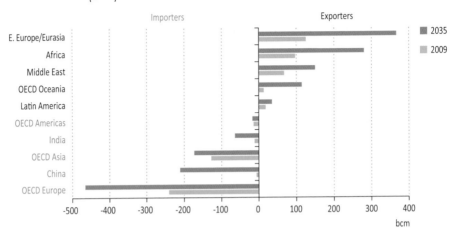

Australia (included in OECD Oceania) becomes a major gas exporter over the projection period, with net exports increasing from 14 bcm in 2009 to 85 bcm in 2020 and 115 bcm in 2035. Six major LNG export projects are underway, including the giant Gorgon project in Western Australia and three projects in Queensland that are the first in the world to be based on coalbed methane. Based on currently operating and sanctioned projects, Australian LNG export capacity could exceed 70 bcm soon after 2015, making it the second-largest global LNG exporter after Qatar.

Even though gas consumption in the Middle East grows rapidly, gas production growth exceeds domestic demand, meaning that the region, led by Qatar, bolsters its standing as a major supplier to global markets. Likewise, net exports from Africa increase rapidly, from just below 100 bcm in 2009 to 280 bcm in 2035, through a mixture of pipeline gas (primarily from North Africa) and LNG. Latin America remains a marginal net exporter.

Europe's requirement for gas imports has increased, compared with last year's *Outlook*, because of the higher projection for natural gas consumption. OECD Europe is projected to import 470 bcm of gas in 2035, almost double the 2009 figure, and the requirement for the European Union rises to 450 bcm in 2020 and 540 bcm in 2035 (86% of total EU gas consumption), up from 310 bcm in 2009. However, the expectation, from a few years ago, that North America would also be a major importer of LNG, has been turned on its head by the increase in regional production of unconventional gas. This means that the North American market does not need to depend on inter-regional imports and has even led to LNG export projects (from North America) moving forward. Our projections suggest that supply and demand of natural gas in North America will be roughly equal over the coming decades (Box 4.2).

The spectre of a gas-supply glut which loomed over the gas market for the past three years has been considerably dissipated with the recovery of demand in 2010. The projections of the New Policies Scenario suggest that the rate of utilisation of inter-regional transportation capacity will be back to pre-crisis levels before 2015.[3] The market is tightening first in the Asia-Pacific region (notably in China and Japan) and the effects could be quickly transmitted to the European market as competition increases for available LNG.

Box 4.2 • **North America: net gas importer or exporter?**

Gas production and demand in North America are currently both some 800 bcm per year. Both are projected to increase to above 900 bcm in 2035. The net trade position, therefore, depends on the balance between two very large numbers. Small shifts on either the supply or the demand side could quickly have relatively large implications for imports into the region or exports from it.

In all three *Outlook* scenarios, the North American market remains broadly in balance throughout the period to 2035 (with the region as a whole a marginal net importer). One of the reasons this balance is maintained is that unconventional supply in the United States is expected to become more responsive to fluctuations in demand, thus helping to maintain a regional equilibrium. Moreover, while higher prevailing prices in Asia and Europe create incentives for export, there may be competitive options within North America, for example, to supply natural gas vehicles or GTL projects. Some LNG export projects are expected to go ahead – for example, the involvement of Asian companies, such as KOGAS, in Canadian gas production could lead to trans-Pacific LNG trade – but our projections do not suggest that North America will assume a major role in global gas trade.

Investment

Projected trends in gas demand and supply call for total cumulative investment of around $9.5 trillion (in year-2010 dollars) in supply infrastructure in the New Policies Scenario for the period 2011 to 2035 (Figure 4.8). Output from currently producing conventional gas fields declines to 1.1 tcm in 2035, supplying less than a third of projected conventional output in 2035. The gross upstream investment requirement for conventional gas production capacity over the projection period is $5 trillion, around three-quarters of which arises in non-OECD countries. Investment needs for unconventional gas production come to $1.8 trillion, but here the regional split is reversed, with three-quarters of this sum being spent within the OECD. Total investment in LNG supply infrastructure over the projection period is estimated at $590 billion, and a further $2.1 trillion is needed for pipeline transmission and distribution systems. A detailed discussion of oil and gas investment and costs is included in Chapter 3.

---

3. Some of the incremental pipeline capacity being built to serve the European market is designed to substitute for, rather than supplement, existing routes to market (and, therefore, is not an indicator of excess supply capacity).

Figure 4.8 ● Cumulative investment in natural gas supply infrastructure by region and activity in the New Policies Scenario, 2011-2035

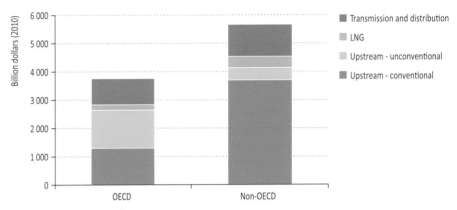

Note: The estimated $80 billion of investment needed worldwide in LNG carriers is not included.

Are we entering a Golden Age of Gas?

The *WEO* special report "Are We Entering a Golden Age of Gas?", released in June 2011, examined the proposition that natural gas might play a significantly greater role in the future global energy mix. It did this on the basis of a Golden Age of Gas Scenario (GAS Scenario) which incorporated plausible but deliberately favourable assumptions about policy, prices and other drivers that positively affect gas demand and supply projections. This section briefly re-caps the main assumptions and findings of this special report and compares the projections of the GAS Scenario with those of the New Policies Scenario in this *Outlook*.

The starting point for the GAS Scenario was the New Policies Scenario from *WEO-2010*.[4] Four major new assumptions were adopted:

- A more ***ambitious policy for gas use*** in China, driven principally by the policy of expanding gas use included in China's 12th Five-Year Plan. China's gas demand in this scenario is also influenced by the assumption of lower gas prices (making gas more competitive versus coal) and of slower growth in nuclear power capacity.

  In the *WEO-2011* New Policies Scenario, we recognise the impact of the 12th Five-Year Plan but adopt a slightly more cautious view of China's gas demand growth, which averages 6.7% per year over the period 2009 to 2035, as opposed to 7.6% in the GAS Scenario (Figure 4.9).

- ***Changing supply and demand fundamentals***, the GAS Scenario incorporates a more optimistic assumption about future gas supply – primarily as a result of the availability

---

4. The GAS Scenario adopted the same assumptions for population and economic growth as the scenarios in *WEO-2010*. However, these have subsequently been updated for the projections in this year's *Outlook* (see Chapter 1).

of additional unconventional gas supplies at relatively low cost – and, accordingly, lower price assumptions. Prices are assumed to be $1.5/MBtu to $2.5/MBtu lower than in the *WEO-2010* New Policies Scenario and around $1/MBtu to $1.5/MBtu lower than in the *WEO-2011* New Policies Scenario.[5] The effect of lower gas prices, both in absolute terms and also relative to other fuels, is to increase gas demand across the board.

The *WEO-2011* New Policies Scenario sees unconventional gas supply rising to over 1.05 tcm in 2035, 22% of total gas production. In the GAS Scenario this supply rises to 1.2 tcm in 2035, or 24% of total output.

- ■ **Greater use of natural gas for transportation**: the GAS Scenario assumes that governments in some countries act vigorously to encourage greater introduction of natural gas vehicles (NGVs); penetration of NGVs is also encouraged by a favourable price differential between natural gas and oil. As a result, the GAS Scenario projects that there will be over 70 million NGVs by 2035 (up from 12 million currently) and that the share of natural gas in the fuel mix for road transportation rises from the current 1% to 5% in 2035.

  In the *WEO-2011* New Policies Scenario, the rise in the number of NGVs is less dramatic, reaching only around 35 million vehicles by 2035, with the share of natural gas in road transportation rising to around 3% at the end of the projection period. This is the main reason for the lower average annual increase in gas use in the transportation sector in *WEO-2011*, at 2.5%, than the 3.6% seen in the GAS Scenario (Figure 4.9).

- ■ **Slower growth of nuclear power capacity**: in the aftermath of Fukushima, the GAS Scenario assumes that the licenses of fewer existing plants are extended and fewer new plants are built, compared with the *WEO-2010* baseline. As a result, the share of nuclear power in primary energy demand rises (from 6% today to 7% in 2035), but more slowly than had previously been projected.

  In the *WEO-2011* New Policies Scenario, we assume a less significant constraint on the growth in nuclear capacity. Whereas the GAS Scenario (developed in early 2011) limited worldwide nuclear generation capacity to a total of 610 GW in 2035, the New Policies Scenario in this year's *Outlook* sees nuclear capacity expanding to over 630 GW (see Chapter 12 for a discussion of the implications of lower nuclear capacity growth).

Taken overall, gas demand growth in the GAS Scenario is 2% per year for 2009 to 2035, compared with 1.7% in the *WEO-2011* New Policies Scenario. Despite the variations arising from different underlying assumptions, the results of our *WEO-2011* New Policies Scenario reinforce the main conclusion of the GAS Scenario – that fundamental factors on both the supply and demand side point to an increased share of gas in the global energy mix. Natural gas resources are abundant, well spread across all regions and recent

---

5. Despite the observed trend of increasing globalisation of natural gas, we assume in both the GAS scenario and in *WEO-2011* that the gas price differentials between the United States, Europe and Japan remain broadly constant. The magnitude of the differences is similar to the costs of transportation between the regions.

technological advances have facilitated increased global trade. Gas is a flexible fuel that is used extensively in power generation, competes increasingly in most end-use sectors and offers environmental benefits when compared to other fossil fuels.

Figure 4.9 ● *Comparison of average annual natural gas demand growth between the New Policies Scenario and the GAS Scenario, 2009-2035*

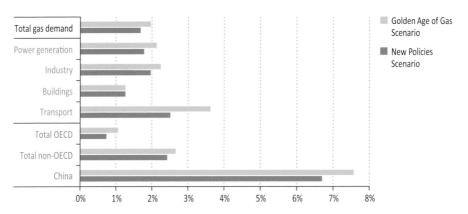

Notes: GAS Scenario growth rates for 2009 to 2035 differ from the figures for 2008 to 2035 that were included in the *Golden Age of Gas* report (IEA, 2011), due to the change in base year. This change permits direct comparison with the *WEO-2011* New Policies Scenario.

This is not to say that a "Golden Age of Gas" is inevitable (Spotlight). Nor should it be thought that increased use of natural gas, in itself, is enough to put the world on a carbon emissions path that is consistent with limiting the average global temperature rise to no more than 2°C. On its own, this is far from enough: CO_2 emissions of around 35 Gt in 2035 in the GAS Scenario put emissions on a long-term trajectory consistent with stabilising the concentration of greenhouse-gas emissions in the atmosphere at around 650 ppm, suggesting a long-term temperature rise of over 3.5°C (this is similar to the trajectory in the *WEO-2011* New Policies Scenario, see Chapter 2).

Ultimately, the extent of the expansion of gas use hinges on the interaction between economic and environmental factors, and policy interventions in the market. In the absence of a price for CO_2, coal is likely to remain cheaper than gas for generating electricity in many regions. However, a cost comparison alone does not reflect the full range of benefits that gas can provide, such as reinforcing the diversity of energy supply, providing flexibility and back-up capability as more intermittent renewable capacity comes online and reducing local pollutants and emissions (when substituting for coal). Natural gas alone cannot provide the answer to the challenge of climate change. Addressing the effects of climate change will require a large shift to low-carbon energy sources, in both the short and long term, increasing energy efficiency and deploying innovative technologies, like carbon capture and storage. But natural gas can play an important part in the transition.

Do all roads lead to a Golden Age of Gas?

There are powerful arguments in favour of a bright future for natural gas. In all the scenarios examined in this *Outlook* (including the GAS Scenario released in June 2011), natural gas has a higher share of the global energy mix in 2035 than it does today. But a golden age of gas could nonetheless be tarnished or cut short by policy decisions or by technological or other developments that reduce the attractiveness of gas relative to other energy sources. While some of these are unlikely or unforeseeable, it is nonetheless worth considering circumstances in which the rise of natural gas might be held back.

A first consideration is that much of the momentum behind the rapid growth of natural gas comes from supportive government policies in non-OECD countries. In China, India and other countries, governments are targeting an increased role for natural gas and are intervening in the market in order to promote the infrastructure and end-uses that will allow this to happen. But if market economics became an absolute priority in choosing the fuel and technology for deployment in power generation in China, for example, gas would have little impact on the power mix, except in regions where transport costs raise coal prices considerably (IEA, 2011). This could, in theory, lead to a reversal or weakening of support for natural gas, leaving it as a niche fuel, rather than a mainstream contributor to non-OECD energy demand growth.

From another angle, a dramatic reduction in the cost of a major renewable technology could have a large impact, as could a significant improvement in the efficiency of carbon capture and storage – although this would have to be somehow weighted in favour of CO_2 capture from coal-fired combustion if it were to tip the scales against natural gas. The role of gas as a back-up to variable renewables would be affected if cost-effective electricity storage were to be developed.

On the supply side, there is already public and political concern about the environmental impact of unconventional gas; if not adequately addressed through the adoption of best practices in gas production, these concerns could become a major constraint in some countries on the expansion of gas output. In markets where alternative supplies are limited, further experience of disruptions to deliveries from an established supplier – as happened in Europe in 2006 and 2009 during the Russia-Ukraine disputes – can quickly raise concerns about gas security that could, in turn, lead to action to temper increased reliance on natural gas consumption.

A setback to a golden age of gas could also come from concerted policy action on CO_2 emissions and efficiency, particularly if deployment of carbon capture and storage is delayed. This could make a reality of the trend that is visible in the 450 Scenario, where gas consumption flattens towards 2030 and then starts a gradual decline.

CHAPTER 5

POWER AND RENEWABLES OUTLOOK
Electrifying solutions?

H I G H L I G H T S

- World electricity demand in the New Policies Scenario is projected to increase from 17 200 TWh in 2009 to over 31 700 TWh in 2035, an annual growth rate of 2.4%, driven by economic and population growth. China and India account for over half of the increase, with OECD countries making up less than one-fifth. Globally, industry remains the largest consuming sector, followed by the residential and services sectors.

- Over the *Outlook* period, 5 900 GW of new capacity is added worldwide in the New Policies Scenario to meet demand growth and replace retired power plants, at an investment cost of nearly $10 trillion (in year-2010 dollars). Renewable energy technologies account for half of cumulative additional capacity and 60% of the investment. Gas and coal each provide one-fifth of new capacity additions. China adds more coal, gas, nuclear, hydro, biomass, wind, and solar capacity than any other country.

- From 2009 to 2035, 44% of the increase in electricity generation comes from renewables. Mainly driven by government policies, generation from non-hydro renewables increases from 3% of the total in 2009 to over 15% in 2035. Coal use increases by almost 50% over the *Outlook* period, remaining the largest source of electricity, but its share of total generation falls by seven percentage points over the *Outlook* period. Gas and hydro broadly maintain their shares in the generation mix. Despite recent announced policy changes, nuclear also retains its share of global electricity generation through 2035, buoyed by expansion in China.

- In the New Policies Scenario, over two-fifths of global investment in the power sector goes to transmission and distribution (T&D), with almost three-quarters of this for distribution networks. Nearly two-thirds of the investment is made in non-OECD countries, to meet growing demand. Two-fifths is to replace ageing infrastructure currently in use. Reinforcing and expanding T&D networks is necessary to integrate more renewables-based generation, accounting for just below 10% of the transmission investment (excluding distribution). Grid integration costs of renewables can be higher for certain regions such as the European Union, where it is 25% of transmission investment. In the 450 Scenario, global grid integration costs rise to over 18% of total investment in transmission.

- Worldwide CO_2 emissions from the power sector in the New Policies Scenario increase by one-fourth between 2009 and 2035, growing more slowly than demand. This is mainly the result of the increased use of renewables and improved plant efficiency that reduce the CO_2 intensity of the power sector by about 30% over the *Outlook* period.

Electricity demand

Demand for electricity grows steadily in each of the three scenarios presented in this year's *Outlook,* continuing the long-term upward trend. In the wake of the economic recession, electricity demand fell by 0.7% in 2009, the first fall since IEA records began in the early 1970s; but it recovered strongly in 2010, growing by 6%. In the New Policies Scenario, global electricity demand is projected to grow by more than four-fifths between 2009 and 2035, from 17 200 terawatt-hours (TWh) to over 31 700 TWh, at an annual growth rate of 2.4% (Table 5.1). It doubles in the Current Policies Scenario, and increases by almost two-thirds in the 450 Scenario. The variation in demand growth across the three scenarios is primarily due to the extent to which more energy-efficient technologies are adopted in end-use sectors in response to changing electricity prices and government measures, and, in the 450 Scenario, to the increased demand from the transport sector, due to the higher penetration of electric vehicles. Differences between scenarios in end-user prices of electricity emerge because of the different mix of fossil-fuel plants, changes in fossil-fuel prices, the impact of carbon pricing in regions where this is introduced, differences in subsidies to renewables and differing assumptions about the phase-out of subsidies over time. Much of the growth in each scenario occurs in countries outside the OECD, driven by their faster economic and population growth, expanding access to electricity and rising per-capita consumption.

Table 5.1 ● Electricity demand* by region and scenario (TWh)

	1990	2009	New Policies Scenario		Current Policies Scenario		450 Scenario	
			2035	2009-2035**	2035	2009-2035**	2035	2009-2035**
OECD	**6 593**	**9 193**	**12 005**	**1.0%**	**12 554**	**1.2%**	**11 343**	**0.8%**
Americas	3 255	4 477	5 940	1.1%	6 119	1.2%	5 612	0.9%
United States	*2 713*	*3 725*	*4 787*	*1.0%*	*4 898*	*1.1%*	*4 505*	*0.7%*
Europe	2 321	3 088	4 028	1.0%	4 244	1.2%	3 802	0.8%
Asia Oceania	1 017	1 628	2 037	0.9%	2 191	1.1%	1 930	0.7%
Japan	*759*	*950*	*1 158*	*0.8%*	*1 225*	*1.0%*	*1 075*	*0.5%*
Non-OECD	**3 492**	**8 024**	**19 717**	**3.5%**	**21 798**	**3.9%**	**16 978**	**2.9%**
E. Europe/ Eurasia	1 585	1 280	1 934	1.6%	2 238	2.2%	1 742	1.2%
Russia	*909*	*791*	*1 198*	*1.6%*	*1 401*	*2.2%*	*1 057*	*1.1%*
Asia	1 049	4 796	13 876	4.2%	15 334	4.6%	11 666	3.5%
China	*559*	*3 263*	*9 070*	*4.0%*	*10 201*	*4.5%*	*7 447*	*3.2%*
India	*212*	*632*	*2 465*	*5.4%*	*2 590*	*5.6%*	*2 117*	*4.8%*
Middle East	190	600	1 393	3.3%	1 525	3.7%	1 264	2.9%
Africa	263	532	1 084	2.8%	1 152	3.0%	1 000	2.5%
Latin America	404	816	1 430	2.2%	1 550	2.5%	1 306	1.8%
Brazil	*211*	*408*	*750*	*2.4%*	*792*	*2.6%*	*675*	*2.0%*
World	**10 084**	**17 217**	**31 722**	**2.4%**	**34 352**	**2.7%**	**28 321**	**1.9%**
European Union	*2 227*	*2 793*	*3 530*	*0.9%*	*3 716*	*1.1%*	*3 351*	*0.7%*

*Electricity demand is calculated as the total gross electricity generated less own use in the production of electricity and transmission and distribution losses. **Compound average annual growth rate.

Around 80% of the growth in electricity demand over the projection period occurs in non-OECD countries, with China and India accounting for nearly two-thirds of it. Although non-OECD annual per-capita electricity consumption increases from 1 450 kilowatt-hours (kWh) in 2009 to 2 750 kWh in 2035, it remains far below the OECD annual average of around 7 500 kWh in 2009. In addition to the lower rates of economic and population expansion, slower growth in OECD electricity demand stems from faster energy efficiency improvements and conservation, prompted in part by higher fuel prices and the adoption of measures such as carbon pricing.

In the New Policies Scenario, industry remains the largest electricity-consuming sector throughout the *Outlook* period, accounting for more than one-third of total electricity demand throughout (Figure 5.1). Electricity demand in the residential sector rises by 88% between 2009 and 2035, outpacing population growth by a factor of over three as access to electricity increases and the use of modern electrical appliances expands. Nevertheless, the residential sector's share remains fairly stable, at slightly over a quarter, over the period. The services sector's electricity consumption also increases steadily, at 2% per year on average throughout the *Outlook* period, though this rate is slower than the 2.5% growth rate in the residential sector, primarily due to energy efficiency measures in OECD countries and an increase in the direct use of renewables for heat both in OECD and non-OECD countries. With an average growth rate of 3.6% per year, transport is the fastest-growing sector, as demand from rail almost triples, though transport still accounts for only 2% of total electricity demand in 2035.

Figure 5.1 ● **World electricity supply and demand by sector in the New Policies Scenario**

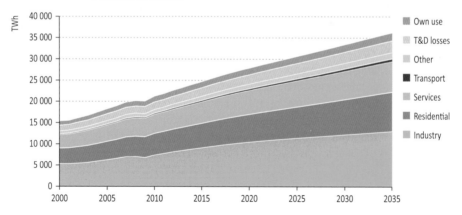

Electricity supply

Global electricity generation is projected to increase by 2.3% per year on average over the *Outlook* period in the New Policies Scenario, from 20 000 TWh in 2009 to more than 36 000 TWh in 2035 (Table 5.2). It grows at an average rate of 2.6% per year in the Current Policies Scenario and 1.8% per year in the 450 Scenario. Supply grows at a slower rate than demand, due to the falling shares of own use and transmission and distribution losses.

Table 5.2 ● Electricity generation by plant type and scenario (TWh)

	1990	2009	New Policies Scenario 2020	New Policies Scenario 2035	Current Policies Scenario 2020	Current Policies Scenario 2035	450 Scenario 2020	450 Scenario 2035
OECD	**7 629**	**10 394**	**11 997**	**13 304**	**12 143**	**13 939**	**11 743**	**12 541**
Fossil fuels	4 561	6 306	6 649	6 165	6 993	7 713	6 133	3 285
Nuclear	1 729	2 242	2 445	2 779	2 389	2 472	2 495	3 463
Hydro	1 182	1 321	1 476	1 592	1 461	1 547	1 505	1 683
Non-hydro renewables	157	525	1 427	2 768	1 300	2 208	1 610	4 110
Non-OECD	**4 190**	**9 649**	**15 884**	**22 946**	**16 426**	**25 429**	**15 092**	**19 683**
Fossil fuels	2 929	7 139	10 944	14 327	11 764	18 463	9 702	7 481
Nuclear	283	454	1 130	1 879	1 105	1 582	1 246	2 932
Hydro	962	1 931	2 904	3 926	2 793	3 597	3 042	4 369
Non-hydro renewables	15	125	905	2 814	764	1 787	1 102	4 900
World	**11 819**	**20 043**	**27 881**	**36 250**	**28 569**	**39 368**	**26 835**	**32 224**
Fossil fuels	7 490	13 445	17 593	20 492	18 757	26 176	15 835	10 765
Nuclear	2 013	2 697	3 576	4 658	3 495	4 053	3 741	6 396
Hydro	2 144	3 252	4 380	5 518	4 254	5 144	4 547	6 052
Non-hydro renewables	173	650	2 332	5 582	2 063	3 995	2 712	9 011

In the New Policies Scenario, coal remains the largest source of electricity generation globally throughout the *Outlook* period, with coal-generated output growing by 48% between 2009 and 2035. Nonetheless, its share of generation falls from 41% to 33% (Figure 5.2). The proportion of electricity produced from oil also drops, from 5% in 2009 to 1.5% in 2035. A marked increase in electricity generation from renewable sources offsets the fall in the shares of coal and oil. The share of generation from non-hydro renewables grows from 3% in 2009 to 15% in 2035, with almost 90% of this increase coming from wind, biomass, and solar photovoltaics (PV). Natural gas, hydro and nuclear all maintain relatively constant shares of electricity generation throughout the period of 22%, about 16% and 13% respectively. In the 450 Scenario, generation from non-hydro renewables rises to 28% of total electricity output in 2035, as a result of policies to enhance energy security and to curb greenhouse-gas emissions.

The change in the mix of technologies and fuels used to produce electricity is driven mainly by their relative costs, which are influenced by government policies. Government targets to reduce greenhouse-gas emissions and local energy-related pollution and, in some countries, the power sector's dependence on imported fuels have a significant impact on technology choices. In the New Policies Scenario, two types of measure have the most significant impact on the generation mix over time: carbon pricing and subsidies to renewables.

Figure 5.2 ● Share of world electricity generation by fuel in the New Policies Scenario

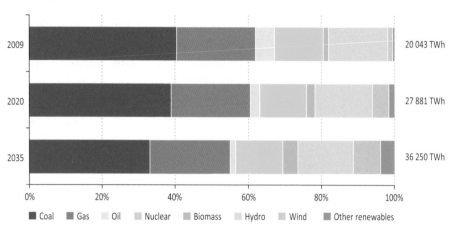

Many countries have introduced or are considering the introduction of some form of carbon price, typically through an emissions-trading scheme, whereby overall emissions are capped and the price that must be paid to emit a tonne of carbon dioxide (CO_2) is set by the market, or through a fixed-rate carbon tax. The carbon price encourages investment in technologies that emit less carbon, such as renewables or nuclear (Figure 5.3), and increases the operating costs of fossil-fuel plants. Both effects increase the cost of electricity to consumers, which lowers overall demand. Carbon pricing increases the absolute cost of gas-fired generation, but as the emissions intensity of gas is lower than that of coal and gas combined-cycle gas turbine plants are more efficient than coal plants, the impact of the carbon price on the cost of gas-fired generation is approximately half that of coal-fired generation. Fossil-fuel plants fitted with carbon capture and storage (CCS) emit significantly fewer carbon emissions, but the relatively high costs of CCS means that its deployment remains limited in the New Policies Scenario. The use of CCS is much more significant in the 450 Scenario (see Chapter 6).

In the New Policies Scenario, it is assumed that carbon pricing, explicit or implicit, is adopted in several OECD countries and in China (see Chapter 1 for details). In the 450 Scenario, the use of carbon pricing is more widespread (all OECD countries, China, Russia, Brazil and South Africa are assumed to adopt it) and prices are higher, resulting in a stronger shift to low-carbon technologies.

Carbon pricing alone does not account for all the growth in renewable electricity generation over the *Outlook* period. A large number of governments have adopted additional policies, including subsidies, designed specifically to stimulate investment in renewable energy technologies. As the cost of renewable energy technologies falls over time, some become fully competitive during the *Outlook* period – particularly in regions where there is a carbon price (see Chapter 14). As is the case for all generation technologies, the economic viability of renewable energy technologies is determined not just by the direct costs of generation, but also by the costs of integrating the capacity into the system. These include the associated grid infrastructure costs (see *implications of increasing renewables-based capacity* in this chapter) and the particular contribution of each technology to system adequacy.

Figure 5.3 ● **Typical levelised cost by plant type and carbon price* in the OECD in the New Policies Scenario, 2020**

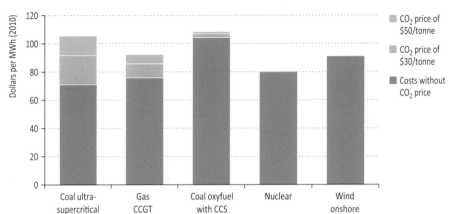

*Levelised cost is cost per unit of electricity generation, taking into account all the costs over the lifetime of the asset, including construction, operation and maintenance, fuel inputs and the cost of capital. In the New Policies Scenario, CO_2 prices range from zero to $30/tonne.

Coal

Electricity output from coal-fired plants worldwide increases from just over 8 100 TWh in 2009 to around 12 000 TWh by 2035 in the New Policies Scenario, though the share of coal in total electricity generation falls from 41% to 33%. Strong growth in non-OECD countries, where coal-fired generation doubles, far outweighs the fall in OECD countries (Figure 5.4). The European Union accounts for almost two-thirds of the fall in coal-fired generation in the OECD, largely as a result of the impact of the European Union Emissions Trading System on the competitiveness of coal relative to other technologies. In the United States, where only a shadow carbon price is assumed in the New Policies Scenario, the use of coal remains comparatively stable, declining by 3% between 2009 and 2035. The biggest growth in coal generation in a single country is in China, but the share of coal in total generation there falls over time. Government targets to reduce local pollution and to increase the deployment of other generation technologies, such as nuclear and wind, reduce the share of coal in China from 79% in 2009 to 65% in 2020. The continuation of these trends further reduces the share of coal to 56% in 2035. In India, coal use is projected to almost triple over the forecast period, eventually displacing the United States as the world's second-largest consumer of coal for power generation. India's increased use of coal for power is driven by strong growth of electricity demand, due to rapid population and economic growth, and coal's strong competitive position (see Chapter 10).

Figure 5.4 • Incremental global coal-fired electricity generation relative to 2009 by region in the New Policies Scenario

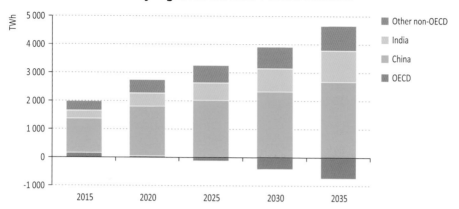

In the New Policies Scenario, the mix of coal-fired generation technologies changes over the *Outlook* period as older plants are retired and more efficient new plants are built, including ultra-supercritical designs and integrated gasification combined-cycle (IGCC) plants (Figure 5.5).[1] Increases in the price of coal, reductions in the capital costs of advanced coal technologies and the introduction of carbon prices all contribute to the shift towards higher-efficiency coal plants. As a result, the average global thermal efficiency of coal plants increases by four percentage points, from 38% in 2009 to 42% in 2035. These more efficient technologies entail higher capital costs than the subcritical and supercritical plant designs that make up most of the current fleet of coal-fired plants, but they use less fuel and, therefore, emit less CO_2 and other emissions for each unit of electricity they generate. Lower fuel use due to higher efficiency plants also helps to moderate import dependence for importing countries.

Figure 5.5 • World coal-fired electricity generation by plant type in the New Policies Scenario

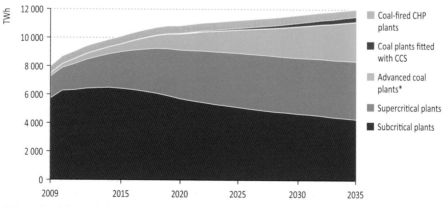

*Advanced coal plants include ultra-supercritical and IGCC plants.

1. See Chapter 10 for a detailed discussion of coal-fired generation technologies.

The deployment of CCS in coal plants is limited in the New Policies Scenario. The cost of CCS falls over the *Outlook* period, but it only becomes competitive with other generation technologies towards the end of the period, in those regions that have adopted a carbon price. In 2035, there are 65 gigawatts (GW) of coal capacity equipped with CCS, contributing only 3% of coal-fired generation and 1% of total electricity generation. In the 450 Scenario, in 2035, generation from coal capacity equipped with CCS accounts for 60% of coal-fired generation and 9% of total electricity generation (see Chapter 6).

Natural gas

Natural gas-fired power plants continue to play an important role in the electricity generation mix in the New Policies Scenario, accounting for a stable 22% of global generation throughout the *Outlook* period. Gas-fired generation grows from around 4 300 TWh in 2009 to a little over 7 900 TWh in 2035. Over three-quarters of the global increase in gas-fired generation over the period occurs in non-OECD countries; China and the Middle East account for one-fifth of the global increase each and one-tenth occurs in India (Figure 5.6). There is also significant growth in generation from gas within the OECD, as further efficiency improvements and carbon pricing enhance the competitiveness of gas-fired plants, relative to coal.

Figure 5.6 ● **Gas-fired electricity generation in selected countries and regions in the New Policies Scenario**

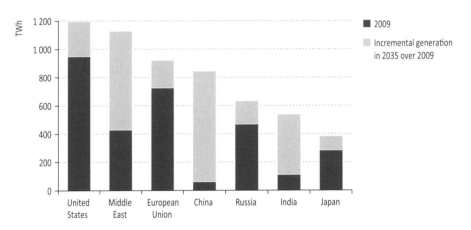

CCGT plants remain the dominant gas-fired generation technology. In the New Policies Scenario, generation from CCGT plants grows from a little under 2 600 TWh in 2009 to almost 4 900 TWh in 2035, accounting for over 60% of the growth in gas-fired generation. Generation from open-cycle gas turbine (OCGT) plants more than doubles, increasing from 370 TWh in 2009 to over 800 TWh. OCGT plants generally operate for relatively few hours each year, generating at periods of particularly high demand. One of the factors driving the growth in this technology is the need for peaking plants, such as these, as the penetration of variable renewables increases. Given their relatively high costs and the nature of the government policies assumed in the New Policies Scenario, deployment of CCS in gas-fired plants is limited; such plants provide less than 1% of the electricity generated from gas in 2035.

Nuclear power

In the wake of the accident at the Fukushima Daiichi nuclear plant in March 2011, several governments are reviewing their policies towards nuclear power (see Chapter 12). Announced changes in policy are taken into account in all the scenarios presented in this year's *Outlook:* early retirement of all nuclear plants in Germany by the end of 2022; no lifetime extensions or new plants in Switzerland; decommissioning of units one to four and no construction of new units at the Fukushima Daiichi site; and delays in capacity additions in China, due to the temporary suspension there of approval for new projects. Due to increased uncertainty, financing may become more difficult to secure, leading to increased cost of capital for nuclear projects. Compared to *WEO-2010,* the expected construction costs of new plants have risen by 5% to 10%. Although the prospects for nuclear power in the New Policies Scenario are weaker in some regions than in last year's projections, nuclear power continues to play an important role, providing baseload electricity. Most non-OECD countries and many OECD countries are expected to press ahead with plans to install additional nuclear power plants, though there may be short-term delays as the safety standards of existing and new plants are reviewed. Globally, nuclear power capacity is projected to rise in the New Policies Scenario from 393 GW in 2009 to 630 GW in 2035,[2] around 20 GW lower than projected last year.[3] Therefore, the share of nuclear in cumulative gross additions is 6% over the *Outlook* period, compared to 7% in *WEO-2010.* In several OECD countries, capacity is projected to grow much less than previously expected, but this is partially offset by an increase in projected non-OECD capacity, which is 8 GW higher than projected last year by 2035, mainly due to improved prospects in China.

In the New Policies Scenario, generation from nuclear power plants worldwide increases by almost 2000 TWh over the *Outlook* period, more than the nuclear output in North America and OECD Europe combined in 2010. This increase comes predominantly from non-OECD countries, with China alone accounting for over two-fifths of the global increase (Figure 5.7). In India, nuclear power generation grows almost ten-fold. In Russia, it grows by two-thirds. About 60% of the nuclear capacity added in the OECD replaces ageing nuclear plants that are retired in the *Outlook* period; in total, capacity increases by only 16%.

---

2. Capacity figures for all technologies are provided in gross terms, therefore including own use. For nuclear, the difference between net and gross capacity figures is generally 5% to 6%.

3. Including units 5 and 6 of the Fukushima Daiichi plant and the Fukushima Daini plant, which are not retired, but are not being used.

Figure 5.7 ● Additions and retirements of nuclear power capacity by region in the New Policies Scenario

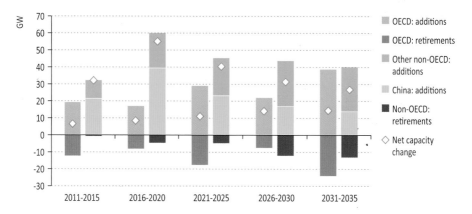

Renewables

The use of renewable energy sources to generate electricity expands significantly in all three scenarios. In the New Policies Scenario, renewables-based electricity generation worldwide almost triples, from 3 900 TWh in 2009 to 11 100 TWh in 2035. This expansion is driven largely by government policies, including subsidies (see Chapter 14), and represents 44% of the growth in total electricity generation over the period. The bulk of this growth comes from four sources: wind and hydro provide approximately one-third each, biomass accounts for about one-sixth and solar PV for one-tenth (Figure 5.8).

Figure 5.8 ● Incremental global renewables-based electricity generation relative to 2009 by technology in the New Policies Scenario

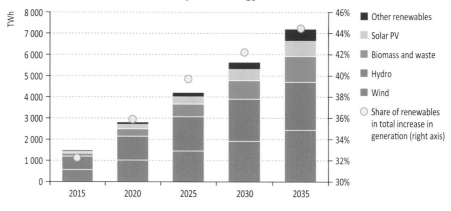

In the New Policies Scenario, over three-quarters of the growth in installed wind capacity and 70% of the growth in solar PV capacity occurs in the United States, European Union, China and India (Figure 5.9). Rapid capacity expansion in China has already seen onshore wind electricity generation increases from just 2 TWh in 2005 to 27 TWh by 2009, and it is projected to reach almost 590 TWh by 2035, making China the world's leading onshore wind power producer. Generation from installed onshore wind capacity increases more than three-fold in the European Union, from 133 TWh in 2009 to 480 TWh in 2035, and more than five-fold in the United States, from 74 TWh to 390 TWh. A steady improvement in the economics of offshore wind power encourages widespread increases in the installed capacity of this technology, which contributes one-fourth of total wind power generation by 2035; output increases from less than 1 TWh in 2009 to 670 TWh in 2035, almost level with generation from solar PV. As with onshore wind, the majority of the growth in offshore wind generation occurs in China, the European Union and the United States.

Figure 5.9 ● Solar PV and wind power capacity by region in the New Policies Scenario

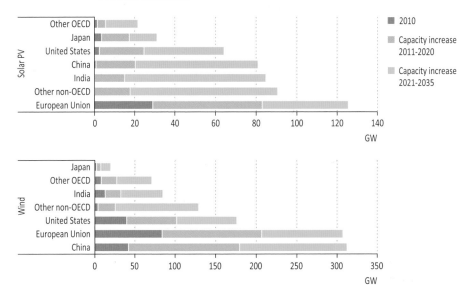

Solar PV electricity generation increases substantially over the *Outlook* period, from 20 TWh in 2009 to 740 TWh in 2035 in the New Policies Scenario, growing at an average rate of 15% per year. The European Union accounted for three-quarters of global solar PV generation in 2010. This has been driven by strong government programmes, particularly in Germany where there has been rapid growth in recent years. Over the early years of the *Outlook* period, Europe continues to exhibit very strong growth in solar PV but, between 2020 and 2035, the increase in solar PV generation in each of China, the United States and India is larger than that in the European Union.

Hydropower has already been developed extensively in many OECD countries and there is limited remaining potential, given the costs and environmental constraints. By contrast, large developments of hydro are expected to take place in many non-OECD countries. These countries account for 85% of total hydro capacity additions in the New Policies Scenario, with China, India and Brazil making up almost 60% of non-OECD hydro additions. In several cases, these resources are located far away from load centres and require significant investment in transmission lines.

Electricity supply from biomass power plants grows at an average annual rate of 6.5%, which results in a five-fold growth in output over the *Outlook* period, from 288 TWh in 2009 to 1 500 TWh in 2035. The bulk of the growth comes from non-OECD countries: the combined growth in demand in China and India is more than one-third of the global total. Other sources of renewable energy – geothermal, concentrating solar power and marine power – gain a small foothold in the power sector by the end of the *Outlook* period.

CO_2 emissions

In the New Policies Scenario, CO_2 emissions from the power sector grow from 11.8 gigatonnes (Gt) in 2009 to 14.8 Gt in 2035 – a rise of 25%. Growth in electricity generation is more than three times larger than this, reflecting a 30% reduction in the CO_2 intensity of the power sector over the *Outlook* period – the amount of CO_2 released into the atmosphere for each unit of electrical energy produced – from 530 grammes of CO_2 per kWh in 2009 to 375 grammes in 2035. This reduction arises mainly from the shift away from coal and oil as fuel sources for electricity generation towards lower-carbon nuclear and renewables-based technologies. Efficiency improvements within the coal and gas generation fleets also make a considerable contribution to improving the carbon intensity of the sector, as older less efficient plants are retired and more efficient plants are installed. The introduction of CCS technology, though only on a relatively small scale, also contributes to driving down the CO_2 intensity of the power sector. These factors yield annual emissions savings of around 7.8 Gt by 2035, compared with the emissions that would have been generated for the projected level of electricity generation had there been no change in the mix of fuels and technologies and no change in the efficiency of thermal generating plants (Figure 5.10). The largest share of these savings comes from more efficient plants (36% of the cumulative reduction), followed by savings attributed to wind, nuclear and hydro, each of which accounts for around 15% of the abatement.

Figure 5.10 ● Global CO_2 emission savings in power generation relative to the 2009 fuel mix* in the New Policies Scenario

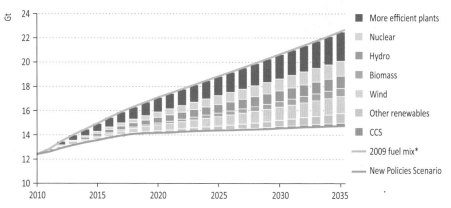

*The emissions savings compared with the emissions that would have been generated for the projected level of electricity generation were there no change in the mix of fuels and technologies, and no change in the efficiency of thermal generating plants after 2009.

In OECD countries, the reduction in carbon intensity of the electricity sector outweighs the growth in generation in the New Policies Scenario, so that CO_2 emissions are more than 20% lower in 2035 than in 2009. In non-OECD countries, demand growth is much stronger and, as a result, emissions increase by more than 50% by 2035.

New capacity additions and retirements

In order to meet growing demand in the New Policies Scenario, global installed electricity generation capacity increases from 5 143 GW in 2010 to over 9 000 GW in 2035 (Figure 5.11). About 2 000 GW of capacity of all types are retired throughout the *Outlook* period (Table 5.3), while 60% of power plants in service or under construction today are still in operation in 2035. This means that 59% of power sector emissions in 2035 are already "locked in", unless future policy changes force early retirement of existing plants or their retrofitting with CCS (see Chapter 6). In the New Policies Scenario, gross capacity additions are about 5 900 GW over the *Outlook* period, equivalent to more than five-times the installed capacity of the United States in 2010 (Table 5.4). Renewable energy technologies account for half the capacity additions, gas and coal for one-fifth each and nuclear power for 6%. Cumulative oil-fired capacity additions, one-third of which are in the Middle East, are less than 2% of total additions.

Figure 5.11 ● Global installed power generation capacity and additions by technology in the New Policies Scenario

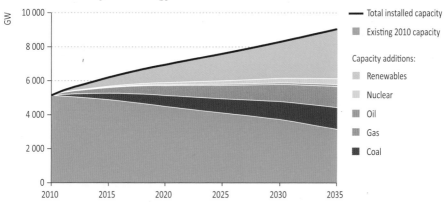

Over the past decade, global capacity additions have been fairly similar across coal, gas and renewables, with comparatively few additions of nuclear and oil-fired power plants. Since 2000, nearly four-fifths of the additional coal-fired power plants have been built in China, while more than half of the gas-fired capacity additions were built in the United States and European Union. Globally, the share of renewables in capacity additions has steadily increased in recent years, reaching about 50% of total additions in 2010. However, as renewables often generate less electricity per unit of capacity installed each year than thermal plants, their contribution to incremental electricity output has been less than their share of incremental capacity.

Table 5.3 ● Cumulative capacity retirements by region and source in the New Policies Scenario, 2011-2035 (GW)

	Coal	Gas	Oil	Nuclear	Biomass	Hydro	Wind	Geo-thermal	Solar PV	CSP*	Marine	Total
OECD	**380**	**212**	**178**	**71**	**42**	**31**	**246**	**1**	**72**	**1**	**0**	**1 235**
Americas	195	129	77	3	16	15	85	1	7	1	0	527
United States	*182*	*120*	*61*	*2*	*12*	*7*	*73*	*0*	*6*	*1*	*0*	*464*
Europe	148	40	53	60	22	14	151	0	55	1	0	544
Asia Oceania	37	43	48	8	5	3	9	0	10	0	0	163
Japan	*13*	*35*	*44*	*7*	*4*	*2*	*4*	*0*	*9*	*0*	*0*	*117*
Non-OECD	**206**	**177**	**114**	**36**	**14**	**49**	**151**	**0**	**19**	**0**	**0**	**766**
E. Europe/Eurasia	80	105	19	32	1	8	4	0	1	0	0	249
Russia	*43*	*84*	*6*	*20*	*1*	*4*	*1*	*0*	*0*	*0*	*0*	*159*
Asia	103	18	30	2	7	26	139	0	15	0	0	340
China	*68*	*1*	*6*	*0*	*1*	*18*	*114*	*0*	*9*	*0*	*0*	*217*
India	*27*	*3*	*2*	*1*	*2*	*3*	*23*	*0*	*4*	*0*	*0*	*65*
Middle East	0	30	39	0	0	1	1	0	1	0	0	72
Africa	20	14	10	0	1	2	3	0	1	0	0	52
Latin America	3	10	16	1	5	12	4	0	2	0	0	52
Brazil	*1*	*1*	*2*	*1*	*3*	*7*	*2*	*0*	*1*	*0*	*0*	*18*
World	**586**	**388**	**292**	**107**	**56**	**80**	**397**	**1**	**92**	**1**	**0**	**2 001**
European Union	*154*	*42*	*53*	*58*	*22*	*10*	*150*	*0*	*55*	*1*	*0*	*545*

*CSP = concentrating solar power.

Table 5.4 ● Cumulative gross capacity additions by region and source in the New Policies Scenario, 2011-2035 (GW)

	Coal	Gas	Oil	Nuclear	Biomass	Hydro	Wind	Geo-thermal	Solar PV	CSP*	Marine	Total
OECD	**260**	**468**	**42**	**125**	**106**	**104**	**686**	**12**	**278**	**31**	**17**	**2 129**
Americas	166	226	10	28	51	41	257	8	77	14	2	880
United States	*159*	*163*	*8*	*20*	*42*	*20*	*209*	*6*	*68*	*11*	*1*	*708*
Europe	67	139	15	48	44	50	390	3	153	15	14	938
Asia Oceania	27	103	16	50	11	12	39	2	48	3	1	311
Japan	*5*	*75*	*16*	*22*	*8*	*7*	*22*	*1*	*36*	*0*	*1*	*193*
Non-OECD	**1 031**	**767**	**71**	**221**	**138**	**578**	**618**	**18**	**275**	**49**	**1**	**3 767**
E. Europe/Eurasia	53	146	1	53	11	30	28	3	6	0	0	331
Russia	*32*	*96*	*0*	*33*	*8*	*18*	*13*	*3*	*2*	*0*	*0*	*204*
Asia	923	348	14	148	103	401	525	12	206	21	1	2 700
China	*555*	*170*	*3*	*114*	*58*	*213*	*384*	*2*	*89*	*16*	*0*	*1 605*
India	*251*	*91*	*2*	*24*	*20*	*79*	*95*	*0*	*89*	*4*	*0*	*657*
Middle East	0	148	40	7	4	14	23	0	21	15	0	271
Africa	47	56	4	6	8	45	19	2	22	11	0	222
Latin America	8	70	11	7	12	88	22	2	20	3	0	243
Brazil	*3*	*40*	*4*	*5*	*7*	*40*	*12*	*0*	*13*	*1*	*0*	*125*
World	**1 291**	**1 235**	**112**	**346**	**244**	**682**	**1 304**	**30**	**553**	**81**	**17**	**5 896**
European Union	*61*	*127*	*13*	*50*	*43*	*32*	*373*	*2*	*152*	*15*	*13*	*882*
Average economic lifetime (years)	*30*	*25*	*25*	*35*	*25*	*50*	*20*	*25*	*20*	*20*	*20*	

*CSP = concentrating solar power.

5

In the OECD, renewables account for about 60% of cumulative capacity additions between 2011 and 2035 in the New Policies Scenario. Wind and solar PV cumulative capacity additions are greater than the sum of additions from gas, coal and nuclear power, though their share of incremental output is generally far lower. In North America, capacity additions of wind and gas are similar (slightly more than 25% each of total additions), with about 220 GW of additional gas capacity (in part because of relatively low gas prices). In OECD Europe, government policies drive wind power capacity additions to nearly three-times those of gas, almost six-times those of coal additions and eight-times those of nuclear power between 2011 and 2035. In Japan, nuclear capacity additions are much lower than both gas and solar PV, and similar to added wind capacity of 22 GW.

In non-OECD countries, fossil-fuel capacity additions (coal, gas and oil) are roughly equal to non-fossil-fuel capacity additions (wind, hydro, solar PV, nuclear, biomass and other renewables) in the New Policies Scenario between 2011 and 2035. Almost all of the coal-fired capacity built in non-OECD countries (and three-quarters worldwide) occurs in Asia. China and India combined account for over 60% of all the coal-fired capacity built during the *Outlook* period. The additional wind and hydro capacity built in non-OECD Asia is also extremely large, greater than that of all OECD countries combined. In Russia and the rest of Eastern Europe/Eurasia, gas-fired plants dominate capacity additions, followed by nuclear, coal and hydro. In Africa, gas, coal, and hydro make the biggest contributions to additional capacity. Latin America relies mainly on hydro and gas for new capacity, but solar PV and wind capacity additions are also significant. In the Middle East, gas and oil account for 70% of new capacity.

Implications of increasing renewables-based capacity

The increasing contribution of renewables to meeting rising electricity demand has important implications for the design of electricity systems, because of the variable nature of the output of several renewables-based technologies, such as wind, solar PV or concentrating solar power (CSP) without storage (see Chapter 10 in *WEO-2010* for further discussion). The integration of these variable renewables into the system results in additional costs for electricity systems to ensure security of supply (Box 5.1).

Our analysis provides a global estimate of the quantity of flexible capacity needed to ensure system adequacy – the ability of the electricity system to meet electricity demand at all times with an acceptably high probability – due to the increasing share of variable renewables in the system in the New Policies Scenario.[4] We find that for every 5 megawatts (MW) of variable renewable capacity installed, about 1 MW of other (flexible) capacity is needed to maintain system adequacy. In 2035 in the New Policies Scenario, this corresponds to around 300 GW, or about 8% of the non-variable capacity additions over the *Outlook* period.[5] This level of additional capacity leads to system adequacy costs ranging from $3 per megawatt-hour (MWh) to $5/MWh of variable renewables generation, depending on the region.

---

4. While ample data are available for the United States and Europe, resources of variable renewables and the hourly electricity demand are not covered with the same quality in all *WEO* regions. The impact of increased penetration of renewables on the rest of the generation mix is an emerging area of research requiring further investigation (IPCC, 2011)

5. Full details of the methodology used by the IEA in this analysis are available at: www.worldenergyoutlook.org/methodology_sub.asp.

Box 5.1 ● Costs of integrating variable renewables into the electricity system

Total system integration costs range from $5/MWh to $25/MWh of variable renewables-based generation. This is made up of three components: adequacy costs, balancing costs and grid integration costs (IEA, forthcoming).

- *Adequacy costs:* Electricity systems must have enough generating capacity available to meet system demand even at peak times. This is known as system adequacy. Variable renewables make a relatively low contribution to system adequacy, because only a small proportion of their potential output is certain to be available at times of peak demand. As a result, other plants are needed on the system to compensate for this variability. Based on our estimates, the costs associated with this additional capacity ranges from $3/MWh to $5/MWh of variable renewables generation.

- *Balancing costs:* The variable nature of supply requires more flexibility from the rest of the power system, typically on short time-scales. Additional services are needed in order to perfectly match supply and demand and to maintain system stability. Balancing services can be supplied by flexible generation, smart grids, strong interconnections between grids or energy storage technologies, such as pumped hydro, compressed-air and large-scale batteries. The costs are largely operational (rather than capital) and range between $1/MWh and $7/MWh of variable generation, depending on the region (IEA, 2011a).

- *Grid integration costs:* Once renewables capacity is built, it must be linked to the existing power grid and existing transmission and distribution (T&D) networks may need to be reinforced to transport the generated electricity to consumers. The costs range from $2/MWh to $13/MWh for variable renewables and are discussed in detail in the T&D section in this chapter.

The additional 300 GW of capacity is required to compensate for the difference between the average power output of variable renewables and the amount of power they can reliably be expected to produce at the times when demand for electricity is highest (the latter measure is known as the capacity credit).[6] This capacity is added to contribute to system adequacy, but runs for significantly fewer hours each year than would be the case in the absence of variable renewables (IEA, forthcoming).

The average power output of variable renewables is higher in most cases than their capacity credit. By the end of the *Outlook* period, globally, wind and solar combined have an average power output of 25% in the New Policies Scenario, higher than their combined

---

6. The capacity credit of variable renewables depends mainly on the time of day (or season) at which the output is available and how well that matches with the time of peak demand. It also depends on the characteristics of the resource (*e.g.* the strength and consistency of the wind), and on the level of deployment of each type of variable renewable.

average capacity credit of 9%.[7] The global capacity credit estimate masks considerable regional variations. In 2035, the combined global capacity credit of wind, solar PV and CSP (without storage) plants in the New Policies Scenario ranges from 5% to 20% among the different regions modelled. It is relatively low in OECD Europe, largely due to the expansion of solar PV in central European countries, such as Germany, where energy output is zero or near zero during times of peak demand, which are generally on cold winter evenings. It is highest in Japan as there is a higher correspondence between the peaks in demand – which are driven in part by air conditioning requirements on sunny afternoons – and the output of solar PV, which make up about 60% of the variable renewables capacity in operation in Japan by 2035.

In the New Policies Scenario, OECD Europe and the United States account for over 40% of the gross capacity additions of wind and solar capacity over the *Outlook* period. In the United States in 2035, installed capacity of wind and solar PV combined is 240 GW, with an average utilisation rate of 27%, resulting in an average power output of around 65 GW. Given their 8% capacity credit, only about 20 GW of this is reliably available at times of peak demand. To compensate for this difference, 45 GW of other types of capacity is required to maintain system adequacy. In OECD Europe, which has a lower capacity credit (5%) and higher variable renewables capacity in 2035 (about 450 GW), the amount of additional capacity required in the system is higher, at around 90 GW (Figure 5.12).

Figure 5.12 ● Capacity of wind and solar PV and their system effects for the United States and OECD Europe, 2035

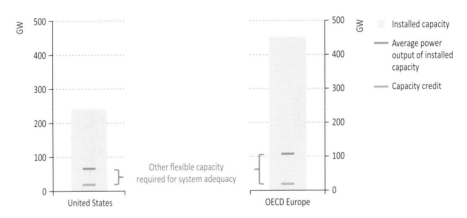

The capacity credit of variable renewables, and therefore the additional capacity required to maintain system adequacy, can be increased in several ways, including wider interconnection between regional grids, which can smooth the variability of wind and solar resources, and may improve alignment between generation and peak demand. Using a more diverse mix

---

7. Based on data from Heide *et al.* (2010), NREL (2010, 2011a, 2011b), and World Wind Atlas (2011).

of renewables can also improve capacity credit because their outputs, being diverse, are more constant overall. Using energy storage technologies that are directly linked to variable renewables is another way to improve reliability in meeting demand.

Investment

In the New Policies Scenario, cumulative global investment in the power sector is $16.9 trillion (in year-2010 dollars) over the *Outlook* period, an average of $675 billion per year (Figure 5.13). New generating capacity accounts for 58% of the total investment, with transmission and distribution (T&D) making up the remaining 42%. Renewables make up 60% of investment in new power plants, led by wind, solar PV and hydro, even though they represent only half of the capacity additions; their larger share of investment reflects their higher capital costs, relative to fossil-fuel power plants. Investment in coal-fired power plants, which account for the largest share of total investment other than renewables as a group, is relatively constant over time. Investment in nuclear power, which is much more capital-intensive, is slightly less than in coal-fired plants while cumulative capacity additions are only half those of coal. Investment in gas-fired plants is lower than coal or nuclear while additional capacity is substantially higher, reflecting far lower capital costs per MW of capacity.

Figure 5.13 ● **Investment in new power plants and infrastructure in the New Policies Scenario**

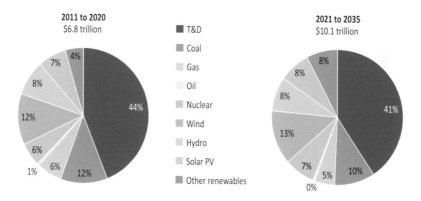

Investment in new power plants in non-OECD countries increasingly outpaces investment in the OECD over time. In the 2011 to 2020 period, the levels of investment in the OECD and non-OECD are similar (less than 20% apart) for all three scenarios. Asia accounts for most of non-OECD investment. In the 2021 to 2035 period, non-OECD investment in new power plants is over 30% beyond the level in the OECD in the New Policies Scenario, with Asia accounting for almost 70% of the investment in non-OECD countries (Table 5.5).

Table 5.5 ● Investment in new power plants in the New Policies Scenario, 2011-2035 ($2010 billion)

	Coal	Gas	Oil	Nuclear	Biomass	Hydro	Wind	Geo-thermal	Solar PV	CSP*	Marine	Total
OECD	**634**	**372**	**27**	**540**	**359**	**272**	**1 210**	**28**	**701**	**129**	**64**	**4 336**
Americas	416	174	7	143	174	105	451	16	189	54	8	1 738
United States	*405*	*131*	*6*	*111*	*157*	*51*	*372*	*11*	*170*	*45*	*5*	*1 463*
Europe	155	115	10	207	155	136	690	8	386	65	51	1 976
Asia Oceania	63	83	10	191	30	31	69	3	126	10	5	621
Japan	*12*	*60*	*10*	*94*	*20*	*18*	*39*	*1*	*97*	*0*	*2*	*354*
Non-OECD	**1 151**	**545**	**59**	**585**	**305**	**1 134**	**932**	**42**	**544**	**156**	**2**	**5 456**
E. Europe/Eurasia	117	128	1	186	29	62	43	7	15	0	0	588
Russia	*75*	*84*	*0*	*116*	*20*	*37*	*19*	*6*	*5*	*0*	*0*	*362*
Asia	943	206	8	331	227	758	794	27	391	66	2	3 752
China	*430*	*87*	*2*	*228*	*137*	*336*	*590*	*7*	*170*	*50*	*1*	*2 037*
India	*347*	*59*	*1*	*68*	*38*	*167*	*138*	*1*	*165*	*14*	*1*	*999*
Middle East	0	130	39	22	8	29	34	0	43	46	0	352
Africa	80	34	4	20	17	90	28	5	48	34	0	360
Latin America	11	47	6	25	24	196	34	4	48	10	0	404
Brazil	*4*	*28*	*2*	*17*	*13*	*93*	*19*	*0*	*30*	*6*	*0*	*211*
World	**1 785**	**917**	**86**	**1 125**	**664**	**1 406**	**2 142**	**71**	**1 245**	**285**	**66**	**9 791**
European Union	*144*	*107*	*8*	*216*	*149*	*92*	*663*	*7*	*382*	*65*	*51*	*1 884*

*CSP = concentrating solar power.

Focus on T&D infrastructure

Robust networks are vital to delivering electricity reliably to consumers. Reinforcement and expansion of network capacity will be needed in the future to accommodate demand growth, integrate greater renewables-based generation, improve access to electricity in developing countries, facilitate the use of electricity in road transport and increase electricity trade across borders. Indeed, transmission and distribution (T&D) infrastructure accounts for 42% of all power sector investment over the *Outlook* period (though it is often under-represented in discussions about the power sector).[8]

Electricity networks can be broadly divided into two types. Transmission grids comprise high-voltage lines and support structures that transport power over large distances from power plants to demand centres and large industrial facilities. Distribution grids deliver power from the transmission level to individual households and businesses. They are denser than transmission grids and account for 90% of the total length of T&D grids worldwide. In 2009, there were about 70 million kilometres (km) of T&D lines globally – more than 1 500 times the circumference of the planet and equivalent to the length of all the world's roads. Most of the features of the technology have remained fundamentally unchanged since the early days of the industry a century ago, but there has been considerable innovation in some areas, including the introduction of digital communication and control technologies. The evolution of such "smart-grids" is expected to continue in coming decades.

Grid expansion

The structure and size of national grids is determined by geographic, political and economic conditions. In industrialised countries, large parts of the electricity grid were built in the 1960s and 1970s. In emerging economies, rapid growth of grids started more recently and is continuing (Figure 5.14). The total length of all grids combined almost tripled between 1970 and 2009.

Since 1970, grid growth has been largest in Asia, led by China. Asia accounted for 60% of the growth in this period, resulting in a more than six-fold increase in the continent's grid length. Grid length in the United States and Europe doubled, but from a higher starting point. One-third of investment in T&D infrastructure in 2009 was made in China, where $62 billion (in year-2010 dollars) was spent on T&D, more than OECD Europe and the United States combined.

T&D assets last on average for about 40 years, but can operate for much longer in some cases. Lifetime varies considerably according to the asset type: poles and lines routinely last longer, while control centres and substations need more frequent refurbishment. The conditions under which the assets are used and the way in which they are maintained are also important determinants of equipment life. For example, poor maintenance during the economic downturn of the early 1990s is expected to shorten considerably the operating lifetimes of network assets in Eastern Europe/Eurasia. By 2035, half of the grid infrastructure in place globally in 2009 will have reached 40 years of age (Figure 5.15).

---

8. Full details of the methodology used by the IEA in this analysis are available at www.worldenergyoutlook.org/methodology_sub.asp.

Figure 5.14 ● Expansion of T&D grids, 1970-2009, and regional shares of global grid investment in 2009

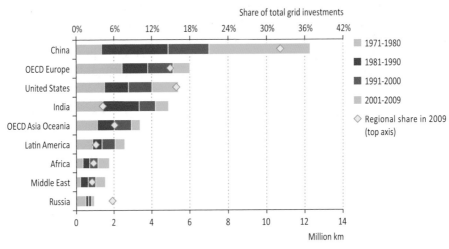

Sources: ABS (2010); ENTSO-E (2011).

The length of the electricity grid per square kilometre (km²) of land area and generation per-capita provide two measures of the maturity of a power system (Figure 5.16). Both tend to increase quickly initially and then begin to level off, reflecting saturation effects. In the early stages of system development, the length of electricity grids usually grows faster than per-capita generation, as distribution networks are expanded to connect more end-users. More advanced stages tend to require only incremental grid extensions, with investment shifting towards strengthening grids to support relatively gradual increases in demand.

Figure 5.15 ● Share of T&D infrastructure in place in 2009 reaching 40 years of age

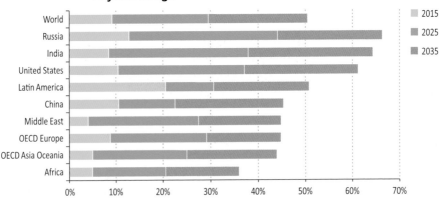

Sources: ABS (2010); IEA analysis.

Increases in grid density in the coming decades will occur primarily in less developed regions. Africa has seen a marked increase in the length of the electricity grid per km² over recent

years and this is set to continue. In the New Policies Scenario, Africa's T&D grid per km²
increases from 10% of the OECD average in 2009 to 14% in 2035. Per-capita generation
growth is slower. In 2009, per-capita generation in Africa amounted to around 600 kWh,
only 7% of the OECD average (about 7 500 kWh). In the New Policies Scenario, per-capita
generation in Africa rises, but is still only 8% of the OECD average by 2035. In the highly
industrialised OECD, grid density and per-capita generation are fairly stable over the *Outlook*
period, due to the power system becoming more efficient and to saturation effects.

Figure 5.16 ● **T&D grid length and per-capita generation for selected regions
in the New Policies Scenario**

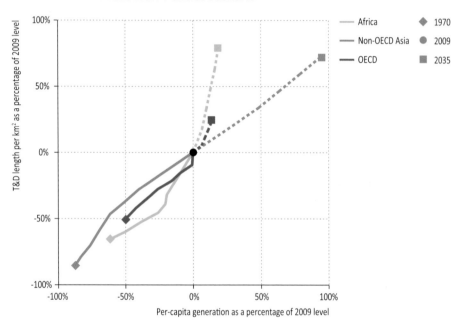

In Asia, both grid density and per-capita generation grow significantly. In India and China,
which have experienced rapid T&D growth in the past, grid density begins to saturate over
the *Outlook* period while per-capita generation continues to grow. The average grid density
in non-OECD Asia is about two-thirds of the OECD value in 2035, compared with 15% in 1970
and 50% in 2009.

T&D investment needs

In the New Policies Scenario, a total of $7.1 trillion (in year-2010 dollars) is invested in T&D
infrastructure over 2011 to 2035. $2.6 trillion, or 36%, of this investment is in the OECD
and $4.5 trillion in non-OECD countries. There are three components of this investment:
additional capacity to meet higher demand, refurbishment and replacement of existing
assets as they reach the end of their technical lifetime and increases due to the integration
of renewables. Additional capacity accounts for 57% of total T&D investment globally,

refurbishment of existing assets for 40%, and grid integration of renewables for 3%. Almost three-quarters of global T&D investment is in distribution lines, which represent about 90% of the total length of T&D networks.

In the OECD, three-fifths of T&D investments are needed for the replacement and refurbishment of existing infrastructure (Figure 5.17). This is due to the age structure of the assets, but also reflects relatively stable electricity demand. Renewables integration costs reach 5% of T&D investments in 2035 in the New Policies Scenario, due to increased renewables deployment. In non-OECD countries, strong growth in electricity demand calls for new transmission and distribution lines to connect many new customers and power plants, with less than one-third of total investment for replacement and refurbishment. Eastern Europe/Eurasia, and Russia in particular, is an exception, as large shares of the infrastructure there need to be refurbished (Table 5.6). In non-OECD countries, the costs of renewables integration amount to about the same as in the OECD in absolute terms, but generally represent only about 3% of the overall investment in 2035.

Figure 5.17 ● **Annual average investment in T&D infrastructure in the New Policies Scenario**

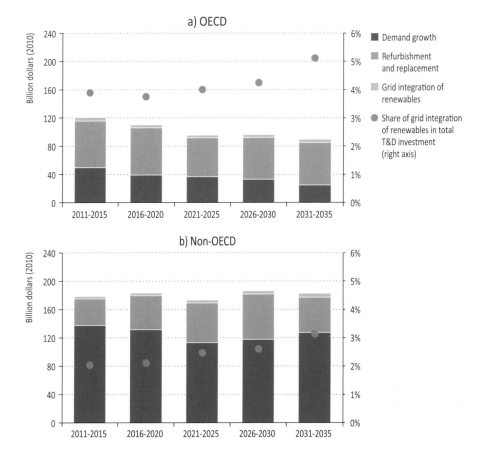

Table 5.6 ● Investment in T&D infrastructure in the New Policies Scenario, 2011-2035 ($2010 billion)

	Total	Transmission				Distribution				Share from renewables
		Total	New	Refurbishment	Renewables	Total	New	Refurbishment	Renewables	
OECD	**2 561**	**626**	**212**	**326**	**87**	**1 936**	**709**	**1 208**	**18**	**4.1%**
Americas	1 271	402	147	217	38	869	283	578	9	3.7%
United States	*1 009*	*312*	*109*	*172*	*31*	*696*	*203*	*486*	*7*	*3.8%*
Europe	915	179	49	87	42	737	297	431	8	5.5%
Asia Oceania	375	45	16	23	7	330	129	200	1	2.1%
Japan	*219*	*24*	*6*	*15*	*4*	*195*	*66*	*128*	*1*	*2.0%*
Non-OECD	**4 530**	**1 214**	**852**	**273**	**89**	**3 316**	**2 288**	**1 008**	**21**	**2.4%**
E. Europe/Eurasia	442	133	47	83	4	308	120	188	1	1.0%
Russia	*252*	*95*	*29*	*65*	*2*	*157*	*54*	*103*	*0*	*1.0%*
Asia	3 267	850	644	137	69	2 417	1 728	673	16	2.6%
China	*1 931*	*623*	*477*	*97*	*49*	*1 308*	*910*	*387*	*11*	*3.1%*
India	*632*	*111*	*76*	*22*	*13*	*521*	*327*	*190*	*3*	*2.6%*
Middle East	231	59	40	13	6	172	137	34	1	3.1%
Africa	278	78	60	13	5	199	156	42	1	2.1%
Latin America	313	93	61	27	5	220	148	71	1	2.1%
Brazil	*193*	*63*	*42*	*19*	*3*	*130*	*89*	*40*	*1*	*2.1%*
World	**7 092**	**1 839**	**1 064**	**599**	**176**	**5 252**	**2 997**	**2 216**	**39**	**3.0%**
European Union	*809*	*160*	*38*	*82*	*40*	*649*	*236*	*406*	*8*	*5.9%*

5

The costs of refurbishing or expanding grids must be recouped by grid operators through their tariffs. Based on global investments in T&D in the New Policies Scenario totalling $7.1 trillion, residential customers around the world will pay between $10/MWh and $20/MWh in 2035 to cover these infrastructure costs, with an average of $12/MWh.[9] The weight of the tariff in the residential end-user price varies across regions (Figure 5.18).

Figure 5.18 ● T&D infrastructure costs as a share of residential end-user price in the New Policies Scenario, 2035

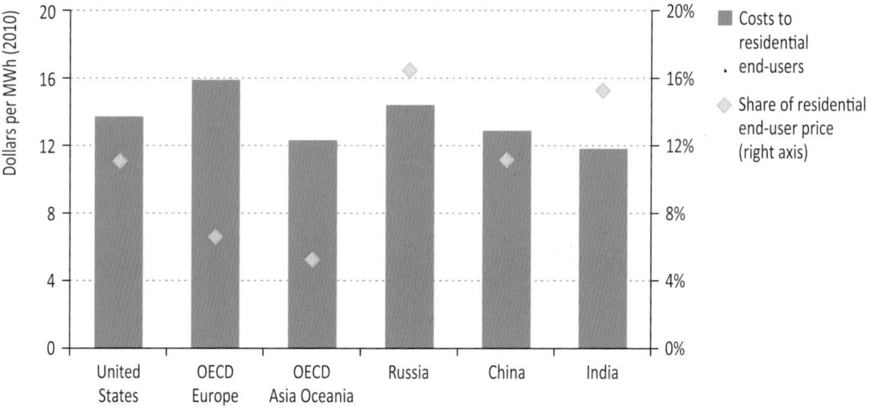

Because T&D infrastructure is considered a natural monopoly,[10] T&D tariffs are normally regulated by public authorities to prevent abuse of monopoly power and excess profits. Grid regulation is crucial to avoid cross-subsidisation, set appropriate tariffs and ensure grid-access rights for new market entrants, an issue especially relevant for renewables generators. Regulations are a key determinant in the efficiency and reliability of T&D infrastructure. Additionally, planning and building new transmission and distribution can take up to ten years or longer and often exceeds the time to build new power plants. Therefore, providing a stable investment framework for grid operators is an important task for regulators.

Grid integration of renewables

The best sites for producing renewables-based electricity are not always located close to centres of demand. To exploit remote renewable energy sources, additional high-voltage transmission lines have to be built and some aspects of existing T&D networks reinforced. The associated capital costs are known as grid integration costs and make up one part of the

9. Our estimates exclude other charges such as system services, losses, congestion and regulatory charges that are often included in, and can add significantly to, transmission tariffs (ENTSO-E, 2011).

10. A natural monopoly in this case means that the economies of scale achieved by a single electricity transmission and distribution network in any territory are so large that it is not feasible for two or more grids to co-exist competitively.

total system integration costs (see Box 5.1). In the New Policies Scenario, over the period 2011 to 2035, grid integration costs of renewables make up 10% of total transmission investment costs worldwide (excluding investment in distribution), 14% of transmission investment in the OECD and 25% of transmission investment in the European Union. The share in global T&D investment costs taken together is 3%.

The grid integration costs of renewables vary considerably by region, driven in part by the proximity of renewable resources to population centres. In the United States, for instance, the areas with the greatest wind potential are located in the middle of the country, far away from the main load centres on the east and west coasts. The situation is similar in China, where high wind speeds can be found close to Mongolia and in the northwest province, while the load centres are located along the coast. In the Europe Union, T&D grids are already well-developed, but the distances between the windiest regions in and around the North Sea and the main centres of demand are still significant, resulting in the cost of grid integration of renewables accounting for 25% of total transmission investment there.

Grid integration costs also vary by technology. Solar PV and CSP entail grid integration costs per unit of capacity similar to those of wind. The location of solar PV, CSP and wind plants is influenced by where sunlight or wind is the strongest and most constant, which can be distant from demand centres. Costs for other renewable energy technologies, such as biomass, are generally lower as they tend to be located closer to existing networks and demand centres. Hydro and geothermal grid integration costs vary greatly across countries, as they are determined by geography. Grid integration costs per unit of output for variable renewables are also higher than those of other types of generators (renewables-based or otherwise) because their average power output is generally lower, making it necessary to spread costs over fewer units of generated power. Due to the variation in the average power output between technologies and their proximity to demand centres, grid integration costs for variable renewables range from $2/MWh to $13/MWh in the New Policies Scenario.

The cumulative global investment in T&D related to the grid integration of renewables is $220 billion (in year-2010 dollars) in the New Policies Scenario, assuming an average cost of $150/kW of renewables capacity.[11] The share of renewables grid integration costs in total transmission investment is above 18% in the 450 Scenario, where significantly more renewables are deployed, or nearly 6% of cumulative T&D investment (Figure 5.19). Higher assumed grid integration costs of $250/kW would increase cumulative investment from 2011 to 2035 to $360 billion in the New Policies Scenario. Even with higher integration costs and significantly more deployment of renewables in the 450 Scenario, the share in global T&D investment costs remains below 10%.

5

---

11. See Figure 5.19 note.

Figure 5.19 ● Renewables grid integration costs as a share of global T&D investment costs in the New Policies Scenario by integration cost, 2011-2035

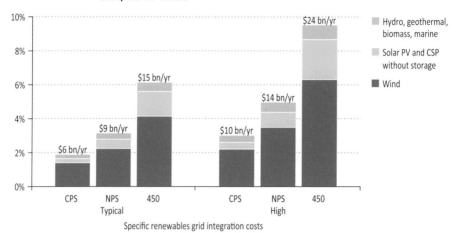

Specific renewables grid integration costs

Notes: Grid integration costs are $150/kW in the New Policies Scenario, and are denoted above as "typical". Additional illustrative examples are provided using a higher grid integration cost of $250/kW. These figures are drawn from detailed studies for Europe and the United States, showing a range from $60/kW to $320/kW (NREL, 2010 and 2011a; ECF, 2010).

Smart grids

A smart grid is an electricity grid that uses digital and other advanced technologies to monitor and manage the transport of electricity from generation sources to meet the varying electricity demands of end-users (IEA, 2011b). Such technologies are expected to play an important role in boosting the flexibility of the system to deal with the challenges of rising demand, shifting load patterns, new sources of supply and the variability of some sources of renewables-based supply.

Smart grids enable electricity generators, end-users and grid operators to co-ordinate their needs and capabilities. This can improve efficiency, reliability and flexibility; lower the cost of electricity supply; and reduce the environmental impacts. Compared with today's systems, smart grids will involve the exchange of more information on all levels. Furthermore, information flows (including the availability of real-time pricing) become bi-directional. With distributed generation and electric vehicles, the consumer can, in principle, feed electricity into the grid. But, for this to occur efficiently, market participants will need access to real-time information about the power system. The potential bi-directional flow of information has created some concerns about data protection and privacy, which will have to be addressed if smart grid technology is to be widely adopted.

Smart grid technologies can increase network efficiency by providing more timely information to operators through wide-area monitoring and integrated communications systems (*e.g.* radio networks, internet, cellular and cable); reducing outage times and maintenance as a result of improved sensing and automation on low-voltage distribution grids; and reducing transmission losses through transmission enhancements (*e.g.* dynamic line rating systems). With the deployment of smart meters and advanced communications systems, smart grids can also increase the flexibility of the power system by enabling end-users (motivated by a suitable financial incentive) to vary their demand according to the needs of the system. Potential areas for such demand-side response include heating and cooling loads in the industrial, commercial and residential sectors. In the future, it could also include supply to the grid from electric vehicles. This enhanced flexibility could alleviate some of the challenges of integrating renewables generation and reduce balancing costs.

Smart grid technologies could facilitate the integration of electricity grids over larger geographic areas. Conceptually, regional electricity grids could be combined and reinforced to interconnect larger regions or even continents, allowing significantly more electricity transport across large areas (known as "super-grids"). There are two main drivers of this potential development: the advantages of physical interconnection of electricity markets, enabling more efficient use to be made of power plants, with higher reliability of supply to consumers, and the need to integrate variable renewables.

A larger grid can facilitate the integration of variable renewable energies by allowing them to serve demand that best matches their variable output, even if the demand is geographically remote. It can also reduce balancing costs by smoothing the cumulative output of variable renewables, because the strength of sunshine or wind is less correlated over longer distances. Based on a simulation of a super-grid spanning the entirety of the United States and another extending across OECD Europe, we estimate that the development of super-grids could reduce system adequacy costs related to variable renewables by up to 25%. System adequacy costs fall because the super-grids facilitate a significant increase in the capacity credit of variable renewables (Figure 5.20). The actual deployment of super-grids will depend on the extent of the other economic benefits that they might provide to the system and their cost relative to other means of managing variability, such as energy storage or the increased installation of flexible generation capacity. They will also be contingent on political and public acceptance of additional transmission lines.

Figure 5.20 ● **Effect of a super-grid on the capacity credit for wind and solar PV for the United States and OECD Europe, 2035**

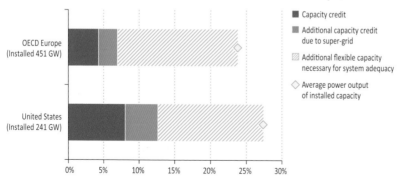

Note: Super-grids were modelled in the United States and Europe as networks that allow unrestricted electricity transport across the entirety of each region.

Sources: Capacity credit calculated based on Heide *et al.* (2010), NREL (2010, 2011a, 2011b) and World Wind Atlas (2011).

CHAPTER 6

CLIMATE CHANGE AND THE 450 SCENARIO
The door is closing, but when will we be "locked-in"?

H I G H L I G H T S

- Global energy-related carbon-dioxide (CO_2) emissions reached 30.4 Gt in 2010, 5.3% above 2009, representing almost unprecedented annual growth. In the New Policies Scenario, our central scenario, CO_2 emissions continue to increase, reaching 36.4 Gt in 2035, and leading to an emissions trajectory consistent with a long-term global temperature increase of more than 3.5°C.

- The 450 Scenario prescribes strong policy action to limit climate change, and results in global CO_2 emissions peaking before 2020 and then declining to reach 21.6 Gt in 2035. The share of fossil fuels in the global energy mix falls from 81% in 2009 to 62% in 2035. Global demand for both coal and oil peak before 2020, and then decline by 30% and 8% respectively by 2035, relative to their 2009 level. Natural gas demand grows by 26%. The 450 Scenario requires additional cumulative investment of $15.2 trillion relative to the New Policies Scenario, but delivers lower fossil-fuel import bills, reduced pollution and health benefits.

- New country-by-country analysis reveals that 80% of the total CO_2 emitted over the *Outlook* period in the 450 Scenario is already "locked-in" by our existing capital stock (*e.g.* power plants, buildings, factories), leaving little additional room for manoeuvre. If internationally co-ordinated action is not implemented by 2017, we project that all permissible CO_2 emissions in the 450 Scenario will come from the infrastructure then existing, so that all new infrastructure from then until 2035 would need to be zero-carbon. This would theoretically be possible at very high cost, but probably not practicable in political terms.

- The long lifetime of capital stock in the power sector means that the sector accounts for half of the emissions locked-in to 2035. This lock-in exceeds the projected emissions for the sector in the 450 Scenario. Delaying action until 2015 would call for early retirement or retrofitting of plants emitting 5.7 Gt in 2035, around 45% of the global installed fossil-fuel capacity. Delaying action is a false economy, because for every $1 of investment avoided before 2020, an additional $4.3 would need to be spent after 2020 to compensate for the increased emissions.

- Carbon capture and storage (CCS) is a key abatement option, accounting for 18% of emissions savings in the 450 Scenario, relative to the New Policies Scenario, but it faces regulatory, policy and technical barriers that make its deployment uncertain. In the Delayed CCS 450 Case, adoption is delayed by ten years, compared with the 450 Scenario, meaning it is widely deployed only after 2030. This increases the cost of the 450 Scenario by $1.14 trillion (8%) and puts unprecedented pressure on other low-carbon technologies, supporting the economic case to invest now in CCS.

Introduction

The last year has not been a good one for those concerned about the contribution of energy-related carbon-dioxide (CO_2) emissions to climate change. Energy-related CO_2 emissions in 2010 increased by 5.3%. Developing countries have continued to play a key role in powering global economic growth, providing modern energy to tens of millions of additional people, but they are doing so, to varying degrees, by using fossil fuels. Many countries have continued to move forward with plans to address climate change but, in many cases, those plans have yet to be implemented and their effectiveness proven. Policy intentions have come under pressure as budgets tighten and priorities change in the face of the uncertain economic outlook. Small steps were taken towards a global climate agreement at the United Nations Framework Convention on Climate Change (UNFCCC) 16th Conference of the Parties (COP-16) in Cancun, Mexico, when a heroic leap was needed to set the world on an emissions trajectory compatible with the stated long-term target of limiting the average global temperature increase to 2°C. Heightened awareness of the vulnerability of oil supply could help or hinder efforts to combat climate change, depending on whether countries see it as a reason to move away from fossil fuels or to invest further in domestic fossil-fuel supply. The events at Fukushima Daiichi in Japan make it more difficult to envisage a substantial and growing component of low-carbon nuclear power in the future energy mix. What do these, and other, developments mean for our efforts to tackle climate change? To what extent are energy-related emissions to 2035 already locked-in, setting the future climate trajectory?

This chapter seeks to reappraise the climate implications of the projected energy future, primarily in terms of our 450 Scenario. Unlike other scenarios presented in this *Outlook*, the 450 Scenario is an output-driven scenario: the constraint set is that the global energy pathway must be compatible with a 50% chance of meeting the goal of limiting the global increase in average temperature to 2°C, compared to pre-industrial levels (Box 6.1). This requires the long-term concentration of greenhouse gases in the atmosphere to be limited to about 450 parts per million of carbon-dioxide equivalent (ppm CO_2-eq). Within the framework of the policy assumptions made in this scenario, in the 450 Scenario an energy future is constructed which produces a level of energy-related emissions compatible with this objective, at the lowest achievable cost to the energy sector. The cost constraint is important: it recognises that there are other ways to achieve such an emissions trajectory, but also that they are likely to be more costly. Where appropriate, comparisons are made between the results of our 450 Scenario and those of our other *WEO-2011* scenarios, in order to demonstrate the divergence between the projected future paths and, particularly, what more needs to be done to achieve the climate goal.

This chapter begins by examining major developments of the last year that have, or may have, a significant impact on global efforts to tackle climate change. It sets out the projections of our 450 Scenario on global energy demand and the consequences for emissions of greenhouse gases and local pollutants, broken down by region and by sector. It then considers the implications of delay in making commitments to the necessary trajectory, in terms of the "lock-in" to a high-carbon infrastructure that would then make meeting the climate targets much more difficult and expensive, or even, in terms of practical politics, impossible.

Box 6.1 • What is special about 2°C?

The expected warming of more than 3.5°C in the New Policies Scenario would have severe consequences: a sea level rise of up to 2 metres, causing dislocation of human settlements and changes to rainfall patterns, drought, flood, and heat-wave incidence that would severely affect food production, human disease and mortality.

Alarmingly, research published since the International Panel on Climate Change's Fourth Assessment Report in 2007 suggests that this level of temperature change could result from lower emissions than those of the New Policies Scenario, due to climate feedbacks (IPCC, 2007a). For example, drying of the Amazon would release CO_2 that would then lead to further warming (Lewis et al., 2011) and rising arctic temperatures would lead to extra emissions from melting permafrost (Schaefer et al., 2011). These feedbacks have not yet been characterised with certainty, but they are expected to be triggered by temperature rises between 2°C and 5°C (Smith et al., 2009). The threshold for large-scale sea level rise may be similar, between 1.8°C and 2.8°C (Lenton et al., 2008; Hansen et al., 2008).

From the perspective of emission scenarios, these feedbacks imply that an increase in emissions can no longer be assumed to result in a pro-rata incremental increase in impacts. Put another way, a decision to relax climate policy and aim for a higher temperature target, such as 2.5°C or 3°C, may not actually allow much room for an increase in emissions, given the likelihood of further emissions and warming being triggered by feedbacks. For example, Schaefer et al. (2011) calculate that under conditions similar to the New Policies Scenario (which stabilises the atmospheric concentration at around 650 ppm CO_2-eq), emissions from melting permafrost would lead to a further increase of 58 to 116 ppm in CO_2 concentrations, resulting in further warming and more feedbacks.

The 450 Scenario, by definition, achieves a long-term atmospheric concentration of 450 ppm CO_2-eq (resulting in average warming of 2°C). Such a temperature increase (even without allowance for additional feedback effects) would still have negative impacts, including a sea-level rise, increased floods, storms and droughts.

The new evidence has led some researchers to conclude that even keeping the temperature rise to 2°C may risk dangerous climate change, and that an even lower temperature threshold and corresponding stabilisation target (such as 350 ppm) should be set (Anderson and Bows, 2011; Hansen et al., 2008; Rockström et al., 2009; Smith et al., 2009). The uncomfortable message from the scientific community is that although the difficulty of achieving 450 ppm stabilisation is increasing sharply with every passing year, so too are the predicted consequences of failing to do so.

Recent developments

In many ways, the pattern of global energy-related CO_2 emissions – the main contributor to global warming – mirrored that of the global economic recovery overall. The anticipated decline in 2009, due to the economic downturn, was only 1.7% compared with 2008, lower than expected. Emissions in almost all OECD[1] countries fell, but increased demand for fossil fuels (and the corresponding increase in emissions) in non-OECD countries partially compensated for this (Figure 6.1). The biggest component of the increase in non-OECD countries was demand for coal, especially in China, where it grew by almost 8% in 2009. India also saw strong growth in emissions due to increasing coal demand – 7% growth in 2009 compared with 2008 – and oil demand, which grew by 9% compared with 2008. In 2010, global emissions returned to growth, reaching 30.4 gigatonnes (Gt). This is despite the fact that OECD emissions remained below their 2008 levels in many cases, including in the United States and European Union.

Figure 6.1 • Energy-related CO_2 emissions by country, 2008-2010

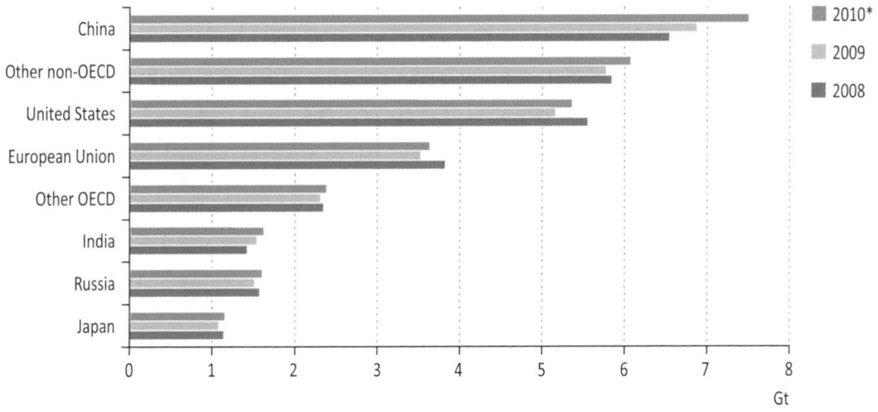

*Estimated.

In contrast to the OECD, energy-related emissions in non-OECD countries as a group were higher in 2010 than ever before, having grown in both 2009 and 2010. China, which became the largest source of energy-related emissions in 2007, continued in this position, with estimated energy-related emissions of 7.5 Gt in 2010. This is 40% higher than the United States, a very significant change over the four years since 2006, when the two countries had virtually the same level of energy-related emissions. Per-capita emissions in China are at 5.6 tonnes, one-third those of the United States. Emissions in India reached just over 1.6 Gt in 2010, India becoming the third-largest emitter on a global level, though per-capita emissions in India remain very low, at just 14% of the OECD average. Russia was one of the few large non-OECD countries to see a drop in emissions in 2009, but strong growth in 2010 meant that energy-related emissions rose slightly higher than pre-crisis levels, approaching levels last seen in the early 1990s (see Chapter 9). Coal was the biggest source of emissions growth in 2010, primarily driven by use in China and India. Natural-gas emissions also increased strongly in 2010, but from a relatively low base. While the increase in emissions

---

1. In this chapter, references to the OECD also include countries that are members of the European Union, but not of the OECD.

was driven by economic growth as the global economy began to recover from the financial crisis, emissions growth was stronger than GDP growth, implying a slight increase in the global CO_2 intensity of GDP.

The substantial rise in emissions was not accompanied by widespread intensification of action to combat climate change. The COP-16 meeting in Cancun resulted in some progress, recorded in two documents, but not the comprehensive, legally binding, global agreement that some still hoped for. The 2°C goal has been made explicit and is for the first time formally agreed; second, the individual emissions reduction pledges made by the countries are made official; and third, there is explicit acknowledgement that in order to limit long-term temperature increase to below 2°C, greater ambition is necessary. The COP explicitly urged countries to go beyond these pledges, opening the door to tighter restrictions on emissions before 2020. Since the COP-16 meeting in 2010, implementation of climate policies across the globe has developed in different directions. Some countries, such as Australia, China and the European Union, seem to be moving forward. Australia has developed plans to tax carbon emissions from the country's largest emitters, starting at 23 Australian dollars per tonne in 2012 before moving to a market-based trading scheme in 2015. This scheme aims to cut national emissions by 5%, compared with 2000 levels, by 2020. A parliamentary vote on the scheme is expected before the end of 2011. China's 12th Five-Year Plan (see Chapter 2) has introduced pilot cap-and-trade schemes, as a lead in to one of the means to achieve China's pledge of 40% to 45% reductions in CO_2 intensity below 2005 levels by 2020. An emissions trading system covering the entire economy and entering into force before 2020 is also under discussion. The European Union is discussing post-2020 targets and policies designed to give investors clear signals of requirements beyond the third trading period of the EU Emissions Trading System. On the other hand, the Arab Spring and the consequences of the Japanese tsunami have switched the attention of energy policy makers away from climate change, with energy security concerns climbing higher up the agenda of many countries.

In any case, the pledges made in association with the Copenhagen Accord and Cancun Agreements are not sufficient to put the world on a climate-sustainable path. Even if country pledges are interpreted in the most stringent possible way, and other policy commitments, such as removal of inefficient fossil-fuel subsidies, are implemented by 2020, action would need to be very significantly stepped-up after 2020 in order to achieve the 450 trajectory described in this chapter. This path differs very little from that set out in the *World Energy Outlook-2010*, which outlined in detail the scale of the effort required after 2020 in order to achieve a 450 trajectory given the lack of ambition of climate pledges for 2020. For instance, *WEO-2010* found that the weakness of the pledges for 2020 would increase the cost of achieving the 450 Scenario by $1 trillion over the period 2010 to 2030, compared with *WEO-2009*'s 450 Scenario, which assumed earlier strong action (IEA, 2010a). Supporting our conclusions, the United Nations Environment Programme (UNEP) emissions gap report found that the level of emissions expected as a result of the Copenhagen pledges is higher than would be consistent with either a "likely" or "medium" chance of staying below 2°C or 1.5°C (UNEP, 2010). Momentum may rise again in the preparation for COP-17 in late 2011, and we make the assumption that action will not cease with the expiry of the Kyoto Protocol first commitment period in 2012; however, very little time is left to get the appropriate policies in place.

Overview of trends in the 450 Scenario

Before examining the main trends and implications of the 450 Scenario, it is important to highlight briefly why the scenario is needed. It is because (as illustrated in Figure 6.2) neither the New Policies Scenario, our central scenario, nor the Current Policies Scenario puts us on a future trajectory for greenhouse-gas emissions that is consistent with limiting the increase in global temperature to no more than 2°C, the level climate scientists say is likely to avoid catastrophic climate change. The 450 Scenario illustrates one plausible path to that objective.

Figure 6.2 ● **World energy-related CO_2 emissions by scenario**[2]

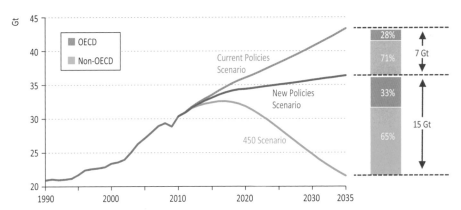

Note: There is also some abatement of inter-regional (bunker) emissions which, at less than 2% of the difference between scenarios, is not visible in the 2035 shares.

In line with practice in previous *World Energy Outlooks,* we have estimated greenhouse-gas emissions from all sources and for all scenarios (Table 6.1). We have then assessed the consequences for long-term concentrations and temperature increases of such emissions trajectories.

The New Policies Scenario, which takes account of both existing government policies and declared policy intentions (including cautious implementation of the Copenhagen Accord and Cancun Agreements), would result in a level of emissions that is consistent with a long-term average temperature increase of more than 3.5°C (see Chapter 2 for energy trends in the New Policies Scenario). The outlook in the Current Policies Scenario, which assumes no change in government policies and measures beyond those that were enacted or adopted by mid-2011, is considerably worse, and is consistent with a long term temperature increase of 6°C or more.

The trends and implications of the 450 Scenario, a scenario based on achieving an emissions trajectory consistent with an average temperature increase of 2°C, are sometimes presented here against the baseline of the New Policies Scenario to help demonstrate what more needs to be done, particularly in terms of carbon abatement. The main changes to the 450 Scenario in *WEO-2011* relate to the policy assumptions, which reflect changes in domestic and international energy and climate policies (Box 6.2). Non-policy assumptions relating to energy and CO_2 prices, GDP and population are presented in Chapter 1.

---

2. In 2009, energy-related CO_2 emissions contributed 61% to total greenhouse-gas emissions.

Table 6.1 ● World anthropogenic greenhouse-gas emissions by scenario ($Gt\ CO_2$-eq)

	2009	New Policies Scenario 2020	New Policies Scenario 2035	Current Policies Scenario 2020	Current Policies Scenario 2035	450 Scenario 2020	450 Scenario 2035
CO_2-energy	28.8	34.4	36.4	36.1	43.3	31.9	21.6
CO_2-other	1.4	1.2	1.1	1.7	1.9	1.0	0.8
CH_4	7.7	7.2	7.1	9.3	10.7	6.4	5.1
N_2O	3.2	3.2	3.2	3.8	4.2	3.0	2.7
F-gases	0.7	0.7	0.9	1.4	2.3	0.5	0.5
LULUCF[3]	5.2	4.3	1.9	4.3	1.9	4.3	1.9
Total	47.1	50.9	50.6	56.5	64.4	47.1	32.6

Notes: F-gases include hydrofluorocarbons (HFCs), perfluorocarbons (PFCs) and sulphur hexafluoride (SF_6) from several sectors, mainly industry. CO_2-other = CO_2 from industrial processes; LULUCF = land use, land-use change and forestry. Peat emissions are not included.

Source: IEA-OECD analysis using OECD Env-Linkages model.

Box 6.2 ● Updates to the 450 Scenario policy framework

To reflect developments over the last year, we have updated the policy framework for the 450 Scenario in the following ways:

● OECD countries: In addition to the emissions trading schemes already in place in the European Union and New Zealand, we assume that Australia introduces a CO_2 price from mid-2012 and an emissions trading scheme from 2015, while Japan and Korea introduce CO_2 pricing in 2020. All other OECD countries are assumed to introduce CO_2 pricing by 2025, either through cap-and-trade schemes or carbon taxes. We assume that trading schemes are linked at a regional level from 2025, when CO_2 prices start to converge (see Chapter 1 for CO_2 price assumptions).

● Non-OECD countries: In addition to China (where a CO_2 price covering all sectors is in place in 2020), Brazil, Russia and South Africa are also assumed to implement domestic CO_2 pricing from 2020, either through cap-and-trade schemes or carbon taxes. All trading schemes have access to carbon offsets.

Under these assumptions, 51% of global energy-related CO_2 emissions are covered by a CO_2 price in 2020 in the 450 Scenario and 36% in 2035.

As in *WEO-2010*, we assume that fossil-fuel subsidies are removed in all net-importing regions by 2020 (at the latest) and in all net-exporting regions by 2035 (at the latest), except for the Middle East where it is assumed that the average subsidisation rate declines to 20% by 2035. We also assume that fuel economy standards are implemented in the transport sector, strict energy efficiency measures are adopted in the buildings sector and that support for CCS and renewables is available.

---

3. Emissions from land use, land-use change and forestry are assumed to remain unchanged in all scenarios. In practice, actions to reduce deforestation and land-use change will reduce these emissions to varying degrees in different scenarios, while increased use of biomass and biofuels will increase them. We assume that these two factors cancel one another out in all scenarios.

Primary energy demand in the 450 Scenario

In the 450 Scenario, global primary energy demand increases by 23% from 2009 to reach nearly 14 900 million tonnes of oil equivalent (Mtoe) in 2035 (Table 6.2). Energy demand increases by an average of 1.4% per year to 2020, but this growth slows to 0.3% per year from 2020 to 2035. This represents a significantly lower level of energy demand growth than in other *WEO-2011* scenarios and principally reflects greater efforts to improve energy efficiency.

The composition of energy demand changes considerably in the 450 Scenario, with the share of fossil fuels in the energy mix declining from 81% in 2009 to 62% in 2035. Global demand for coal peaks around 2016 and then declines by 2.7% per year on average, to reach 3 310 million tonnes of coal equivalent (Mtce) in 2035, 30% lower than 2009. In the 450 Scenario, global demand for oil also reaches a peak around 2016 before declining to 78.3 million barrels per day (mb/d) in 2035, 6.9% lower than 2009. This peak in global oil demand is brought about by additional policy intervention, such as stringent fuel economy standards for passenger vehicles and strong support for alternative fuels. While oil continues to hold the largest share of the energy mix throughout the *Outlook* period, its share drops from 33% in 2009 to 25% in 2035. Natural gas demand grows steadily, at 1.2% per year through 2030, stabilising thereafter and reaching almost 3 900 billion cubic metres (bcm) in 2035.

Table 6.2 ● World primary energy demand by fuel in the 450 Scenario

	Demand (Mtoe)			Share		
	2009	2020	2035	2009	2020	2035
Coal	3 294	3 716	2 316	27%	26%	16%
Oil	3 987	4 182	3 671	33%	29%	25%
Gas	2 539	3 030	3 208	21%	21%	22%
Nuclear	703	973	1 664	6%	7%	11%
Hydro	280	391	520	2%	3%	3%
Biomass and waste	1 230	1 554	2 329	10%	11%	16%
Other renewables	99	339	1 161	1%	2%	8%
Total	**12 132**	**14 185**	**14 870**	**100 %**	**100 %**	**100 %**

In the 450 Scenario, the overall share of low-carbon fuels in the energy mix doubles from 19% in 2009 to 38% in 2035. Demand for all low-carbon fuels grows strongly. Electricity generated from nuclear power increases by around 140% over the *Outlook* period, with almost 70% of this increase being in non-OECD countries (for the implications of a slower nuclear expansion for the achievability of the 450 Scenario, see Chapter 12 and Box 6.4). China, in particular, sees strong growth in nuclear power, its share of primary energy demand growing from 1% in 2009 to 13% in 2035. Electricity generated from hydropower increases globally by an average of 2.4% per year over the *Outlook* period, to stand at just over 6 000 terawatt hours (TWh) in 2035. In 2035, hydropower accounts for nearly 70% of the electricity generated in Latin America, almost 25% in Africa and nearly 20% in developing Asia. Global demand for biomass nearly doubles, reaching 2 330 Mtoe in 2035. Biomass for power generation and biofuels in transport account for 44% and 29% of the growth respectively. Other renewables, including wind, solar, geothermal and marine power, grow more than ten-fold over the *Outlook* period, albeit from a low base, and their share of the energy mix grows from 1% in 2009 to 8% in 2035. Electricity generation from wind power experiences a large increase, going from 1% of global electricity generation in 2009 to 13% in 2035, with China, the European Union and the United States being the largest markets. Solar, meanwhile, increases its share from less than 1% in 2009 to nearly 7% in 2035. The largest proportional increases are seen in India, where solar reaches 8% of total electricity generation, and North Africa, where solar reaches almost a quarter of total generation.

Energy-related emissions and abatement[4]

In the 450 Scenario, energy-related CO_2 emissions peak at around 32.6 Gt before 2020 and then decline to 21.6 Gt by 2035, almost back to 1990 levels. Global CO_2 emissions from coal peak around 2016 and then decline at an ever more rapid pace to reach 5.4 Gt in 2035, about 40% of the level of 2010 (Figure 6.3). Global CO_2 emissions from oil peak at 11.3 Gt around 2016, before declining to 9.6 Gt in 2035. In contrast, CO_2 emissions from natural gas, which emits the lowest level of CO_2 of all fossil fuels when combusted, peak around 2025, after which they stabilise and then fall slightly as carbon capture and storage (CCS) for gas is more widely deployed, reaching around 6.6 Gt in 2035, 8% higher than 2010. CO_2 intensity – the amount of emissions per unit of GDP – declines at a rate of 3.5% per year from 2010 to 2020, and 5.5% per year from 2020 to 2035. This rate is more than six-and-a-half times greater than the annual intensity improvements achieved in the last ten years.

Figure 6.3 • **World energy-related CO_2 emissions by fossil fuel in the 450 Scenario**

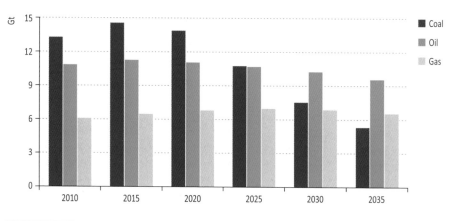

4. Abatement analysis in the 450 Scenario is calculated against a baseline of the New Policies Scenario, our central scenario.

Energy efficiency measures account for half the cumulative CO_2 abatement achieved in the 450 Scenario, relative to the New Policies Scenario, over the *Outlook* period (Figure 6.4). The scale of this reduction underlines the importance of strong policy action to ensure that potential efficiency gains are realised, in such forms as building standards, vehicle fuel economy mandates and insistence on widespread use in industry of the best-available technologies (Box 6.3). After the cheaper energy efficiency measures are exploited early in the *Outlook* period, more expensive abatement options take a larger share, and the annual share in abatement of efficiency measures falls to 44% in 2035. The increased adoption of renewable energy (including biofuels) is the second-most important source of CO_2 abatement, relative to the New Policies Scenario, growing from a combined 19% in 2020 to 25% in 2035, or a cumulative 24% over the period as a whole. Nuclear power grows rapidly in importance and accounts for a cumulative 9%, while CCS also accounts for an increasing share, growing from only 3% of total abatement in 2020 to 22% in 2035, or a cumulative 18%.

Figure 6.4 ● **World energy-related CO_2 emissions abatement in the 450 Scenario relative to the New Policies Scenario**

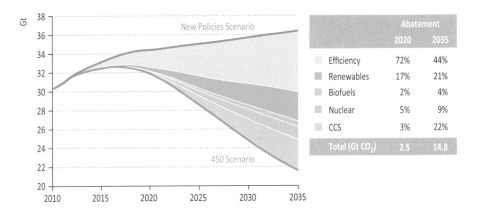

	Abatement	
	2020	2035
Efficiency	72%	44%
Renewables	17%	21%
Biofuels	2%	4%
Nuclear	5%	9%
CCS	3%	22%
Total (Gt CO_2)	2.5	14.8

Box 6.3 ● **Reaping abatement through efficiency in the 450 Scenario**

In the 450 Scenario, energy efficiency policies and measures are the cheapest abatement option available and the most important source of abatement. Efficiency is responsible for half of cumulative global abatement relative to the New Policies Scenario, or 73 Gt, between 2011 and 2035. The role of energy efficiency varies by country, according to the remaining potential and energy pricing. In OECD countries, despite the strong efficiency improvements already occurring in the New Policies Scenario, efficiency in the 450 Scenario is responsible for almost 42% of abatement relative to the New Policies Scenario. This share rises to 54% in non-OECD countries, where efficient energy-producing and -using technologies are in general less widely deployed. This is due to both their higher costs relative to less efficient technologies, and because energy subsidies often present in these countries do not encourage energy efficiency.

Efficiency policies are by far the largest source of abatement in end-use sectors relative to the New Policies Scenario. They are responsible for 91% of *buildings* sector direct abatement. To reap this abatement, we assume that strict energy efficiency standards are set for new buildings, incentives are put in place for refurbishment of existing buildings, and increasingly high efficiency standards are set for appliances, alongside higher electricity prices due to pricing of CO_2 in many countries. In the *transport* sector, where efficiency measures are responsible for 64% of cumulative abatement relative to the New Policies Scenario over the projection period, we assume the widespread implementation of fuel economy standards for cars. We also assume that the level of fuel taxes is adjusted to maintain the price of oil-based fuels for vehicles (which would otherwise fall as demand falls) at a level similar to the Current Policies Scenario, so as to avoid a rebound effect which would lead to more energy use. Efficiency gains in the *industry* sector, which are responsible for 59% of abatement in this sector, are driven by the implementation of CO_2 pricing in OECD countries, China, Russia, Brazil and South Africa. In order to avoid international relocation (referred to as carbon leakage) in those sectors that could face it, we have assumed that countries not introducing CO_2 pricing would enter into international sectoral agreements setting minimum efficiency standards.

Full energy efficiency potential will not be realised while end-user prices are too low. Fossil-fuel subsidies are a major barrier to energy efficiency, as the artificial reduction in energy costs leads to higher than optimal demand for energy (see Chapter 14). Removal of fossil fuel subsidies in the 450 Scenario accounts for a cumulative 7.9 Gt of abatement from 2010 to 2035, relative to the New Policies Scenario.

Energy efficiency measures have associated benefits, in terms of energy security and reduced local pollution. Despite this, and their sound economic rationale, energy efficiency measures are more difficult to implement than one might suppose. For example, landlords may not be willing to invest in double-glazed windows, as they do not reap the benefits of reduced energy bills or a driver may prefer the experience of driving a more powerful car than a more efficient one. To help governments overcome such obstacles, the IEA has made recommendations to the G-8 in 2006 and subsequent years to improve energy efficiency. The IEA estimates that if these were implemented globally without delay, they would result in annual CO_2 savings of as much as 7.6 Gt by 2030 (or 70% of the savings between the New Policies and 450 Scenarios in that year). In 2011, the IEA published a progress report on the implementation of these recommendations by IEA countries, which found that 70% of the recommendations can be categorised as fully or partially implemented, or that implementation is underway (IEA, 2011). To conform with the 450 trajectory, full implementation of those recommendations would be required in all countries, coupled with subsidies removal.

Policies aimed at reducing energy-related CO_2 emissions in the 450 Scenario also have the effect of reducing emissions of other air pollutants that have a negative impact on human health and the environment, such as sulphur dioxide (SO_2), nitrogen oxides (NO_x) and particulate matter ($PM_{2.5}$) (Figure 6.5). By 2035, SO_2 emissions are almost 42% lower than 2009 levels in the 450 Scenario, a reduction of 38 million tonnes (Mt). Most OECD countries already have sulphur control measures in place, so the majority of these reductions (25 Mt) occur in other countries and result mainly from reduced fossil-fuel consumption. Global emissions of NO_x, which causes acidification of rain and ground-water and contributes to ground-level ozone formation, are 27% lower in the 450 Scenario in 2035 than in 2009, as a result of reduced use of coal and oil. Emissions of $PM_{2.5}$, which together with NO_x is the main cause of smog formation and the subsequent deterioration of urban air quality, are 10% lower globally by the end of the *Outlook* period. In OECD countries, $PM_{2.5}$ emissions in 2035 are almost identical to 2009 levels, as a result of greater use of biomass in the residential sector; but emissions of $PM_{2.5}$ in non-OECD countries decrease by 4 Mt over the *Outlook* period, with the largest reductions occurring in China and India.

Figure 6.5 ● Emissions of major air pollutants by region in the 450 Scenario

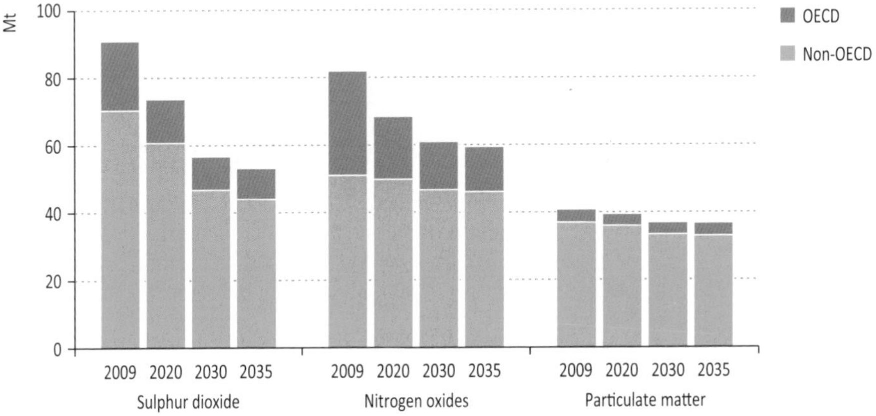

Note: The base year of these projections is 2005; 2009 is estimated by IIASA.
Source: IIASA (2011).

Regional energy-related CO_2 emissions and abatement

Energy-related CO_2 emissions in the OECD decline by 50% between 2009 and 2035 in the 450 Scenario, to reach 6 Gt, and their share of global emissions falls from 42% in 2009 to 28% in 2035. In 2035, the OECD accounts for 31% of global CO_2 abatement, relative to the New Policies Scenario, equal to 4.6 Gt. CO_2 emissions in non-OECD countries fall by a much smaller 9% over the *Outlook* period, to reach 14.3 Gt in 2035, though this is still a substantial 10.0 Gt of CO_2 abatement, relative to the New Policies Scenario, in 2035. Around three-quarters of the global abatement of CO_2 emissions in the 450 Scenario, relative to the New Policies Scenario, occurs in six countries or regions, highlighting their importance in achieving an emissions trajectory consistent with 2°C (Figure 6.6).

China's energy-related CO_2 emissions peak at around 9.1 Gt just before 2020 and then decline to just under 5.0 Gt in 2035, 28% lower than 2009. Despite this decline, China remains the largest CO_2 emitter throughout the *Outlook* period (although in cumulative terms from 1900 to 2035, China's energy-related emissions remain below those of both the United States and European

Union in the 450 Scenario). China achieves cumulative CO_2 abatement in the 450 Scenario relative to the New Policies Scenario of 49.5 Gt over the *Outlook* period, more than any other country, and accounts for 34% of the global cumulative abatement. China's CO_2 abatement comes mainly from efficiency measures, which are responsible for half of the total, and from fuel-switching to renewables (including biofuels), which accounts for 18% of the total. Adoption of CCS also plays an important role, but is more prominent towards the end of the *Outlook* period. In 2035, China is by far the largest market for renewables and electric vehicles (EVs). China's per-capita emissions increase to exceed those of the European Union before 2020, but by the end of the projection period strong policy action means that they have fallen again and, at 3.6 tonnes per person, are only marginally higher than those of the European Union, and well below the level of Japan, the Middle East, Russia and the United States (Figure 6.7).

Figure 6.6 ● **Energy-related CO_2 emissions in the 450 Scenario and abatement relative to the New Policies Scenario by region, 2009 and 2035**

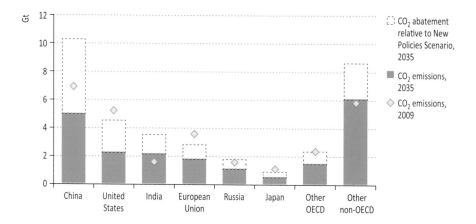

In the 450 Scenario, energy-related CO_2 emissions in the United States decline by 56% from 2009 to reach 2.3 Gt in 2035. A shift away from coal in favour of natural gas in power generation is central to achieving these reductions. Demand for renewables grows throughout the projection period, with generation from renewables increasing from around 11% of total generation in 2009 to 17% in 2020 and then more sharply to 38% by 2035. In the transport sector, strong improvements in fuel economy, increased use of biofuels and a shift to electric, hybrid and plug-in hybrid electric vehicles (PHEVs) are also important. By 2035, biofuels account for a quarter of fuel use in transport and electric vehicles and PHEVs account for 54% of passenger light-duty vehicle (PLDV) sales.

India's energy-related CO_2 emissions grow steadily to 2.2 Gt in 2035, 39% above the 2009 level. India's emissions-intensity target of improvements of 20% to 25% by 2020, compared with 2005, is exceeded in the 450 Scenario, mostly thanks to investment to generate offset credits (either from domestic sources for sale to other countries seeking to meet their own emissions-reductions targets or by foreign companies and governments seeking to meet their own obligations) and direct financial transfers from developed countries (without offset credits). Efficiency and renewables play a big role in abatement. Government policies such as the new trading scheme for energy-efficiency or 'white' certificates in industry and feed-in tariffs for renewables (assumed to be implemented in all scenarios, but with the most stringent targets in the 450 Scenario) are expected to be key instruments in the transformation. India's per-capita emissions remain very low at only 1.4 tonnes by 2035, 43% below the global average, as lack of access to modern energy persists (Spotlight).

Figure 6.7 ● Energy-related CO$_2$ emissions per capita in the 450 Scenario by region

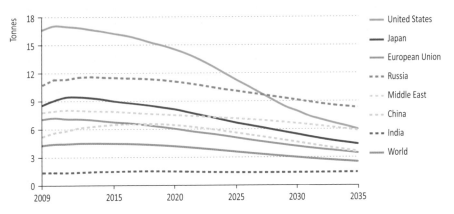

In the European Union, energy-related CO$_2$ emissions are 1.8 Gt in 2035, 49% below the level in 2009. Renewables account for 23% of primary energy demand in 2020, exceeding the target of 20%.[5] Discussion is ongoing in the European Union to define objectives for 2030 and beyond on the path to an 80% cut in emissions by 2050, relative to 1990 levels, and strong action is expected to continue in the region. Renewables continue to grow strongly across the period, reaching a share in the total energy mix of 35% in 2030 and 43% by 2035. Biomass represents 60% of renewables use in the European Union in 2035, being widely used for heating, power generation and transport. By 2035, wind represents the largest single share of installed electricity generating capacity, 29% higher than that of natural gas. CO$_2$ emissions from new PLDVs reach on-road levels of 50 gCO$_2$/km in 2035, 70% below current levels.

Russia's energy-related CO$_2$ emissions peak before 2015 in the 450 Scenario, falling from their 2009 level of 1.5 Gt to 1.1 Gt in 2035. Russia's target of reducing emissions by 25%, relative to 1990 levels, is exceeded in the energy sector in the 450 Scenario, where emissions in 2020 are 29% below their 1990 level. This is partly due to the sharp decline which occurred in Russia's emissions after the collapse of the Soviet Union, but policies to improve energy efficiency and support the deployment of renewables and nuclear power also play a role. Later in the period, the implementation of a CO$_2$ price encourages efficiency and

5. The European Union uses a slightly different methodology to calculate energy statistics than that of the IEA. These figures are calculated using the EU methodology.

The International Year of Sustainable Energy for All: can universal access be achieved without increasing CO$_2$ emissions?

2012 has been named the "International Year of Sustainable Energy for All" by the United Nations General Assembly. This year's *Outlook*, again addresses the issue of energy poverty, focussing on financing in the Energy for All Case (Chapter 13). But the analysis in that chapter is seen in the context of the New Policies Scenario: it is valuable also to examine the additional costs imposed by attempting to meet as much as possible of the electricity needs of the poor through low-carbon sources.[6] Today 1.3 billion people lack access to electricity, 84% of whom live in rural areas. Electrification of the urban poor through low-carbon sources would be extremely difficult, as, broadly speaking, generation for the national grids in poor countries is largely high-carbon throughout our projection period, and it is unrealistic to assume all new electrification would come entirely from low-carbon options due to the high cost. In fact, it is also unnecessary, as climate goals can be achieved through abatement in countries other than the poorest.

However, rural electrification offers more scope for low-carbon solutions and they are indeed often more suitable than conventional sources to meet off-grid demand (due, for example, to cost-effectiveness at small scale). Achieving universal rural electrification by 2030 with only low-carbon off-grid options, such as solar photovoltaics, small and mini-hydro, biomass, and wind, would require additional generation from these sources of 670 TWh in 2030.

The cumulative investment needed to bring this low-carbon electricity to all rural populations is estimated to be $586 billion between 2010 and 2030. About 60% of this investment is needed in sub-Saharan Africa and over 20% in India. In our Energy for All Case, we estimate the costs of rural electrification with a mix of fossil-fuel-based grid connection and off-grid options to be around $480 billion between 2010 and 2030. In that case, total additional CO$_2$ emissions amount to 240 Mt in 2030, with rural electrification accounting for around 38% of those additional CO$_2$ emissions. For an extra $5 billion a year, rural electrification could be achieved while saving over 90 Mt of CO$_2$ emissions in 2030.

6

strengthened support for low-carbon technologies also contributes. Per-capita emissions fall only gradually across the period and, at 8.3 tonnes per person in 2035, they are among the highest of any country, over three-times the global average (see Chapter 9).

Japanese energy-related CO$_2$ emissions reach around 520 Mt in 2035 in the 450 Scenario, less than half the 2009 level and 43% below the level in the New Policies Scenario. Measures to drive efficiency improvements are the biggest source of this abatement, delivering nearly

---

6. Since much of the population without access to modern energy sources relies on kerosene for lighting and cooking at present, the baseline against which the change in emissions as a result of switching to electricity use is not zero-emissions. Although we cannot quantify this effect, the net result is to reduce the additional emissions resulting from additional grid connections.

40% of cumulative abatement, relative to the New Policies Scenario. Renewables account for almost a quarter of abatement, as their share in generation increases to 31% by 2035, and CCS is also important, at 19% of cumulative abatement. Despite a fall in its share in the early years of the period, nuclear power remains very significant in the Japanese energy mix, accounting for 17% of abatement relative to the New Policies Scenario and reaching 46% of electricity generation by 2035.

Ten policies account for 42% of abatement, relative to the New Policies Scenario, in 2020 and 54% in 2035 (Table 6.3). These policies are implemented in just five regions – the United States, the European Union, China, India and Russia. Rapid and stringent implementation of policies in these countries is vital to the success of the 450 Scenario. Chinese power generation is, due to its sheer scale and reliance on coal, the largest source of abatement in both 2020 and 2035, with Chinese industry not far behind. Abatement in these sectors partly reflects government policies to rebalance the economy, but also ambitious capacity targets for low-carbon energy sources. Economy-wide CO_2 pricing, coming into effect from 2020, drives further abatement later in the period. CO_2 pricing in the United States, European Union and Russia also plays an important role by 2035, with reductions achieved in the power and industry sectors, both domestically and through the purchase of international offset credits.

Sectoral emissions and abatement

The power sector, currently the source of 41% of global energy-related CO_2 emissions, is the biggest source of abatement, in the 450 Scenario relative to the New Policies Scenario. Emission reductions due to fuel-switching to less carbon-intensive generation, power sector efficiency and reduced electricity demand reach a cumulative 99.0 Gt by 2035, or two-thirds of total abatement from all sectors (Figure 6.8). Abatement from power generation is so substantial that by 2035 transport, which currently emits only just over half of the amount that is emitted by power generation, becomes the largest sectoral source of emissions. Abatement from buildings and industry, where there is generally less low-cost abatement potential, accounts for 5% and 10% each of total cumulative abatement from all sectors.

Figure 6.8 • World energy-related CO_2 abatement by sector in the 450 Scenario compared with the New Policies Scenario

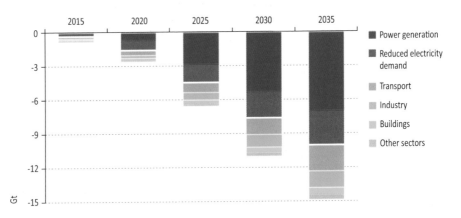

The power generation sector plays a crucial role in the 450 Scenario, with emissions declining by 60% from 2009 levels to reach 4.8 Gt in 2035. The sector's share of total energy-related emissions falls from 41% in 2009 to 22% by the end of the *Outlook* period. The 450 Scenario reflects a strong policy push towards low-carbon technologies in power generation, resulting in their share of global electricity generation increasing from one-third in 2009 to three-quarters in 2035. This transformation occurs mainly thanks to a combination of:

- The introduction of CO_2 pricing in all OECD countries and several major non-OECD countries.
- Enabling policies for deployment of low-emissions technologies, including nuclear and CCS.
- Higher support for renewables-based electricity generation technologies.

Of the total cumulative global abatement in power generation, relative to the New Policies Scenario, around 66 Gt comes from power plant efficiency and fuel-switching to lower-carbon energy sources and technologies, while the remaining 33 Gt of savings are attributable to reduced demand for electricity from final end-use sectors (Figure 6.9).

Figure 6.9 ● **Change in world energy-related CO_2 emissions from the power generation sector in the 450 Scenario compared with the New Policies Scenario**

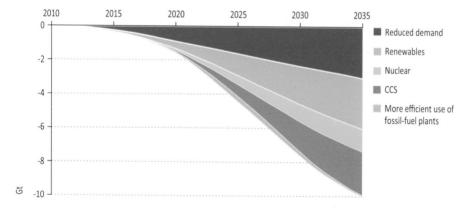

In the 450 Scenario, emissions from transport peak before 2020 at around 7.3 Gt and then gradually decline to reach 6.6 Gt in 2035. Improved fuel efficiency is the most significant factor in reducing emissions, accounting for more than half of the reduction observed in 2035, relative to the New Policies Scenario. In the 450 Scenario, electric vehicles and plug-in hybrids account for 6% of total light-duty vehicles sales in 2020 (up from less than 0.01% today) and this share increases to 37% by 2035 (see Chapter 3 for more on electric vehicles and hybrids). Cumulatively, electric vehicles and plug-in hybrids constitute 14% of all vehicle sales during the *Outlook* period. Biofuels are the second-largest contributor to transport emissions savings in the 450 Scenario, with their share in total transport fuels increasing from about 2% in 2009 to 13% by 2035, or 7.8 mb/d. In 2035, more than 60% of all biofuels consumed are advanced biofuels derived from ligno-cellulosic or other non-food crops, highlighting the need for early investment in research and development to bring these fuels to commercial viability.

Table 6.3 • Top ten sources of abatement in the 450 Scenario relative to the New Policies Scenario by policy area

	2020				2035		
	Sector	Measures	Abatement (Mt CO_2)		Sector	Measures	Abatement (Mt CO_2)
China	Power generation	More ambitious government targets for low-carbon generation, including wind, nuclear and solar.	198	China	Power generation	Domestic CO_2 price covering all sectors from 2020. Support to renewables and nuclear. Deployment of CCS.	2 477
China	Industry	Rebalancing of the economy and efficiency improvements.	175	United States	Power generation	Domestic CO_2 price from 2020. Extended support to renewables and nuclear.	1 344
China	Buildings	50% of building stock has improved insulation to reduce energy consumption per unit area by 65% vs. 1980 level. 50% of appliances stock meets highest-available efficiency standards.	143	China	Industry	Domestic CO_2 price from 2020. Rebalancing of the economy and use of CCS.	929
India	Power generation	Feed-in tariff to reach 15% of renewables installed capacity by 2020. Support and regulation for more efficient coal.	94	China	Buildings	Domestic CO_2 price from 2020. More stringent implementation of standards. Mandatory energy efficiency labels. 100% of buildings stock has improved insulation to reduce energy consumption per unit area by 65% vs. 1980 level.	712
European Union	Buildings	Enhanced efficiency standards for existing buildings. Zero-carbon footprint for all new buildings as of 2018.	84	India	Power generation	Support to renewables, nuclear and efficient coal. International offset projects.	680

Table 6.3 ● Top ten sources of abatement in the 450 Scenario relative to the New Policies Scenario (continued)

	2020			2035		
	Sector	Measures	Abatement (Mt CO$_2$)	Sector	Measures	Abatement (Mt CO$_2$)
European Union	Power generation	EU Emissions Trading System. Support for renewables and CCS.	81	Power generation	EU Emissions Trading System. Extended support to renewables and CCS.	371
India	Industry	Efficiency improvements in iron, steel and cement due to international offset projects.	73	Transport (road)	Passenger light-duty vehicle fuel economy standards, biofuels incentives and incentives for natural gas use in trucks.	358
United States	Buildings	More stringent mandatory building codes. Extension of energy efficiency grants. Zero-energy buildings initiative.	61	Power generation	CO$_2$ pricing implemented from 2020. Support for nuclear power and renewables.	293
India	Buildings	Mandatory energy conservation standards and labelling requirements for all equipment and appliances. Increased penetration of energy efficient lighting.	54	Buildings	More stringent mandatory building codes. Extension of energy efficiency grants. Zero-energy buildings initiative.	277
China	Transport (road)	More stringent on-road emissions targets for PLDVs to gradually reach 100 gCO$_2$/km by 2035. Enhanced support to alternative fuels.	42	Buildings	Zero-carbon footprint for all new buildings as of 2018.	241
Total			**1 005**			**7 682**
Share of total abatement			**42%**			**54%**

6

In the industry sector, energy demand in the 450 Scenario continues to increase throughout the *Outlook* period, with average growth of 1.2% per year. This slowdown relative to the New Policies Scenario is accompanied by fuel-switching away from coal and oil and towards electricity (generated in an increasingly decarbonised power sector) and renewables. As a result, direct emissions from the industry sector decline by 0.2% per year over the *Outlook* period, to reach 4.7 Gt in 2035.

Global direct emissions from the buildings sector peak in the 450 Scenario at 3.0 Gt around 2012 and then decline gradually to 2.6 Gt in 2035. The emissions profile in the buildings sector is influenced heavily by developments in OECD countries, where emissions are expected to peak in 2010 and decline by 1.1% per year thereafter. In non-OECD countries, emissions do not start to decline in this sector until around 2020, and then fall on average by 0.2% per year. This is because energy demand in the sector in non-OECD countries is under countervailing upward pressure from increasing population, increasing average incomes (driving residential energy demand) and the shift towards tertiary sectors in the growing economies (driving up services sector demand).

Investment in the 450 Scenario

The global energy sector will require very substantial investment over the next 25 years, regardless of the path followed; the total investment differs by scenario, but total investment in energy supply is much larger than the differences between scenarios. In the 450 Scenario, investment in energy supply, including in coal, oil and gas extraction and transportation, biofuels production and electricity generation, transmission and distribution, amounts to a cumulative total of $36.5 trillion (in year-2010 dollars) from 2011 to 2035. This is around 4% less than is required in the New Policies Scenario, with investment in fossil-fuel supply in the 450 Scenario falling by around $4.6 trillion in total compared with the New Policies Scenario. This supply-side reduction in investment relative to the New Policies Scenario is, however, more than offset by an increase in investment in low-carbon technologies and efficiency measures, both on the supply side and on the demand side. This includes additional investment by consumers in more energy-efficient and lower-emitting vehicles and appliances. Overall, additional investment in low-carbon technologies and energy efficiency in the 450 Scenario, relative to the New Policies Scenario, amounts to a cumulative $15.2 trillion over the period (Figure 6.10).

Average annual investment in the 450 Scenario, over and above that needed in the New Policies Scenario, grows fast over time, from just over $160 billion per year in the decade 2011 to 2020 to $1.1 trillion annually by the end of the *Outlook* period. The increase is due to the fact that the abatement achieved in 450 Scenario up to 2020, even with action arising from the Cancun Agreements, leaves much to be accomplished in the later period and at a higher capital cost per unit of CO_2 saved. As a result, only 11% (or $1.6 trillion) of total additional investment is incurred before 2020.

Of the cumulative additional investment in the 450 Scenario relative to the New Policies Scenario, $6.3 trillion, or over 40%, is needed in the transport sector. Most of this is directed towards the purchase of more efficient or alternative vehicles. The building sector is the second-largest recipient of additional investment, amounting to $4.1 trillion (or 27% of the total).[7] Refurbishment of buildings in OECD countries and solar PV installations account for most of the investment. The decarbonisation of the power sector requires a net additional $3.1 trillion. About two-thirds of total investment in electricity generation goes to renewable-based technologies, 14% to nuclear, 8% to plants fitted with CCS and 12% to fossil-fuel plants not fitted with CCS. Industry invests an additional $1.1 trillion, almost a third of it directed to CCS.

Figure 6.10 ● Cumulative energy sector investment by scenario, 2011-2035

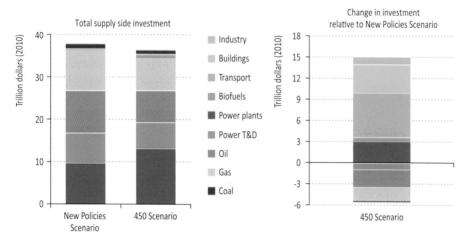

Notes: Investment in solar PV in buildings is attributed to power plants in supply-side investment. Elsewhere, it is attributed to the buildings sector. T&D = transmission and distribution.

Within power generation, there is some avoided investment in electricity transmission and distribution lines, totalling about $930 billion. The lower level of electricity demand in the 450 Scenario – achieved through the $2.7 trillion investment made in buildings and industry in improving efficiency of electricity end-use – leads to a reduction in grid infrastructure investment of around $1.1 trillion. The increased usage of renewable energy, which requires greater investment in transmission and distribution than other energy sources (see Chapter 5), adds nearly $165 billion in the 450 Scenario, partially offsetting the savings due to lower demand.

Mirroring their importance in global abatement relative to the New Policies Scenario, China and the United States need the largest additional investment – $3.2 trillion and $2.8 trillion respectively. Non-OECD countries account for almost half of the total cumulative additional investment relative to the New Policies Scenario, with their share increasing towards the end of the period in line with their share of abatement.

Other spending in the 450 Scenario: fuel costs and subsidies

The changes in the energy sector to achieve the 450 Scenario have an impact on fuel expenditure, relative to the New Policies Scenario, as lower international fuel prices interact

---

7. It is important to note that this investment finances not only the direct abatement from the buildings sector reported in Figure 6.8, but also a proportion of the indirect abatement through electricity demand reduction due to investment in more efficient end-use equipment.

with policies aimed at addressing climate change, such as carbon pricing, fossil-fuel subsidy removal and support for low-carbon energy sources, and the resulting changes in the structure of demand. These various forces act together to varying degrees to produce differing effects on consumer fuel expenditure in different regions. At the global level, the net effect is a reduction in consumers' expenditure on energy, with total cumulative savings, relative to the New Policies Scenario, of $690 billion (in year-2010 dollars) over the period 2011 to 2035.

However, not all regions see savings. Lower international fuel prices and lower demand might be expected, other things being equal, to lead to lower fuel expenditure by consumers. But, a number of factors cut across this. We assume that end-use fuel-prices in the transport sector are kept at a level similar to the Current Policies Scenario through taxation, in order to avoid a rebound effect, as lower demand would otherwise drive end-use prices down, thus discouraging efficient behavior. This reduces potential savings to consumers (although increasing importing countries' governments' revenues). Pricing of CO_2, which is intended to reduce demand for carbon-intensive technologies, is assumed to be passed onto consumers, increasing fuel costs relative to the New Policies Scenario. At the same time, removal of fossil-fuel subsidies also increases expenditure by end-users (although simultaneously reducing expenditure by governments). In the Middle East, for example, expenditure on energy is a cumulative $2.5 trillion higher in the 450 Scenario than in the New Policies Scenario – subsidies on fossil fuels amount to $4 trillion across the period, around $2 trillion less than in the New Policies Scenario. China and India, on the other hand, see total combined savings on energy costs of more than $1.4 trillion over the *Outlook* period, $840 billion in China and $600 billion in India.

Global spending on fossil-fuel subsidies is a cumulative $6.3 trillion over the period 2011 to 2035 in the 450 Scenario, $4.1 trillion lower than in the New Policies Scenario. Subsidies for renewables, meanwhile, are a cumulative $550 billion higher in the 450 Scenario than in the New Policies Scenario. Again, there is significant regional variation, with China spending a cumulative $366 billion more on renewable-energy subsidies in the 450 Scenario, compared with the New Policies Scenario, and making around $62 billion in savings on fossil-fuel subsidies, while Russia spends around $28 billion more on renewables subsidies over the period and saves just over $340 billion on fossil-fuel subsidies. As mentioned, the greatest change is in the Middle East, with an additional $39 billion spent on renewables subsidies, far outweighed by the $2 trillion reduction in fossil-fuel subsidies.

Benefits of the 450 Scenario

The 450 Scenario, apart from leading to lower adaptation costs than the other scenarios presented in *WEO-2011* (see, for example, Parry *et al.*, 2009; IPCC, 2007b; Stern, 2006), gives rise to a number of other benefits, notably in terms of import bills, local pollutants and health impacts. In the 450 Scenario, crude oil, steam coal and natural gas import prices are much lower than in the other scenarios, reflecting lower demand. In real terms, the IEA crude oil import price needed to balance supply and demand reaches $97/barrel (in year-2010 dollars) in 2020 and remains stable at that level thereafter. The oil price in 2035 is $23/barrel lower than in the New Policies Scenario. Likewise coal and natural gas prices are much lower in the 450 Scenario than in the New Policies Scenario, with the greatest change being to coal prices, which by 2035 are $42/tonne, or 38%, lower in the 450 Scenario (Table 6.4). Lower demand for oil and gas are also likely to reduce uncertainty in upstream investment and diminish the volatility in oil and gas markets compared with the New Policies Scenario[8] (see Chapter 3 for the implications of deferring investment in oil and gas), while energy security is also enhanced by diversification of the energy mix which reduces import dependence.

Table 6.4 ● **Fossil-fuel import prices in the 450 Scenario** (2010 dollars per unit)

	Unit	2010	2020	2035	Change from NPS in 2035
IEA crude oil imports	barrel	78.1	97.0	97.0	-19%
OECD steam coal imports	tonne	99.2	93.3	67.7	-38%
Natural gas - Europe	MBtu	7.5	9.8	9.4	-22%
Natural gas - Pacific	MBtu	11.0	12.0	12.1	-15%
Natural gas - North America	MBtu	4.4	6.5	7.8	-8%

Note: NPS = New Policies Scenario.

Lower oil-import requirements and lower international oil prices significantly reduce import dependence, and reduce oil-import bills in the 450 Scenario, compared with the New Policies Scenario. Over the whole period, the five-largest importers – China, the European Union, the United States, India and Japan – collectively spend around $7.4 trillion, or more than one-fifth, less than in the New Policies Scenario, while all importing countries as a group spend $9.1 trillion less. These savings increase over time as the impact of efficiency and diversification measures grows and as the difference between oil prices in the different scenarios increases.

In some OECD importing countries, oil-import bills are actually lower in 2035 than in 2010. The oil-import bill in the United States peaks in 2014, at around $350 billion, and declines to some $130 billion in 2035, less than half 2010 levels. The savings for the United States are also very large relative to the import bill in the New Policies Scenario – more than $140 billion in 2035. Among OECD countries, the proportionate impact on the import bill is highest in the United States, but the reduction in other countries is also marked (Figure 6.11). In the European Union, import bills peak around 2015, at $340 billion, and decline steadily to just over $230 billion in 2035, again lower than the 2010 level, while import bills in Japan in the 450 Scenario are also well below their 2010 level by 2035. Nonetheless, the revenues of oil-exporting countries are projected to grow in the 450 Scenario: and OPEC countries' revenues are projected to be almost three-times higher in the period 2011 to 2035 than they were in the period 1985 to 2010, at a cumulative $25 trillion over the period 2011 to 2035. This compares to cumulative revenues of $32 trillion in the New Policies Scenario.

---

8. Lower levels of international fossil-fuel prices do not always imply lower end-user prices compared to the New Policies Scenario. Subsidies removal, the introduction of CO_2 prices, and the shift towards more costly electricity generation options tend to increase end-user prices, cancelling or tempering the decline in international prices at the end-user level.

Figure 6.11 • Oil-import bills in selected regions by scenario

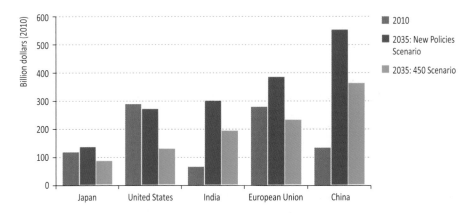

Lower fossil-fuel consumption in the 450 Scenario reduces local pollution. Noxious emissions from the burning of fossil fuels, and particularly fine particles (or particulate matter) have a significant health impact – exposure to these emissions induces increased respiratory disease and shortens life expectancy. A commonly used measure is the number of years of life lost, obtained by multiplying the number of people exposed to the pollutants by its impact in terms of the reduced life expectancy, measured in years. On this basis, fine particles caused the loss of about 1.3 billion life years in China and 570 million in India in 2009. In the 450 Scenario, by 2035 this number drops by 18% in China. In India it increases by 81%, but is still 21% lower than in the New Policies Scenario. This has implications for health-related costs, often borne by government sources, as well as for productivity (IIASA, 2011). The global cost of pollution control, including of particulate matter, reaches around $430 billion in 2035, or just over $125 billion lower than the cost in the same year in the New Policies Scenario. Over the period as a whole, this represents a saving of almost $1.3 trillion, relative to the New Policies Scenario (Table 6.5).

Table 6.5 • Cost of pollution control by region and scenario ($2010 billion)

		New Policies Scenario		Current Policies Scenario		450 Scenario	
	2010	2020	2035	2020	2035	2020	2035
United States	68	86	91	90	100	85	67
European Union	77	107	109	110	121	103	92
Other OECD	51	66	69	67	75	63	52
Russia	7	13	17	13	18	12	13
China	39	87	115	90	133	82	86
India	5	11	28	12	31	11	21
Middle East	10	20	29	21	34	19	22
Other non-OECD	40	74	98	76	106	69	75
World	**296**	**465**	**555**	**479**	**617**	**443**	**428**

Source: IIASA (2011).

Implications of delayed action

This section discusses the implications for the energy sector if action to intensify policy measures and deploy low-carbon technology is delayed, despite adherence to the long-term objective of limiting temperature increase to 2°C. The present slow pace of international negotiations towards a global architecture to tackle emissions makes it important to understand the implications of each year of delay. Most important is the way investment in the built-in infrastructure of energy assets locks in the situation for years to come. Lock-in is a problem that is widely discussed, but not yet widely understood, although increasing attention is being paid to quantification of this issue.[9] Through a sector-by-sector analysis, covering power generation, industry, transport and buildings, we have quantified the emissions projected to come from the energy infrastructure now existing and that under construction – the lock-in.[10] For the first time, we present a quantification of the costs and implications of delaying action. We then analyse the implications for achieving a 450 Scenario of a ten-year delay in the roll-out of CCS. Both analyses point to a clear conclusion: actions being currently taken are insufficient, costs are rising and time is running out very fast.

Lock-in in the energy sector

In the 450 Scenario presented so far, we assume that an international climate agreement is reached quickly and that governments vigorously enact policies – such as CO_2 pricing and support for low-carbon technologies – which are strong enough to steer the energy sector onto a steep decarbonisation path. This new, intensive action is implemented as early as 2013 in all OECD countries and within this decade for other large emitters. Pricing emissions, as in the 450 Scenario, leads to rapid transformation across all sectors and, in some sectors – like the power and industry sectors – the CO_2 price leads to actions such as early retirement, refurbishment, or retrofitting, which are economic under the new circumstances.

The current pace of international negotiations is unlikely to produce an early international agreement, and only a few countries have put in place a clear framework adequate to incentivise investment in low-carbon infrastructure. In this situation, investment decisions will tend to reinforce a carbon-intensive infrastructure which, when CO_2 and other negative externalities are not priced, is often cheaper. The result will be a higher level of emissions from existing infrastructure over a longer period than the 450 Scenario suggests.

Emissions that will come from the infrastructure that is currently in place or under construction can be thought of as "locked-in", because they cannot be avoided without stringent policy intervention to force premature retirements, costly refurbishment and retrofitting or letting capacity lie idle to become economic. They are not unavoidable, but avoiding them does not make economic sense in the current policy context. To examine the consequences, we pose the following questions, sector-by-sector: What quantity of

---

9. A quantification of the extent of lock-in in the energy sector (but over a different time period and under differing assumptions) has been undertaken by Davis et al., (2010).

10. The analysis also covers emissions from fossil-fuel combustion in the agriculture, non-energy use and other energy sectors.

emissions is currently locked-in by existing infrastructure? What quantity of additional emissions will be locked-in for every year of delay in agreeing to an international climate framework? What are the additional costs of this delay?

A country-by-country and sector-by-sector analysis of the capital stock in place and under construction shows that energy-related CO_2 emissions of 23.9 Gt in 2020 (or nearly 80% of 2010 emissions) come from power plants, buildings, factories and vehicles already in place or under construction today (Figure 6.12). Looking further out, the options are, of course, more open, but still notably constrained: by 2035, this figure is 16.1 Gt (or 53% of 2010 emissions). This is under the assumption that over the *Outlook* period the existing infrastructure is allowed to operate as in the New Policies Scenario, *i.e.* in the absence of strong new government interventions stemming from an international climate agreement.[11] That is, once built, infrastructure is allowed to operate for the term of its economic life according to the conditions prevailing in place in the New Policies Scenario.

Cumulative emissions over the *Outlook* period from infrastructure in place and under construction amount to more than 590 Gt. This is 80% of the total emissions from the energy sector over the *Outlook* period consistent with a 450 trajectory. Installations in non-OECD countries account for 57% of these locked-in emissions.

Figure 6.12 ● World energy-related CO_2 emissions from locked-in infrastructure in 2010 and room for manoeuvre to achieve the 450 Scenario

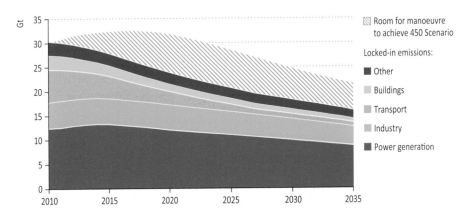

If infrastructure investments continue in line with the New Policies Scenario beyond today, the picture would look even bleaker. If strong action to move away from the New Policies Scenario trajectory were not implemented until 2015, and all infrastructure in place or under construction in the New Policies Scenario at that point were allowed to operate as in that scenario, emissions from locked-in infrastructure would represent almost 95% of the

11. This means that policy interventions going beyond what is assumed in the New Policies Scenario could reduce these emissions. For instance, changes in CO_2 prices, early retirements, retrofitting, refurbishment, mothballing, fuel-switching (*e.g.* from coal to biomass), among others, would reduce the actual emissions below the projected level. Some of these actions would be expected to take place under the 450 Scenario, leaving greater space for emissions from newer, more efficient infrastructure. All analysis of locked-in emissions in this section is based on the assumptions of the New Policies Scenario.

cumulative budgeted emissions of the 450 Scenario to 2035. If action were to be delayed until 2017, all the available headroom for emissions through 2035 would be taken up by the plants and equipment already existing or under construction at that date (Figure 6.13). Expressed another way, no new investment could be made after 2017 in new power generation, industrial capacity additions, new buildings, passenger and commercial vehicles, appliances, space heating and agricultural equipment unless it were zero-carbon. Any new emitting plant or facility installed after this point would require the early retirement of some existing plant or facility to create headroom for the new investment.[12]

Every year of delay of introduction of a global framework with the sufficiently powerful economic incentives to direct investment to follow the path of the 450 Scenario has two consequences:

- It increases the amount of capital stock that will need to be retired early, mostly in the power and industry sectors.

- It limits dramatically the amount of more carbon-intensive infrastructure that can be added in the future.

Figure 6.13 ● World energy-related CO_2 emissions in the 450 Scenario and from locked-in infrastructure in 2010 and with delay

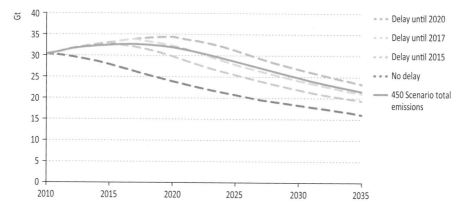

The power generation and industry sectors have the greatest lock-in, due to the long lifetime of power plants and industrial installations.[13] They account respectively for 50% and 21% of cumulative locked-in emissions. The high-carbon infrastructure already in place or under construction will continue to be a source of emissions far into the future unless there are very strong economic incentives for retrofitting or early retirement. Even once such incentives are in place, there would be some lag before there would be an impact on power sector emissions, both because such policies take time to implement and because fossil fuel-fired plants under construction would still come on-stream as planned.

---

12. This is in addition to the early retirements necessary in the 450 Scenario even without delay.

13. This analysis focuses on direct emissions from end-use sectors. Indirect emissions due to electricity demand in end-use sectors are attributed to the power sector, since they are determined by the generation mix. This reduces the rate of lock-in attributed to the end-use sectors, particularly the buildings sector.

If power plants were allowed to operate as in the New Policies Scenario, we estimate that in 2020 around 12.1 Gt of emissions would come from power plants that are already in place today or are currently under construction. The level of locked-in emissions would fall to 8.8 Gt by 2035. Cumulative emissions over the *Outlook* period, just from power plants existing and under construction today, would exceed all the budgeted emissions for the power sector in the 450 Scenario. Delaying the implementation of stringent new measures to 2015 would make returning to a 450 trajectory after that date more costly and difficult, as emissions in the early part of the period would exceed the 450 trajectory laid out in this *Outlook*, necessitating greater abatement later on, meaning even more retirements of power plants before the end of their economic lifetime (Figure 6.14). In order to regain a trajectory consistent with a long-term atmospheric stabilisation at 450 ppm CO_2-eq, investors would have to retire early or retrofit plants emitting 5.7 Gt by 2035 – around 45% of the world's fossil-fuel power generation capacity.

Figure 6.14 • **World energy-related CO_2 emissions in the 450 Scenario and from locked-in infrastructure in 2010 and with delay to 2015 in the power sector**

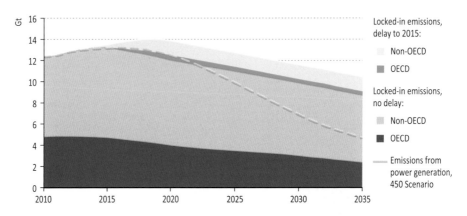

Globally, coal plants account for nearly three-quarters of the emissions locked-into the power sector, and gas-fired power plants for most of the remainder. Installations in non-OECD countries account for 67% of these locked-in emissions. Annual locked-in emissions in China are 4.0 Gt in 2020 (20% above 2009 levels) and they decline slightly to 3.2 Gt in 2035 (or 96% of current emissions). The profile is similar for India – a country, like China, where most of the recent capacity additions have been coal-based – though at an absolute level around one-quarter that of China. Although the efficiency of plants recently installed, particularly in China, is typically higher than the average level of new installations in the advanced economies, the quantity of emissions projected to come from these new plants over the projection period is nonetheless high because of the scale of recent construction. Plants in non-OECD countries are relatively young as most have been built to respond to heightened demand growth during the past two decades. About 63% of installed fossil-fuel capacity in those countries is less than 20 years old, so most of it will not be reaching the end of its technical lifetime before 2035. Reflecting the rate of growth of electricity demand in these countries, 330 GW of fossil-fuel plants are currently under construction. In contrast, plants in OECD countries are ageing, particularly the coal plants that have long provided baseload generating capacity. About 30% of the installed fossil-fuel capacity in 2010 in OECD countries will be approaching the end of its lifetime in the next ten to fifteen years.

Industry is the second-largest source of locked-in emissions. In 2020, annual emissions from capacity currently in place or under construction will be 5.2 Gt (similar to today's levels). By 2035 locked-in emissions will be around 25% lower. Delaying the implementation of new policy action to 2015 broadly increases emissions across the period by around 1 Gt per year (Figure 6.15). Taking either 2011 or 2015 as the point up to which capacity is locked-in, the proportion of locked-in emissions declines only very slowly over the period, due to the long lifetime of capital stock in industry – up to 80 years in some cases, and with an average of around 50 years. This means only capacity installed more than 25 years ago would, in normal circumstances, to be retired over the course of the projection period.

Figure 6.15 • World energy-related CO_2 emissions in the 450 Scenario and from locked-in infrastructure in 2010 and with delay to 2015 in industry

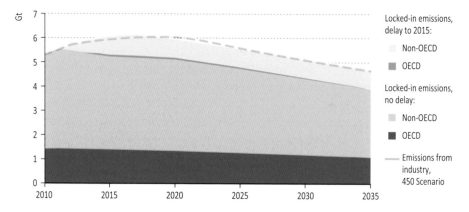

As in the power sector, most of the locked-in emissions arise in non-OECD countries and almost all additional emissions resulting from delay in introducing more stringent standards arise in those countries. This is due to the fact that over the next five years, almost no growth in projected installed industrial capacity occurs in the OECD. By contrast, growth in non-OECD countries is very substantial and is expected to continue after the end of our projection period (IEA, 2010b). We do not, at present, envisage a need for large-scale early retirements in industry in order to achieve the 450 Scenario (although we do expect that refurbishment and retrofitting of existing capital stock, particularly with CCS, would take place in order to improve efficiency and reduce emissions). But because of the long lifetime of capital stock in the industry sector, any delay now could have very severe consequences for the continued cuts in emissions necessary after 2035. Indeed, our analysis suggests that delay until 2015 would mean that installations in place or under construction in 2015 would give rise to emissions equal to the total emissions in the 450 Scenario, leaving no room for manoeuvre in this sector. Since the global location of industrial activities is relatively flexible, the consequence of exhausting the national "budget" of emissions from this sector could well be that activities would simply be switched to countries without CO_2 restrictions. This points to the need for a global sectoral approach when addressing the issue of emissions control in the industry sector.

In the OECD countries, the chemical sub-sector has the highest rate of lock-in and accounts for 34% of the cumulative emissions from industry over the *Outlook* period which are already locked-in. By 2035, chemical plants currently in place or under construction will emit 470 Mt (or 80% of their current level of emissions). Cement has the second-highest rate of lock-in in OECD countries, accounting for one-fifth of cumulative locked-in emissions. In non-OECD countries, cement accounts for 37% of cumulative locked-in emissions. Iron and steel accounts for 21%. By 2035, emissions from iron and steel plants currently existing or under construction in non-OECD countries will be 750 Mt (or 80% of current emissions). Pulp and paper also has a very high lock-in rate, both at the global and regional level, with emissions from plants currently existing and under construction actually higher in 2035 than the current level of emissions from these plants in non-OECD countries. This is due to increased production from these plants in these countries. However, this sub-sector accounts for only 6% of total industry emissions over the period. China has the highest average lock-in across sectors, due to strong growth in recent years. India, by contrast, has relatively low lock-in for most industrial sectors. However, due to the current rapid pace of economic development taking place in India, there is the potential for lock-in to increase rapidly in the next few years, as new capacity additions continue to be made. Other developing Asian economies are expected to have higher levels of lock-in than India, but less than China, mainly because strong growth in production is not expected until later in the period.

Passenger and other road vehicles (rather than long-lived stock like railways) are the main source of CO_2 emissions in the transport sector. Because of their shorter lifetime, locked-in emissions from road transport, which stand at 2.7 Gt in 2020, decline to 850 Mt in 2035. Of the cumulative energy sector emissions over the period which are already locked-in, road transport accounts for 12%. Even with a delay to 2015, there is still substantial room for manoeuvre. However, for the transport sector, the major lock-in is the manufacturing and fuel retail infrastructure in place, and as such it is difficult to capture in this kind of analysis.

The extent of lock-in in the buildings sector is largely dependent on the building shell, as installed equipment, like boilers and heaters, has a significantly shorter lifetime. Thus, the level of locked-in emissions declines steeply initially, leaving a residual level of emissions which changes slowly due to the slow depreciation of building stock. And for this reason, the non-OECD regions, although locked-in initially at high levels of energy intensity, provide an excellent window of opportunity: current rates of development and growth suggest that much investment in buildings and their associated equipment will be made in the near future in these countries. But, by the same token, every year of delay leads to greater lock-in in the basic infrastructure. Emissions from the building sector account for 7% of the cumulative emissions over the period from the energy sector that are already locked-in. As is the case for transport, however, there is a degree of lock-in that exists in urban planning and infrastructure – such as the natural gas distribution networks – that will commit emissions for a very long time, but which is not captured in this analysis.

The cost of lock-in

If action is delayed until 2015, emissions from the power sector will overshoot the trajectory of the 450 Scenario in the early years of the projection period. These additional emissions must be offset by reductions later in the period. The additional abatement could, theoretically, occur in the power sector or elsewhere. Our analysis shows that in most countries the power sector still offers the cheapest abatement. We therefore assume that the additional reductions will be made in this sector. Postponing the compensating abatement until later than 2035 means that emissions would have to become negative, *i.e.* widespread use of biomass generation with CCS would be needed. As this technology is not proven at commercial scale, and therefore cannot be counted upon, we assume that the majority of the additional power sector abatement takes place, over the *Outlook* period, by other means.

Given a delay in action to 2015, containing cumulative emissions from the power sector to a level compatible with stabilisation of the atmospheric concentration of greenhouse gases at 450 ppm CO_2-eq over the *Outlook* period would require the following actions, from the cheapest to the most expensive:

- Retrofitting plants with CCS, when this is more economic than early retirement. In this case, the associated capital cost is the cost of retrofit (and there may be some additional operating costs).

- Shut down of plants that are beyond their economic lifetime, but still safe and profitable to operate, *i.e.* plants for which the initial investment has been repaid. This will reduce revenues compared to the New Policies Scenario.

- Retirement of plants before the end of their economic lifetime, *i.e.* before the upfront investment has been fully recovered. In this case there will be a lost sunk cost, as well as a loss of revenues.

- Additional investment in low-carbon generation.

Delay in action results in some financial savings in the early years of the projection period, relative to the 450 Scenario. While there is increased investment in fossil fuel-based generation, particularly in cheaper, inefficient plants, there would be a reduction in investment in highly efficient and low-carbon plants. Over the decade 2011 to 2020, we estimate that the net effect would be to avoid $150 billion of investment, relative to the 450 Scenario. After 2020, the additional abatement to compensate for higher emissions earlier in the period means that more low-carbon plants and equipment need to be installed, relative to the 450 Scenario, with a net effect of adding $650 billion to investment over the period 2021 to 2035 (Figure 6.16). In other words, for every $1 of avoided investment between 2011 and 2020, either through reduced low-carbon investment or adoption of cheaper fossil-fuel investment options, an additional $4.3 would need to be spent between 2021 and 2035 to compensate.

Figure 6.16 ● **Change in investment in power generation by technology in the Delayed 450 Case, relative to the 450 Scenario**

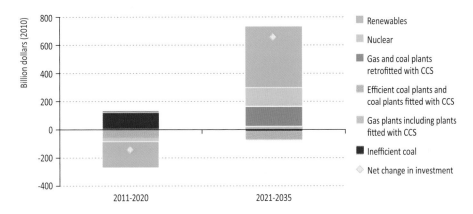

What if CCS does not deliver?

The 450 path set out in this *Outlook* is just one way of achieving an emissions path consistent with long-term stabilisation of the atmospheric concentration of CO_2-eq at 450 ppm. There are, of course, other paths that the energy sector could take in the future that would have the same outcome. What matters for a given climate goal is not the particular choice of policy measures, nor the investment cost – nor, even the specific emissions profile (at least in the absence of major overshooting, which could lead to the kind of feedback effects discussed in Box 6.1) – but the cumulative level of emissions over a certain period. One way of viewing this is to regard the world as having a total emissions budget. At what rate and how this budget is used up does not matter from a climate change viewpoint. What matters is that the budget is not exceeded. A budget of a cumulative 1 000 Gt of CO_2 emitted between 2000 and 2049 would, if respected, give a 75% chance of keeping the global average temperature increase to 2°C or less (Meinshausen *et al.*, 2009). In the 450 Scenario, this budget is exceeded by 2035.

Reducing the probability of success to 50% increases the budget to 1 440 Gt. Since a total of 264 Gt of emissions have already been emitted between 2000 and 2009, 1 176 Gt more can be emitted from 2010 to 2049. The implications for global energy markets of this budget are profound, as it implies that less than half of the remaining proven fossil-fuel reserves can be used over the next 40 years (Figure 6.17). More of the reserves could be used without exceeding the emissions budget if CCS technology becomes available. However, CCS technology has not yet been proven at commercial scale and the level of commitment to demonstration plants leaves doubts about how fast this can be achieved. We examine in this section what it would mean for the energy sector's contribution to attaining the 2°C goal if this crucial technology is not deployed on a large scale until ten years later than we envisage in the 450 Scenario. This would mean no commercial application in the power sector until 2030, leaving only five years of operation within the projection period; and in the industry sector, CCS only beginning to be introduced at any scale by 2035.

Figure 6.17 ● Potential CO₂ emissions from remaining fossil-fuel reserves and in the 450 Scenario, compared with the emissions budget to achieve 2°C

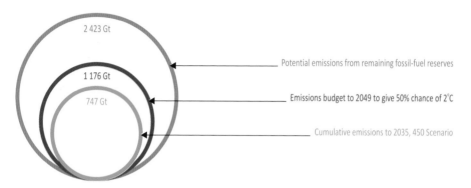

While CCS in 2020 accounts for only 3% of the abatement, relative to the New Policies Scenario to reach the 450 Scenario, by 2035 this share increases to 22% (or 18% across the period). This makes CCS the third biggest source of cumulative abatement, after energy efficiency measures and renewables. With the delay assumed above, CCS would have a much more limited application before 2035, reducing the cumulative abatement contribution from this source to 1% over the period, and leaving 17%, or around 25 Gt, to be achieved by other means.

Delayed CCS 450 Case – Implications for the energy sector

Of the total global capacity of 1 270 GW of coal-fired generation in place in 2035 in the 450 Scenario, 32% of it is CCS-equipped. For gas, it is a much smaller, but still significant, 10% of a total of 2 110 GW of capacity.[14] Around 6 Gt, or 40%, of cumulative abatement in the industry sector over the *Outlook* period comes from CCS. Delay in the commercial availability of CCS would make necessary either a shift to other low-carbon generating technologies or a decrease in energy demand through greater efficiency (or a combination of the two) in order to keep cumulative emissions over the period at the level of the 450 Scenario. In this sensitivity analysis, both turn out to be the case, although energy demand decreases only marginally, with total demand 0.06% lower than the 450 Scenario in 2020 and 1.5% lower in 2035, a cumulative difference of around 1 870 Mtoe over the period. This puts much more pressure on other low-carbon generation options.

In the 450 Scenario, CCS accounts for 18% of cumulative abatement globally. With a ten-year delay in the roll-out of this technology, this falls to under 1%, or around 1.6 Gt. As we assume that the total level of emissions is unchanged, reduced abatement from CCS has to be made good elsewhere. This happens in power generation largely through a shift to renewables (both in power plants and through additional solar photovoltaic generation in the buildings sector) and nuclear. Abatement due to renewables, including biofuels, increases by almost 45% compared with the 450 Scenario, accounting for 50 Gt of the cumulative abatement across

---

14. Due to the higher carbon content of coal, a CO₂ price acts to make CCS-equipped coal plants commercially viable before this is true of gas-fired plants.

the period as a whole. Total abatement from efficiency increases only very marginally, as the 450 Scenario already absorbs almost all of the available efficiency potential in most regions. For comparison, the Low Nuclear 450 Case (Box 6.4 and Chapter 12) sees a smaller increase in the contribution of renewables including biofuels, from a share of 23% of cumulative abatement in the 450 Scenario to 32%. The remainder of the shortfall in this case is made up by expansion of CCS, which delivers more than a fifth of abatement in the Low Nuclear 450 Case (Figure 6.18). In both of these sensitivity cases, cumulative additional investment, relative to the New Policies Scenario, is substantially higher than in the 450 Scenario, increasing by $1.14 trillion in the Delayed CCS 450 Case and by $1.5 trillion in the Low Nuclear 450 Case. This underlines the importance of these two technologies in achieving the 450 Scenario at the lowest possible cost. Investment in the Delayed CCS 450 Case is discussed below.

Box 6.4 ● The implications of less nuclear power for the 450 Scenario

The accident at the Fukushima Daiichi power station in early 2011 has led to a re-evaluation of the risks associated with nuclear power and to greater uncertainty about the future role of nuclear power in the energy mix. In order to assess the impact on our projections of a reduced role for nuclear, we include in this year's *Outlook* a chapter outlining two special cases. The first is a variant of the New Policies Scenario, which we call the Low Nuclear Case, in which nuclear plays a much smaller role in global energy supply. The second examines the implications of those same low nuclear assumptions for achievement of the 2°C global climate objective – the Low Nuclear 450 Case, a variant of the 450 Scenario. The results of those analyses are presented in Chapter 12.

To contain energy sector emissions within the CO_2 budget of the 450 Scenario with a much lower component of nuclear power is a formidable challenge. It would require much greater deployment of other low-carbon technologies. Renewables-based electricity generation capacity in 2035 would need to be almost 20% higher than its already very high level in the 450 Scenario, while CCS-equipped coal- and gas-fired capacity would need to be more than a third higher than in the 450 Scenario. There would be a need for additional cumulative investment of $1.5 trillion over the period 2011 to 2035. In practice, it is by no means certain whether these projections could actually be realised, given the scale of the imposed commercial losses and practical limitations on deploying low-carbon technologies on such a large scale and so quickly.

At a regional and country level, abatement in the Delayed CCS 450 Case mostly follows a similar pattern to the global changes compared with the 450 Scenario. OECD countries see an increase in the importance of nuclear for abatement of about 50% while abatement from renewables including biofuels increases by 33%. The abatement in major emerging markets from these sources also increases substantially, with China's abatement from nuclear rising by nearly 70%. Other developing countries, with much less nuclear to begin with, see a relatively small increase in abatement from this source, instead relying on a large increase in renewables.

It is important to note that the results of the Delayed CCS 450 Case assume extremely rapid roll-out of renewable energy technologies, nuclear generation and electric vehicles. For example, it would require an average annual addition of some 90 GW of wind power between 2011 and 2035, which is 50% more than the total installed capacity in non-OECD countries in 2010 (Table 6.6). Sales of hybrids, plug-in hybrids and electric vehicles would need to be 75% of total sales of PLDVs in 2035, requiring a significant transformation of the infrastructure used to fuel/recharge the cars. Every year, some 27 GW of new nuclear reactors need to be built, necessitating widespread acceptance of such technology. Other technologies which will be called upon, such as large hydro and biofuels, also carry with them issues of social acceptability due to their potential environmental consequences and sustainability issues. It is also worth noting that, given long lead-times in constructing nuclear power plants in particular, the fact that CCS would not be available as planned in the 450 Scenario would have to be apparent early in the period in order for nuclear to reach the necessary level in a timely manner, a degree of foreknowledge and forward planning that might not be available in the real world. Given the fact that achieving the 450 Scenario would already be extremely challenging, it is highly optimistic to assume the feasibility of rolling out technology even faster, and transforming both individual behaviour and urban planning.

Figure 6.18 ● Cumulative share of abatement relative to the New Policies Scenario in the 450 Scenario, Delayed CCS 450 Case and Low Nuclear 450 Case

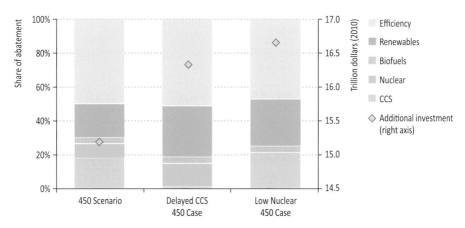

If changes on this scale were to be achieved, demand for individual fuels would change very significantly. Coal demand peaks in the 450 Scenario at around 3 880 Mtoe in 2016, falling to 2 315 Mtoe by 2035; in the Delayed CCS 450 Case it peaks in the same year at the same level, but falls much more steeply to under 1 700 Mtoe by 2035. Although coal is the fuel which – as would be expected – suffers the most dramatic change, gas demand also falls significantly and is 4% lower in 2035 than in the 450 Scenario (Figure 6.19). The biggest increase among the low-carbon fuels is nuclear, demand for which is just over 15% higher in 2035 in the Delayed CCS 450 Case than in the 450 Scenario, while demand for renewables increases by 8%. Since much of this increase in demand for renewables comes from a shift to renewables-based generation in the power sector, the Delayed CCS 450 Case actually sees an increase in coal and gas capacity without CCS. This is because variable renewables require additional flexible capacity, partially provided by coal and gas plants (for which CCS is not available) (see Chapter 5).

Table 6.6 ● Consumption, capacity and stock of selected technologies by scenario

	2010	2035		Average annual additions, 2006-2010	Average annual additions, 2011-2035		
		New Policies Scenario	450 Scenario	Delayed CCS 450 Case		New Policies Scenario	450 Scenario	Delayed CCS 450 Case
Hydro (GW)	1 027	1 629	1 803	1 976	40.0	27.3	34.2	41.2
Wind (GW)	195	1 102	1 685	2 059	32.1	52.1	75.8	90.7
Solar PV (GW)	38	499	901	1 030	10.0	22.1	38.5	43.7
Nuclear (GW)	393	633	865	998	1.7	13.8	21.9	27.2
EV stock (million)*	0	13	140	173	0.0	0.6	6.3	7.6
PHEV stock (million)**	0	18	280	285	0.0	0.8	13.4	13.7
Biofuels consumption (mb/d)	1.3	4.4	7.8	8.4	0.2	0.1	0.3	0.3

*EV= electric vehicle. **PHEV= plug-in hybrid electric vehicle.

Figure 6.19 • Change in global energy demand by fuel in the Delayed CCS 450 Case compared with the 450 Scenario

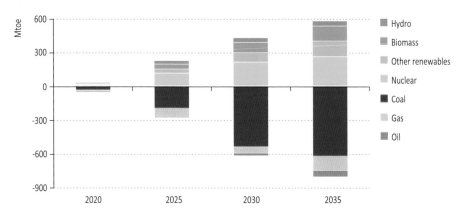

The cost of delaying CCS

Overall, the cost of living within the cumulative emissions budget of the 450 Scenario would be increased by a ten-year delay in the roll-out of CCS by $1.14 trillion (or 8%) over the period 2011 to 2035 (Figure 6.20). These additional costs are weighted towards the power generation and transport sectors because, in the absence of CCS, these sectors are the most cost-effective sources of abatement. Overall, investment in industry actually falls relative to the 450 Scenario because, due to the long lifetime of industrial capacity and the CO_2 intensity of industrial processes, retrofitting with CCS in this sector is by far the largest source of abatement, and was expected to have accounted for around one-third of investment in the 450 Scenario. Without CCS, other abatement options in industry are limited and are costly. The greatest proportional increase in overall investment occurs in the last five years of the period, when investment is 12% higher than in the 450 Scenario.

Figure 6.20 • Additional investment in the Delayed CCS 450 Case

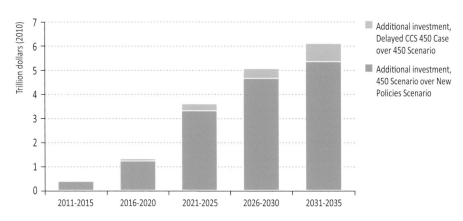

This analysis demonstrates the importance of CCS to cost-effective avoidance of dangerous climate change. Though the effectiveness and commercial acceptability of CCS have yet to be proven (and hard-headed reappraisal will be needed year-by-year for several years yet), intensive investment and effort to demonstrate the commercial viability of CCS is the rational course of action for governments seriously intent on restricting the average global temperature rise to no more than 2°C. Successful demonstration will need to be followed by equally rapid and widespread deployment. This judgement holds even after allowing for the heavy discounting appropriate to more distant investment in alternative solutions.

Regulatory and policy support will be critical to successful CCS deployment. CCS generates no revenues or other market benefits if there is no carbon price. It is both costly to install and, once in place, increases operating costs. To overcome these barriers, effective, well-designed policy is essential.

In addition, the ongoing analysis for the Fifth Assessment Report of the International Panel on Climate Change shows that the majority of new emissions profiles compatible with 2°C would see energy-related emissions becoming negative around 2070.[15] This presumes, in practice, widespread use of biomass with CCS, underlining even further the importance of quickly deploying this technology at a large scale. Any delay in action is likely to make CCS even more important, as retrofitting coal- and gas-fired power plants and industrial installations is the only economically viable way to avoid locking-in these emissions.

---

15. See www.pik-potsdam.de/~mmalte/rcps/.

OUTLOOK FOR RUSSIAN ENERGY

PREFACE

Part B of this Outlook looks at one of the key players in the global energy economy, the Russian Federation.

Chapter 7 starts with a brief overview of energy in Russia and the role of Russian energy in the world. It then sets out the assumptions underpinning our projections, including the macroeconomic and policy landscape, the critical issue of domestic energy pricing and the potential for energy efficiency savings. It also considers the main policy issues and challenges that will affect the development of the Russian domestic energy market and sets out our projections for Russian energy demand to 2035.

Chapter 8 goes into detail on Russian resources and supply potential, focusing primarily on hydrocarbons, but also covering the potential for nuclear power, hydropower and other renewable sources of energy. It assesses the costs and challenges of developing these resources and, on the basis of our projections to 2035, identifies the potential role of Russia as a global energy supplier.

Chapter 9 examines the national and global implications of the trends identified in the previous two chapters. It looks at the role of energy in Russian national economic development and also integrates the outlook for Russia into the broader picture by examining the contribution of Russia to regional and global energy security, the potential development of new trading relationships, notably with China, and the impact of activity in the Russian energy sector on environmental sustainability.

CHAPTER 7

RUSSIAN DOMESTIC ENERGY PROSPECTS

Energy: master or servant of the Russian economy?

H I G H L I G H T S

- Russia is a key player in the global energy economy and the energy policy choices made by the authorities in the coming years will shape not only the prospects for Russia's own economic development but will also have major implications for global energy security and environmental sustainability.

- The energy intensity of Russian GDP has improved in recent years but, even allowing for Russia's industrial structure and harsh climate, energy use in Russia is still highly inefficient. Raising efficiency in each sector of the economy to the levels of comparable OECD countries would save more than 200 Mtoe of primary energy per year, 30% of total demand and an amount similar to the energy used in a year by the United Kingdom.

- Stronger energy efficiency policies and price reforms in the New Policies Scenario start to tap into this energy savings potential, dampening overall increases in energy demand. As a result, the energy efficiency gap between Russia and OECD economies narrows, but it remains significant: energy savings potential in 2035, relative to OECD efficiencies, would still be 18% of total primary consumption.

- Total energy demand rises 28% to reach 830 Mtoe in 2035, at an annual average rate of 1%. Energy use for transportation, grows at the fastest pace, followed by the industry and power sectors. Demand growth in the 450 Scenario flattens after 2015 and averages only 0.4% per year, while in the Current Policies Scenario growth is higher at 1.3% per year.

- Higher domestic natural gas and electricity prices encourage efficiency, but on their own do not lead to major changes in the energy mix: the share of gas falls only slightly from 54% in 2009 to 52% in 2035. Gas demand rises by 0.8% per year on average and reaches 530 bcm in 2035. Oil consumption edges higher to 3.2 mb/d in 2035 from 2.7 mb/d in 2009. Coal use stays within a range of 155 to 175 Mtce per year.

- Electricity output grows to 1 440 TWh in 2035, an average increase of 1.5% per year. Natural gas remains the most important fuel in the power sector even though nuclear and, from a low base, renewables grow more quickly. The ability of a liberalised electricity market to deliver adequate and timely investment will be a key measure of its success; the power sector requires total investment of $615 billion (in year-2010 dollars) during the period to 2035.

- Reform efforts have been directed at the electricity market, but have yet to touch the district heating sector in the same way. Demand for district heating grows only at 0.3% per year as it struggles to compete with own-use boilers (and CHP) installed by industry and with decentralised space heating in apartments and private homes.

Introduction[1]

Russia's energy sector operates on a grand scale. In 2010, Russia was the largest oil producer, the largest producer and exporter of natural gas and the fourth-largest energy consumer (after China, the United States and India). It has exceptional reserves of natural gas, oil, coal, uranium, metals and ores. It has major potential for hydropower and other renewables and, in Siberia, one-fifth of the world's forests. The sheer size of the country and its resources mean that energy policy choices made by the Russian authorities in the coming years will help to shape not only the prospects for national economic development in Russia, but also global energy security and environmental sustainability.

Strong rates of gross domestic product (GDP) growth from 2000 to 2008, supported by the rise in international energy prices, have helped Russia recover from the precipitous economic decline of the 1990s (Table 7.1). Measured by purchasing power parity (PPP), Russia now has the sixth-largest economy in the world. Among the leading emerging economies (the so-called BRICS - Brazil, Russia, India, China and South Africa), it has by some distance the highest GDP per capita. With fossil-fuel prices high and projected to remain so and Russia's standing among resource-holders enhanced by the uncertainty facing some other key producers (see Chapter 3), a bright future for Russia's energy sector might appear almost guaranteed. Yet the challenges facing Russia's energy sector are, in many ways, no less impressive than the size of its resources.

Table 7.1 ● **Key energy-related indicators for Russia**

	Unit	1991	2000	2010*	2000-2010**
GDP (MER)	$2010 billion	1 300	919	1 465	4.8%
GDP (PPP)	$2010 billion	1 973	1 395	2 223	4.8%
Population	million	148	147	142	-0.4%
GDP (PPP) per capita	$2010 thousand	13.3	9.5	15.7	5.1%
Primary energy demand	Mtoe	872	620	687	1.0%
Primary energy demand per capita	toe	5.9	4.2	4.8	1.4%
Primary energy demand per GDP	toe/$1 000 ($2010, MER)	0.67	0.67	0.47	-3.6%
Net oil export	mb/d	4.4	3.9	7.5	8.5%
Net gas export	bcm	177	185	190	0.3%
Energy-related CO_2 emissions	Mt	2 168	1 492	1 604	0.7%

*2009 is the base year for the projections in this *Outlook* and the last year for which an energy balance for Russia was available at the time of writing, but it was an anomalous year because of the impact of the financial and economic crisis. Preliminary energy data for 2010, where available, are incorporated in the projections: they are also presented in this table and elsewhere in Part B where appropriate.

**Compound average annual growth rate.

Notes: MER = market exchange rate; PPP = purchasing power parity.

The oil and gas fields in Western Siberia that account for the bulk of current Russian production are entering, or are already in, a period of long-term decline. In the natural gas sector, this consideration applies particularly to two super-giant, low-cost gas fields

---

1. The analyses in Chapters 7, 8 and 9 benefitted greatly from discussions with Russian officials, industry representatives and experts, notably during an IEA workshop held in Moscow in April 2011.

(Urengoy and Yamburg) that have made a preeminent contribution to the gas balance over the last two decades. The task of compensating for declines from existing fields is complicated by the fact that new production areas are higher-cost, more difficult technically and often even more remote. Uncertainties over the fiscal regime and the pace of demand growth in different markets further complicate decisions on the timing and extent of Russia's investment in new production and transportation capacity.

On the domestic front, Russia needs to modernise its ageing infrastructure and tackle the inefficiency of its energy use. These are recognised by the Russian authorities (Box 7.1) as major policy priorities, but the scale of the task is huge. For the moment, it is unclear to what extent Russia will realise its potential for energy savings and whether it will create domestic energy markets that deliver strong commercial incentives both for investment and for efficient operation and end-use. For the next few years, strong energy exports will be essential to overall economic growth but, looking further ahead, there are competing paths for the Russian economy and the role of its energy resources: is Russia to remain predominantly an exporter of raw materials, highly dependent on the oil and gas sectors for economic growth; or will Russia foster a more broadly based and diversified economy, served – but not dominated – by a market-driven energy sector? The answer to these questions will determine not only the future of Russian energy but also, to a significant extent, the rate of economic growth in Russia and how the benefits of growth are spread among the population.

Trends in energy demand and supply

Recent trends in energy demand reflect the economic upheaval of the 1990s and the subsequent recovery (Figure 7.1). A deep slump in industrial and economic activity in the early part of the 1990s meant a similarly sharp decline in energy use: between 1991 and 1998, Russian GDP declined by 40%, while in the same period domestic energy demand fell by almost one-third. Following the 1998 financial crisis, the Russian economy started a prolonged period of growth which halted, and then reversed, the decline in energy use.

Figure 7.1 ● Primary energy demand in Russia by fuel and GDP, 1990-2009

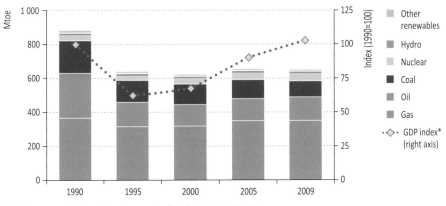

*GDP is measured at market exchange rates (MER) in year-2010 prices.

Figure 7.1 also shows significant divergence between the paths of economic growth and energy consumption growth since 2000. Between 2000 and 2009, GDP grew by more than 50% while energy demand increased by less than 5% as economic output shifted towards less energy-intensive sectors. This represents an improvement of one-third in the energy intensity of Russia's GDP over this period. However, the amount of energy required to produce a unit of GDP (in PPP terms) is, in Russia, still more than twice the average in OECD countries.

On the supply side, Figure 7.2 shows the bumpy ride taken by Russian oil and, to a lesser degree, coal production since the early 1990s, compared with the much more stable levels of output from the gas sector. Continuity in gas supply was facilitated by the commissioning of key super-giant fields brought on stream in the 1980s and by the transformation in 1989 of the Soviet-era Ministry of the Gas Industry into the new corporation, Gazprom. Oil output, which was already stagnating in the late Soviet period, was depressed further in the early 1990s by the fragmentation of productive capacity into multiple state entities, many with only regional scope, that were often plagued by poor management and under-investment. Oil production declined by more than 40% before starting to recover around the turn of the century, as the industry was consolidated and partially privatised.

Box 7.1 ● Policy making and regulation in the Russian energy sector

Given the importance of the Russian energy sector to the national economy, most key decisions about energy policy are taken at the highest levels of government. Below this level, multiple ministries and other executive offices work on the development of energy sector policy proposals and different aspects of policy making.

The Ministry of Energy takes the lead on day-to-day regulation and supervision of the energy sector, but is by no means the only body with responsibilities in this area. Among the other ministries, the Ministry of Natural Resources and Environment is responsible for regulating upstream activities, awarding field licenses and monitoring compliance. The Ministry of Finance answers for fiscal policy, a critical component of the investment climate. The Ministry for Economic Development holds sway over tariff policy in the gas and electricity sectors and co-ordinates energy and energy efficiency policies with the overall priorities for national economic development.

Other important agencies for the energy sector are the Federal Anti-Monopoly Service, responsible for competition policy, the Federal Tariff Service, which sets transportation, transmission and other regulated tariffs and the Russian Energy Agency, which is responsible, under the Ministry of Energy, for promoting and monitoring implementation of energy efficiency policies and measures.

Figure 7.2 ● Energy production in Russia by fuel, 1990-2009

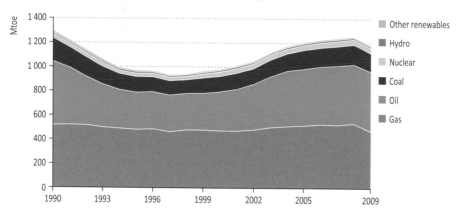

Russia produces far more energy than it consumes internally. Indeed, energy dominates Russia's overall export mix, accounting for around two-thirds of exports by value. Russia exported an average of 7.5 million barrels per day (mb/d) of oil in 2010, around two-thirds as crude oil or natural gas liquids (NGLs) and the rest as refined products. Net exports of natural gas were 190 billion cubic metres (bcm) and net coal exports were 82 million tonnes of coal equivalent (Mtce).[2] The bulk of these exports were transported westward along well-established routes to neighbouring countries and to European markets; however, a small but increasing share of oil, gas and coal is also being delivered to China and the Asia-Pacific region. This incipient shift in favour of fast-growing Asian markets is expected to gain momentum throughout the projection period.

The presentation in this *Outlook* of projections for Russia as a whole should not disguise the fact that there are strong variations between the Russian regions in terms of resources, prices and policy challenges. Energy production is heavily concentrated in specific parts of the country, notably in the Urals Federal District, which includes the autonomous regions of Khanty-Mansisk (almost 60% of oil production) and Yamalo-Nenets (close to 90% of gas production), and in Siberia, which includes the main coal-producing Kuzbass and Krasnoyarsk regions as well as most of Russia's hydropower production and potential (Figure 7.3). By contrast, the Central Federal District, which includes Moscow, has negligible resources and production of fossil fuels. The 37 million inhabitants of this District, in common with others in European regions of Russia, rely on supplies from other parts of the country, delivered over long distances. Transportation costs lead to large differentials between energy prices in different regions, as well as variations in the actual and potential fuel mix.

---

2. Preliminary data for 2010 are provided here to give a more representative picture of current export volumes.

Figure 7.3 ● Share of Russia's population, energy consumption and fossil-fuel production by federal district, 2009

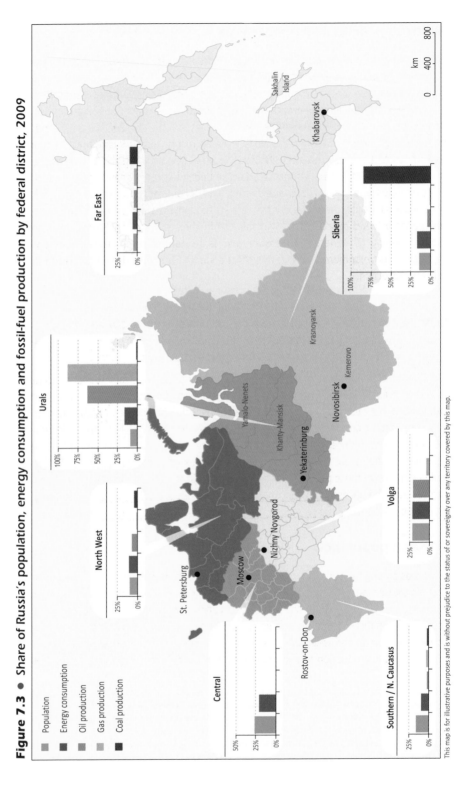

This map is for illustrative purposes and is without prejudice to the status of or sovereignty over any territory covered by this map.

Trends in policies and governance

The experience of economic turbulence in the 1990s has been reflected in trends in energy-sector policies and governance for much of the last decade. With an emphasis often on stability, rather than far-reaching reform, the state maintained or, in some cases, increased its involvement in many key parts of the national economy. However, this trend has not been uniform – the liberalisation of the electricity sector in the mid-2000s is an exception – and it is being challenged.[3] There is a widely shared recognition in Russia that the country needs to strengthen the quality of both state and market institutions if it is to realise its full potential. Survey data[4] suggests that Russia is still perceived as prone to serious problems with corruption and bureaucratic interference which contribute to the costs and risks of doing business. Doubts persist about the respect for contracts and private sector property rights, a factor which discourages investment. A further challenge for Russia, as for many resource-owning countries, is to ensure that large natural resource revenues do not prejudice the development of other parts of the economy or of effective national institutions and good governance.

Key assumptions for the Russian energy outlook

As elsewhere in this *Outlook,* our energy projections for Russia are contained in three scenarios: the Current Policies Scenario, the New Policies Scenario and the 450 Scenario (see Chapter 1 for a full description of the assumptions underlying each scenario). Detailed results are presented only for the New Policies Scenario. This scenario takes into account both existing policies and Russia's policy intentions including, where available, targets for the energy sector and the environment that are set out in national strategy documents and sector programmes, even where the relevant measures or instruments for their implementation are not yet in place. We are cautious in assessing the prospects for full implementation of these policies and targets, bearing in mind the difficulties that often arise with securing the necessary budgetary and financial support and, crucially, in ensuring the optimal performance of the relevant institutions and administrative mechanisms.

GDP and population

Russia was among the countries most affected by the recent global economic and financial crisis: an 8% decline in GDP in 2009 highlighted the risks associated with Russia's exposure to downturns in international commodity prices and energy demand. The economy is now recovering from this sharp shock: a large fiscal stimulus, equivalent to 9% of GDP, cushioned the initial effects of the recession in some sectors and higher oil prices since then have helped to boost production and employment. However, in the view of the International

---

3. An example is the proposal to divest part of the state holding in companies such as the oil giant, Rosneft, the power company, InterRAO, the hydropower company, Rushydro, and possibly even the oil pipeline company, Transneft (although it is not yet clear when, or whether, these plans will come to fruition).

4. Russia ranks 123rd out of 183 countries in the World Bank's Ease of Doing Business index (World Bank, 2011) and 154th among 180 surveyed in Transparency International's Corruption Perceptions Index (Transparency International, 2010).

Monetary Fund (IMF), Russia needs strengthened policies to avoid growth tapering off to less than 4% per year in the medium term, significantly less than the rate of growth seen prior to the financial and economic crisis (IMF, 2011a).

Our GDP growth assumptions for the period to 2015 are drawn from the IMF World Economic Outlook (IMF, 2011b); this means real GDP is assumed to grow at an average annual rate of 4.3% for the period from 2009 to 2015. After 2015, the rate of economic growth is assumed to slow down progressively in the longer term, such that the average for the entire 2009 to 2035 period is 3.6% (Table 7.2). The contribution of the services sector to GDP is assumed to increase steadily over time, continuing the shift seen since 1990. Although Russia could surprise in either direction, longer-term growth is assumed to be constrained from rising higher by a decline in the population, a weak banking sector, relatively low rates of investment in new or upgraded productive capacity and continued concerns over the investment climate and governance. Nonetheless, the annual average figure of 3.6% is higher than the 3% assumption in last year's *Outlook* for the period 2008 to 2035.

Table 7.2 ● **Indicators and assumptions for population and GDP in Russia**

	Population			GDP ($2010, PPP)			GDP per capita ($2010, PPP)		
	2009 (millions)	1991-2009* (%)	2009-2035* (%)	2009 ($ billion)	1991-2009* (%)	2009-2035* (%)	2009 ($)	1991-2009* (%)	2009-2035* (%)
Russia	142	-0.2%	-0.3%	2 138	0.4%	3.6%	15 069	0.7%	3.9%
World	6 765	1.3%	0.9%	70 781	3.2%	3.6%	10 463	1.9%	2.6%
European Union	501	0.3%	0.2%	14 911	1.9%	1.9%	29 755	1.5%	1.7%

*Compound average annual growth rate.

A key medium-term challenge will be to mobilise the investment required to upgrade and renovate Russia's industrial base. Gross fixed capital formation (that is, spending on physical assets such as machinery, land development, buildings, installations, vehicles or technology) is around 20% of GDP (having reached a high point of 22% in 2008), a relatively low figure compared to other emerging economies. Foreign investment has played a relatively minor role: capital inflows rose steadily in the 2000s, bolstered by Russian capital returning from abroad, but – even at its peak of $75 billion in 2008 – foreign investment did not exceed 5% of GDP (MER).[5]

Our GDP assumptions, which are constant across all three scenarios, are generally lower than those that have been used to develop Russia's own strategy documents and development plans for the energy sector. For example, the baseline scenario of Russia's Energy Strategy to 2030 assumes average annual GDP growth of almost 5% per year (Institute of Energy Strategy, 2010). The difference in GDP assumptions naturally has a major impact on energy demand. To allow for a more direct comparison between our projections

---

5. Despite limitations on participation in oil and gas projects, a significant share of foreign investment has nonetheless come to the energy sector: Rosstat data suggest that 30% of foreign investment was in oil, gas and electricity in the period 2004 to 2010.

and those in Russian strategy documents, we have run a high GDP case in line with the more optimistic outlook for GDP growth from the Russian Ministry of Economic Development (the "favourable" scenario for economic growth). The results are presented in Chapter 9.

Demographic trends are a source of political concern within Russia and of uncertainty with regard to the future. Russia's population has fallen by around 4% since 1990, from 148 million to 142 million, although Rosstat data shows that the rate of decline has slowed in recent years. In our scenarios, the Russian population is assumed to continue to decrease, but at a slower rate, reaching 133 million in 2035. This falls between the "low" and "medium" scenarios from Rosstat, which show figures for 2030 of 128 and 139 million, respectively. Over the period to 2035, the share of Russia's population living in urban areas is expected to increase from 73% to 78%. The results of the 2010 Census will enrich understanding of Russian demographics; provisional figures, released in second-quarter 2011, put the total population at 142.9 million in 2010. They also suggest some strong regional variations in trends within Russia, with the largest decline in population seen in the Far East.

GDP per capita (PPP) was $15 100 in 2009, more than twice as high as the figure for China and half the level of the European Union. Projected average growth in GDP per capita to 2035 is 3.9% per year, higher than the headline rate for GDP growth because of the expected decline in the population. As living standards rise, so do our assumptions about residential space per capita, which increases from an average of 22 square metres (m^2) in 2009 to 38 m^2 in 2035, and indicators such as car ownership, which rises from 220 passenger vehicles per 1 000 people in 2009 to 390 vehicles per 1 000 people in 2035.

Energy and climate policies

The Energy Strategy to 2030 (Government of Russia, 2009) provides a detailed overarching framework of long-term policy priorities for the energy sector. The Strategy is supplemented and, in some cases, modified by development programmes for the oil, gas and coal sectors and a similar document for the power sector, adopted in 2008 and then amended in 2010. Investment, efficiency, security and reliability are recurrent themes in the Energy Strategy, which foresees three main changes to the Russian energy balance in the period to 2030: a reduction in the share of natural gas in the primary energy mix to under 50%; an increase in the share of non-fossil fuels in primary energy consumption to 13% to 14% (from 10% today); and a reduction in the energy intensity of GDP. The Energy Strategy adopts a multi-scenario approach to test the implications of different future levels of demand and supply. A comparison with the projections of this *Outlook* is included in Chapter 9.

Market reform has proceeded at various speeds in different sectors of the Russian energy economy. Domestic coal, oil and wholesale electricity markets have been liberalised, while reforms to other parts of the electricity sector (retail and capacity markets) are at an earlier stage, as are reform efforts in the domestic gas sector and in heat supply. The interplay between sectors with different market and regulatory structures is an element of complexity and uncertainty in the Russian domestic energy outlook. We assume that the domestic coal, oil and wholesale electricity markets will remain commercially competitive, with gradual reform in the other areas. We assume that the upstream structure of the oil and gas industry remains dominated by Russian companies (state and private) throughout the projection period (see Chapter 8).

Russia's policies on energy efficiency and intensity and, to a lesser extent, its commitments on greenhouse-gas emissions are important in shaping the outcomes examined in this *Outlook* (Table 7.3). Russia's pledge to the Copenhagen Accord was a 15% to 25% reduction in emissions by 2020, relative to a 1990 baseline.[6] Implementation of this pledge is included in our assumptions, the lower figure, which is considered as a business-as-usual trajectory, in the Current Policies Scenario, a 20% target in the New Policies Scenario and the higher figure, 25%, in the 450 Scenario. These are all achieved, with room to spare, in our projections so they do not impose additional policy constraints in the period to 2020 (see Chapter 9). By contrast, the aim to reduce Russia's energy intensity by 40% to 2020, compared to that of 2007, sets a much higher level of ambition. This target was announced by President Medvedev in 2008 and its achievement would have substantial implications for energy use.

Table 7.3 ● **Main assumptions for Russia by scenario**

	New Policies Scenario	450 Scenario
Electricity and natural gas prices	- Gas prices for industry reach "netback parity" by 2020; gradual above-inflation increase in residential electricity and gas prices.	- As in the New Policies Scenario, but higher rates of increase in residential electricity and gas prices.
Power generation	- State support to the nuclear and hydropower sectors; support mechanisms for non-hydro renewables introduced from 2014.	- Domestic emissions trading scheme post-2020 for power generation. - Stronger support for nuclear and renewables.
Industry	- Efficiency measures driven by prices. - Reduction in share of GDP value-added in favour of the services sector.	- Domestic emissions trading scheme post-2020.
Transport	- Accelerated development of natural gas vehicles.	- Introduction of fuel-efficiency standards.
Buildings	- New building codes, meter installations and refurbishment programmes lead to efficiency gains in space heating 50% above Current Policies Scenario. - Efficiency standards for appliances. - Per-capita residential space increases from 22m² to 38m².	- As in the New Policies Scenario, but a higher rate of efficiency gains in space heating, 150% above Current Policies Scenario. - Stricter efficiency standards for buildings and appliances.
CO₂ emissions	- 20% reduction in 2020 compared with 1990.	- 25% reduction in 2020 compared with 1990.
Oil and gas supply *(all scenarios)*	- The tax regime for oil and gas will succeed in mobilising the necessary investment to allow an appropriate level of exploitation of the oil resources in each region, according to the economic possibilities (see Chapter 8). - The target for 95% utilisation of associated gas, *i.e.* a reduction in gas flaring, is reached in 2014.	

6. The target within this range depends on the extent to which the role of Russia's forests as a carbon sink will be taken into account and whether all major emitters adopt legally binding obligations.

Another target adopted by the Russian authorities for 2020 is to increase the share of renewable energy resources in the electricity mix to 4.5%. For the moment, this is not backed up by legislation or economic incentives and this target is not met in the New Policies Scenario, though we do assume that renewables support mechanisms are in place from 2014, leading to faster growth in renewable energy use later in the projection period. In the 450 Scenario, we see a more concerted effort to promote lower-emissions technologies both in power generation and in other energy-intensive sectors. Many of the policies considered in the 450 Scenario come from the Climate Doctrine Action Plan (Government of Russia, 2011), which was adopted in April 2011.[7] This plan sets out a range of measures for different sectors of the Russian economy, including economic instruments for limiting greenhouse-gas emissions in industry and power generation, which accounts for our introduction of a domestic emissions trading system in the 450 Scenario after 2020.

Energy pricing

Russia has made substantial changes to pricing in both the electricity and gas sectors in recent years, improving the incentives for investment and efficiency. In the New Policies Scenario, we assume that industrial electricity prices remain liberalised and that industrial gas prices follow a path consistent with reaching parity with gas export prices, minus the export duty and transportation costs, in 2020. We assume that regulated residential prices for both gas and electricity increase at a rate above the inflation rate, reducing, but not entirely removing, subsidies to households.

In the electricity sector, full liberalisation of the wholesale market took place in January 2011, ensuring cost-reflective wholesale pricing for industrial consumers.[8] Pricing of electricity for all residential consumers remains government-controlled: regulated electricity tariffs are set by regional energy commissions, within overall boundaries set by the Ministry of Economy and Federal Tariff Service. The system of regulation has been under challenge in 2010/2011 because of significant increases in distribution network charges, leading to the imposition of additional end-user price caps in 2011. While this particular imposition was, arguably, justified, undue and persistent price capping would distort market operation and undermine the prospects for efficient investment.

In the gas sector, the Federal Tariff Services sets wholesale tariffs for natural gas destined for industrial and power sector use; tariffs to residential and municipal customers, as for electricity, are established on a local basis by regional energy commissions. Gazprom is

---

7. The Action Plan is not supported by funding and, in most areas, has more of the character of an agenda for policy research (and possible future implementation), rather than a specific declaration of policy goals that could be considered for inclusion in the New Policies Scenario.

8. The wholesale market does not cover the whole of Russia: there are "non-price zones" (around 5% of total consumption) where, because of limited competition, all prices remain regulated. North Siberia is not part of the unified power network and is excluded from wholesale trade (Solanko, 2011).

required by law to supply pre-negotiated volumes of gas to customers at regulated prices, regardless of profitability. Additional gas can be purchased from Gazprom or independent producers at higher prices; this deregulated sector has grown and now accounts for around one-third of domestic gas supply.

Average natural gas prices to Russian industry have increased consistently in recent years, from $0.4 per million British thermal units (MBtu) in 2000 to $2.8/MBtu in 2010. The stated aim, since 2007, has been to increase industrial prices to "netback parity", a level equivalent to the Gazprom export price to Europe, minus export duty and transportation costs. This is a moving target and in some years even quite significant domestic price increases have not kept pace with the faster rise in the reference price, which is in part linked to oil prices through the contractual conditions of long-term supply contracts to Europe. The initial target date to reach netback parity was 2011, but this has been postponed to 2014-2015 and we assume that a further postponement will occur. Indeed, it is questionable whether the target of netback parity, conceived in an era of lower oil prices, will remain the formal objective of gas pricing policy.

One alternative to netback parity would be a regulated price ceiling for the domestic market. Another possibility, in the event of further domestic gas sector reform, would be a transparent market-driven price, set at a gas trading exchange or hub. A price set on such an exchange would tend to reflect the long-run marginal cost of Russian gas supply (and an efficient regulated price would be set at similar levels). The assumed marginal cost of Russian supply varies, depending on which region the marginal gas is sourced from and according to the applicable tax rate, but our analysis suggests that such an approach would produce an average domestic gas price around $5.5/MBtu (in year-2010 dollars) in 2020. In practice, this is only slightly below the $6.4/MBtu figure produced by a calculation of the netback value, taking the OECD Europe gas import price in 2020 as the starting point. For the purposes of our analysis, we assume that higher average industrial gas prices will remain an objective of Russian policy and we have retained netback parity as a ceiling for future price increases. Average Russian industrial gas prices are assumed to move on a path consistent with reaching this level, *i.e.* $6.4/MBtu, in 2020, following which the pace of real price increases slows to less than 1% per year.

The amount spent on energy in all sectors of final consumption in Russia, expressed as a share of GDP, rose from around 4% of GDP in 2000 to an estimated 11% in 2011 (Figure 7.4). Comparing levels of spending on energy across countries can be misleading because of differences in climate and the structure of GDP, but a comparison of trends over the last ten years can nonetheless be instructive. Russia's energy spending as a share of GDP has converged towards the figures for the European Union and China and, since 2007, has been higher than the figure for the United States, highlighting the impact of price increases for gas and electricity over this period. Moreover, the fact that Russia – still with relatively low average end-user prices – expends more than 10% of its GDP on energy is a pointer to the extent of wasteful energy consumption. This indicates that greater energy efficiency needs to complement further price reforms, in order to mitigate the impact on household budgets and industrial costs.

Figure 7.4 ● Total energy costs as a percentage of GDP

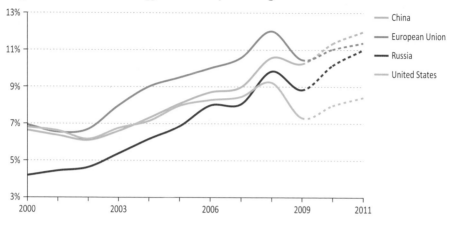

Note: 2010 data are preliminary; 2011 data are estimates.

Energy savings potential

There is greater scope to use energy more efficiently in Russia than in almost all other countries. How Russia uses this potential over the coming decades will shape its energy balance, helping to determine the requirement for upstream investment and the availability of resources for export. The energy intensity of Russian GDP – the amount of energy used to produce a unit of Russian output – has declined since reaching a peak in 1996. However, as the Russian government has recognised (Government of Russia, 2010), this improvement in energy intensity has been due mainly to structural changes in the economy, *i.e.* a drop in the share in GDP of energy-intensive output. Only a relatively small part of the change since 2000, one-fifth, was derived from actual improvements in the efficiency of energy use and, despite this limited improvement, Russia's energy intensity remains among the highest in the world. The country's size, its long, harsh winters and its industrial structure provide a partial explanation for the high intensity of Russia's energy consumption but, even taking these factors into account, the potential for efficiency gains is still enormous.

If Russia used energy as efficiently as comparable OECD countries in each sector of the economy in 2008, it could have saved more than 200 million tonnes of oil equivalent (Mtoe) from its primary energy demand – equal to 30% of its consumption that year and an amount similar to the total primary energy used by the United Kingdom. This snapshot of a more energy-efficient Russia emerges from a detailed and disaggregated analysis of Russia's energy consumption in 2008 relative to OECD benchmarks[9] (Figure 7.5). With

9. 2009 was an anomalous year because of the recession and so 2008 was chosen as the base year for analysis. The same calculations for 2009 showed that the volume of potential savings was lower (slightly less than 200 Mtoe), but the percentage saving, at 30%, was the same. Benchmarking was against OECD Europe in all sectors and sub-sectors except those affected by climate, *i.e.* higher heating needs for buildings, where Finland and Canada were used as points of comparison. The comparison was facilitated by reference to IEA (2010, 2011), UNIDO (2010), CENEF (2008). Analysis by the World Bank (2008) and the Russian Academy of Sciences (2009) measured Russia's energy savings potential relative to best-available energy technologies (rather than against OECD countries who themselves can improve their efficiency in various sectors), and so found even higher savings potentials.

these savings, Russia's energy intensity would still be about 60% higher than the OECD average (or 85% higher than the European Union), due to Russia's more energy-intensive industrial structure and the large share of its population living in areas with high heating requirements.[10] These savings would, though, bring Russia's energy intensity very close to the figure for Canada, which is the OECD country most similar to Russia in terms of average annual temperatures and of the share in GDP of energy and heavy industry. An alternative way of reading this analysis is to conclude that Russia's current level of energy consumption could in practice support a considerably larger economy *i.e.* future economic growth need not be accompanied by increases in energy demand if effective policies promoting energy efficiency are implemented.

Figure 7.5 ● Primary energy savings potential in Russia based on comparable OECD efficiencies, 2008

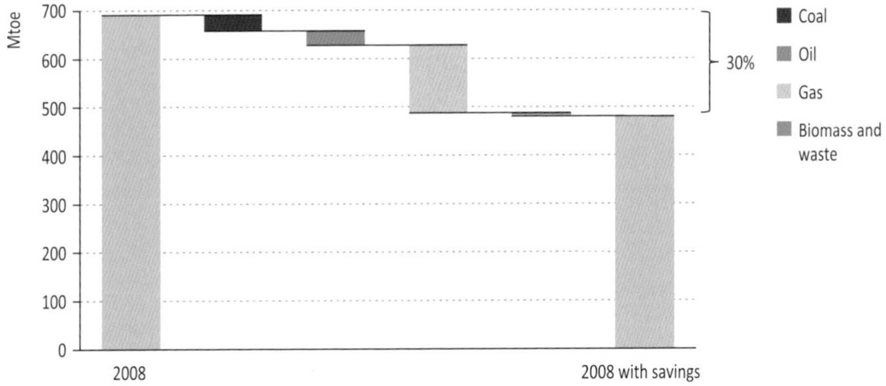

Among primary fuels, the savings potential is equal to almost 180 bcm of gas, 600 kb/d of oil and oil products and more than 50 Mtce of coal. The current international market value of these saved primary resources is about $70 billion, which is 46% of Russian domestic spending on energy in 2008. Final electricity consumption would be 170 terawatt-hours (TWh) lower than today's levels, a reduction equivalent to the annual output from about 75 combined-cycle gas turbine (CCGT) plants.[11] Among final consumers, the greatest potential savings are in the buildings sector (including residential end-uses and services), followed by industry and then transport (Figure 7.6). Large savings of primary fuels are available in the power and heat sectors, as a result of greater efficiency of conversion in generation, more efficient transmission and distribution and also because less electricity and heat need to be generated in order to supply more efficient final uses.

Among the fossil fuels, the largest share of savings is in natural gas. These savings arise mainly from improvements in the efficiency of power and heat generation, and from lower demand for electricity in a more efficient economy, although the buildings sector also

---

10. Comparisons of current energy intensity are expressed as total primary energy demand divided by GDP expressed in PPP terms; future trends are presented using GDP at market exchange rates (MER).

11. Plants of 400 megawatts (MW) each operating at an assumed utilisation rate of 65%.

accounts for a significant share (Figure 7.7). Given the huge scale of Russian gas demand, policies affecting gas use have important implications for the requirement for investment in Russian gas supply and for the availability of gas for export. Potential savings of almost 180 bcm, including those from reduced gas demand and those from reduced gas flaring, are equivalent to the planned plateau output from the three-largest fields on the Yamal peninsula combined (Bovanenkovo, Kharasaveysk and South Tambei). It is also close to Russia's net export of natural gas in 2010 (or 120% of its net export in 2009).

Figure 7.6 ● Energy savings potential in Russia by sector, 2008

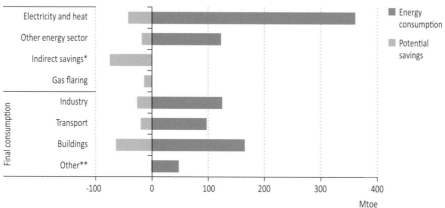

*Indirect savings reflect the additional decrease in primary energy use that accompanies savings in final consumption, *e.g.* 1 MWh of saved electricity in the buildings sector results in savings of the primary fuels that would have been used to generate that unit of electricity.

**Includes agriculture and non-energy use.

Despite some selective investment in more efficient energy use in Russia, notably in export-oriented industrial sectors, thus far the progress made in reducing the wasteful energy practices inherited from Soviet times has been slow. Prices have yet to reach a level sufficient to generate widespread efficiency improvements, meaning in many cases that while the technical potential for savings is there, energy efficiency investments face long payback periods and uncertain rates of return. A related issue is the poor availability of data and insufficient communication, leaving households and companies either unaware of the potential gains from investing in efficiency or underestimating their value. When suitable energy efficiency investments are identified, Russian capital markets are often non-responsive to the opportunities. There is also a relative scarcity of energy efficiency expertise, both within energy-using institutions and to support growth in the fledgling energy services sector.

The policy and regulatory framework to support energy efficiency improvements has evolved rapidly since 2009. Prior to then, and with the exception of a brief burst of national activity in the late 1990s, there were no systematic efforts to implement energy efficiency policies at national level. There were some initiatives taken by municipal and regional authorities, but energy efficiency measures by industry and households that were implemented were motivated in part by rising gas and electricity prices or, more often, by the imperative to replace older pieces of machinery and equipment as they reached the end of their useful life. However, President Medvedev's definition of energy efficiency as a strategic priority and, in particular, the target to reduce energy intensity by 40% in 2020 (relative to 2007), has given new impetus to national policy.[12]

Figure 7.7 ● **Natural gas savings potential in Russia, 2008**

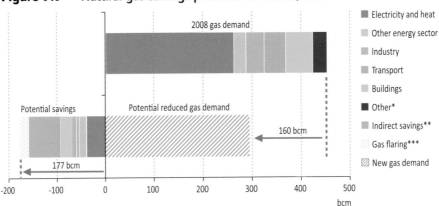

*Includes agriculture and non-energy use.

**Indirect savings reflect the additional decrease in primary gas use that accompanies savings in final consumption.

***Gas flaring, a non-productive use of gas, is not a component of gas demand and therefore reductions in this area are reflected only in the total potential savings.

The main measures promoting more efficient energy use are contained in the 2009 framework law on energy efficiency and the State Programme for Energy Saving to 2020 (Government of Russia, 2010), which was adopted in late 2010 and is in the early stages of implementation. Although there are still gaps in policy as well as in the institutional capacity to implement policies effectively, there are now measures in place or under development covering compulsory energy metering by industry and households, energy efficiency standards for appliances, energy efficiency building codes and standards, compulsory energy audits for large energy consumers and mandatory reductions in specific energy consumption in public buildings. Moreover, there is committed federal support for the development and implementation of regional energy efficiency programmes and a system of federal guarantees for energy efficiency programmes put in place by large enterprises.

Where mechanisms are in place to realise its provisions, the energy efficiency law is considered in the Current Policies Scenario. Partial implementation of the State Programme is taken into account in the New Policies Scenario and fuller implementation in the 450 Scenario. Aspects of the energy efficiency programme that rely on regulation, such as

---

12. The IEA prepared energy efficiency policy recommendations across 25 fields of action for the G-8 and evaluated the progress of G-8 countries against these benchmarks (IEA, 2009). Russia was assessed to have "full and substantial implementation" in only 10% of these areas and in around one-third there was no sign that implementation was planned or underway. However, an updated assessment conducted by Russia's Centre for Energy Efficiency (Bashmakov, 2011) showed significant progress by Russia since 2009 in adopting the relevant regulatory acts.

stricter building codes, are considered more likely to have an additional impact on energy demand than those parts (for example in relation to industrial energy use), which rely on limited state funds to act as a catalyst for mobilising much larger amounts of private investment.[13] In calibrating our assumptions about the rate of efficiency improvement in the New Policies Scenario, we take into account the institutional capacity and expertise required for full implementation of an energy efficiency programme, as well as the record of falling short of targets set by a previous energy efficiency programme.[14]

Russian domestic energy outlook

Overview

Russian total primary energy demand expands progressively both in the New Policies Scenario (at an average pace of 1% per year from 2009 to 2035) and in the Current Policies Scenario (at a faster rate of 1.3% per year). By the end of the projection period in 2035, Russia's energy consumption in the Current Policies Scenario exceeds the amount used in its first year of independence (910 Mtoe versus 872 Mtoe in 1991). However, at 830 Mtoe, energy consumption in the New Policies Scenario in 2035 is well below the 1991 figure. In the 450 Scenario, total primary energy demand increases much more modestly, at only 0.4% per year (Figure 7.8), reaching 720 Mtoe in 2035, 20% lower than in the Current Policies Scenario.[15]

Figure 7.8 ● **Total primary energy demand by scenario**

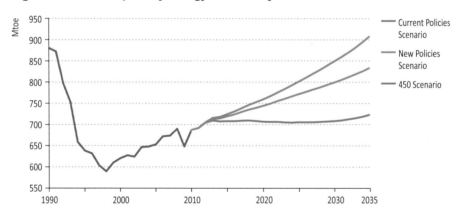

13. Total federal funding for the energy efficiency programme for the period 2011 to 2020 is envisaged at 70 billion roubles ($2.5 billion), out of a total estimated cost of 9 500 billion roubles ($340 billion), less than 1% of the total. Regional budgets are expected to contribute a further 6.5% of the total; but the overwhelming majority (the remaining 93%) is expected to come from non-budgetary funds, *i.e.* commercial or international lending or investment by industry and households.

14. The "Federal Targeted Program for an Energy Efficient Economy for the Period 2002-2005 with an Outlook to 2010" that was adopted in 2001.

15. Growth rates are lower if the period is calculated from 2010 to 2035 so as to exclude the demand effects of the 2009 recession; in this case, the annual average growth in energy demand falls to 0.8% in the New Policies Scenario, 1.1% in the Current Policies Scenario and 0.2% in the 450 Scenario.

Combined with the projected gradual decline in the Russian population, these trends produce notable differences in energy demand per capita in the different scenarios: starting from 4.6 tonnes of oil equivalent (toe) per capita in 2009, the 450 Scenario shows a slight increase in per-capita consumption over the projection period to reach 5.4 toe in 2035; the New Policies Scenario is higher at 6.3 toe in 2035; while in the Current Policies Scenario the level reaches 6.8 toe over the same period. For comparison, the respective indicator for the European Union in 2009 is 3.3 toe per capita, a figure that remains stable throughout the period to 2035.

Energy savings

In the New Policies Scenario, higher energy prices, an improved policy and regulatory framework and progress with implementation of the State Programme for Energy Saving to 2020 are all expected to hold back energy demand growth. The results vary by sector,[16] but cumulative savings of energy over the projection period in this scenario, relative to the Current Policies Scenario, are 715 Mtoe, more than one year of current energy consumption. These savings also lower the overall energy bill for Russian consumers (Box 7.2).

Box 7.2 ● Counting the benefit of increased energy efficiency

Properly designed energy efficiency measures save both energy and money. The efficiency measures taken into account in the New Policies Scenario, even if they do not realise Russia's full potential for energy saving, lead to a substantial reduction in the energy bills paid by Russian industry and households. Cumulative spending on energy in Russia in all end-use sectors is $230 billion less (in year-2010 dollars) in the New Policies Scenario than in the Current Policies Scenario from 2010 to 2035. The annual savings rise through the projection period; by 2035 they amount to over $25 billion per year, almost 1% of projected GDP (MER). These savings arise despite the assumption that domestic energy prices rise more slowly in the Current Policies Scenario. The largest share of the benefit goes to households, services and agriculture, which together reap more than $100 billion; industry saves a further $90 billion and there are also savings from greater efficiency in energy used for transport.

At the same time, the impact of the policies and measures considered in the New Policies Scenario is not sufficient to take advantage of all the potential efficiency gains. A sector-by-sector analysis of the efficiency of Russia's energy use in 2035 in the New Policies Scenario, compared with the levels achievable if efficiency levels matched those projected for OECD countries, reveals that, in 2035, Russia would still have potential savings of almost 150 Mtoe, or 18% of projected primary energy consumption (Figure 7.9). The level of achievement in the New Policies Scenario is a clear improvement on the situation in 2008, when savings relative to OECD efficiencies were estimated at 30% of primary energy consumption, but it falls short of Russia's aspirations.

---

16. Details on policies and projections by sector (power and heat, industry, transport and buildings) are included later in this chapter.

Figure 7.9 ● Primary energy savings potential in Russia based on comparable OECD efficiencies in the New Policies Scenario, 2008 and 2035

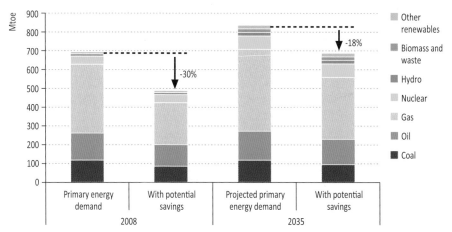

Notes: The data for 2008 are as presented in Figure 7.5. The data for 2035, based on the New Policies Scenario, are a simplified version of the same analysis, *i.e.* using OECD benchmarks for most sectors except those directly influenced by climatic factors, where indicators from Canada and Finland were used for comparison.

The relatively modest pace of energy efficiency improvement projected in the New Policies Scenario is due in part to the assumed rate of GDP growth, which does not allow for a rapid renewal or replacement of the Russian capital stock. It is also due to some remaining gaps in policy. Despite the ambition of some of Russia's policy objectives – notably the 40% reduction in energy intensity that is targeted by 2020 – all the measures that would be required to reach them have yet to be put in place or even, in some cases, identified. An example of the latter is the transport sector, which has seen the fastest pace of energy demand growth in recent years but where there are, for the moment, no standards for fuel efficiency or other efficiency measures under consideration. Our analysis suggests that to realise a greater share of Russia's energy efficiency potential, including achievement of the 2020 target for energy intensity, would require further market-driven reforms, both inside and beyond the energy sector, and enhanced efforts to more fully reflect costs and externalities in energy prices.

An even more substantial challenge is one shared by many countries around the world: to see that policies are implemented in an effective way. Russia has put in place a policy and regulatory framework for energy efficiency at a quick pace since 2009, but there are no guarantees that the returns on this investment will be as rapid. Experience from OECD countries suggests that the different regulatory, institutional, financial and behavioural obstacles to energy efficiency improvements are not easily removed (IEA, 2009). Despite efforts to develop Russia's institutional capacity and expertise on energy efficiency, notably in the Russian Energy Agency, this process is still at a relatively early stage and will require a sustained commitment of human and financial resources. Early evidence suggests that some important aspects of the energy efficiency strategy, for example, the regional energy efficiency programmes and the industrial energy audits, are making progress but they are running behind the schedule originally envisaged. Monitoring and evaluation of policies, a crucial element of any successful energy efficiency strategy, is hindered by gaps in the energy data.

The speed with which Russia meets its energy intensity targets does not depend only on energy prices or policies. The rate of economic growth and the rate of structural change in the economy away from energy-intensive industries will both have a profound effect on the ratio of energy demand to GDP. As noted above, this was the main reason for the fall in energy intensity observed in the period 2000 to 2008. Our assumption that GDP growth will be slower in the years ahead than in 2000 to 2008 also holds back the pace of this structural change. This puts greater weight on actual improvements in the efficiency of use in order to deliver the desired reduction in energy intensity. In the New Policies Scenario, the 40% reduction in energy intensity that is targeted by the Russian authorities for 2020 is achieved in 2028 (Figure 7.10); it is reached in 2025 in the 450 Scenario.[17] While the projected decline in Russian energy intensity narrows the gap with OECD countries and the European Union, the rate of the improvement in Russia is slower than the rate anticipated for the other main emerging economies (the BRICS countries). Russia cuts the energy intensity of its GDP by half between 2009 and 2035, while, over the same period, the reduction in energy intensity in the other BRICS is 56%.

Figure 7.10 ● **Primary energy intensity in Russia and other selected regions in the New Policies Scenario**

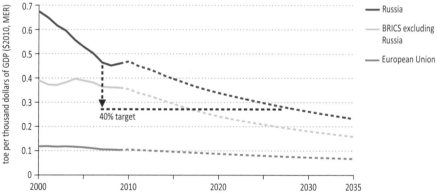

Note: The BRICS countries are Brazil, Russia, India, China and South Africa.

Domestic energy trends by fuel

Fossil fuels are by far the most important source of domestic energy supply in Russia and this is projected to remain the case throughout the New Policies Scenario. In 2009, fossil fuels accounted for 90% of total primary energy supply, down marginally from 93% in 1991;

---

17. This target would be achieved sooner with faster expansion of less energy-intensive sectors such as light industry and the services sector. The implications of higher GDP growth for the energy sector are examined in Chapter 9.

this decline is projected to continue during the period to 2035, reaching 85% in 2035 as the shares of nuclear and renewables gradually increase (Figure 7.11).

Figure 7.11 ● Primary energy demand in Russia by fuel in the New Policies Scenario

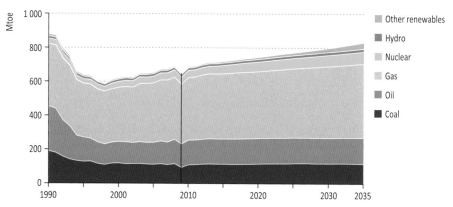

Among the fossil fuels, a major strategic issue highlighted in the Energy Strategy to 2030 is the desirable share of natural gas in the primary energy mix. It increased from 43% in 1991 to 54% in 2009 and Russia's domestic gas consumption has rebounded almost to the levels of the early 1990s, while domestic oil and coal demand remain at just over half of the 1991 figures. Gas demand growth in the 1990s was encouraged by domestic pricing policies that kept gas prices low, in large part to manage the social and industrial impact of the post-Soviet recession, while those for coal and oil were liberalised. Increases in natural gas prices since 2000 were designed to reign in the rise in gas consumption, while also supporting the necessary supply side investments in new gas production. However, natural gas consumption has remained buoyant in Russia, increasing by an average of 1.4% per year from 2000 to 2010. The target set in the 2009 Energy Strategy is to reduce the share of gas in the fuel mix from 54% to between 46% and 47% in 2030.

In the New Policies Scenario, natural gas consumption grows at the more modest rate of 0.8% per year on average over the period from 2009 to 2035, reaching 530 bcm in 2035. Annual demand for coal remains within a range from 155 Mtce to 175 Mtce over the projection period, while demand for oil moves higher to 3.2 mb/d. Comparing the shares by fuel in the energy mix (using our projections for 2030, not 2035) with those targeted in the Energy Strategy to 2030, the share of natural gas in our *Outlook*, at 53% in 2030, is substantially higher than the figure targeted by the Energy Strategy. As a consequence, the shares of both coal and oil are lower than their respective targets in the Energy Strategy: 19% for oil (versus a target of 22%) and 15% for coal (versus 18% to 19%).

Our analysis suggests that, even as natural gas prices rise, gas will continue to have a strong competitive position in Russia (Box 7.3) and this underpins its continued prominence in the fuel mix. The shift in relative prices would need to be larger in order for coal or other competing fuels to make significant inroads into the market share of gas. Gas has the advantage of being the incumbent fuel across large parts of European Russia, where most

Russian industrial and residential demand arises. Where price differences are marginal, gas remains the preferred fuel for new equipment in industry and power generation because of its flexibility and environmental performance. Coal is not considered a plausible competitive fuel for residential use and there are tough constraints on the competitive position of coal as an alternative fuel for industrial use or power generation, most notably the long distance and the high transportation costs between the main areas of coal production in Siberia and the main centres of demand in the European part of Russia. To an extent, the same consideration applies to hydropower, where the largest share of potential is also concentrated in Siberia (see Chapter 8).

Box 7.3 ● Higher gas prices, efficiency and fuel switching

Rising industrial gas prices in Russia will have implications for the efficiency of Russian gas use and will affect also the attractiveness of gas compared to other fuels. At certain price levels, it would make sense for the big consumers of gas – in industry and power generation – to switch to alternative fuels: the main competing fuel is domestic coal. The average industrial gas price in Russia in 2010 was $2.8/MBtu. In the New Policies Scenario, this price increases to $6.4/MBtu in 2020 ($230 per thousand cubic metres; all prices in year-2010 dollars) as it reaches an equivalent level to the assumed Russian export price, minus export duty and transportation costs. Real price increases after 2020 are much more modest, at less than 1% per year. The average delivered price of steam coal to industry remains at between $50 to $60 per tonne throughout the projection period, less than half the average post-2020 gas price on an energy-equivalent basis.

While this would appear to suggest that coal will be in a good position to supply a portion of the domestic market currently using natural gas, a closer look reveals that substitution of gas by coal may in practice be more limited. The present average price of steam coal is kept low by the predominance of coal deliveries over relatively short distances from the main areas of coal production in Siberia. But if coal is to take appreciable market share from gas, it needs to compete further away in the European part of Russia that has the largest concentrations of population and industry (the Central and Volga Federal Districts in Figure 7.3). Transporting this coal across the country pushes up the delivered cost substantially. With incremental transportation costs by rail of at least $30/tonne, the price of steam coal to industry or power plants in the heart of European Russia rises to $80 to $90/tonne.

At these price levels, the benefits of choosing coal over natural gas are less evident, noticeably in the power sector, as coal-fired power generation achieves a lower conversion efficiency than natural gas. Analysing the long-run marginal costs of Russian electricity generation, we estimate that, at $85/tonne of steam coal in European Russia, gas prices would have to rise to $7.5/MBtu for coal-fired power to be competitive. In the New Policies Scenario, average natural gas prices in Russia do not reach this level at any point in the projection period. As a result, there is little fuel switching from gas to coal in the power sector (and no sign of this in the end-use sectors).

The current contribution of non-fossil fuels to the Russian energy balance is dominated by nuclear power, which meets 7% of primary energy demand; the share of renewable energy sources is small, at 3%, most of which consists of hydropower (2%). Over the period to 2035, all of these energy sources are projected to grow at faster rates than those of the fossil fuels. Nuclear power expands at an average of 2% per year and use of renewable energy grows at 4% per year, with non-hydro renewables growing quickly from a very low base. As a result, the combined share of nuclear and renewables in the New Policies Scenario rises from 9% today to 15% in 2035 (13.5% in 2030, in line with the 13% to 14% targeted by the Energy Strategy).

Domestic energy trends by sector

The increase in Russian energy demand since 2000 has varied strongly by sector. Annual consumption in the buildings sector (including residential and services) fell by an average of 0.9% per year over the last ten years; energy use in industry was flat over the decade; the largest contribution to overall demand growth, by some distance, came from the transportation sector, where consumption grew by 3.1% per year.[18] Electricity and heat production also showed divergent trends; electricity output increased by almost 2% per year, while district heating supply fell by an annual average of 1%.

In the New Policies Scenario, energy demand in the transport sector is projected to continue to grow at the fastest rate, albeit at a slower pace, averaging 1.3% per year (Figure 7.12). This is followed by energy use for industry and for power and heat generation. The slowest rate of energy demand growth is in the buildings sector (including residential and services), reflecting the very high potential for energy saving in this sector and the impact, in part, of new efficiency policies.

Figure 7.12 ● **Incremental energy demand by sector and fuel in the New Policies Scenario, 2009-2035**

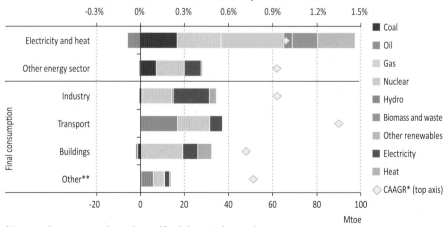

*Compound average annual growth rate. **Includes agriculture and non-energy use.

---

18. Figures are provided for the period 2000 to 2010, using preliminary data for 2010; growth rates for the period 2000 to 2009 are lower because of the impact of the economic and financial crisis.

Electricity and heat

The electricity sector in Russia is the fourth-largest in the world after the United States, China and Japan and, together with the extensive heat supply network, constitutes the backbone of the Russian economy. The electricity and heat systems, linked through the widespread installation of combined heat and power (CHP) plants, face some similar challenges with ageing infrastructure but have, in practice, followed quite different paths over the last two decades. Electricity demand has returned almost to the levels of 1991 and the structure and operation of the industry have been transformed by a major, albeit incomplete, market-driven liberalisation. Production of district heating, by contrast, is around 40% below 1991 levels, growth prospects are tepid and, although most CHP plants are now in private hands, there has been relatively little progress in designing or implementing reforms to meet the particular challenges of the heat sector.

The reform of the Russian electricity market, launched in 2003, is expected to have a substantial impact on Russia's energy sector and longer-term economic performance. This was one of the most ambitious electricity sector reforms undertaken anywhere, and involved restructuring the RAO UES electricity system and executing the largest electricity sector privatisation to date. Generating capacity of 100 gigawatts (GW) was sold to new owners, including major Russian companies (Gazprom, SUEK, Lukoil) and foreign ones (Enel, E.ON, Fortum). If Russia can make progress with effective regulation and relieve some pressing constraints on inter-regional transmission, there is the possibility of a competitive wholesale spot market covering European Russia, the Urals and most of Siberia.

Achieving and maintaining a competitive environment for power generation will be a challenging task. Although liberalisation put a large share of thermal generation capacity into private hands, the state still owns or controls more than 60% of total capacity and this figure has crept up in recent years. Nuclear and most hydropower assets are state-owned and an additional share of thermal power plants is being brought under the control of majority state-owned companies, notably Inter RAO and Gazprom.[19] This balance could be affected by new privatisation efforts, for example, in the hydropower sector. However, the current trend towards consolidation within state-owned or state-controlled entities makes independent, objective regulation and supervision indispensable if a really competitive wholesale market structure is to flourish.

Another challenge for Russian policy makers over the projection period lies in the cross-linkages between different parts of the energy market that are at different stages of development, including the links between wholesale and retail electricity markets, heat supply, arrangements for new generation capacity, and the markets for different fuels (primarily natural gas). Further development of the electricity sector could be a catalyst for wider energy-sector reforms, but this outcome is far from certain. The market orientation of the wholesale electricity sector stands out from the approach to other sectors of the

---

19. This trend was reinforced in mid-2011 with the announced merger of Gazprom's power assets (36 GW), which include the Mosenergo company supplying the capital city, with those of privately held IES Holding (16 GW). IES will get a 25% stake in Gazprom Energoholding, whose post-merger 52 GW would represent one-third of Russia's thermal generation capacity.

energy economy and the government could be tempted to intervene in the power market in ways that could damage the incentives for efficient investment, operation and end-use in the longer term.

The total installed capacity for electricity generation in Russia is around 225 GW, of which over two-thirds are thermal power plants, a further 21% are hydropower and 11% are nuclear. Slightly more than half of the thermal plants are CHP, although precise classification of the capacity shown in Figure 7.13 is complicated by the fact that most electricity-only plants also produce and market small amounts of heat.[20] Gas-fired power plants (electricity-only and CHP together) make up 44% of total capacity.

Figure 7.13 ● **Breakdown of installed electricity and CHP capacity in Russia, 2009**

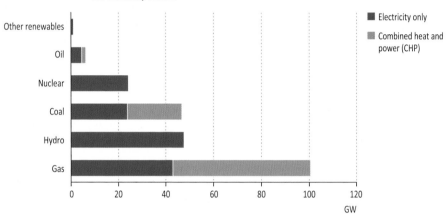

In the New Policies Scenario, generation capacity increases to 280 GW in 2035 and total electricity generation in Russia is projected to grow by an average of 1.5% per year between 2009 and 2035, reaching 1 440 TWh (Figure 7.14). Current annual final electricity consumption per capita, at around 5 megawatt-hours (MWh), is close to current levels in the European Union (5.5 MWh) and is projected to grow at around 1.9% per-year, overtaking per capita consumption in the European Union in 2017. Natural gas remains the most important fuel for power generation; gas-fired power output rises from 470 TWh in 2009 to 630 TWh in 2035, a 44% share of the total. Coal-fired power output stabilises at around 225 TWh. The use of fuel oil for power generation tails off almost entirely.

The fastest pace of growth in electricity output comes from nuclear power and also, later in the projection period, from non-hydro renewables. Considering figures for 2030 (to allow for comparison with the goals set by the Energy Strategy to 2030), the shares of nuclear power (19%), hydropower (15%) and other renewables (4%) in electricity generation in 2030 are close to the objectives set by the Energy Strategy (in percentage terms rather than absolute

---

20. IEA statistics classify almost all thermal capacity as CHP because of its heat output, but for the purposes of modelling the sector we have adopted the split between power, *i.e.* electricity-only plants, and CHP shown in Figure 7.13, based on the type of technology used at the different plants.

levels of output; see Chapter 8 for more detail). As a result, the share of non-fossil fuels in total power output rises from 34% to 38% in 2030, in line with the Strategy's target figure.

Figure 7.14 ● **Electricity generation by fuel in Russia in the New Policies Scenario, 1990-2035**

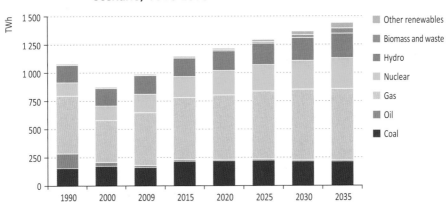

The ability of the electricity system to deliver efficient investment will be essential to its success. The post-1990 decline in power demand meant that there has been relatively little new thermal or nuclear capacity brought online in Russia over the last twenty years (Figure 7.15). This has left Russia with a stock of power generation assets that is significantly older than that in OECD Europe (and radically different to that of China, where rapid population and economic growth mean that the majority of generation capacity has been built within the last ten years). The high average age of Russian installed capacity also means low average efficiency: according to IEA data, the average thermal efficiency of gas-fired power generation in Russia (excluding CHP) is 38%, compared to an average of 49% in OECD countries and up to 60% for a new CCGT plant, which is the best-available technology. From another perspective, the fact that a lot of power generation capacity needs to be replaced is an opportunity for Russia, given a supportive regulatory framework, to improve rapidly the efficiency and environmental performance of the sector. In this sense, Russia's "room for manoeuvre" is much greater than that of many other leading industrial economies (see Chapter 5).

New investment is currently secured to 2018 through a contractual obligation placed on purchasing parties under the privatisation process. This administrative approach was a justifiable way to ensure stable operation of the system, particularly in the absence of well-developed financial markets and experienced regulators. However, it lacks flexibility and reduces the scope for innovation. The government is now reviewing the capacity regime and examining options for securing investment beyond 2018, with the intention of moving to a more market-based approach that will deliver efficiently timed, sized and well located generation investment at least cost. This will require additional improvements to the regulatory framework. Policy makers will also need to consider the implications of a changing profile of annual electricity demand over the next decades, in part because of a likely increase in demand for summer cooling (Box 7.4).

Figure 7.15 • Age profile of installed thermal and nuclear capacity in Russia, comparison with selected countries and regions, 2010

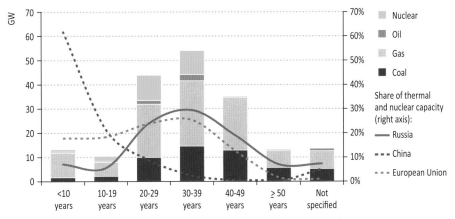

Sources: Platts World Electric Power Plants Database, December 2010 version; IAEA (2011).

Box 7.4 • Keeping Russia cool: heat waves and demand for air-conditioning

The summer of 2010 was the hottest on record in Russia: in the European part of Russia and the southern Urals the variation from average summer temperatures was over 6°C (RosHydroMet, 2011). While not quite as hot, 2011 has continued the run of summers well above trend temperatures. As Russia gets wealthier and as a changing climate potentially pushes average summer temperatures higher, demand for cooling is likely to grow. Ownership of air conditioners is still relatively low, at six per 100 households, according to Rosstat data for 2009, but the market has been increasing rapidly.

The effects on seasonal electricity consumption of increased demand for cooling can be seen across the countries of southern Europe, all of whom have experienced a large increase in summer electricity demand over the last decades. In Italy, ownership of air conditioners rose by 15% per year between 2001 and 2009, from 11 to 34 units per 100 households (sales were also given a boost in Italy by an especially intense heat wave in 2003). This has contributed to a summer peak in electricity demand that is now regularly higher than peak demand in the winter.

It is highly unlikely that the seasonal profile in Russia would change to this extent, electricity demand during the Russian winter is around 30% higher than during the summer months. Nonetheless, even a slight flattening of the demand profile through the year would have implications for the operation of the electricity system. Russia would tend to need more baseload generation relative to other capacity and schedules for plant repair and maintenance, traditionally conducted in the summer months, would need to be reviewed. Demand for cooling would also contribute to a more general trend, the increased share of electricity demand, relative to heat demand, that will affect decisions on new capacity and limit the scope for new CHP plants.

In the New Policies Scenario, we estimate that Russia will require total investment of $615 billion (in year-2010 dollars) in the power sector between 2011 and 2035, of which more than $250 billion (40%) is in transmission and distribution and the rest, $360 billion, is in generation (Figure 7.16). Investment in generation is driven mainly by the anticipated retirement or renovation of existing capacity: based on the age of existing plants, we assume that over 80% of the existing thermal electricity generation capacity will be replaced or refurbished during the projection period, including 64 GW of thermal electricity-only capacity and 68 GW of CHP. A competitive power market is assumed to deliver new thermal generation capacity at efficiency levels close to the best-available technology and this is projected to result in a significant improvement in the overall efficiency of the Russian electricity sector: electricity output from thermal power plants in 2035 is more than 30% higher than in 2009, but the fossil fuel inputs required to produce this electricity rise by only 5% over the same period.[21] Compared with other sectors of the Russian energy economy, the electricity sector, and electricity-only generation in particular, does the most to narrow the gap with projected efficiencies in OECD countries over the period to 2035.

In the transmission and distribution sector, two-thirds of the total investment requirement is to replace or refurbish old infrastructure, while one-third is to meet growing electricity demand. As in the generation sector, there has been little investment in transmission and distribution networks since Soviet times – especially in distribution systems which are ageing and in urgent need of critical maintenance. We estimate that, as of 2009, 3.7% of the Russian transmission and distribution network reached 40 years of age, significantly higher than the global average of 1.6% or the 1.8% figure for OECD Europe (see Chapter 5). This is one of the reasons why around 11% of the electricity generated in Russia is lost in the transmission and distribution system; these losses have increased from 8% in 1990 and are now considerably higher than the average figure of 6% in OECD countries, even after allowing for the increased transmission distances across Russia.

Figure 7.16 ● **Cumulative power sector investment in Russia by type in the New Policies Scenario, 2011-2035**

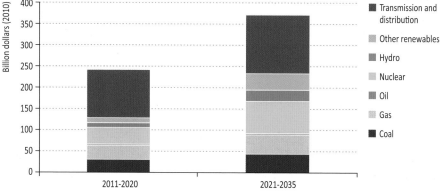

---

21. This includes fossil-fuel inputs to electricity-only plants as well as an estimated share of the fuel inputs to CHP plants that are used for electricity generation.

Reform efforts directed at the electricity market have yet to touch the heat sector in the same way, making the future of district heating in Russia a major uncertainty (Spotlight). Consumption of heat produced by centralised plants – CHPs and large boiler plants – fell during the 1990s and, in contrast to the rebound in electricity demand after 2000, continued to decline thereafter (Figure 7.17). Disconnections from the district heating system and reductions in consumption by existing consumers – particularly by large industrial users – outweighed increased demand due to new housing construction. We project that heat supply will remain relatively flat over the projection period to 2035 in the New Policies Scenario, rising only by an average of 0.3% per year.

Figure 7.17 ● District heat supply by sector in the New Policies Scenario, 1990-2035

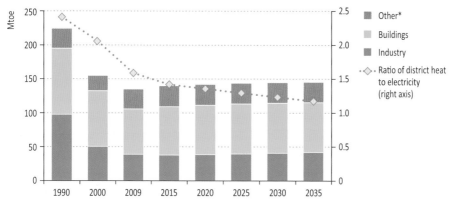

*Agriculture, non-energy use and other energy sector, including losses in heat supply.

The heat sector is one of the biggest product markets in Russia and, as the consumer of around one-third of Russia's primary energy, it is a critical sector from an energy saving and environmental perspective. Data on the heat sector is generally poor, but IEA data puts total deliveries from centralised heat supply sources in 2009 at 5 650 petajoules (PJ) (135 Mtoe or 1 570 TWh), around 1.6 times higher in energy terms than the output from the electricity sector. More than 40% of this heat is produced by around 500 CHP plants, 50% by heat-only boilers and the balance by industrial and other sources.[22]

We estimate that between 70% and 80% of the housing stock is at present covered by district heating and that almost 3 billion square metres of heated floor area are connected to centralised systems; this is currently the primary source of heat for around 100 million people. However, the market for centralised heat supply is being squeezed by industry moving away from urban areas and by the rising popularity of new single-family homes away from city centres. A combination of poor levels of service and increasing prices (particularly for industry) is also driving away the sector's existing customers: in Chelyabinsk, for example,

---

22. Heat generated by industry and households for their own use does not appear as heat production in IEA statistics, only heat generated and sold is included.

over the period 1992 to 2002, more than 660 MW of heat load was detached from the district heating system, as industrial customers installed their own boilers. Some better-off residential customers are also opting for decentralised space heating in apartments. These trends exacerbate a number of the problems facing the district heating sector. Suppliers are left with a relatively poor customer base that is even less able to cover the costs of much-needed investment. For CHP plants, it reduces the efficiency of their output by cutting the heat load relative to electricity demand.

······················· S P O T L I G H T ······················

What future for district heating in Russia?

A large share of district heating and CHP plants is often considered a hallmark of efficient energy use, but this is not the case in Russia. The sector currently supplies inflexible and generally low quality heating services at increasing tariffs (although, in most cases, still below cost recovery) and with a high level of wasted energy. One objective of the Russian Energy Strategy to 2030 is to develop an advanced heat supply system, with CHP expanding its share of the market at the expense of heat-only boilers. This would be in line with trends in the rest of the world. However, in practice, heat supply from CHP plants in Russia has fallen substantially since 1991 and has yet to show signs of recovery. Although CHP offers great potential for efficiency, compared with separate heat and electricity production, market conditions are not supportive and the sector as a whole remains starved of investment. Small-scale and micro-CHP units are likely to be widely deployed by individual enterprises or isolated communities, but a continued divergence in performance and prospects between a liberalised power market and an unreformed heat sector is unlikely to favour the commissioning of new large-scale CHP plants.

A first priority for Russian policy is to install the meters and controls that will provide accurate information on heat production and consumption and allow for a real calculation of costs and expenditures (as well as allowing consumers to regulate their heat use). The creation of a commercially oriented district heating sector will also require movement on the sensitive issue of tariffs and tariff methodologies so as to allow them to reflect the full costs of efficient supply and to remove cross-subsidies. But an equally significant challenge to reform of the district heat sector (which has moved ahead with the adoption of a law on heat supply in 2010) is the absence of a single ministry or federal entity having overall responsibility for an industry consisting of thousands of heat suppliers, distributors and local municipalities.

There are large potential savings in Russia's heat supply systems, both in heat generation at CHP plants and boilers and, in particular, in the heat supply networks. The efficiency of Russia's CHP plants and heat boilers is well below that of the best technology plant used internationally. According to the Ministry of Energy, 80% of Russian boilers are over 30 years old (20% are over 50 years old) and there are high, and increasing, losses from an ageing network of almost 200 000 km of heat supply pipes across the country. These losses are difficult to assess with any accuracy, because of poor metering and data, but are put at 19% of total heat production in the Energy Strategy to 2030. Losses tend to be even higher for deliveries to households, because of the longer distribution network, reaching 30% of heat supply and beyond, compared with 5% to 15% in Finland.

Our projection for centralised heat supply in the New Policies Scenario is based on the assumption that reform efforts in the heat sector will continue to be hesitant and subject to strong regional variations. In the aggregate, investment is sufficient to keep urban district heating systems operational, but the marginal increases achieved in the efficiency of heat supply are not expected to win back customers or to reverse the decline in market share – the trend towards decentralised heat provision in industry and the residential sector continues apace. This means a growing discrepancy between the heat and power sectors, which would be felt first and foremost by the regional electricity companies that supply both products. As shown in Figure 7.17, the ratio of district heat to total electricity demand is projected to fall substantially over the period to 2035, implying a contracting share of CHP in Russia's electricity production.

Industry

Russian industry consumes around 125 Mtoe of energy per year, 29% of total final energy consumption, with gas, electricity and heat accounting for 76% of this amount. The main industrial energy users in Russia are producers of iron and steel (29%), chemicals (23%), non-metallic minerals (12%), such as cement and non-ferrous metals (5%), such as aluminium. The energy intensity of Russian industrial production has halved since reaching a peak in the mid-1990s. Increased use of productive capacity, closure of the most inefficient plants and implementation of some energy efficiency measures have all contributed to this improvement, but a sector-by-sector analysis reveals that Russian industry still uses a great deal more energy per unit of production than its international counterparts.

The iron and steel industry is a case in point. This sector has undergone significant structural changes over the last two decades: for example, the share of Russian steel production from open-hearth furnaces, the most inefficient technology, declined from over 50% in 1990 to under 10% in 2009 – a faster shift away from this technology than seen in Ukraine (World Steel Association, 2000, 2010). Even so, we estimate that the energy savings potential in Russia remains significant, at 5.3 gigajoules (GJ) per tonne of steel produced (IEA, 2010).[23] Raising the efficiency level of Russian iron and steel production to OECD levels would result in savings of 5.3 Mtoe from total energy consumption in this sector of around 35 Mtoe. For the industrial sector as a whole, the estimated savings relative to OECD efficiencies are 27 Mtoe, or 21% of current consumption.

In the New Policies Scenario, industrial energy consumption grows by 27% over the period 2009 to 2035, at an average rate of 0.9% per year. This is less than the rise in overall industrial output, an intensity improvement that is helped by a gradual shift towards lighter manufacturing. The structural shift, together with a swing in relative gas and power prices in favour of power,

---

23. This is based on a comparison with best-available technologies; the savings considered for our energy savings analysis, relative to OECD benchmarks, are less.

pushes up industrial demand for electricity, which grows at an average of nearly 2% per year, becoming by 2035 the largest single direct energy input to the sector (Figure 7.18). Industry's use of district heat is expected to remain relatively flat, as manufacturers build more on-site co-generation plants and boilers (for which gas remains the preferred fuel).

Figure 7.18 ● Industry energy demand by fuel in Russia in the New Policies Scenario, 2000-2035

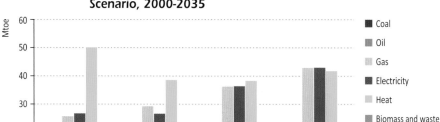

Energy efficiency improvements in the industrial sector are expected to be driven by the introduction of new machinery and technologies, as existing industrial assets reach the end of their useful life. The speed of this turnover will depend in part on energy prices, particularly for those industrial sectors that are competing on export markets. However, in many cases the share of energy in overall production costs is relatively low and it is not axiomatic that price increases alone will be sufficient to lead to more widespread investment in new technologies. This is the logic behind the provision in energy efficiency legislation for mandatory energy audits for all industrial enterprises spending more than 10 million roubles per year on energy ($350 000 at mid-2011 exchange rates), a provision that applies to around 150 000 companies. The audits and the energy "passports" issued as a result are intended to raise awareness of energy saving opportunities and propel companies into adopting efficiency projects but, thus far, implementation is proceeding relatively slowly.[24] The energy services market, which could help companies take advantage of energy saving opportunities, is still relatively undeveloped, held back by some technical and legislative barriers as well as poor access to longer-term finance.

Transport

The transport sector in Russia accounts for 21% of final energy consumption, 90 Mtoe. It has been the fastest growing of all the end-use sectors, with energy demand increasing at an average of 3.4% per year from 2000 to 2008, before declining by 8% in 2009. Overall energy use in transportation is divided between passenger and freight transportation (65% of the total) and the energy used for oil and gas transportation. In the New Policies Scenario,

---

24. The first round of energy audits for eligible industrial companies should be completed by the end of 2012. This requirement also applies to all utilities, energy suppliers and public bodies and authorities. The Russian Energy Agency is making efforts to accelerate this process, but given a relatively slow start and the large number of eligible companies the deadline may need to be pushed back.

energy demand in the transport sector grows by 1.3% per year (Figure 7.19). The sub-sectors responsible for the largest amounts of the additional energy consumed are pipelines and road transportation (33% and 28%, respectively), but the fastest rates of growth in demand come from railways (3.5% per year) and domestic aviation (2.9% per year).

Figure 7.19 ● **Energy consumption in the transport sector by type in the New Policies Scenario, 2000-2035**

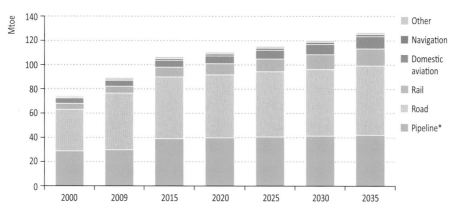

*Pipeline energy consumption refers to the energy used for oil and gas transportation.

The Russian car market has grown very quickly over the last ten years, particularly taking into account the decline in the population. According to Rosstat, the number of privately-owned passenger vehicles increased from under 20 million in 2000 to over 31 million in 2009 (from under 10 million in 1991). Russia was one of the fastest growing light-duty vehicle markets in the world before the market was hit hard by the economic crisis: sales were down by almost 50% in 2009, compared with the previous year. A recovery is underway, but the prospects for consistently high rates of market growth are constrained by long driving distances between Russian cities and the relatively slow pace of infrastructure development; this is expected to hold back increases in energy demand for road transportation after 2015.

Russia aims to double the length of its roads by 2030 and to expand the rail network and domestic aviation and to make more efficient use of inland waterways (Government of Russia, 2008). The current density of the national road network in Russia is low by international standards, because of the country's huge size; at 776 000 km, the total length of Russian paved roads is 20% less than the figure for Japan, even though Russia's land area is 45 times larger. There is also a growing disparity between vehicle sales growth and infrastructure development: the size of the Russian vehicle fleet increased by 60% in the years from 2000 to 2009, but the road network increased by only 3% over the same period. In the New Policies Scenario, car ownership rises steadily through the projection period, to around 390 cars per 1 000 people in 2035, but the increase is slower than during the period from 2000 to 2009, when the number of cars per 1 000 people increased from 140 to 220 (Table 7.4).

Table 7.4 ● Passenger light-duty vehicle ownership in selected countries in the New Policies Scenario (vehicles per 1 000 people)

	2000	2009	2020	2035	2009-2035*
Russia	140	220	300	390	2.2%
China	4	30	110	300	9.5%
European Union	430	475	520	560	0.7%

*Compound average annual growth rate.

We estimate that the average on-road consumption of the Russian car fleet is over 13 litres per 100 km. If Russia were to raise efficiency levels to match those of OECD Europe (where the average efficiency is about 8 litres per 100 km), this would result in an annual saving of around 12 Mtoe (nearly 240 kb/d) in oil products. In the New Policies Scenario, in the absence of specific policies to encourage fuel efficiency or renewal of the vehicle stock, Russia moves relatively slowly to tap these potential gains. We project a gradual increase in the average efficiency of the Russian car fleet, reaching 8 litres per 100 km by 2035. The improvement is due to broader international trends in car design and efficiency, transferred to Russia in the shape of imported cars and foreign brands for local assembly,[25] that influence also the fuel economy of Russian brands. The trend towards greater efficiency is helped by a shift away from premium high-performance vehicles towards cheaper, more efficient models for larger-volume markets.

The use of natural gas vehicles in Russia is projected to increase rapidly, helped by support from Gazprom and regulations that keep the price of compressed natural gas (CNG) below the gasoline price. There are currently around 100 000 natural gas vehicles in Russia and 250 CNG filling stations. Gazprom is supporting a steady expansion of the CNG infrastructure and this is expected to boost natural gas use in road transportation by an average rate of 13% per year. In the New Policies Scenario, sales of natural gas passenger vehicles rise to more than 200 000 per year by 2035 (10% of total Russian sales). Starting from a low base, overall natural gas consumption in road transportation is projected to reach 3 bcm of gas per year in 2035, amounting to a share of around 5% of the total energy used for road transportation.

Energy demand for oil and gas pipeline transportation is projected to increase gradually over the projection period by around 1.4% per year, driven by a rise in gas production and exports and the resultant growth in the overall length of the pipeline transportation system. We estimate that potential savings from greater efficiency in gas transportation are around 6 Mtoe (7 bcm). Gazprom has been investing in replacing inefficient equipment and compressor stations (more than $5 billion in the period 2006 to 2009, according to Gazprom's annual reports). However, the available data does not show that the amount of energy used to transport Russia's gas is declining: for the period 2000 to 2009, the amount of energy required to move 1 million cubic metres along 1 km of pipeline has remained stable,

---

25. According to Rosstat, non-Russian brands account for more than a third of the Russian vehicle fleet and over half in the two largest cities, Moscow and St. Petersburg.

at around 20 kilogrammes of oil equivalent. This suggests that while Gazprom's investment may be counteracting the natural deterioration in performance as existing equipment gets older, it is not yet increasing the efficiency of the system as a whole.

In the gas sector, much of the attention and investment is focused on the high-pressure transmission network, but a greater concern, from the viewpoint of efficiency and environmental impact, is the much longer gas distribution system. Gas distribution losses are around 6 bcm, according to IEA data (but the real figure may be considerably higher). Based on earlier research (IEA, 2006), we make a conservative estimate that up to 12 bcm per year could be lost, due to leakages, across Russia's gas network as a whole. In the distribution network the investment that would be required to reduce leakages is difficult to justify on economic grounds, even with carbon credits, since the leaks occur across more than half-a-million kilometres of pipes.

Buildings

The buildings sector (including residential and services) accounts for 35% of Russian final consumption, a total of 147 Mtoe in 2009, making it the largest energy end-use sector. Consumption of heat makes up the largest share of energy use (45%), followed by natural gas (30%) and electricity (16%). Most of the energy is used for space and water heating. The combined share of energy used for cooking, appliances and lighting is estimated to be less than 20%. Energy use in buildings has fallen since 1991, at the rapid rate of 2.3% per year, although the rate of decline has slowed since 2000. Even so, the building stock remains one of the main sources of potential energy efficiency savings available to Russia. Comparing Russian residential energy use with Canada, the OECD country with the closest average temperatures to Russia, we found that more than twice as much energy is used to heat a square metre of residential space in Russia (Figure 7.20).[26] This finding matches evidence from projects implemented in Russia, for example a World Bank project in Cherepovets during the late 1990s that retrofitted 650 buildings and reduced heat demand by 45% (World Bank, 2010). Total potential savings in the buildings sector are 61 Mtoe, of which almost 47 Mtoe are in space heating.

A challenge for Russia in raising the efficiency of its residential housing stock is to accelerate the refurbishment rate of existing buildings. This rate has been slowing in Russia over recent decades. In the 1970s to 1980s, major repairs were carried out on 3% of total housing stock each year, but this indicator had fallen to 0.6% by 2009. In multi-family buildings it is difficult to co-ordinate building-wide efficiency initiatives among the multiple owners and then very hard for homeowners' associations to obtain credit for capital improvements: easing access to finance for such improvements in the buildings sector will be vital to achieve Russia's efficiency goals.

A range of new policies and measures, not all yet backed up by mechanisms for implementation, are assumed in the New Policies Scenario to moderate the increase in

7

---

26. The calculation was adjusted for the different composition of the housing stock in the two countries, the majority of residential buildings in Russia are multi-family homes rather than single-family, whereas the opposite is the case in Canada.

energy use in buildings and could – if extended and implemented in full – have a profound impact on the Russian buildings sector.[27] These include stricter building codes for new buildings, new appliance efficiency standards and the phase-out of incandescent lamps, requirements to renovate the existing building stock and incorporate energy efficiency improvements, and the obligation to install meters in all buildings for electricity, heat, water and natural gas. There is a specific policy emphasis on efficiency in the public sector; as of 2010, for example, all state entities have been obliged to reduce their specific energy consumption of energy and water by at least 3% per year, for the five years to 2015.[28] In some areas, these initiatives at federal level follow in the footsteps of existing measures enacted by individual regions or cities (Box 7.5); in most cases, though, they represent a substantial new direction for Russian policy.

Box 7.5 ● As efficient as... Moscow?

The construction boom in Moscow over the last decade has added over 3 million m² of residential space to the housing stock each year; the rate of renovation of the existing building stock has also been higher than elsewhere in the country. The capital has not only grown, it has also become richer: over the period from 2000 to 2009, Moscow's real GDP increased by over 6% per year (8.5% per year excluding 2009). Yet this increase in wealth and living space is not reflected in the indicators for energy and water use. Water consumption fell over the same period by an average of almost 4% per year, while hot water consumption fell by 1.4% per year. Residential demand for heat was flat and gas consumption by households declined marginally by 0.3% per year. Of all the main indicators, only residential electricity use showed an increase.

The performance of Moscow stands out for two reasons: because of the concentration of construction activity and wealth in the capital and also because Moscow, along with regions such as Tatarstan and Chelyabinsk, has been among the front-runners in promoting more efficient energy use in Russia in the buildings sector. Acting ahead of federal legislation, Moscow introduced a new building code in 1994, tightened the requirements again in 1999 and was one of the first movers in promoting the installation of electricity meters. As well as providing incentives for efficiency improvements, metering revealed, in some cases, that actual consumption was well below the level that had been assumed and billed.

Measures such as these are now being adopted more widely – as of 2009, 53 regions of Russia, out of 83, had introduced building codes stipulating energy efficiency standards – and policy efforts are now underpinned by stricter federal regulation as well. Successfully replicating and surpassing Moscow's efforts across the country would be a notable achievement for energy efficiency in Russia.

---

27. We assume that these measures are implemented in part (or in some cases, with a delay) in the New Policies Scenario (Table 7.3).

28. This type of measure raises the broader question about the design of some of Russia's energy efficiency measures, whether they find the right balance between administrative measures and financial incentives to encourage more energy efficient choices.

In the New Policies Scenario, policy action is not strong enough to continue the decline in energy use in the buildings sector experienced since 1991. Overall energy demand for buildings (including residential and services) grows at 0.7% per year on average, pushed higher by demand in the services sector, which expands at 1% per year, and by the assumed increase in living space per capita. There is an improvement in the efficiency of energy use in the residential sector: the average amount of energy required to heat a square metre of floor space goes down from 0.023 toe in 2009 to 0.018 toe in 2035 (Figure 7.20). However, the projected pace of this improvement still leaves Russia using 50% more energy to heat a square metre of floor space in 2035 than Canada used in 2009.

Figure 7.20 ● **Efficiency of energy consumption for space heating in the residential sector, 2009-2035**

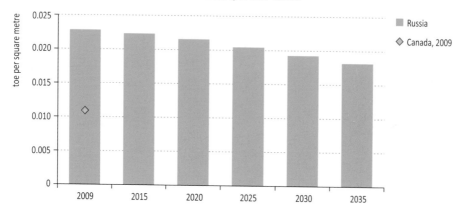

CHAPTER 8

RUSSIAN RESOURCES AND SUPPLY POTENTIAL

Raising the next generation of super-resources?

H I G H L I G H T S

- Russia has world-class energy resources sufficient to underpin its continuing role as a major producer and exporter throughout the period to 2035 and beyond. Our projections of fossil fuel supply are not limited by availability of resources, but rather by the long time scales and technical challenges involved in developing new fields in remote areas, while production from existing fields is declining.

- In the New Policies Scenario, oil production plateaus around 10.5 mb/d for the coming five years before starting a slight decline that takes output to 9.7 mb/d in 2035. Oil exports decline from 7.5 mb/d in 2010 to 6.4 mb/d in 2035. There is an important shift to new, higher-cost production areas in Eastern Siberia, the Caspian and the Arctic. Natural gas liquids also play a growing role.

- We assume that the tax regime will provide sufficient incentives both for the development of new production areas and for continued investment in the core region of Western Siberia. However, if the prevailing effective tax rate in the traditional producing regions were maintained, our projections for production (and export) would be around 1.8 mb/d lower in 2035.

- Gas production increases from 637 bcm in 2010 to 860 bcm in 2035. Net gas exports rise substantially from 190 bcm to close to 330 bcm in 2035. Production from the Yamal peninsula becomes the new anchor for Russian gas supply, helping to offset the expected declines in other parts of Western Siberia and to meet demand growth, alongside output from the Barents Sea and Eastern Siberia. A larger share of gas output is expected to come from companies other than Gazprom, but both gas and oil production remain dominated by Russian state and private companies.

- Coal production rises to 270 Mtce in the mid-2020s and then starts a slow decline. Domestic demand for coal is flat over the period and export levels start to tail off in the latter part of the *Outlook* period, as global coal demand plateaus and Russian coal struggles to compete.

- Russia has large-scale plans to expand the role of nuclear power and hydropower. Although our projections for new nuclear capacity are lower than these official plans, output from nuclear plants still expands by two-thirds over the period to 2035, at a rate of 2% per year. Large hydropower increases at 1% per year. The contribution of non-hydro renewables increases at the fastest rate, but remains small relative both to other fuels and to the large potential.

Overview

Russia has world-class energy resources: 13% of the world's ultimately recoverable resources of conventional oil, 26% of gas and 18% of coal (Figure 8.1), as well as clear potential to increase its output from renewable sources of energy. Although often located in remote regions with harsh climates, these resources underpin Russia's continuing role as a major energy producer and exporter throughout the projection period and beyond. Production is not expected to be limited by access to capital (though this could be improved), but by the long time scales and technical challenges involved in developing new fields in remote areas at a time when production from existing fields will be declining.

Figure 8.1 ● *Russian share in global resources, production and export of fossil fuels, 2010*

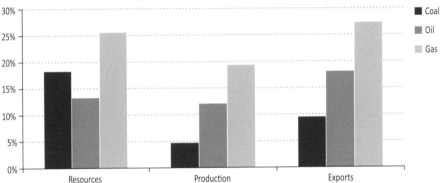

Notes: Only conventional oil and gas resources have been taken into account. If unconventional gas resources are included, Russia's share of gas resources drops to 22%. Data for production and exports is from 2010; the figure for exports is a share of net exports from all WEO regions.

Source: IEA databases.

In the New Policies Scenario, Russian oil production plateaus around 10.5 million barrels per day (mb/d) for the coming five years and then declines slowly to 9.7 mb/d in 2035 (Figure 8.2). Net exports of crude and refined products are reduced as internal demand gradually increases, particularly from the transport sector; this means that the high water mark for Russia's net oil exports is projected prior to 2015. Over the projection period, the centre of gravity of Russian oil production shifts further east, away from the traditional production areas in Western Siberia. There is an accompanying re-orientation of export flows in favour of fast-growing Asian markets.

Gas production increases from 637 billion cubic metres (bcm) per year in 2010 to 690 bcm in 2020 and 860 bcm in 2035 (Figure 8.3).[1] Moderate growth in domestic demand, as Russia starts to use gas more efficiently, allows for an increase in net exports to close to 330 bcm per year by 2035. The major change in exports over the projection period is the start of gas trade with China, anticipated towards 2020, which increases rapidly after 2020 to reach 75 bcm per year in 2035. Exports to Russia's long-standing export markets in Europe increase more slowly, reaching around 235 bcm in 2035, compared with 200 bcm in 2010.[2]

---

1. For an explanation of the way that IEA presents gas production volumes from different countries, refer to Box 8.3.

2. Gazprom imports gas from the Caspian region, so the total volume of gas exported by Russia is higher than the figure for net exports.

Figure 8.2 ● Russian oil balance in the New Policies Scenario

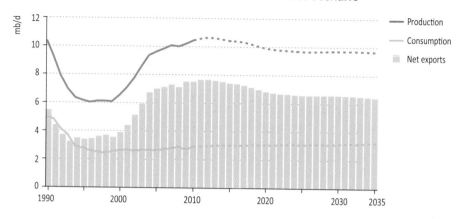

Figure 8.3 ● Russian gas balance in the New Policies Scenario

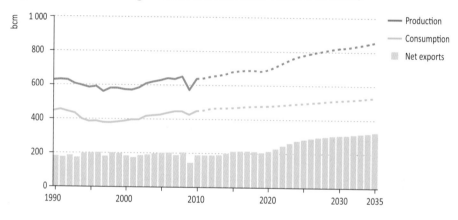

Figure 8.4 ● Russian coal balance in the New Policies Scenario

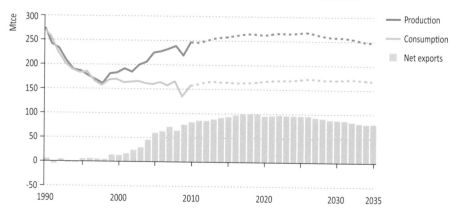

Coal production rises to close to 270 million tonnes of coal equivalent (Mtce) in the mid-2020s and then decreases slowly thereafter (Figure 8.4). Domestic coal demand remains flat, as efficiency improvements in coal use offset the effects of economic growth. Although exports have expanded strongly in recent years, the competitive position of Russian coal, both on domestic and international markets, is held back by high transport costs. As European demand for imported coal declines, so the focus for Russian coal export switches to the east; China's relative proximity to the main Russian coal reserves in Siberia makes it the main export destination. But, overall, Russia struggles to compete with other exporters as global demand for coal decreases slowly after peaking in 2025 in the New Policies Scenario.

Alongside its wealth of fossil-fuel resources, Russia also has a major nuclear industry and huge potential both for hydropower and for harnessing other renewable sources of energy. The contribution of nuclear power and renewables is projected to increase steadily over the projection period, with their combined share in Russia's primary energy supply rising from 10% in 2009 to 15% in 2035. The largest increases come from nuclear power, which grows at an average of 2% per year, and hydropower at 1% per year. However, the fastest rates of growth come from non-hydro renewables, which rise from a very low base at an average of 7% per year, with the increases heavily weighted towards the latter part of the projection period, as support mechanisms are put in place and technology costs come down (Figure 8.5).

Figure 8.5 ● **Russian nuclear and renewables output in the New Policies Scenario**

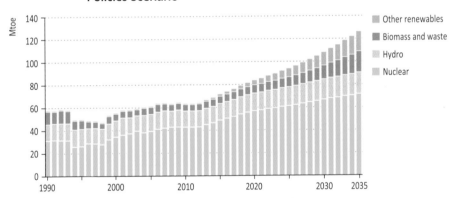

The lion's share of production from nuclear power and renewables is consumed domestically, although there are projects underway or under consideration that have an eye to export markets, such as hydropower projects near the Chinese border or the proposed nuclear power plant in the Russian Baltic enclave of Kaliningrad. Future renewables-based electricity and biomass projects could also be oriented towards export to the European Union, where they could help to meet policy targets for increased use of renewable energy resources. The primary actors in the Russian nuclear power industry are state-owned companies and there has also been a consolidation of hydropower assets within the majority state-owned Rushydro over recent years, although some hydropower plants are controlled by private

Russian industrial groups (and there are also indications that the government may reduce its stake in Rushydro over the next few years). Growth in the renewables sector is expected to open up opportunities both for new entrants and for the existing electricity utilities.

The main industry players in oil production are all Russian companies: two majority state-owned companies, Rosneft and Gazprom (including its primarily-oil-producing subsidiary GazpromNeft), and seven that are privately owned or publicly traded: Lukoil, TNK-BP, SurgutNefteGaz, Tatneft, Bashneft, Russneft and Slavneft (controlled jointly by TNK-BP and GazpromNeft). Together, these produce about 90% of the country's oil (Figure 8.6). Smaller private Russian oil companies and a sprinkling of international oil companies (ExxonMobil, Shell, Total, Statoil, Wintershall, ENI[3]) contribute the rest under various arrangements (production sharing agreements in the cases of ExxonMobil in Sakhalin-1 and Total in Kharyaga; minority ownership, with or without operatorship, for the others). Even where privately owned, the large Russian oil companies all retain strong ties with the government. Long distance oil pipelines are a monopoly of state-owned Transneft.[4]

Figure 8.6 ● Estimated Russian oil and gas production by type of company, 2010

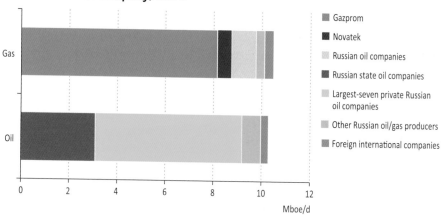

Notes: TNK-BP is classified as Russian. Data are in million barrels of oil equivalent (Mboe) to allow for comparison by energy content; the 10.5 Mboe/d of gas production is equivalent to 637 bcm per year.

There are about 80 gas producing companies in Russia (Henderson, 2010). Most are small and regional in scope and gas production is dominated by Gazprom (80%). There are growing contributions from private gas producer Novatek (6%) and from oil companies producing associated gas (10%), with the rest being divided between foreign companies and small local independents. Gazprom's hold on the Russian gas industry is secured by its monopoly over long-distance gas transport pipelines and gas exports, although its dominant position in Russian production could be challenged to a degree by the rise of Novatek and other Russian producers.

---

3. Wintershall and ENI (together with ENEL) are partners in gas fields with production of condensate.

4. The only exception is the Caspian Pipeline Consortium (CPC) pipeline that crosses from the northern Caspian to the Russian Black Sea port of Novorossiysk. Transneft is a minority shareholder in this pipeline. Around 80% of CPC capacity is used for transportation of Kazakhstan crude oil.

We assume that the structure of the oil and gas industry remains dominated by Russian state and private companies throughout the period. Despite intermittent signs from the government of a desire to open the Russian oil and gas industry to foreign investment, history suggests this is likely to be a slow process. A possible exception in the medium term could be the participation of Chinese (and perhaps Indian) companies in Eastern Siberia and the Far East, possibly as providers of loans rather than providers of equity. Previous efforts by Chinese companies to secure roles in the Russian upstream have yielded moderate results,[5] but China's growing role in the worldwide energy industry and its position as a market for Russia's eastern resources would make this a natural evolution, albeit a politically sensitive one. Participation of western international companies is likely to be limited mainly to the Arctic or deep offshore resources, which are very capital intensive and require the most recent technologies, as exemplified by the participation of Total and Statoil in the Shtokman project, of Total and other foreign companies in the Yamal-LNG project and the agreement between ExxonMobil and Rosneft for exploration in the deep waters of the Black Sea and the Kara Sea in the Arctic. Participation by western companies in a large gas-to-liquids (GTL) plant could also be a possibility. On the other hand, the provision of oilfield services is likely to come increasingly from independent service companies, as the service arms of previously vertically integrated oil and gas companies continue to be spun-off, bringing increased competition and efficiency.

Russia's coal industry was slower than the oil industry to restructure in response to the post-Soviet downturn. But the sector has now been fully liberalised and production increased steadily after 1999 until it was hit by the 2009 economic crisis. Of the ten major producers with output greater than 10 Mtce per year, SUEK has emerged as the largest. Several of the key players are part of large diversified groups, with activities including power generation, steel production, mineral extraction and metals fabrication, e.g. Mechel, Evraz, Severstal, which tend to use their coal production within the group rather than market it. SUEK is the number one exporter, with 30 Mtce out of about 80 Mtce total net exports in 2010. We assume that the industry will continue to rationalise.

Oil

Resources

We estimate proven reserves of oil in Russia to be about 77 billion barrels. This estimate is close to the numbers quoted by the BP annual statistical review (BP, 2011) and by BGR (BGR, 2010). It matches well with the sum of the numbers published by the main Russian oil companies. As all the major Russian oil companies have their reserves credibly audited (both according to the rules of the US Securities and Exchange Commission (SEC) and the recommendations of the Petroleum Resources Management System [PRMS]), this proven reserves number is well established. Ultimately recoverable resources (URR) are less well established. Based on the 2000 United States Geological Survey (USGS) assessment and

---

5. Sinopec has a joint-venture with Rosneft (Udmurtneft) producing oil in the Volga-Ural basin, and China National Petroleum Corporation has an exploration joint-venture with Rosneft in Eastern Siberia (Vostok Energy). Also of note is the Chinese loan for the construction of the Eastern Siberia-Pacific Ocean pipeline in exchange for preferential future supply.

subsequent updates,[6] IEA analysis puts the level of ultimately recoverable resources at around 480 billion barrels (crude and natural gas liquids), of which about 144 billion have already been produced.

Due to Russia's size and diversity, we have considered and modelled the main resource-rich areas of Russia individually in this *Outlook*; these eight basins (the same for both oil and gas) are shown in Figure 8.7. The bulk of these resources are in the historical producing regions of Western Siberia and Volga-Urals, which together are estimated to hold almost 65% of the total remaining recoverable conventional oil (Table 8.1). Two factors might make this an overstatement of the dominance of the traditional areas:

■ Attributing too large a reserve growth to these regions; on the one hand they are very mature; on the other hand, small increases in recovery rates in their huge historical fields would yield large additional reserves.

■ Fairly small resources estimates in the USGS study for the newer regions. Indeed other sources give larger numbers for ultimately recoverable resources in Eastern Siberia (Efimov, 2009), in the Sakhalin area, and in the Arctic offshore continental shelf (Government of Russia, 2009; Kontorovich, 2010; Piskarev, 2009), including the Barents Sea, the Kara Sea and the largely unexplored Laptev, East Siberian and Chukchi Seas. The deep-offshore Black Sea was not assessed by USGS, but is generally credited with 7 billion barrels of recoverable resources.

Table 8.1 ● **Conventional oil resources in various Russian regions, end-2010** (billion barrels) (crude + NGLs)

	Proven reserves*	Ultimately recoverable resources	Cumulative production	Remaining recoverable resources		
				Total	Share	Share per ABCD**
Western Siberia	48	266	80	186	55%	55%
Volga Urals	16	81	51	29	9%	10%
Timan Pechora	4	28	5	22	7%	7%
Eastern Siberia	5	21	0	21	6%	14%
Sakhalin	2	9	1	7	2%	3%
Caspian	2	25	5	20	6%	5%
Barents Sea	0	18	0	18	5%	3%
Other offshore Arctic	0	30	0	30	9%	3%
Others	0	2	1	0	0%	0%
Total Russia	**77**	**480**	**144**	**336**	**100%**	**100%**

*Proven reserves are approximately broken down by basin based on company reports.**This column is an IEA estimate based on the Russian ("ABCD") classification system (Box 8.1) taking into account recovery factors and the probabilities of the various categories to estimate a mean value.

Sources: USGS; data provided to the IEA by the US Geological Survey; IEA databases and analysis.

---

6. We have used: updates to USGS resource estimates published since the 2000 assessment; cumulative production as per IEA databases; and some simplifying assumptions on the worldwide distribution of the "reserves growth" component of the USGS assessment.

Figure 8.7 ● Oil and gas basins in Russia

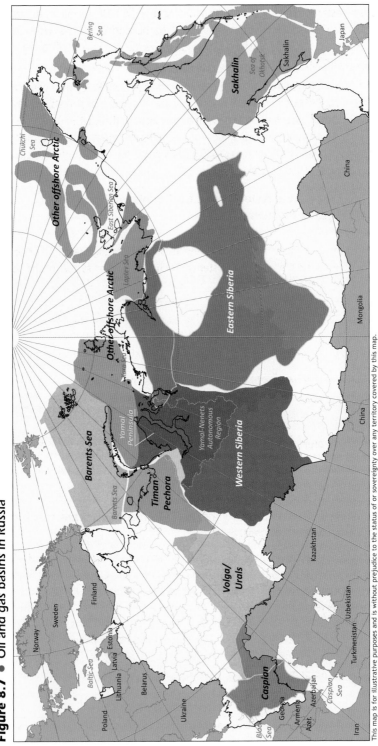

This map is for illustrative purposes and is without prejudice to the status of or sovereignty over any territory covered by this map.

Notes: "Western Siberia" includes only onshore; offshore (Kara Sea) is included in "Other offshore Arctic". "Eastern Siberia" includes the northern part of the Krasnoyarsk administrative region (the location of the Vankor field); although this area is geologically related to the Western Siberian basin, it is reported as Eastern Siberia, to which it belongs administratively. Sakhalin includes Sakhalin Island and the Sea of Okhotsk. Caspian includes the Russian sector of the Caspian Sea and the pre-Caspian basin: the North Caucasus region is included here as well, both because the fields in the North Caucasus are geologically related to the pre-Caspian basin and because production and reserves in the region are considered state secrets, making it difficult to establish separate projections.

Our USGS-based estimates have been compared to the "ABCD" resource estimates published by the Russian government (Table 8.2) and, to the extent that the two systems can be compared, the results are consistent (Box 8.1 details the Russian classification of oil and gas resources and how it compares to other international systems).

Addition of the proven, probable and possible reserves of the main Russian oil companies (as audited under the PRMS scheme) gives a total of about 150 billion barrels. This takes into account only already licensed fields and does not include the vast regions that are unlicensed and poorly explored, including most of the Arctic continental shelf. So the total value for ultimately recoverable resources of 480 billion barrels, which forms the basis of our projections for future production, is plausible.

Box 8.1 ● Russian reserves system versus the Petroleum Resources Management System

Although the reserves of Russian oil and gas companies are increasingly reported under the Petroleum Resources Management System (PRMS) classification system or the US Securities and Exchange Commission rules for reserves reporting, documents from the Russian government and academic institutes generally use the Russian system of reserves classification. This system uses alphabetical classes (A, B, C1, C2, D1, D2 and D3) that reflect decreasing certainties for technically recoverable resources:

- "A" are reserves that have been fully ascertained through drilling and production.

- "B" are reserves that have been established through well-testing.

- "C1" are estimates for established fields, including parts that may not have been drilled and tested yet, but for which geophysical information is available.

- "C2" represents preliminary estimates based on exploration.

- "D1" represent estimates of hydrocarbon potential, based on surface seismic.

- "D2" are possible resources in new areas, based on studies of regional geology.

- "D3" represent a prognosis for the hydrocarbon resources of new basins, based on general geological considerations.

There is no unique correspondence between the Russian classification and the PRMS one. But in general, industry experts assume that A, B and C1 reserves ("ABC1") lie in between proven and proven + probable; C2 and part of C1 correspond to probable and possible reserves; while D1, D2 and D3 resources are closer to estimates of undiscovered hydrocarbons. Further, the D1, D2 and D3 resources are normally quoted as oil-in-place, rather than recoverable oil. C1 and C2 are typically technically recoverable, not necessarily economically recoverable. A and B, as they are based on an approved development plan, are normally both technically and economically recoverable.[7]

---

7. For gas, it is traditional in Russian assessments to assume a recovery factor of 100%, which is one of the reasons why PRMS assessments give smaller values than "ABC1".

Table 8.2 ● Conventional hydrocarbon resources in various Russian regions, end-2009, according to the Russian system of classification

	Oil (billion barrels)		NGLs (billion barrels)		Gas (trillion cubic metres)	
	ABC1,C2	C3,D1,D2	ABC1,C2	C3,D1,D2	ABC1,C2	C3,D1,D2
Western Siberia	111	208	16	30	41	71
Volga Urals	28	23	0.6	0.9	0.9	2
Timan Pechora	15	31	0.5	3	0.7	4
Eastern Siberia	19	79	3	20	8	40
Sakhalin	3	15	0.7	4	1	8
Caspian	4	15	8	4	6	8
Barents Sea	3	9	0.5	12	5	12
Other offshore Arctic	0.1	28	0.0	11	4	18
Others	0.4	5	0.0	0.0	0.5	0.2
Total Russia	**183**	**414**	**29**	**85**	**68**	**163**

Notes: For oil, "ABC" reserves are considered state secrets, so there is uncertainty on the total and the regional values. D resources are published by the Ministry of Natural Resources, as well as all categories for gas and gas-condensates. This table summarises the key values by region, including IEA estimates for oil "ABC" numbers. C3, a previously used category of undiscovered resources, has been merged into D1 in the latest version of the classification system, which comes into full force in January 2012.

Production

More than 1 000 different fields produce oil or condensate in Russia, though about 35 major fields contribute 50% of total production. Production peaked during Soviet times at around 11.5 million barrels per day (mb/d) in the 1980s and then went through a trough, around 6 mb/d, in the mid-1990s, before recovering in the 2000s to exceed 10 mb/d in 2007. The Energy Strategy to 2030 (Government of Russia, 2009) foresees a gradual increase in production over the coming decades, to between 530 and 555 million tonnes in 2030 (around 11 mb/d). The General Scheme for Development of the Oil Industry to 2020 (Ministry of Energy of the Russian Federation, 2011) is oriented more towards keeping production constant, through a combination of fiscal incentives, efficiency improvements and increases in exploration and drilling activities. We draw from these strategy documents the assumption, underpinning our supply projections, that policy (including fiscal policy) will be designed and adjusted through the projection period with the aim of encouraging production around current levels, *i.e.* at or around 10 mb/d, for as long as possible.

In the New Policies Scenario, we project that Russian oil production will stabilise around 10.5 mb/d for the next few years and then start a slight decline, while remaining above

9.6 mb/d throughout the period to 2035 (Figure 8.2). This results from a balance between decreases in old (Volga-Urals) or ageing (Western Siberia) fields, on the one hand, and new field developments, often in remote areas. The slight overall decline comes despite a significant projected increase in production of natural gas liquids (NGLs), which is discussed more fully in the gas section of this chapter and Box 8.4.

The recent history of oil production in Russia has two main phases (Figure 8.8). The first saw the resurgence of supply from Western Siberia, driven by the application of more modern field management technologies, until around 2006. The second phase, over the last five years, has seen production coming online in newer oil provinces – Timan Pechora, Sakhalin and, more recently, Eastern Siberia – while production from Western Siberia flattened out or even decreased. Of note is the remarkable resilience of production from the Volga-Urals basin throughout the period – in spite of this basin having already produced more than 60% of its estimated recoverable resources in the seventy years since production started. This is also due to the application of modern technology, a process that started later than in Western Siberia.

Given its dominant position in Russian oil production, accounting for around two-thirds of the total, the evolution of production in Western Siberia is a key element in any projection of future oil production. The flattening of production there since 2006 might indicate that all the easy wins from modern technology have been realised already. At the same time, the large volume of remaining recoverable resources suggests that production could be maintained or even increased, with suitable investment. Existing tax holidays and other incentives generally favour investment in new fields in regions such as Eastern Siberia, so a critical issue for Western Siberia – and therefore for Russian output as a whole – is whether the fiscal system will be restructured in a way that promotes investment in small fields that have not been developed so far or in enhanced recovery at existing fields.

Figure 8.8 ● **Recent evolution of Russian oil production by region**

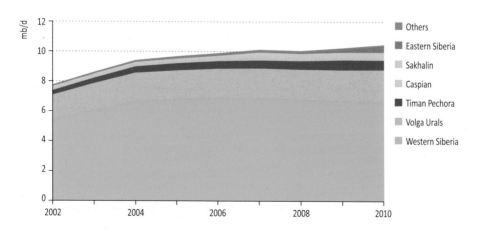

As noted, our projections for the various regions in the New Policies Scenario are based on the view that Russian policy will adjust as necessary to encourage total production at (or as close as possible to) current levels. For our modelling of Russian supply, this means in practice an assumption that the tax regime will succeed in mobilising the necessary investment to allow an appropriate level of exploitation of the oil resources in each region, according to the economic possibilities. This is similar to the graduated approach that is currently taken by the fiscal authorities to encourage new field developments in Eastern Siberia, but would be an innovation in the traditional production areas of Western Siberia and the Volga Urals (Box 8.2).

Box 8.2 ● A balancing act of tax and economics[8]

Investment in oil production in Russia, as elsewhere, is very sensitive to changes in the fiscal regime. The current taxation system for oil fields has three main parts: a Mineral Extraction Tax (MET) for each barrel produced;[9] an export tax for each barrel exported;[10] and a corporate profit tax rate of 20%.[11] As they stand, this makes it very difficult to achieve profitability in new field development. A project with lifting costs of $6/barrel, capital costs of $6/barrel and transport costs of $6/barrel has an internal rate of return (IRR) of less than 5% (at an $80/barrel oil price for Urals blend). This is insufficient to justify the investment, particularly given the political and logistical risks in Russia. In this example, the effective tax rate for exported oil amounts to 84%.

This is why the Russian government has granted a number of tax exemptions or tax holidays, until a certain level of production is reached, to some of the greenfield projects in Eastern Siberia, the Yamal-Nenets region and also, from 2012, the Black and Okhotsk Seas. This is a deliberate policy to promote the development of new production areas and to support regional economic development. It is also possible to apply for specific exemptions on a project-by-project basis. By adjusting the tax rate, the Russian government is believed to be ready to allow the oil companies an IRR of about 15%, if they can demonstrate that the economics are especially difficult.

This case-by-case approach has its merit but it leaves oil companies uncertain about the stability of the tax regime. There has been, and continues to be, a lively debate among the various ministries and companies about the need to reform the tax system. Any such reform would need to find the right balance between preserving state revenue levels (given the importance of oil tax revenue to the state budget) and providing sufficient

8. This discussion focuses on oil, but similar issues exist also for gas projects, although the existing tax burden for gas projects is significantly lighter (on an energy equivalent basis). Year-on-year tax increases for the gas industry are planned in 2012, 2013 and 2014.

9. The MET is calculated in $/barrel as 0.22*(P-15), where P is the market price of the Ural Blend in $/barrel. For example, for a (Ural) oil price of $80/barrel, MET is $14.24/barrel. Older fields get a discount of 30% to 70% on MET, depending on their degree of depletion.

10. Export taxes are calculated in $/barrel as 4+(0.65*(P-25)). For example, for an (Ural) oil price of $80/barrel, it amounts to $39.75/barrel. A decrease of the 0.65 coefficient to 0.60 is under discussion, as part of what is called the "60-66" rebalancing of export taxes between crude and refined products.

11. Profit tax is regionally determined: regional administrations can reduce it to a minimum of 15.5%.

incentives for companies to invest in new projects (thereby promoting economic development in new regions), while also safeguarding the incentive to reduce costs through the application of new technologies.

Designing the tax system to provide sufficient incentives to at least maintain production in the core region of Western Siberia will be essential, as a shortage of investment there could lead to a rapid decline in total Russian oil production. The currently prevailing effective tax rate in the traditional producing regions of Western Siberia and Volga Urals is around 75% for exported oil. We estimate that, if this rate were to be maintained throughout the projection period, total Russian oil production (and exports) would drop by a further 1.8 mb/d in 2035. Although the tax revenue would be about the same (with higher taxes offsetting lower production), an accompanying significant loss of economic growth would be likely, given the importance of the oil sector in Russia's GDP.[12]

With this assumption about the fiscal regime, the projections for oil production in each of the main basins (Table 8.3) track, to a degree, the proportionate distribution of remaining regional resources, given in Table 8.1. Declines in the Volga Urals region are delayed, but accelerate after 2020; the Timan-Pechora and Sakhalin regions remain at or close to plateau; while Eastern Siberia and the Caspian basin realise their potential for significant production increases. Crucially, production in Western Siberia remains relatively buoyant, falling slightly in the period to 2025, but rebounding towards the end of the projection period, helped by the anticipated increase in NGLs production from the gas sector. The offshore Arctic continental shelf also has large potential, but logistical challenges are assumed to prevent it from becoming a major production area until the closing years of the projection period.

Table 8.3 ● **Projections for oil* production by main basins** (mb/d)

	2010	2015	2020	2025	2030	2035
Western Siberia	6.75	6.33	5.87	5.80	5.98	6.19
Volga Urals	2.10	2.07	1.90	1.72	1.46	1.15
Timan Pechora	0.60	0.62	0.59	0.57	0.58	0.59
Eastern Siberia	0.43	0.72	0.76	0.75	0.74	0.72
Sakhalin	0.31	0.30	0.31	0.31	0.32	0.29
Caspian	0.23	0.27	0.34	0.35	0.38	0.41
Barents Sea	0.00	0.00	0.01	0.05	0.10	0.13
Other offshore Arctic	0.00	0.00	0.00	0.00	0.01	0.01
Others	0.02	0.11	0.11	0.11	0.16	0.18
Total Russia	**10.45**	**10.42**	**9.89**	**9.68**	**9.72**	**9.66**

*Includes crude oil, NGLs, and unconventional oil.

Notes: "Others" include projections for additives (IEA, 2010a, Chapter 4 for a definition of additives) and GTL for all of Russia, which explains its growth. 2010 data do not include additives, for lack of historical data.

12. For the purpose of this analysis of sensitivity to tax rates, we assumed GDP growth was unchanged.

Looking at Western Siberia in more detail, production is stabilised at around 5.9 mb/d in the 2020s. Prospects in this region depend on multiple fields (Figure 8.9), of which some of the major producers are: [13]

- *Samotlor*, a super-giant field opened in 1965, with original recoverable resources of the order of 28 billion barrels. Production peaked in 1980 at more than 3 mb/d. It still contains about 7 billion barrels of reserves and production has recently stabilised at around 500 thousand barrels per day (kb/d), a level TNK-BP plans to maintain through steady investment.

- *Priobskoe*, one of the "younger" giant fields in the region, with about 7 billion barrels of reserves. It is divided in two parts by the Ob River. The left bank started production in 1988, the right bank in 1999. Production has been increasing in the last ten years to reach about 800 kb/d, the largest producer in Russia.

- *Krasnoleninskoe*, actually a group of fields, with oil-in-place comparable to Priobskoe. Although production started in the 1980s, it is a complex reservoir that is being developed only recently on a large scale. It currently produces about 150 kb/d.

- *Urengoy*, better known for its gas production, is also a giant oil and condensate reservoir, with about 1 billion barrels of oil reserves and 4 billion barrels of condensate. It currently produces about 70 kb/d of condensate and less than 10 kb/d of crude.

- *Lyantor* and *Federovskoe* are examples of old, declining giants, with ultimately recoverable resources in the order of 10 billion barrels. They still represent the best part of Surgutneftegas' declining Western Siberian production, with production of more than 150 kb/d each.

- *Tevlinsko-russkinskoe, Povkhovskoe, Vateganskoe* are also ageing giants, forming the basis of Lukoil's production in Western Siberia. With remaining reserves of about 2.5 billion barrels, they have recently declined by about 6% per year, to about 400 kb/d in 2010. However Lukoil has indicated that this was partly due to limitations on electrical power availability to lift the increasingly water-rich production, so production could stabilise after additional investment.

- *Uvat* is an interesting case of a greenfield development, demonstrating that the Western Siberian basin still contains significant untapped resources.[14] Located in the south of the basin, far from the existing infrastructure, its development is only just starting. Recoverable resources are estimated to be close to 2 billion barrels, with planned plateau production around 150 kb/d. There are other fields in the same part of the basin that could make use of the infrastructure developed for Uvat.

---

13. As much as possible, we follow Russian GOST (state standard) 7.79-2000 for the transliteration of Russian field names, except when there is a widely used name in English.

14. Another example of the potential for new large fields in Western Siberia is the Gydan peninsula in the Yamal-Nenets district, close to the border of the north of the Krasnoyarsk district.

Figure 8.9 • Major oil fields and supply infrastructure in Russia

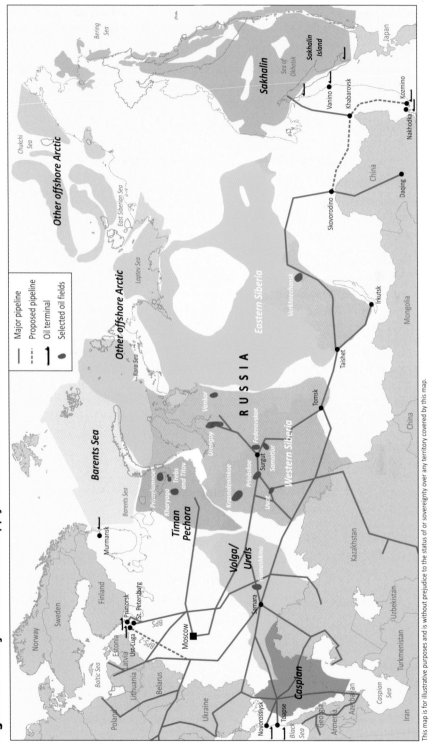

This map is for illustrative purposes and is without prejudice to the status of or sovereignty over any territory covered by this map.

In spite of its remarkable performance over the last ten years, production in the Volga-Urals region is projected to start an inexorable decline after 2016, due to resource depletion. The region features some old giants, such as:

■ *Arlanskoye* in Bashkiria which, with more than 8 000 wells, produces with more than 93% water content. With initial reserves of about 2 billion barrels, it is now largely depleted and produces about 60 kb/d of oil.

■ *Romashkino* in Tatarstan is the grandfather of Russian super-giant fields. Opened in 1948 with 18 billion barrels of estimated original recoverable resources, it still contains about 2 billion barrels of proven reserves. Tatneft has been maintaining steady production of around 300 kb/d for a number of years.

The Timan Pechora basin contains many recently developed fields. They are generally of medium size and we are projecting production to remain at around 0.6 mb/d to 2035. Of note are:

■ *Prirazlomnoe*, in the Pechora Sea (not to be confused with the onshore field of the same name in Western Siberia). It is the first offshore development in the Russian Arctic continental shelf and, as such, receives a lot of attention as a precursor of possible future developments. It reuses the decommissioned North-West Hutton production platform from the North Sea. Production is expected to start in 2012 and to reach a plateau of 120 kb/d.

■ *Kharyaga*, operated by Total under a production sharing agreement, with current production of 25 kb/d and proven and probable reserves in excess of 1 billion barrels.

■ The *Trebs* and *Titov* fields, together one of the last known super-giant fields to be licensed. With proven and probable reserves estimated around 1 billion barrels, these fields were licensed to Bashneft at the end of 2010. Development is planned in partnership with Lukoil.

Figure 8.10 ● **Changes in Russian oil production by region in the New Policies Scenario**

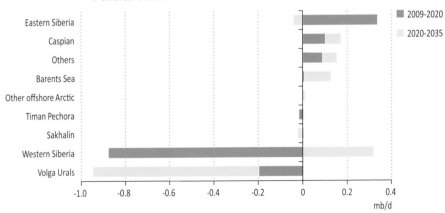

Note: "Others" include projections for additives and GTL for all of Russia, which explains its growth.

Production in Eastern Siberia is projected to make the largest contribution to incremental Russian supply (Figure 8.10) during the period to 2035, with supply rising by more than 300 kb/d in the next five to eight years to reach a total of almost 0.8 mb/d. Production then remains near this level throughout the projection period, with declines in the recently developed fields being offset by gradual expansion of production from new fields farther from existing infrastructure. Overall, Eastern Siberia has the most recently discovered fields:

- *Vankor*, with reserves estimated at 3 billion barrels, came on stream in 2009/2010 and has largely been responsible for the increase in Russian oil production since then. It reached 250 kb/d in 2010 and 315 kb/d in July 2011. Plans to increase production to 500 kb/d have not been fully confirmed and may depend on negotiations with the government on the applicable tax regime. Although belonging administratively to the Krasnoyarsk region, and therefore usually reported as an Eastern Siberian field, Vankor is related geologically to the Western Siberian basin and located very close to the Yamal-Nenets region, so it can tap into the transport infrastructure existing there. Rosneft has constructed a new 500 kilometre (km) pipeline, linking Vankor to the Transneft system (at Purpe), that creates the link between the Yamal-Nenets region and the Eastern Siberia – Pacific Ocean (ESPO) pipeline and eastern markets. This enhanced access to market removes a major obstacle to the development of other fields in the region.

- *Verkhnechonskoe* is in the Irkutsk region.[15] With more than 1 billion barrels of reserves, its development was made possible by the construction of the ESPO pipeline, which passes close to the field, ensuring easy transportation in an otherwise remote region. Currently at 60 kb/d, production is being increased towards an expected peak of 150 kb/d.

- *Talakanskoe*, with 800 million barrels of reserves, is located close to Verkhnechonskoe and therefore to the ESPO pipeline. It is a key element of the expansion of Surgutneftegas into Eastern Siberia. Production, at 40 kb/d, is in the ramp-up phase.

Commercial production in the Russian sector of the Caspian Sea began in 2010. This area is set to play a small but increasing role in Russia's oil production, with output increasing gradually from the current 230 kb/d to 400 kb/d in 2035 (including the onshore pre-Caspian and North Caucasus parts of the basin):

- The *Yuri Korchagin* field is the first large offshore field to be developed in the Russian sector of the Caspian. Operated by Lukoil, it contains more than 200 million barrels of proven reserves and produced first oil in 2010.

- The *Filanovskiy* field, the largest offshore field discovered in the Russian Caspian Sea with more than 1 billion barrels of recoverable resources, is the next planned offshore Caspian development.

The main developments on Sakhalin Island are producing significant quantities of liquids and there are substantial recoverable resources in the region. However, while a number of

8

---

15. Verkhnechonskoe, and a few similar fields in Eastern Siberia, is unusual in that it produces from pre-Cambrian rocks; this feature, together with the absence of nearby source rock, had been used by proponents of the existence of abiotic oil as evidence for their theory. However, more recent work has fully established the biotic origin of the oil and confirmed it migrated from source rocks located a significant distance away (Everett, 2010).

additional Sakhalin projects are under discussion (Sakhalin-3, -4, -5...), we project that high costs will delay developments, with production essentially flat, near 300 kb/d, until 2035.

- *Chayvo, Odoptu* and *Arkutun-Dagi* are the three main oil producing fields, operated by ExxonMobil as part of the Sakhalin-1 project. With proven reserves of 300 million barrels, Chayvo produced about 100 kb/d in 2010, a 50% decline compared to its peak in 2007. This decrease does not necessarily indicate early onset of decline, as production is limited by the absence of a gas export facility. Production from Odoptu, on the other hand, is growing, compensating for the reduced production in Chayvo.

- *Piltun-Astokhskoe* and *Lunskoye*, part of Sakhalin-2, primarily a gas and liquefied natural gas (LNG) project also contain large recoverable oil resources of the order of 700 million barrels. They currently produce about 150 kb/d of oil.

The Barents Sea is primarily a gas province but development of the Shtokman field, assumed for the period after 2020, will provide a sizeable volume of NGLs.

Investment and costs

The levels of oil production that we project in the New Policies Scenario imply an overall requirement for upstream oil investment of $740 billion (in year-2010 dollars) in Russia over the period from 2011 to 2035, or an average of over $29 billion per year. This investment is needed to compensate for the decline in production from existing fields, since crude oil output in 2035 from fields that were in production in 2010 drops from 9.8 mb/d to only 3 mb/d, a fall of 70% (Figure 8.11). On this basis, there is a need, over the *Outlook* period, to add a total of 5 mb/d of conventional capacity, in order to meet the projected level of total output. Around two-thirds of the crude oil produced in 2035 will come from fields that have already been found, but a further one-third will need to come from new fields that have yet to be proven or discovered. We anticipate that most of these new fields will be in Siberia, both Eastern and Western, in the Caspian region and, to a lesser extent, in the Timan Pechora basin.

Figure 8.11 ● Russian oil production by type in the New Policies Scenario

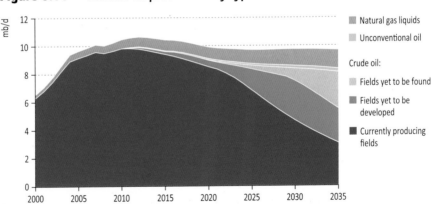

World Energy Outlook 2011 - OUTLOOK FOR RUSSIAN ENERGY

Estimating decline rates at existing Russian oil fields is complicated by the fact that production has been affected by many factors unrelated to geology over the last 20 years. To start with, as noted in Box 8.2, production rates vary in response to changes in the tax regime. Most old fields have also seen an increase in production in the period 2000 to 2010 due to application of new technologies (hydraulic fracturing, horizontal drilling and pump optimisation) and the development of satellite fields, located around the main producing zones, or of layers originally by-passed above or below the main production zones. The swings in gas output linked to the 2009 economic downturn also affected NGL output and therefore total liquids production. However, there are a few examples of fields showing a clear decline trend, at a rate of about 5% to 6% per year, indicating the likely future trend of the large fields in the traditional producing regions.[16]

Costs of production are low in the mature producing regions of Western Siberia and Volga-Ural, with lifting costs (including marketing and other general costs) estimated to be of the order of $4/barrel to $8/barrel. However they are on the increase, with the ageing fields producing more and more water along with the oil, the growing costs of electricity in the liberalised electricity market and the worldwide inflation in service costs. Greenfield developments – whether in parts of Western Siberia away from the main producing infrastructure or in new regions (Caspian, Eastern Siberia, Sakhalin, Timan-Pechora basin) – have higher lifting costs, ranging from $6/barrel to $10/barrel due to their remoteness and the limited infrastructure. Offshore developments, such as Sakhalin-1 and Prirazlomnoe (due to start production at the end of 2011 in the Pechora Sea), have even higher lifting costs, probably in the order of $15/barrel.

Capital costs for greenfield projects are in the range of $5/barrel to $10/barrel. Although new developments in new regions could be expected to be much more capital intensive, because of the requirement for new infrastructure, in practice the development of new regions is likely to be undertaken step-by-step, reducing capital costs significantly: the first projects will be those closest to existing infrastructure, with subsequent projects building on the incremental advances made by previous ones. Capital costs in these cases are similar to those needed to maintain production in old fields, as exemplified by the investment budget of TNK-BP in Samotlor: at $4.6 billion over five years to maintain production at around 500 kb/d, this works out at a capital cost of about $8/barrel. Offshore fields have higher capital requirements, with Prirazlomnoe estimated to be on the order of $10/barrel to $12/barrel. Because of the huge distances from most fields to market, transportation costs also play a key role in the economics of oil (and gas) production, with $5/barrel to $10/barrel being typical for most oil exports (with the lower part of the range applying to Volga-Ural exports to Europe, and the upper range for Western Siberia exports to China via the ESPO pipeline).

Exports

In the New Policies Scenario, we project that oil exports (crude and refined products) will decline slowly, from a peak of 7.7 mb/d in 2012, just slightly above the 2010 value, to

---

16. This is in line with the findings of IEA (2008), which gives 5.5% as the post-plateau decline rate for onshore giant fields.

6.4 mb/d in 2035, as crude production falls and domestic demand for transport fuel continues to grow. Exports will continue to go through a variety of routes: existing pipelines to Europe, shipments from northern, eastern and Black Sea ports, and completion and extension of the ESPO pipeline to the Pacific Coast and China. We expect continued expansion of westward oil flows through the Russian export terminals at Primorsk and Ust-Luga on the Baltic Sea, in part to avoid transit through third countries; but the major anticipated change over the projection period will be the expansion of eastward connections to China and Asia-Pacific markets.

Russian oil exports to the east have been slower to develop than commercial logic might suggest: it has taken time to reconcile commercial and state interests in a manner capable of moving projects forward, both within Russia (where Yukos' early leadership in pipeline discussions with China was lost after the company's bankruptcy) and between Russia and the other main regional actors, China and Japan. China's loan agreements in 2009 with Rosneft ($15 billion) and Transneft ($10 billion) facilitated decisions both on the pipeline route and on volumes (since the loans were to be repaid in oil supply).

Phase I of the ESPO project, completed in December 2009, consisted of a 2 700 km pipeline from Taishet to Skovorodino with a capacity of 600 kb/d. From January 2011, pipeline deliveries of 300 kb/d began via a southward trunk line from Skovorodino to Daqing in China; the remaining volumes are currently transported by rail to Kozmino Bay on the Russian Pacific Coast. Phase II of the ESPO, already underway, will extend the pipeline from Skovorodino to the coast, a further 2 100 km, and increase its overall capacity to 1 mb/d. There is also the possibility of doubling the capacity of the spur to China (to 600 kb/d). We assume that the expansion to 1 mb/d will be completed in 2013 and that a further expansion of the system will bring total capacity to 1.6 mb/d by the early 2020s. This, together with the potential to further reinforce eastward export routes later in the projection period in response to rising oil import demand in Asian markets, will give Russia the opportunity to balance its exports between East and West (see Chapter 9).

The split between the export of crude oil and of refined products will be determined largely by fiscal policy. Over the last ten years, the Russian government has made various attempts to promote exports of refined products at the expense of crude. The favourable tax regime for product exports was originally intended to provide incentives for much-needed modernisation of Russian refineries and thereby to capture more value-added within Russia. Thus far, this policy has failed: although in 2010 around a third of Russian exports were in the form of refined products, most were in the form of low-end, low-value-added fuel oil, as Russian refineries were still not equipped to supply high-end gasoline products. Anticipated changes in taxation in 2011,[17] coupled with more stringent gasoline specifications on the internal market, are intended to realise the original objective.

---

17. Such as the much-discussed "60-66" rebalancing of export taxes which would see the marginal tax rate on crude oil exports decreased from 65% to 60%, while that on exports of some refined products would be set at 66% of the export tax on crude. At the time of writing, the latter part has been adopted, while the former is still awaiting decision.

Natural gas

Resources

Proven reserves of natural gas in Russia are generally quoted to be about 45 trillion cubic metres (tcm).[18] This matches well with the sum of the numbers published by the main producing companies, using the "ABC1" reserves under the Russian classification system. Under the PRMS, however, proven reserves are closer to 26 tcm. Gazprom, for example, reports 33 tcm as "ABC1" reserves, but only 19 tcm as proven under PRMS. The lower PRMS figure takes into account actual recovery factors in the large traditional fields and the fact that the development of new large fields, such as Yamal or Shtokman, is still under study.

Table 8.4 ● Conventional gas resources in various Russian regions, end-2010 (tcm)

	Proven reserves*	Ultimately recoverable resources	Cumulative production	Remaining recoverable resources		
				Total	Share	Share per ABCD**
Western Siberia	22	59	18	41	39%	53%
Volga Urals	1	5	1	4	3%	1%
Timan Pechora	1	3	1	2	2%	2%
Eastern Siberia	1	7	0	7	7%	18%
Sakhalin	1	3	0	3	3%	3%
Caspian	1	7	1	6	6%	7%
Barents Sea	0	23	0	23	21%	7%
Other offshore Arctic	0	20	0	20	19%	9%
Others	0	1	0	1	1%	0%
Total Russia	**26**	**127**	**21**	**106**	**100%**	**100%**

*Proven reserves are approximately broken down by basin based on company reports. **This column is an IEA estimate based on the Russian ("ABCD") classification system (Box 8.1), taking into account recovery factors and the probabilities of the various categories to estimate a mean value.

Sources: USGS; data provided to the IEA by the US Geological Survey; IEA databases and analysis.

Using a methodology similar to that in this chapter's analysis of oil, based on USGS publications, we estimate the level of ultimately recoverable resources at close to 130 tcm, of which 21 tcm have already been produced. The breakdown of our estimated figure for ultimately recoverable resources between the main basins (Table 8.4) shows the predominance of Western Siberia, where all the largest Gazprom fields, either producing (Urengoy, Yamburg, Zapolyarnoe) or under development (Yamal peninsula) are located, as well as most of the gas associated with oil fields produced by oil companies. Although not producing today, the Barents and Kara seas are considered very strong prospects.

As in the case of oil, the methodology based on the USGS assessment could be overestimating reserve growth in Western Siberia and underestimating the resources in the poorly explored regions of Eastern Siberia and the Arctic offshore continental shelf (other

---

18. 45 tcm in BP, 2011; 48 tcm in O&GJ, 2010; 46 tcm in Cedigaz, 2011; 48 tcm in Government of Russia, 2009.

than the Barents Sea). Overall, our USGS-based numbers are conservative compared with the values under the Russian classification system, even taking the different definitions into account (Table 8.2). The Russian "ABCD" numbers show, in particular, more resources than our USGS-based analysis in absolute terms, as well as higher percentages for Western and Eastern Siberia.[19]

Box 8.3 ● What's in a bcm?

A bcm (billion cubic metres) of natural gas is a commonly used measure of gas production and trade, but what that "bcm" represents depends on how it is measured and how much energy it contains. The IEA standard is to report gas volume as actual physical volumes, measured at 15°C and at atmospheric pressure. This means that a bcm of Russian gas, in terms of energy content, can have a different value from a bcm of gas from another country. For example, an average Russian bcm (at 15°C) contains 38.2 petajoules (PJ) of energy (according to the conversion factors used by IEA) compared with 41.4 PJ for a bcm from Qatar. In the case of Russia, there is additional scope for confusion because the Russian standard reports gas volumes measured at 20°C and atmospheric pressure, slightly different to the IEA.[20]

Negotiating a way through the multiple pitfalls of different calorific values and conversion factors can be difficult and there are different ways of doing so.[21] The IEA approach is to keep the underlying balances for each country on an energy basis (rather than a volume basis) and to maintain a database of the different energy content of gas imports, exports, production and consumption for each country (IEA, 2010b). For the figures presented in bcm units in this *Outlook*, 1 bcm of Russian gas equals 0.82 Mtoe; 1 bcm is also equivalent to 1.017 bcm reported according to the Russian standard, allowing for the different temperatures at which the volumes are measured.

Production

In the New Policies Scenario, total Russian gas production increases from 637 bcm in 2010 to 690 bcm in 2020, 820 bcm in 2030 and 860 bcm in 2035 (Box 8.3 explains the IEA presentation of gas volumes). Production of 820 bcm in 2030 puts our projection around 6% below the range targeted in the Russian Energy Strategy to 2030; this is due in part to our lower projected demand numbers for the domestic market. As the indications for remaining

---

19. Some of the Russian data includes the Kara Sea in Western Siberia, accounting for part of the larger number for Western Siberia and the lower number for other offshore Arctic.

20. A further complication is that the energy content of hydrocarbons can be reported on a gross calorific value basis (GCV) or on a net calorific value basis (NCV); for gas, NCV is approximately equal to 90% of GCV. Russia reports on an NCV basis; IEA uses GCV when reporting energy in joules, but NCV when reporting energy in Mtoe (to facilitate comparison with other fuels).

21. Other organisations, *e.g.* BP in their yearly statistical publication or Cedigaz, report volumes on an energy-equivalent basis, *i.e.* they use a "standard" gas cubic metre with a gross calorific value of 41.87 megajoules (MJ)/m³ (BP) or 40 MJ/m³ (Cedigaz, 2011). This is the same approach as expressing oil production in tonnes of oil equivalent, which is actually an energy unit rather than a mass unit.

recoverable resources suggest, Russian output continues to be concentrated in Western Siberia. However, the overall share of this region in total Russian production is projected to decline from around 90% in 2010 to 78% in 2035, due to rapid increases in output from Eastern Siberia and the Barents Sea.

A breakdown of recent production by region (Figure 8.12) underlines the preeminent position of Western Siberian fields in the overall Russian picture. Within Western Siberia, three fields stand out: the stalwarts, Urengoy and Yamburg,[22] which have provided the backbone of Russian production for the past two decades, and Zapolyarnoe, which started production in 2001. However, the contribution of other regions to the Russian gas balance is growing, notably with the development of gas production and export from Sakhalin.

Figure 8.12 ● Recent gas production trends by region

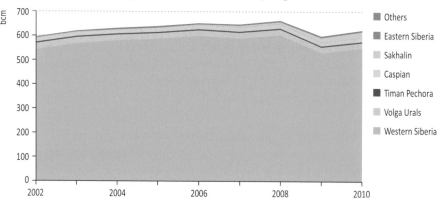

In recent years, Russia has been able to satisfy domestic gas demand and its export commitments through the development of smaller satellite fields in Western Siberia, together with drawing on a steady increase in production by private companies, such as Novatek, and oil companies' associated gas output. Demand reduction, due to the economic and financial crisis, eased pressure on supply and drastically reduced the need for large-scale gas imports from Central Asia. But while relatively small supply increments have proved to be sufficient to meet Russia's needs in the recent past, a swift rebound in global gas demand and the need for Russia to compensate for production declines at the traditional Urengoy and Yamburg fields (at a current rate of 50 bcm every four to five years) will impose a need for new upstream development.

A strategic question for the Russian gas industry is the extent to which Russia will rely on Gazprom and Gazprom-led mega-projects, such as Yamal and Shtokman, to meet these future production needs; or, seen from another perspective, whether a larger share of output could come from multiple smaller fields and from other Russian gas producers such as Novatek and the Russian oil companies, who own some significant and under-exploited

---

22. These are two of the traditional "big three" of Russian gas production in West Siberia. The third is normally the Medvezhe field. However, production from Medvezhe was less than 15 bcm in 2009, so the "big three" label is no longer accurate (unless the third of the three is considered to be Zapolyarnoe).

gas assets (Spotlight). Our assumption is that the structure of the Russian gas market will change slowly over the projection period, with Gazprom's super-dominance of gas production, transportation and sales somewhat reduced over the projection period, but not dismantled. Even with relatively marginal improvements in access to transportation capacity and exchange-based gas trading[23] (and through joint gas marketing ventures with Gazprom), non-Gazprom producers are expected to play an expanded supporting role in the overall Russian gas balance, displacing to a large extent the previous reliance on gas from Central Asia.

Our projection for gas supply by region (Table 8.5) shows a gradual increase in output from Western Siberia, from 564 bcm in 2010 to about 665 bcm in 2035. Behind this figure there is a change in the geographical focus of production within Western Siberia, away from the Nadym-Pur-Taz region, the location of the largest concentration of producing fields. Declines in the Nadym-Pur-Taz region have so far been offset, in part by the development of nearby satellite fields and deeper horizons in the major fields, but mainly by the development of Zapolyarnoe since the early 2000s (now at plateau, around 110 bcm per year). Over the projection period, there is a gradual shift to the Yamal peninsula, thanks to the new transport infrastructure built for the development of the Bovanenkovo super-giant field there. Bovanenkovo is now expected to come on stream in 2012, with plateau production in the first phase planned at 110 bcm per year, with other neighbouring fields following later.

Table 8.5 ● Projections for gas production by main basins (bcm)

	2010	2015	2020	2025	2030	2035
Western Siberia	564	604	604	630	646	665
Volga Urals	24	20	16	14	11	10
Timan Pechora	3	3	2	2	2	2
Eastern Siberia	5	7	24	61	67	77
Sakhalin	23	25	25	26	27	28
Caspian	17	18	17	17	17	17
Barents Sea	0	1	2	27	50	58
Other offshore Arctic	0	0	0	1	1	1
Others	1	1	1	1	1	1
Total Russia	**637**	**679**	**692**	**779**	**822**	**858**

The other large Yamal Peninsula project, the Yamal LNG plant proposed by Novatek, is targeting production and liquefaction of 20 bcm per year by 2016, with gas produced from the giant South Tambei and neighbouring fields. The economics of this project received a major boost when production from Yamal gas fields earmarked for LNG was exempted from the Minerals Extraction Tax (Box 8.2) until cumulative gas output reaches 250 bcm (and condensate output 20 million tonnes).

---

23. A pilot Russian gas exchange functioned from 2006 to 2008; since then, there have been various draft proposals to reintroduce it on a permanent basis, but no decision as yet on its form or scope.

The last of the mega-projects?

Two gas mega-projects are under way in Russia: the Yamal development was launched in 2008 and production is expected to start in 2012; the Shtokman project, though it has experienced numerous delays, is assumed to start production at the end of the current decade.[24] By any reckoning, these two developments, particularly Yamal, are set to play an important role in the Russian supply picture. Yet these projects look less indispensable to the Russian supply outlook than they once did. The reason for this change is the growing availability of other supply options, in particular from non-Gazprom producers. An increased contribution from these producers is anticipated in the government's Energy Strategy, which foresees their share of output rising to above 25% by 2030, from around 20% today. A detailed, asset-by-asset analysis (Henderson, 2010) puts the potential for non-Gazprom gas production at even higher levels, growing from about 150 bcm in 2010 to more than 300 bcm in 2020 and 370 bcm in 2030 (close to 45% of total projected output). A part of this growth is already provided for in committed company investment plans and it could be realised in full with easier access to Gazprom pipelines and to the more lucrative parts of the domestic market.

Growth of this magnitude would fulfil our entire projected Russian gas production increase in the New Policies Scenario, leaving Gazprom with the task of keeping its production about constant at today's levels. In this case, development of the Yamal peninsula and Shtokman field would still be needed to offset decline in current Gazprom fields, but output, particularly for Shtokman, could be built up more slowly than currently envisaged.

Reliance on a larger number of smaller projects with shorter lead times (alongside a greater focus on the efficiency of domestic gas use) would be a coherent strategic response by Russia to the uncertainties over the pace of gas demand growth in Europe. To a degree, this is already reflected in Gazprom's own plans: after Bovanenkovo (with envisaged peak production of 110 bcm per year) and Shtokman (70 bcm to 90 bcm per year), output from the next largest fields envisaged for development drops to 30 bcm to 40 bcm per year. With infrastructure in place on the Yamal peninsula, Gazprom will have more flexibility to pace its investment in additional smaller fields in response to market developments in Europe and elsewhere.

But this does not mean that we have seen the last of such mega-projects. Over the projection period, demand from faster-growing markets outside Europe is a more likely foundation for such major upstream developments: Gazprom is already considering marketing Shtokman LNG to India; and Novatek is seeking markets for Yamal LNG in the Asia-Pacific region. Another generation of Russian mega-projects, albeit on a smaller scale to Yamal and Shtokman, is likely to emerge in East Siberia for export to Asia.

8

---

24. At the time of writing, the development decision has not yet been taken by the partners in the project.

According to our projections, developments in Western Siberia play the most important role in meeting overall increases in output, joined towards 2020 by new fields in other areas, such as the offshore Shtokman development in the Barents Sea. Plans for Shtokman are still fluctuating, primarily due to uncertainty over where and how the gas will be marketed. The earliest date for first production is around 2017, but we assume that production begins at the end of the decade. In our projections, output from the Barents Sea rises to close to 60 bcm in 2035. As well as uncertainty over the start-date, there is also upside potential for this projection, since preliminary plans from Gazprom forecast a plateau capacity from Shtokman of 70 bcm.

Two large fields in Eastern Siberia, Kovykta, with resources of the order of 2 tcm, and Chayandin, with more than 1 tcm of gas resources, are assumed to play a key role in feeding the much-discussed gas pipeline to China. Kovykta's ownership has been hotly contested in the last few years, holding back its development. At the beginning of 2011, it was acquired by Gazprom and could provide up to 40 bcm per year. A development challenge for many Eastern Siberian gas fields, including Kovykta, is their high helium content. Although a valuable product with growing worldwide shortages, it is hard to transport helium economically from the very remote Eastern Siberian fields to the market. As Russian regulations rightly prevent venting this valuable resource, project developers are investigating a number of approaches, including underground storage, to enable production to begin. Based on our assumption that Russia and China will find mutually acceptable terms for the start of gas trade, we project that output from Eastern Siberia increases from 5 bcm in 2010 to more than 60 bcm by the mid-2020s and to over 75 bcm by the 2035, providing the second-largest contribution, after Western Siberia, to overall supply growth (Figure 8.13). Even without agreement on a pipeline to China, there would be opportunities for sizeable growth in Eastern Siberian gas production, for export as LNG to Asian countries, particularly Japan, Korea and, again, to China.

Further to the east, in the Sakhalin area, the Sakhalin-2 project includes the first LNG plant in Russia, with a capacity of 14 bcm. This is a joint venture between Shell, Gazprom, Mitsui and Mitsubishi, developing the Lunskoe and Piltun-Astokhskoe fields. Expansion of the plant by constructing another train of 7 bcm is under discussion, with gas possibly coming from the Sakhalin-1 project or other gas fields in the Sakhalin area. As well as gas for export, a part of Sakhalin output (and of that from Eastern Siberia) is earmarked for domestic use; the government is promoting the construction of new transmission and distribution networks to make gas more widely available for local industrial and residential use, initially in Vladivostok (via the newly built connection to Sakhalin via Khabarovsk). Our projection is for a very slight increase in overall gas production, which reaches 28 bcm per year in 2035.

The other major gas producing region in Russia is the Caspian basin, which is home to another super-giant gas field, the Astrakhan field, with more than 3 tcm of recoverable resources. Production, at about 10 bcm per year, is small for a field of this size, because of the technical challenges and the additional costs created by the high hydrogen sulphide content. Production could increase with progress in the technology needed for such very sour gas fields. The region also features a couple of medium size offshore prospects in the Caspian Sea: the Tsentralnoe field and the Khvalynskoe field. The Energy Strategy to 2030 (Government of Russia, 2009) foresees that gas production from offshore Caspian Sea fields will rise gradually to 21 bcm to 22 bcm per year by 2030; our projection is more conservative at 17 bcm.

Figure 8.13 ● Changes in Russian natural gas production by region in the New Policies Scenario

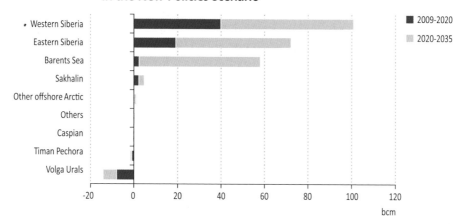

Box 8.4 ● The curious case of the missing natural gas liquids...

A feature of Russia's gas output, compared to other large producers such as Qatar, is the relatively small content of natural gas liquids (NGLs). The explanation is that traditional gas production from the fields situated onshore in Western Siberia is mostly from the uppermost, "cenomanian" layer. This contains very dry gas, almost pure methane. This is also the case for new projects, such as Bovanenkovo and Shtokman, which will produce from the same layer. As this layer has been the mainstay of Russian gas production for many years, Russia historically had relatively little production of NGLs and so limited capacity in gas processing plants. In fact, NGLs are poorly reported, with different numbers being reported by different sources. IEA estimates put Russian production of NGLs in 2010 at about 650 kb/d, half of which is field condensate and the other half gas plant liquids.

More recently, the deeper layers ("valenginian" and "Achimov") of the traditional fields, containing wetter gas, are being brought into production, with the potential of producing more NGLs. New regions, such as Eastern Siberia and Sakhalin, also have wetter gas. Reduction of flaring of associated gas, a naturally rich gas, contributes further to production of NGLs. Investment in gas processing plants to recover the NGLs has been slow, so only a small part of the ethane and about half of the butane and propane is currently marketed. But further development of the gas processing facilities is an important part of the strategy of the Ministry of Energy, with a dozen new or refurbished plants planned for the next ten years, so we project increasing recovery and use of NGLs throughout the period, leading to a doubling of NGLs production by 2035. The oil price and tax regime is currently very favourable for NGLs and has already prompted companies like Novatek to invest in gas processing facilities and to export NGLs. This not only makes a very positive contribution to the economics of their gas fields, but also provides access to export markets, while they are restricted to the domestic market for gas.

Overall, Russian gas production will gradually move north and east, with the Yamal peninsula, the Barents Sea and Eastern Siberia accounting for one-third of production by 2035. In the traditional regions of Western Siberia, the focus will be on deeper, less productive layers, using the existing infrastructure wherever possible. There will be accompanying efforts to monetise NGLs, the production of which is projected to double between 2010 and 2035, reaching over 1.3 mb/d (Box 8.4).

Investment and costs

The projections in the New Policies Scenario imply total investment in the upstream gas sector of over $730 billion (in year-2010 dollars) between 2011 and 2035. A major share of this investment will be required in order to compensate for declining production at existing fields. Decline rates cannot be easily estimated from the production data, not least because varying the rate of production in the large fields of Western Siberia has been used by Gazprom as a means of matching supply to demand (as demonstrated during the 2009 economic crisis – although some of the oil companies also saw their access to Gazprom pipelines reduced during this period). Overall, we estimate that Russia will need to bring on 640 bcm of new capacity by 2035 in order to meet the projection for supply in the New Policies Scenario.

Costs vary significantly between traditional onshore Western Siberian projects and the new greenfield projects requiring new infrastructure, such as Yamal, Shtokman or Kovykta. Traditional projects have very low capital costs, of the order of $4 per thousand cubic metres (kcm), as exemplified by some of the recent developments by Novatek, while the new projects have capital costs as high as $30/kcm to $60/kcm. Similarly, operating costs run from about $5/kcm for the traditional onshore projects up to $50/kcm for the future Arctic LNG projects. Transportation costs probably vary between $10/kcm and $50/kcm, although for pipeline transport, with Gazprom's monopoly, costs are not necessarily closely reflected in prices.

The tax regime for gas is currently more favourable than that for oil, with export tax being no higher than 30% and the Mineral Extraction Tax (MET) being about ten times less than that for oil, on an energy equivalent basis. However, the Russian government has already announced significant increases in the MET for gas over the next few years, starting in 2012, and reform of the export tax is also on the agenda. Even with the current tax regime, the new greenfield projects represent attractive investments only if they benefit from tax breaks, which are considered on a project-by-project basis (as, for example, with Yamal LNG). Projects in the traditional production areas have, in theory, attractive economics, but these are tempered by the Gazprom pipeline and export monopoly, which means reduced sales prices for other gas producers.

Exports are subject to a duty of 30%, paid on the realised export price. There are some exemptions currently in place: for gas exported via the Blue Stream pipeline across the Black Sea to Turkey; for some of the gas exports to neighbouring countries; and, thus far, for all LNG export projects. The fact that LNG exports are zero-rated, while pipeline-oriented projects generally are subject to 30% duty, is an important consideration for the Shtokman project as the partners weigh the different options for marketing this gas.

Flaring

The rate of utilisation of associated gas produced by oil companies has been steadily improving, with the amount of gas produced (and not flared) by oil companies steadily increasing, both in absolute terms and when considered relative to oil production. The numbers reported for utilisation of associated gas vary between 50% and 95%, depending on the companies, with an average around 75%, but some scepticism has been expressed concerning these values. The state companies, Rosneft and GazpromNeft, accounting together for more than half of the gas flared, have the worst reported performance.

Since 2002, the rise in the output of associated gas has been larger than the growth in oil production, *i.e.* the ratio of associated gas production to oil production has increased (Figure 8.14). This is consistent with the view that the amount of gas being flared has decreased in recent years. Nonetheless, the regulatory authorities' aim to reach a 95% utilisation rate of associated gas by 2012 is unlikely to be met and there are indications that it might be achieved only in 2014.

Figure 8.14 ● **Production of associated gas, expressed in volumes and as a ratio to oil production, 2002-2010**

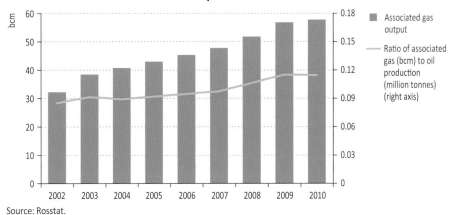

Source: Rosstat.

The exact amount of gas being flared is uncertain, as more than 50% of the flares do not have meters. Different ministries and officials give estimates ranging from 16 bcm to more than 20 bcm in 2010. The Global Gas Flaring Reduction Partnership (GGFRP) gives an estimate, based on satellite measurements, of 35 bcm in 2010 (GGFRP, 2010), but the methodology and calibration for satellite measurements is still being refined. Indeed, the GGFRP data shows a large reduction in flaring in Russia from 2009 to 2010, while Russian statistics indicate an increase, due to the start of new fields in remote regions of Eastern Siberia. Other estimates (PFC, 2007) provide figures of around 30 bcm (after adjustment to take into account reductions in flaring since the time of the study).

To promote utilisation of associated gas, the government plans to increase 100 times from its current (very low) value the fee paid by companies for flaring, and to apply an additional fee for flares that are not metered. However, improved access to Gazprom pipelines and further gas price increases in the internal market would likely be a more effective way to promote a reduction. If the 95% utilisation were to be achieved, and taking into account the likely extent of current under-reporting, we estimate that additional volumes of up to 20 bcm per year of gas would become available.[25]

Transportation and storage

All production, processing, storage and distribution facilities are integrated in the unified supply gas system, owned and operated by Gazprom. Twenty-five sub-surface gas storage facilities, with a working storage capacity of 65 bcm, were in operation at the end of 2009. Their maximum daily output was 0.62 bcm/d. Gazprom is investing in additional storage capacity to match its expectation of growth in domestic demand: 87 bcm of working storage capacity and 1 bcm/d of possible output are planned by 2030, together with further development of the distribution system to deliver gas to regions that currently have limited gas supply (particularly in the Russian Far East).

Prospects for natural gas export flows

Net gas exports increase by almost 75% from 190 bcm in 2010 to close to 330 bcm by 2035 (Figure 8.3). Total gas exports grow from around 215 bcm to 360 bcm over the same period, with a modest rise in exports to Europe accompanied by a much faster projected rate of increase in exports to Asia, as LNG and by pipeline.

Pipelines will continue to provide the route to market for the bulk of Russian exports throughout the projection period (Figure 8.15). In addition to the existing pipelines to Europe through Belarus and Ukraine, the Nord Stream pipeline through the Baltic Sea will provide new capacity and increased flexibility in export routes. Nord Stream Phase 1, with 27 bcm per year capacity, is due to come into operation at the end of 2011. Its capacity is due to double in the second phase, expected in late 2012. No firm decision on the proposed South Stream pipeline across the Black Sea is yet known. If built according to the schedule envisaged by Gazprom and its project partners, with first deliveries in 2015, South Stream's 63 bcm of additional annual transport capacity would change the pattern of Russian export flows considerably (see Chapter 9). By partially substituting for the existing export channels, the pipeline would reduce the perceived risks of transit through Ukraine, as well as the transit fees. However, South Stream remains an expensive project; the preliminary capital cost estimate of $22 billion (South Stream, 2011) is considerably more than the estimated $3 billion cost of rehabilitating the Ukrainian pipeline system (and this estimate does not include the additional infrastructure that would be required within Russia to bring larger quantities of gas to the start of the South Stream pipeline on the Russian Black Sea coast).

---

25. Savings of 17 bcm from reduced gas flaring are included in the calculation of potential energy savings available to Russia in Chapter 7.

Figure 8.15 ● Major gas fields and supply infrastructure in Russia

Legend:
- Selected gas field
- Existing gas pipeline
- Pipeline planned/under const.
- ★ Existing LNG export terminal
- ☆ Planned LNG export terminal

This map is for illustrative purposes and is without prejudice to the status of or sovereignty over any territory covered by this map.

Construction of a new gas pipeline to China is dependent on agreement on the pricing of future deliveries. The overall package of agreements to launch bilateral gas trade could include Chinese loans or pre-payment for gas but is thought unlikely at this stage to include Chinese participation in Russian upstream gas developments (as was the case for China's gas agreements with Turkmenistan). Two routes for a new pipeline are under discussion, with the Altai, or Western, route being favoured by Russia in the first phase (Figure 8.15). A second phase could involve a more eastern pipeline route, closer to markets in Manchuria and northern China. Gas for China would come initially from Western Siberia, tying in to the existing infrastructure there, and subsequently from new developments in Eastern Siberia, such as the giant Kovykta field. If agreement on prices is reached in 2011, the pipeline could start operation as soon as 2016. We project that Russian gas exports to China by pipeline start towards 2020 and reach 75 bcm in 2035.

Russia is also aiming to diversify gas exports through the development of LNG export capacity. This goal was articulated at a time when North America appeared to be a promising market, but expectations in this direction have been undercut by the boom in North American unconventional gas production and a consequently much reduced import requirement. While eastern projects (Sakhalin and Vladivostok) could serve growing Asian LNG import markets, the market outlook for Arctic LNG projects (Yamal and Shtokman) is less clear, adding to uncertainty over the timing of Russia's emergence as a global LNG supplier.

In addition to the currently operating Sakhalin-2 plant, several new LNG projects have been proposed, the main ones being:

■ The addition of a third train to the Sakhalin-2 plant, with a capacity of 7 bcm per year. This could possibly receive gas from the Sakhalin-1 fields, eliminating some of the oil production bottlenecks there.

■ The Vladivostok LNG plant, proposed by Gazprom, with a capacity of 14 bcm per year, to be fed either from Eastern Siberian fields through a new pipeline, or with gas from Sakhalin. A preliminary agreement, already reached with a consortium of Japanese companies, for a feasibility study of this $7 billion project suggests that a target start date of 2017 is feasible.

■ Novatek's Yamal LNG project, with a capacity of 20 bcm per year. The project includes plans to send LNG across the Northern route to Asian markets (Box 8.5). Completion is proposed for 2016, although this date is expected to be pushed back.

■ The Shtokman project in the Barents Sea would feature both pipeline deliveries and, in a second step, an LNG plant of 10 bcm per year capacity. The tentative schedule for first production is 2017, but a decision has not been confirmed yet. We anticipate that this project will start at the end of the decade.

Although there are uncertainties surrounding all these projects and a likelihood of delays in execution of the Arctic projects, given the harsh climate, we project a progressive expansion of LNG capacity from the current 14 bcm per year to 33 bcm per year in 2020 and 70 bcm per year in 2035. LNG would represent about 20% of total Russian exports by the end of the projection period. The required investment in LNG infrastructure is estimated at close to $80 billion (in year-2010 dollars).

Box 8.5 ● The "Northern Route" to market

The "Northern Route" (or Northeast Passage) from the Atlantic to the Pacific across the Russian Arctic seas has fascinated mariners and traders for centuries, though it has never been more than a marginal route for global trade because of the very harsh environment. The gradual reduction in Arctic ice cover, due to climate change, is however likely to increase the number of ice-free sailing days in the summer and to lessen the need to call on ice-breaking assistance. This has attracted the attention of potential producers of LNG in Russia's remote Arctic north, not least because demand for imports in their initial target market, North America, appears to have evaporated. A demonstration of this renewed interest was Novatek's deliveries of condensate from Murmansk to China across the Russian Arctic in the summer of 2010 and again in July 2011, the latter being one of the earliest ever summer shipments along this route.

If passable, the Northern Route provides by far the quickest sea route from Europe to Northeast Asia: the Novatek shipments took around 22 days, half the time required for the next best itinerary via the Suez canal. For LNG suppliers, the Northern Route would mean a sea journey from Murmansk to Chinese LNG terminals only around 30% longer than the journey from Qatar. But this route is still a long way from reliable year-round operation and its regular use will involve significant additional expense: it would require special ice-strengthened LNG carriers, presumably with lower capacities than typical carriers because of size constraints to negotiate the narrow straits,[26] and the use of accompanying icebreakers (whose services also come at considerable cost), if only for insurance purposes. Seasonal contracting would see the Asian market favoured during the summer months, when the Northern Route is open, and sales into Europe or the Atlantic basin at other times. To avoid interrupting production because of transport unreliability, expensive additional storage would be required.

8

Unconventional resources

Because of the large conventional resources of oil and gas, unconventional resources have received comparatively little attention in Russia and the extent of the existing unconventional resources is generally poorly known. Nonetheless, Russia has considerable potential in these areas; as technology improves and costs come down – mainly due to investments in other regions of the world – so the opportunities for Russia to develop its unconventional potential will grow.

Extra-heavy oil and bitumen

Bitumen and extra-heavy oil resources are known to be extensive in Russia, but there are significant discrepancies between the published estimates. BGR gives 345 billion barrels

---

26. The size restrictions are a maximum draft of 12.5 metres (m) and maximum beam of 30m (Ragner, 2008); this compares to 12m draft and 50m beam for a Q-Flex LNG carrier and a similar scale (12m and 53m) for a Q-Max.

for recoverable resources (BGR, 2010), while Russian sources are more conservative, at around 250 billion barrels, as discussed in the *World Energy Outlook*-2010 (IEA, 2010a). Recent Russian government data is even more prudent, giving more than 120 billion barrels, around one-third in Tatarstan, half in Eastern Siberia and some around St. Petersburg. There have been several pilot projects with steam-based thermal methods of recovery, such as steam-assisted gravity drainage, both in Tatarstan (Tatneft) and in the Timan Pechora region (Lukoil). However, large-scale developments are very much in their infancy. For example, the General Scheme for Development of Oil Industry to 2020 calls for only modest developments in Tatarstan, with a capacity of the order of 40 kb/d in the 2020s. As a result we project output of close to 100 kb/d by 2035, though the resources suggest there is potential for a significantly higher figure.

Kerogen shales

Kerogen shales, also known as oil shales, are poorly known in Russia. BGR and USGS estimates (IEA, 2010a) for near-surface resources are around 290 billion barrels, though the recoverable amount is not known. The better studied deposits are near the Baltic Sea and in the Volga-Urals basin. Some deposits are also known in Eastern Siberia. The Baltic deposits were historically exploited as solid fuel for power plants (as in nearby Estonia), but were abandoned in favour of gas. There is currently no plan for exploitation of near-surface kerogen shales in Russia.

Of note is the Bazhenov shale, the source rock underlying all the Western Siberia reservoirs (IEA, 2010a). Probably the most extensive shale formation in the world, it contains both some light tight oil, similarly to the Bakken shale in the United States, and very large remaining amounts of kerogenic matter. Technology to economically produce the latter, which lies at a depth of 3 000 metres, would be a breakthrough, potentially extending the life of the Western Siberian infrastructure by many years. We have not included any production for kerogen shales in our projections to 2035.

Gas-to-liquids and coal-to-liquids

Gas-to-liquids (GTL) could be an attractive way for Russia to exploit gas fields located far from pipelines, while also hedging against decoupling of gas and oil prices. Some projects have been proposed, for example in the Yakutia region, to produce either diesel and naphtha or methanol. However, no project seems to have passed the conceptual stage so far. Taking into account possible technological developments in small-scale GTL (IEA, 2010a), we project GTL production in Russia to start in the 2020s and to grow to close to 120 kb/d in 2035. A large project based on Arctic gas, as an alternative or in addition to some of the planned LNG plants, could be a viable way to extract value from some expensive Arctic gas fields; such a development is not included in our projections.

Coal-to-liquids (CTL) is also potentially attractive, as Russia has large coal resources located far from markets. There have been some reports of preliminary discussions about CTL projects – China's Shenhua group has expressed interest – but in the absence of specific information about project developments, we have not included any CTL production in our projections to 2035.

Coalbed methane

Coalbed methane (CBM) exploitation is the most advanced among the unconventional hydrocarbon resources in Russia. Pilot projects are already operated by Gazprom in the Kemerovo region and there are concrete plans to move to large-scale production of 4 bcm per year by 2016 and up to 20 bcm per year in the longer term. We project this interest will continue, with production growing to 38 bcm per year by 2035. Recoverable resources are estimated to be 17 tcm, ample to sustain this level of production.

Shale gas and tight gas

Shale gas resources in Russia are very poorly known, with the pioneering work of Rogner still being the basis for most estimates (Rogner, 1997). IEA analysis estimates about 4 tcm of recoverable shale gas in Russia (IEA, 2009). Russia's shale formations are in regions that are not as forbidding as the Arctic, such as the Volga Ural region, the Baltic region or even the Moscow region, so this resource would seem to deserve more attention. Similarly, there is a shortage of information about other sources of tight gas (gas contained in very low permeability formations) in Russia; if such formations are defined as those requiring the use of hydraulic fracturing technology to achieve economical production, we estimate current tight gas production in Russia to be about 20 bcm per year. In the New Policies Scenario, we project that production from gas shales and other tight gas formations will grow slowly, to around 30 bcm per year in 2035.

Methane hydrates

Although the total amount of methane in hydrate deposits around the world is the object of widely different estimates, there is no doubt that it is extremely large (IEA, 2009) and that a substantial part of these resources is located in the Russian Arctic, both onshore in permafrost and offshore on the continental platform of the Arctic seas. Although it is often reported that the Messoyakha field in the north of Western Siberia produced gas from methane hydrates, this appears to be somewhat by chance, with a conventional gas reservoir happening to be recharged from above by depressurisation of methane hydrates. Even this interpretation is not fully established.

Although the resources are extremely large, the lack of established technologies to produce methane from hydrates and the large conventional gas resources available in the same Russian Arctic regions, account for the absence of a concrete pilot production project in Russia. In fact, methane hydrates present in permafrost are more often seen as a safety hazard in northern drilling than a resource, quite apart from being a potential environmental threat (Box 8.6). We do not project any production from methane hydrates during the period of this *Outlook*.

Box 8.6 ● Methane hydrates and climate risks

Methane hydrates are not only potential gas resources; they are also a possible major contributor to climate change. Global warming may trigger the dissociation of methane hydrates, potentially releasing massive amounts of methane into the atmosphere. Since methane has a greenhouse gas effect 25 times that of CO_2 on a mass basis over 100 years, such a release could trigger a catastrophic feedback loop.

The hydrates located in the Arctic are considered most at risk, in this sense, due to the recent rapid warming in the Arctic and the shallow depth of the seas or low-lying permafrost coastal areas where they are located. In particular the Eastern Siberian Sea region is considered to be the most vulnerable. Recent, preliminary measurements (Shakhova and Semiletov, 2010) suggest methane hydrate dissociation rates far higher than previously thought. Given the very large amount of methane thought to exist as hydrates in the region, if such rates were confirmed, there could be a sudden surge of interest in producing the methane before it is naturally released. For example, at a long-term price of CO_2 of $50 per tonne of CO_2 equivalent, the value of preventing methane release to the atmosphere would be a staggering $890/kcm ($25/MBtu).

Coal

Russia has vast coal resources. Ultimately recoverable resources are estimated to be of the order of 4 trillion tonnes, ranking third in the world after the United States and China. About two-thirds of this amount is hard coal and the rest is brown coal. Proven reserves amount to about 160 billion tonnes under the PRMS classification (190 billion tonnes under the Russian "ABC1" classification), of which some 70 billion tonnes are hard coal and the rest brown coal (BGR, 2010).

Although there are coal deposits in many regions of Russia, the Kuznets basin (Kuzbass) in the Kemerovo region in Siberia alone accounts for 60% of production. The second-largest basin is the Kansko-Achinsk, in the nearby southern part of the Krasnoyarsk region, with about 15% of production, primarily lignite. The rest comes mostly from various parts of Eastern Siberia and the Far East and, to a lesser extent, from the Timan-Pechora basin and the Russian part of the Donetsk basin (Donbass), near the border with Ukraine. Many of the coal basins suffer from a harsh climate and tend to be very remote, with limited coal transport infrastructure. The dominance of the Kuzbass and Kansko-Achinsk basins results both from their vast resources and from their location near the trans-Siberian railway, in the south of Siberia. Remoteness is responsible for the vast resources of Siberia's Tungusk basin or the Lena basin being hardly exploited.

Production has steadily increased over the last ten years (Figure 8.16), except in 2009 when the industry suffered as a result of the economic crisis. This pattern of increase has been driven by increased exports rather than growth in domestic demand. Russia has become the third-largest coal exporter in the world (after Australia and Indonesia). The main export market has been the European Union, which absorbs more than 50% of Russia's exports, but our projections in the New Policies Scenario indicate that the focus will switch to the east, to China in particular. This dovetails with the expectation in Russian strategy documents that the main areas of growth in coal production will be in the east of the country, notably the Kansko-Achinsk basin and Eastern Siberia. In these circumstances, as in the oil and gas sectors, increased Chinese involvement in the Russian coal sector would be logical: alongside discussion of possible loans for coal sector development, there are already strong indications of interest from Chinese companies in direct participation in ventures to develop Russia's coal resources. Coal imports to Russia have been steady at around 20 Mtce per year. They come primarily from Kazakhstan, as a number of power plants in Russia are linked to their traditional supply source from Soviet times.

Figure 8.16 ● Russian coal production and exports

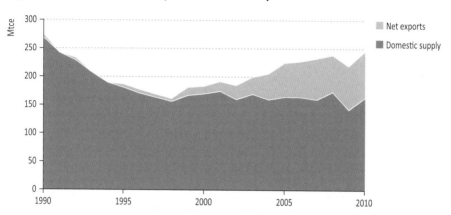

The Russian Energy Strategy sets some ambitious goals for increased coal production over the coming decades, with output targets of between 315 Mtce and 375 Mtce by 2030, depending on the scenario. The implementation of the Strategy detailed in the long-term development programme for the coal sector similarly sets the lower range for future output at 320 Mtce in 2030. Our projections are more conservative (Table 8.6), as transport costs put a limitation on the ability of Russian coal to compete in export markets and internal demand is gradually reduced by improved efficiency. We project coal production in Russia to be about 250 Mtce per year in 2035, after peaking just below 270 Mtce in the mid-2020s.

Table 8.6 ● Coal production in Russia by type in the New Policies Scenario (Mtce)

	2009	2015	2020	2025	2030	2035
Coking coal	58	61	62	60	59	58
Steam coal	126	158	162	167	161	154
Brown coal	35	38	38	39	37	36
Total	**219**	**258**	**262**	**267**	**257**	**248**
Net exports	77	94	96	96	88	80

Despite an existing policy objective to increase coal use in the domestic market, the share of coal in Russian primary energy consumption has continued to fall, from 22% in 1990 to 19% in 2000 and 15% in 2009. Coal use in Russia has faced formidable competition in the shape of low-priced and readily available domestic gas supplies. As discussed in Chapter 7, the price advantage of gas in the Russian domestic market is gradually being reduced as gas prices increase. This creates an opportunity for coal to increase its share in the Russian energy mix – particularly in the event that the expansion of nuclear power capacity does not proceed as planned. However, while the mining costs of Russian coal are relatively low by international standards, the distance from the main coal production areas in Siberia to the main consumption centres in the European part of Russia is a constraint on pricing coal competitively with gas in the domestic market – all the more so as the western-most Russian reserves in the Pechora fields and the Donbass are being depleted quickly.[27] There are also logistical and, potentially, environmental issues to overcome before coal could be used more widely to generate power and heat for the largest cities of Russia. This applies especially to combined heat and power (CHP) plants, which tend to be located in residential areas.

The prospects for exports in the New Policies Scenario are determined by the shifting supply/demand balances projected for the European and Chinese markets. The main current export destination is the European Union: deliveries in this direction are expected to decline as the European Union import requirement falls from the peak of over 190 Mtce, reached in 2008, to 155 Mtce in 2020 and 110 Mtce in 2035. Net Chinese import demand (around 125 Mtce in 2010) increases over the current decade, reaching around 190 Mtce in 2018, before tailing off to 80 Mtce in 2035, but is vulnerable to small changes in the Chinese domestic supply and demand balance. The result for Russia is that total net exports rise to around 100 Mtce in 2018, before decreasing to less than 80 Mtce in 2035 as China's thirst for imports declines.

The competitive position of Russian coal on international markets, as within Russia, is affected by relatively high transportation costs, which account for a high proportion of export costs. Russia is at the top of the international cash-cost curve for internationally traded steam coal (see Figure 11.5 in Chapter 11), leaving it with the smallest margins on international sales and meaning that Russia is likely to be among the exporters first affected by any downturn in international demand. A key strategic issue, identified in the Russian authorities' draft programme for the coal sector, is adequate rail and port capacity as well as efficient management of the logistics. Charges at Russian ports, for example, can add up to $10/tonne to costs, compared with typical charges of $2/tonne to $5/tonne elsewhere.

Nuclear

The Russian nuclear industry languished for many years in the wake of the nuclear catastrophe at Chernobyl, but its fortunes have improved over the last decade, with increased political support, new state funding for the domestic construction programme and a greatly enhanced role for nuclear in Russia's energy strategy. Rosatom – the state nuclear energy corporation – and the Russian government have plans in place that would more than double Russia's nuclear capacity over the next two decades. To date, these plans have

---

27. See Box 7.4 in Chapter 7. A possible alternative would be to locate coal-fired power plants closer to the reserves in Siberia and then transmit electricity rather than transport coal. Thus far this has been held back by insufficient cross-country transmission capacity, a constraint which is expected to be eased (but not removed) by new investment in the electricity grid during the projection period.

not been affected by any change of policy after the Fukushima accident and Russia has the raw materials, and industrial and technological foundation to rebuild its nuclear industry. However, thus far, implementation has been held back by high capital costs, financial resource constraints and lengthy commissioning periods. We expect these constraints to remain, at least in the period to 2020. Russia's commitment to a number of international nuclear projects is also likely to slow the pace of growth at home.

In the New Policies Scenario, Russian nuclear capacity rises from the current 24 gigawatts (GW) to 31 GW in 2020 and 37 GW in 2035, an average increase of 1.7% per year over the projection period (Figure 8.17). This is more than 25% below the lower end of the range targeted in the Energy Strategy to 2030 and in other development plans for the electricity and nuclear sectors. However, since electricity demand in the New Policies Scenario is also considerably lower than the level anticipated by Russian strategy documents, the share of nuclear power in overall electricity generation still increases, from the current 17% to 19% in 2035 (and 18.6% in 2030, meeting the increase targeted by the Energy Strategy to 2030). Nuclear power plays a more prominent role in the 450 Scenario, pushed by the assumed introduction of a domestic emissions trading scheme after 2020. In this scenario, nuclear capacity rises to 50 GW and electricity output to over 365 terawatt-hours (TWh) in 2035.

Figure 8.17 ● Installed nuclear capacity and share of electricity generation in the New Policies Scenario

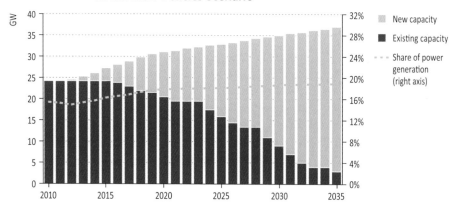

Note: The schedule for retirement of existing capacity is based on the assumption that all current plants receive a 15-year extension to their original 30-year license period.

As of 2011, Russia has 32 nuclear reactors at ten power plants. All are operated by the state nuclear power generation company, Rosenergoatom (part of Rosatom, the company with second-largest nuclear generation capacity in the world, after EdF).[28] Three reactors have

---

28. Of the reactors in operation, sixteen use VVER technology (a Soviet technology similar to the pressurised water reactor technology in OECD countries), fifteen use RBMK technology (a Soviet graphite-moderated design, of the type used in Chernobyl, and now considered obsolete) and there is one fast-breeder reactor. The standard generating unit for new-build reactors is an updated VVER design, providing 1 200 MWe.

been commissioned in the last decade, Rostov-1 (also known as Volgodonsk-1) in 2001, Kalinin-3 in 2004 and Rostov-2 in 2010. Two others started operation in the 1990s. All of the rest (27 out of 32) were commissioned in the 1970s and 1980s, for an initial licence period of thirty years. Rosatom has already granted, or is considering, the extension of the operating lives of these reactors, normally by an additional fifteen years.

The output of Russia's nuclear plants rose from 130 TWh in 2000 to 164 TWh in 2009, primarily because of an increase in the load factor to close to 80% in 2009. Electricity production from nuclear power is projected to reach 270 TWh in 2035. Given that almost all of the existing units are scheduled for retirement by the end of the projection period, this means a major acceleration in the commissioning of new reactors. Of the additional capacity, 12 GW is expected to come from the completion of previously stalled constructions; the rest would be new reactors, both at existing sites and at twelve new power plants around Russia.[29]

Rapid development of Russia's nuclear power generation capacity will be very capital-intensive and costs could be pushed even higher if the Fukushima incident leads to additional safety requirements. Our estimate of the total cost of the nuclear capacity additions foreseen in the New Policies Scenario is $115 billion (in year-2010 dollars), an average of $4.6 billion each year. This is broadly in line with figures announced by Rosatom for 2010, in a total investment programme of 163 billion rubles ($5.2 billion), of which around three-quarters was dedicated to new build. As of 2010, Rosatom was building ten new reactors on six sites, as well as a floating nuclear power plant designed for remote, Arctic locations.[30]

Along with consistently high levels of state financing, the nuclear expansion programme will require sufficient and timely commitment from Russian industry, for example, to build the huge pressure vessels housing the reactors. It will also require a high number and level of nuclear specialists: as in other countries this is no small task, given that there were few qualified young engineers attracted to the nuclear industry in the 1990s. The available industrial and human resources will need to be sufficient not only for the domestic nuclear programme, but also for a growing number of international projects. Atomstroyexport, also part of Rosatom, has been an increasingly active competitor for business abroad: Russia has nuclear power projects underway or in the planning stage in fifteen countries around the world, with the firmest prospects in China (where two reactors were completed in 2006/2007 at Tianwan), India, Turkey, Belarus, Vietnam, Armenia and Ukraine. All of these factors lead us to be cautious when assessing the prospects for a very swift increase in Russian nuclear capacity in the domestic market.

---

29. Not all of the new capacity would be for the Russian market: the first of a project for two 1 200 MWe VVER units in the Kaliningrad region of Russia, close to the Lithuanian border, is provisionally scheduled to start operation in 2016 and is aiming to sell more than half of its power output to Germany, Poland and the Baltic states.

30. The launch of the world's first floating nuclear power plant, the Akademik Lomonosov, equipped with two 35 MW reactors, took place in 2010. It is planned to begin to provide power at an offshore Arctic drilling site from 2012. Small modular reactors are a new technology gaining considerable interest; however, there are question marks about the economics and potential risks of this type of small nuclear plant.

Russia has significant uranium resources to support its ambitions for nuclear power: these are estimated at 648 thousand tonnes (kt) of which 100 kt are proven reserves, around 4% of the global total (BGR, 2010).[31] The figure for proven reserves is 284 kt under the Russian "ABC1" classification. Most of these are in the Sakha Republic in Russia's Far East and along the border with Mongolia, particularly in the Zabaikalsky region. Russia produced 3.6 kt in 2009, or 7% of global output. Unlike fossil fuels, mined production of uranium is less than consumption, since commercial and military inventories account for around a quarter of global supply; but increased uranium demand for nuclear power, both in Russia and abroad, is expected to tighten the global balance.

The main current sources of Russian supply are in the Zabaikalsky region, but Russia has plans to increase uranium production by exploiting the more remote reserves of the Sakha Republic. The major project in this area is the huge Elkon development, where production is expected to start around 2015 and to reach 5 kt by the mid-2020s, more than Russia's entire output today. Major challenges, as with many of the remaining deposits, are the distance from existing infrastructure and a severe climate. Several foreign companies from Europe, Japan, India and China participate in, or are interested in, joint projects for uranium mining in Russia, in partnership with state-owned AtomRedMetZoloto (ARMZ). ARMZ is also expanding its operations outside Russia, having taken a controlling stake in Canada's Uranium One (which operates several projects in Kazakhstan). Russia has extensive conversion, enrichment and nuclear fuel fabrication capacity, providing fuel and services both to Russian plants and to foreign nuclear operators. Russia has also created the International Uranium Enrichment Centre in Angarsk, Siberia, a multilateral initiative that aims to ensure guaranteed supplies of uranium products to member countries (aside from Russia, these are currently Armenia, Kazakhstan and Ukraine) as an alternative to development of their own enrichment capacities.[32]

Hydropower and other renewables

Russia has a well-established hydropower sector, developed largely during Soviet times. However, Russia has far from exhausted its hydropower potential and it has not used more than a fraction of the possibilities for other, non-hydro renewables. The current contribution of renewable energy to the Russian energy balance is anchored in 48 GW of installed hydropower capacity, which accounted for 18% of total electricity output in 2009 (Figure 8.18). The other main contribution comes from biomass, mostly firewood used for heating in rural areas. The use of modern renewable technologies is marginal.

---

31. Unlike other energy resources, reserves and resources of uranium are categorised according to production cost; reserves figures from BGR are <$80/kg.

32. The initiative also now incorporates the idea of a "fuel bank", under an agreement with the International Atomic Energy Agency, which establishes a reserve of uranium that would be available to nations that face supply disruptions unrelated to technical or commercial reasons.

Figure 8.18 ● Share of renewable energy in Russian total primary energy demand, electricity and heat production, 2009

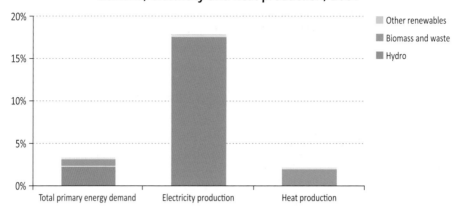

Hydropower

Russian hydropower, like the nuclear industry, is emerging from a period of dramatic slowdown in activity in the 1990s that saw construction of many planned new plants halted for lack of funding. As with the nuclear sector, the hydropower sector has some ambitious plans for expansion in the coming decades, but there are uncertainties over how quickly they will be realised. There was a tragic setback in 2009 when an accident at the 6 400 megawatt (MW) Sayano-Shushenskaya plant in Siberia destroyed three of the ten turbines at Russia's largest hydropower facility and damaged the rest.

In 2009, total production from Russia's hydropower and pumped storage plants was slightly below the post-1990 peak of 177 TWh, reached in 2007. The largest hydropower company in Russia is the majority state-owned Rushydro; other assets belong to regional energy companies and some are also linked to energy-intensive industries and industrial groups. Output from the Krasnoyarsk HPP (6 GW), for example, goes almost entirely to the huge Krasnoyarsk aluminium smelter.

We project that total production from hydropower will increase slowly over the initial part of the projection period, but then more rapidly after 2020, to reach 218 TWh in 2035. Its share in the overall electricity mix falls from 18% in 2009 but stays at around 15% after 2020. Total capacity increases from today's 48 GW to 51 GW in 2020 and then to 61 GW in 2035. Additional capacity in the early part of the projection period comes mainly from the completion of unfinished projects, many of which are in the North Caucasus region. Later on, growth is expected in parts of European Russia (predominantly in the form of pumped storage), but the main locations of capacity growth are anticipated to be Siberia and – to a lesser extent – the Far East.

The investment challenge for hydropower in Russia is not related to a shortage of resources: overall, Russia uses 20% of its economically exploitable hydropower potential.[33] The largest

---

33. This figure for Russia is 852 TWh per year (WEC, 2010) and is defined as amount of the gross theoretical capability that can be exploited within the limits of current technology under present and expected local economic conditions.

obstacle, as discussed for the coal sector, is the distance to market. Most of the remaining hydropower potential is in Siberia, far from the main electricity consumption centres. Use of Russia's hydropower potential is expected to remain constrained by limited cross-country transmission capacity, although investment in the electricity network eases this constraint somewhat in the latter part of the projection period. An alternative solution to the problem of limited local demand would be an expansion of cross-border electricity trade with China; bilateral declarations and commercial interest have yet to be turned into specific initiatives, with the main barrier being the lack of long-distance transmission capacity.

The sector also faces a major challenge with ageing equipment and infrastructure: of 510 functioning hydropower plants across the country, 72% are more than thirty years old and 38% are already older than fifty years. Part of future investment must therefore go to upgrading existing facilities, with a short-term imperative, for Rushydro, to reconstruct the Sayano-Shushenskaya plant. Repair works at this plant are scheduled for completion by 2014, at an estimated total cost of 33 billion rubles ($1.2 billion).

Other renewables

Aside from large hydropower, modern renewables currently occupy only a small niche in the Russian energy mix, a position well below their potential. In the New Policies Scenario, the role of non-hydro renewables is expected to increase to 2035; their share of total primary energy supply reaches 4% in 2035 from 1% in 2009, a very large increase in percentage terms, but still constituting only a small part of Russia's overall energy mix.

The use of renewables for power generation is confined at present to a small number of local and regional projects, none of which have been built into the integrated network. The cumulative capacity of small hydropower projects (defined as less than 25 MW) is around 250 MW. There are geothermal stations in the Far East (Kamchatka), providing another 80 MW of capacity; and around 16 MW from pilot projects for wind power. There is also an experimental 1.7 MW tidal project in the Barents Sea that dates back to 1968. Traditional biomass (firewood) is widely used for space heating in rural areas, providing 2% of Russia's heat supply,[34] although this is not necessarily sustainable and the firewood is often burned at low efficiencies.

A sharp increase in the uptake of renewable technologies is an objective set in a number of Russian strategy documents.[35] The most prominent of the various targets is to achieve a 4.5% share for renewables (excluding large hydro) in electricity generation by 2020: this was accompanied by intermediate targets for 2010 (1.5%, which has already been missed) and 2015 (2.5%). Based on electricity demand projections, we estimate that the 2020 target would require 55 TWh of electricity to come from renewables (excluding large hydro). This implies the addition of nearly 15 GW of renewable electricity generation capacity over

---

34. Estimates of the share of biomass in Russia's heat production are often higher, around 4%; use of firewood for fuel is often under-reported in national statistics.

35. Expanding the share of renewable energy is identified as a priority in government documents: Energy Strategy to 2030 (Government of Russia, 2009); Concept of Long-term Social and Economic Development to 2020 (Government of Russia, 2008); the Russian Climate Doctrine and related Action Plan (Government of Russia, 2011); as well as the State Programme on Energy Saving to 2020 (Government of Russia, 2010).

the next nine years, or 1.6 GW per year.[36] For comparison, China added over 37 GW of renewables capacity in 2009 alone.

The investment required to meet this target in 2020 is estimated at $26 billion (in year-2010 dollars), or $3 billion per year. For the moment, there are no supportive policies or incentives in place to attract this sort of capital to the renewables sector. Various schemes are under discussion but, against a background of concern about high end-user electricity prices, the government appears reluctant for the moment to add higher costs into the wholesale mix by agreeing to include a premium in tariffs for new renewables-based electricity. We assume in the New Policies Scenario that support mechanisms will be put in place in Russia by 2014 (see Chapter 7). With implementation of support schemes only from that date, the 4.5% share of renewables in electricity generation is not achieved by the official target date of 2020 (Figure 8.19). In the 450 Scenario, we assume a much more concerted effort to promote low-carbon technologies, leading to their earlier and faster deployment.

Figure 8.19 ● **Share of renewables in power generation in Russia in the New Policies Scenario** (excluding large hydropower)

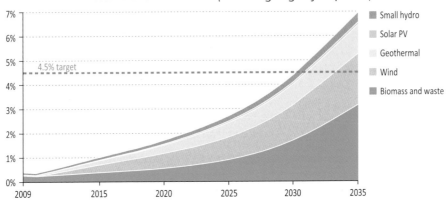

There are opportunities for renewable energy in Russia that are less contingent on state support. Chief among these is the provision of energy to isolated communities. Large parts of Siberia and the Far East are very sparsely populated and are not connected to the unified grid.[37] Power often comes from small diesel generators and the supply of fuel to these communities is very expensive – in some cases having to be brought in by helicopter. Small-scale renewable technologies, with back-up from the existing generators, can be a very competitive proposition in these areas and would not necessarily need financial incentives. With improved access to financing on commercial terms, better information and the removal of administrative and other barriers, there could be a significant expansion of off-grid renewables, primarily wind, geothermal and biomass.

---

36. Our estimate is based on the mix of renewable technologies that Russia is projected to have when electricity output from renewables reaches 55 TWh per year (in 2028).

37. These are excluded from wholesale trade (and all electricity is sold at regulated prices); support schemes designed for the wholesale market – as currently under discussion – would in any case have no impact on renewables projects in isolated areas.

The adoption of renewable technologies in Russia may be held back initially by the limited industrial capacity to supply components or products (aside from small hydropower plants and, to an extent, geothermal energy). The government is taking steps to address this issue, but it will take time to develop a domestic manufacturing base in the renewables sector (and, in the meantime, there are relatively few strong domestic voices – with the exception of Rushydro – pushing for stronger policies). Nonetheless, with the progressive removal of this constraint and the introduction of support schemes for the wholesale market, we project more rapid growth for a range of renewable technologies. The opportunities for small-scale hydropower are mainly in the North Caucasus and parts of Siberia; geothermal potential is concentrated in the Far East and in the North Caucasus; and, although most of Russia's wind potential is along the northern and Pacific coasts, areas of very low population density, there is the potential for increased generation of electricity from wind in the southern part of European Russia (IEA, 2003, and Popel et al., 2010).

There are also extensive opportunities for Russia to use forestry residues and other biomass for power and heat generation. Examples already exist in Russia of heat boilers that have been switched to use biomass (wood pellets), but only 1 600 of the 66 000 residential heat supply plants in Russia are fired by renewable sources (Ministry of Natural Resources and the Environment of Russia, 2010). The areas with the richest sources of this raw material are in the northwest and in Siberia. In northwest Russia, the estimated annual resource of residues from the forestry industry is more than 30 million m^3, based on actual production from the timber industry (Gerasimov and Karjalainen, 2011). This is equivalent to over 5 Mtoe, or 60 TWh of energy, only a tiny proportion of which is used. In neighbouring Finland, solid biomass use of 7 Mtoe provides almost 20% of primary energy supply; in northwest Russia, the share of all renewables is around 2%.

There is scope to increase biomass use through co-firing or conversion of existing boilers and biomass-based CHP could also contribute to meeting the Russian 4.5% renewables target. In many eastern parts of the country, greater use of biomass for heat and combined heat and power could be an efficient alternative to the gasification programme that is currently planned. In western areas, there is also considerable scope to develop renewables projects primarily with a view to exporting electricity to the European Union. In addition to possible exports of the biomass itself, any electricity generated from renewable energy for export to the European Union would contribute to meeting member country targets.[38]

8

---

38. The European Union Directive 2009/28 on promotion of renewables aims to encourage joint energy projects between European Union members and third countries; electricity imported from new renewables projects in neighbouring countries can count towards an European Union country's renewables target. This would require, in many cases, the expansion of cross-border interconnection capacity.

CHAPTER 9

IMPLICATIONS OF RUSSIA'S ENERGY DEVELOPMENT
Who depends on whom?

H I G H L I G H T S

- Although Russian energy use becomes more efficient and reliance on oil and gas in the national economy declines, the pace of these changes in the New Policies Scenario is more modest than Russia's aspirations. The share of the oil and gas sectors in Russian GDP decreases from an estimated 24% in 2011 to 20% in 2020 and 15% in 2035. Faster implementation of energy efficiency measures could help to accelerate the modernisation of the Russian economy and thereby reduce more quickly the risks to the Russian economy from excessive reliance on the oil and gas sectors.

- Russia will need a cumulative investment of over $2.5 trillion (in year-2010 dollars) to meet the supply requirements of the New Policies Scenario, with the largest share of this in the gas sector (43%), followed by oil (31%) and then power (25%); average annual investment needs are over $100 billion, 7% of Russia's current GDP (MER).

- China becomes a major contributor to Russia's revenues from the export of fossil fuels, its share increasing from 2% to 20% over the *Outlook* period while that of the European Union falls from 61% to less than half. Total revenues from the export of fossil fuels rise from $255 billion in 2010 to $420 billion (in year-2010 dollars) in 2035. The domestic gas market is an increasingly important source of value as domestic natural gas prices are raised.

- At 9.7 mb/d of production in 2035, Russia is the largest non-OPEC oil producer and the second-largest global producer, underlining its vital role in oil markets even as oil exports decline over the projection period. New pipeline routes to both east and west create a more diverse and flexible oil export system, as well as opportunities to enhance Russia's role as a transit country for Caspian oil.

- By 2035, Russia provides more than 30% of the gas imported both by the European Union (over 170 bcm) and by China (75 bcm), underlining Russia's central position in Eurasian and global gas security. Although not at the levels of westward Russian gas export to Europe, the Russia-China relationship is set to become one of the main arteries of global gas trade.

- Russia benefits from greater diversity of gas export markets, which create a degree of competition between Europe and Asia for positions in Russian supply. However, the changing dynamics of global gas markets and pressure on traditional pricing models for gas are creating competitive challenges for Russia as it moves to higher-cost sources for incremental gas output and looks to expand its position in LNG markets.

Energy and national economic development

Energy use in Russia is changing. A large part of Russia's industrial and social infrastructure – its factories, power plants, buildings, networks – needs to be rebuilt or replaced over the coming decades: much of it is reaching the end of its useful life. Given the low efficiency of the existing stock compared to the average technologies available on the market, all three scenarios examined in this *Outlook* (and almost any plausible scenario for Russian economic development) bring an improvement in the overall efficiency of energy use. The speed and depth of this change will depend on Russian policy choices. Driving these choices will be Russia's aspiration to create a more efficient, dynamic and broad-based economy, less dependent on the oil and gas sectors (Box 9.1).

Box 9.1 ● Oil and gas in the Russian economy

It is often said that Russia relies too heavily on oil and gas, but it is surprisingly difficult to pin down exactly how much these sectors contribute to the Russian economy. There are some useful indirect indicators, for example the share of oil and gas in Russia's overall export earnings at around two-thirds (Figure 9.1), and the share of oil and gas revenues in federal budget income at almost half.[1] But there are different figures on the size of the oil and gas sector in Russia's gross domestic product (GDP).

According to the Federal State Statistics Service (Rosstat), oil and gas production and related services represent a relatively small share of national output, around 6% to 8% as indicated by data for 2004 to 2009. United Nations statistics likewise show that the entire extractive sector, including oil and gas, accounts for 9% of Russian GDP. This compares to the equivalent figure of over 20% in Norway and around 50% in Saudi Arabia.

Investigating why the figure for Russia is lower than for other oil and gas exporters, different studies (World Bank, 2005, and Kuboniwa *et al.*, 2005) have pointed to the share of Russia's oil and gas activity that is undertaken by trading companies which are related to, but separate from, the production entities. These trading companies often sell the oil and gas on domestic and international markets and, in the statistics, this has the effect of moving profits and value-added from the oil and gas (manufacturing) sector to the trade (services) sector.

The Russian Ministry of Economic Development has put the share of oil and gas in Russian GDP at 18.7% for 2007 (Government of Russia, 2008) and Gurvich (2010) has also reviewed the data to come up with higher estimates. For the purposes of this *Outlook*, we calculated value-added from the oil and gas sector based on revenue from domestic and foreign oil and gas sales, minus relevant production, transportation and intermediate costs. We estimate that the oil (excluding refining) and gas sectors provided 17% of GDP in 2007, 24% in 2008 and – after falling back in 2009 – 21% in 2010.

---

1. A 43% share of budget revenues in 2010 from the oil and gas sectors (3.6 trillion rubles out of 8.3 trillion) includes mineral extraction tax, VAT and excise taxes, and export duties.

Figure 9.1 ● Structure of Russian exports by value, 2009

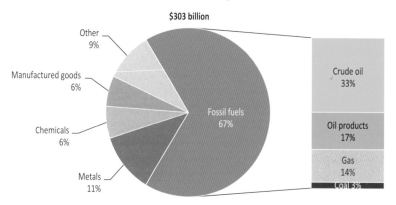

Note: 2009 was an anomalous year in terms of the volume of exports, but in terms of the value structure of Russian exports fossil fuels held a similar share of the total in 2007 (66%) and 2008 (69%).

Sources: Ministry of Natural Resources and the Environment of Russia; Central Bank of Russia.

Improvements in energy efficiency are critical to the modernisation of the Russian economy (OECD, 2011): efficiency improvements in energy production and use would be accompanied and sustained by a broader process of technological change and innovation, as the Russian economy adjusts to the demand for more efficient equipment and related energy services. Industries that produce tradable goods, and particularly those seeking export markets, benefit from lower production costs and increased competitiveness. Energy savings by households and industry release resources that can be used for productive investment. Energy efficiency also reduces the investment required for domestic energy supply (or, alternatively, frees up additional resources for export), as well as improving environmental outcomes.

The efficiency gains projected in the New Policies Scenario are relatively modest compared with Russia's potential, but some of these benefits are nonetheless visible and quantifiable from the analysis conducted in Chapter 7.[2] Cumulative spending on energy is $230 billion less (in year-2010 dollars) than in the Current Policies Scenario (see Box 7.2); the investment requirement in upstream oil, gas and coal is reduced by a total of $130 billion; emissions of greenhouse gases and major air pollutants are significantly lower.

By improving Russia's medium-term prospects for economic development, the resources and income released by energy efficiency also help to address a second vulnerability facing Russia, the dependence of the national economy on oil and gas (Box 9.1), underlined by the large share of fossil fuels in the value of Russian exports (Figure 9.1). As brought home in 2009, the main risk of such high reliance on energy is the way that it ties Russian economic

---

2. Total primary energy demand in 2035 in the New Policies Scenario is 830 million tonnes of oil equivalent (Mtoe), compared with over 900 Mtoe in the Current Policies Scenario. However, Russia taps into a relatively small proportion of its potential efficiency gains in the New Policies Scenario. Our analysis in Chapter 7 suggests that potential savings from increased energy efficiency, relative to comparable OECD countries, are equal to 30% of Russian energy consumption in 2008 and remain at 18% in 2035 in the New Policies Scenario.

and fiscal fortunes to movements in international commodity prices. Russia was among the countries worst affected by the global financial and economic crisis – in large part because of the plunge in oil prices and sharp contraction of global energy demand. We estimate that total Russian revenues from oil and gas exports fell by 40% in 2009 compared to the record levels of 2008. The direct effect of the oil price shock was compounded by similar falls in the prices of other important export commodities, such as base metals, as well as a sharp swing in capital flows as investors drew back from commodity plays and investments perceived as high risk, including emerging market assets. As a result, Russia suffered a larger decline in GDP (8%) than any other major economy in 2009 and, after budget surpluses in every year since 2000, the budget deficit in 2009 was 6.3% of GDP. High reliance on the oil and gas sectors also creates potential pitfalls for Russia that have been documented in the literature on the "resource curse", *i.e.* the risk that an abundance of natural resources can actually hinder broader economic growth and human development in the longer term.[3]

Based on our GDP assumptions and projections for oil and gas production and export, we project that the contribution of oil and gas to Russia's GDP will decline gradually from the estimated 2011 level of 24%, reaching 20% in 2020 and 15% in 2035 in the New Policies Scenario (Figure 9.2). The relative contributions of oil and gas converge over the projection period, with the share attributable to the gas sector growing from 5% in 2010 to a high point of 7% in the mid-2020s, before settling back to 6% in 2035, while that of oil falls steadily from a peak of 19% in 2011 to 9% in 2035. The pace of change projected in this *Outlook* is significantly slower than the rate of change targeted by Russia, in part because of different assumptions about GDP growth: the Concept for Social and Economic Development to 2020 (Government of Russia, 2008) envisages reducing the share of the oil and gas sector to 12.7% as early as 2020.

Figure 9.2 ● Estimated share of oil and gas in Russian GDP in the New Policies Scenario

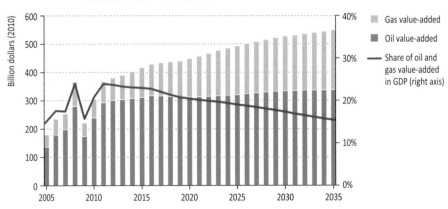

3. High resource export earnings can strengthen the exchange rate and discourage production in other sectors of the economy, although this risk is mitigated in part in Russia through two funds (the Reserve Fund and the National Welfare Fund), managed by the Ministry of Finance, which accumulate revenues from oil and gas export duties when oil revenues exceed a certain threshold.

Taken together, our assumption for Russia's GDP growth, the gradual projected decrease in the role of oil and gas in the economy and the moderate progress with energy efficiency improvements examined in Chapter 7 imply some continuing risks to Russia's national economic development. In particular, the trajectory anticipated in this scenario suggests that Russia could remain relatively exposed to external shocks, such as that in 2009, with any fall in international commodity prices having a substantial impact on overall economic activity.

·············· S P O T L I G H T ··················

What would higher GDP growth imply for the energy sector?

Russia's existing strategy documents for the energy sector, including the Energy Strategy to 2030, are based on a more optimistic set of GDP assumptions than those used in this *Outlook.* The differences are particularly evident in the period after 2015, when growth is assumed to be in the 5% to 6% range (as in the "favourable" scenario developed by the Ministry of Economic Development) before tailing off after 2025. This compares to average rates of growth in *WEO-2011* scenarios of 3% to 4% over the same period.

To allow for more direct comparison, we ran a "high GDP" case in line with the favourable scenario from the Ministry of Economic Development.[4] We kept most assumptions constant from the New Policies Scenario, but assumed a faster pace of energy efficiency improvement – in line with the logic that moving Russia on to a sustained trajectory of higher GDP growth will require a more far-reaching modernisation of the Russian economy and, in turn, that energy efficiency is an indispensable component of such a strategy. The main results include:

- Total primary energy demand rises to over 1 090 Mtoe in 2035, increasing by an average of 2% per year compared with 1% in the New Policies Scenario. Electricity demand reaches 1 870 terawatt-hours (TWh) in 2035, 75% higher than current levels.

- Russia meets its target for a 40% improvement in energy intensity, relative to a 2007 baseline, in 2023, considerably earlier than in the New Policies Scenario (when it is met in 2028). This underlines that, if Russia is to get close to the 40% reduction target by 2020, it will need to combine high GDP growth and a concerted effort to implement energy efficiency policies.

- While energy intensity targets are met earlier, Russia's targets for reducing greenhouse-gas emissions become more difficult to achieve as higher GDP growth pushes up overall demand and emissions. CO_2 emissions in 2020 are about 1 920 million tonnes (Mt), only 12% below 1990 levels, compared to 23% below in the New Policies Scenario.

9

4. In this high-GDP analysis, average GDP growth in the period 2009 to 2035 was assumed to be 5.3% per year compared with 3.6% in the regular scenarios.

A steeper trajectory for GDP, as targeted by policy makers as part of a revised strategy for social and economic development to 2020, would bring GDP projections back towards the levels used to underpin the Energy Strategy to 2030 (Spotlight). For the energy sector, we assume that high GDP growth in the medium term would require accelerated efforts to implement energy efficiency policies and raise the technical efficiency of Russia's capital stock, as well as putting in place the market structures necessary to ensure efficient investment, operation and end-use.[5]

Investment

Although one aim of the modernisation process is to diversify the Russian economy away from oil and gas, the need for innovation and efficiency applies to these parts of the energy industry as much as to the energy sector and the economy as a whole. With their very large investment needs over the projection period, there is an opportunity for the oil, gas and power sectors to play an important, even a leading, role in the technological transformation of the Russian economy. Russia will need a total of over $2.5 trillion in cumulative investment (all figures in year-2010 dollars) in order to meet the energy supply requirements of the New Policies Scenario over the period 2011 to 2035 (Figure 9.3).

Figure 9.3 ● **Cumulative investment requirement in coal, oil, gas and power supply in the New Policies Scenario, 2011-2035**

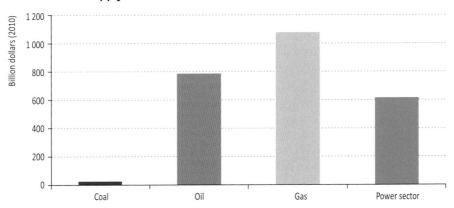

The largest share goes to the gas sector with over $1 trillion, of which $730 billion is in the upstream (including delivery of the gas to the existing transmission infrastructure), $250 billion is for maintaining the transmission and distribution network and $80 billion is

---

5. The link posited here between increased energy efficiency and high GDP growth means a difference in emphasis compared with the Russian Energy Strategy to 2030: the Strategy's "energy efficiency" scenario, which includes greater efforts to reduce greenhouse-gas emissions, assumes a lower rate of GDP growth compared to the baseline (by one year's growth in the period to 2030). While we recognise that a transformation of energy production and use at global level, as required by the 450 Scenario, could reduce global GDP by the equivalent of one year's growth to 2035, there are strong counter-arguments – not least the significant potential costs to GDP from a changing climate. This is why GDP is assumed not to change between the main scenarios analysed in this *Outlook*. In the Russian case, the downside risks to GDP would appear to be highest in a business-as-usual scenario.

in infrastructure for liquefied natural gas (LNG). The investment needs for the oil sector are $790 billion, most of which is in the upstream. The power sector will need $615 billion, 60% of this for generation and 40% for transmission and distribution. The investment needs of the coal sector are relatively modest, but still amount to $24 billion. The average annual investment required in the energy sector as a whole is over $100 billion, 7% of Russia's current GDP at market exchange rates (MER).

In the short term, the economic and financial crisis has given Russia some breathing space for energy investment by holding back energy demand both at home and on its major export markets. Even so, the current level of investment is not so distant from the average level required over the projection period. In the oil and gas sectors, for example, we estimate that total investment in upstream and midstream oil and gas projects was around $50 billion in 2010, against our annual average requirement in the New Policies Scenario of more than $70 billion for these sectors.[6] Nonetheless, it will be a challenge to mobilise this level of investment throughout the projection period. Success will depend not only on price levels and the fiscal regime, but also on Russia's choice of economic model for the next stage in its development, in particular the extent to which this investment will be expected to come from a limited number of state-owned or state-directed companies, rather than from multiple market players, both state and private, operating in a broadly non-discriminatory and competitive environment.

Revenues

Revenue from the energy sector will continue to be an important driver of growth and source of national wealth, particularly in the near term. We estimate that annual revenue from export sales of fossil fuels, including oil, gas and coal, will increase from $255 billion in 2010 to $420 billion in 2035 in the New Policies Scenario (in year-2010 dollars). Oil continues to account for the largest share of export revenue, although this share falls from 79% to 65% over the projection period as export volumes fall. The share of gas increases from 17% to 33%, while that of coal remains relatively small. The most significant shift is in the geographical sources of this revenue (Figure 9.4). While, in 2010, China accounted for a small share of the overall figure (2%), by 2035 this increases to 20%. The European Union remains the largest source of Russian fossil-fuel export earnings, but there is a reduction in its share from 61% to 48% in 2035.

A second notable shift over the period to 2020 is the rising value of the domestic market for natural gas, compared with export markets. This is due to the continuing process of gas price reform from the mid-2000s that put Russia on a path to equalise the domestic price paid by industry with the European export price (minus differential transportation costs and export duty). Gazprom estimates that it lost up to $50 billion over the last decade on price-restricted domestic sales, but reported a profit on domestic sales for the first time in 2009. Other natural gas producers in Russia, who are denied access to export markets, are obliged to sell only on domestic markets.

---

6. See tables 3.7 and 3.8 in Chapter 3 for data on oil and gas industry investment in Russia.

Figure 9.4 ● Sources of revenue from fossil fuel export sales, 2010 and 2035

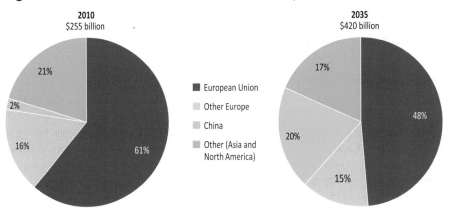

Note: Revenue is in year-2010 dollars.

We estimate that the domestic gas market produced around one-fifth of total Russian gas sales revenue in 2000 and that this share rose to 40% in 2010. In our projections, this share will continue to increase, as domestic prices rise, to account for half of total gas sales revenue in 2020 before declining in the latter part of the projection period (Figure 9.5). The share of the European Union in total Russian gas sales revenue has fallen from 60% in 2000 to under 40% in 2010. Although revenue from gas sales to the European Union rises in real terms over the projection period, expressed as a percentage of total revenue it continues to fall, to less than 30% in 2035, because of the value of domestic sales and then the increasing importance of Russian exports to China and other Asian economies.

Figure 9.5 ● Estimated share of Russian gas sales revenue from domestic and international gas sales in the New Policies Scenario

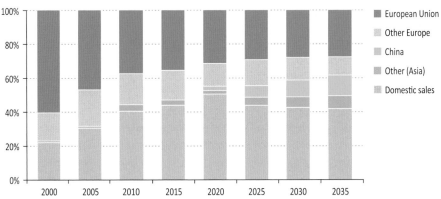

Eurasian and global energy security

Russia's weight in fossil-fuel resources, production and export give it a critical role in ensuring the adequacy and reliability of global energy supply. As of 2011, it is the largest oil producer and its role in oil markets would be underscored by any deferment of investment in the Middle East (see Chapter 3). Russia is also set to remain, throughout the projection period, the largest global gas producer, the largest external gas supplier to the European market and an increasingly important provider of gas – and, to some extent, also of coal – to China and the Asia-Pacific region.

However, there is a range of risks, both short term and longer term that attach to Russia's position in the global energy economy. The perceived short-term risks concern, particularly, the possibility of unexpected disruption to supply, whether caused by technical failures, accidents or politics; the longer-term factors relate to the adequacy of investment to enable Russian supply to keep up with demand. These risks are magnified, both for consumers and producers, in cases where there is high dependence on a single supplier, route, or export market, which is why diversity in these areas features strongly among policies designed to promote energy security. Some key issues related to Russia's contribution to global energy security are: the incentives offered for investment in supply (see Chapter 8); the strength and geographical sources of demand for Russian fossil-fuel exports; the reliability and diversity of the routes to market for Russian oil and gas; Russia's role in providing access to the market for other oil and gas producers (notably in the Caspian region); and the share of Russian energy (particularly natural gas) in individual export markets. As examined in Chapter 7, the policy framework for the domestic market and the efficiency of domestic consumption will also have a major impact on the availability of resources for export and in reducing the risk of demand running ahead of deliverability.

Regional energy relationships and oil and gas transit

Russia has a complex set of energy relationships with neighbouring countries. There is some evidence that ties across the former Soviet space are being normalised, as prices for Russian gas exports converge towards the equivalent of international prices. Yet there remain important open questions about future trade and transit flows. The first of these relates to Russia's gas relationship with Central Asia. Until 2009, large-scale Russian gas imports from Central Asia – and from Turkmenistan in particular – were a significant component of the Russian gas balance. But Russia now has a reduced need for gas imports, as a result of the effects of the economic crisis, more efficient gas use on the domestic market and rising production from non-Gazprom producers (see Chapter 8), while the commissioning of the Turkmenistan-China pipeline in late 2009 means that Central Asian gas is no longer available to Russia at a steep discount to international prices, removing the attractive possibilities for arbitrage trade. Under these circumstances, we assume that total Russian imports from Central Asia will remain under 40 billion cubic metres (bcm) in the New Policies Scenario.[7]

A second question concerns Ukraine and its role in the transit of Russian gas exports to the main European markets, after two disputes in 2006 and 2009 that led to interruptions in

---

7. As described in the *WEO-2010* focus on the Caspian region (IEA, 2010), higher Russian import levels are likely only if Central Asian gas is either available at a lower price or if there is a particular strategic decision to increase them, for example an attempt to forestall the development of alternative export routes from Central Asia to Europe. It is also possible that the call on Central Asian gas could rise in the event of a delay to major upstream developments in Russia, such as the gas projects on the Yamal peninsula, but non-Gazprom production within Russia would in all probability be a cheaper way to fill any gap.

gas deliveries to many parts of Europe. Despite some progress towards a more transparent, commercial basis for gas supply and transit, political considerations continue to bear on this relationship, as indicated by the gas supply agreement of April 2010, which provided Ukraine with temporary relief on the price of imported gas at the same time as an agreement was signed extending Russian rights to base its Black Sea Fleet in Crimea. This apparently irreducible element of politics, alongside the slow progress being made by Ukraine in tackling the inefficiency of its own gas consumption (so as to mitigate its high dependency on Russian gas imports) perpetuates the perception of risk associated with this route to market.

The share of Russia's export flows transiting Ukraine has decreased from over 90% in the 1990s to around 70% with the launch of new pipelines through Belarus (Yamal-Europe) in 1999 and across the Black Sea to Turkey (Blue Stream) in 2003. This figure will fall further with the commissioning of Nord Stream in 2011/2012. The planned South Stream pipeline across the Black Sea represents an additional threat to Ukraine's role as a transit country (Figure 9.6). If South Stream were to be built according to the schedule announced by the project sponsors then, with exports from Russia to Europe as projected in the New Policies Scenario, there would be a major shift in the pattern of gas flows, involving lower utilisation of the existing routes through Ukraine (and therefore Slovakia and the Czech Republic) and possibly also through Belarus (and therefore Poland). This picture could change if gas demand in Europe is higher[8] and/or if Russia were to gain a higher share of European gas imports. For example, if Russia had a share of European Union gas imports of 40% throughout the projection period, then exports would be 20 bcm higher in 2020 and 45 bcm higher in 2035 than indicated in Figure 9.6.

Figure 9.6 • Projected gas flows from Russia to Europe and potential growth in gas-export pipeline capacity

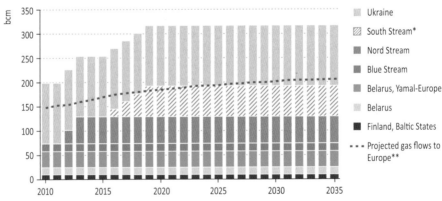

*The dates for commissioning of South Stream are the planned dates indicated by the project consortium and are not IEA projections. **Projected gas flows are from the New Policies Scenario and include exports to the European Union, to other OECD Europe and southeast European countries, but exclude Ukraine and Belarus.

Notes: Some of these future flows could also come via LNG. Pipeline capacities are assumed constant; this would require consistent investment in maintenance that may not happen in practice, particularly for any capacity that is not being used.

8. See, for example, the Low Nuclear Scenario (Chapter 12) and the Golden Age of Gas Scenario (IEA, 2011b).

There is similar evidence in the oil sector of Russia's desire to minimise its exposure to transit routes. The commissioning of the Baltic Pipeline System to the terminal at Primorsk near St. Petersburg in 2001 has reduced oil flows to non-Russian ports and changed the pattern of flows along other transit pipelines. Primorsk has already become the largest export outlet for Russian crude, taking the place of the Druzhba pipeline network to Central and Eastern Europe and allowing Russia also to ease congestion in the Turkish Straits by diverting exports away from the Black Sea port of Novorossiysk.

The expansion of the East Siberia Pacific Ocean (ESPO) pipeline to the east, together with the completion of the second Baltic Pipeline System (BPS-2) to Ust-Luga, will give Russia a further 2.6 million barrels per day (mb/d) of oil export capacity (130 million tonnes per year), without any corresponding increase in export levels. As oil exports are projected to decline marginally over the coming decades, the looming excess capacity in the Russian pipeline network has implications for Kazakhstan (Box 9.2) as well as, potentially, for the role of Russian in international markets.

Box 9.2 ● The Russian route to market for Caspian oil

Despite the commissioning in the 2000s of new export routes from the Caspian region to both the east (Kazakhstan-China) and the west (Baku-Tbilisi-Ceyhan), Russia still provides the route to market for more than half of non-Russian oil exports from the Caspian region, including more than three-quarters of the oil exported from Kazakhstan, the region's largest producer. The decision to expand the Caspian Pipeline Consortium pipeline from the Kazakhstani port of Atyrau on the North Caspian coast to the Russian Black Sea port of Novorossiysk will reinforce the predominance of export routes through Russia in the period to 2015.

The trajectory for the expansion of Kazakhstan's production beyond 2015 is unclear, with the start date for the second phase of the Kashagan project a major uncertainty (IEA, 2010). In the New Policies Scenario, we project that Kazakhstani output will rise from 1.6 mb/d in 2010 to 2.3 mb/d in 2020 and 3.9 mb/d in 2035. The bulk of this will be available for export and, along with an enlarged Kazakhstan-China pipeline, the intention has been to provide for future export growth through a new large-capacity trans-Caspian export route (known as the Kazakhstan Caspian Transportation System or KCTS) that would see Kazakhstani oil arrive at Black Sea or Mediterranean ports via the South Caucasus. However, though Kazakhstan is likely to be wary of too high a level of dependence on any single transit country, Russia is expected to have ample westbound capacity in its oil pipeline system by 2020 and could, if it wished, make a competitive offer to transport incremental volumes of its Caspian neighbour's oil.

Russia in global oil markets

The share of Russia in global oil production decreases over the projection period from 12% in 2009 to 10% in 2035 in the New Policies Scenario (Figure 9.7). Nonetheless, at 9.7 mb/d of production in 2035, Russia is still the largest non-OPEC producer and the second-largest

9

global producer, behind only Saudi Arabia (13.9 mb/d). It remains well ahead of the next largest producers, Iraq and the United States. As emphasised in Chapter 8, while there is no shortage of resources in Russia to meet these levels of production, this projection is sensitive to decisions on fiscal policy that will determine the attractiveness of the necessary investments.

From the late 1990s until around 2005, Russia was the provider of the "incremental barrel", meaning that Russia contributed much of the production growth necessary to keep pace with rising global demand. This will no longer be the case in the future, as exports decline steadily from 7.5 mb/d in 2009 to 6.4 mb/d in 2035. Nonetheless, Russia retains a critical role in the global oil balance as the largest non-OPEC producer and, moreover, a producer with the resources and strategic goal to keep oil output at a consistently high level (Box 9.3).

Figure 9.7 ● *Oil production in Russia and selected countries in the New Policies Scenario*

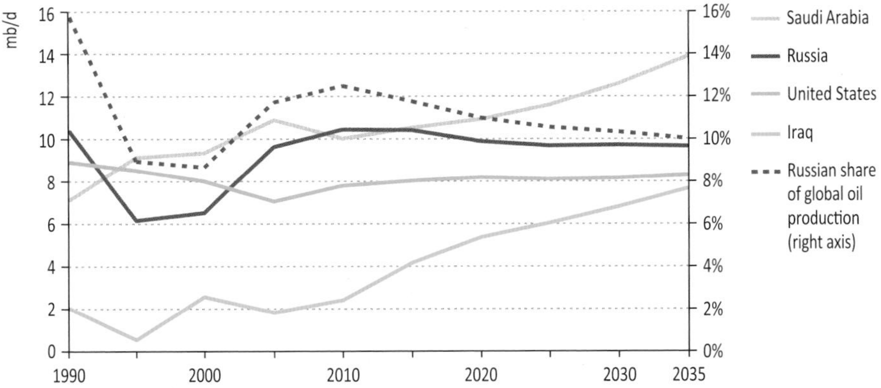

Russia's role in oil markets will also be shaped by the way that it gets oil to market. As noted, Russia will have an increasing amount of spare capacity in its oil pipeline network through the projection period and, in theory, this will give Russia more flexibility in managing its oil export flows, overcoming to a degree the rigidity of its traditional pipeline choices. To the extent that Russian exporters and traders have options available, this will increase their bargaining power relative to their potential purchasers, particularly those dependent on Russian deliveries by pipeline such as some refineries in Central and Eastern Europe. Additional spare capacity could also facilitate differentiation between the quality of different streams of crude oil passing through the Transneft system, allowing producers of higher-quality crudes (including, potentially, producers in Kazakhstan) to capture more of their value. This flexibility could offer Russia the possibility of adapting its export strategies quickly, in response to changing market needs; but the scope for short-term flexibility may not be great – Asian markets are projected to be the main source of global demand growth but direct eastward routes to these markets will be constrained by the capacity of the East Siberia-Pacific Ocean pipeline system.

Another issue for Russia is the balance between the export of crude oil and refined products. Boosting the share of refined products has been a long-standing Russian policy goal, in line with the overall aim to decrease the share of raw materials in Russia's export mix. Higher fuel specifications on the domestic market and, in the longer term, a fiscal advantage attached to the export of higher-grade refined products are expected to stimulate further refinery investment in Russia. Export to China is likely to remain dominated by crude, but increasing demand for imported products in other markets, notably diesel in Europe, could stimulate additional product export from Russia. Europe has a structural shortage of diesel output to meet increasing demand and Russia was the largest single supplier of Europe's net 1.1 mb/d imports in 2010. This import requirement is expected to increase to 1.4 mb/d already in 2015 (IEA, 2011a).

Box 9.3 ● Russian role in co-operation among oil and gas producers

There have been occasions – most recently following the precipitous decline in international oil prices in late 2008 – when Russia has expressed the wish to co-ordinate production levels with OPEC members. However, Russia has no spare oil production capacity and, even though the largest companies on the domestic market are either state-owned or are subject to a degree of state direction, it is not easy to see how Russia would acquire the flexibility in output levels that is, at least in theory, required of OPEC members and their national oil companies. Russia's large revenue needs and the important role of the oil sector in meeting them also lessen the likelihood of any willingness to restrain output.

In the natural gas sector, Russia has assumed a leading role among global gas producers and was instrumental in the creation of the Gas Exporting Countries' Forum (GECF), which became a fully-fledged international organisation in 2008. The GECF is focusing on analytical issues and information-sharing and has no role in market management that would justify a comparison with OPEC (despite the "Gas OPEC" tag that has been attached to the organisation). Even if some members were to push the Forum in this direction, it would be difficult for the GECF to co-ordinate cutbacks in production, given volume commitments in long-term contracts and the relative ease with which other fuels can substitute for gas. An alternative, about which Russian officials have spoken in the past, is a role for the GECF in co-ordination of investment programmes; but it is not yet clear how this could be managed and whether such co-ordination, if indeed it were to take place, would occur among the group as a whole or, more informally, between individual GECF members on a bilateral basis.

Russia in global gas markets

After briefly ceding its position as the largest producer of natural gas to the United States in 2009, in the New Policies Scenario Russia consolidates its place as the leading global producer and exporter of natural gas in the period to 2035. Between 2009 and 2035, the increase in gas production in Russia is greater than that in any other country, accounting for 17% of global gas supply growth (Figure 9.8). Russia remains the largest global exporter of natural gas in all scenarios; the volume of gas produced varies widely from 970 bcm in the

Current Policies Scenario to 635 bcm in the 450 Scenario in 2035 because of the different policies affecting global gas demand and greater or lesser efficiency of gas use within Russia. Russia's position would be enhanced further if gas demand in key global markets were to be higher than in the New Policies Scenario, as posited in the Low Nuclear Case (see Chapter 12) and also in the Golden Age of Gas Scenario (IEA, 2011b). Russia has considerable scope, given a supportive market structure, to increase output from non-Gazprom producers, as well as from Gazprom itself, and so would be in a position to take up a significant share of incremental demand in both Europe and Asia in these "higher-gas" scenarios.

Figure 9.8 ● *Gas production in selected countries in the New Policies Scenario, 2009 and 2035*

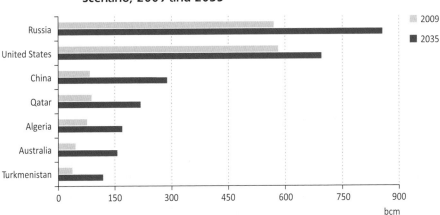

Developments in the Russian gas industry have indirect impacts on gas markets around the world, but the direct energy security implications are concentrated in those markets where Russia is a supplier. The main Eurasian gas trade flows, derived from our modelling results in the New Policies Scenario, are shown in Table 9.1. Europe is set to remain the largest export market for Russia, but Asian markets become increasingly important, with eastward exports (mainly to China, Japan and Korea) projected to rise above 100 bcm in 2035, almost 30% of Russia's total gas exports.

China plays a critical role in the orientation of Eurasia's gas export flows. From 4 bcm in 2010, exports from Russia and producers in the Caspian region to China grow to account for 16% of Eurasia's total gas exports in 2020 – mainly due to exports from Turkmenistan. This figure rises to close to 30% in 2035, as exports from Russia and the Caspian region increase. This shift is a natural response to China's growing import needs: China is projected to account for over one-third of the total growth in inter-regional gas trade over the projection period (see Chapter 4).

While our projections for production and export are lower in absolute terms than those in Russian strategy documents, the thrust of the export trend does correspond to Russia's strategic priorities, namely, to maintain its position on the European market while decreasing its proportional dependence on European customers by increasing the share of its exports going to Asia.

Table 9.1 ● Main gas trade flows from Russia and the Caspian region in the New Policies Scenario (bcm)

	2010*	2020	2035
Caspian net exports	*42*	*100*	*135*
to China	3	37	57
to Europe via Southern Corridor	6	23	42
to Russia (a)	*24*	*34*	*31*
Russia net exports (b)	*190*	*214*	*328*
Total Russia exports *(a+b)*	214	248	359
to Europe	201	225	237
to China	1	13	75
to other OECD countries	12	10	35
to other non-OECD countries	0	0	12
Total Russia and Caspian to China	**4**	**50**	**132**
Total Russia and Caspian to Europe	**207**	**248**	**279**
Russia to European Union	*118*	*158*	*171*

*Preliminary 2010 data.

Notes: Figures for the Caspian region include data for Azerbaijan, Kazakhstan, Turkmenistan and Uzbekistan; residual exports from the Caspian region to Iran are not shown. Exports from Russia to other OECD countries are mainly to Japan and Korea; exports to other non-OECD countries are mainly to India.

In the New Policies Scenario, Russia's share of imports to the European Union market no longer declines steadily, as it has over the last decade (from almost 50% in 2000 to 38% in 2009 and 34% in 2010). Russia's share remains at 35% in 2020, before falling gradually to 32% in 2035 (Figure 9.9). Expressed as a share of EU total gas consumption, Russia's share recovers slightly from 23% in 2009 to 27% in 2020, remaining at this level until 2035. The share of Russian gas in the Chinese import mix is projected to grow quickly after 2015, reaching 10% in 2020 and 35% in 2035. This represents 15% of total Chinese gas consumption in 2035.

Figure 9.9 ● Russian share of natural gas imports and consumption in the European Union and China in the New Policies Scenario

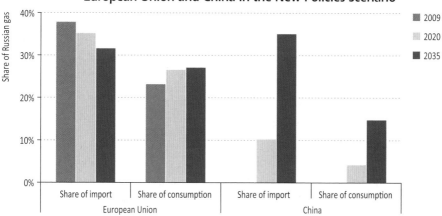

Although its share of imports into the European Union drops slightly by 2035, Russia's increased share of overall gas consumption in both the European Union and China highlights its central position in global gas security and, especially, the importance of Gazprom, as the dominant exporter of Russian gas. Russia is set to have a significant degree of market power in Europe, while becoming less reliant, proportionally, on revenue from gas sales to Europe. Spare pipeline capacity on routes to Europe would give Russia scope to re-direct export flows in response to a gas supply emergency or other contingency, an energy security bonus for customers contemplating a hypothetical shortage of gas supply, but a situation which also reinforces Russian market influence. To the east, alongside increased exports to Japan and Korea, gas supply from Russia to China is set to become one of the main arteries of global gas trade, providing Russia with diversity of markets and revenues, and China access to the large and as yet underdeveloped gas resources of Eastern Siberia and the Russian Far East. Some will see the picture presented in Table 9.1 as a healthy balance in gas trade relationships across Eurasia; others may see it as presaging a degree of competition between Asia and Europe for positions in Russian gas supply, including investment (Box 9.4).

Box 9.4 ● **Europe versus Asia: competing for Russian gas?**

As Russia opens up gas trade with China it gains greater diversity of gas export markets and some strategic benefits, but the way in which this brings China into competition with Europe for Russian gas can be overstated or misunderstood. Russia will continue to supply the majority of its gas via pipelines, anchored by long-term supply contracts; under these circumstances, the contractual scope for Gazprom to vary supplies by destination is small. Moreover, in relation to switching supplies between Europe and China, the physical scope for abrupt change is even narrower. With the exception of the proposed 30 bcm Altai pipeline that would link existing fields in Western Siberia to China, there is unlikely to be any scope for Gazprom to make discretionary choices about the direction of pipeline exports between east and west.

Most of the fields earmarked for export to China are in Eastern Siberia and the Far East, thousands of kilometres from the existing westbound infrastructure. These fields are remote and, as noted in Chapter 8, will be expensive to develop. With this in mind, Gazprom has reason to be concerned about its own bargaining position for, once fixed infrastructure from these remote fields to China is in place, Gazprom's options and therefore its leverage are limited. This helps to explain Gazprom's preference to supply China initially from Western Siberia (via the so-called Western route, the Altai pipeline): since it also supplies existing European markets from Western Siberia, it can keep pressure on China to bid for this gas at a price closer to the European export price.

While geography and the structure of the gas trade make it unlikely that China and Europe will compete for incremental Russian supply on a short-term basis, this is not to say that there will be no competition over the longer term. But they will be competing to influence the investment priorities and strategies of Russian companies, primarily but not exclusively those of Gazprom. Political factors will influence some of the Russian investment decisions, but more significant will be the commercial factors: the extent and reliability of demand, the price available on the respective markets and, perhaps, reciprocal investment opportunities.

At the same time, the position of Russia on the Chinese market is not yet secured, and its position on the European market is not entirely secure. As noted, Gazprom's share of European Union gas imports has declined markedly to 34% in 2010, 16% below its import share in 2000. The most recent drop in this share, by 4% in 2010, was the result of a Gazprom strategy to make minimal changes to its oil-indexed pricing formulas for gas export even as they came under pressure from cheaper LNG in Europe. This may have optimised revenues as oil-indexed prices rebounded during the year, but it meant that Gazprom progressively lost market share to other suppliers with pricing formulas more sensitive to gas-to-gas competition. Gas demand growth in Europe is becoming increasingly concentrated in the power sector, calling into question the viability of an export strategy based on indexation to a commodity – oil – that is no longer used for electricity generation in Europe.[9] A similar insistence on maximising export prices holds risks also in relation to supply into China, where pricing remains the main stumbling block in negotiations over gas (even though it seems clearer than in the past that China may be ready to pay the prices necessary to ensure development of East Siberian gas resources).

Developments within the European gas market will affect the degree of dependence of some customers on Russian gas. As of 2010, thirteen countries rely on Russia for more than 80% of their total gas consumption and a total of seventeen receive more than 80% of their gas imports from Russia.[10] Some of these countries in central and southeastern Europe have a realistic perspective to gain access to alternative pipeline supplies from the Caspian region as well as, in some cases, LNG. This part of Europe – for example in Poland – could see some development of unconventional gas production. The European Union has also established a strategic priority to develop a more inter-connected European natural gas grid, through inter-connector and reverse flow projects; this would also have a potential impact in southeast Europe through the Energy Community Treaty that aims to integrate this region into a broader European internal energy market. To the extent that these developments provide viable alternatives to Russian supply, they will reduce reliance on Russian gas and diminish Gazprom's pricing power in Europe. In the longer term, seriously enhanced efforts in Europe to de-carbonise its energy system would have a significant impact on the trajectory of European demand for imported gas, compared with the New Policies Scenario.[11]

These uncertainties and competitive pressures are set to increase at a time when Russia is obliged to move to higher-cost sources for incremental gas supply (as compared with existing output from Western Siberia). Plans to increase its presence on the international LNG market form one element of the Russian response and this is already paying dividends

---

9. Less than 3% of the electricity generated in OECD Europe comes from oil, this figure has halved over the period 2000 to 2009.

10. According to IEA data, countries relying on Russia for more than 80% of their gas consumption in 2010 were Armenia, Belarus, Bosnia and Herzegovina, Bulgaria, Czech Republic, Estonia, Finland, Latvia, Lithuania, FYROM, Moldova, Serbia and Slovak Republic; in addition to these, Croatia, Romania, Poland and Ukraine also relied on Russia for more than 80% of their gas imports.

11. In line with the objective to decrease greenhouse-gas emissions by 80% in 2050, relative to a 1990 baseline; this is modelled in the 450 Scenario which shows a gradual decline in the European Union's gas import requirement from the early 2020s; European Union gas import demand in the 450 Scenario is 360 bcm in 2035, compared with 540 bcm in the New Policies Scenario.

in Asian markets with the start of LNG shipments from Sakhalin. But there are questions about LNG export projects in the Russian Arctic, which are set to rely, at least in part, on markets in the Atlantic basin. Given that the North American market is now not expected to require imported gas, this raises the question of how future Russian LNG supply may affect Russian pipeline deliveries to European markets. Russian strategy will need to be responsive to the changing dynamics of global gas markets.

Russia in global coal markets

Russian exports of coal have increased sharply since 2000. In the New Policies Scenario, Russia's overall contribution to global exports is expected to be maintained – the projections anticipate that Russian exports of hard coal will remain within a range of 75 million tonnes of coal equivalent (Mtce) to 100 Mtce throughout the period to 2035, keeping Russia as one of the main coal exporting countries behind only Indonesia, Australia and Colombia.

But the direction of Russia's coal exports is expected to change. At present, more than 50% of Russian coal exports go to the European Union; but the European requirement for coal imports is projected to fall by almost 40% over the period to 2035, as demand for coal drops. To a degree, the Chinese market presents a viable alternative export market for Russia but here, too, the market opportunity narrows after 2020 in the New Policies Scenario (see Chapters 10 and 11). Russia is a relatively high-cost supplier and faces increasingly strong competition in Asian markets from other producers, primarily Indonesia and Australia but increasingly also Mongolia (although Mongolia relies in part on Russian transit to reach non-Chinese markets). In our projections, this constrains Russia's contribution to the global balance and exports decline slowly after 2020.

Environment and climate change

The dramatic economic decline in Russia in the early 1990s brought some environmental benefits and emissions reductions, but at a huge social cost. Energy-related carbon-dioxide (CO_2) emissions fell by 35% between 1990 and 1997, but this was due entirely to a collapse in Russian industrial output and fossil-fuel use. The rebound in economic activity since 2000 has been less carbon intensive, such that, while GDP surpassed the 1990 figure in 2007 (at market exchange rates), CO_2 emissions in 2009 were still 30% below 1990 levels. Nonetheless, Russia remains the fourth-largest emitter of CO_2 and one of the most carbon-intensive economies in the world: the production of each unit of Russian GDP releases three times more CO_2 than the equivalent in the European Union. Expressed on a per-capita basis, annual emissions are almost 11 tonnes of CO_2 (tCO_2) compared with around 7 tCO_2 in the European Union (where GDP per capita is twice as large).

Russia has the opportunity over the coming decades to achieve significant additional environmental improvements if policies are introduced to support more efficient energy production and use, and the deployment of low-carbon technologies. The projections in this *Outlook* show a wide range in the level of energy-related CO_2 emissions, depending on the scenario (Table 9.2). Higher fossil energy use pushes up emissions fastest in the Current

Policies Scenario, so that they are close to the 1990 level by 2035. In the New Policies Scenario, emissions growth is more moderate at 0.6% per year; and in the 450 Scenario, emissions decline after 2015, pushed down by more concerted efforts at energy saving and, after 2020, by the assumed introduction of a domestic emissions trading scheme for power generation and the most energy-intensive industrial sectors. The carbon intensity of Russia's GDP decreases in all scenarios, but most quickly in the 450 Scenario. In terms of emissions per-capita, the 450 Scenario is the only scenario that registers a decline, from the current 11 tCO_2 to 8 tCO_2, while this figure rises to above 13 tCO_2 per-capita in the New Policies Scenario and to above 15 tCO_2 in the Current Policies Scenario.

Table 9.2 ● Energy-related CO_2 emissions in Russia by scenario (million tonnes)

	1990	2009	2020	2035	2009-2035*
New Policies Scenario	2 179	1 517	1 687	1 787	0.6%
Current Policies Scenario	2 179	1 517	1 732	2 046	1.2%
450 Scenario	2 179	1 517	1 551	1 102	-1.2%

*Compound average annual growth rate.

Notes: Energy-related CO_2 data from the Russian national inventory, reported to UNFCCC, differ from those of IEA. The Russian data on emissions are up to 6% lower than IEA data for all years except 1990 and 1991, when they are around 10% higher; thus the Russian figure for 1990 is 2 287 million tonnes and for 2009 is 1 387 million tonnes.

Russia's emission pledge for 2020, submitted to the UNFCCC following the Copenhagen Accord, is a 15% to 25% reduction, relative to 1990. The target within this range depends on the extent to which the role of Russia's forests as a carbon sink will be taken into account and whether all major emitters adopt legally binding obligations.[12] The minimum target of a 15% reduction in 2020, applied to energy-related emissions, sets a limit of 1 852 million tonnes of CO_2 emitted in 2020: this is achieved with room to spare in all three scenarios. A 25% reduction would set the limit at 1 634 million tonnes: this would require additional policy efforts, since it is met only in the 450 Scenario. Our analysis suggests that Russia could comfortably afford to adopt a limit at the more ambitious end of the proposed range; indeed, it would need to be at or beyond 25% if it is to provide a meaningful stimulus to national policy in the period to 2020.

Moving Russia towards a lower-emissions trajectory will require much more concerted efforts to improve energy efficiency and to deploy low-carbon fuels and technologies (Figure 9.10). Improvements in energy efficiency contribute half of the cumulative abatement in emissions in the transition to the 450 Scenario. Increased deployment of renewables and power plants equipped with CCS technology, particularly after 2020, accounts for the bulk of the remainder.

9

---

12. As a member of the G-8, Russia is also associated with the longer-term goal of achieving at least a 50% reduction in global emissions by 2050 (2008 Hokkaido G-8 Summit) and the goal of developed countries to reduce emissions by 80% or more by 2050 compared to 1990 or more recent years (2009 L'Aquila G-8 Summit).

Figure 9.10 • Energy-related CO_2 emissions abatement in Russia by source in the 450 Scenario compared with the New Policies Scenario

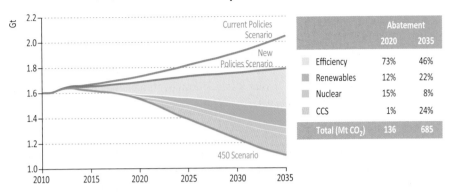

An environmental scenario does not just bring benefits to the global climate through reductions in CO_2. Sulphur dioxide (SO_2), nitrogen oxides (NO_x) and particulate matter $(PM_{2.5})$ all have negative effects on human health and the environment. Policies aimed at decreasing consumption of fossil fuels and reducing emissions of CO_2 also have the effect of lowering emissions of these pollutants (Table 9.3). In the New Policies Scenario, lower energy demand means that emissions of SO_2 fall by more than 500 000 tonnes compared with the Current Policies Scenarios, a reduction of 10%; emissions of NO_x in 2035 are also 8% lower. In both of these cases, there is a decline in the absolute volume of emissions, relative to 2009. The improvements are due to the assumed implementation of measures to control air pollution from power generation, industrial processes and transportation. However, these measures are not stringent enough to produce a decline in emissions of particulate matter, which remain flat in the New Policies Scenario while increasing slightly relative to 2009 in the Current Policies Scenario. Emissions are even lower in the 450 Scenario, further improving air quality and reducing negative health impacts.

Table 9.3 • Emissions of major air pollutants in Russia by scenario (thousand tonnes)

	New Policies Scenario			Current Policies Scenario			450 Scenario		
	2009	2020	2035	2009	2020	2035	2009	2020	2035
Sulphur dioxide (SO_2)	6 019	4 223	4 353	6 019	4 317	4 858	6 019	3 977	3 477
Nitrogen oxides (NO_x)	4 797	3 653	3 348	4 797	3 718	3 636	4 797	3 442	2 732
Particulate matter $(PM_{2.5})$	1 301	1 305	1 302	1 301	1 327	1 387	1 301	1 281	1 214

Notes: Estimates are based on assumed implementation of a range of pollution control measures in place or under consideration in Russia.[13] The regulatory regime is not assumed to change by scenario; variations are due to differences in the level and structure of energy use.

Source: IIASA, 2011.

13. Pollution control measures include: large combustion plants are equipped with moderate control measures like in-furnace control of SO_2 emissions or combustion modification for NO_x; measures for mobile sources are based on Russian plans to implement European standards, i.e. Euro IV by 2015; controls on process sources from the non-ferrous metals industry (important sources of SO_2 and dust) have been assumed according to programmes presented by the industry; medium-to-high efficiency electrostatic precipitators are introduced to control emissions of $PM_{2.5}$ from large stationary combustion and process sources.

Comparing this *Outlook* with Russian scenarios and objectives

How do the results of this *Outlook* compare with the scenarios and targets set out in Russian strategy documents? Care is needed when comparing projections from various sources since they may have different underlying assumptions. But it is, nonetheless, instructive to look at five key objectives for 2030 from Russian strategy documents alongside the corresponding outputs from the *WEO-2011* (Table 9.4).

Table 9.4 ● *WEO-2011* projections (in 2030) compared with selected forecasts from the Russian Energy Strategy to 2030

	WEO-2011			Energy Strategy to 2030 (range)
	New Policies Scenario	Current Policies Scenario	450 Scenario	
Total primary energy demand (Mtoe)	799	849	708	963 - 1 096
Electricity demand (TWh)	1 351	1 514	1 219	1 740 - 2 164
Gas production (bcm)	822	888	657	870 - 925
Oil production (mb/d)	9.7	10.2	8.5	10.6 - 10.7
Coal production (Mtce)	257	309	132	282 - 381
CO_2 emissions (Mt)	1 756	1 915	1 232	2 048 - 2 288

Notes: The range for electricity demand in the power sector development plan (General Scheme), as amended in 2010, is lower, at 1 553 TWh to 1 860 TWh; electricity demand in this table and in the Spotlight is calculated as production minus net exports; the Energy Strategy's figures for gas production are adjusted to the IEA standard bcm (Box 8.3).

The differences in total primary energy demand and electricity demand between Russian scenarios and the *WEO-2011* are driven largely by different assumptions about GDP growth. If we adjust for this factor, our results approach the range included in the Energy Strategy to 2030 (see Spotlight earlier in the chapter), though oil, gas and coal production in the New Policies Scenarios are all below the lower end of the targets included in the Energy Strategy, in part because of lower projections of domestic demand.

Higher official expectations in Russia about total primary energy demand and electricity demand have important implications for energy policy and investment planning, in particular in the electricity sector (see Chapter 7). Russian projections of electricity demand were revised sharply downwards in 2010 in the wake of the economic crisis (note to Table 9.4) and a correction made to the General Scheme for the sector. Nonetheless, pending the introduction of a fully market-based approach to new investment in the power sector, there is still a risk of imposed plans providing for levels of demand that may not materialise, with consumers or taxpayers ultimately paying for capacity that is not efficiently timed, sized or well located.

PART C
OUTLOOK FOR COAL MARKETS

PREFACE

Continuing the practice of shining a searchlight on different issues each year in the WEO, Part C (Chapters 10 and 11) focuses on coal, the fuel which has accounted for nearly half of additional global energy consumption over the decade to 2010. Chapter 10 deals with demand and Chapter 11 with supply and investment. Recent developments in the sector are reviewed and the prospects for coal in the three scenarios are presented.

Chapter 10 explains the forces which determine demand for coal and the extent of competition between coal and other fuels in different markets. It looks at the environmental aspects of coal use and developments in coal-based technology. It concludes with a detailed regional analysis of the main coal-consuming areas of the world.

Chapter 11 carries out a similar supply-side analysis, detailing global coal resources, the cost of exploiting them by different mining techniques, production, and investment prospects and international trade, again concluding with a series of country studies.

CHAPTER 10

COAL DEMAND PROSPECTS
Treading water or full steam ahead?

H I G H L I G H T S

- Coal accounted for nearly half of the increase in global energy use over the past decade. Its prospects vary markedly across the three scenarios in this *Outlook*, demonstrating the significance of government policies towards energy, according to the balance they strike between social, economic and environmental effects. Much depends too on the pace of investment into existing clean coal technologies and the development of carbon capture and storage.

- In the New Policies Scenario, global coal use rises through to the early 2020s and then remains broadly flat, above 5 850 Mtce, throughout the rest of the projection period; one-quarter higher than in 2009. Coal continues to be the second-largest primary fuel globally and the backbone of electricity generation. In the Current Policies Scenario, demand carries on rising after 2020, increasing overall by nearly two-thirds to 2035. But in the 450 Scenario, coal demand peaks before 2020 and then falls heavily, declining around one-third between 2009 and 2035.

- China, responsible for nearly half of global coal use in 2009, holds the key to the future of the coal market with an ambitious 12th Five-Year Plan to reduce energy and carbon intensity through enhanced energy efficiency and diversifying the energy mix. In the New Policies Scenario, China accounts for more than half of global coal demand growth, with its demand growing around one-third by 2020 and then declining slightly before remaining broadly stable, above 2 800 Mtce, through to 2035.

- India also plays an increasingly important role. By more than doubling its coal use by 2035 in the New Policies Scenario, India displaces the United States as the world's second-largest coal consumer by 2025. Over 60% of the rise comes from the power sector, reflecting the enormous latent demand for electricity in India: in 2009 about one-quarter of the nation's population still lacked access to electricity. Bringing electricity access to all the world's population by 2030, could entail more than half of the resultant increase in on-grid generation capacity coming from coal, compared with the New Policies Scenario.

- Power generation remains the main driver of global coal demand over the projection period, accounting for at least three-quarters of the increase in both the New and Current Policies Scenarios. Stronger uptake of existing clean coal technologies and carbon capture and storage, could boost the long-term prospects for coal use. If the average efficiency of all coal-fired power plants were to be five percentage points higher than in the New Policies Scenario in 2035, such an accelerated move away from the least efficient combustion technologies would lower CO_2 emissions from the power sector by 8% and reduce local air pollution.

Chapter 10 - Coal demand prospects

353

Overview

Over the decade to 2010, coal accounted for close to half the global increase in fuel use (Figure 10.1). The dip in consumption of coal during the global economic crisis was much smaller than that in other fossil fuels (Box 10.1). Even in percentage terms, growth in coal use over the decade outpaced growth in renewables, although this was a time of take-off for many renewables (under the stimulus of widespread government support) and they started from a low base. As a result, the importance of coal in the global energy mix is the highest since 1971, the first year for which IEA statistics exist. Globally, coal is the most important fuel after oil. According to preliminary estimates, coal use accounted for 28% of global primary energy use in 2010, compared with 23% in 2000. Coal is the backbone of global electricity generation, alone accounting for over 40% of electricity output in 2010. Even in the OECD, coal fuelled more than one-third of electricity generation in 2010. In non-OECD countries, where coal resources are often abundant and low cost, coal is the most important fuel: it accounted for 35% of total primary energy use, 36% of total industry consumption and nearly half of total electricity generation in 2010 (Box 10.2).

Figure 10.1 ● **Incremental world primary energy demand by fuel, 2000-2010**

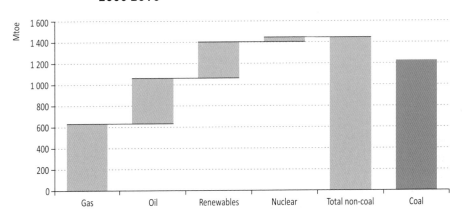

Note: IEA estimates for 2010.

The coal demand prospects outlined in this chapter diverge markedly across our three scenarios. This is because coal is a carbon-intensive fuel and the environmental consequences of its use can be significant, especially if it is used inefficiently and without effective emissions and waste control technologies. Government decisions reflecting their judgements as to the balance between the relevant social, economic and environmental considerations, particularly their success in encouraging the development and deployment of clean coal technologies, are crucial to the future pattern of coal demand; and different intensities of government intervention in energy markets are what characterise our scenarios (see Annex B). One factor, however, is constant: non-OECD Asia, especially China and India, dominate the global picture of future coal use, whatever its level (Figure 10.2).

Box 10.1 ● A decade of booming coal use

For all the talk about natural gas and renewables, coal unquestionably won the energy race in the first decade of the 21st century. Global coal demand accelerated strongly in the 2000s, following minimal growth in the 1990s. On average, between 2000 and 2010 (a year for which only preliminary estimates are available), primary coal demand grew by 4.4% per year – well above the average of 2.7% for natural gas and 1.1% for oil. In 2010, world coal demand was almost 55% higher than in 2000 – a bigger increase in both volume and percentage terms than for any other fuel category, including renewables. In energy terms, the increase in global coal demand amounted to 1 750 million tonnes of coal equivalent, equal to 75% of China's coal demand in 2010. In other words, coal accounted for nearly half of the increase in global primary energy demand during the past decade. Coal demand was relatively robust during the global economic crisis: it declined by only 0.7% in 2009, as a result of a 4% rate of growth in non-OECD countries, which offset a drop of nearly 10% in the OECD. In contrast, world demand for oil and natural gas fell by 1.9% and 2.1%, respectively.

In the New Policies Scenario, which assumes cautious implementation of the policy commitments and plans that have been announced by countries around the world, global coal use rises through to the early 2020s and then remains broadly flat to 2035, at above 5 850 million tonnes of coal equivalent (Mtce).[1] By then, demand is one-quarter higher than in 2009 and 17% higher than in 2010, based on preliminary estimates for that year. The prospects are very different in our other two scenarios. In the Current Policies Scenario, which assumes no change in government policies relative to mid-2011, demand continues rising through to 2035, reaching a level nearly one-third higher than in the New Policies Scenario. In the 450 Scenario, which assumes much stronger policies to achieve the goal of limiting the long-term global temperature increase to 2°C, coal demand peaks before 2020 and then falls sharply; by 2035, demand is nearly 45% lower than in the New Policies Scenario and back to the level of the early 2000s. Should nuclear generation end up developing at a slower pace than we have assumed in the New Policies Scenario, global demand could be nearly 300 Mtce higher in 2035 in that scenario – a volume equivalent to over half of global inter-regional steam coal trade in 2009 (see Chapter 12). Similarly, if the United Nations "International Year of Sustainable Energy for All" in 2012 should result in a successful commitment to bring electricity to all the world's people by 2030, more than half of the resultant increase in on-grid electricity generation capacity is expected to be coal-fired, increasing global coal consumption marginally in that year (see Chapter 13). On the other hand, stronger policies supporting the use of natural gas and increased supplies from unconventional sources, as described in the Golden Age of Gas Scenario, would lower coal demand by nearly 400 Mtce in 2035 compared with the New Policies Scenario (see Chapter 4).

10

---

1. A tonne of coal equivalent is defined as 7 million kilocalories, so 1 tonne of coal equivalent equals 0.7 tonnes of oil equivalent.

Figure 10.2 ● World primary coal* demand by region and scenario

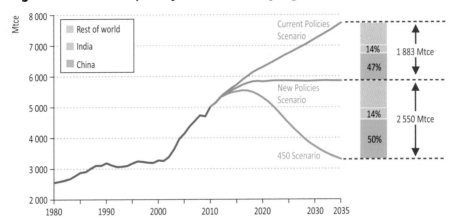

*Includes hard coal (coking and steam coal), brown coal (sub-bituminous coal and lignite) and peat.

As noted, the overwhelming bulk of any future growth in coal demand will come from non-OECD countries. Their share of world coal demand rises from 70% in 2009 to around 80% in all three scenarios as a consequence of faster population and economic growth and the availability of significant low-cost local and regional coal resources, particularly in non-OECD Asia (Table 10.1).

Box 10.2 ● The shift to Asia

Over the last decade, more than 80% of the global increase in coal demand came from China alone. The country's share of global coal demand rose from 27% in 2000 to 47% by 2010; its coal use more than doubling to 2 350 Mtce. This increase corresponds to twice the coal consumption in the United States in 2010, the world's second-largest consumer of coal. Even during the period 2008 to 2010, overshadowed by the global economic crisis, China's coal demand grew by around 80% of the total coal used in the European Union in 2010. India accounted for a further 11% of the global coal demand growth from 2000 to 2010, consolidating its position as the world's third-largest coal consumer. Demand growth in OECD countries was much slower up to 2007 and then fell for two consecutive years as a result of economic troubles; by 2010, demand was 6% below peak use in 2007, and nearly back to 2000 levels. In 2010, almost half of OECD coal demand came from the United States, where prospects for future use are threatened by expanding supplies of unconventional natural gas and renewables, and tighter air pollution restrictions.

In the New and Current Policies Scenarios, China and India together account for well over two-thirds of the projected increase in demand over the *Outlook* period (Figure 10.3). By contrast, there is little prospect of long-term growth in coal demand in the OECD, where coal demand comes under much greater pressure from natural gas and renewables as a result of the imposition of a carbon price and other government policies that improve the economics of low-carbon energy sources. Even in the Current Policies Scenario, OECD consumption reaches a plateau by 2020 and falls back close to the level of 2010 by 2035, while in the 450 Scenario OECD coal demand is 60% lower than 2010 by the end of the projection period. In the 450 Scenario, China is the biggest contributor to the global fall in coal demand relative to the New Policies Scenario, reflecting its dominant position in global coal use.

Table 10.1 ● Coal demand by region and scenario (Mtce)

	1980	2009	New Policies Scenario		Current Policies Scenario		450 Scenario	
			2020	2035	2020	2035	2020	2035
OECD	1 380	1 476	1 494	1 146	1 609	1 588	1 400	623
United States	537	693	705	599	751	773	698	326
Europe	663	415	383	264	431	400	333	151
Japan	85	145	158	115	165	156	141	60
Non-OECD	1 179	3 229	4 339	4 713	4 699	6 154	3 908	2 685
China	446	2 179	2 863	2 820	3 069	3 709	2 596	1 535
India	75	399	619	883	699	1 148	531	521
Russia	n.a.	136	166	168	173	203	150	96
World	2 560	4 705	5 833	5 859	6 308	7 742	5 309	3 309
Share of non-OECD	46%	69%	74%	80%	74%	79%	74%	81%
Share of China	17%	46%	49%	48%	49%	48%	49%	46%
Share of India	3%	8%	11%	15%	11%	15%	10%	16%

Figure 10.3 ● Incremental world primary coal demand by region and scenario

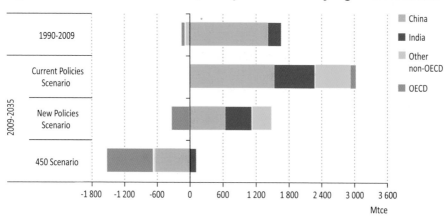

The power sector remains the main driver of global coal demand over the projection period in all three scenarios. Power generation accounts, respectively, for just over 80% and around 75% of the increase in world coal demand in both the Current and New Policies Scenarios, keeping the share of this sector in total coal demand around or above two-thirds

(Table 10.2). In the 450 Scenario, the power sector accounts for almost all of the fall in global coal demand, driving down the sector's share to half. Coal use in industry and transformation processes such as blast furnaces and coke ovens, by contrast, nearly peaks in all three scenarios around 2020. The price differential between oil and coal prices fosters increased use of coal as feedstock in coal-to-liquids plants (CTL) in all three scenarios, overtaking that of the buildings sector, but this use nonetheless remains small relative to power generation or industry.

Table 10.2 ● World coal demand by sector and scenario (Mtce)

	1980	2009	New Policies Scenario		Current Policies Scenario		450 Scenario	
			2020	2035	2020	2035	2020	2035
Power generation	1 242	3 063	3 811	3 920	4 225	5 547	3 394	1 661
Industry	604	915	1 172	1 092	1 204	1 235	1 104	928
Transformation*	300	447	530	461	543	508	510	388
Coal-to-liquids	3	27	60	178	67	226	55	159
Buildings	339	178	178	129	186	143	164	99
Other	72	75	82	79	84	83	81	75
Total	2 560	4 705	5 833	5 859	6 308	7 742	5 309	3 309
Share of power generation	49%	65%	65%	67%	67%	72%	64%	50%
Share of industry	24%	19%	20%	19%	19%	16%	21%	28%

*Primarily blast furnace and coke oven transformation, and own use.

Understanding the drivers of coal demand

Current patterns of coal use are the result of a multitude of factors, most important of which are the overall demand for energy in stationary uses (since coal is rarely directly used in transport) and the level of economic activity driving that demand, the extent of resource endowments and the cost of producing and delivering them to market, the competitiveness of coal against other fuels, technological developments affecting the use of coal and other energy sources, and the geopolitical, policy and regulatory environment. The prospects for coal demand in each sector and region – and the different types of coal demanded[2] – depend, therefore, on how these different factors change (Table 10.3). The coal demand projections for the three scenarios in this *Outlook* rest on assumptions about each of these drivers. This section assesses how sensitive coal demand is to each factor and how predictable they are, in order to shed light on the degree of uncertainty surrounding the outlook.

---

2. In practice, coal is very heterogeneous, with big differences in the characteristics across deposits. Specific qualities of coal referred to as coking coal, generally with low ash content and high calorific value, are required for the production of coke for steel production. A wider range of types of anthracite and bituminous coal, generally referred to as steam coal, are used in power generation and steam-raising in industry. Brown coal – sub-bituminous coal and lignite – which has a low calorific value, is also used for power generation (see Box 11.1 for classifications and definitions of coal types).

Table 10.3 ● Summary of the main drivers of coal demand by sector

	Economic activity	Price	Policies	Technology
Power generation	Electricity demand is strongly correlated with industrial output, commercial activity and household incomes.	Competitiveness of coal-fired plant is sensitive to changes in relative fuel prices (especially for new plants and relative to natural gas).	Policies for low-carbon technologies, including carbon pricing and air pollution, strongly affect decisions to build and operate coal-fired plants.	Clean coal technologies, including more efficient plant, co-firing with biomass and carbon capture and storage (CCS) could support further growth.
Industry	Industrial output, especially iron, steel and cement, is the main underlying driver.	Iron and steel coal demand is relatively insensitive to prices; for steam-raising, demand is highly price sensitive relative to natural gas and oil.	Strongly influenced by air pollution and other environmental regulations (slag and ash disposal). Carbon pricing can raise industry production costs significantly.	Some remaining scope for efficiency gains in iron and steel and steam-raising; CCS unlikely to be commercially viable in most cases for many years.
Buildings	Personal income and cost determines demand for space and water heating.	Normally coal is only used where the price is competitive with other fuels.	Local pollution regulations and subsidies are the main factors.	Important technological advances that would boost demand are not expected.
Coal-to-liquids (CTL)	Indirect impact through demand for mobility and therefore liquid fuels.	Viability of new projects is highly sensitive to the ratio of coal to oil prices; can be an attractive way to monetise remote coal deposits.	Policies on energy security and use of CCS with CTL will be vital. Local water scarcity can hamper development.	Technological advances crucial, particularly CCS; combined with biomass, CCS could achieve negative carbon emissions.
Other uses	Weak correlation.	Feedstock demand for coal in natural gas and chemical production (in some countries) is highly price sensitive.	More stringent environmental policies would depress demand in most applications.	Technological cost reduction breakthroughs in coal-to-gas might boost demand.

Economic activity

At the global level, coal demand is strongly correlated with economic activity: other things being equal, faster economic growth normally results in faster growth in coal demand. Therefore, uncertainty about the rate of growth in gross domestic product (GDP) in the medium term is a major source of uncertainty for global coal demand. This is especially true of the rate of economic growth in China, the world's second-largest economy, which accounted for close to half of global coal demand in 2010. The correlation between coal demand and GDP is much stronger in some regions, depending on the importance of coal in the fuel mix and other factors. The intensity of coal use in a given economy is measured by the amount of coal consumed per unit of GDP at market exchange rates (MER). On this measure, China and India have by far the most coal-intensive economies, with China in 2009 consuming five times as much coal per dollar of GDP as the world as a whole (Figure 10.4).

Coal intensity in all regions is expected to fall significantly over the projection period, in line with the long-term downward trend in energy intensity and stronger penetration of low-carbon sources. In the New Policies Scenario, global average coal intensity is nearly halved between 2009 and 2035. The economies of India and China remain much more coal intensive than the rest of the world, though the gap narrows substantially.

Figure 10.4 ● **Primary coal intensity by region as a percentage of 2009 world average in the New Policies Scenario**

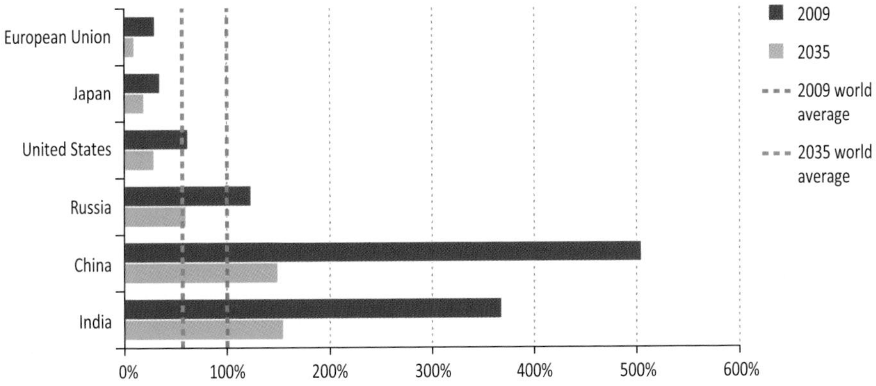

Note: Calculated based on GDP expressed in year-2010 dollars (MER).

Globally, growth in electricity demand is highly correlated to GDP growth, with a coefficient of close to one (*i.e.* every 1% increase in GDP is accompanied by a 1% increase in electricity demand). Though high initially, the link between coal-fired power generation and GDP is expected to weaken in the future as the role of other fuels and advanced technologies expands. This is particularly the case in the 450 Scenario, where the share of coal in power generation drops from 40% in 2009 to 15% in 2035, compared with 33% in the New Policies Scenario.

The relationship between coal use and GDP growth is likely to remain strong in the industrial sector, partly because of difficulties in finding commercially viable replacements for coking coal in existing technologies in iron and steel production. The iron and steel sector accounts for around 40% of total industrial coal use in 2009. Historically, iron and steel demand is highly correlated with many factors – urbanisation, development of infrastructure and growth of the heavy-industry sector – which results in a strong indirect correlation between these factors and coal use, mainly coking coal. In the New Policies Scenario, a slowdown in the rate of growth in global crude steel output, continuing efficiency gains (spurred in part by higher prices) through technological improvement and fuel substitution towards electricity and natural gas result in coal demand for iron and steel peaking around 2020 (Figure 10.5). While the coal intensity of steel production worldwide is projected to drop by one-third by 2035 in the New Policies Scenario, technological innovation (*e.g.* using hydrogen and plastic wastes) could further reduce coal use in iron and steel production and, therefore, lead to a faster decoupling of industrial coal use from GDP.

Figure 10.5 ● World crude steel production, and iron and steel coal use versus GDP in the New Policies Scenario

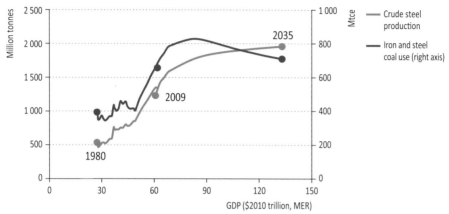

Sources: IEA analysis; World Steel Association.

Coal prices and inter-fuel competition

In principle, coal can be replaced by at least one other fuel in most applications (though the ease with which this can happen varies), making the price of coal relative to other fuels a vital determinant of the competitiveness of coal and, therefore, the share of coal in the overall energy mix in any given market. In the short term, with a fixed stock of energy-related equipment, fuel switching – to or from coal – can occur only if unutilised capacity or multi-firing facilities are available. Data on fuel-switching capability is generally poor, but due to inherent technological aspects, it is seen to be greatest in the power sector. For example, in the United States, there is a significant amount of old, relatively inefficient coal-fired capacity (around half of its coal-fired power generation capacity is subcritical and over 30 years old) that is held in reserve, generally to meet peak load, but can be brought online at other times when coal prices are low relative to other fuels, especially natural gas.

In the long term, opportunities for fuel-switching are greater as new equipment capable of using different fuels can be installed. However, most often, power stations and industrial boilers are built without multi-firing capabilities and it can be impractical or expensive to install such equipment or to modify the installation to use another fuel at a later stage. Decisions about the type of capacity to be built may have long-term repercussions for coal use, particularly in the case of power plants which have operating lives spanning many decades. Thus, the use of coal at the end of the projection period will be influenced both by the delivered price of coal relative to other fuels at that time, as well as the price that will prevail between now and then. A period of relatively low coal prices followed by a steep increase would result in significantly higher coal demand in 2035 than if prices rise sharply in the near term and then remain high.

Coal price trends

Around 17% of global hard coal production was traded internationally in 2009. While the price for a large proportion of coal output is relatively stable, internationally traded coal prices tend to fluctuate over short periods (roughly in line with oil and natural gas prices), reflecting the dynamics of inter-fuel competition (supported by financial trading markets), the importance of oil in coal mining operations and the cost of transporting coal over long distances (see Chapter 11). Although coal prices weakened relative to both oil and natural gas prices in the early part of the 2000s, they have strengthened since 2005 in response to differences in market conditions. Because coal markets are regionalised, reflecting the significant cost of transportation and inland bottlenecks in the coal supply chain, prices vary markedly across countries and regions. They are at times related to local weather conditions.

The prices of internationally traded coal have fluctuated significantly in recent years, surging to record highs in 2008. To a large extent, this movement was part of a broad surge in energy and other commodity prices worldwide, which was driven by booming demand in China and elsewhere; other factors, including a shortage of ships (which caused freight rates to surge), rail transportation constraints in South Africa and Russia, and flooding in Australia, also played a role. Prices fell back sharply in late 2008, due to the financial and economic crisis. They have since rebounded strongly, reflecting continuing strong demand in China, which became a net importer of coal for the first time in 2009, and renewed flooding in Queensland, Australia in 2010/2011, strikes in Colombia and increasing costs of production in all the major exporters. For example, European steam coal import prices peaked at $210 per tonne in September 2008, plunged to $60/tonne early in 2009 and then recovered to around $120/tonne by mid-2011. Prices have increased even more in Asia, resulting in a shift in South African exports away from Europe to Asia. Moreover, rapidly strengthening demand for seaborne coal imports has tended to produce a relatively more global market for both steam and coking coal.

Differentials between the prices of various qualities of coal, notably steam and coking coal can also oscillate markedly. The price of internationally traded coking coal has tended to increase relative to the steam coal price in recent years. In mid-2011, the price differential had reached its highest level ever, both in absolute and relative terms. This reflected exceptionally strong demand for coking coal for steel making, the relatively small volume of internationally traded coking coal, the existence of fewer suppliers and the inherently higher energy content compared to steam coal. While the quality differential is expected to persist, supply constraints are likely to diminish as investment shifts from steam to coking coal. As long as the differential remains high, coking coal will retain preferential access to transportation infrastructure and certain cross-category coals will be more utilised for metallurgical purposes.

The prospects for international coal prices are uncertain. The most critical factor is the outlook for imports into China, which could account for a large share of international coal trade, even though such imports would remain a fairly small fraction of domestic demand. Chinese import needs are sensitive to small variations between the very large volume of coal domestically produced and the total volume required on the domestic market. A relatively small shift in China's supply and demand balance could lead to a sudden shift in import needs and significant swings in international prices (see Spotlight in Chapter 2). Chinese utilities with plants located on the coast are adept at arbitraging the domestic and international markets and, over the longer term, policy decisions in China and other major consuming countries, particularly India, will be central to the evolution of coal demand and, hence, prices. In the New Policies Scenario, we assume that the average OECD steam coal import price – a proxy for international prices – reaches $110/tonne (in year-2010 dollars) or just under $200/tonne in nominal terms by 2035 (Figure 10.6).[3] Coal prices rise more quickly in the Current Policies Scenario, as global coal demand and international trade grow more strongly, but fall heavily in the 450 Scenario, as a result of far-reaching policy action to reduce demand, which in consequence reduces markedly the volume of international coal traded.

Figure 10.6 ● Average OECD steam coal import price by scenario

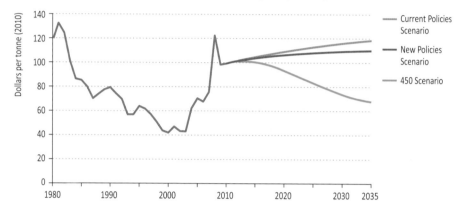

In the New Policies Scenario, coal prices are assumed to increase by significantly less than oil and natural gas prices in percentage terms, both because production costs are expected to increase less rapidly and because global coal demand levels off around 2025 (Figure 10.7). A detailed analysis of coal supply costs in each of the main exporting countries can be found in Chapter 11, while more details on the fuel price assumptions for each scenario in this *Outlook* can be found in Chapter 1.

Figure 10.7 ● Ratio of average OECD steam coal import price to average regional natural gas and IEA crude oil import prices in the New Policies Scenario

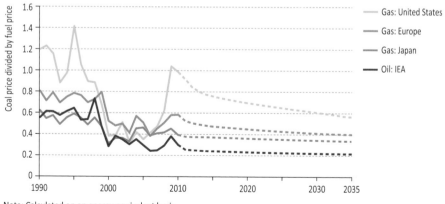

Note: Calculated on an energy-equivalent basis.

---

3. While various price indices can be chosen as a proxy for international prices, the average OECD steam coal import price reflects a significant part of global trade that takes place both in the Atlantic and Pacific Basin markets by encompassing such importing regions as Japan, Korea and Europe.

Fuel choice in power generation

Fuel choice in power generation involves both short-term decisions about which type of plant to run at any point in time given, a fixed set of generating assets, and periodic long-term decisions about investment in new generating capacity. In competitive markets, short-term decisions are largely a function of the prevailing prices and availability of different fuels, which account for most of the operating costs of power generation. Long-term decisions about the choice of fuel for new capacity are based on a broader set of factors which together determine the full, or levelised, cost of generation over the whole operating lifetime of the plant. These factors include assumptions about the future price of fuel inputs and about capacity factors (how often the plant runs) together with assumptions about the capital costs, and non-fuel operating and maintenance costs (including any actual or potential penalties for emissions) for each type of plant. These costs are influenced by construction lead times, plant efficiencies and discount rates. Other relevant factors, which are not directly financial but have clear financial implications, include the difficulty of obtaining planning permission, technical risks and operational versatility. Taken overall, at low coal prices, subcritical coal-fired plants – the cheapest type of coal plant – typically have the lowest levelised cost of all the different generating options (Box 10.3). At higher coal prices, more efficient supercritical or ultra-supercritical plants may have lower levelised costs than subcritical plants, but they may then not be competitive with other energy sources. In practice, the costs of different generating options vary significantly across countries.

On an energy-content basis, coal prices have traditionally been much lower than the equivalent prices of oil and natural gas and they are assumed to remain so in all three scenarios in this *Outlook*. Consequently, coal-fired plants generally find themselves high up in the merit order for dispatch from existing plants (though usually below renewables-based or nuclear capacity, for which short-term marginal costs are generally extremely low as fuel cost plays a negligible role or none at all). As a result, the utilisation rates for coal-fired plants are normally high, despite their generally lower conversion efficiencies compared with natural gas-fired plants (especially combined-cycle gas turbines [CCGTs]). But, when it comes to new capacity, the relative attractiveness of a coal plant *vis-à-vis* a natural gas plant is reduced by its typically higher construction costs, longer construction time and, for the most part, lower operational flexibility (CCGTs are among the most flexible plants). As a result, the levelised cost of coal-fired plants in some cases can be higher than that of other fuels and technologies. A decision to build a coal-fired plant also has to take into account the commercial and financial risk that carbon penalties or other tougher pollution controls may be introduced at some point in the future, requiring costly retrofitting, and that incentives may be given to alternative low-carbon technologies, affecting the cost advantage of coal and potentially reducing the utilisation of the coal-fired plant later in its operating life. For example, the existing EU Emissions Trading System and the currently envisaged more stringent air pollution controls in the United States are discouraging investment in coal-fired capacity and may continue to do so for some time.

Coal-fired capacity accounted for one-third of all the generating capacity additions worldwide over the period 1990 to 2010. Nearly 70% of these coal-fired capacity additions were in China. The bulk of the coal-fired capacity that has been built over the last two decades is subcritical – a large proportion of it in China (Figure 10.8).

Box 10.3 ● Coal-fired power generating technologies

Different types of coal-fired power generation technologies in operation today, or under development, have markedly different characteristics and costs. The main types are:

● *Subcritical*: Conventional boiler technology – the most commonly used in existing coal-fired plants – in which water is heated to produce steam at a pressure below the critical point of water (the point at which water reaches a pressure of 22.06 megapascals [absolute] and a temperature of 374°C). A water separator (or drum) facility must be installed in order to separate water and steam. Thermal efficiency is typically below 40% (gross terms and lower heating value).

● *Supercritical*: Steam is generated at a pressure above the critical point of water, so no water-steam separation is required (except during start-up and shut-down). Supercritical plants are more efficient than subcritical plants (normally above 40%), but generally have higher capital cost.

● *Ultra-supercritical*: Similar technology to supercritical generation, but operating at an even higher temperature and pressure, achieving higher thermal efficiency (can exceed 50%). Although there is no agreed definition, some manufacturers classify plants operating at a steam temperature in excess of 566°C as ultra-supercritical.

● *Integrated gasification combined-cycle (IGCC)*: Involves the production of a flue gas by partially combusting coal in air or oxygen at high pressure. Electricity is then produced by burning the flue gas in a combined-cycle gas plant. Thermal efficiency can exceed 50%.

In order to reduce carbon-dioxide (CO_2) emissions from coal-fired power plants, there is much development underway in the field of carbon capture and storage (CCS). This technology involves the capture of CO_2 produced in the power plant, its transportation and long-term storage underground.

Sources: IEA (2010a); IEA (2010b).

But the shares of both supercritical and advanced technologies, such as ultra-supercritical and IGCC, have grown, especially in the last five years. No commercial large-scale, integrated coal-fired power stations with carbon capture and storage (CCS) have yet been built, though several demonstration plants are planned. In 2010, roughly three-quarters of coal-fired capacity worldwide was subcritical, compared with close to 85% in 1990; another 20% was supercritical and only 3% were advanced technologies.

In the Current and New Policies Scenarios, coal remains the backbone of power generation, despite the projected share of coal in global power output falling from 40% in 2009 to 33% in 2035 in the latter scenario. In the New Policies Scenario, coal-fired plants are expected to account for around 27% of the total new additions to generating capacity worldwide between 2011 and 2020, and around 22% between 2011 and 2035. Of the coal-fired capacity additions, the share of more efficient plants is expected to continue to rise, driven by improvements in technology, growth in operational experience and rising coal and carbon prices, which improve the competitiveness of more efficient technologies. Between 2021 and 2035, just over 15% of new capacity is projected to be subcritical, while the share of advanced coal-fired plants is projected to rise to 40% (Figure 10.9). Reflecting the expected pace of development of CCS technology, coal plants fitted with CCS account for 10% of total coal capacity additions over the same period, with half being built in OECD countries (see Chapter 5).

Figure 10.8 ● World coal-fired generating capacity by type and major region

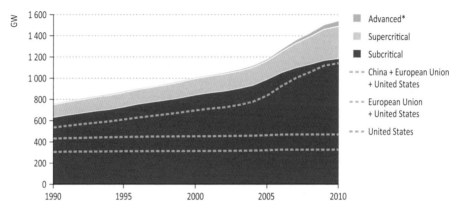

*Includes ultra-supercritical and IGCC.
Note: Excludes coal-fired generation from combined heat and power (CHP) plants.
Source: Platts World Electric Power Plants Database, December 2010 version.

Figure 10.9 ● New additions of coal-fired electricity generating capacity by technology and region in the New Policies Scenario

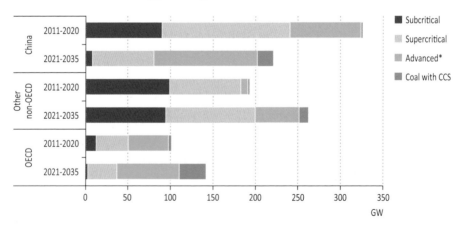

*Includes ultra-supercritical and IGCC.
Note: Excludes coal-fired generation from CHP plants.

The projections described above, and those for the other scenarios, are derived from a detailed analysis of the levelised costs of all the main coal-fired generating technologies, together with assumptions about government policies. The results by 2020 in the New Policies Scenario for the three main coal-consuming countries – China, the United States and India – and for selected key generating technologies are shown in Figure 10.10 (the detailed economic and technical assumptions[4] underlying this analysis can be found in Table 10.4). Based on these assumptions, coal is the most competitive fuel option in China and India, while in the United States, natural gas is much more competitive. The levelised costs of subcritical plants are, in most cases, cheaper than supercritical plants, but decisions about which type of plant to build will take account of expectations about higher coal prices in the future, which may tip the balance in favour of supercritical, ultra-supercritical or even IGCC plants. The assumed introduction of a shadow CO_2 price in the United States and a CO_2 price in China (see Chapter 1) reduces the competitiveness of coal-fired technologies, notably subcritical plants.

The type of coal-fired power generation technology that is deployed over the projection period for a given output of coal-generated electricity makes a difference to the total amount of coal inputs required and consequent emissions. If the average efficiency of all coal-fired power plants were to be five percentage points higher than in the New Policies Scenario in 2035, such an accelerated move away from the least efficient combustion technologies would lower CO_2 emissions from the power sector by 8% and reduce local air pollution. Opting for more efficient technology for new power plants would require relatively small additional investments, but improving efficiency levels at existing plants would come at a much higher cost. More efficient plants would also substantially reduce the generation efficiency penalty related to future CCS retrofitting. In reality, the penetration of the most efficient coal-fired power generation technologies is constrained by barriers such as financing (less efficient plants are often the cheapest way of generating power), the absence of adequate pricing of the external environmental costs of coal-fired power (including CO_2 emissions) and technical considerations (Spotlight).

10

·············· S P O T L I G H T ··················

What is impeding the deployment of more efficient coal-fired generation?

The growth in CO_2 emissions from coal-fired power generation can be curbed by choosing the most efficient plant designs for new plants and by replacing old plants with more efficient ones. CCS technology imposes an efficiency penalty and additional costs, making the underlying plant efficiency of the utmost importance. Improving the efficiency of new coal-fired plants needs to be a high priority where fleets are being rapidly expanded, notably in China and India.

---

4. The projections of power generation by fuel and type of technology are very sensitive to the assumptions made about future capital and operating costs, operational performance and financial parameters. Relatively small changes in these can have a significant impact on the relative financial attractiveness of the different coal-fired technologies and alternative means of generation, and, therefore, the share of coal in total generation and the volume of coal consumed in power stations in absolute terms.

In practice, however, power companies do not always opt for the most efficient plant designs, because they are often relatively expensive on a levelised cost basis. The thermal efficiency of an ultra-supercritical plant is typically up to 50% higher than that of a conventional subcritical plant. But the capital and maintenance costs are higher, which can make the subcritical plant the cheaper option at low coal prices, where there is no penalty for CO_2 emissions or when regulated electricity rates result in losses for the producer. In China, recent data indicate that a typical ultra-supercritical plant today has capital costs only 15% higher than those of a supercritical plant, yet subcritical plants continue to make up a substantial proportion of new coal-fired capacity, especially in non-OECD countries. In China, for example, 31% of all the coal-fired capacity brought online in 2010 was subcritical. But the share of supercritical and ultra-supercritical plants has been growing as the construction cost differences diminish, delivered coal prices increase and policy signals that encourage their deployment strengthen. Where coal is relatively expensive, ultra-supercritical technology would normally be the option of choice. For example, in Japan such plants now account for 28% of coal-fired capacity.

Other factors affect decisions about which technology to use for new plants, including:

- Fuel quality: local coal may have high moisture or ash content, or other impurities, such as chlorine.

- The technical capability and experience required to construct and operate such plants.

- The longer planning and construction lead times of more complex plants.

- The size of the unit: larger plants tend to be more efficient than smaller ones.

- Ambient conditions, such as the availability of cold seawater, or indeed water availability in any sense, which may mean less efficient air cooling may have to be used.

Taken overall, the best-available technology may not be the cheapest or most practical solution. However, in an environment where coal prices are rising, policy makers need to be wary of "locking-in" less efficient technology (see Chapter 6). The presence of a carbon price provides a powerful incentive for utilities to opt for more efficient plants and maximise the efficiency of existing plants. The cost of avoiding emissions through more efficient coal-fired generation can be low.

In the medium to longer term, newer technologies, such as IGCC, promise even higher efficiency. But these advanced technologies require further development to reduce costs before they can be deployed commercially on a large scale.

Table 10.4 ● Levelised electricity generating cost assumptions in the New Policies Scenario, 2020

| | Coal | | | | | Gas | Nuclear | Wind |
	SUB	SC	USC	IGCC	Oxyfuel with CCS	CCGT	power	onshore
United States								
Capacity factor	75%	75%	75%	75%	75%	55%	90%	28%
Thermal efficiency (gross, LHV)	39%	44%	49%	50%	40%	61%	33%	100%
Capital cost ($/kW)	1 800	2 100	2 300	2 600	4 000	900	4 600	1 750
Construction lead time (years)	5	5	5	5	5	3	7	1.5
Economic plant life (years)	30	30	30	30	30	25	35	20
Unit cost of fuel (various*)	55	55	55	55	55	7	3	-
Non-fuel O&M costs ($/kW)	45	63	69	91	120	23	104	26
China								
Capacity factor	75%	75%	75%	75%	75%	60%	85%	25%
Thermal efficiency (gross, LHV)	37%	42%	48%	49%	39%	59%	33%	100%
Capital cost ($/kW)	600	700	800	1 100	1 700	550	2 000	1 500
Construction lead time (years)	4	4	4	4	4	3	6	1.5
Economic plant life (years)	30	30	30	30	30	25	35	20
Unit cost of fuel (various*)	70	70	70	70	70	11	3	-
Non-fuel O&M costs ($/kW)	21	28	32	50	68	18	70	23
India								
Capacity factor	75%	75%	75%	75%	75%	65%	80%	23%
Thermal efficiency (gross, LHV)	34%	39%	44%	45%	35%	58%	33%	100%
Capital cost ($/kW)	1 200	1 500	1 700	1 900	2 975	700	2 800	1 550
Construction lead time (years)	5	5	5	5	5	3	7	1.5
Economic plant life (years)	30	30	30	30	30	25	35	20
Unit cost of fuel (various*)	60	60	60	60	60	11	3	-
Non-fuel O&M costs ($/kW)	42	60	68	86	119	25	105	23

*Natural gas in $/MBtu; coal in $/tonne; nuclear in $/MWh.

Notes: All costs are in year-2010 dollars. Unit fuel costs do not include CO_2 prices. The weighted-average cost of capital is 7% for China and India, and 8% for United States. Investment costs are overnight costs (see Chapter 5). For coal and nuclear, capacity factors are estimated averages for baseload operation, and mid-load operation for natural gas. SUB = subcritical; SC = supercritical; USC = ultra-supercritical; IGCC = integrated gasification combined-cycle. O&M = operation and maintenance; LHV = lower heating value.

The choice of power generation technology, and in particular the choice of coal-fired technology in China, will be of enormous importance for global CO_2 emissions from coal use. Policy decisions determining the role of nuclear power and renewables may shape the residual role for fossil-fuel based plants. When competing against natural gas for new capacity on a purely commercial basis, the relative financial attractiveness of coal-fired generation will hinge critically on relative coal and natural gas fuel prices, as well the extent and cost of requirements to install pollution-control equipment at coal plants, notably flue-gas desulphurisation (FGD).

Figure 10.10 ● Levelised electricity generating costs by component for selected technologies and countries in the New Policies Scenario, 2020

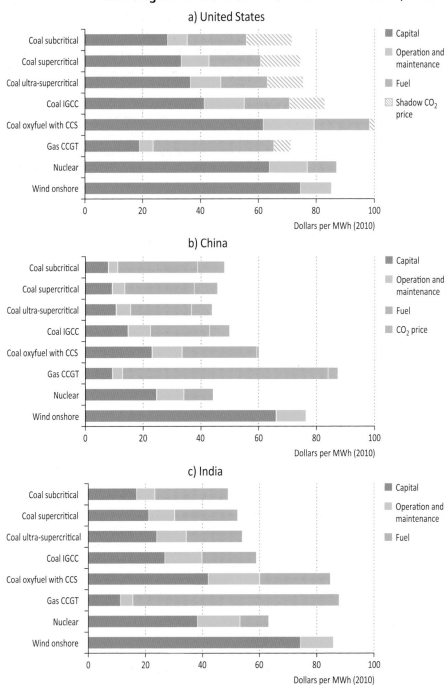

a) United States

Coal subcritical
Coal supercritical
Coal ultra-supercritical
Coal IGCC
Coal oxyfuel with CCS
Gas CCGT
Nuclear
Wind onshore

■ Capital
■ Operation and maintenance
■ Fuel
▨ Shadow CO$_2$ price

Dollars per MWh (2010)

b) China

Coal subcritical
Coal supercritical
Coal ultra-supercritical
Coal IGCC
Coal oxyfuel with CCS
Gas CCGT
Nuclear
Wind onshore

■ Capital
■ Operation and maintenance
■ Fuel
■ CO$_2$ price

Dollars per MWh (2010)

c) India

Coal subcritical
Coal supercritical
Coal ultra-supercritical
Coal IGCC
Coal oxyfuel with CCS
Gas CCGT
Nuclear
Wind onshore

■ Capital
■ Operation and maintenance
■ Fuel

Dollars per MWh (2010)

Notes: The CO$_2$ price in the United States is a shadow price, *i.e.* the price that investors assume will prevail on average over the lifetime of the plant. No actual CO$_2$ price is assumed to be introduced in the United States in the New Policies Scenario.

Fuel choice in industry

Coal has different uses in industry and can be substituted by other fuels with varying degrees of ease. About 40% of all the coal used in industry at present is for iron and steel production; non-metallic minerals production, such as cement and glass, account for one-quarter. Steel is produced through a dozen or so processing steps, which can take various configurations, depending on the product mix, available raw materials, energy supply and investment capital (IEA, 2010b). In principle, any fossil fuel source can be used instead of coal to make steel, but in practice coking coal is normally the preferred source (Figure 10.11), as it is most easily and cheaply converted into the coke (almost pure carbon), which is needed to reduce iron ore (most often in a blast furnace) and then into steel (in a basic oxygen furnace). A wider range of coal types, including steam coal, can be used to process iron ore using pulverised coal injection technology, which involves the injection of coal directly into the blast furnace to provide the carbon for iron fabrication, displacing some of the coke required for the process.

Figure 10.11 ● **World iron and steel sector energy consumption by type in the New Policies Scenario**

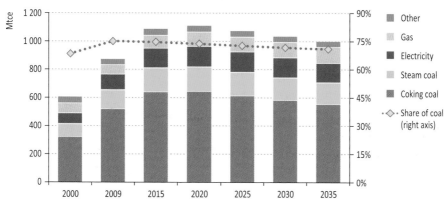

The choice of steel making technology depends, among other things, on the relative prices of coking and steam coal. Steel can be produced using less coal (or other sources of carbon) in an electric-arc furnace, in which the feedstock includes a large proportion of scrap steel, or be replaced by natural gas in the direct reduced iron process. In the New Policies Scenario, coal remains the principal energy source for steel making worldwide, at over 70%, though the shares of electricity and, especially natural gas, increase over the projection period (partly as the availability of scrap steel increases and alternative production processes are adopted).

In the cement industry, any source of energy can be used to make clinker – lumps or nodules produced by sintering limestone and alumina-silicate at high temperatures – which is then ground up to make cement. But coal is generally the preferred fuel where it is cheaper and its use is less restricted by environmental regulations. Coal use for cement manufacturing has been soaring in China and other emerging economies, driven by booming construction of basic infrastructure. In the New Policies Scenario, while cement production continues to grow over the projection period, use of coal for cement and other non-metallic minerals peaks around 2015, at 270 Mtce, and then declines throughout the rest of the projection period, partly as a result of substitution by natural gas. In industry as a whole, the share of coal in overall energy use drops from 28% in 2009 to 23% by 2035 in the same scenario, as lighter, more highly specialised industrial activities, where coal is less suitable for process energy needs and more stringent environmental regulations tip the balance in favour of other forms of energy. Electricity overtakes coal's contribution to industrial energy demand by 2035.

Fuel choice in other uses

Coal is no longer widely used in buildings (services and residential sectors) for space and water heating, for which the primary energy inputs in most cases are natural gas, oil products and electricity. Nonetheless, in some countries, notably China and India, a significant number of poor households rely on the direct use of coal. In the New Policies Scenario, global coal use in buildings is projected to level off within the next decade and then to decline, as rising incomes, infrastructure development and health concerns encourage switching to cleaner and more practical fuels. By 2035, the share of coal in energy consumption in buildings drops to 2%, from 4% in 2009.

Coal-to-liquids (CTL) plants have emerged as a new and potentially important market for coal. CTL, a process involving the production of liquid hydrocarbons from coal, has a long history. First used industrially in Germany during the Second World War, it was then extensively applied in South Africa. Since 1955, Sasol's plants, with a combined capacity in 2010 of 160 thousand barrels per day (kb/d), have produced more than 1.5 billion barrels of synthetic liquid fuel. In these plants, the coal is gasified to create "syngas", a mixture of hydrogen and carbon monoxide. This is similar to the old "town gas" that was used before natural gas became widely available (the same process of gasification is used in IGCC power plants, albeit in different types of gasifiers). In a second step, the syngas is turned into a liquid hydrocarbon, typically high quality diesel, using the Fischer-Tropsch (FT) technique, with an iron or cobalt catalyst. An alternative "direct" route involves reacting coal with hydrogen in the presence of suitable catalysts to produce liquid oil that can be used in a standard refinery to produce commercial hydrocarbon products. This is the technology used by Shenhua Group from China in its 24 kb/d plant in Inner Mongolia, which is being brought up to full production at present.

Around 20 CTL projects have been announced in the past five years, the majority in the United States and China, with a few in Indonesia, India, Australia, Canada, and South Africa. All of the announced projects are expected to incorporate CCS technology, but the uncertainty surrounding the regulatory framework for CCS is thought to be a key reason for the slow pace of development of these projects. Apart from water scarcity, future coal and oil prices are another key factor for CTL projects: the higher the oil price relative to the coal price, the better the economics of CTL production.

Because no large CTL plant has been built recently, there is a wide range of estimates for the capital costs associated with CTL technology, from $80 000 to $120 000 per barrel per day of capacity. Syngas/FT plants offer significant economies of scale, falling within this range of capital costs only for capacities above 50 kb/d. The capital costs of plant using the methanol and direct routes are less dependent on size. The oil price required to make CTL economic is in the range $60 to $100/barrel, depending on the location of the projects (China being in the lower part of the range) and the cost and quality of the feedstock. These prices include CCS, which typically adds only modestly to the cost and will probably be a requirement in most cases, as CO_2 emissions from CTL plants without CCS are very high (IEA, 2010c). CTL is economic at the assumed oil price trajectories in all three scenarios. In the New Policies Scenario, coal demand for CTL production rises from 27 Mtce in 2009 to 60 Mtce in 2020 and around 180 Mtce in 2035, with resultant oil production increasing from 0.2 mb/d to 0.4 mb/d and 1.2 mb/d, respectively (Figure 10.12). There is also growing interest in using coal as a feedstock for the production of petrochemicals and methanol, which is increasingly used as a blend stock for gasoline in China and some other countries.

Figure 10.12 ● Coal-to-liquids inputs by country in the New Policies Scenario

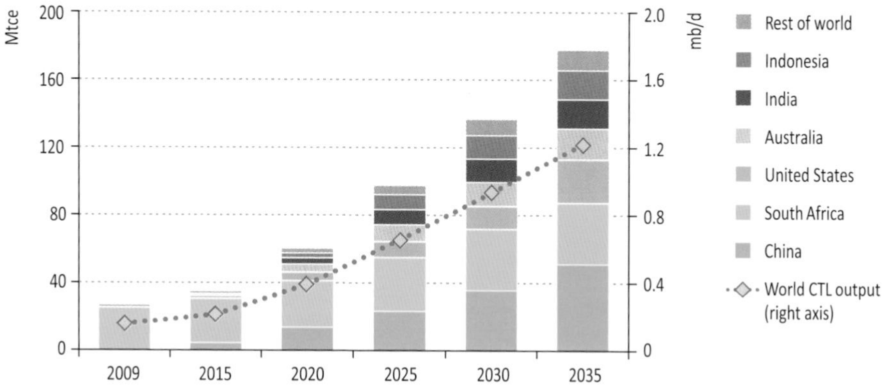

Energy and environmental policies

Government energy policies can have a decisive influence over coal use. Its use may be deliberately encouraged for economic or social reasons (to provide a market for local production and support employment and regional development, or to keep down energy costs to households and businesses) or for energy security reasons (to reduce reliance on energy imports). Policies may be designed to encourage switching away from coal to more environmentally benign or lower carbon sources, for example, air quality regulations or carbon penalties (taxes or cap-and-trade schemes). The structure of energy taxation can often favour the use of coal, relative to the other two fossil fuels. However, other measures, such as subsidies for renewables, may skew the competitive playing field against coal and other fossil fuels.

Environmental attitudes are especially important for coal use, as burning coal gives rise to airborne emissions of various pollutants, CO_2 and other greenhouse gases (though specialised equipment can be installed to lower or eradicate such emissions). As the most carbon-intensive fossil fuel, coal is the leading source of CO_2 emissions globally, accounting for 43% of all fuel-combustion-related CO_2 emissions in 2009. In the New Policies Scenario, global coal-related emissions climb by 20%, though their share of global emissions drops to 41% (Table 10.5). All of the growth in coal-related CO_2 emissions comes from non-OECD countries; emissions in the OECD fall. By contrast, in the 450 Scenario, coal-related CO_2 emissions fall by 57% globally. According to differences in policies and regulations, coal-related emissions of local air pollutants show divergent trends. In the New Policies Scenario, global emissions of nitrogen oxides and particulate matter grow only marginally through the projection period, while emissions of sulphur dioxide (SO_2) fall (see Chapter 2). Coal is the leading source of SO_2 emissions, but contributes much less than other fuels to NO_x and particulate emissions.

Table 10.5 ● CO_2 emissions from coal combustion by region and scenario (gigatonnes)

	1980	2009	New Policies Scenario		Current Policies Scenario		450 Scenario	
			2020	2035	2020	2035	2020	2035
OECD	3.6	4.0	4.0	2.8	4.3	4.1	3.7	0.6
Non-OECD	3.0	8.5	11.4	12.1	12.4	16.1	10.2	4.8
World	6.6	12.5	15.4	14.9	16.7	20.2	13.9	5.4
Share of coal in world CO_2	36%	43%	45%	41%	46%	47%	44%	25%

The differences in emissions and the amount of coal consumed between each scenario result mainly from fuel-switching in power generation, and lower electricity use, which reduces the need to burn coal in power stations (Figure 10.13). These two factors account for more than three-quarters of the fall in coal use in 2035 in the 450 Scenario *vis-à-vis* the New Policies Scenario. The other contributing factors are efficiency gains in the power sector and in other end-use sectors as well as fuel-switching.

Figure 10.13 ● Reduction in world primary coal demand by sector and scenario

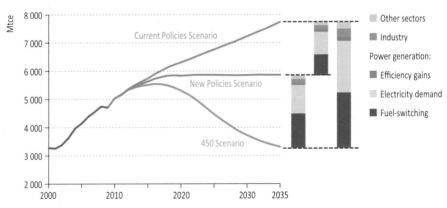

In the 450 Scenario, OECD coal demand falls most, to less than half of the 2009 level. Coal use would be even lower in that scenario were it not for increased deployment of CCS, especially in the OECD countries; by 2035, roughly one-third of global coal is consumed in power stations and industrial facilities equipped with CCS (mainly in China and the United States), compared with only 2% in the New Policies Scenario (Figure 10.14).

Figure 10.14 ● **World primary coal demand by region and scenario**

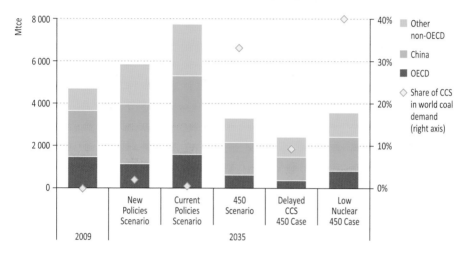

Note: See Chapter 6 for assumptions behind the Delayed CCS 450 Case and Low Nuclear 450 Case.

In some countries, the government subsidises the consumption of coal, typically through price controls that keep the price of coal below the level that would prevail in a competitive market (Figure 10.15). However, consumption subsidies to coal worldwide are small and diminishing, compared with those to oil products, natural gas and electricity, accounting for only 0.7% of these subsidies (see Chapter 14). Coal consumption subsidies may be linked to subsidies to production, where the aim is to protect output and jobs in indigenous mining activities (see Chapter 11). The objective may also be to keep energy costs down to improve the international competitiveness of domestic industry and the commercial sector, or to protect the purchasing power of households for social reasons. In monetary terms, China has the biggest coal consumption subsidies (measured as the differences between actual prices and the full economic cost of supply)[5], amounting to around $2 billion in 2010, though the rate of subsidisation (measured as a percentage of the full economic price) is less than 5%. In principle, the Chinese government lifted the remaining formal price controls on coal in 2007, but intervention in coal pricing by the central and provincial governments does still occur. For example, the National Development and Reform Commission imposed a price freeze on coal sold by domestic producers in June 2010, as part of a package of measures to tame inflation.

---

5. Reference prices used to calculate consumption subsidies have been adjusted for coal quality differences, which affect the market value of the fuel.

Figure 10.15 ● Coal consumption subsidies for selected countries, 2010

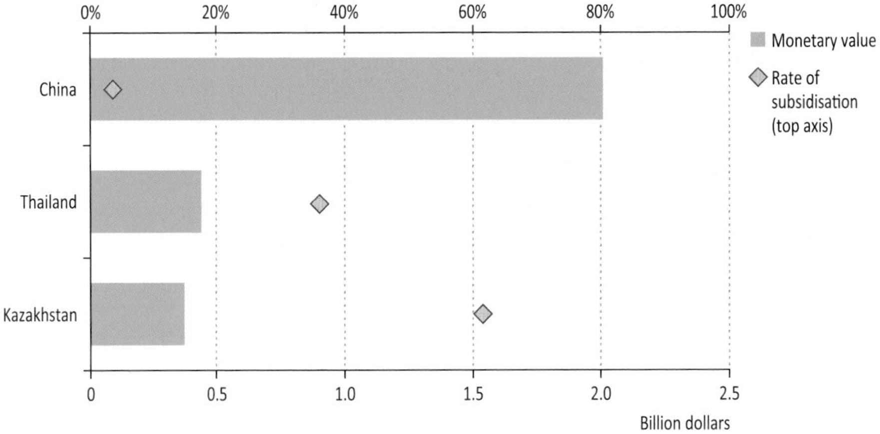

Notes: Measured using the price gap approach (see Chapter 14). The rate of subsidisation is the subsidy expressed as a percentage of the full economic price.

Technological innovation

Future choices of energy technology will be determined by a combination of economic and policy factors. Those choices, particularly in producing power, will have a major effect on coal consumption worldwide. Technological innovation could lead to large changes in the operational and economic characteristics of the different coal-fired technologies and in competing technologies, with potentially important consequences for coal demand, especially in the longer term. Experience suggests that technological change is likely to accelerate, increasing the uncertainty involved in investment decisions. The prospects for advances in coal technology, and their consequences, are discussed briefly by sector in this section.[6]

Coal-fired generating technologies

Almost all coal-fired generation today is based on the pulverised coal combustion (PCC) method, which in a boiler creates steam and then uses the steam to drive a turbine to produce electricity. Most existing plants operate under subcritical steam conditions, with a maximum efficiency of about 39%. However, an increasing share of new plants are capable of operating under supercritical or ultra-supercritical conditions, with efficiencies of up to 47%. Advances in materials have paved the way for larger unit capacities, with single PCC units of 1 000 megawatts (MW) now in commercial operation. Further developments in materials are underway, involving the use of nickel-based super alloys, which should permit the use of steam temperatures of 700°C and higher in the next generation of coal plants, potentially boosting efficiencies to over 50%.

There is also potential for improving fluidised-bed combustion (FBC) technology, which is currently used in niche applications, notably where flexibility is required to adapt to varying fuel quality: FBC can be designed to work well with low-quality coals, biomass and general waste. Since FBC plants operate at lower combustion temperatures than conventional

---

6. This section draws on the findings of the 2010 edition of the IEA's *Energy Technology Perspectives* (IEA, 2010b).

PCC plants, they are said to reduce emissions of nitrogen oxides easily. They also offer the potential for integrated in-bed sulphur reduction. Today, there are several hundred FBC plants in operation, most of which are small. Manufacturers hope to be able to scale up designs to offer units within the range of 500 MW to 800 MW.

Integrated gas combined-cycle (IGCC) technology, which fell out of favour in the late 1990s because of problems with reliability and high costs, could make a come-back with design improvements. IGCC plants have the potential to achieve levels of efficiency as high as those of PCC plants. Today, only four plants are still in commercial operation – two in Europe and two in the United States – with capacities of 250 MW to 300 MW each and efficiencies of 40% to 43%. Future designs may achieve efficiencies of over 50%, possibly with the addition of a water-gas shift reactor, whereby additional hydrogen can be produced and carbon monoxide converted to CO_2 for capture and storage. Given the high efficiency and flexibility of an IGCC plant, significant effort is being devoted to the development of IGCC technologies, particularly in China, the United States, Japan and Europe.

Carbon capture and storage

Carbon capture and storage (CCS) technology, if widely deployed, could potentially reconcile the continued widespread use of coal with the need to reduce CO_2 emissions. While the technology exists to capture, transport and permanently store CO_2 in geological formations, it has yet to be deployed on a large scale in the power and industrial sectors and so costs remain uncertain. The experience yet to be gained from the operation of large-scale demonstration projects will be critical to the prospects for widespread deployment of CCS. The demonstration phase, which is only just beginning, is likely to last for over a decade. At the end of 2010, a total of 234 active or planned CCS projects had been identified across a range of technologies, project types and sectors, but only eight projects (of which five are considered full CCS projects, in that they demonstrate the capture, transport and permanent storage of CO_2 utilising sufficient measurement, monitoring and verification systems and processes to demonstrate permanent storage) are currently operating – most of them at demonstration-scale (GCCSI, 2011). The likelihood is that there will be, at best, no more than a dozen large-scale demonstration plants in operation by 2020.

CCS raises many legal, regulatory and economic issues that must be resolved before it can be widely deployed. Several initiatives have been taken by the IEA and other bodies, such as the Global CCS Institute and the Carbon Sequestration Leadership Forum, to develop the policy and regulatory framework to facilitate commercial deployment of CCS on a large scale, but much remains to be done.

The main challenges to successful full-scale demonstration and commercial deployment of CCS include (IEA, 2009):

- High construction costs (assuming an average project cost for a CCS plant of $3 800 per kilowatt (kW), equates to around $2 billion for a 500 MW plant) and difficulties in financing large-scale projects, particularly in the absence of any carbon penalty.

- Higher operating and maintenance costs and the reduced thermal efficiency of plants fitted with CCS, relative to similar coal-fired power plants without CCS.

significant. These emissions are different from those of a coal-fired power plant. The direct CTL process involves converting coal into diesel or gasoline by adding hydrogen and making it react with the coal to form hydrocarbon chains. CO_2 emissions arise primarily from generating hydrogen. In the indirect route, it is intrinsic to the syngas generation process that the energy comes from the coal itself. CO_2, which is produced with the syngas, must be separated from the syngas prior to the FT (or methanol) process. So the bulk of the CO_2 is, in any case, captured. This is why CCS is a relatively inexpensive addition to indirect CTL production processes: only compression, transport and storage need to be added and these are normally much less expensive than capture. Estimates for the cost of adding CO_2 purification to a CTL plant, as required for sequestration, range from \$3/barrel to \$5/barrel of oil produced (IEA, 2010c). Nonetheless, since CTL with CCS has yet to be demonstrated commercially, the efficacy of the technology and its cost are uncertain.

Another potential area of technological progress in CTL is in the use of underground coal gasification (UCG) to produce the syngas, with a Fischer-Tropsch plant to transform the syngas into liquid hydrocarbon. In principle, UCG can provide the syngas at much lower capital costs and allows deeper, un-mineable, coal beds to be exploited. UCG has been piloted in Australia, Canada, China, Spain and South Africa, with mixed success. The world's first UCG-based liquids production demonstration plant was commissioned by Linc Energy in 2008 in Queensland, Australia and the company is planning a first commercial plant in South Australia. Widespread success with UCG could pave the way for more CTL production in countries rich in coal resources that are located far from markets.

Regional Analysis

In the New Policies Scenario, which is the focus of this section, global primary coal demand rises through the early 2020s and then remains broadly flat throughout the rest of the projection period.[9] By 2035, global coal use is one-quarter higher than in 2009 and it remains the second-largest fuel in the primary energy mix and the backbone of electricity generation.

The next two-and-a-half decades see divergent trends in coal use across the main regions. Primary coal demand in non-OECD countries, which already accounts for close to 70% of world coal demand, continues to grow, albeit at a decelerating rate, reaching around 4 715 Mtce in 2035 – an increase of 45% on 2009 levels (Table 10.6). Demand expands in all non-OECD regions and just three countries – China, India and Indonesia – account for over 80% of this increase; China and India account for at least one-third each (Figure 10.17).

Over the projection period, China continues to account for nearly every second tonne of coal consumed globally and, around 2025, India overtakes the United States as the world's second-largest consumer of coal. By contrast, OECD demand is projected to decline to just under 1 150 Mtce, a drop of one-third, compared to peak consumption levels in 2007. Coal use in OECD Europe and the United States falls the most in absolute terms – in total, by around 250 Mtce between 2009 and 2035 – but the United States, though dropping to the third-largest in the world after China and India, remains by far the biggest coal consumer in the OECD. While

---

9. See Annex B for a summary of policies and measures assumed in the New Policies Scenario.

coal demand rebounds in the short term, along with economic recovery and partly to alleviate shortfalls related to reduction in nuclear output, long-term coal demand declines in all OECD regions, mainly due to replacement by natural gas and renewables in power generation.

Table 10.6 ● Primary coal demand by region in the New Policies Scenario (Mtce)

	1980	2009	2015	2020	2025	2030	2035	2009-2035*
OECD	**1 380**	**1 476**	**1 567**	**1 494**	**1 406**	**1 280**	**1 146**	**-1.0%**
Americas	573	743	798	763	738	690	636	-0.6%
United States	*537*	*693*	*738*	*705*	*684*	*644*	*599*	*-0.6%*
Europe	663	415	420	383	346	299	264	-1.7%
Asia Oceania	145	318	350	348	322	291	246	-1.0%
Japan	*85*	*145*	*166*	*158*	*144*	*131*	*115*	*-0.9%*
Non-OECD	**1 179**	**3 229**	**4 067**	**4 339**	**4 457**	**4 576**	**4 713**	**1.5%**
E. Europe/Eurasia	517	276	310	304	303	301	299	0.3%
Russia	*n.a.*	*136*	*164*	*166*	*171*	*169*	*168*	*0.8%*
Asia	573	2 775	3 548	3 812	3 921	4 037	4 184	1.6%
China	*446*	*2 179*	*2 749*	*2 863*	*2 839*	*2 823*	*2 820*	*1.0%*
India	*75*	*399*	*519*	*619*	*701*	*778*	*883*	*3.1%*
Indonesia	*0*	*44*	*67*	*87*	*107*	*127*	*146*	*4.8%*
Middle East	2	2	2	3	3	3	3	1.9%
Africa	74	151	170	179	184	185	180	0.7%
South Africa	*68*	*141*	*152*	*158*	*161*	*162*	*160*	*0.5%*
Latin America	14	26	37	41	46	49	46	2.3%
Brazil	*8*	*16*	*24*	*23*	*23*	*21*	*20*	*1.0%*
World	**2 560**	**4 705**	**5 634**	**5 833**	**5 863**	**5 856**	**5 859**	**0.8%**
European Union	*n.a.*	*381*	*371*	*326*	*282*	*233*	*200*	*-2.5%*

*Compound average annual growth rate.

Figure 10.17 ● Incremental primary coal demand by region in the New Policies Scenario

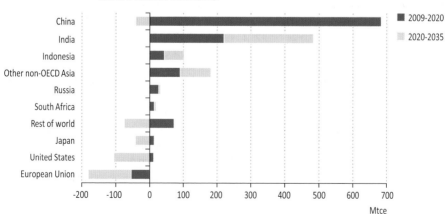

China

China will remain far and away the largest coal consumer in the world. Based on preliminary estimates for 2010, China accounted for nearly half of global coal use. Coal demand in China amounted to around 2 350 Mtce in 2010 – equal, in energy terms, to twice the oil consumption of the United States, the world's biggest oil consumer. Chinese coal consumption grew by 50% alone, between 2005 and 2010 to satisfy surging demand in power generation and heavy industry; in volume terms, this increase was equivalent to more than total coal demand in the United States in 2010.

In the New Policies Scenario, China's coal demand is projected to grow by about 30% from 2009 to over 2 850 Mtce by 2020, and then decline slightly, before stabilising for the rest of the projection period above 2 800 Mtce (Figure 10.18). The share of coal in China's total primary energy demand declines to around 50% by 2035, from just over 65% in 2009. The power sector continues to drive demand and, towards the end of the projection period, accounts for two-thirds of China's total coal demand. The increase in power-sector coal demand between 2020 and 2035 is largely offset by a fall in demand in industry, where coal demand peaks before 2020. The share of coal in both sectors declines through 2035 and, while coal still remains the principal source of energy for power generation, coal is overtaken by electricity use in industry. These projections are highly sensitive to the underlying assumptions about economic growth and industrial output, which are the primary drivers of electricity demand and coal needs for both generation and industry. GDP growth is assumed to decelerate through the projection period as the economy matures (see Chapter 1).

Figure 10.18 ● Coal demand in China by sector in the New Policies Scenario

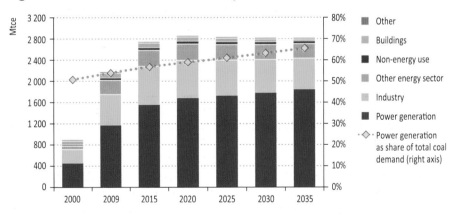

As elsewhere, government policies will have a major influence on coal demand in China (and, therefore, globally) in the longer term. For energy diversification and environmental reasons, China has been targeting a large expansion of nuclear power, natural gas and renewables, as well as major improvements in the efficiency of energy use. The 12th Five-Year Plan for 2011 to 2015, which was adopted in March 2011, sets targets for reducing energy intensity and carbon intensity, which are in line with China's Copenhagen pledge (Box 2.2). The share of non-fossil-fuel consumption in the primary energy mix is to be increased from 8.3% in 2010 to 11.4% by 2015. Nuclear, renewables and natural gas are expected to be aggressively promoted. Many of the policies reflected in the 12th Five-Year Plan are expected to be taken forward through more detailed targets. In some cases, this *Outlook* anticipates such targets, based on credible reports at the time of writing, in order to provide further depth to the analysis. The energy targets outlined by officials show coal consumption reaching around 3.8 billion tonnes in 2015. Coal demand could turn out to be higher if the targets are missed or if the Chinese economy grows faster than assumed in the plan. On the other hand, more ambitious targets might be set in future plans, leading to lower coal use over the *Outlook* period.

Power sector

The recent surge in demand for coal for power generation mainly results from soaring electricity demand, growth of which has averaged around 12% per year since 2000. China's electricity generation reached an estimated 4 220 terawatt-hours (TWh) in 2010, up by almost 15% since 2009 and more than 200% since 2000. It has continued to grow strongly, with output expected to reach above 4 500 TWh in 2011; when China will almost certainly become the world's biggest power producer, overtaking the United States. China's expansion of generating capacity has been impressive, more than doubling from around 455 GW in 2005 to nearly 1 000 GW in 2010, with coal-fired plants contributing more than 60% of the increase. Demand has grown so quickly that power shortages have been common and seem likely to continue in the near future, especially in hotter months when demand peaks. Part of the reason for the shortages is that electricity prices remain under tight governmental control as a means of limiting inflation. However, coal prices have been largely deregulated and have increased significantly since 2007. This has led to power companies incurring losses, discouraging investment in new plants and worsening shortages.

10

Growth in electricity consumption has come principally from the industrial sector, which currently accounts for over half of total electricity use. Industrial electricity demand grew very quickly to 2007, slowed substantially in 2008 as export-destined manufacturing was affected by the global economic crisis, and has since resumed its strong growth. On average, it grew by around 13% per year between 2000 and 2010. Urbanisation and a nationwide programme of electrification have boosted power use in the buildings sector by 12% annually over the same period. Even so, annual per-capita use, at around 3 000 kWh, remains low compared with the OECD, at 8 200 kWh. Electricity demand in all sectors will continue to grow, though policy interventions could temper the pace of growth, particularly in the industrial sector. In the New Policies Scenario, Chinese electricity demand in total is projected to grow on average by 8% per year between 2009 and 2015, and then slow to an average of 3% per year over the remainder of the projection period as economic growth slows and electricity use becomes more efficient. By 2035, China's total electricity demand is nearly equal to that of the OECD in 2009.

Coal is abundant, in absolute terms and relative to other fossil fuels, and low cost in most regions of China, and will remain the backbone of the country's power sector for some time to come. However, mining costs have been rising; they now fall into an average range of $55/tonne to $70/tonne for steam coal at the mine mouth. If users and utilities are located far from producing areas, transport costs, which may include both rail and seaborne freight, can add substantially to this cost. In mid-2011, competing imported steam coal was trading at around $120/tonne delivered to coastal regions in China. Considering all of these factors, it appears that even relatively expensive natural gas at $10 per million British thermal units (Mbtu) to $12/MBtu may be competitive in some locations in some applications in the power sector (Figure 10.19). At present, coal prices are low enough for coal to be generally competitive with natural gas in most locations. But higher coal prices and tighter environmental regulations could increasingly favour natural gas-fired capacity over coal. The share of natural gas in total generation increases from a mere 2% in 2009 to 8% in 2035 in the New Policies Scenario. The rate of penetration of natural gas in the Chinese power sector makes a sizeable difference to fuel needs: an additional 10 GW of coal-fired power replaced by natural gas would save around 20 Mtce, while requiring around 10 billion cubic metres (bcm) of additional natural gas supply.

Figure 10.19 ● **Breakeven price of coal versus natural gas for power generation in China, 2020**

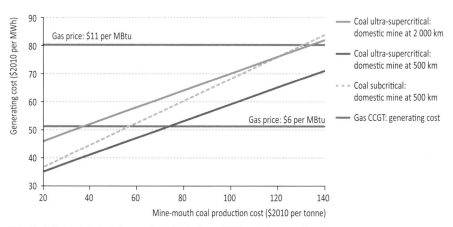

Note: Coal plant costs include flue-gas desulphurisation and NO_x emission control equipment.

Taking all factors into account, the size of coal's share in overall power generation use in China is expected to decline substantially, as a result of diversification policies and steady improvement in the thermal efficiency of the stock of coal-fired plants. In the New Policies Scenario, around 330 GW of coal-fired generating capacity is added during the current decade, accounting for 40% of total capacity additions in the period (Figure 10.20). Of this coal-fired capacity, 90 GW is already under construction, around 70% of which is relatively high efficiency supercritical or ultra-supercritical plant; the remaining is less efficient, subcritical plants. Over the whole projection period, coal plants account for only one-third of the total net increase in China's generating capacity, as other technologies, such as wind, hydro, natural gas and nuclear, gain prominence.

Figure 10.20 ● New additions of power-generating capacity in China by type in the New Policies Scenario

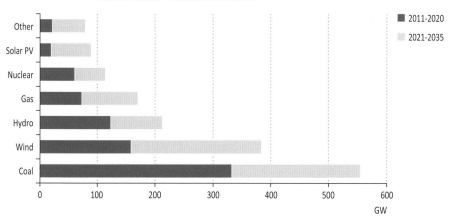

In the longer term, the importance of coal in China is expected to decline even though it will undoubtedly remain the cornerstone of the electricity mix throughout the *Outlook* period (Figure 10.21). In the New Policies Scenario, the share of coal in electricity generation drops from 79% in 2009 to 65% in 2020 and 56% in 2035. However, coal consumption in the power sector rises briskly, from 1 175 Mtce in 2009 to around 1 700 Mtce in 2020, and then grows more gradually to 1 850 Mtce in 2035. The main reasons for the slowdown in coal needs for power is slower electricity demand growth after 2020, more efficient coal-fired capacity additions and the increased penetration of other fuels and technologies. Natural gas-fired generation increases particularly rapidly: by 2035, it is fourteen-times higher than in 2009, contributing 8% to total electricity generation. The share of low-carbon power generation in China — nuclear power, CCS-fitted coal- and natural gas-fired plants, hydropower and other renewables — expands rapidly and by 2035 these sources make up almost 40% of total generation, compared with around 20% in 2009. As a result of this shift and efficiency gains, the carbon intensity of China's power generation falls significantly, from around 800 grammes of CO_2 per kWh in 2009 to 500 g CO_2/kWh in 2035.

10

Figure 10.21 ● Electricity generation in China by type in the New Policies Scenario

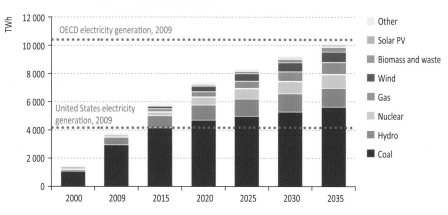

Industry and other sectors

The recent rapid expansion of coal use in China's non-power sectors is set to reverse over the projection period (Figure 10.22). Coal demand in industry – the main non-power sector – grew at an annual rate of 9% between 2000 and 2009. The bulk of industrial coal use is in the steel and cement sectors. In the New Policies Scenario, industrial coal demand slows markedly from 2015 and then begins to fall steadily by around 2020, at an average rate just below 1.5% through 2035. This results from a combination of factors: progressively slower growth in industrial production (notably steel), increasing energy efficiency in industry and displacement of coal by electricity and natural gas. The share of coal in total industrial energy consumption drops from around 60% in 2009 to just below 40% by 2035. Three-quarters of the growth in energy use in industry is met by electricity, which increases its share of final demand in the sector to 44% by 2035; the sectoral share of natural gas rises from 2% to 9%. The projected change in the industrial energy fuel mix is driven by macroeconomic factors, notably a shift in the economy towards lighter industry, inter-fuel competition, stricter environmental regulation and already announced targets to decarbonise the economy. Natural gas increasingly displaces coal in the production of process heat. The restructuring of the economy towards lighter industry, including high-technology products and services, favours the use of electricity.

Figure 10.22 ● Non-power generation coal demand in China by sector in the New Policies Scenario

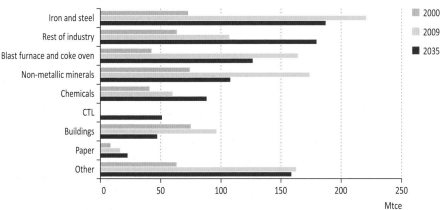

Coal use in steel making, the single most important industrial sector, is set to drop significantly in the longer term. Annual crude steel production in China almost quintupled between 2000 and 2010, reaching around 630 million tonnes. This is almost six times higher than that of the second-largest producer, Japan. This led to an unprecedented increase in the sub-sector's need for coal: coal-based energy use in steel making (including blast furnace and coke oven activity) increased from around 115 Mtce in 2000 to 385 Mtce in 2009, an average annual rate of growth of 14%. The share of coal in the steel industry's total energy consumption increased from 83% to 86% over the same period, mostly at the expense of oil (the share of electricity also increased). Coking coal accounted for around 85% of the increase in coal use, which more than quadrupled between 2000 and 2009 to nearly 300 Mtce. In the New Policies Scenario, total coal demand in iron and steel production is projected to increase by around 75 Mtce by 2015. Thereafter, the peaking of China's crude steel production around 2020 and the introduction of more efficient techniques lead to a decline in coal consumption, averaging 2% per year from 2015 until the end of the projection period. A similar pattern is expected in the cement and other non-metallic minerals.

One new sub-sector that is expected to see significant growth in coal use is coal-to-liquids. China recently brought on stream its first commercial CTL facility – the 24 kb/d Shenhua Group plant in Inner Mongolia, which uses the direct CTL route. Several other projects, including the joint-venture project between ExxonMobil and Jincheng Anthracite Mining Group that uses the former's coal-based methanol-to-gasoline technology, are in the start-up phase. Coal use for CTL is expected to reach around 5 Mtce in 2015 and 50 Mtce in 2035, as higher oil prices make new investments in this technology more profitable. This sub-sector accounts for around 2% of China's total primary coal demand by the end of the *Outlook* period.

The use of coal in the buildings sector in China, mostly for space and water heating, still made up 15% of total final energy demand in buildings in 2009. In the New Policies Scenario, that share declines to 5% by the end of the projection period, pushing coal consumption down to about 50 Mtce from 100 Mtce in 2009. Natural gas, and especially, electricity for water heating and other appliances account for most of the increase energy demand in buildings.

India

India – currently the third-largest coal user worldwide behind China and the United States – is likely to see continued rapid expansion in coal demand in the absence of radical policy change. Coal demand increased by about 80% between 2000 and 2010, reaching 420 Mtce, with growth accelerating to more than 7% per year on average since 2005, compared with 5% in the first half of the decade. Booming demand is being driven by rapid economic growth, which is pushing up energy needs for power generation and in industry, where coal – mainly domestically sourced – is the main fuel. In the New Policies Scenario, India becomes the world's second-largest consumer of coal by around 2025, with demand more than doubling to 880 Mtce by 2035 (Figure 10.23). Consequently, India is poised to become the world's biggest importer of coal soon after 2020, as rapid demand growth outstrips domestic supply (see Chapter 11). The share of coal in India's total primary energy demand increases in parallel over the next decade, to nearly 46%, before decreasing to the level of 42% from 2009 towards the end of the projection period.

More than 60% of the projected rise in Indian coal demand in the New Policies Scenario comes from the power sector, to meet surging demand. There is enormous latent demand for electricity: across the country, an estimated 300 million people still lack access to electricity; by 2030, this number is expected to fall to around 150 million in the New Policies Scenario – a significant improvement, but still 10% of the population. Coal is expected to remain the primary source of power generation, mainly because it is the most economically competitive generating option, in most cases, for both existing plants and new capacity. The power sector's share of total coal use nonetheless dips, from 72% in 2009 to 67% in 2035, as demand in industry and other transformation processes grows even faster. The rapid increase in industrial coal demand – averaging 4% per year between 2009 and 2035 – stems from a continuing boom in crude steel production, as well as from other manufacturing and processing industries. Industry's share of total coal demand rises from 16% in 2009 to 21% by 2035.

Figure 10.23 ● Coal demand in India by sector in the New Policies Scenario

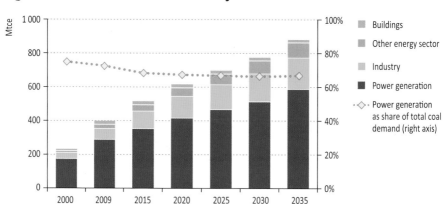

Power sector

In the New Policies Scenario, coal use in power generation in India grows from around 290 Mtce in 2009 to 590 Mtce in 2035. Electricity demand almost quadruples, as a result of the rising population and economic growth. Around 2015, India becomes the world's third-largest consumer of electricity, behind China and the United States. Coal remains the main source of electricity generation, although its share declines from around 70% in 2009 to 53% in 2035. In absolute terms, coal-fired generation increases by over 1 100 TWh between 2009 and 2035 – an increase greater than that of any other energy source.

Net additions of coal-fired generating capacity total around 250 GW during the projection period, accounting for 40% of the total additions, while combined capacity of natural gas, hydro and nuclear expands by nearly 200 GW. Coal plants have dominated new build over the past decade, with around 25 GW of capacity having been added. But coal capacity will rise even faster in the next few years, as around 100 GW more is under construction and about as much again is planned. Of the capacity being built, about 40 GW is known to be supercritical or ultra-supercritical, which will increase sharply the share of those technologies in overall coal capacity. Supercritical or ultra-supercritical plants account for more than half of all the new coal capacity that is projected to be built in the second-half of the projection period (Figure 10.24). India also has ambitious plans to add 20 GW of solar photovoltaic capacity by 2022.

Figure 10.24 ● Coal-fired generating capacity in India by type in the
New Policies Scenario

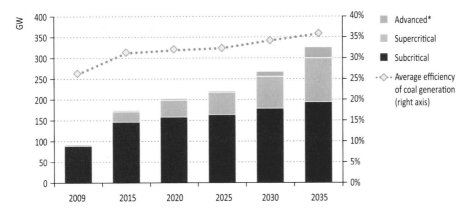

*Includes ultra-supercritical and IGCC.

Delivering power to its vast population has always been a challenge for India. Although output rose from around 290 TWh in 1990 to 900 TWh in 2009, this has not been enough to provide electricity to the entire population. Despite governmental efforts, the grid has not been extended to all key rural areas and many dwellings in urban areas of informal housing are still not connected to the grid. As a result, annual per-capita electricity consumption is only around 620 kWh, compared with almost 3 000 kWh in China, where almost all the population has access to electricity. Power shortages in India are endemic, largely due to under-investment in generation and transmission. Technical and non-technical losses in transmission and distribution are extremely high by international standards. The recent involvement of the private sector in generation and distribution has improved the situation somewhat, but there are still wide differences between different states.

Industry and other sectors

Total coal demand in the industrial sector in India is projected to increase almost three-fold, from 65 Mtce in 2009 to almost 190 Mtce in 2035 in the New Policies Scenario – the result mainly of strong demand from steel making, cement and other heavy industries. Iron and steel accounted for 70% of total industrial coal use in 2009 and the sector is expected to remain a primary user through the projection period. India is the fifth-largest steel producer in the world, just behind Russia. In 2010, steel production reached around 70 million tonnes (Mt), an increase of 6% on 2009 and an average of 10% per year since 2000. The Indian iron and steel industry is unique because of the high share of production that relies on feeding direct reduction of iron into electric furnaces – about 40%. This technique offers several advantages, including lower plant construction costs and the fact that it does not require the use of high-quality coking coal (any kind of carbonaceous fuel can be used). This makes the process especially attractive for India, which has very limited reserves of high-quality coking coal. Nonetheless, Indian steel making is still comparatively energy inefficient. Around 40% more energy is used per tonne of crude steel production in India compared with the world average, but new technologies and polices are being introduced to close the gap. Total coal consumption in steel making, including the use of coal in blast furnaces and coke ovens, is projected to grow by 4% per year on average, from around 45 Mtce in 2009 to nearly 140 Mtce in 2035. The other big industrial user of coal is the cement and other non-metallic minerals sector, production from which has been growing rapidly in recent years. India, with production of 220 Mt in 2010, is now the second-largest cement producer in the world, after China. There is limited scope for saving energy in Indian cement production, because of low clinker-cement ratios and the increasing use of dry-process kilns with pre-heaters. As domestic cement demand remains strong, this will lead to increasing energy demand, most of which is expected to be met by coal, as it is the cheapest energy source. Coal use in the cement and non-metallic minerals sector is projected to double, to around 20 Mtce in 2035.

The remaining industrial sub-sectors, as a group, see continuing growth in demand for coal too, their total consumption reaching nearly 30 Mtce by the end of the projection period. Coal use outside the power and industrial sectors, in buildings and agriculture, is relatively small and is projected to remain broadly flat. The CTL sub-sector is one exception: higher oil prices making investments in CTL worldwide more profitable, and in response to rising oil imports, India is projected to consume around 20 Mtce of coal in CTL by 2035, resulting in a production of 125 kb/d of synthetic oil, 1.5% of India's oil demand.

United States

Based on preliminary estimates for 2010, the United States was the world's second-largest coal-consumer, with consumption equal to 30% that of China. Coal accounts for around one-fifth of the country's primary energy use – a larger share than for most other OECD countries, primarily due to abundant indigenous resources of relatively low-cost coal (the country is a major producer and a net exporter of coal). In 2009, about 90% of coal demand in the United States was in the power sector. However, environmental concerns, including local air quality and greenhouse-gas emissions, are expected to severely constrain this use of coal in the coming decades (Figure 10.25).

Figure 10.25 • Coal demand in the United States by sector in the New Policies Scenario

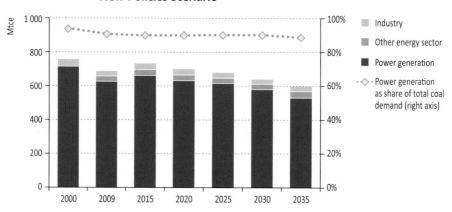

In the New Policies Scenario, coal demand rebounds from the level of 2009 in the short-term, but never recovers to pre-recession levels, meaning that coal consumption in the world's second-largest consumer peaked in 2005 at a level of nearly 800 Mtce. Between 2015 and 2035 coal demand is projected to decline by 1% per year, reaching 600 Mtce by 2035 – 25% lower than in 2005.

The projected long-term decline in coal demand in the United States is explained mainly by the dwindling use of coal for power generation, even though new coal-fired power stations currently under construction will add about 8 GW of capacity (over end-2009 levels) by the end of 2012 and additional stations are expected to be built before the end of the current decade, some of them replacing older stations that will be mothballed or retired. In many states, coal is currently the cheapest option for new baseload capacity, but the cost advantage over natural gas is small. However, coal faces increasing public opposition, as well as mounting air pollution restrictions, and, in the longer term, competition from renewables-based generation and nuclear power, thanks to government support for those alternatives, as well as from natural gas. The competitive position of coal is undermined, in part, by the assumption in the New Policies Scenario that the power sector takes account in its investment decisions of a shadow CO_2 price of $15/tonne (in year-2010 dollars) as of 2015 and that the price reaches a level of $35/tonne by 2035.[10] Growth in natural gas-fired generation is supported by relatively low natural gas prices, compared with the rest of the world (thanks to abundant resources of unconventional natural gas, especially from shale) and low capital and non-fuel operating costs. Between 2009 and 2035, total generating capacity increases for all fuels except coal without CCS, which falls by about 10%, and oil; wind power capacity increases the most in absolute terms. Increased natural gas, nuclear and renewables-based generation displaces coal generation and consequently lowers the average load factor for coal-fired plants, which results in the share of coal in total power generation in the United States dropping from 45% in 2009 to 35% in 2035 (Figure 10.26). Coal use would be even lower if it were not for cumulative additions over the projection period of 20 GW of plants fitted with CCS.

Traditional use of coal in industry, notably for steam generation and coke production for iron and steel fabrication, also declines throughout the projection period, mainly as a result of efficiency improvements that reduce the need for process steam and an expected decline in smelting and increased use of electric-arc furnaces in steel making, which reduces coking coal needs. In the New Policies Scenario, total industrial coal demand falls by 10% between 2009 and 2035, reaching just under 30 Mtce. Lower demand for coal in power generation and in industry is partially offset by increased use of coal as a feedstock for liquids production in CTL plants (in some cases, blended with biomass feedstock), especially towards the end of the projection period. Around ten CTL plants have been proposed, though none has yet been given the green light, mainly because of uncertainties about oil prices and penalties for CO_2 emissions. Up from zero in 2009, coal inputs to CTL plants are projected to reach around 25 Mtce in 2035 and result in an output of 180 kb/d of oil products.

---

10. The CO_2 price in the United States is a shadow price, *i.e.* the price that investors assume will prevail on average over the lifetime of the plant. No actual CO_2 price is assumed to be introduced in the United States in the New Policies Scenario.

Figure 10.26 ● Electricity generation in the United States by type in the New Policies Scenario

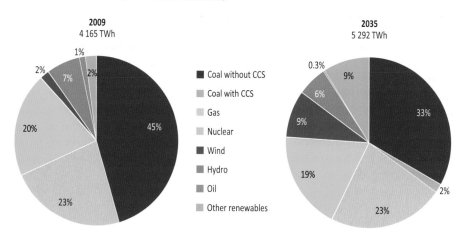

2009
4 165 TWh

2035
5 292 TWh

- Coal without CCS
- Coal with CCS
- Gas
- Nuclear
- Wind
- Hydro
- Oil
- Other renewables

Other non-OECD Asia

In the New Policies Scenario, primary coal consumption in non-OECD Asia outside China and India more than doubles, from around 200 Mtce in 2009 to 480 Mtce 2035 – an average annual rate of growth of 3.5%. Growing demand in the electricity and industrial sectors drives this increase in coal use.

Indonesia is the biggest single contributor to this demand growth – it is already the largest coal consumer in non-OECD Asia after China, India and Chinese Taipei – with consumption in 2009 of around 45 Mtce. Demand in the world's fourth most populated nation is expected to grow strongly in the coming decades, driven by rapid economic growth and underpinned by a large resource base: Indonesia is already the world's largest steam coal-exporter, the second-largest hard-coal exporter and the fifth-largest coal producer. The government plans to give priority over export sales to meeting domestic coal needs. In the New Policies Scenario, the country's domestic demand triples to nearly 150 Mtce by 2035, a rate of growth of 5% per year. This exceeds substantially the rate of growth in demand of any major coal-consuming country or region, including India. By the end of the *Outlook* period, Indonesia moves from being the fourteenth-largest coal consumer globally to the sixth-largest, overtaking Japan. The power sector drives most of the growth in demand, with coal-fired capacity more than quadrupling, to just over 50 GW by 2035.

Eastern Europe/Eurasia

Russia is the largest coal consumer in the Eastern Europe/Eurasia region, accounting for around half of the region's primary demand. Yet coal plays a relatively modest role in Russia's energy mix, meeting only 15% of its total primary energy needs and fuelling about one-fifth of its electricity generation. Despite Russia's ample indigenous resources, the share of coal in total primary energy demand is expected to remain at around 2009 levels over the projection period, as the share of nuclear and renewables in the fuel mix increases. In the New Policies Scenario, after rebounding to pre-recession consumption levels by 2015, demand remains fairly flat, at around 165 Mtce, throughout the projection period, as energy efficiency gains temper excessive electricity growth and other fuels mainly cover increased electricity demand. In absolute terms, coal use for power generation remains broadly flat, and coal's share in the generating mix decreases from 17% in 2009 to 16% in 2035 (Figure 10.27). However, there is scope for significantly higher coal use for power generation, if the programme of building new nuclear reactors does not proceed as currently planned (see Chapter 7).

Figure 10.27 ● **Electricity generation in Russia by type in the New Policies Scenario**

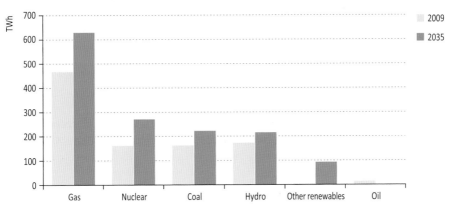

The Caspian region, mainly Kazakhstan, is another big coal consumer in Eastern Europe/ Eurasia. Nearly two-thirds of the Caspian's primary use of coal is in power generation. In the New Policies Scenario, coal demand is expected to grow modestly overall, mainly because incremental energy needs are principally met by increased use of natural gas. The planned construction of additional nuclear power, which is assumed to be commissioned towards the end of the projection period, also limits the scope for increased use of coal for power generation. A significant portion of increased electricity demand could also be met by reducing losses in transmission and distribution, which are very large (IEA, 2010c).

OECD Asia Oceania

Based on preliminary estimates for 2010, Japan – the world's third-largest economy, the largest coal importer and the fourth-largest coal user – accounted for almost half of coal demand in the OECD Asia Oceania region. But this share drops in the New Policies Scenario as a result of slow economic growth (in line with a declining population) and a correspondingly modest rise in its electricity needs, as well as a major increase in power supply from renewable sources and natural gas. Japan's primary coal demand is projected to drop by 20%, from 145 Mtce in 2009 to about 115 Mtce in 2035, a level last witnessed in the

1990s. Demand declines in all sectors, including industry, but is most marked in the power sector (already today Japan has one of the most efficient coal-fired power generations fleets in the world, partly due to its import dependency). The overall fall in coal use, is nonetheless, much less marked than was projected in last year's *Outlook*, mainly because of a downward revision to our projections for nuclear power as a result of events at the Fukushima Daiichi nuclear power plant (see Chapter 12). As a result, less coal-fired capacity is expected to be retired and coal-fired plants achieve higher load factors, especially in the medium term, because of the loss of baseload nuclear capacity.

In the rest of the OECD Asia Oceania region – Korea, Australia and New Zealand – coal demand is also projected to decline in the New Policies Scenario. Their combined primary demand drops from around 175 Mtce in 2009 to 130 Mtce in 2035, a fall of 25%. As in Japan, falling demand in the power sector is the main reason for lower coal demand in Korea, Australia and New Zealand. Increased use of nuclear power, renewables and natural gas in power generation largely explains the projected 45% fall in coal demand in Korea, to 52 Mtce by 2035.

Abundant resources of cheap and high quality coal have made Australia the world's ninth-largest coal consumer, just behind Korea. Some 76% of power generation is coal-fired, using hard coal in New South Wales and Queensland and lignite in Victoria. Gas is more important in other mainland states. In Queensland and New South Wales, coal production for domestic power tends to be driven by export-focused operations: the coal is often low cost, but lower quality as well, with higher ash and moisture, although usually low sulphur. The domestic coal-fired plants that use this coal and Victorian lignite were built mostly in the 1970s and 1980s, so much of the capacity is subcritical; low coal prices have discouraged investment to increase efficiency. Only around 10% of capacity is supercritical, all of it built in the last decade. As a result of this high coal dependence in the power sector (and lack of low-carbon options such as hydro and nuclear), Australia has relatively high per-capita emissions of greenhouse gases. However domestic coal consumption is projected to decline slowly over the projection period, as natural gas and renewable sources (especially wind) increase their share. This change is underpinned by significant policy initiatives, responding to growing public demand, including an expanded mandatory renewable energy target and carbon pricing.

OECD Europe

Coal demand in Europe has been declining steadily in recent years, as a result of a combination of factors: sluggish economic growth; the accompanying modest rise in electricity demand; an expansion of natural gas and renewables-based generating capacity; the phase-out of subsidies to indigenous coal production; the introduction of carbon pricing in the European Union; and increasingly stringent local environmental regulations. These factors are set to continue to depress coal demand over the *Outlook* period, though the rate of decline will depend on the strength of government energy and environmental policies. In the New Policies Scenario, Europe's primary coal demand plunges by over one-third, from around 415 Mtce in 2009 to 265 Mtce in 2035. Most of this decline results from reduced coal burning for power generation. Load factors at existing plants are set to fall, with increased competition from other baseload and must-run plants, and few new coal plants are likely to be built, given the strong preference for natural gas plants, the unfavourable economics of coal (taking account of CO_2 prices and support for renewables), the threat of heavier carbon penalties in the future and more stringent environmental controls, and strong local opposition to coal. Industrial coal use also falls – the result of declining production in heavy industry, switching to natural gas and electricity and increased energy efficiency.

Africa

South Africa, a large coal resource holder and producer, currently accounts for over 90% of Africa's total coal consumption. It is expected to account for most of the growth in the continent's coal demand in the coming decades. In the New Policies Scenario, total primary coal demand in South Africa grows slowly to a peak of around 165 Mtce by 2025 and then falls back slightly. Coal use in the next few years will be boosted by the entry into service of several new coal-fired power plants in South Africa, where rapidly rising electricity demand has led to severe power shortages in parts of the country.

The use of coal for CTL production in South Africa is also set to rise in the longer term. The country already has the two largest CTL plants in the world, with a combined output capacity of 160 kb/d. These plants account for about one-fifth of the country's total primary coal consumption. More capacity is expected to be added, boosting coal use in this sector to around 35 Mtce in 2035 – an increase of 40% on current levels.

Rest of the world

Coal demand in the rest of the world is very small: Canada, Latin America, Mexico and the Middle East together making up less than 2% of world demand in 2009. While Canada, Colombia and Venezuela are endowed with sizeable coal resources, the majority of their future coal production is projected to be exported in the New Policies Scenario. Considering also that Canada, Latin America, Mexico and the Middle East already rely heavily on indigenous oil and natural gas resources, there is, accordingly, little projected coal demand growth in these regions as a whole in the New Policies Scenario. By 2035, these regions still account for less than 2% of world coal demand.

10

CHAPTER 11

COAL SUPPLY AND INVESTMENT PROSPECTS
The sweet and the sour of Asian trade

H I G H L I G H T S

- Globally, coal is the most abundant fossil fuel, with reserves totalling 1 trillion tonnes, or some 150 years of production in 2009 – in energy terms, around 3.2 and 2.5 times larger than those of natural gas and oil, respectively. The coal resource base is much larger and geographically diverse, and as market conditions change and technology advances, more resources can become reserves.

- In the New Policies Scenario, the lion's share of the nearly 20% growth in global coal production between 2009 and 2035 occurs in non-OECD countries. China contributes more than half of the increase in global supply to 2035; the bulk of the rest comes from India and Indonesia. Australia is the only major OECD producer to increase production to 2035; output in the United States falls around 2020, while European output continues its historical decline.

- Rising prices of inputs to coal mining and the opening of more expensive mines since 2005 have driven up supply costs on a weighted-average basis by around 12% per year. Yet, international coal prices have risen much more, boosting operating margins and the profitability of investments. Continued depletion of economically attractive seams and the need to shift new investment to deposits that are less easy-to-mine and/or more distant from existing infrastructure is expected to drive supply costs further upwards, especially if extraction rates keep rising rapidly. Increasingly stringent environmental, health and safety legislation as well as changing tax regimes will also add to cost pressures.

- In the New Policies Scenario, inter-regional trade in hard coal grows rapidly to 2020, with volumes stabilising thereafter. By 2035, 18% of world coal production is traded inter-regionally and the pattern of trade continues to shift towards the Pacific Basin markets. Australia and Indonesia command nearly 60% of inter-regional hard coal trade in 2035, while new supplies from Mongolia and Mozambique gain prominence.

- The international coal market will become increasingly sensitive to developments in China, where marginal variations between very large volumes of coal production and demand will determine China's net trade position. Hard coal net imports would fall to zero by 2015 (compared with our projection of around 185 Mtce) if China's output rises by just 1.2% per year higher than in the New Policies Scenario.

- India is poised to become the world's biggest importer of hard coal soon after 2020, as rapid demand growth outstrips the rise in indigenous production and India's inland transport capacity. Projected imports reach nearly 300 Mtce in 2035 – about 35% of India's hard coal use and 30% of inter-regional trade in the New Policies Scenario.

Overview of projections

The outlook for coal supply is intrinsically linked to the prospects for prices and sectoral coal demand presented in Chapter 10; consequently, production by type varies markedly across the three scenarios in this *Outlook* (Box 11.1). Compared to an average annual growth rate of 2.3% between 1980 and 2009, global production grows by 0.7% on average per year to 2035 in the New Policies Scenario, but in the Current Policies Scenario it expands at more than double the rate, 1.8%, while in the 450 Scenario it *declines* by 1.5%. Like demand, coal production continues to be dominated by China and other non-OECD countries, especially India and Indonesia, with their combined share of global production reaching around 80% in all three scenarios, compared to 72% in 2009 (Table 11.1).

Box 11.1 ● Classification and definition of coal types

Coal is an organic sedimentary rock formed from vegetable matter, particularly during the Carboniferous period, and consolidated in seams between strata of non-organic rock. The IEA classifies coal as hard coal and brown coal, and also includes peat.

Hard coal: coal with a gross calorific value greater than 5 700 kilocalorie per kilogramme (kcal/kg) (23.9 gigajoules per tonne [GJ/tonne]) on an ash-free but moist basis and with a mean random reflectance of vitrinite of at least 0.6. It includes anthracite and bituminous coal. Hard coal is calculated as the sum of coking and steam coal:

- Coking coal: Hard coal with a quality that allows the production of coke suitable to support a blast furnace charge (see footnote for the term metallurgical coal[1]).

- Steam coal: All other hard coal not classified as coking coal. Also included are recovered slurries, middlings and other low-grade coal products. Coal of this quality is sometimes referred to as thermal coal.

Brown coal: non-agglomerating coal with a gross calorific value less than 5 700 kcal/kg (23.9 GJ/tonne) containing more than 31% volatile matter on a dry mineral matter free basis. Brown coal is the sum of sub-bituminous coal[2] and lignite:

- Sub-bituminous coal: Non-agglomerating coal with a gross calorific value between 4 165 kcal/kg and 5 700 kcal/kg.

- Lignite: defined as non-agglomerating coal with a gross calorific value less than 4 165 kcal/kg.

Peat: a solid formed from the partial decomposition of dead vegetation under conditions of high humidity and limited air access. It is available in two forms for use as a fuel, sod peat and milled peat. Milled peat is also made into briquettes for fuel use.

---

1. The terms coking and metallurgical for coal are, strictly, not synonymous. Metallurgical coal includes all types of coal used in the metals sectors. Although coking coal accounts for most of this, it also includes some high-quality steam coal.

2. For the following countries, IEA includes sub-bituminous coal in steam coal: Australia, Belgium, Chile, Finland, France, Iceland, Japan, Korea, Mexico, New Zealand, Portugal and the United States.

Table 11.1 ● Coal* production by type and scenario (Mtce)[3]

	1980	2009	New Policies Scenario		Current Policies Scenario		450 Scenario	
			2020	2035	2020	2035	2020	2035
OECD	**1 385**	**1 403**	**1 421**	**1 197**	**1 534**	**1 640**	**1 297**	**608**
Steam coal	847	971	934	805	1 013	1 163	902	351
Coking coal	305	219	298	266	306	290	231	185
Brown coal	233	213	190	126	216	187	164	71
Non-OECD	**1 195**	**3 525**	**4 412**	**4 662**	**4 774**	**6 102**	**4 012**	**2 701**
Steam coal	868	2 866	3 649	3 945	3 985	5 274	3 319	2 139
Coking coal	196	505	556	483	570	537	547	444
Brown coal	131	155	207	234	218	292	145	118
World	**2 579**	**4 928**	**5 833**	**5 859**	**6 308**	**7 742**	**5 309**	**3 309**
Share of steam coal	*66%*	*78%*	*79%*	*81%*	*79%*	*83%*	*80%*	*75%*
Share of coking coal	*19%*	*15%*	*15%*	*13%*	*14%*	*11%*	*15%*	*19%*
Share of brown coal	*14%*	*7%*	*7%*	*6%*	*7%*	*6%*	*6%*	*6%*

*Includes hard coal (coking and steam coal), brown coal (sub-bituminous coal and lignite), and peat. See Box 11.1 for classifications and definitions of coal types.

In the New Policies Scenario, production of steam coal grows faster than that of coking coal, so the share of steam coal in global production rises by three percentage points to 81% in 2035. The share reaches 83% in the Current Policies Scenario, due to even stronger demand from the power sector, but in the 450 Scenario the share drops three percentage points below the 2009 level of 78% as coal-fired power generation declines much more than coking coal use in industry. The share of brown coal, mainly utilised in the power sector, declines slightly from the 2009 level of 7% in all three scenarios.

Regardless of the scenario, China – which produced 45% of global coal output in 2009 – remains the key producing country over the projection period, making the largest contribution to the growth of global production in the Current and New Policies Scenarios, and to the *decline* in the 450 Scenario; 44%, 58%, and 40% respectively (Figure 11.1). Chinese coal production increased from 2008 to 2009 by almost 6%, more than offsetting the drop in production in the rest of the world as a whole due to the global economic crisis. In each of the three scenarios, the other main producers include India, Indonesia and Australia, though they are driven by different objectives. India increases production to slow down the pace of growing imports, Indonesian production expands to satisfy both growing domestic and international demand, while Australian output changes, for the most part, in line with shifts in international demand across the three scenarios.

The share of global hard coal output which is traded between *WEO* regions has more than doubled since 1980 to 17% in 2009, as growing domestic needs in various parts of the world could not be sufficiently (or, at times, economically) satisfied by local output. Across the three scenarios, the volume of coal traded inter-regionally varies markedly, reflecting the role of trade, economics and prices in balancing local needs (Table 11.2). In the New

11

---

3. A tonne of coal equivalent is defined as 7 million kilocalories, so 1 tonne of coal equivalent equals 0.7 tonnes of oil equivalent.

Policies Scenario, 18% of global hard coal production is inter-regionally traded in 2035 and in the Current Policies Scenario 22%, driven primarily by rising imports into non-OECD Asia. The global share declines to 15% in the 450 Scenario, as the need to import coal is reduced because indigenous resources are generally able to meet more of the lower level of demand. The opposite is true in the Current Policies Scenario and, by 2035, global trade in volumetric terms is more than three-times as big as in the 450 Scenario. As coking coal resources are less abundant than steam coal in key projected demand centres and there are fewer alternatives for coking coal, the share of global coking coal output which is inter-regionally traded increases in all three scenarios.

Figure 11.1 ● Incremental coal production by scenario and region, 2009-2035

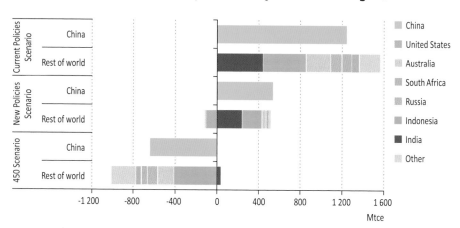

Box 11.2 ● *WEO-2011* coal-supply modelling enhancements

For this year's *Outlook,* the coal supply module in the IEA's *World Energy Model (WEM)* has been improved to incorporate more detail at the level of coal type and country, and to provide for modelling for a wider range of factors, including resource availability, production and transportation costs, international coal prices and infrastructure bottlenecks. In our projections, the international coal prices assumed (and presented for each scenario in Chapter 1) reflect our judgment as to the levels needed to bring forward the investment required to ensure sufficient supply to meet projected demand and international trade. More details on the IEA's *WEM* coal supply module, as well as the assumptions and methodologies used to project energy demand by fuel type in the IEA's *WEM,* can be found at www.worldenergyoutlook.org/model.asp.

Trade patterns are expected to continue gravitating towards Asia and away from Atlantic Basin markets. Within Asia, China and India are set to increase their dominance of trade, as the importance of Japan and the European Union diminishes. The international coal market will remain very sensitive to developments in China, which became a net importer in 2009 (Figure 11.2). Marginal changes in its very high levels of production and demand will determine China's future net trade position, with profound implications for global international trade (see Spotlight in Chapter 2).

Table 11.2 ● World inter-regional* hard coal trade by type and scenario (Mtce)

	1980	2009	New Policies Scenario 2020	New Policies Scenario 2035	Current Policies Scenario 2020	Current Policies Scenario 2035	450 Scenario 2020	450 Scenario 2035
Hard coal	173	753	1 056	1 017	1 212	1 571	721	480
Steam coal	64	560	740	720	878	1 232	491	261
Coking coal	109	192	316	298	334	339	231	220
Share of world production								
Hard coal trade	8%	17%	19%	18%	21%	22%	14%	15%
Steam coal trade	4%	15%	16%	15%	18%	19%	12%	10%
Coking coal trade	22%	27%	37%	40%	38%	41%	30%	35%

*Total net exports for all WEO regions, not including trade within WEO regions.

Figure 11.2 ● China's coal trade balance, 2000-2011

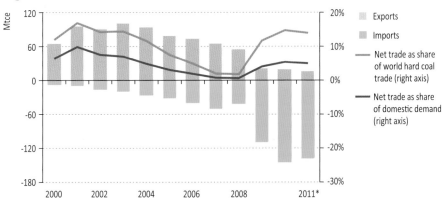

*Preliminary estimates.

Understanding the drivers of coal supply

Future trends in coal supply will be determined by a combination of factors related essentially to resource endowment, the economics of production – the interaction between the cost of supply (mining and transportation) and international prices – and policy considerations. Not all coal resources can be produced profitably, due to high production costs, unfavourable geology (poor quality, thin and deep deposits), infrastructure bottlenecks and long haulage distances, or lack of affordable labour. Local opposition, safety and environmental hazards, as well as emissions penalties can also affect the economic viability of mining, for example, if fugitive methane emissions are included in emissions control schemes.

Globally, a vast amount of coal remains in the ground but, over the past century, easy-to-mine and high-quality coal deposits have often been extracted, driving investment to less attractive deposits or locations further from existing transport infrastructure or demand centres. On the other hand, international coal prices continue to rise and technological improvements, such as longwall mining and coal-washing, have maintained or boosted productivity so as to help keep unit costs down. The majority of coal produced today still serves to meet domestic demand, because of the high cost of transportation per unit of energy; only the highest quality coals, or those that can be mined very cheaply, are traded internationally on a large scale. This is likely to remain the case over the coming decades, especially for steam coal.

Resources and reserves

At the end of 2009, world coal reserves – the part of resources estimated to be economically exploitable with current technology (Box 11.3) – amounted to 1 trillion tonnes, equivalent to 150 years of global coal output in 2009 (BGR, 2010). In terms of energy content, these reserves are approximately 3.2 and 2.5 times larger than those of natural gas and oil, respectively, and are more widely dispersed geographically. Total coal resources are many times greater. Globally, coal resources beneath land are estimated at around 21 trillion tonnes but, as market conditions change and technology advances, more coal will be "proven" over time. Around 90 countries are known to have coal resources, but 95% of the global endowment is concentrated in the regions of North America, Asia-Pacific and Eastern Europe/Eurasia (Table 11.3). Nearly 40% are located in the Asia-Pacific region alone, a region crucial for future coal demand, production and trade prospects.

Table 11.3 ● Coal resources and reserves by region and type, end-2009
(billion tonnes)

	Hard coal		Brown coal		Total	
	Reserves	Resources	Reserves	Resources	Reserves	Resources
North America	232	6 652	33	1 486	265	8 138
Asia-Pacific	309	6 861	76	1 075	385	7 936
E. Europe/Eurasia	124	2 891	108	1 324	232	4 215
Europe	17	467	55	282	72	748
Africa	30	78	0	0	30	79
Latin America	9	28	5	20	15	48
Middle East	1	40	-	-	1	40
World	723	17 017	278	4 187	1 001	21 204

Notes: World excludes Antarctica. Classifications and definitions of hard and brown coal can differ between BGR and IEA due to different methodologies.

Source: BGR (2010).

Just under three-quarters, or nearly 725 billion tonnes, of global coal reserves are hard coal, more than half of which is found in just two countries: 31% in the United States and 25% in China. Hard coal production is much higher in China (it accounted for almost half of the world's output in 2009, compared with a 16% share for the United States), so China's hard coal reserves-to-production ratio is lower – at 70 years, compared with 250 years for the United States. Most of the rest of the world's hard coal reserves are in India, Russia and Australia (Figure 11.3). Global brown coal reserves total just under 280 billion tonnes and are distributed somewhat differently to those of hard coal, with four countries holding the bulk: Russia, Germany (the world's biggest brown coal producer in 2009), Australia and the United States.

Box 11.3 ● Coal resources and reserves definitions

Conventions for classifying and defining fossil energy resources vary across fuels, countries and international bodies. In the interest of consistency (and because of the high reputation of the institution), the coal resource and reserve data presented in this chapter come from the German Federal Institute for Geosciences and Natural Resources (BGR, 2010).

Resources include hydrocarbons in the ground "which are either proved but are at present not economically recoverable, or which are not demonstrated, but can be expected for geological reasons". This concept of resources is equivalent to the term "hydrocarbons in place" used to describe the total amount of oil and natural gas in the ground, not "ultimately recoverable resources", the measure of long-term oil and natural gas production potential used in the *Outlook*. For this reason, direct comparisons of the numbers given for coal and hydrocarbon resources can be misleading.

Reserves are the portion of resources that is known in detail and can be recovered economically, using current technologies. Accordingly, the amount of reserves depends on current international prices, as well as the state of technological progress. This concept of reserves for coal is equivalent to proven reserves of oil and natural gas (see Chapters 3 and 4).

11

Similar to oil and natural gas, coal resources and reserve data will evolve over time as more detailed appraisals are made, and present data therefore provide only a partial indication of where coal production growth is likely to occur. This will be determined less by the absolute size of the coal resource or reserve base and more by, the geological characteristics of the deposits (affecting the cost of mining and the quality of the coal produced), potential environmental restrictions, the cost of transportation and relative economics. These vary widely from one region to another. Nonetheless, globally, it is clear that the overall size of the coal resource base will not be a constraint for many decades to come. The rising cost of supply, however, coupled with a likely tightening of environmental regulations and other policies affecting coal demand, might be.

Figure 11.3 ● Coal reserves by country and type, end-2009

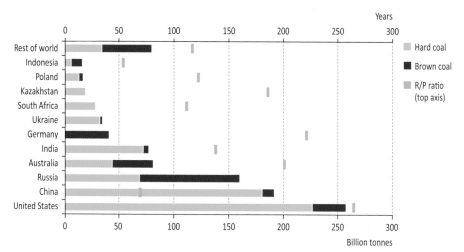

Notes: R/P ratio stands for reserves-to-production ratio. Not shown in the figure, Ukraine and Russia have R/P ratios of 625 and 525 years, respectively.

Source: BGR (2010).

Cost trends and technology

Coal mining is a much less capital-intensive business than oil and natural gas extraction, with operational costs making up most of the overall cost. In this section, though we make reference to changing investment costs in the coal supply chain, we review coal cost trends largely in terms of the free-on-board (FOB) cost per tonne.[4] This is the cash cost to the producer of their operations. Capital costs and profit are recovered through the margin between FOB cash costs and the FOB price.

Within the category of capital expenditure, transport infrastructure can be a significant part of total investment in the coal supply chain, with construction lead times for long-haul domestic transport infrastructure often longer than the timescale for the mine itself. To take one example, investment in Australia averages $30 to $40 per tonne of annual incremental port capacity and a similar amount is required for inland transport (depending on haulage distances), compared with an average of $90 or $150/tonne per year for investment in coal mining capacity (opencast and underground, respectively).

In FOB terms, cash costs for internationally traded steam coal have risen globally on a weighted-average basis by around 12% per year since 2005 – a cumulative increase of around 70%. Despite a small dip in 2009 in the wake of the economic crisis, all the main steam coal exporters have seen significant rises in their supply costs, due to changes in the input factors contributing to operational costs (Figure 11.4). Nonetheless, supply costs have

---

4. Free-on-board is a shipping term whereby the seller pays for the transportation of the goods to the port of shipment, plus loading costs. The buyer pays the cost of marine freight transport, insurance, unloading and transportation from the arrival port to the final destination.

generally risen much less than international coal prices, widening operational margins and boosting the profitability of investments – especially for coking coal, whose international price has risen the most.

Figure 11.4 ● **Change in average FOB supply cash costs relative to 2005 for internationally traded steam coal by selected country and component**

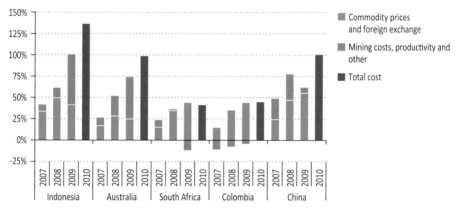

Sources: IEA analysis based on data from IEA Clean Coal Centre and the Institute of Energy Economics at the University of Cologne.

As in the past, future coal supply costs will depend to a large extent on trends in the cost of materials, equipment, labour and fuel, as these are major input factors into mining and shipping operational costs (Meister, 2008). Also, foreign exchange rate effects can have a major impact (outside the United States, a weaker dollar automatically increases the cost of local inputs expressed in dollars). In mature production regions, the depletion of economically attractive seams and the need to shift investment towards less easy to mine coal deposits can be expected to drive upwards operating mining costs (and capital costs), especially if extraction rates keep rising rapidly. Moreover, moving production to new coal deposits, even those with favourable geology and good coal quality (*e.g.* Mongolia, Mozambique), can increase overall supply costs (including capital recovery) because of the lack of existing infrastructure (or remoteness from it). In Australia, for example, the next batch of export mines to be brought into production are in less developed or completely undeveloped basins, like Surat or Galilee, where rail freight distances can be as much as 500 kilometres (km) (ABARES, 2011). Increasingly stringent environmental, health and safety legislation may also add to cost pressures. Rising costs for marginal sources of supply would underpin future international coal prices, within the limits of coal's ability to compete with alternative fuels. Uncertainty about the above factors and the risk that new government policies might curb the growth in coal demand (with an especially marked impact on traded coal markets) could discourage mining companies from investing in new mining capacity and transport infrastructure.

Mining

As in recent editions of the *Outlook,* we present here the results of analysis of the recently prevailing FOB cash costs to key exporters of producing and transporting steam coal to local ports for onward shipment (Figure 11.5). Cash costs include the variable costs of production, such as labour, materials, inland transportation and port handling fees, but exclude the depreciation of capital assets and corporate overheads. The weighted-average FOB cash cost of mining internationally traded steam coal in nominal terms reached an estimated $56/tonne in 2010, up from $43/tonne in 2009. This cost increase was driven by rising input prices at existing operations, but also higher operational costs at new mines. Yet international FOB steam coal prices rose even more, to around $90/tonne on average in 2010, yielding healthy margins and providing a strong incentive for new investment.

Figure 11.5 ● **Average FOB supply cash costs and prices for internationally traded steam coal, 2010**

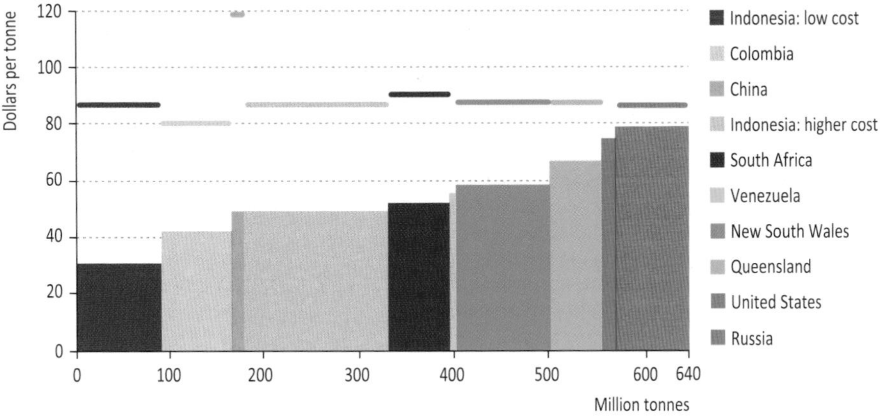

Notes: Prices, costs and volumes are adjusted to 6 000 kcal/kg. Boxes represent FOB costs and bars show FOB prices.
Sources: IEA Clean Coal Centre analysis partly based on Marston, IHS Global Insight and Wood Mackenzie.

Costs vary markedly across countries but, as in 2009, Indonesia and Colombia had the lowest FOB cash costs of all the main steam coal exporters in 2010. Indonesia, the world's largest steam coal exporter, is by a wide margin the cheapest producer of steam coal, with some mines achieving FOB cash costs of $30/tonne or less. Higher cost mines in Indonesia produce on average at $50/tonne FOB, but they still remain at the front of the global steam coal supply cost curve, alongside Colombia, China and South Africa. Due to the dependence of Indonesian mining companies on diesel fuel in their opencast truck-and-shovel and inland barging operations, the removal of diesel subsidies in 2005, together with worsening geological conditions and coal qualities, means that Indonesian mining costs have been rising significantly. For example, some collieries have faced FOB cost increases of up to $15/tonne, shifting more of Indonesia's production up the global supply curve.

Supply costs in South Africa, Australia (New South Wales and Queensland) and the United States have also increased significantly since 2009, mostly due to the rising costs of labour and commodity inputs, such as explosives, chemicals, diesel fuel, electricity and tyres, as well as exchange rate fluctuations. Environmental levies and taxes have also driven up costs. South African steam coal now costs on average $50/tonne FOB to produce, while costs in Australia and the United States have moved up to the $60 to $75 range. The United States, as the swing supplier in the Atlantic Basin market, exports only when FOB prices exceed $75/tonne and domestic coal prices are lower (which has recently been the case, due to competition from lower natural gas prices, spurred by the increase in unconventional natural gas output). Historically, export-mining operations in the United States were mainly located in the Appalachians, but geological conditions, coal qualities and productivity have been deteriorating, due to the depletion of favourable coal deposits. Increasing amounts of coal originating from the Illinois and Powder River basins in the United States are now finding their way onto the export market. Russian steam coal for export remains the most expensive, mainly due to the extremely long inland transport distances to the Baltic and Pacific ports (between 4 000 and 6 000 km).

In the coking coal market, global demand has been strong coming out of the economic crisis, due to rising steel production, mainly in Asia. Hard coking coal prices between Australian mines and Japanese steel mills were well above $300/tonne FOB during the first half of 2011, due to major floods in Queensland, which affected coking coal supplies in particular. For similar reasons to steam coal, despite rising supply costs, coking coal margins have increased sharply to, on average, over $100/tonne in 2010 (Figure 11.6). In response, coking coal operations are being expanded in the key exporting countries, Australia, the United States and Canada, while brand new mines, notably in Mongolia and Mozambique, are being developed.

Figure 11.6 ● Average FOB supply cash costs and prices for internationally traded coking coal, 2010

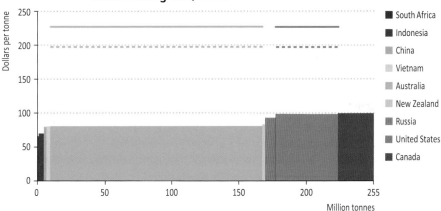

Note: Boxes represent FOB costs for all metallurgical coal qualities, while dashed and solid bars show lower and higher quality coking coal FOB prices, respectively.

Sources: IEA Clean Coal Centre, citing data from Wood Mackenzie; IEA analysis.

The relative importance of each input factor in total supply cash costs varies greatly between individual mines and countries, due to geological conditions and mining techniques (Box 11.4). The mining method is particularly important. In general terms, underground mining uses less diesel per tonne of coal mined, so countries with a large share of output from underground mines, such as South Africa, the United States and Russia, have seen smaller increases in fuel costs, as their production has been less exposed to rising international oil prices (Rademacher, 2011).

Box 11.4 ● Coal mining techniques

Depending on the geological conditions of a deposit, coal can be either mined through underground or opencast methods.

Opencast mining is economically feasible only if coal seams are close to the surface and fulfil certain other requirements. This method can recover 90% or more of the coal in-situ and can be very low-cost, as labour costs are low due to use of large-scale mining equipment. Typically, the overburden (rock and soil between the surface and the coal seam) is fractured by explosives and then removed. Once the coal seam is exposed, it can be drilled, fractured and then mined in three different ways:

● Truck and shovel: large capacity electric-powered mining shovel and diesel-powered hydraulic excavator are used to strip waste material and recover the coal. A truck or conveyor belt then transports the coal to a preparation plant or directly to the final point of use, typically a power station. This method is used in almost all major coal-producing countries and it can be used in most opencast mines regardless of geological characteristics; but it requires large amounts of diesel fuel.

● Dragline: is more capital-intensive than truck and shovel and essentially involves dragging a large bucket, suspended from a boom via wire ropes, over the overburden or seam surface. A dragline usually operates on electricity and can move at times more than 400 tonnes of material in one cycle.

● Bucket wheel-excavator: is among the largest machinery in the world and some are capable of moving up to 240 000 tonnes of coal or waste rock per day. The machine has a large, rotating wheel with a configuration of scoops that is fixed to a boom and is capable of pivoting. This continuous mining method is mainly employed in brown coal production in Germany, Russia and Australia, and the coal is immediately transported, via conveyor belts, to a stacker or power plant.

Underground mining, which accounts for about 60% of world coal production, is used where coal seams are too deep to be mined economically using opencast methods. There are two principal underground mining methods:

● Room and pillar: involves cutting a network of chambers in the seam, while up to 40% to 50% of the seam remains as pillars to support the mine roof. This method can be started up quickly and is less capital-intensive than longwall mining. On the other hand, it requires more roof support materials and related equipment.

● Longwall: employs a mechanical shearer that is able to mine coal from a seam face up to 300 metres long. Self-advancing, hydraulically-powered shields temporarily hold up the roof while coal is extracted, and later the roof is allowed to collapse. Over 75% of the coal in the deposit can be extracted with this method, but capital expenditure is high.

Source: IEA based on World Coal Association (2005).

Underground mining methods (longwall, room and pillar) are comparatively more labour-intensive than opencast methods (dragline, truck and shovel), where larger mining equipment can be utilised (Figure 11.7). Underground mining also depends more on steel, machinery and electricity. Unsurprisingly, explosives, tyres and diesel costs are significant in truck and shovel operations, while draglines are less diesel intensive, but electricity is a more important cost factor (Trüby and Paulus, 2011). Globally, prices for diesel, explosives, tyres, steel and machinery parts have increased dramatically since 2005. Consequently cash costs for opencast mining operations have increased sharply, due to their very high exposure to these input factors, notably in Indonesia, where mining is exclusively carried out using the truck and shovel method. The production of explosives and chemicals requires oil and natural gas as feedstock, so the future prices of these commodities will have an indirect, as well as direct impact on coal mining costs.

Figure 11.7 ● **Share of key input factors in coal mining costs by technique, 2009**

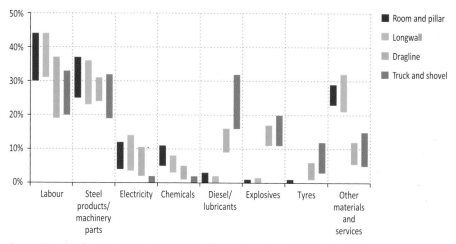

Source: IEA analysis based on data from the Institute of Energy Economics at the University of Cologne.

Coal mining productivity, measured as average production per employee per hour, has declined substantially over the past five years in major producing countries, such as Australia and the United States (Figure 11.8). While various factors (such as changes in international coal prices or the size of the workforce) contribute to changes in productivity, significant productivity deterioration can occur as existing coal deposits are exhausted, seams become thinner and mining operations move deeper. As more overburden has to be removed or trucks have to move coal over longer distances from the mine mouth, the cost of mining increases. For example, productivity in opencast mines in the major Australian mining state of Queensland fell, on average, by around 6% per year from 2003 to 2009, as existing deposits became depleted; nonetheless, Australian mines are still among the most productive in the world. During the same time period, productivity, on average, fell by around 4% per year in the United States, with notable falls in Appalachian underground mines – a region that has seen extensive coal mining since the 19th century.

Figure 11.8 ● Coal mining productivity in Australia and the United States, 2004-2009

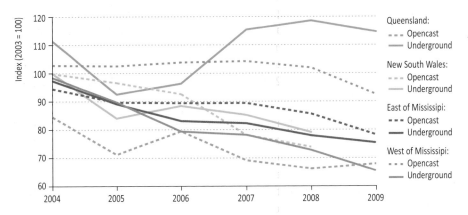

Note: Productivity measured as average production in tonnes per employee per hour.

Sources: US DOE/EIA (2011); NSW Department of Primary Industries (2011); Queensland Department of Mines and Energy (2011).

Nevertheless, there is still scope for productivity gains in some key countries. For example, the Chinese mining industry is currently undergoing consolidation, with the closure of small, inefficient and dangerous provincial or community mines and the development of several large-scale coal-power bases, which involve building coal-fired power plants next to mining operations with higher productivity and safety standards (Rui, Morse and He, 2011). Furthermore, technological improvements may help to offset increasing mining costs. These include the introduction of larger and more efficient machinery (*e.g.* larger haul trucks, larger capacity shovels and draglines), and innovative materials-handling systems (*e.g.* in-pit crushing and conveying systems).

Transport

Operational transport costs can make up a sizeable part of the total cost of supplying coal. This includes inland transport and, in the case of internationally traded coal, coal terminal fees and sea haulage rates for bulk carriers. In certain cases, there may be significant additional costs associated with delivering the coal to final users once the coal is off-loaded at its port of destination.

The importance of inland transportation depends on the distance to the final user or export terminal. Typically, coal is moved by railway, truck or river barge, though conveyor belts are sometimes used if the mine is located close to the port. Due to the availability of river transport and/or the requirement to move coal over only short distances, coal is cheap to transport in Indonesia and Colombia, around $4 to $8/tonne. Australia and South Africa have inland transport costs of $8 to $15/tonne, as distances are longer, so coal has to be moved by railway. In Russia, known for its long railway haulage distances for domestic coal output, inland transport costs have almost doubled in the course of the last five years, to around $35/tonne. Inland transport costs are expected to rise in many countries, as mining operations move further inland, due to the depletion of deposits near the coast. Furthermore, bottlenecks in railway infrastructure have become a major issue in recent years, not least in South Africa and Russia. For example, the railway line linking the Russian Kuzbass mining region with the Pacific terminals is operating at near full capacity (VDKI, 2010). In similar cases, expanding exports or domestic deliveries will require substantial new investment in transportation infrastructure. Coal turnover at export terminals comprises the second main link in the coal transport chain. While actual handling costs are relatively low, $2 to $5/tonne, limitations to port throughput capacities can become critical bottlenecks.

In recent years, utilisation of port infrastructure in several of the major exporting countries has been high and in 2010 it increased in most cases (Figure 11.9). Australian ports have recently been operating at close to capacity, with the queue to load coal averaging more than 100 bulk carriers in 2010 (effective capacity in Australia was temporarily reduced by capacity expansion works and weather-related incidents). By contrast, export terminals in countries like Poland and the United States, that used to be major exporters but now have less competitive supply costs, still have plenty of spare port capacity. Apart from supply economics, another reason for the low utilisation of ports in the United States is the difficulty of effective co-ordination between rail operators and the port authorities responsible for loading schedules, due to limited coal storage space, especially in eastern ports which export high volumes of coking coal, each type of which needs to be stored individually.

Figure 11.9 ● Coal export port utilisation rates for selected countries

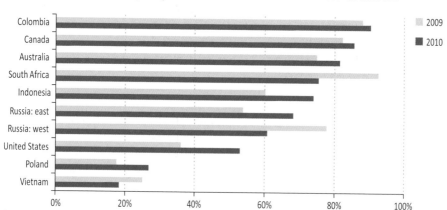

Note: South Africa includes export capacity from the port of Maputo, in Mozambique.

Sources: IEA analysis; McCloskey (2007-2011); DnB NOR (2011); ABARES (2011).

Port utilisation rates remain high in Indonesia and Australia, despite major port expansions in recent years (Schiffer, 2011). Australia increased port capacity to a level of about 380 million tonnes (Mt) per year in 2011 and 80 Mt/year of capacity is expected to be added over the next five years (ABARES, 2011). With other countries also planning to expand coal export capacities, a total global expansion of around 240 Mt/year, or nearly 20%, is envisaged by 2016 (Table 11.4). Port bottlenecks should ease if these capacities come online as scheduled, though lack of railway capacity can limit their use. For example, in South Africa port capacity was expanded significantly in 2009 and again in 2011 to 108 Mt/year; however, lack of railway capacity has kept effective export capacity to less than 70 Mt.

Table 11.4 ● **Coal export port capacities for selected countries** (Mt/year)

	2009	2010	2011	2009 to 2011	2013	2016	2011 to 2016
Australia	350	365	379	8%	448	459	21%
Indonesia	304	304	315	4%	337	366	16%
United States	121	121	129	6%	155	193	50%
South Africa	75	97	108	45%	108	108	-
Colombia	76	76	77	1%	88	104	35%
Russia: west	85	85	89	5%	89	89	-
Russia: east	45	46	47	4%	55	55	16%
Canada	35	39	41	17%	41	41	-
Vietnam	34	34	34	-	34	34	-
Poland	20	20	20	-	20	20	-
Mozambique	3	5	6	100%	6	16	167%
Venezuela	11	11	11	-	11	11	-
Total	**1 158**	**1 203**	**1 256**	**8%**	**1 391**	**1 495**	**19%**

Note: South Africa includes export capacity from the port of Maputo, in Mozambique.

Sources: IEA analysis; McCloskey (2007-2011); DnB NOR (2011); ABARES (2011).

On average, transport costs make up a much bigger share of the cost of supplying internationally traded coal than is the case for oil. Dry bulk carriers are used to ship hard coal over long distances and, occasionally, maritime shipping can be the biggest cost component in the coal-supply chain. Shipping costs, which can be very volatile, have fallen recently as a result of new vessels becoming available, yet they still account for a significant share of the total cost of supply to some markets, for example to northwest Europe (Figure 11.10).

In 2008, bulk freight rates hit all-time highs, reaching up to $40/tonne and accounting for up to 45% of total coal delivered costs on certain routes. This spike in freight rates led to increased investment in new shipping capacity but, due to the lead times in shipping construction, the new vessels have been coming online since 2009 and 2010, just when freight demand stabilised. The result has been a collapse in freight rates. For example, freight rates for the Australia to northwest Europe route have averaged $18/tonne over the last two years, compared with around $40/tonne in 2007 and 2008 (McCloskey, 2007-2011). It is likely that there will continue to be ample spare shipping capacity in the next two to four years, keeping freight rates low (Figure 11.11). Ship order books are still at very high levels, with almost 2 000 new Panamax and Capesize bulk carriers due to be commissioned over the period to 2015.

Figure 11.10 ● **Steam coal supply cash costs to northwest Europe by component for selected exporters, 2007 and 2009**

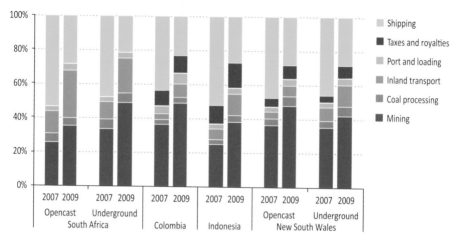

Sources: IEA Clean Coal Centre analysis citing data from Marston; McCloskey (2007-2011); IEA analysis.

Figure 11.11 ● **Dry bulk carrier market evolution, 2011-2015**

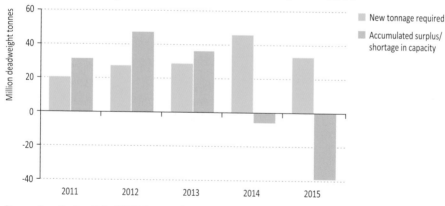

Sources: Barry Rogliano Salles (2011), DnB NOR (2010); IEA analysis.

Government policies

Governments intervene in energy markets in many different ways, with direct or knock-on effects on both coal demand and supply. Energy and environmental policies, including those aimed at mitigating climate change, mainly affect supply through changes in demand, but some measures directly affect supply, either positively (*e.g.* via subsidies to local production) or negatively (*e.g.* restrictions on mining for local environmental reasons). Supply-side government coal policies can be important in determining the costs and volumes of coal produced and, therefore, market prices.

Generally, exploration or exploitation of mineral resources requires a permit. The procedure to gain the permit, together with the requirements on the exploration or exploitation of the deposit, including royalties, can be critical in fostering or hampering mining activity. While technology exists to minimise their environmental impact, noise, dust, water use, liquid and solid wastes, and significant temporary (and, to a lesser extent, permanent) modifications to the landscape may all inhibit mining operations. Local opposition and the stringency of regulatory requirements relating to these externalities have a substantial impact on financial risk, cost and, ultimately, investment in mining.

In a growing number of countries, including China, waste management from mining, particularly reducing the risk of groundwater contamination, and requiring site-rehabilitation, has improved enormously in recent decades and is expected to improve in the future. But such essential improvements unquestionably increase costs. The extent and nature of the environmental impact and the rigour of mineral resource permitting are closely linked in practice. Mining is sometimes prohibited in areas that are protected for environmental reasons. Access to land is also affected by the nature of the ownership of mineral resources: when a coal resource is owned by a private landowner, the incentive to lease the land (for mineral exploitation) may be stronger, as the owner benefits directly.

Government policies on energy taxation generally also affect coal production. For example, a carbon cap-and-trade scheme not only pushes up the effective price of coal to end-users, but can also have a significant impact on production costs and supply, as diesel and electricity (the prices of which are affected by the carbon price) are important cost components in coal mining and transportation. The inclusion of fugitive emissions of methane from coal mines in any carbon cap-and-trade scheme would have a highly significant impact on coal-production costs. In general, mining regulation is expected to become progressively more stringent.

Some countries continue to support coal mining directly (Table 11.5). Usually this is for reasons of local employment and regional development. Coal production is often a labour-intensive industry (especially in developing countries), both through direct employment in the mines and in auxiliary industries located nearby, such as transport, equipment repair and material and equipment supply. In some regions, coal production is overwhelmingly the main economic activity, lending the authorities to intervene to support mining activities when costs are rising or market prices are falling. Energy security is also sometimes cited as a rationale for subsidising indigenous coal production in order to displace oil, natural gas or coal that would otherwise have to be imported. The rationale for coal production subsidies – whether for social, regional or energy-policy reasons – is often poorly defined and little overt attempt is made to measure this effectiveness.

Coal production can be supported in a variety of ways. Most commonly, subsidisation takes the form of guaranteeing prices to mining companies and/or guaranteeing demand (usually from local power stations). Other forms of financial support include royalty and tax exemptions and credits, grants for investment, cheap loans, preferential rail-transportation tariffs (usually where the rail company is publicly owned) and the imposition of tariffs or quotas on imported coal. Assistance may also be offered to retired miners (which does not directly affect current production) and to cover the costs associated with land reclamation.

Table 11.5 ● Government support to coal production in OECD countries

	Type of support
Germany	The cost of producing hard coal in Germany (lignite mining is not subsidised) is normally far higher than the price of imported coal; the difference is made up by a subsidy to Ruhrkohle AG (RAG), the dominant producer. The cost of these subsidies peaked at $8.5 billion in 1996 and in 2011 was around $2 billion, even though production had been declining for many years. In mid-2007, the federal government, the governments of the states with mines, the unions and RAG agreed on a detailed road map to end all subsidies by 2018. Under the deal, production is being gradually scaled back. Subsidies will continue to be paid jointly by the federal and state governments until 2014, after which time the federal government will pay all subsidies. Liability costs that remain after the closure of the pits will primarily be paid out of a fund which will be financed by the proceeds of a public sale of shares in RAG. Another programme, in place since 2001, provides older coal miners with early retirement payments until they become eligible for regular pension payments.
Korea	Support to producers of anthracite coal has been in place for several decades, involving price support, subsidies for acquiring capital equipment, subsidies for exploration and support of a more general nature. The price-support component was repealed at the end of 2010. The government also provides support to the production of anthracite briquettes, mainly by setting the price below cost (to protect low-income households) and paying the difference to producers. Support is due to be phased out progressively and terminated by the end of 2020, though a scheme to provide vouchers to subsidise consumption is expected to be expanded to offset the impact of higher prices.
Poland	Most of the costs currently associated with aiding the restructuring of the hard coal industry are associated with historic liabilities. Since 2007, the costs of mine closures have been met by a dedicated fund, established for this purpose by the remaining mining enterprises. Coal sales are not subsidised and state aid is no longer given to support operating costs or to maintain access to already exploited coal reserves.
Spain	Transfer payments are made by the government to private coal companies to compensate them for the difference between their operating costs and the prices at which they sell their output to local power plants (which are negotiated directly). Under the National Plan for Strategic Coal Reserves 2006-2012, operating aid is to be reduced progressively and production is due to fall from 12.1 Mt in 2005 to 9.2 Mt in 2012. Aid is available to pay benefits to former miners, to cover the costs of mine closures, for alternative industrialisation projects and for developing infrastructure in the affected mining regions.
Turkey	The government subsidises directly hard coal mining by paying the difference between the price at which TTK (the monopoly state-owned hard coal producer) sells the coal to power generators (linked to the import price) and the actual cost of production. The government plans to phase out this subsidy with the planned restructuring of TTK.

Source: OECD/IEA databases and analysis.

Information about government support to coal production is patchy, especially for countries outside the OECD. A handful of OECD countries, mostly in Europe, still support coal production, though the level of support has in most cases been declining for several years. For example, Germany still has hard coal production subsidies, but is due to phase them out completely by 2018. All support to coal production in Europe is subject to European Union (EU) rules on state aid.

11

Coal market and industry structure

International markets

International coal trade takes place largely within two distinct geographical zones: the Atlantic and Pacific Basins. In the Atlantic Basin, Colombia, Russia and South Africa are the main exporters of steam and coking coal, mostly to Europe, though South African coal also moves eastwards, mainly to India. In the Pacific Basin, the main exporters are Australia and Indonesia, supplying their traditional markets in Japan, Korea and Chinese Taipei, but recently also China and India. Exceptions to these general export patterns do exist: some Australian and Indonesian coal goes to Europe and the United States exports steam coal to Europe and coking coal to other markets. Canada is also an exporter of coking coal in the Pacific Basin and Mongolia is joining its ranks. Mozambique will soon be exporting coking and steam coal in both the Atlantic and Pacific Basins. China and India have become active players in the coal import market. India, which is facing the prospect of increasing import needs in coming years, has established strong relations with Indonesia and South Africa in particular. China's position is more complex as import volumes depend heavily on arbitrage opportunities between domestic and international prices.

The Atlantic Basin market for steam coal has matured and become highly liquid: paper trade has grown rapidly over the last four years and in 2010 was thought to be around ten-times as big as the physical market. There is a wide range of financial instruments for hedging and managing risks. Liberalisation of electricity markets, a shift from long-term contracts to spot index pricing and increased price volatility have helped to drive this development. Flexibility on both the demand side (particularly concerning the types of coal used) and on the supply side (including the ability to diversify coal quality through blending facilities) is likely to increase in the future.

By contrast, the Pacific Basin has not seen the same increase in the sophistication of trading practices, probably because of the continued preference on the part of Japanese and Korean buyers for long-term contracts, for security-of-supply reasons, which are related to steel production to support domestic manufacturing. The Pacific Basin has been dominated by long-term contracts with annual price negotiations, with spot purchases accounting for only a small share of total trade. The share of long-term contracts is thought to be from 85% to 90% in Japan, 80% to 85% in Chinese Taipei and 75% to 80% in Korea. India is moving from spot pricing to an equal mix of spot and long-term contracts. The slow pace of electricity market deregulation, aversion of some exporters to hedging and difficulties in establishing reliable indices in the complex Chinese and Indian markets, have impeded the development of a more liquid market. Nevertheless, there are signs that the situation is changing in some parts of Asia, with paper and financial trading playing a growing role. For example, the first Indonesian futures contract for steam coal was launched in 2010.

In the global coking coal market, trade has recently moved from mostly annual contracts to quarterly contracts and pressure from the producers to move to monthly pricing is growing. The existence of iron ore and coke derivatives and the launch of the first coking coal derivatives, which together allow hedging along the whole steel-value chain, are likely to lead to a rapid change in the way coking coal is traded. It is unclear to what extent this may be affected by any reduction in the differential between steam and coking coal prices – a prospect that is looking increasingly likely, given recent large investments in new coking coal mines.

Industry concentration

The structure of the coal industry worldwide has been undergoing radical change over the past decade or so, involving increased merger and acquisition activity (both internationally and within domestic industries) and increased diversification on the part of the mining companies, both in terms of the types of commodities they mine and the extent of overseas activities. In general, units within the industry have become larger, while most of the big companies have become more international. In China and some other developing countries, mining has been put on a more commercial footing, with some liberalisation of domestic markets, greater private-sector involvement and modernisation of operations.

Nonetheless, the coal industry remains far-less concentrated than other fuel industries. The 30 leading coal companies (whose share has risen slightly since 2008) accounted for around 40% of global production and global trade in 2010, far smaller shares than in the oil or natural gas sectors. The share of the four big diversified mining companies (Anglo American, BHP Billiton, Rio Tinto and Xstrata) in global production actually declined from 7% in 2008 to 5% in 2010, though this was mainly because the three-largest Chinese producers (Shenhua Group, China Coal Group and Datong Coal Group) expanded their market share from 7% to 9%. The recent strong growth in output in China has resulted from the ongoing consolidation of small-scale mining operations into larger, more efficient and safer mining complexes.

The concentration of global coal mining is likely to become more marked in the medium term, partly as a continuation of recent trends (in line with global trends across most industrial sectors) and because production growth is expected to come mainly from countries where the industry is already dominated by a small number of large companies. The trend towards concentration will be accelerated if prices and investment returns remain high, as they will enhance the ability of big companies to seek out new acquisitions. The increased financial and technical risk associated with greenfield mining and related transport infrastructure will also tend to favour greater concentration, though joint-venture operations can spread this risk. Nonetheless, there may be opportunities for small companies to build market share where policies are put in place to increase production by opening mining blocks to competition.

International trade in hard coal is more concentrated. The four big diversified mining companies, together with Drummond and Peabody Energy in the United States, SUEK and Kuzbassrazrezugol in Russia, and Bumi in Indonesia, accounted for nearly 40% of inter-regional coal trade in 2010. The share of these companies was just under 45%

in 2008. As Australia and Indonesia account for more than half of the world inter-regional hard coal trade, mining companies in these countries naturally also account for a large share. Like production, trade is likely to become even more concentrated in the future, though, as demonstrated by the recent increasing importance of Indonesian mining companies, the barriers to market entry by smaller players are not insurmountable.

Current investment trends[5]

Despite rising costs, investment in coal mining resumed its strong upward path in 2010, following a temporary slowdown in growth in 2009 in the wake of the global financial and economic crisis. The 30 leading coal companies worldwide, accounting for around 40% of world coal production, invested a total of $16 billion in 2010 – an increase of $1.5 billion, or 10%, on the 2009 level (Table 11.6). The jump in investment by these companies was particularly marked in China and the United States, where their investment had stalled in 2009, due to financing difficulties and a slump in demand. On the assumption that the level of investment per tonne of coal produced by the 30 leading coal companies was matched by all other coal companies, world coal mining investment totalled about $40 billion in 2010. Recently, an increasing share of investments has been directed towards coking coal operations, because of the widening differential between steam and coking coal prices. Spending on acquisitions of coking coal companies (not included in our investment estimates) has also increased.

The four big diversified mining companies – Anglo American, BHP Billiton, Rio Tinto and Xstrata – invested a total of $4.6 billion in 2010, marginally higher than in 2009. Investment by Xstrata increased sharply, with the development of six new coal mines, with a total capacity of more than 31 Mt per year, which are due to enter production by 2013. This increase was largely offset by a drop in capital expenditure by BHP Billiton. The combined spending of the two big Chinese coal-mining companies (Shenhua Group and China National Coal Group) surged by about 70% to $5.2 billion in 2010, driven by the continuing rapid growth in Chinese coal demand. Consequently, their share in the total investments of the 30 leading coal companies increased from just over one-fifth to one-third. In 2010, investment by the leading Indonesian companies dropped, partly as a result of unusually high capital expenditures in 2009; but it was still 85% higher than in 2007.

The 30 leading coal companies combined have a high reserves-to-production ratio, averaging about 60 years. But the ratio differs widely among the companies; for instance four companies, including the largest Chinese producer Shenhua Group, have a remaining reserve lifetime of less than 25 years. The average reserve lifetime of the four big diversified mining companies, at around 40 years, is relatively long. Eight companies have reserves covering more than 50 years of production at current rates, three of which are based in the United States. Companies like RWE Power in Germany and China National Coal Group report very large coal reserves. However, reserve statements have to be treated with care due to different classification systems and accounting standards between countries.

---

5. Investment figures presented in this section exclude expenditure on mergers and acquisitions.

Table 11.6 ● Key figures for the 30 leading coal companies

Name (corporate base)	Production (Mt) 2010	R/P* (Years)	Exports (Mt) 2010	Investment ($ million) 2008	Investment ($ million) 2009	Investment ($ million) 2010
Coal India (India)	431	50	n.a.	600	629	513
Shenhua Group (China)	352	21	n.a.	2 090	1 169	2 626
Peabody Energy (US)	198	41	19	264	261	557
Datong Coal Mining Group (China)	150	n.a.	n.a.	n.a.	n.a.	n.a.
Arch Coal (US)	146	27	6	497	323	315
China National Coal Group (China)	138	134	n.a.	1 142	1 874	2 564
BHP Billiton (UK-Australia)	104	61	63	938	2 438	1 534
Shanxi Coking Coal Group (China)	101	n.a.	n.a.	1 132	1 732	628
RWE Power (Germany)	99	350	n.a.	331	459	241
Anglo American (UK-South Africa)	97	28	48	832	496	491
SUEK (Russia)	89	66	29	449	351	429
Cloud Peak Energy (US)	85	11	3	138	120	92
Xstrata (UK-Switzerland)	80	45	58	1 204	1 111	1 998
Alpha Natural Resources (US)	77	49	9	331	339	345
Rio Tinto (UK-Australia)	73	27	32	515	512	609
Consol Energy (US)	66	31	6	446	580	733
PT Bumi Resources (Indonesia)	59	31	53	567	484	287
Kuzbassrazrezugol (Russia)	50	n.a.	24	667	126	336
Banpu (Thailand)	43	49	n.a.	120	82	42
Sasol (South Africa)	43	n.a.	n.a.	121	170	232
PT Adaro Indonesia (Indonesia)	42	20	n.a.	226	141	290
Kompania Węglowa (Poland)	40	n.a.	n.a.	371	316	265
Massey Energy (US)	34	63	n.a.	737	275	n.a.
Drummond (US)	32	n.a.	32	n.a.	n.a.	n.a.
Patriot Coal (US)	28	58	n.a.	121	78	123
Mitsubishi Development (Japan)	28	n.a.	n.a.	n.a.	n.a.	n.a.
Alliance Resource Partners (US)	27	23	1	177	328	290
PT Kideco Jaya Agung (Indonesia)	24	n.a.	n.a.	n.a.	n.a.	n.a.
Teck Cominco (Canada)	23	32	n.a.	118	69	355
International Coal Group (US)	14	70	1	173	88	107
Total	2 775	60	384	14 304	14 551	16 002

*Reserves-to-production (R/P) ratios are based on the last year of available data for production and the sum of probable and proven reserves. Includes IEA estimates on reserve data, which may not conform to the Australasian Joint Ore Reserves Committee (JORC) standards.

Sources: Company reports and IEA analysis.

11

Focus on the New Policies Scenario

Production prospects

In the New Policies Scenario, global coal production increases by nearly 20%, from about 4 930 million tonnes of coal equivalent (Mtce) in 2009 to a plateau of 5 860 Mtce in 2025, before remaining broadly flat throughout the rest of the projection period (Table 11.7). The majority of the net growth in world production occurs in non-OECD countries, while production falls markedly in the OECD (particularly in Europe and the United States), at an average rate of 0.6% per year. Non-OECD production over 2009 to 2035 expands by almost 1 150 Mtce, or 1.1% per year on average, with roughly 50% of the increase coming from China and the bulk of the rest from India and Indonesia. Australia is the only major OECD producer to see a rise in production (12% between 2009 and 2035), but growth is mainly confined to the period to 2020.

Table 11.7 ● **Coal production by region in the New Policies Scenario** (Mtce)

	1980	2009	2015	2020	2025	2030	2035	2009-2035*
OECD	1 385	1 403	1 462	1 421	1 358	1 281	1 197	-0.6%
Americas	673	810	856	833	799	750	697	-0.6%
United States	*640*	*757*	*794*	*769*	*740*	*699*	*652*	*-0.6%*
Europe	609	249	218	192	169	146	118	-2.8%
Asia Oceania	103	343	388	396	390	386	382	0.4%
Australia	*74*	*338*	*382*	*391*	*384*	*381*	*377*	*0.4%*
Non-OECD	1 195	3 525	4 172	4 412	4 505	4 575	4 662	1.1%
E. Europe/Eurasia	519	364	407	408	406	393	382	0.2%
Russia	*n.a.*	*219*	*258*	*262*	*267*	*257*	*248*	*0.5%*
Asia	568	2 873	3 423	3 634	3 725	3 805	3 903	1.2%
China	*444*	*2 197*	*2 563*	*2 675*	*2 691*	*2 710*	*2 739*	*0.9%*
India	*77*	*349*	*399*	*441*	*488*	*537*	*589*	*2.0%*
Indonesia	*0*	*238*	*338*	*380*	*406*	*415*	*429*	*2.3%*
Middle East	1	1	1	1	1	1	1	0.9%
Africa	100	207	238	254	251	255	256	0.8%
South Africa	*95*	*202*	*218*	*224*	*216*	*216*	*214*	*0.2%*
Latin America	8	80	103	115	122	121	120	1.6%
Colombia	*4*	*68*	*90*	*101*	*109*	*107*	*107*	*1.8%*
World	2 579	4 928	5 634	5 833	5 863	5 856	5 859	0.7%
European Union	*n.a.*	*238*	*201*	*171*	*142*	*117*	*89*	*-3.7%*

*Compound average annual growth rate.

Trade prospects

In the New Policies Scenario, total hard coal trade between *WEO* regions is projected to rise from about 750 Mtce in 2009 to a plateau of around 1 050 Mtce by 2020, before falling back slightly to nearly 1 020 Mtce by 2035 (Table 11.8). The average annual value of this hard coal trade over the period 2011 to 2035 amounts to about $140 billion (in year-2010 dollars); although this appears large, it is far below the average annual value of trade in oil ($1.9 trillion) and natural gas ($385 billion). Patterns of coal trade have been shifting in recent years, with a growing share taken by non-OECD Asian imports, and this trend is set to continue. Consequently, the centre of gravity in international coal trade continues to move to the Pacific Basin market in the New Policies Scenario (Figure 11.12).

Table 11.8 ● **Inter-regional* hard coal net trade by country in the New Policies Scenario** (Mtce)

	2009		2020		2035	
	Mtce	Share of primary demand**	Mtce	Share of primary demand**	Mtce	Share of primary demand**
OECD	**-120**	**9%**	**-72**	**6%**	**50**	**5%**
Americas	38	5%	70	9%	61	9%
United States	*32*	*4%*	*65*	*9%*	*53*	*8%*
Europe	-181	65%	-191	72%	-147	78%
Asia Oceania	23	7%	48	13%	136	37%
Australia	*256*	*81%*	*310*	*83%*	*301*	*83%*
Japan	*-144*	*100%*	*-158*	*100%*	*-115*	*100%*
Non-OECD	**178**	**5%**	**72**	**2%**	**-50**	**1%**
E. Europe/Eurasia	87	30%	103	31%	83	26%
Russia	*77*	*42%*	*96*	*43%*	*80*	*38%*
Asia	-19	1%	-178	5%	-281	7%
China	*-88*	*4%*	*-188*	*7%*	*-81*	*3%*
India	*-61*	*16%*	*-178*	*30%*	*-294*	*34%*
Indonesia	*191*	*100%*	*288*	*99%*	*278*	*94%*
Middle East	-1	43%	-2	54%	-2	55%
Africa	56	27%	75	29%	75	29%
South Africa	*63*	*31%*	*66*	*29%*	*54*	*25%*
Latin America	54	70%	74	66%	74	63%
Colombia	*63*	*93%*	*95*	*94%*	*99*	*93%*
World	**753**	**17%**	**1 056**	**19%**	**1 017**	**18%**
European Union	*-156*	*62%*	*-155*	*69%*	*-110*	*76%*

*Total net exports for all *WEO* regions, not including trade within *WEO* regions. **Share of production in the case of exporting regions.

Notes: Positive numbers denote exports; negative numbers imports. The difference between OECD and non-OECD figures in 2009 are due to stock change.

11

Figure 11.12 ● World inter-regional* hard coal net trade by major region in the New Policies Scenario

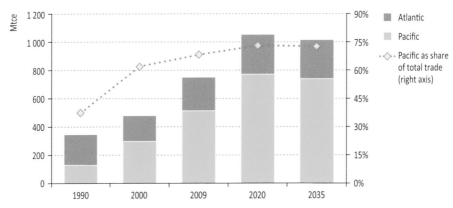

*Total net exports for all *WEO* regions, not including trade within *WEO* regions.

Whereas Japan and the European Union have long been the world's largest coal importers, China and India are now emerging as equally important. In particular, India is set to rely increasingly on imports, as its production fails to keep pace with booming domestic demand. India's hard coal imports were around 60 Mtce in 2009 and by 2035 are projected to rise five-fold to 300 Mtce – by far the largest volume of any single country (Figure 11.13). International coal price movements will increasingly be driven by Indian and Chinese import needs, which have already had a profound impact on the prices paid by coal-importing OECD countries. Falling demand in the European Union drives down net imports by almost 50 Mtce between 2009 and 2035, to 110 Mtce; most of the change happens after 2020.

China is expected to become an increasingly important net importer in the medium term, but in line with a projected slowdown in domestic demand growth and expected improvements in domestic coal infrastructure, its trade in coal returns to balance by the end of the projection period. However, this projection is very sensitive to projected rates of demand and production growth, since trade is the difference between very large volumes and, therefore, a modestly faster pace of demand growth or slightly slower output growth would lead to much higher import requirements (and *vice versa).* Assuming all else was equal, hard coal net imports would fall to zero by 2015 (compared with our projection of around 185 Mtce) if China's output rises by just 1.2% per year above our projection in the New Policies Scenario.

Among the coal exporters, by far the biggest increase in volumes traded occurs in Indonesia and Australia, although their growth is concentrated in the period to 2020. These two countries, which together already accounted for about 60% of global hard coal trade in 2009, are projected to contribute nearly half of the increase in global inter-regional trade to 2035 in the New Policies Scenario. Most of their exports go to India, China, Japan and Korea. Exports from South Africa, the United States and Russian initially follow the growth in global coal trade in the New Policies Scenario to 2020, before declining towards 2035 as global trade decreases and competition from the lowest and/or largest exporters intensifies.

Colombia, along with new entrants such as Mongolia and Mozambique, is projected to contribute substantially to incremental growth in inter-regional trade, increasing diversification of sources for importers (especially of coking coal in the case of Mongolia and Mozambique).

Figure 11.13 ● **Major hard coal importers in the New Policies Scenario**

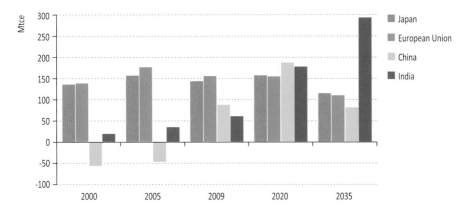

Investment outlook

In the New Policies Scenario, global cumulative investment requirements in the coal industry over the period 2011 to 2035 amount to about $1.2 trillion (in year-2010 dollars), or $47 billion per year. Around half of this capital expenditure is in the period to 2020 and China alone accounts for over half of the cumulative investments over the projection period (Figure 11.14). Nearly 95% of projected investment goes to mining, including existing and new operations, and the rest to ports and shipping. While far from negligible, coal capital expenditures, at only 3% of global energy-supply investment requirements, are small relative to other sectors.

The average annual projected rate of global investment is around 15% higher than the investment estimated to have been made in 2010. In effect, the diminishing need for new capacity, as demand growth slows and saturates by the early 2020s in the New Policies Scenario, is offset by the assumed rise in capital costs per unit of output. As seen above, after a short period of respite following the onset of the global economic crisis, capital expenditure in mining and transportation per unit of output is on the rise again. The global weighted-average investment between 2008 and 2010 is estimated at $7.8/tonne of annual mine capacity (in year-2010 dollars), though costs vary significantly across countries and basins. We assume that investment unit costs in real terms will continue to rise over the projection period due to continuing upward pressures on labour, steel, equipment, materials and energy costs, albeit at a slower rate than in the recent past – the global weighted-average investment is assumed to be $9.3/tonne of annual mine capacity over the projection period.

11

Figure 11.14 ● Cumulative coal-supply investment by region in the New Policies Scenario, 2011-2035

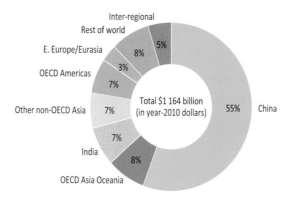

Regional analysis

China

Mining

China has the second-largest coal reserves in the world, after the United States, with around 180 billion tonnes of hard coal and 10 billion tonnes of brown coal. Reserves are concentrated in the north-central and western provinces of the country, mainly in Shanxi, Shaanxi, Inner Mongolia and Xinjiang. Based on preliminary data, China produced close to 2 300 Mtce of coal in 2010, of which 83% was steam coal and 17% coking coal. Coal production in China more than doubled between 2000 and 2010. The Chinese mining industry is divided into large state-owned mining enterprises, provincial mining companies and small-scale community operations. While most state-owned enterprises, like Shenhua Group, have efficient and safe mining operations, where modern equipment is employed, the smaller-scale operations at the provincial or community level are often less efficient and less safe. Production from state-owned enterprises makes up around half of total coal production, with the majority of the remainder coming from small-scale operators (VDKI, 2010).

Due to the rapid exploitation of coal deposits in recent years, the Chinese coal-mining industry faces two key challenges. First, existing operations are moving deeper underground, with more than 60% of coal reserves now found at depths greater than 1 000 metres (Minchener, 2007). Underground mining already makes up 90% of total output and, with greater depth, mining costs are increasing, further raising the costs of supply (which are already increasing due to transportation distances). Second, new coal-mining operations are moving westward into Inner Mongolia, Shaanxi and Xinjiang, as more eastern deposits, closer to the main markets, are exhausted. The production targets that the government has set for the Xinjiang are challenging: 1 billion tonnes by 2020, up from an estimated 75 Mt in 2009. Achieving this will require the construction of a new railway line, dedicated to freight, to connect the coal-fields with the inland provinces of Gansu and Qinghai. However, transport costs to the coast will be very high, due to the long distances involved.

Between 2000 and 2009, the average cost of production of the major state-owned enterprises in Shanxi province broadly tripled (Figure 11.15). The labour productivity of newer coal-mining operations in China is similar to that in the Appalachian region of the United States or in South Africa, thanks to the recent industry consolidation and efforts to increase mechanisation, but, worsening geological conditions have outweighed productivity gains in recent years. Levies and taxes have been introduced or increased, mostly aimed at diminishing the environmental impact of mining, while the coal resource tax has also contributed to higher costs.

Figure 11.15 • Average coal production costs of major state-owned enterprises in Shanxi province, 2000-2009

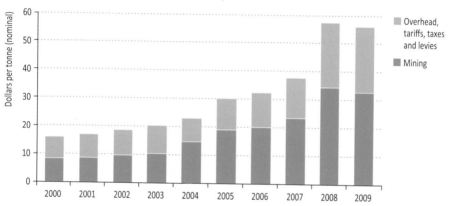

Sources: Wang and Horii (2008); Tu (2010); IEA analysis.

The massive restructuring of the Chinese coal-mining industry that has been underway for several years – aimed at consolidating mining operations, increasing efficiency and improved health and safety – has contributed to constraining growth in production, as small inefficient mines have been shut down. As a result, production has failed to keep pace with demand, causing imports to rise and exports to drop in the short term, though the restructuring is expected to reinforce the ability of the industry to attain higher production in the longer term. However, there are concerns about the adequacy of the incentives to invest, given the pressure on the coal industry from power generators to hold down prices. Power generators have incurred big financial losses recently as a result of the government capping electricity prices (to combat inflation and protect household incomes) to levels that do not cover the rising cost of coal. The government attempted to address this problem in the 11th Five-Year Plan by planning large "coal-power bases" – the development of combined mining operations and coal-fired power plants at the same location. The electricity so produced is to be transmitted via ultra-high-voltage direct current lines to demand centres. This approach can reduce transport costs and also allows for larger, more efficient and safer mining operations. Chinese officials hope that the distribution of business risk resulting

11

from this growing integration of the power and coal sectors will reduce the conflict between coal and power producers. Currently, fourteen coal-power bases are under development, controlled by state-owned enterprises and the central government. Each will produce and burn more than 100 Mt of coal per year.

Inland transportation

The greater part of Chinese demand for coal and energy generally is concentrated along the coast, mainly in the industrial centres around Beijing, Shanghai, and Guangdong (Figure 11.16). Consequently, coal has to be hauled long distances from the western coal-mining areas, sometimes exceeding 1 300 km. As a result, domestic transportation makes up a significant part of the total cost of supply. It can cost around $35/tonne to move coal from the central coal provinces to the coast at present (Table 11.9). Most of the country's coal-fired power stations are located near the coastal demand centres, as building plants close to mines was previously hindered by a regulatory scheme that discouraged electricity-grid operators from building transmission lines. The share of coal transported by railway has rapidly dropped in recent years, but rail is still the dominant mode of transportation, accounting for 44% of total coal production in 2009 (NBS, 2010). Because of inland transportation constraints, growing volumes of domestic coal are shipped by sea using bulk carriers, from Qinhuangdao and other ports in the north to major demand centres in the south, and a smaller fraction of the coal produced is moved by road and river barges.

Table 11.9 ● Major railway-to-port coal routes in China

	Distance (km)	Costs ($/tonne)		Capacity (Mt/year)	
		2006	2010	2006	2010
From Northern Shanxi/Shaanxi/Inner Mongolia to:					
Qinhuangdao	650	16	26	193	193
Tianjin	500	13	20	90	108
Jintang	600	15	24	16	46
Hunaghua	600	15	24	90	90
From Central/Southeast Shanxi to:					
Qingdao	900	23	36	15	15
From Southwest/Southeast Shanxi to:					
Rizhao	900	23	36	27	45
Lianyungang	950	24	38	17	20

Sources: Tu (2010); Morse and He (2010); Paulus and Trüby (2011); IEA estimates and analysis.

In total, coastal freight amounts to around 500 Mt/year and is expected to increase in the short term. Coal-transportation capacity has increased rapidly: the capacity of the three major railway links connecting the western coal-bearing provinces with the northern coal ports has grown from 550 Mt/year in 2005 to around 1 000 Mt/year in 2010. Even so, rail bottlenecks have occurred and road transportation has been hindered by massive and common traffic jams on the roads to eastern demand centres. Rising rail freight rates have also driven up coal-supply costs: the railway network is under the control of a state monopoly, which is able to extract monopoly rents from coal producers. The government's policy of encouraging coal-power bases is expected to reduce the need for coastal transport, relieve the pressure on the rail-transport system and limit the market power of the state rail company over future coal power projects.

Figure 11.16 ● Major coal production centres and transport routes in China, 2009

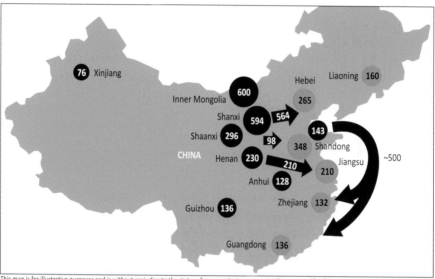

This map is for illustrative purposes and is without prejudice to the status of or sovereignty over any territory covered by this map.

Notes: All numbers are in million metric tonnes. The black circles indicate production and the green circles are deliveries to major demand centres. Arrows depict major railway lines or coastal shipping. Note that only major production and demand centres of (more than 100 Mt) have been included, except Xinjiang.

Sources: NBS (2010b); Tu (2010); IEA analysis.

Trade

China swung from being a net exporter to a net importer of coal in 2009, with a net import balance of 88 Mtce. In 2010, net imports continued to climb to an estimated 126 Mtce, making China the world's second-largest coal importer after Japan. China continues to export some coal, but imports are now much bigger. Exports have been controlled by the government through export quotas since 2004, mainly because of concerns about over-exploitation of domestic resources and the prospect of growing dependence on imported coal. Only four mining companies (China Coal Company, Minmetals, Shanxi Group and Shenhua Group) hold export licenses. Export quotas are set on an annual basis and have been reduced progressively from 80 Mt in total in 2005 to 38 Mt in 2011. The government also levies a tax on coal imports, currently 17% of the FOB price. While subject to oscillation and consequently arbitrage, transportation bottlenecks and high rail-freight tariffs for coal moved from the northwest of China have led to an increase in domestic delivered coal prices and reduced the competitiveness of Chinese production compared with imports (Figure 11.17). Average coal prices at Qinhuangdao, the main export terminal, increased from $71/tonne in 2007 to $115/tonne in 2010, and were higher again in 2011.

Figure 11.17 ● China's steam coal import volumes from selected countries

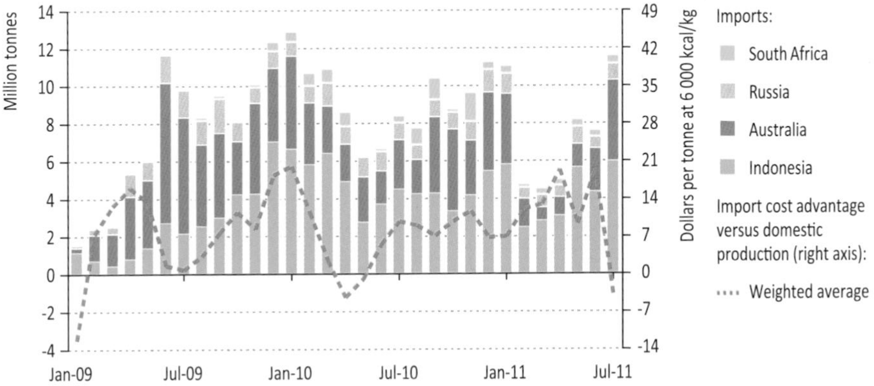

Source: Based on data from IEA Medium-Term Coal Market Report (2011), forthcoming.

Outlook

Our projections for China's production reflect the expected trends in mining and transport infrastructure outlined above and the underlying coal demand trends presented in Chapter 10. Many of the policies contained in China's 12th Five-Year Plan are expected to be taken forward through more detailed targets. In some cases, this *Outlook* anticipates such targets, based on credible reports at the time of writing, in order to provide further depth to the analysis. The energy targets outlined by officials show coal consumption reaching around 3.8 billion tonnes in 2015, which would call for average annual growth of production of only around 5% from 2011 to 2015, compared with almost 10% between 2000 and 2010. In the New Policies Scenario, total coal production in China is projected to increase by around 540 Mtce between 2009 and 2035, reaching 2 740 Mtce. Most of the increase – about 90% – occurs by 2020, production growing at an annual rate of 1.8% from 2009 to 2020 (Figure 11.18). All of the increase to 2020 is in steam coal, which reaches a level of nearly 2 300 Mtce. After 2020, coal production remains almost stable until the end of the projection period, as industrial energy demand slows, with coal increasingly substituted by natural gas and electricity, and other fuels (like natural gas, nuclear and renewables) become more important in power generation. As a result of fuel substitution and due to saturation of steel use, coking coal production peaks before 2020 and declines through the *Outlook* period from 380 Mtce in 2009 to around 300 Mtce in 2035.

In the New Policies Scenario, Chinese net coal imports increase from around 90 Mtce in 2009 to 185 Mtce by 2015 (Figure 11.19). Over this period, lower economic growth rates, improved energy efficiency and a shift of focus away from coal towards natural gas, renewables and nuclear energy gradually dampen coal demand growth and slow the growth in coal imports. In addition, the opening up of new, highly cost-competitive coal mines in the west of China, together with increased transmission of electricity rather than transport of coal should increase the competitiveness of domestic supply, compared to imports. Between 2015 and the end of the projection period, net imports decline, though slowly, as a result of stagnating domestic Chinese coal demand. By 2035, China's net imports are around 3% of the nation's demand – a level comparable to 2009.

Figure 11.18 ● **Cumulative coal production in China by period and type in the New Policies Scenario**

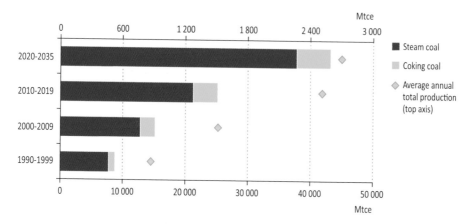

Figure 11.19 ● **China's hard coal net trade by type in the New Policies Scenario**

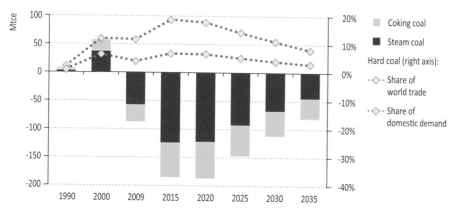

Nevertheless, with the import levels projected in the medium term in the New Policies Scenario, China would remain a crucial player in global coal trade, up to 2020. China's share of total inter-regional hard coal trade would remain just below 20%, China becoming the world's biggest importer in volume terms; later on, China would be overtaken by India. On the other hand, Chinese net imports would still meet only around 7% of total Chinese coal demand around 2020. As Chinese demand is so large, a relatively small variation in domestic coal demand or supply would have major implications for international markets (see Spotlight in Chapter 2). In this context, prices could be highly volatile, as short-term fluctuations in the Chinese domestic market amplify price movements on the international market.

United States

Mining

Reserves of hard and brown coal in the United States amounted to 227 billion tonnes and 31 billion tonnes respectively at end-2009 (BGR, 2010). Together, these reserves are sufficient to maintain 2009 production levels for 265 years. The United States accounts for around a quarter of the world's coal reserves. Reserves are concentrated primarily east of the Mississippi River, notably in Illinois, Kentucky and West Virginia, although output from west of the Mississippi River now accounts for more than half of national output. All of the sub-bituminous coal, which makes up 37% of the reserve base, is west of the Mississippi River, mostly in Montana and Wyoming. Low-value lignite, which accounts for about 9% of reserves, is located mostly in Montana, Texas and North Dakota. Anthracite, the highest-quality coal, makes up only 1.5% of the reserve base and is found in significant quantities only in northeastern Pennsylvania.

Environmental policies will be crucial to the prospects for coal production, through their impact on access to resources and on the cost of complying with regulations aimed at minimising the risks and damage associated with mining activities. Mining regulations are likely to continue to get tighter. Regulation could also act to cut coal demand. Air pollution regulations are expected to severely constrain the use of coal in the industry and power sectors in this decade by leading to a substantial shutdown of old plants. Additionally, a big concern on the part of the coal-mining industry is that action to curb the growth in greenhouse-gas emissions, possibly through a cap-and-trade scheme, will limit the scope for using coal or increase the cost of doing so. For these reasons, mining companies are increasingly looking at the possibility of expanding export sales to compensate for any loss of domestic sales. A number of projects underway are designed to expand US coal-export capacity (subject to regulatory and legal constraints), such as the Asia-focused Gateway Pacific Terminal being built in Washington state.

Outlook

Apart from rebounding in the short term from the dip due to the economic crisis, coal production in the United States is projected to decline steadily through the projection period, from 757 Mtce in 2009 to just above 650 Mtce in 2035, an average rate of decline of 0.6% per year (Figure 11.20). The main reason for this drop in production is the weak outlook for domestic demand (see Chapter 10), resulting from strong competition from low-cost natural gas and growing renewable energy sources in power generation, and the assumption that from 2015 investment decisions in the power sector will take into account a "shadow-price" for carbon. The United States has large remaining resources of coal, but much of the resource is relatively expensive to extract and transport to market, especially for areas east of the Mississippi. However, due to strong global demand and high prices, there has been a rebound in exports over the past few years, especially of coking coal. In the New Policies Scenario, hard coal net exports are projected to rise from 32 Mtce in 2009 to 53 Mtce in 2035, with most of the growth occurring in the first decade of the projection period and consisting predominantly of coking coal.

Figure 11.20 ● Coal production in the United States in the New Policies Scenario

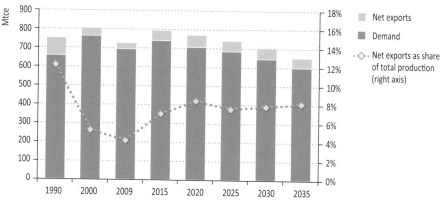

Much of the projected fall in production is likely to occur in the higher-cost Appalachian region, which in 2009 accounted for around one-third of all the coal produced in the United States. Appalachian coal is mined mainly from large underground mines and most is supplied to markets within the region (mainly power stations), though growing volumes are exported. The region, especially the central areas, has been extensively mined and production is relatively costly. In addition, concerns about the environmental impact of the practice of mountaintop mining in central Appalachia will probably constrain output; a freeze on surface-mining permits was recently announced and a review of permitting is underway.

Production is expected to remain strong in the western United States, where coal is mined primarily from very large opencast mines, with generally lower costs than those in the Appalachian and the interior region – the other main producing area, located in Illinois. Low-cost supplies of coal from the west will increasingly meet demand from coal-fired power plants east of the Mississippi River and will supply most of the coal needed for the new coal-to-liquids (CTL) plants that are expected to be built during the projection period. Production in the interior region, which has been in decline in recent years, could be boosted by investment in new mines to exploit the substantial reserves of mid- and high-sulphur bituminous coal found in Illinois, Indiana and western Kentucky.

India

Mining

Although coal is widely used in every state in India, coal deposits are heavily concentrated, with 80% of both reserves and resources located in just four states: Chattisgarh, Jharkand, Orissa and West Bengal. Roughly 95% of the country's 77 billion tonnes of coal reserves are hard coal (BGR, 2010). Indian hard coal is generally of poor quality, with high levels of ash and organic impurities, though the sulphur content is low. Even though Indian power plants are designed to be able to handle coal with ash content in a range from 25% to 55%, the high ash content causes technical problems and gives rise to higher operating costs at power plants. As a result, coal-washing facilities are increasingly being installed. Calorific values are generally low, in a range of 3 100 kcal/kg to 5 100 kcal/kg and averaging around 4 100 kcal/kg (IEA-CCC, 2007). Only around 10% of domestic coking coal readily meets quality standards; the remainder requires intensive washing to make it suitable for coke production.

Coal continues to play a central role in underpinning rapid economic development in India, though the country increasingly has to import coal to meet its booming energy needs. Indian hard coal production jumped from around 200 Mtce in 2000 to 330 Mtce in 2010 (based on preliminary data), growing at an average annual rate of 5%. Brown coal output has grown from 8 Mtce to 13 Mtce over the past decade.

As in China, a mismatch between the geographical locations of coal supply and demand requires coal to be transported over long distances. More than 50% of the coal traffic, including imports into India, is moved by rail; another 20% is shipped by road and the rest is consumed (mainly) in mine-mouth power plants. Rail shipments increased from 300 Mt in 2005 to 400 Mt in 2009; government targets for production imply an increase to 700 Mt by 2025.

The Indian coal industry is highly concentrated with several state-controlled mining companies dominating the market. One state-owned company, Coal India, controls around 80% of total Indian coal output. Private companies account for less than 10%. Most coal comes from opencast operations, as over 60% of coal resources are found at a depth of less than 300 metres (Mills, 2007) and opencast mining is usually more efficient and less capital intensive than underground mining. Indian mining companies have in recent years increased the share of output from opencast operations from less than 80% in the late 1990s to almost 90% in 2009. Nonetheless, productivity is very low: the output per employee year of Coal India was around 1 100 tonnes, compared with, on average, around 9 000 tonnes in Australian opencast mines. New opencast mines are more difficult to develop, as many of the reserves are in environmentally protected areas and permits can be difficult to obtain from the various administrative authorities. There is also a lack of coal-washing capacity. Further increases in production will hinge on the introduction of appropriate policies and investment to address these concerns.

Trade

A growing share of Indian demand is met by imports, as demand has outstripped production for several years. Indian hard coal imports grew from 20 Mtce in 2000 to 75 Mtce in 2010. Of this, roughly one-third was coking coal. As domestic coal often has to be hauled over long distances, domestic coal supplies are not always competitive with imported coal, even at inland demand centres. This has increased the incentive to import. Indian coal imports are very likely to increase further: in February 2010, the Indian coal ministry announced that, in the fiscal year 2011/12, Indian hard coal imports could increase to 142 Mt, an increase of around 70%, partly due to regulatory hurdles that are impeding growth of domestic production.

Indonesian and South African steam coals constitute the largest part of Indian steam coal imports. Imported coal is also used for blending with low-quality domestic coal to bring it up to acceptable standards. Some existing coal-fired plants depend exclusively on imported coal. The limited amount of domestic coking coal reserves and their relatively poor coking capabilities mean that India in 2009 imported almost half of its coking coal needs, mostly sourced from Australia.

Outlook

In the New Policies Scenario, Indian hard coal production grows at an annual average rate of 2%, rising from 340 Mtce in 2009 to 570 Mtce in 2035 (Figure 11.21). India's share of total world production increases steadily over the projection period and reaches 10% in 2035, consolidating India's position as the world's third-largest coal producer. By 2035, India's total coal production is only 10% lower than the projected production level of the United States, compared to about 55% lower in 2009. To maintain India's production growth in the near-term, it will be necessary to relieve domestic transport-infrastructure bottlenecks and open up new mining capacity. The Planning Commission for the 11th Five-Year Plan has already had to revise coal output projected for 2012 down from 420 Mtce to 390 Mtce and the latest estimates suggest that even this new figure might not be met, with output in 2012 reaching only 370 Mtce.

Figure 11.21 ● **Coal production in India by type in the New Policies Scenario**

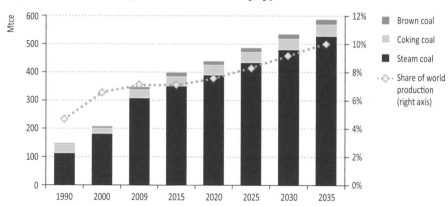

Despite continuing strong growth in demand in the Indian steel sector, coking coal production is projected to grow at a moderate pace of 1.3% annually until 2035. This is mostly due to the relatively limited resource base, particularly of good-quality coking coal. Brown coal production is projected to grow more rapidly, at around 2.5% per year on average over the projection period, but from a lower base, production almost doubles from 11 Mtce in 2009 to 20 Mtce in 2035. Nonetheless, the share of brown coal in total Indian production still reaches only 3.5% in 2035.

In the New Policies Scenario, net imports of hard coal are projected to rise by more than 10% per year between 2009 and 2015, reaching 120 Mtce. Thereafter, imports rise at a slower pace, as domestic production picks up, transportation infrastructure expands and demand growth moderates. They increase to just under 300 Mtce in 2035, growing at a rate of around 4.5% per year from 2015 (Figure 11.22). Dependence on coking coal imports increases most, reaching nearly 70% by 2035 (up from 47% in 2009). Overall dependence on hard coal imports rises from around 15% to around 35% over the same period.

Figure 11.22 ● Hard coal net imports in India by type in the New Policies Scenario

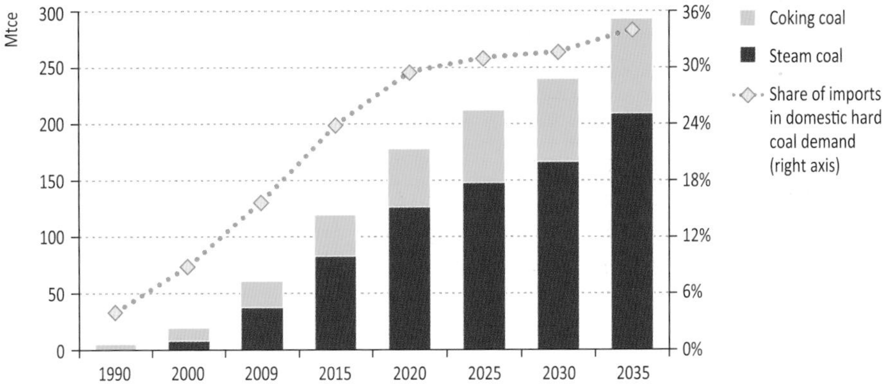

Indonesia

Mining

Indonesia had 5.6 billion tonnes of hard coal reserves and 10.1 billion tonnes of brown coal, at end-2009 (BGR, 2010). Most are found on the islands of Sumatra and Kalimantan. Kalimantan is home to a large part of Indonesia's coal reserves, which are mostly of sub-bituminous or bituminous quality in the range of 5 100 to 6 100 kcal/kg. Indonesian coals are usually low in ash and sulphur, but have a high moisture and volatile matter content. Reserves in Sumatra are mostly low-quality lignite and sub-bituminous coals, with a calorific value of less than 5 100 kcal/kg, making them expensive to transport. As a result, coal is increasingly being upgraded by drying or is consumed at the mine-mouth in power stations, which are linked by high-voltage transmission lines to the major demand centres.

Indonesia's coal production has risen sharply over the last decade, from 65 Mtce in 2000 to 282 Mtce in 2010 (based on preliminary data) – a remarkable average growth rate of 16% per year. Most production is steam coal, but output in 2010 included 46 Mtce of brown coal, mainly used for domestic power generation. The government has given priority to boosting coal production, partly to compensate for the decline in crude oil production (which led the country to suspend its OPEC membership in 2008).

In recent decades, the Indonesian government operated a so-called "coal contracts of work" (CCOW) mine-licensing system, whereby contract holders were granted the right to exploit a given coal deposit for 30 years in return for payment of a royalty of 13.5% of the price of the coal sold at the mine-mouth. The contractors were also obliged to offer Indonesian investors at least 51% of the mining stock after a ten-year operating period (Baruya, 2009). In 2001, this provision forced two foreign investors – Rio Tinto/BP and BHP Billiton – to cede shares. At the beginning of 2009, a new Indonesian mining law was enacted which gives foreign as well as Indonesian investors eight years to carry out exploration and another 23 years to build and operate the mines. Foreign investors are no longer obliged to offer stock to Indonesian investors. The general objective of the policy is to promote mining development by simplifying licensing, improving the planning of mining areas and clarifying responsibilities between central, provincial and district authorities. The government has also established a minimum floor price for coal as a means of increasing state revenues from royalties, which might be otherwise be based on artificial transfer prices.

Trade

Indonesia, which became the world's largest exporter of steam coal in 2006, exported 233 Mtce of steam coal in 2010, nearly 85% of its total coal production. Steam coal production all goes to the export market. There is hardly any trade in brown coal, as it is uneconomic to ship it far due to its low energy content. The main destinations for Indonesian exports are Korea, Japan and Taiwan. Exports to India and China have been growing very fast since 2007, reaching 32 Mtce and 40 Mtce respectively in 2009.

Practically all of the hard coal that is exported comes from south and east Kalimantan. A network of navigable rivers allows many mining operations to haul coal shipments via river barges to offshore terminals, where coal is loaded onto larger bulk carriers. This system avoids for the most part the need for domestic railway transport, which is common in other major exporting countries, such as Australia and South Africa. This has helped Indonesian producers to keep costs down and also to increase exports very quickly, as the supply chain is not so constrained by domestic physical transport infrastructure where bottlenecks can occur.

The Indonesian government plans to meet rapidly-rising domestic power demand through a massive expansion of coal-based power generation. Therefore, the government has decided to give the domestic market priority over increasing exports. In order to meet this objective, the government has implemented a range of measures, such as setting a minimum percentage of coal production that has to be sold to domestic customers. In addition, in early 2011, the government announced its intention to ban exports of low-quality coal, starting in 2014. This timeframe coincides with the projected completion of about 7.2 GW of coal-fired power generation capacity by 2014 (half of which will be brought online by 2012).

Outlook

In the New Policies Scenario, Indonesian production continues its rapid growth in the short term, with 6% annual growth rate until 2015, when output reaches nearly 340 Mtce (Figure 11.23). This projection takes into account new projects that have already been launched by mining companies operating in Indonesia. After 2015, production growth slows to 1.2% per year on average, as investors are forced to turn to the lower-quality remaining coal deposits and as global steam coal trade reaches a plateau. These projections are about 15% lower than the production target set by the Indonesian government of around 460 Mtce by 2025. By 2035, brown coal production will make up 30% of total volumes – 10% higher than the share in 2009.

Figure 11.23 ● **Coal production in Indonesia in the New Policies Scenario**

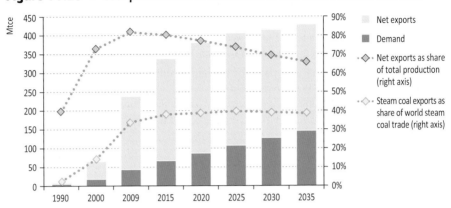

In the New Policies Scenario, the continuation of the recent surge in Indonesian coal production underpins growth in exports of almost 4% per year until 2020, when they reach around 290 Mtce. Thereafter, rising domestic demand and a plateau in global steam coal trade result in exports remaining almost stable until the end of the projection period. This projection is consistent with the government's recent interventions to ensure that more coal supplies are made available to domestic power stations. Coal exports as a share of total production fall from around 82% in 2009 to just over 65% in 2035. Nonetheless, Indonesia remains the world's largest exporter of steam coal throughout the projection period and increases its share of this market from 34% in 2009 to nearly 40% in 2035.

Australia

Mining and transport

Coal is Australia's largest fossil-energy resource, with reserves of hard coal at end-2009 of 44 billion tonnes, placing Australia in fifth position globally behind the United States, China, India and Russia (BGR, 2010). Australia also has considerable hard coal resources, requiring further exploration to prove up, totalling around 1 575 billion tonnes. Furthermore, Australia's resources of brown coal are the world's seventh-largest. Much of the country's coal resource is of high quality (low ash and sulphur) and low cost, since it can be mined by opencast operations (which account for around 80% of output in 2009) or efficient longwall underground mining. Coal resources are also located close to demand centres and actual or potential routes to export.

These resources already underpin a large coal-mining industry. Production totalled 355 Mtce in 2010 (based on preliminary estimates), made up of 188 Mtce (53%) of steam coal, 148 Mtce (42%) of coking coal, and 19 Mtce (5%) of brown coal. At these production rates, total coal reserves would last around 200 years. Exploration spending directed at expanding coal reserves is currently at record levels, most of it taking place in Queensland and New South Wales. The last time comparable levels of exploration took place was in the early 1980s, in response to a rapid increase in global coal demand, and they resulted in a significant increase in Australia's coal reserves, notably in the Bowen Basin in central Queensland, which has gone on to become a major export region.

The Australian coal industry is gearing up for a massive export-focused expansion over the near to medium term. Already coal output is expanding rapidly, with 20 new mines entering production in 2008 to 2010, increasing annual output by some 54 Mtce. This involved capital spending of some $3.9 billion. Over the same period, spending on rail and port infrastructure totalled $3.1 billion, boosting export capacity by around 80 Mt per year. In the period to 2015, plans are advanced to lift infrastructure capacity still further by between 100 Mt and 160 Mt. There is little doubt that the coal resource can support this level of development and the mining sector is capable of raising the necessary capital and of increasing output to these levels. However, in the past, matching and coordinating timely development of adequate transport infrastructure has proved challenging.

The growth in Australia's coal industry is taking place amid a general resources boom, including natural gas for export as liquefied natural gas (LNG) (see Chapter 4) and ores such as iron, nickel and copper. While such a cyclical boom is not unknown in Australia, this cycle appears to be deeper, more widespread and longer-lasting than previous ones. Managing the impact of such a large development on a small economy is proving difficult, as evidenced by project delays and cost over-runs. These pressures may become particularly severe where developments overlap. One such case is in the Gladstone area, where three large, first-of-a-kind LNG plants, based on coalbed methane, are under construction and where there is already a heavy concentration of construction activity, including expansion of the coal port and other activities. Successfully completing such complex infrastructure projects will require careful co-ordination across industry, state and federal governments, as well as co-operative initiatives to increase the supply of skilled labour.

11

The issue of how best to obtain a suitable return to the community from coal-export earnings, while continuing to encourage investment, has proven difficult. The federal government attempted to introduce a secondary tax on mining profits in 2010, but subsequently modified this to include only some iron-ore and coal mines. A second challenging issue is addressing greenhouse gases through carbon pricing. The government announced in July 2011 an ambitious proposal for an initial carbon tax, starting in 2012 at 23 Australian dollars per tonne, which would rise on an annual basis before being replaced by an emissions trading scheme in 2015. Fugitive emissions from coal mining would be included in the scheme – a first for a large coal-producing and exporting country. The government has earmarked some $1.2 billion over five years in assistance for the most emissions-intensive mines and $64 million to implement carbon-abatement technologies. Carbon pricing would also have a major impact on the power sector and the proposal includes several transition and compensatory measures to assist power generators and other coal users to adjust to higher prices.

Trade

In common with Indonesia, but unlike most other major coal-exporting countries, Australia exports the bulk of the coal produced. Based on preliminary estimates, in 2010, nearly 80% of output was shipped abroad. Annual coal output has risen around 4% per year since 2000, but domestic use by only 1%, resulting in over 60% of growth in exports in the past decade. Australia's total coal exports in 2009 amounted to 256 Mtce. Around 48% of exports are coking coal, the rest is steam coal. Coking coal production is concentrated in central Queensland's Bowen Basin and exports are made via five ports, including Hay Point and Gladstone; steam coal export is centred on the port of Newcastle in New South Wales, with output from the Hunter Valley hinterland. Newcastle is the world's largest coal export-terminal, with annual capacity in 2010 around 115 Mt. Its capacity is set to increase to 170 Mt by 2013.

In the case of coking coal, Australia's exports totalled 143 Mtce in 2010, putting the country in a leading position in inter-regional world trade with over 50% of the market. Exports go mainly to the Asia-Pacific region, including Japan, Korea, China, and India, though some coal is exported elsewhere (for example, 22 Mtce were shipped to OECD Europe in 2008). In the case of steam coal, Australia is less dominant, with exports of 140 Mtce in 2010 – around one-fifth of world steam coal trade. Nonetheless, Australia's exports of steam coal have nearly doubled since 2000 and (unlike coking coal) continued growing strongly even during the recent global economic downturn, expanding by some 18% between 2008 and 2009. Despite being the world's second-largest steam coal exporter behind Indonesia, Australia's volumes meet only 3% of global steam coal demand, since the majority of global steam coal continues to be consumed relatively close to where it is produced, and only a small portion is internationally traded. The main export destination for Australia's steam coal is Japan, which accounted for more than 60% of purchases in 2009. However, sales to other Asian markets, especially China and India, have grown rapidly, more than 40% in 2009 over 2008.

Outlook

Growth in coal use early in the projection period, especially in non-OECD countries, is expected to create significant opportunities for additional Australian exports of both coking and steam coal. Demand from India, where coking coal resources are limited, and China is expected to remain strong. Beyond 2015, the outlook for continuing growth in exports depends highly on the long-term prospects for global coal demand. In the New Policies Scenario, Australia's coal production is projected to increase to a peak of about 390 Mtce around 2020, before slightly declining to just under 380 Mtce in 2035 (Figure 11.24). Exports follow a similar trend, peaking at just over 310 Mtce around 2020, before falling to 300 Mtce in 2035.

Prospects for Australian coal exports are uncertain in the long term, hinging particularly on demand and supply trends in China and India. There is considerable uncertainty surrounding the extent to which both countries will moderate their coal demand growth through greater energy efficiency and diversifying fuel use in their power sector. Similarly there is uncertainty about how far their domestic coal industries will be able to keep pace with demand and the extent to which coal imports will become a structural feature of their markets. Additionally, over the past few years, Chinese and Indian companies have been making large-scale investments in existing or new coal mining operations in Australia, pointing to a likely continued strong participation of their economies in the international coal trade market. India looks a more likely market for increasing Australian imports of both coking and steam coal, especially given transport issues in the south of the country; but China may be less certain. Australia's coal exports to European countries seem likely to fall, as the impact of carbon pricing, and renewable and energy efficiency mandates takes effect, though early closures of nuclear plants may slow this decline.

Figure 11.24 ● Coal production in Australia in the New Policies Scenario

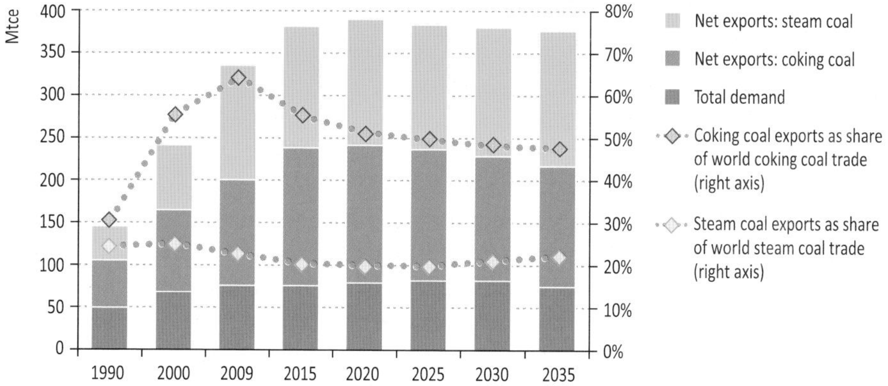

South Africa

Coal has been the cornerstone of the South African economy, providing it with revenues, employment, low-cost electricity and liquid fuels for decades. However, growing concerns about greenhouse-gas emissions from coal use, if translated into firm action, will curb the growth in demand and, therefore, the need to expand production capacity. South Africa has not yet finalised its climate change mitigation policy, but it has pledged to cut emissions below a business-as-usual baseline. In addition, bottlenecks in rail transportation are likely to continue to constrain export sales.

South Africa's coal reserves amount to 28 billion tonnes, equal to about 4% of the world total: no data are available on resources (BGR, 2010). Some 96% of reserves are bituminous coal, 2% anthracite and the balance coking coal. Total reserves are equal to around 110 years of production levels in 2009. There are no significant brown coal resources. The bulk of the reserves and the mines are in the Central Basin, which includes the Witbank, Highveld and Ermelo coal-fields. Other coal reserves in the Limpopo Province are also being explored, with a focus on coking coal.

A large proportion of the steam coal for export and domestic markets is produced from eight mega-mines, each of which produce more than 12 Mtce per year; seven of them are in the Central Basin. Five companies – Anglo-American, Exxaro, Sasol, BHP Billiton and Xstrata – account for about 80% of total South African coal production. Production costs are on the low side, on average, as most coal seams are thick and close to the surface; a quarter of South Africa's bituminous coal is between 15 to 50 metres below the surface and much of the remainder between 50 to 200 metres (Eberhard, 2011). Consequently, roughly half of production comes from opencast mines. Labour productivity is low by international standards, but this factor is outweighed by low wages. Ash content is generally high and sulphur content is typically between 0.6% and 0.7%. Export-grade coal generally requires washing to reduce ash content below 15%. Average heating values are declining, as the better quality reserves in the Central Basin are depleted. Steam coal used for domestic power and synfuel production have much lower calorific values, and higher ash content and are supplied mostly from screened run-of-mine production (although about a third of the coal supplied for electricity production derives from the middlings fraction from coal washing).

The Central Basin has been producing coal for many decades and, with the best seams now depleted, production is expected to reach a peak within the next few years and then begin to decline. This decline is likely to be offset by higher production from the northern Waterberg coal-field, though significant investment in mining, processing and transport infrastructure will be needed.

The main constraint on raising production for domestic uses and, especially, for export is the country's ageing rail infrastructure. Almost all exported coal is transported by rail along a dedicated 580 km line from the Central Basin coal-fields to the Richards Bay terminal on the east coast. The line is owned and operated by the state-owned rail monopoly, Transnet. The line is fully utilised, though more efficient loading operations could, in principle, raise capacity. Even so, any significant increase in supply from the Waterberg field would require the construction of a new line close to 500 km in length, to link up with the existing Central Basin-Richards Bay line. The existing Transnet freight network is capable of moving only small quantities of Waterberg coal. As a result of these constraints, the Richards Bay terminal, which can handle up to 91 Mt per year following an expansion in 2010, has been operating at well below capacity. Coal exports through Richards Bay peaked at 69 Mt in 2005, but declined to 63 Mt in 2010 – less than 70% of capacity. Transnet has targeted an increase in deliveries to 70 Mt in 2011. As a result of the constraints, South Africa's global ranking as a coal exporter has declined, from second behind Australia in 2000 to fifth in 2009, overtaken by Indonesia, Russia and Colombia.

In the longer term, South African coal production will be driven primarily by domestic demand for coal for power generation. This will hinge on the type of capacity that will be built by Eskom – the national power utility – to meet soaring electricity demand. Eskom recently restarted three coal plants that had been shut for a decade, and two new plants are under construction. In 2009, electricity generated from coal represented 94% of South Africa's total power output. In the New Policies Scenario, coal is projected to continue to play a dominant role in the power generation mix, but its share gradually declines to make way for new investments in natural gas-fired plants, nuclear power and renewables. Coal production is projected to rise to a peak of around 230 Mtce around 2020 and then falls back to 210 Mtce by 2035 (Figure 11.25). As the projected growth in demand exceeds that of supply, South Africa's coal exports gradually decline to around 55 Mtce in 2035 from 65 Mtce in 2009.

Figure 11.25 ● Coal production in South Africa in the New Policies Scenario

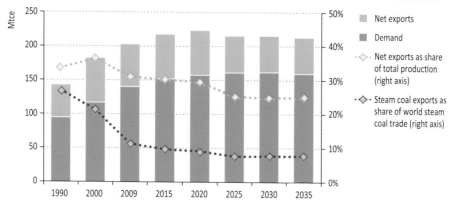

Russia

Russia's coal resources stood at 3.9 trillion tonnes, end-2009, ranking it third in the world after the United States and China (BGR, 2010). Hard coal accounts for about two-thirds and over 40% (70 billion tonnes) of Russia's resources and reserves, respectively. Increased export demand has driven a steady increase in Russia's coal production over the past decade and the country is now the world's third-largest coal exporter. The main export market has been the European Union, which absorbs around 60% of Russia's exports, but there are signs that the focus for future exports may be switching to the Pacific market – to China in particular (see Chapter 8 for a broader analysis of the outlook for Russian coal supply).

The Russian Energy Strategy contains some ambitious targets for increases in coal production over the coming decades, with output targets of between 315 and 375 Mtce by 2030. Our projections are more conservative, as transport costs put a limitation on the ability of Russian coal to compete in the export market and internal demand is gradually reduced by improved efficiency. In the New Policies Scenario, coal production rises from 220 Mtce in 2009 to around 265 Mtce by the early 2020s, and then slowly declines to just below 250 Mtce in 2035. Domestic demand for coal is fairly stagnant over the second-half of the projection period and exports, after an initial increase to around 100 Mtce by 2020, start to dwindle back to 2009 levels of around 80 Mtce, as global coal trade levels out and Russian coal struggles to compete with other lower cost exporters.

Rest of the world

The countries covered in detail above accounted for almost 90% of world coal production in 2010, based on preliminary data. Around 40% of the remaining production comes from OECD Europe. Production there has been in steady decline for many years, with cuts in subsidies and the closure of high-cost mines. At 240 Mtce in 2010, production was about 40% of its 1980 level and 25% lower than in 2000. In 2010, the two key main producers were Poland and Germany, which together accounted for 60% of the region's total

11

coal output. Some European coal producers continue to subsidise their production for social and energy-security reasons, though EU rules on state-aid require subsidies to be phased out by 2018. Turkey also plans to phase out its coal subsidies. The biggest drop in production is expected to occur in Germany, where a deal negotiated in 2007 provides for all subsidies to current production to be phased out by 2018, which will result in most hard coal mines being closed (brown coal mines are generally profitable). Mining operations in some countries – notably Poland and the United Kingdom – are already cost-competitive with imported coal or close to being so. In total, OECD Europe coal production is projected to fall to around 190 Mtce in 2020 and 120 Mtce in 2035 – almost one-half of output in 2009.

Colombia is currently the fifth-largest coal exporter worldwide, but has the potential to become an even more important player. Even though coal deposits are plentiful and mining costs are generally low, exports of around 60 Mtce have remained significantly below their maximum potential during recent years, due to limitations in port capacity. Coal production is mostly controlled by large, vertically-integrated global mining companies which own production, rail and port capacity. Alongside South Africa, Colombia has been the dominant supplier in the Atlantic Basin, exporting mostly steam coal to US and European power generators. In the first half of 2010, Colombia exported around 6 Mtce to southeast Asia, due to strong demand and favourable freight rates; but Colombia is not at present in an competitive position to serve the growing Asian markets. The eastern route to Chinese ports is almost double the length of the route South African coal shipments have to traverse. The Pacific route is 4 500 miles shorter but, even disregarding high transit fees, the Panama Canal has been a bottleneck. In the New Policies Scenario, Colombian coal exports sharply increase from about 60 Mtce in 2009 to 85 Mtce in 2015 (Figure 11.26). This increase is driven by comparatively low mining costs, strong private investment in mining and export infrastructure and governmental interventions further to expand coal exports. Additionally, Colombia's transport disadvantage to supply Asian coal demand is expected to diminish: the expansion of the Panama Canal is projected to be completed by 2014. This will allow bulk carriers up to double the maximum current size to pass through the canal, allowing coal to be shipped westwards at significantly lower cost. Also, China and Colombia agreed in February 2011 to build a 220 km railway line to allow goods, including coal, to be moved from the Colombian port city of Cartagena to the Pacific Ocean, providing an alternative outlet for coal shipments. In the second-half of the New Policies Scenario, export growth saturates in line with developments in global demand and by 2035 Colombia's exports are just below 100 Mtce.

Venezuela is the other significant coal producer in Latin America. There are two major mining operations: Mina Norte and Paso Diablo. The Venezuelan mining minister announced in 2009 that the concessions granted for Mina Norte (which expires in 2011) and Paso Diablo (which expires in 2013) will not be renewed. Granting of new concessions has been suspended for environmental and political reasons, and the government will henceforth work under operating contracts or joint ventures between the state and private partners. Production is not projected to increase substantially in the New Policies Scenario, unless there is a significant shift in policy and investment is stepped up.

Figure 11.26 ● Hard coal net exports from selected smaller exporters in the New Policies Scenario

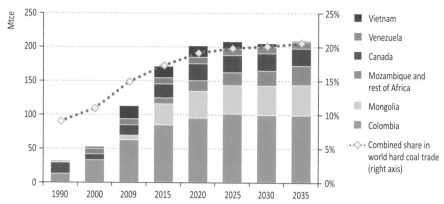

Mongolia has seen rapid growth in its brown and coking coal industry, with production increasing more than seven-fold since 2000, to almost 20 Mtce in 2010 (based on preliminary estimates). Production rose sharply, by 9 Mtce, in 2010 alone. The country has large resources and, given its close proximity to strong Chinese demand, it could further increase production, both for export and to meet growing domestic needs, mainly for power generation. The government has indicated that exports to China could reach 50 Mt by 2015, but such an expansion hinges on large investments in developing new mines and building transportation infrastructure. Access to water, which is scarce in areas where the bulk of resources are located, could be a major constraint. By 2035 Mongolian, coal production is projected to reach 50 Mtce. Mongolia together, with Mozambique, become two new important suppliers of coal (especially coking coal) to the international market.

Mozambique is a promising new player in international hard coal trade. Generally, Mozambique, which enjoys favourable geology, is expected to produce relatively low cost coal, of high quality (particularly coking coal). Three projects Benga, Moatize and Zambeze, all located near Tete in the northwest of the country, are currently being developed. First coal deliveries from the Benga and Moatize opencast mines are already scheduled for 2011/2012. These two projects together are planned to reach a capacity of more than 30 Mt/year when fully operational. Coal will be transported via the 665 km long Sena railway line, which links the Tete coal fields to the port of Beira. This port is currently expanded to handle 6 Mt/year and the government is planning a new terminal at Beira that should be able to handle between 18 Mt/year to 24 Mt/year. From a freight cost perspective, Mozambique is ideally located to serve coal demand in both, the Atlantic and Pacific Basin markets, (especially India). With new operations reaching full capacity over the coming decade, Mozambique's coal exports increase steeply until 2020 in the New Policies Scenario. With the projected slow-down in global coking coal demand thereafter, Mozambique's exports grow at a lower rate in the longer run. Yet, Mozambique's export and production growth rates remain among the highest in the world.

11

PREFACE

Part D deals with three disparate issues, each of major significance.

Nuclear energy is going through a new period of reappraisal. Though the assumptions embedded in the main scenarios of *WEO-2011* already allow for declared revisions of government policy in the wake of the accident at Japan's Fukushima Daiichi nuclear power plant, it pays to consider the implications if there were to be a much more severe cut in the nuclear component of energy supply. That is the focus of Chapter 12. It examines in detail what a smaller role for nuclear power would mean, including the implications for achieving the 450 Scenario, which depends on nuclear energy as a component of low-carbon electricity generation.

Chapter 13 develops further an ongoing theme of the *WEO* since 2002 – bringing modern energy to all the world's people, in terms of their access to electricity and modern cooking facilities. Starting from what is already being done, the chapter evaluates the scale of the remaining challenge, on the basis of the projections of the New Policies Scenario, and the cost of achieving universal access by 2030. It then concentrates on matching the investment needs of different aspects of the problem to the actual and potential funding sources.

Chapter 14 looks at an entirely different aspect of energy funding – subsidies to fossil fuels and renewables. The sums are much greater than those needed to alleviate energy poverty, outlined in Chapter 13. Subsidies to fossil fuels often fail to meet their stated purpose or could be applied much more effectively in different ways and there is a G-20 commitment to phase out inefficient fossil fuel subsidies. So the chapter examines what is happening and how subsidy reform might best be implemented. Financing renewable energy development is seen in a different light. Subsidies in this sector exist to overcome technological and market barriers, with a view to their eventual phase-out as the technology becomes established in the market. The extent of this support and its prospective evolution is measured.

CHAPTER 12

THE IMPLICATIONS OF LESS NUCLEAR POWER
How would it affect energy markets and climate trends?

H   I   G   H   L   I   G   H   T   S

- The Low Nuclear Case examines the implications for global energy balances of a much smaller role for nuclear power. The lower nuclear component of electricity supply is not a forecast, post Fukushima, but an assumption adopted for the purpose of illustrating a global energy outlook in such a low nuclear world.

- In the Low Nuclear Case, the total amount of nuclear power capacity falls from 393 GW at the end of 2010 to 335 GW in 2035, just over half the level in the New Policies Scenario. The share of nuclear power in total generation drops from 13% in 2010 to just 7% in 2035, with implications for energy security, diversity of the fuel mix, spending on energy imports and energy-related CO_2 emissions.

- In 2035, coal demand increases by 290 Mtce compared with the New Policies Scenario, over twice the level of Australia's current steam coal exports. The increase in gas demand of 130 bcm is equal to two-thirds of Russia's natural gas exports in 2010. The increase in renewables-based generation of about 550 TWh is equal to almost five-times the current generation from renewables in Germany.

- Energy import bills rise as higher coal and natural gas imports drive up international energy prices. In 2035, net-importing regions collectively spend an additional $90 billion, or 12%, on gas and coal imports over the New Policies Scenario. Energy-related CO_2 emissions rise with increased use of fossil fuels in the power sector, ending up some 2.6% (0.9 Gt) higher in 2035 than in the New Policies Scenario.

- These aggregate numbers mask more significant impacts in countries with limited indigenous energy resources that were projected to rely relatively heavily on nuclear power, such as Belgium, France, Japan and Korea. Similarly, such a slowdown in the growth of nuclear power would make it more challenging for emerging economies, particularly China and India, to satisfy their rapidly growing electricity demand.

- The Low Nuclear 450 Case, a variant of the 450 Scenario, draws a plausible trajectory which achieves the climate goal of limiting temperature increase to 2°C with a lower nuclear component. Additional investment needs in 2011 to 2035 increase by $1.5 trillion (or 10%), with renewables in particular being called upon to make a larger contribution to abating CO_2 emissions. Following this trajectory would depend on heroic achievements in the deployment of emerging low-carbon technologies, which have yet to be proven. Countries that rely heavily on nuclear power would find it particularly challenging and significantly more costly to meet their targeted levels of emissions.

Why the Low Nuclear Case?

The devastating earthquake and resulting tsunami that struck Japan in March 2011 have disrupted the country's energy sector and had repercussions for energy markets around the world. In Japan, many oil refineries as well as oil, gas and electricity distribution networks were damaged. Several power plants in the northeast were shut down, while others were damaged and could not be restarted quickly. The worst damage occurred at the Fukushima Daiichi nuclear power plant, located some 200 kilometres (km) northeast of Tokyo (Box 12.1). In the hours and days that followed the earthquake and tsunami, four of the plant's six reactors were severely damaged, resulting in off-site releases of radioactivity.

Apart from the immediate and medium-term effects on Japan's ability to meet its electricity needs, the incident has raised new doubts about the safety risks associated with nuclear energy, inevitably raising questions about the future role of nuclear power in the global energy mix. Opposition to nuclear power in Japan has increased, casting doubt over the existing plans to expand nuclear capacity. A few countries have already changed their policies, either abandoning previous steps towards building new nuclear plants, as in Italy, or accelerating or introducing timetables for the phase-out of nuclear plants, as in Germany and Switzerland. These factors, alongside expectations of lower natural gas prices, have led us to make a downward revision to the projected growth in nuclear power in the New Policies Scenario compared with *WEO-2010*.

With efforts continuing to make the Fukushima plants safe and to appraise the full implications of what occurred when the tsunami struck, any assessment of the long-term consequences can only be provisional and policy conclusions should be confined only to the actions necessary in the short term. Nonetheless, the prospects for nuclear power are inevitably now much more uncertain. Additional countries may act to reduce their dependence on nuclear power. Investors may demand higher risk premiums on lending or require stronger guarantees and government incentives, making the financing of new plants more challenging. Tighter safety regulations could push up the cost of new reactors and necessitate unforeseen investment in the existing fleet. At a minimum, investment in new nuclear capacity may be delayed. Some existing nuclear plants may have a shorter operating lifespan either as a result of early retirement or increased reluctance on the part of the authorities to grant life extensions. Greater public resistance to nuclear power can be expected in countries that have been trying to revive dormant nuclear programmes or in those trying to deploy nuclear power for the first time. These conditions are more likely to arise in OECD countries, where existing plants are concentrated, but they may also slow the expansion of capacity in non-OECD countries.

In view of these uncertainties, we have prepared a special case – the Low Nuclear Case – which does not attempt to predict international reactions, but to make them better informed by analysing the possible implications for global energy balances of a much smaller role for nuclear power than that projected in any of the three scenarios presented in this *Outlook*. This chapter looks first at the role of nuclear power in the energy mix today. It then reviews the immediate impact of Fukushima on global energy markets and on nuclear policy in particular, and considers how recent events may alter the economics of nuclear power.

On the basis of a necessarily arbitrary assumption about the extent of the loss of nuclear capacity, compared to that in the New Policies Scenario, the chapter then presents detailed projections for the cost of energy, the diversity of the fuel mix and CO_2 emissions in the Low Nuclear Case, compared with the New Policies Scenario. The final section considers what the Low Nuclear Case would mean for the trajectory to be followed in order to meet the climate goal of limiting the average global temperature increase to 2° Celsius, as in our 450 Scenario, making some preliminary comments on its feasibility.

Box 12.1 ● **The Fukushima Daiichi nuclear power station**

On 11 March 2011, a magnitude-9.0 earthquake on the Richter scale – the most powerful in Japan's recorded history – occurred off the country's northeast coast, triggering a deadly tsunami. There were some 26 000 casualties, and many cities, towns and villages were devastated. All eleven reactors that were operating in the northeast region shut down as designed. However, the massive tsunami engulfed the Fukushima Daiichi nuclear power station, prompting a series of events that led to a failure of three of the plant's six reactors and a leak of radioactive material.

The tsunami flooded the entire Fukushima Daiichi plant – which had already been cut off from the external power grid by the earthquake – destroying the back-up electricity system that powered pumps to cool the nuclear fuel rods. The reactor cores in three of the units began to overheat and water began to evaporate, increasing pressure and lowering the water level inside the reactor vessels, thereby exposing the fuel rods and resulting in the production of hydrogen. The hydrogen subsequently escaped from the reactors and primary containment vessels where it reacted with oxygen, resulting in explosions that damaged the outer buildings. In a fourth unit that had been in cold shutdown prior to the earthquake, the cooling system for the spent fuel pool was damaged, also leading to a hydrogen explosion. These events led to off-site releases of radioactivity.

Some 78 000 people living within a 20 km radius of the plant were evacuated and those within 30 km were asked to stay indoors. Radiation levels in the vicinity of the plant soared initially, but then declined. At the time of writing, efforts to cool the crippled reactors and stem leakages of radioactive material were continuing. They are scheduled to be brought to "cold shutdown" by early 2012.

12

The role of nuclear energy today

The development of nuclear power began almost 60 years ago with the commissioning of the Obninsk Nuclear Power Plant near Moscow in 1954 (Figure 12.1). This signalled the start of a growth period for the global nuclear industry, with an average of seven reactors being built annually up to 1965. Construction accelerated in the mid-1960s, reaching a peak of 37 construction start-ups in 1968 and 1970, and gained further momentum from the first oil shock of 1973/1974, as countries sought to reduce dependence on oil-fired power. The nuclear industry then entered a major downturn in the late 1970s, triggered by soaring costs and delays, coupled with safety concerns following the Three Mile Island accident in 1979 in the United States. This accident sparked public protest, slowed the regulatory process and led many planned projects to be cancelled or suspended. In 1986, the Chernobyl nuclear power plant accident in Ukraine – the most serious in the history of the industry – further depressed activity and prompted several countries to impose restrictions on existing or new nuclear power plants.

Figure 12.1 ● **Nuclear reactor construction starts, 1951-2011**

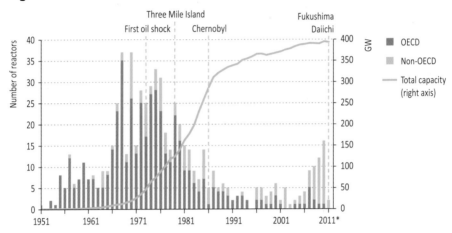

*Data as of 31 Aug 2011.

Outside Japan and Korea, there was little growth in nuclear capacity in the 1990s due to concerns about public acceptance and safety, construction delays, higher than expected construction costs at some nuclear plants and a return to lower fossil-fuel prices. However, since the mid-2000s, global nuclear capacity had been on an upward trend, largely because of rapid development in China, but also power uprates (a process which enables the power output of a reactor to be increased) and life extensions at existing sites in OECD countries. Construction began on 16 new plants in 2010, the largest number since 1980; all but one were in non-OECD countries. This renewed interest in nuclear power stemmed from the need to cost-effectively satisfy rapidly growing electricity demand in the emerging economies, as well as efforts to achieve energy and environmental policy objectives, including mitigating greenhouse-gas emissions and providing a secure, diversified and low-cost electricity supply.

In 2010, nuclear power plants supplied 13% of the world's electricity, down from a peak of 18% in 1996. At the start of 2011, a total of 30 countries around the world operated 441 nuclear reactors, with a gross installed capacity of 393 gigawatts (GW) (374 GW net) – 83% being in OECD countries. Another 17 countries have announced their intention to build reactors. New construction is now overwhelmingly centred in non-OECD countries, where 55 of 67 new reactors are being built (Table 12.1). China alone accounts for 63%

of the construction starts in 2010, followed by Russia, with 13%. The average age of the world's, operating nuclear power plants was 26 years at the end of 2010 and has been steadily increasing with the slowdown in new construction and as more reactors have been given life extensions.[1]

Table 12.1 ● **Key nuclear power statistics by region, end-2010**

	Operational reactors	Installed gross capacity (GW)	Average fleet age (years)	Share of total generation	Reactors under construction
OECD	**343**	**326**	**27**	**21%**	**12**
United States	104	106	31	19%	1
France	58	66	25	75%	1
Japan	54	49	25	27%	2
Germany	17	21	28	23%	0
Korea	21	19	17	31%	5
Canada	18	13	26	15%	0
United Kingdom	19	11	29	16%	0
Other	52	40	28	24%	3
Non-OECD	**98**	**68**	**21**	**4%**	**55**
Russia	32	24	28	15%	11
Ukraine	15	14	22	48%	2
China	13	11	8	2%	28
India	19	5	17	3%	6
Other	19	14	24	9%	8
World	**441**	**393***	**26**	**13%**	**67**

*393 GW of gross capacity is equivalent to 374 GW of net capacity.

Sources: International Atomic Energy Agency Power Reactor Information System; IEA databases.

What might cause expansion of nuclear capacity to slow?

Possible changes in policy concerning nuclear power

While the long-term implications of Fukushima Daiichi remain very uncertain, it is clear that this is a critical point for the industry as many governments are now reassessing their plans for the use of nuclear power (Table 12.2). Risks associated with the safety of nuclear energy have been brought to the fore and within weeks of the incident most countries with nuclear programmes had announced comprehensive safety reviews and some had temporarily suspended approvals for new projects. While most of these reviews were continuing at the time of writing, they are

---

1. The fleet is generally older in OECD regions (average 27 years) than in non-OECD regions (average 21 years) because the latter account for most of the reactors built in recent years.

expected to lead at least to regulatory changes that slow plans for expansion and mandate additional investment to improve safety at existing plants. Before the safety reviews have been completed, some plants, deemed particularly at risk because of concerns about their age or location, have been closed. A few countries have decided that they will completely abandon the use of nuclear energy. Some countries that were working towards introducing nuclear power for the first time, or to revive dormant nuclear programmes, have announced that they will no longer do so or that plans will be delayed. On the other hand, several countries have re-affirmed the importance of nuclear power in their energy mix, including some that have firm plans to substantially increase nuclear capacity in order to meet rising baseload demand or to reduce their dependence on fossil-fuel imports and use.

Table 12.2 ● Recent announcements by selected countries regarding nuclear power

	Capacity (GW)	Mid-2011
European Union	127	Announced plans to "stress test" all 143 plants in its 27 countries.
United States	106	Continues to support nuclear power while stressing safety as paramount concern.
France	66	Continues to support nuclear power while carrying out European Union stress test and looking to increase the role of renewables.
Japan*	46	Announced a review of the existing plan for nuclear power to account for 53% of electricity output by 2030.
Russia	24	Affirmed plan to double nuclear capacity by 2020; undertaking comprehensive safety review.
Korea	19	Affirmed plan to continue expansion of the nuclear industry and to conduct safety checks.
Germany**	13	Immediately shut reactors operational before 1980 and announced that all other reactors would be closed by 2022, effectively reversing a decision taken in 2010 to delay a previous phase-out plan agreed in 2001.
China	12	Temporarily suspended approval of new nuclear reactors, but affirmed 12th Five-Year Plan target to start construction of an additional 40 GW of nuclear capacity in 2011 to 2015.
United Kingdom	11	Affirmed commitment to nuclear power by announcing plans to build eight new reactors by 2025.
India	5	Affirmed plans to boost nuclear capacity to 63 GW by 2032 and to review safety.
Czech Republic	4	Affirmed plans to build two new units at its Temelin nuclear power station.
Switzerland	3	Announced plans to close its five nuclear reactors by 2034.
Turkey	0	Affirmed no change to plans to commission the first of four planned reactors of 1.2-GW by 2018.
Italy	0	A referendum in June 2011 imposed a permanent ban on the reintroduction of a nuclear power programme.
Poland	0	Affirmed plans to commission its first reactor by 2020.
Indonesia, Thailand	0	Delayed, or is considering delaying, their first nuclear power plant projects until after 2020.
Saudi Arabia, UAE, Vietnam	0	Each affirmed no change to plans to build their first nuclear power plants.

*Incorporates the permanent shutdown of Fukushima Daiichi units 1-4. **Incorporates 9 GW of capacity closed in the first half of 2011.

SPOTLIGHT

How did Fukushima Daiichi impact Japanese and global energy markets ?

The devastating earthquake and resulting tsunami that struck Japan in March 2011 led to the Fukushima Daiichi accident and resulted in the immediate loss of 9.7 GW of nuclear capacity, which was automatically shut down. It also led to the temporary loss of 12 GW of coal-, gas- and oil-fired capacity. In total, this amounted to about 8% of the country's total power production capacity. In early-May 2011, the 3.6 GW Hamaoka nuclear power plant was also temporarily shut down, until safety measures could be implemented to protect the plant from tsunamis and earthquakes. Efforts to bring the remaining plants (other than the Fukushima Daiichi units 1-4, which are permanently shutdown) back online are continuing.

The shortfall in baseload nuclear power capacity necessitated additional oil- and gas-fired generation, which increased Japan's short-term reliance on fossil-fuel imports. The incremental demand for oil in Japan's power sector in 2011 is estimated to be 230 thousand barrels per day, while demand for liquefied natural gas (LNG) is expected to rise by 10 billion cubic metres (bcm).

The need for additional imports into Japan has been limited, since many factories and other parts of the infrastructure were out of operation following the earthquake and tsunami, thus removing a significant portion of energy demand. The scarce availability of spare coal-, oil- and gas-fired capacity was a constraint and the authorities stepped up efforts to encourage energy conservation and, in some cases, rationed supplies to cope with shortages. Businesses and households in certain regions have been encouraged to cut electricity consumption by as much as 15% in order to reduce the strain on power grids.

The impact on global oil and gas was limited, as the volumes involved were not large relative to global supplies: the increase in Japanese imports equalled only about 0.2% of global oil supply, and 0.4% of natural gas supply. International oil, gas and coal prices increased in the immediate aftermath of the incident, but fell back later. Fukushima Daiichi no doubt contributed to this upward pressure on energy prices, but other factors – including strong demand growth in emerging economies and supply problems, notably the loss of Libyan oil production – played a more major role.

In *Japan*, Fukushima Daiichi has thrown the long-term role of nuclear power into doubt. In 2010, Japan was the world's third-largest producer of electricity by nuclear power and the government had ambitious plans to expand the nuclear component of the country's energy mix. The pre-crisis official target calls for nuclear to provide 53% of the country's total power supply by 2030 (up from about 27% in 2009) and, before Fukushima, plans were in place to construct nine new reactors by 2020 and another five by 2030. Public resistance has now hardened against these plans and the government has announced that it will revise them. It has announced immediate measures to boost nuclear safety and plans to undertake a stringent safety assessment at each reactor to check its capacity to withstand extreme natural events. Only 11 of Japan's 54 nuclear reactors were in operation as of mid-

September 2011. All reactors are required to be shut down at least once every thirteen months for regular safety inspections and there is a risk that the number online could fall further if delays occur. The government has indicated that its long-term plans may include deploying more renewable energy, stepping up measures to improve energy efficiency and to encourage cleaner use of fossil fuels.

Outside Japan, the most significant impact of Fukushima Daiichi has been in *Germany*, where 17 reactors were operating in 2010 with a total gross capacity of 21 GW, which provided 23% of the country's electricity. Within days of the accident, the government ordered the suspension of operations at seven of its older nuclear plants, removing 7 GW of capacity and it was also decided that another older plant, which had already been shut down, should not be re-started. There followed a decision in May 2011 to completely abandon nuclear power, in a step-by-step process that is to culminate in 2022, reversing a decision taken in 2010 to deploy a previous phase-out programme. This makes Germany the largest user of nuclear power to discontinue its use and close reactors before the end of their economic life.

Responses in other countries have varied. *Switzerland* has announced its intention to phase out nuclear power by 2034. In *Italy*, a referendum in June 2011 resulted in the rejection of a proposal to lift the indefinite ban on nuclear power. And *Thailand* has announced a three-year delay in plans to commission its first reactor, which is now due to come online by 2023.

Elsewhere, several countries, while announcing plans to review safety regulations that may delay the introduction or expansion of new capacity, have also re-affirmed their overall commitment to nuclear power:

- In the *United States*, the world's biggest nuclear power producer, the Nuclear Regulatory Commission has launched a comprehensive review of the country's nuclear facilities to identify lessons that need to be applied as a result of the accident in Japan. The first part of the safety review, aimed at identifying immediate changes needed to maintain safety during emergencies such as earthquakes, hurricanes or power outages, was completed in June 2011. The second part will examine other changes that may be needed.

- The *European Union* decided in March 2011 to conduct stress tests on all of its 143 reactors. The tests will cover "extraordinary triggering events, like earthquakes and flooding, and the consequences of any other initiating events potentially leading to multiple losses of safety functions requiring severe accident management." The results of the stress tests are expected in 2012, when individual member states will have to decide how to respond should any reactors fail the tests. It is expected that any such reactors will be shut down and decommissioned if upgrades prove to be technically or economically impractical. The European Union has also asked neighbouring countries to commit to implementing the same stress tests on their own nuclear plants.

- In *France,* the world's second-largest nuclear-power producing country (and the most nuclear dependent country with from 75% to 80% of its power generation from nuclear), the Nuclear Safety Authority has been charged with carrying out safety assessments of the country's 58 reactors. The government has confirmed its intention to increase the share of renewables in the electricity generation mix. In July 2011, EdF announced that its Generation III European Pressurised Reactor (EPR), being built in Flamanville, would be delayed by two years, stemming in part from the need to carry out new safety tests.

■ In the *United Kingdom*, an interim report by the country's chief inspector of nuclear installations concluded that there is no need to alter the operation of its nuclear plants or change plans for adding new capacity. The final report will be published in third-quarter 2011. In June 2011, the government announced a list of eight sites deemed suitable for new nuclear plants to be built by 2025.

■ *Russia* has announced that it will not be altering its nuclear power expansion plans while instructing Rosatom, the state-owned nuclear corporation, to review nuclear plant safety (see Chapter 8 for our projections for nuclear power in Russia).

■ *China,* which has the world's most ambitious nuclear expansion plans, with 28 reactors under construction in 2010, initially froze approvals of new projects and ordered safety checks on existing plants and those under construction. In June 2011, all of China's operating nuclear reactors were reported to have passed their safety inspections.

■ *India,* which is actively promoting the role of nuclear power in meeting its growing electricity demand, has ordered emergency safety checks to be carried out on all nuclear plants. India has signalled that there will be no change to its target of quadrupling nuclear capacity to 20 GW by 2020 and reaching 63 GW of installed capacity by 2032.

Possible changes to the economics of nuclear power

Electricity utilities choose between various types of plants – nuclear reactors, fossil-fuel thermal power stations and renewable energy technologies – on the basis of the relative lifetime costs at which they can generate electricity, evaluating this within their particular portfolio of plants and taking into account relative risks. In general, the factors that influence the lifetime costs include the required level of investment, financing costs, construction times, expected utilisation rates, operational flexibilities, fuel costs, penalties for CO_2 emissions and the cost for plant decommissioning. The cost of generating electricity from newly built power plants varies significantly across countries and regions according to the characteristics of each market. Relative costs among the different technologies tend to fluctuate over time. Worldwide, no single technology currently holds a consistent economic advantage:

■ Nuclear plants are characterised by very large up-front investments, technical complexity, and significant technical, market and regulatory risks, but have very low operating costs and can deliver large amounts of baseload electricity while producing almost no CO_2 emissions. Typical construction times are between five and eight years from first concrete poured.

■ Coal-fired plants have moderate capital cost and relatively low operating costs as coal is typically cheaper on an energy equivalent basis, compared to other fossil fuels. The competitiveness of coal plants is quickly reduced if a price on carbon is in place and starts rising (see Chapter 10).

- Gas-fired plants typically have the lowest capital costs, short construction times and high operational flexibility; but, unless gas supplies are available at low cost, they have high operating costs. Gas plants have the lowest CO_2 emissions of all fossil-fuel based technologies, typically half of those of coal-fired plants.

- The competitiveness of various renewables-based electricity generating technologies typically depends on favourable local conditions, *e.g.* wind or solar availability, and policy support. Certain technologies often have low load factors, which mean less electricity is generated from a given capacity. Rising fossil-fuel and CO_2 prices, technological advances and economies of scale with wider deployment are expected to make renewables-based systems increasingly cost-competitive in coming decades.

Post-Fukushima Daiichi, the relative economics of nuclear power compared with other generating technologies may deteriorate. Finance providers may demand tougher financing conditions, driving up the cost of capital, and some may decide to discontinue investing in nuclear projects altogether. More stringent safety regulations may lengthen lead times for construction and increase construction and operating costs, as could more vigorous action by opponents of nuclear power.

In the New Policies Scenario, overnight costs of new nuclear power plants in OECD countries range from \$3 500 to \$4 600 per kilowatt of capacity.[2] The wide range in costs reflects the importance of local conditions, including the structure of the industry. At the lower end of these estimates, the overnight cost is almost five-times that of building a natural gas plant. Moreover, financing costs, which are not included in overnight costs, account for a significant share (sometimes as much as half) of the total cost of building a nuclear plant. Financiers of nuclear plants demand interest rates sufficient to compensate for the risk that the utility may not be able to repay loans in full, because of delays in construction, abandonment of the project prior to completion or political or regulatory factors that lead to higher costs or early closure. The cost of financing is often significantly lower in emerging economies, as plants are usually built by publicly-owned utilities with access to cheap government-backed finance. Such projects are usually centrally planned, which may further reduce construction and operating risks. This is the case in China. In the United States, the federal government recently offered loan guarantees for two new reactors at Georgia Power's Vogtle Plant.

The cost of building a nuclear power plant is sensitive to the construction time: the longer the project takes to complete, the longer interest payments accrue without any offsetting positive cash flow. The two evolutionary power reactors (EPRs) currently being built at Taishan in China are expected to take around four years to build. By comparison, the Flamanville reactor in France is expected to take some eight years. Construction delays, which are common in new nuclear programmes or when building non-standard designs, can greatly increase costs. In liberalised markets, it may not always be possible to recoup the cost increases through higher tariffs.

---

2. The overnight cost is the cost that would be incurred if the plant could be built "overnight", so excludes any financing costs.

Our projections of future nuclear power plant construction are based primarily on assumptions about the capital costs of new plants and the cost of capital. Electricity generating costs are particularly sensitive to changes to these key parameters. For example, relative to a base case nuclear plant that generates electricity at a LRMC of $66/MWh, increasing the overnight capital cost by $1 000/kW, would push up the LRMC by 24% (Figure 12.2). Similarly, an increase of two percentage points in the cost of capital would push up the LRMC by 19%. Cost increases of such a magnitude could certainly switch investment to other forms of electricity generation, especially in liberalised markets.

Figure 12.2 • **Sensitivity of long-run marginal cost of nuclear generation to various parameters**

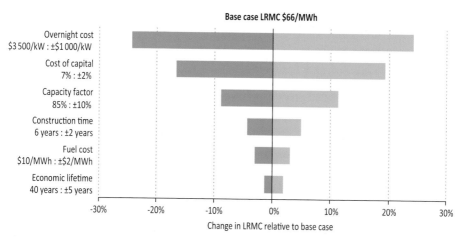

Source: IEA analysis.

Lifetime extensions often provide a means of maintaining nuclear power capacity at relatively low cost. When first built, most of today's nuclear power plants were expected to operate for between 25 and 40 years. However, licence renewals to extend operating lifetimes have become commonplace. For example, of the 104 reactors in the United States, 66 have had their operating lives extended from the original 40 years to 60 years (WNA, 2011). In France, three of the country's reactors have received lifetime extensions over the last decade, from 30 to 40 years, and many are now seeking extensions to 50 years. Experience of lifetime extensions in the emerging economies is more limited, as fewer reactors have yet come to the end of their design life. Extending the life of a nuclear plant or modifying it to increase output is usually more economic than building a new facility, as the upfront capital costs have already been depreciated and output increases can often be achieved at relatively low marginal cost, even though utilities typically have to make significant investments to upgrade plant safety and replace worn equipment to secure lifetime extensions. In the United States, for example, between 2000 and 2008 nuclear output increased 5%, although no new reactors entered service. But such extensions may become uneconomic if regulators make them contingent on very stringent conditions.

Implications of the Low Nuclear Case for the global energy landscape

It is still early to arrive at a definite judgment on the extent of any reduction in nuclear power generation which might result from Fukushima Daiichi. The Low Nuclear Case attempts to make no such judgment. Rather, it is intended to illustrate how the global energy landscape would look with a lower component of nuclear supply. The assumptions about the extent of the lost nuclear capacity are necessarily arbitrary. We have modelled the impact of the following assumptions about nuclear power, keeping all other assumptions the same as in the New Policies Scenario (Table 12.3):

■ In OECD countries, no new reactors are built beyond those already under construction.

■ In non-OECD countries, only 50% of the capacity additions projected in the New Policies Scenario proceed as planned, although all those already under construction are completed.

■ Reactors built prior to 1980 are retired after an average lifetime of 45 years (50 years in the New Policies Scenario).

■ Reactors built from 1980 onwards are retired after a lifetime of 50 years on average (55 years in the New Policies Scenario).

Table 12.3 ● **Key projections for nuclear power in the New Policies Scenario and the Low Nuclear Case**

	Low Nuclear Case			New Policies Scenario		
	OECD	Non-OECD	World	OECD	Non-OECD	World
Gross installed capacity (GW)						
in 2010	326	68	393	326	68	393
in 2035	171	164	335	380	252	633
Share in electricity generation						
in 2010	21%	4%	13%	21%	4%	13%
in 2035	9%	5%	7%	21%	8%	13%
Gross capacity under construction (GW)*	14	54	69	14	54	69
New additions in 2011-2035 (GW)**	6	84	91	111	167	277
Retirements in 2011-2035 (GW)	176	42	218	71	36	107

*At the start of 2011. **Includes new plants and uprates, but excludes capacity currently under construction.

Power sector

In the Low Nuclear Case, the total amount of nuclear power capacity drops from 393 GW in 2010 to 335 GW in 2035 – a fall of 15% – as a result of the slower rate of new construction and a bigger wave of retirements (Figure 12.3). This contrasts with an increase to 633 GW in the New Policies Scenario. In other words, nuclear capacity is little more than half that projected in the New Policies Scenario. The disparity begins to widen just after 2020 and then accelerates. Consequently, the share of nuclear power in total power generation drops from 13% in 2009 to 12% by 2020 and to 7% by 2035. Capacity grows by 97 GW, or over 140%, in non-OECD regions, but this is more than offset by the big fall in OECD regions of 155 GW, or almost 50%. Retirements of nuclear power capacity amount to 218 GW over the *Outlook* period, equal to 55% of the currently installed capacity. Some 11% of existing nuclear capacity in the OECD is retired by 2020 and over 50% by 2035. The fall in capacity, compared with the New Policies Scenario, results in a corresponding fall in nuclear power generation, from around 2 700 terawatt-hours (TWh) in 2009 to around 2 450 TWh in 2035 (compared with around 4 660 TWh in the New Policies Scenario).[3]

Figure 12.3 ● **Nuclear power capacity in the Low Nuclear Case**

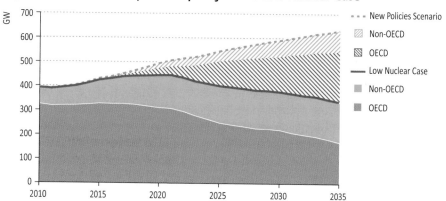

The lower level of nuclear generating capacity in the Low Nuclear Case is offset by a corresponding increase in the combined capacity of plants based on coal, natural gas and renewable energy (Figure 12.4). Globally, in 2035 the share of coal-fired generation reaches 36%, compared with 33% in the New Policies Scenario (from 41% in 2009), while that of gas reaches 24% compared with 22% in the New Policies Scenario (and 21% in 2009). These figures reflect the installation of some 80 GW of additional coal-fired electricity generation capacity and 122 GW of additional gas-fired electricity generation capacity. In keeping with announcements made by some governments since Fukushima Daiichi, it is assumed that more ambitious policies to promote renewables are implemented to offset part of the shortfall in nuclear power. This results in the share of renewables in the generation mix rising to 32% in 2035, compared with 31% in the New Policies Scenario (and 19% in 2009), reflecting the installation of over 260 GW of additional capacity, leading to an increase of about 550 TWh of renewables generation. The use of oil remains very small, with little difference in 2035 compared with the New Policies Scenario.

3. For modelling purposes, we have assumed that global electricity demand is the same as in the New Policies Scenario throughout the *Outlook* period. In reality, the increase in electricity prices that would result from a lower share of nuclear power in the generation mix is likely to lead to slightly lower demand.

Will Fukushima Daiichi affect the industry as severely as Three Mile Island and Chernobyl?

In considering the long-term implications of Fukushima Daiichi, accidents at Three Mile Island in 1979 and Chernobyl in 1986 are obvious points of reference. These accidents profoundly affected the trajectory of nuclear power, derailing new builds globally over the following decades. New construction starts fell from an average of 26 per year in the 1970s to just 7 in the 1980s and 1990s. Will nuclear power development after Fukushima Daiichi experience a similar slowdown?

The downturn for nuclear power following the Three Mile Island and the Chernobyl accidents can be partly attributed to the two accidents. Low energy prices throughout much of the 1980s and 1990s also reduced the financial incentive to accept the risks of the higher capital investment needed to build nuclear plants, whereas the first oil price shock had provided considerable impetus for the soaring level of construction in the 1970s. Moreover, the liberalisation of power markets in many OECD countries transferred the financial risk associated with energy sector investment from the public sector to private industry. Given the prevailing market conditions and without government support, utilities generally preferred lower-risk options (such as coal- and gas-fired plants), which have lower upfront costs, take less time to build and recover costs more quickly, despite their higher unit operating costs.

Today, the key drivers and players expected to underpin the growth in nuclear power are significantly different. Emerging economies, led by China and India, dominate future prospects. They need to utilise all options to meet their rapidly growing electricity demand, and there is no indication that they are contemplating ruling out nuclear power. The return of high energy prices and heightened geopolitical risks in the Middle East and North Africa may increase the value of generating solutions which are independent of large-scale and continuing fuel imports, provided they are safe. Climate change is also a driving force behind nuclear power development, since it is one of the few mature low-carbon technologies that can be widely deployed in large increments. Bearing in mind the stringency and comprehensiveness of the nuclear safety regulations and measures implemented over the past three decades, it remains to be seen how far new requirements, post Fukushima, might affect the economics of nuclear generation.

Nonetheless, in the aftermath of Fukushima Daiichi, trust in the nuclear industry has suffered greatly and will take years to re-build. There have now been three major nuclear incidents over the past three decades, the latest of which occurred in one of the most advanced nuclear nations, albeit that the events at Fukushima Daiichi were triggered by the combination of an exceptionally powerful earthquake and tsunami in a zone exceptionally exposed to such risks.

Figure 12.4 ● Power generation by fuel in the New Policies Scenario and Low Nuclear Case

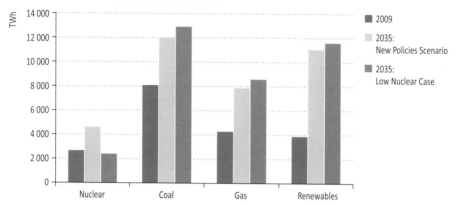

Note: Oil-fired power generation is not shown due to its small share of total output.

Overall investment needs in the power-generation sector increase slightly in the Low Nuclear Case, as lower investment in nuclear plants is more than offset by increased expenditure on coal-fired, gas-fired and, especially, renewables-based plants. Cumulative global investment in nuclear plants over 2011 to 2035, in real 2010 dollars, falls by about $675 billion (or 60%), compared with the New Policies Scenario, while investment in coal-fired plants increases by about $170 billion (or 10%), gas-fired plants by about $104 billion (or 11%) and in renewables-based capacity by about $500 billion (or 9%). The net increase is about $110 billion (or 1%).

International fuel markets

With the projected increase in coal- and gas-fired electricity generation in the Low Nuclear Case, global demand for both fuels inevitably increases over the *Outlook* period, compared with developments in the New Policies Scenario. Demand for coal increases most, reaching 6 150 million tonnes of coal equivalent (Mtce) in 2035, or 5% higher than in the New Policies Scenario. This additional demand is equivalent to over twice Australia's current steam coal exports. Natural gas demand in 2035 is 3% higher, reaching almost 4 900 bcm. The increase is equivalent to two-thirds of Russia's natural gas exports in 2010. OECD countries, where nuclear capacity declines the most, are responsible for about a 70% and 60% of the additional demand for coal and gas, respectively.

Higher demand leads to in greater upward pressure on prices in international markets, especially for natural gas. Coal prices are almost 2% higher in 2035 than in the New Policies Scenario and gas prices are between 4% and 6% higher, depending on region. These higher fossil-fuel prices put upward pressure on the overall cost of generation and electricity prices increase as a result. Greater coal and natural gas-import requirements and higher international energy prices push up spending on imported coal and gas. The additional spending on imports, relative to the New Policies Scenario, rises over time, as the share of nuclear power in the generation mix progressively falls. In 2035, aggregated across all net-importing countries, spending on gas imports is up by around $67 billion, or 11%, while spending on coal imports is up by around $22 billion, or 17% (Figure 12.5). For countries that rely heavily on nuclear power and have limited indigenous energy resources (such as Belgium, France, Japan and Korea), the impact will be more pronounced than the aggregate numbers suggest.

Figure 12.5 • Global primary coal and gas demand and annual spending on imports in the Low Nuclear Case

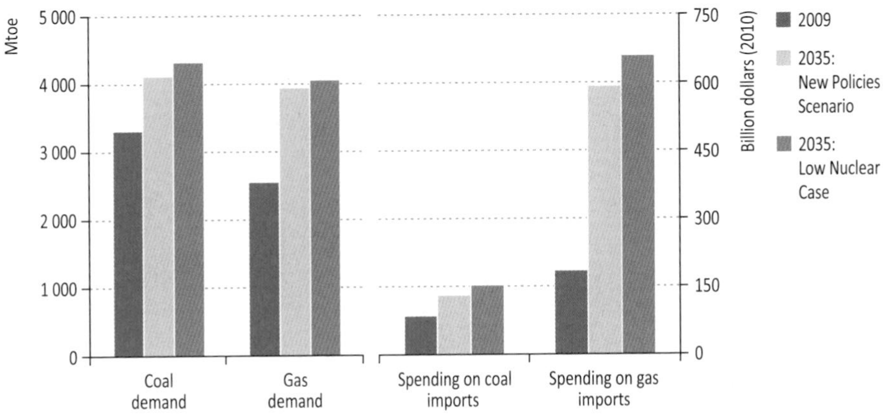

Note: Calculated as the value of net imports at prevailing average international prices.

The additional demand for natural gas and coal in the Low Nuclear Case has important implications for energy security. Although the share of generation coming from renewables increases, the diversity of the power-generation mix declines. The prospect of a limited number of producing regions increasingly dominating global gas supply and trade would raise concerns about the risk of supply disruptions as well as the risk that some countries might seek to use their dominant market position to force prices even higher.

CO_2 emissions

One of the major advantages of nuclear power compared with electricity generated from fossil fuels is that it does not directly generate emissions of carbon dioxide (CO_2) or other greenhouse gases.[4] If the 13% of global electricity production that came from nuclear power plants in 2010 had instead been generated equally from natural gas and coal (based on current average efficiency levels) we estimate that global CO_2 emissions from the power sector would have been 2.1 gigatonnes (Gt), or 17%, higher.

As a result of the increased use of fossil fuels in the Low Nuclear Case, energy-related CO_2 emissions are higher than in the New Policies Scenario. At the global level, the increase in energy-related CO_2 emissions is roughly 2.6% in 2035. Cumulative CO_2 emissions in the

---

4. There are some CO_2 emissions linked to the use of fossil fuels in the nuclear fuel cycle, such as uranium mining and enrichment, but these are at least an order of magnitude lower than the direct emissions from burning fossil fuels.

period 2011 to 2035 are higher by 10.2 Gt, or 1.2%, adding to the rising concentration of greenhouse gases in the atmosphere and making it harder and more expensive to combat climate change. These aggregate numbers mask more dramatic increases in countries that rely more heavily on nuclear power.

CO_2 emissions from power plants reach 15.7 Gt in 2035 in the Low Nuclear Case, 0.9 Gt, or 6.2%, higher than in the New Policies Scenario (Figure 12.6). Compared with the New Policies Scenario, 34% of the increase comes from power plants in non-OECD countries.

Figure 12.6 ● **Energy-related CO_2 emissions from the power sector in the New Policies Scenario and the Low Nuclear Case**

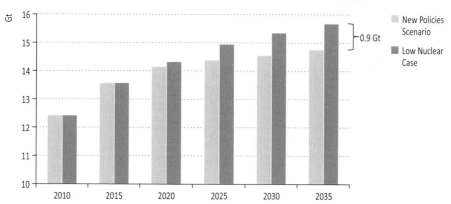

Box 12.2 ● **Human capital and the nuclear industry**

The development, maintenance and growth of a nuclear power programme require a well-trained and experienced work force. Consequently, various stakeholders – including governments, industry, academia and intergovernmental organisations – invest significant resources into human resource management and development.

One area of concern in recent years has been a looming shortage of people with the necessary skills, due to the ongoing retirement of workers and the low number of new entrants (IEA, 2010). In the United States, the nuclear industry work force numbered 120 000 people in 2009; approximately 40% of this work force will be eligible to retire by 2015 and, in order to maintain the current numbers, the industry will need to hire about 25 000 more workers by then (NEI, 2010). In Korea, where the nuclear industry is expected to expand in the coming decade, the Ministry of Knowledge Economy estimates that an additional 23 900 nuclear workers will be needed by 2020.

In order to address the possible skills shortage, governments and industry have taken various initiatives to attract students to the nuclear industry, including increasing the availability of training programmes on nuclear energy, offering financial scholarships and providing internship opportunities to university and graduate students. In the United States, for example, since 2007 the Nuclear Regulatory Commission, the US Department of Energy and Center for Energy Workforce Development have funded expansion of nuclear energy degree programmes through the Nuclear Uniform Curriculum Program. Between 2003 and 2008, the number of students receiving B.S. and M.S. degrees in the United States increased by 288 (or 173%) and 128 (or 97%) respectively. The trend is expected to continue. Between 2008 and 2009 enrolment in nuclear engineering programmes increased at both undergraduate and graduate levels by around 15% and 5% respectively (ORISE, 2010).

The large-scale phase-out of nuclear power assumed in the Low Nuclear Case would lead to a fewer opportunities to gain practical experience in all phases of the nuclear programme. Such experience complements academic training and its absence would result in a deterioration of technical skills. Given the design lifetime of nuclear reactors (typically 40 years) and the expectation of life extensions (typically 20 years), the nuclear work force will need to be maintained for a long period of time. A combination of public opposition to nuclear power and policy uncertainty about its future role in the energy mix could discourage students from pursuing a career in the nuclear industry and could result in a critical shortage.

Meeting the global climate goal with less nuclear power generation

In the Low Nuclear Case, energy-related CO_2 emissions grow faster than in the New Policies Scenario, leading to likely global average temperature increases of over 3.5°C. Our 450 Scenario, unlike our other scenarios does not project the outcome of a given set of assumptions. Rather it describes a plausible path to the pre-determined objective of realising a future in which government policies transform the energy system in such a way as to meet the 2°C goal (see Chapter 6). But in that scenario, the role of nuclear power is greater than in the New Policies Scenario and much greater than in the Low Nuclear Case.

If there is less nuclear power in the global energy mix, then other changes are essential if the objective of limiting the long-term rise in the global average temperature to 2°C is to be achieved:

- The share of renewables needs to be greater even than that projected in our 450 Scenario.
- Overall primary energy needs must be reduced further, through even greater improvements in energy efficiency.
- Carbon capture and storage needs to be very widely deployed by 2035.

In short, to the extent that energy demand in the 450 Scenario cannot be reduced to offset any loss of nuclear supply, all the replacement supply must come from low-carbon technologies, so that there is no net increase in global emissions. This will be extremely challenging and costly.

The Low Nuclear 450 Case

In order to quantify the implications of less nuclear power being available to meet the 2°C goal, we have analysed another case, called the Low Nuclear 450 Case. Since the trajectory of energy-related CO_2 emissions in the 450 Scenario is, by definition, the one which must be achieved to realise the climate change objective, energy-related emissions must be held to that same overall trajectory in the new circumstances. This means that as the 450 Scenario is modified to incorporate the assumptions on nuclear power used in the Low Nuclear Case, much stronger government policy actions have to be imposed in order to cut demand further and ensure yet wider deployment of other low-carbon technologies. Compared with the New Policies Scenario, overall emissions in the Low Nuclear 450 Case are 2.4 Gt, or 7%, lower in 2020 and 15 Gt, or 41%, lower in 2035 (Figure 12.7). Compared with the 450 Scenario, the use of fossil fuels is somewhat higher, but emissions are the same, as the increase in demand for fossil fuels is assumed to come solely from fossil-fuel-based power plants that are equipped with carbon capture and storage (CCS). Increased reliance on renewables compared with the 450 Scenario make up the difference. Overall CO_2 emissions broadly follow the same trajectory as in the 450 Scenario, however, their geographic distribution is slightly different: emissions are slightly higher in the OECD countries and correspondingly lower in the non-OECD countries.

Figure 12.7 ● **World energy-related CO_2 emissions abatement in the Low Nuclear 450 Case relative to the New Policies Scenario**

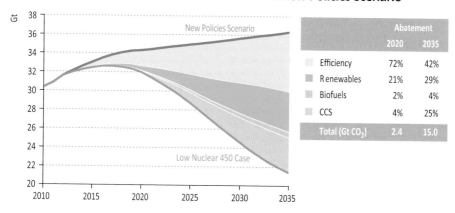

In the Low Nuclear 450 Case, as in the 450 Scenario, energy efficiency measures are the largest source of emissions abatement, accounting for about 47% of the cumulative emissions savings over the period 2011 to 2035 and underlining the importance of very vigorous policy action in this area. The shares of total abatement arising from renewables in power generation and CCS, are substantially higher than in the 450 Scenario. Renewables account for 29% of abatement in 2035 – eight percentage points higher than in the 450 Scenario – and CCS for 25%, three percentage points higher. The wider deployment of CCS allows for an 8% increase in primary coal use and a 3% increase in the use of natural gas in 2035, compared with the 450 Scenario. Renewable energy use is 9% higher, while nuclear power production is 63% lower.

In the Low Nuclear 450 Case, the transformation required in the power sector is even more ambitious than in the 450 Scenario, due to the narrower portfolio of abatement options. The share of fossil-fuel-fired power stations in total generation in 2035 drops to just 38%, from 67% in 2009 (Figure 12.8). Total renewables-based capacity reaches 6 000 GW in 2035, over 970 GW, or one-fifth, more than in the 450 Scenario. By 2035, the capacity of plants equipped with CCS exceeds 820 GW, an increase of more than 210 GW, or a third, more than in the 450 Scenario. In practice, it is by no means certain that these projections could actually be realised, as it is still unclear that low-carbon technologies can be deployed on this scale (the implications for meeting the 2°C climate goal of slower CCS deployment than that in the New Policies Scenario are discussed in Chapter 6).

Figure 12.8 ● **Share of world power generation by source in the 450 Scenario and Low Nuclear 450 Case**

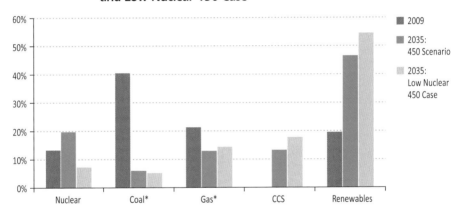

*Excludes plants fitted with carbon capture and storage (CCS).

With reduced abatement coming from nuclear power, the cost of achieving the same cumulative emissions as the 450 Scenario increases by some $1.5 trillion in real 2010 dollars, compared with the 450 Scenario (or 10%) over the period 2011 to 2035 (Figure 12.9). Countries that were expected to rely heavily on nuclear power as a source of CO_2 abatement would find the financial (and practical) burden particularly heavy.

Of the additional investment required in the Low Nuclear 450 Case, only 10% (or $150 billion) is incurred before 2020, the remaining 90% (or $1.3 trillion) arising over the last fifteen years of the projection period, reflecting the profile of fewer additions and additional retirements of nuclear power plants. The power sector absorbs the vast bulk of the net incremental investment ($1.2 trillion). This figure provides for incremental investment of $2.1 trillion for generation from renewables, which need greater nameplate generating capacity than other generating options because of variable resource availability (see Chapter 5). The cost of the new generating capacity is partially offset by a decrease in incremental investment of $1.3 trillion for generation from nuclear.

Figure 12.9 ● **Incremental energy-related investment in the Low Nuclear 450 Case relative to the 450 Scenario**

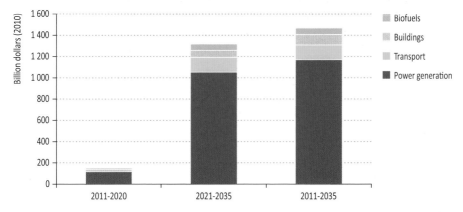

CHAPTER 13

ENERGY FOR ALL

Financing access for the poor

- Modern energy services are crucial to human well-being and to a country's economic development; and yet over 1.3 billion people are without access to electricity and 2.7 billion people are without clean cooking facilities. More than 95% of these people are either in sub-Saharan Africa or developing Asia and 84% are in rural areas.

- In 2009, we estimate that $9.1 billion was invested globally in extending access to modern energy services, supplying 20 million more people with electricity access and 7 million people with advanced biomass cookstoves. In the New Policies Scenario, our central scenario, $296 billion is invested in energy access between 2010 and 2030 – an average of $14 billion per year, 56% higher than the level in 2009. But, this is not nearly enough: it still leaves 1.0 billion people without electricity (more than 60% of this is in sub-Saharan Africa) and, despite progress, population growth means that 2.7 billion people will remain without clean cooking facilities in 2030. To provide universal modern energy access by 2030, cumulative investment of $1 trillion is required – an average of $48 billion per year, more than five-times the level in 2009.

- The $9.1 billion invested in extending energy access in 2009 was sourced from multilateral organisations (34%), domestic government finance (30%), private investors (22%) and bilateral aid (14%). To provide the $48 billion per year required for universal access, we estimate that around $18 billion per year is needed from multilateral and bilateral development sources, $15 billion per year from the governments of developing countries and $15 billion per year from the broad range of actors that form the private sector.

- Private sector investment needs to grow the most, but significant barriers must first be overcome. Public authorities must provide a supportive investment climate, such as by implementing strong governance and regulatory reforms. The public sector, including donors, needs also to use its tools to leverage private sector investment where the commercial case is marginal. At present, energy access funding tends to be directed primarily toward large-scale electricity infrastructure. This does not always reach the poorest households. Access to funding at a local level is essential to support initiatives that cater effectively for local needs, building local financial and technical capacity and stimulating sectoral development.

- Achieving universal access by 2030 would increase global electricity generation by 2.5%. Demand for fossil fuels would grow by 0.8% and CO_2 emissions go up by 0.7%, both figures being trivial in relation to concerns about energy security or climate change. The prize would be a major contribution to social and economic development, and helping to avoid the premature death of 1.5 million people per year.

Introduction

Energy is a critical enabler. Every advanced economy has required secure access to modern sources of energy to underpin its development and growing prosperity. While many developed countries may be focused on domestic energy security or decarbonising their energy mix, many other countries are still seeking to secure enough energy to meet basic human needs. In developing countries, access to affordable and reliable energy services is fundamental to reducing poverty and improving health, increasing productivity, enhancing competitiveness and promoting economic growth. Despite the importance of these matters, billions of people continue to be without basic modern energy services, lacking reliable access to either electricity or clean cooking facilities. This situation is expected to change only a little by 2030 unless more vigorous action is taken.

Developing countries that import oil are today facing prices in excess of $100 a barrel when, at a comparable stage of economic development, many OECD countries faced an average oil price of around $22 a barrel (in 2010 dollars). In little over a decade, the bill of oil-importing less developed countries[1] has quadrupled to hit an estimated $100 billion in 2011, or 5.5% of their gross domestic product (GDP) (Figure 13.1). Oil-import bills in sub-Saharan Africa increased by $2.2 billion in 2010, more than one-third higher than the increase in Official Development Assistance (ODA) over the year.[2] In contrast, oil exporters in sub-Saharan Africa, such as Nigeria and Angola, are benefitting from the oil price boom and tackling energy poverty is, financially at least, within their means. We estimate that the capital cost of providing modern energy services to all deprived households in the ten-largest oil and gas exporting countries of sub-Saharan Africa[3] would be around $30 billion, equivalent to around 0.7% of those governments' cumulative take from oil and gas exports.

Figure 13.1 • **Oil-import bills in net-importing less developed countries**

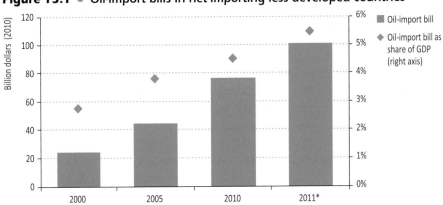

*Estimated, assuming an average oil price of $100 per barrel.

Notes: Calculated as the value of net imports at prevailing average international prices. Oil-import bills as a share of GDP are at market exchange rates in 2010 dollars.

---

1. The group includes India and the oil-importing countries within the United Nation's classification of least developed countries (available at www.unohrlls.org). This group has a combined population of 1.8 billion people and accounts for 65% of those lacking access to modern energy.

2. Data available from the OECD/DCD-DAC database at www.oecd.org.

3. These countries include: Angola, Cameroon, Chad, Democratic Republic of Congo, Equatorial Guinea, Gabon, Ivory Coast, Mozambique, Nigeria and Sudan.

International concern about the issue of energy access is growing. While the United Nations Millennium Development Goals[4] (MDGs) do not include specific targets in relation to access to electricity or to clean cooking facilities, the United Nations has declared 2012 to be the "International Year of Sustainable Energy for All". Other strategic platforms to discuss the link between energy access, climate change and development include the "Energy for All Conference" in Oslo, Norway in October 2011 and the COP17 in Durban, South Africa in December 2011. These issues are also expected to be addressed at the United Nations Conference on Sustainable Development (UNCSD) in Rio de Janeiro, Brazil in June 2012. That conference will aim to secure renewed political commitment to sustainable development, to assess progress to date and to address new and emerging challenges. It will bring to centre-stage in the international debate the need to reconcile environment, development and poverty eradication issues such as energy access.

The *World Energy Outlook (WEO)* has focused attention on modern energy access for a decade, providing the international community with quantitative, objective analysis. This year our analysis tackles the critical issue of financing the delivery of universal modern energy access.[5, 6] The chapter begins by providing updated estimates of the number of people lacking access to electricity and clean cooking facilities, by country. It offers, to the best of our knowledge, for the first time in energy literature, an estimate of the total amount of investment taking place globally to provide access to modern energy services and provides details on the sources of financing. The chapter then examines what level of modern energy access might be achieved by 2030, in relation to the projections in the New Policies Scenario, our central scenario, and the level of investment involved (the New Policies Scenario takes account of both existing government policies and cautious implementation of declared policy intentions). While the *Outlook* period for *WEO-2011* is 2009 to 2035, analysis in this chapter is based exceptionally on the period 2009 to 2030. This period has been adopted to be consistent with the key goal proposed by the United Nations Secretary-General of ensuring universal energy access by 2030 (AGECC, 2010). Since the level of projected investment in the New Policies Scenario is not nearly enough to achieve universal access to modern energy services by 2030, we then estimate the level of additional investment that would be required to meet this goal – as defined in our Energy for All Case.[7] The chapter then examines the main sources of financing, and the types of financing instruments that appear to be the most in need of scaling-up in order to achieve the Energy for All Case (Figure 13.2). This is derived from a bottom-up analysis of the most likely technology solutions in each region, given resource availability, and government policies and measures. Throughout, we have drawn on experience from existing programmes using different financing and business models to provide modern energy access.

4. See www.un.org/millenniumgoals for more information.

5. This chapter benefitted from a workshop held by the IEA in Paris on 13 May 2011, and was presented for the first time to a meeting of government leaders and international institutions hosted by the government of Norway in Oslo on 10-11 October 2011.

6. Due to the focus of this chapter on financing, some elements of *WEO* analysis on energy access, such as the Energy Development Index (EDI), are not included here but have been updated and will be made available online at www.worldenergyoutlook.org.

7. Referred to in *WEO-2010* as the Universal Modern Energy Access Case. See Box 13.1 for the definition of modern energy access used in this analysis.

13

Figure 13.2 ● Financing modern energy access

Technology solutions		Financing instruments		Financing sources
■ Electricity: on-grid, isolated off-grid, mini-grid ■ Cooking: liquefied petroleum gas (LPG), biogas, advanced cookstoves		■ Grants ■ Equity ■ Loans ■ Insurance ■ Subsidies ■ Guarantees		■ Multilateral organisations ■ Bilateral Official Development Assistance (ODA) ■ Developing country governments ■ Private sector

Current status of modern energy access

We estimate that in 2009, 1.3 billion people did not have access to electricity, around 20% of the global population, and that almost 2.7 billion people relied on the traditional use of biomass for cooking, around 40% of the global population (Table 13.1).[8] This updated estimate reflects revised country-level data, where available. More than 95% of the people lacking access to modern energy services (Box 13.1 includes our definition of modern energy access) are in either sub-Saharan Africa or developing Asia and 84% live in rural areas. Sub-Saharan Africa accounts for only 12% of the global population, but almost 45% of those without access to electricity. Over 1.9 billion people in developing Asia still rely on the traditional use of biomass for cooking, with around 840 million of these in India and more than 100 million each in Bangladesh, Indonesia and Pakistan. In sub-Saharan Africa, Nigeria also has over 100 million people without access to clean cooking facilities.

Despite these sobering statistics, some countries have made notable progress in recent years in improving access to electricity and reducing the number of people relying on the traditional use of biomass for cooking. In India, the most recent data show that expenditure on electricity was reported by 67% of the rural population and 94% of the urban population in 2009 (Government of India, 2011), up from 56% and 93% respectively when surveyed in 2006 (Government of India, 2008).[9] In Vietnam, the electrification rate (the share of the population with access to electricity) has increased in the last 35 years from below 5% to 98%. Bangladesh and Sri Lanka have seen progress on access to both electricity and clean cooking facilities, but more so on increased electrification. Angola and Congo both have seen the share of the population with access to modern energy services expand considerably in the last five years, mainly in urban areas.

---

8. While throughout this analysis we focus on the number of people relying on the traditional use of biomass for cooking, it is important to note that some 0.4 billion people, mostly in China, rely on coal. Coal is a highly polluting fuel when used in traditional stoves and has serious health implications (United Nations Development Programme and World Health Organization, 2009).

9. See www.mospi.nic.in.

Table 13.1 ● People without access to modern energy services by region, 2009 (million)

	People without access to electricity	Share of population	People relying on the traditional use of biomass for cooking	Share of population
Africa	587	58%	657	65%
Nigeria	76	49%	104	67%
Ethiopia	69	83%	77	93%
DR of Congo	59	89%	62	94%
Tanzania	38	86%	41	94%
Kenya	33	84%	33	83%
Other sub-Saharan Africa	310	68%	335	74%
North Africa	2	1%	4	3%
Developing Asia	675	19%	1 921	54%
India	289	25%	836	72%
Bangladesh	96	59%	143	88%
Indonesia	82	36%	124	54%
Pakistan	64	38%	122	72%
Myanmar	44	87%	48	95%
Rest of developing Asia	102	6%	648	36%
Latin America	31	7%	85	19%
Middle East	21	11%	n.a.	n.a.
Developing countries	**1 314**	**25%**	**2 662**	**51%**
World*	**1 317**	**19%**	**2 662**	**39%**

*Includes countries in the OECD and Eastern Europe/Eurasia.

Box 13.1 ● Defining modern energy access

There is no universally-agreed and universally-adopted definition of modern energy access. For our analysis, we define modern energy access as *"a household having reliable and affordable access to clean cooking facilities, a first connection to electricity and then an increasing level of electricity consumption over time to reach the regional average"*.[10] By defining access to modern energy services at the household level, it is recognised that some other categories are excluded, such as electricity access to businesses and public buildings that are crucial to economic and social development, *i.e.* schools and hospitals.

Access to electricity involves more than a first supply connection to the household; our definition of access also involves consumption of a specified minimum level of electricity, the amount varies based on whether the household is in a rural or an urban area. The initial threshold level of electricity consumption for rural households is assumed to be 250 kilowatt-hours (kWh) per year and for urban households it is

13

---

10. We assume an average of five people per household.

500 kWh per year.[11] In rural areas, this level of consumption could, for example, provide for the use of a floor fan, a mobile telephone and two compact fluorescent light bulbs for about five hours per day. In urban areas, consumption might also include an efficient refrigerator, a second mobile telephone per household and another appliance, such as a small television or a computer.

Once initial connection to electricity has been achieved, the level of consumption is assumed to rise gradually over time, attaining the average regional consumption level after five years. This definition of electricity access includes an initial period of growing consumption as a deliberate attempt to reflect the fact that eradication of energy poverty is a long-term endeavour. In our analysis, the average level of electricity consumption per capita across all those households newly connected over the period is 800 kWh in 2030. This is comparable with levels currently seen in much of developing Asia.

This definition of energy access also includes provision of cooking facilities which can be used without harm to the health of those in the household and which are more environmentally sustainable and energy efficient than the average biomass cookstove currently used in developing countries. This definition refers primarily to biogas systems, liquefied petroleum gas (LPG) stoves and advanced biomass cookstoves that have considerably lower emissions and higher efficiencies than traditional three-stone fires for cooking. In our analysis, we assume that LPG stoves and advanced biomass cookstoves require replacement every five years, while a biogas digester is assumed to last 20 years. Related infrastructure, distribution and fuel costs are not included in our estimates of investment costs.

Current status of investment in modern energy access

For the billions of people currently deprived, the lack of access to modern forms of energy tends to go hand-in-hand with a lack of provision of clean water, sanitation and healthcare. It also represents a major barrier to economic development and prosperity. The importance of modern energy access is being recognised increasingly by many organisations that provide development funding. We estimate that capital investment of $9.1 billion was undertaken globally in 2009 (Box 13.2 describes our methodology) to provide 20 million people with access to electricity and 7 million people with advanced biomass cookstoves ($70 million of the total). An incomplete set of past observations suggests that this is the highest level of investment ever devoted to energy access.[12]

---

11. The assumed threshold levels for electricity consumption are consistent with previous *WEO* analyses. However, we recognise that different levels are sometimes adopted in other published analysis. Sanchez (2010), for example, assumes 120 kWh per person (600 kWh per household, assuming five people per household). As another point of reference, the observed electricity consumption in India in 2009 was 96 kWh per person in rural areas and 288 kWh in urban areas *on average* over all people connected to electricity, implying a lower consumption for those that have been connected more recently (Government of India, 2011).

12. There are currently no comprehensive data available, and those that do exist employ varying methodologies. Our estimate is constructed from a variety of sources and includes some necessary assumptions. It is to be hoped that this shortcoming in the data receives greater attention in future.

Box 13.2 ● Measuring investment in modern energy access

Our estimate of investment in modern energy access is based on the latest data available and has several components. The estimate is of the capital investment made to provide household access to electricity (both the cost of the provision of first connection and the capital cost to sustain an escalating supply over time) and the cost of providing clean cooking facilities to those who currently lack them. Operating costs, such as fuel costs and maintenance costs, are not included. Broader technical assistance, such as policy and institutional development advice, is also not included. In the case of on-grid and mini-grid solutions for electricity access, the estimate does not include the investment required in supportive infrastructure, such as roads.

Our estimate is based on an average of high and low estimates of investment data from several sources:

● *Bilateral Official Development Assistance* (ODA) – In line with the Multilateral Development Banks' Clean Energy Investment Framework methodology,[13] our estimate of total ODA for energy access includes the investment flows for electricity generation, transmission and distribution in countries eligible for International Development Association (IDA) funding, *i.e.* the poorest countries. We have also included financing for off-grid generation and transmission for those countries eligible for International Bank for Reconstruction and Development (IBRD) funding (countries which, while not among the poorest, still have difficulty accessing commercial credit markets).

● *Multilateral organisations* (development banks,[14] funds, etc.) – This estimate is based on the organisations' own data when available,[15] or the same methodology as ODA where data is not available.

● *Domestic governments in developing countries* – This estimate includes investments made both directly by the governments and through state-owned utilities. It includes investment independently conducted by the governments as well as government investment leveraged through multilateral funding. In IDA countries, it is estimated that for every $1 spent in aid on energy access, it is matched by an additional equal amount from either the private sector or developing country governments. Countries eligible for IDA funding account for 82% of the total population lacking access to electricity, so the same leverage factor has been applied to all countries.

● *Private sector* – The broad range of private sector actors makes this the most challenging category for which to produce a comprehensive estimate. In constructing this estimate, which is based on data on private sector investment in infrastructure, including public-private partnerships (PPP), sourced from the World Bank PPI database,[16] we have assumed that the private sector component of PPP-funded projects is around 50% and that between 5% and 20% of the total investment goes towards energy access, depending on the region.

---

13. See www.worldenergy.org/documents/g8report.pdf.
14. Multilateral Development Banks are a channel for funds from bilateral sources and from bond markets.
15. Publicly available sources supplemented with bilateral dialogue.
16. See www.ppi.worldbank.org.

We estimate that bilateral ODA accounted for 14% of total investment in extending energy access in 2009, only slightly more than 1% of total bilateral ODA in the same year (Figure 13.3). Multilateral organisations, such as international development banks and funds, accounted for more than $3 billion of such investment in energy access, around 34% of the total. This was just over 3% of total multilateral aid in the same year. An estimated 30% of investment in energy access was sourced from domestic governments in developing countries. This included investments made directly by the governments and through state-owned utilities. The private sector is estimated to have accounted for 22% of the total investment in energy access. In the case of investment in energy access by domestic governments and the private sector, the share of total investment directed to energy access is estimated to be less than 1% of the gross fixed capital formation in these countries in 2009. While sources of investment are referred to separately here, in practice two or more often operate in conjunction to deliver an energy access project. Blending funds from different sources can bring important benefits, such as reducing funding risks and securing buy-in from project participants. Multilateral development banks generally enter into partnerships with developing country governments and/or the private sector to deliver projects, such as the Asian Development Bank's biogas programme in Vietnam.

Figure 13.3 ● **Investment in energy access by source, 2009**

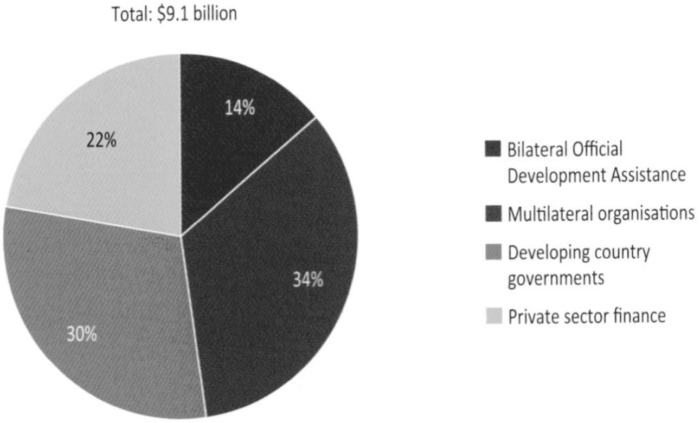

Total: $9.1 billion

Legend:
- Bilateral Official Development Assistance
- Multilateral organisations
- Developing country governments
- Private sector finance

Outlook for energy access and investment in the New Policies Scenario

In the New Policies Scenario, our central scenario, we project that total cumulative investment in extending access to modern energy is $296 billion from 2010 to 2030, an average of $14 billion per year.[17] The projected annual average investment required is therefore 56% higher in the New Policies Scenario than the level observed in 2009. All sources of finance increase their investment in absolute terms to meet this requirement in the New Policies Scenario. Domestic finance in developing countries and multilateral developing banks are the largest sources of finance. But private sector finance is close behind and actually sees the most growth.

---

17. We focus primarily on the average level of investment per year over the projection period, as a better illustration of ongoing investment activity than the cumulative total.

In the New Policies Scenario, around 550 million additional people gain access to electricity and 860 million are provided with clean cooking facilities from 2010 to 2030. The increase in access to modern energy services is driven largely by rapid economic growth in several developing countries, accompanied by rapid urbanisation in some cases, but population growth acts as a countervailing force. For example, in the case of China, the 12th Five-Year Plan (covering the period 2011 to 2015) provides for rapid urbanisation, with plans to create 45 million new urban jobs and an expectation that the urbanisation rate will increase to 52% by 2015, the date by which the country also expects to achieve full electrification.

In several countries, national targets to increase electricity access succeed in delivering improvements over the projection period, but only on a limited scale: many such targets will not be achieved unless robust national strategies and implementation programmes are put in place. Access to clean cooking facilities has in the past often received less government attention than electricity access, with the result that there are fewer related programmes and targets in place at a national level. At an international level, an important step forward was taken in September 2010 when the UN Foundation launched the "Global Alliance for Clean Cookstoves". The Alliance seeks to overcome market barriers that impede the production, deployment and use of clean cookstoves in the developing world, so as to achieve the goal of 100 million households adopting clean and efficient stoves and fuels by 2020.[18]

Access to electricity

In the New Policies Scenario, around $275 billion of investment goes toward providing electricity access from 2010 to 2030. This represents annual average investment of $13 billion to connect around 26 million people per year. The capital-intensive nature of electricity generation, transmission and distribution means that this investment accounts for over 90% of total investment to deliver modern energy services over the projection period. The average annual level of investment in electricity access increases by almost 45%, compared with that observed in 2009. While the share of the global population lacking access to electricity declines from 19% in 2009 to 12% in 2030, 1.0 billion people are still without electricity by the end of the period (Table 13.2). The proportion of those without access to electricity in rural areas was around five-times higher than in urban areas in 2009, and this disparity widens to be around six-times higher in 2030. There are examples of progress in increasing rates of rural electrification, such as in Angola and Botswana, but this is often from a low base.

Annual investment to increase on-grid electricity access averages $7 billion in the New Policies Scenario. The main sources of investment for on-grid access are domestic government finance and the private sector. Almost 55% of total private sector investment is estimated to be in on-grid solutions. Over 40% of the investment made by multilateral development banks is also estimated to be in on-grid solutions. Investment in mini-grid and off-grid electricity generation together averages around $6 billion annually in the New Policies Scenario.[19] Private sector investment represents a significantly smaller share of the total for such projects, reflecting the obstacles to developing commercially viable projects.

13

---

18. See www.cleancookstoves.org.
19. Mini-grids provide centralised generation at a local level. They operate at a village or district network level, with loads of up to 500 kW. Isolated off-grid solutions include small capacity systems, such as solar home systems, micro-hydro systems, wind home systems and biogas digester systems.

Table 13.2 ● People without access to electricity by region in the New Policies Scenario (million)

	2009			2030		
	Rural	Urban	Share of population	Rural	Urban	Share of population
Africa	466	121	58%	539	107	42%
Sub-Saharan Africa	465	121	69%	538	107	49%
Developing Asia	595	81	19%	327	49	9%
China	8	0	1%	0	0	0%
India	268	21	25%	145	9	10%
Rest of developing Asia	319	60	36%	181	40	16%
Latin America	26	4	7%	8	2	2%
Middle East	19	2	11%	5	0	2%
Developing countries	1 106	208	25%	879	157	16%
World*	1 109	208	19%	879	157	12%

*Includes countries in the OECD and Eastern Europe/Eurasia.

At a regional level, the number of people without access to electricity in sub-Saharan Africa *increases* by 10%, from 585 million in 2009 to 645 million in 2030, as the rate of population growth outpaces the rate of connections. The number of people without access to electricity in sub-Saharan Africa overtakes the number in developing Asia soon after 2015. This increase occurs in spite of pockets of progress, such as the government electrification programme in South Africa, which has provided 4 million households with access to electricity since it was launched in 1990 and aims to achieve complete access nationally by 2020. Table 13.3 provides examples of national electrification programmes. While the adoption of national targets and programmes for modern energy access is important, in practice it has been relatively commonplace for initial ambitions to be downgraded subsequently.

The number of people without access to electricity in developing Asia is projected to decrease by almost 45%, from 675 million people in 2009 to 375 million in 2030. Around 270 million people in rural areas are given access to electricity but, despite this, the rural population still constitutes the great majority of those lacking access in 2030. China has provided 500 million people in rural areas with electricity access since 1990 and is expected to achieve universal electrification by 2015. In India, the Rajiv Gandhi Grameen Vidyutikaran Yojana Programme is making progress towards a goal of electrifying over 100 000 villages and providing free electricity connections to more than 17 million rural households living below the national poverty line. Our projections show India reaching a 98% electrification rate in urban areas and 84% in rural areas in 2030. In the rest of developing Asia, the average electrification rate reaches almost 93%. The difference in trajectory between developing Asia and sub-Saharan Africa is clear, with an improving situation in the former and a worsening one in the latter. In developing Asia, India accounts for much of the increased access to electricity, while in sub-Saharan Africa a more mixed story within the region does not, in aggregate, overcome the deteriorating picture, driven primarily by population growth.

Table 13.3 ● **Major programmes and targets for improving access to electricity in selected countries**

	Programme name	Description	Financing arrangements
Bangladesh	Master Plan for Electrification – National Energy Policy of Bangladesh 1996-2004	Electricity for all by 2020.	Loans and grants from donors are passed on, under a subsidiary agreement, to the Rural Electrification Board. Domestic government funds cover all local costs of construction.
Brazil	Light for All	Launched in 2003, extended in 2011 to 2014. So far the programme has connected more than 2.4 million households and it aims for full electrification.	Funded largely by the extension of a Global Reversion Reserve tax incorporated into electricity rates. The scheme also benefits from an investment partnership of federal government, state agencies and energy distributors.
Ghana	National Electrification Scheme – Energy Plan 2006-2020	Electricity access for all by 2020.	Funded through grants and loans by donors and $9 million per year in domestic government budgetary support.
India	Rajiv Gandhi Grameen Vidyutikaran Yojana	Electrify 100 000 villages and provide free electricity connections to 17.5 million households below the poverty line by March 2012.	Total funds of $5.6 billion disbursed between 2005 and 2011. A government subsidy of up to 90% of capital expenditure is provided through the Rural Electrification Corporation. Those below the poverty line receive a 100% subsidy for connection.
Indonesia	Rural electrification programmes – National Energy Management	Electricity access for 95% of the population by 2025.	Investment costs are covered by cross subsidies by the state-owned power utility (PNL) and other costs are funded by donors.
Nepal	Rural Electrification Program – National 3-Year Interim Plan	Electricity access for 100% of the population by 2027.	A Rural Electrification Board administers specific funds for electrification of rural areas.
Philippines	Philippines Energy Plan, 2004-2013	Electrification of 90% of households by 2017.	Funded by grants and loans from a National Electrification Fund and PPPs.
South Africa	Integrated National Electrification Programme	Electricity access for 100% of the population by 2020.	Government funding disbursed by the Department of Energy to Eskom (state-owned utility) and municipalities.
Zambia	Rural Electrification Master Plan	Electricity access for 78% in urban areas and 15% in rural areas by 2015.	The government has created a Rural Electrification Fund that is administered by the Rural Electrification Authority.

13

Outside Asia and Africa, there are at present smaller, but significant, numbers of people without access to electricity in Latin America, but near-universal access is achieved there by 2030 in the New Policies Scenario. In Brazil, Luz para Todos (light for all) is a government programme, operated by a majority state-owned power utility company, and executed by electricity concessionaires and co-operatives. The project promotes renewable energy as the most practical solution in remote areas, with the government providing funding to help cover the costs for renewable energy projects in these areas.

Access to clean cooking facilities

In the New Policies Scenario, $21 billion is invested in total from 2010 to 2030 to provide 860 million people with clean cooking facilities. This is equivalent to an average annual investment of $1 billion to provide facilities to an average of 41 million people per year. After an initial increase, the number of people without clean cooking facilities drops back to 2.7 billion, the level of 2009, in 2030 (Table 13.4). The proportion of people globally without clean cooking facilities declines from 39% in 2009 to 33% in 2030.

Table 13.4 ● **People without clean cooking facilities by region in the New Policies Scenario** (million)

	2009			2030		
	Rural	Urban	Share of population	Rural	Urban	Share of population
Africa	480	177	65%	641	270	58%
Sub-Saharan Africa	476	177	78%	638	270	67%
Developing Asia	1 680	240	54%	1 532	198	41%
China	377	46	32%	236	25	19%
India	749	87	72%	719	59	53%
Rest of developing Asia	554	107	63%	576	114	52%
Latin America	61	24	19%	57	17	14%
Middle East	n.a.	n.a.	n.a.	n.a.	n.a.	n.a.
Developing countries	2 221	441	51%	2 230	485	43%
World*	2 221	441	39%	2 230	485	33%

*Includes countries in the OECD and Eastern Europe/Eurasia.

Over the projection period, almost 60% of the investment in clean cooking facilities is expected to be made in biogas solutions, with advanced cookstoves and LPG solutions each accounting for around 20%.[20] Private sector operators in parts of Asia have already made

---

20. Advanced biomass cookstoves, with significantly lower emissions and higher efficiencies than the traditional three-stone fires, are assumed to cost $50. An LPG stove and canister is assumed to cost $60. In the analysis, we assume that LPG stoves and advanced biomass cookstoves require replacement every five years, but only the cost of the first stove and half of the cost of the second stove is included in our investment projections. This is intended to reflect a path towards such investment becoming self-sustaining. The assumed cost of an average-sized biogas digester varies by region. Based on 2010 data provided by SNV, the Netherlands Development Organisation, the cost is $437 for India, $473 in China, $660 in Indonesia, $526 in other developing Asia, $702 in Latin America and $924 in sub-Saharan Africa. Related infrastructure, distribution and fuel costs are not included in the investment costs.

significant progress in establishing profitable markets for biogas solutions. In 2010, China led the market with 5 million biogas plants installed, while the next three largest Asian markets (India, Nepal and Vietnam) had another 0.2 million units collectively (SNV, 2011). In the case of LPG stoves, multilateral development banks and governments are often the source of the initial capital investment, but the private sector may subsequently be involved in fuel distribution. Advanced biomass cookstoves receive relatively more funding from bilateral and multilateral donors. Much of this goes to indirect subsidies intended to establish local, self-sustaining cookstove markets and to increase the demand for advanced cookstoves. Examples of how such funds are applied include the training of stove builders and information campaigns on the health and other benefits of more efficient stoves. Expenditure of this kind is not included in our calculation of the estimated investment cost of access.

In the New Policies Scenario, the number of people in sub-Saharan Africa without clean cooking facilities *increases* by nearly 40%, to reach more than 900 million by 2030, despite a fall in the proportion of population without access. Almost 65% of the increase in number occurs in rural areas. By 2030, one-third of the people without clean cooking facilities globally are in sub-Saharan Africa, up from one-quarter in 2009.

In developing Asia, the number of people without access to clean cooking facilities declines from 1.9 billion in 2009 to around 1.7 billion in 2030. In the New Policies Scenario, the number of people without clean cooking facilities in India peaks before 2015 and then declines, but India still has nearly 780 million people lacking them in 2030. India previously had the "National Programme for Improved Chulhas" (1985 to 2002), and has recently launched the National Biomass Cookstoves Initiative (NCI) to develop and deploy next-generation cleaner biomass cookstoves to households. The government is piloting the demonstration of 100 000 cookstoves during 2011 and 2012 – providing financial assistance for up to 50% of the cost of the stoves – and this will be used to formulate a deployment strategy for India's next five year plan (2012 to 2017).

The number of people without clean cooking facilities in China maintains a declining trend and stands at around 260 million in 2030. China, like India, builds on previous national programmes, such as the National Improved Stove Program, to distribute cookstoves to rural areas. Together, China and India account for all of the fall in the number of people lacking clean cooking facilities in the region. Across the rest of developing Asia, the number of people without access increases by 4.5% to reach 690 million.

Investment needed to achieve modern energy access for all

The remainder of this analysis focuses on the investment required to achieve the goal of universal access to electricity and clean cooking facilities by 2030 – referred to here as the Energy for All Case – and the methods of financing that may be the most appropriate to support this. We have calculated the cost of achieving this goal to be $1 trillion (in year-2010 dollars). This estimate includes the $296 billion reflected in the New Policies

Scenario. Achieving modern energy access for all by 2030 would therefore require more than three-times the expected level of investment in the New Policies Scenario, growing from $14 billion per year to $48 billion per year (Figure 13.4).[21] This means that an additional $34 billion is needed every year, over and above investment already reflected in the New Policies Scenario. The total required is more than five-times the estimated level of actual investment in 2009. Nonetheless, the total investment required is a small share of global investment in energy infrastructure, around 3% of the total.

Figure 13.4 ● **Average annual investment in modern energy access by scenario**

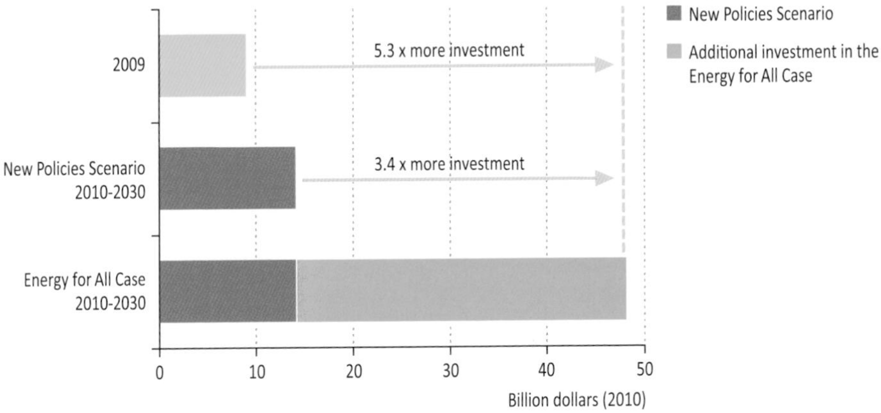

Legend:
- New Policies Scenario
- Additional investment in the Energy for All Case

2009 — 5.3 x more investment
New Policies Scenario 2010-2030 — 3.4 x more investment
Energy for All Case 2010-2030

Billion dollars (2010)

Investment in electricity access

In the Energy for All Case, the additional investment required to achieve universal access to electricity is estimated to be around $640 billion between 2010 and 2030 (Table 13.5).[22] To arrive at this estimate, it was first necessary to assess the required combination of on-grid, mini-grid and isolated off-grid solutions. To identify the most suitable technology option for providing electricity access in each region, the Energy for All Case takes into account regional costs and consumer density, resulting in the key determining variable of regional cost per megawatt-hour (MWh). When delivered through an established grid, the cost per MWh is cheaper than that of mini-grids or off-grid solutions, but the cost of extending the grid to sparsely populated, remote or mountainous areas can be very high and long distance transmission systems can have high technical losses. This results in grid extension being the

---

21. The estimated additional investment required is derived from analysis to match the most likely technical solutions in each region, given resource availability and government policies and measures, with financing instruments and sources of financing.

22. For illustrative purposes, if we instead adopted the assumed minimum consumption threshold of 120 kWh per person in Sanchez (2010), together with our own assumption of five people per household, i.e. a threshold electricity consumption level of 600 kWh per household, this would increase the additional investment required in the Energy for All Case by 4%, taking the total additional investment required to $665 billion to 2030.

most suitable option for all urban zones and for around 30% of rural areas, but not proving to be cost effective in more remote rural areas. Therefore, 70% of rural areas are connected either with mini-grids (65% of this share) or with small, stand-alone off-grid solutions (the remaining 35%). These stand-alone systems have no transmission and distribution costs, but higher costs per MWh. Mini-grids, providing centralised generation at a local level and using a village level network, are a competitive solution in rural areas, and can allow for future demand growth, such as that from income-generating activities.

Table 13.5 ● **Additional investment required to achieve universal access to electricity in the Energy for All Case compared with the New Policies Scenario** ($2010 billion)

	2010-2020	2021-2030	Total, 2010-2030
Africa	119	271	390
Sub-Saharan Africa	118	271	389
Developing Asia	119	122	241
India	62	73	135
Rest of developing Asia	58	49	107
Latin America	3	3	6
Developing countries*	243	398	641
World	243	398	641

* Includes Middle East countries.

More than 60% of the additional investment required is in sub-Saharan Africa, with the region needing the equivalent of an extra $19 billion per year to achieve universal electricity access by 2030. There is greater dependency here on mini-grid and isolated off-grid solutions, particularly in countries such as Ethiopia, Nigeria and Tanzania, where a relatively higher proportion of those lacking electricity are in rural areas. Developing Asia accounts for 38% of the additional investment required to achieve universal electricity access. Achieving universal access to electricity by 2030 requires total incremental electricity output of around 840 terawatt-hours (TWh), and additional power generating capacity of around 220 gigawatts (GW) (Box 13.3 discusses the potential role of hydropower).

In the Energy for All Case, mini-grid and off-grid solutions account for the greater part of the additional investment, $20 billion annually. The annual level of investment is expected to increase over time, reaching $55 billion per year towards 2030 (Figure 13.5). This growth over time reflects the escalating number of additional connections being made annually in the Energy for All Case, going from 25 million people per year early in the projection period to more than 80 million by 2030, and the increasing shift in focus to mini-grid and off-grid connections. It also reflects the gradually increasing level of capital cost associated with the higher level of consumption expected from those households that are connected earlier in the period.

Figure 13.5 ● Average annual investment in access to electricity by type and number of people connected in the Energy for All Case

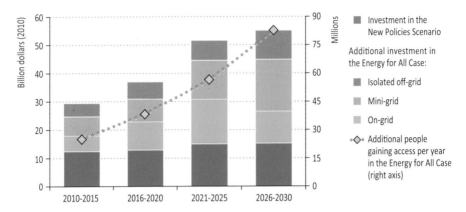

Box 13.3 ● What is the role of hydropower in increasing energy access?

Renewables play a large role in the Energy for All Case. As a mature, reliable technology that can supply electricity at competitive costs, hydropower is one part of the solution to providing universal access to electricity. It has a place in large on-grid projects and in isolated grids for rural electrification. The global technical potential for hydropower generation is estimated at 14 500 TWh, more than four-times current production (IJHD, 2010), and most of the undeveloped potential is in Africa and in Asia, where 92% and 80% of reserves respectively are untapped.

Water basins can act as a catalyst for economic and social development by providing two essential enablers for development: energy and water. Large hydropower projects can have important multiplier effects; creating additional indirect benefits for every dollar of value generated (IPCC, 2011). However, they may have adverse environmental impacts and induce involuntary population displacement if not designed carefully.

The Nam Theun 2 hydropower plant in Laos is an example of a project that has advanced economic and social goals successfully. While managing to achieve this, there are still lessons to be learned in terms of how governments, private developers and multilateral development banks partner to deliver projects more simply and efficiently. Small-scale, hydropower-based rural electrification in China has had some success. Over 45 000 small hydropower plants (SHPs), representing 55 GW, have been built and are producing 160 TWh per year. While many of these plants form part of China's centralised electricity networks, SHPs constitute one-third of total hydropower capacity and provide services to more than 300 million people (Liu and Hu, 2010).

In the Energy for All Case, hydropower on-grid accounts for 14% of additional generation, while SHPs account for 8% of off-grid additional generation. Overall, additional investment in hydropower amounts to just above $80 billion over the period 2010 to 2030. Successfully raising this investment will depend on mitigating the risks related to high upfront costs and lengthy lead times for planning, permitting and construction. Projects that provide broader development benefits and arrangements to tackle planning approval and regulatory risks are important to achieve the required level of investment for hydropower development.

Investment in access to clean cooking facilities

In the Energy for All Case, $74 billion of additional investment is required to provide universal access to clean cooking facilities by 2030, representing nearly four-times the level of the New Policies Scenario. Of this total, sub-Saharan Africa is estimated to need $22 billion. While the largest share of additional investment in the region is for biogas systems, a significant proportion (around 24%) is needed to provide advanced biomass cookstoves to 395 million people in rural areas. Developing Asia accounts for almost two-thirds of the total additional investment required for clean cooking facilities, the largest element ($26 billion) being for biogas systems, principally in China and India (Figure 13.6).

Figure 13.6 ● *Average annual investment in access to clean cooking facilities by type and region, 2010-2030*

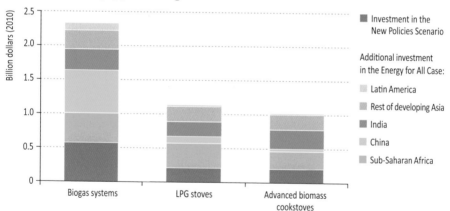

We estimate that to provide over 250 million households worldwide with advanced biomass cookstoves an additional cumulative investment of $17 billion will be needed to 2030 (Figure 13.7). Additional investment of $37 billion is required in biogas systems over the projection period, providing access to around 70 million households.[23] An estimated

---

23. Infrastructure, distribution and fuel costs for biogas systems are not included in the investment costs. Due to an assumed 20-year lifecycle, we assume one biogas system per household in the period 2010 to 2030, thus replacement costs are not included (see footnote 20 for cost assumptions for each technology).

Figure 13.7 ● Average annual investment required by region and technology in the Energy for All Case, 2010-2030

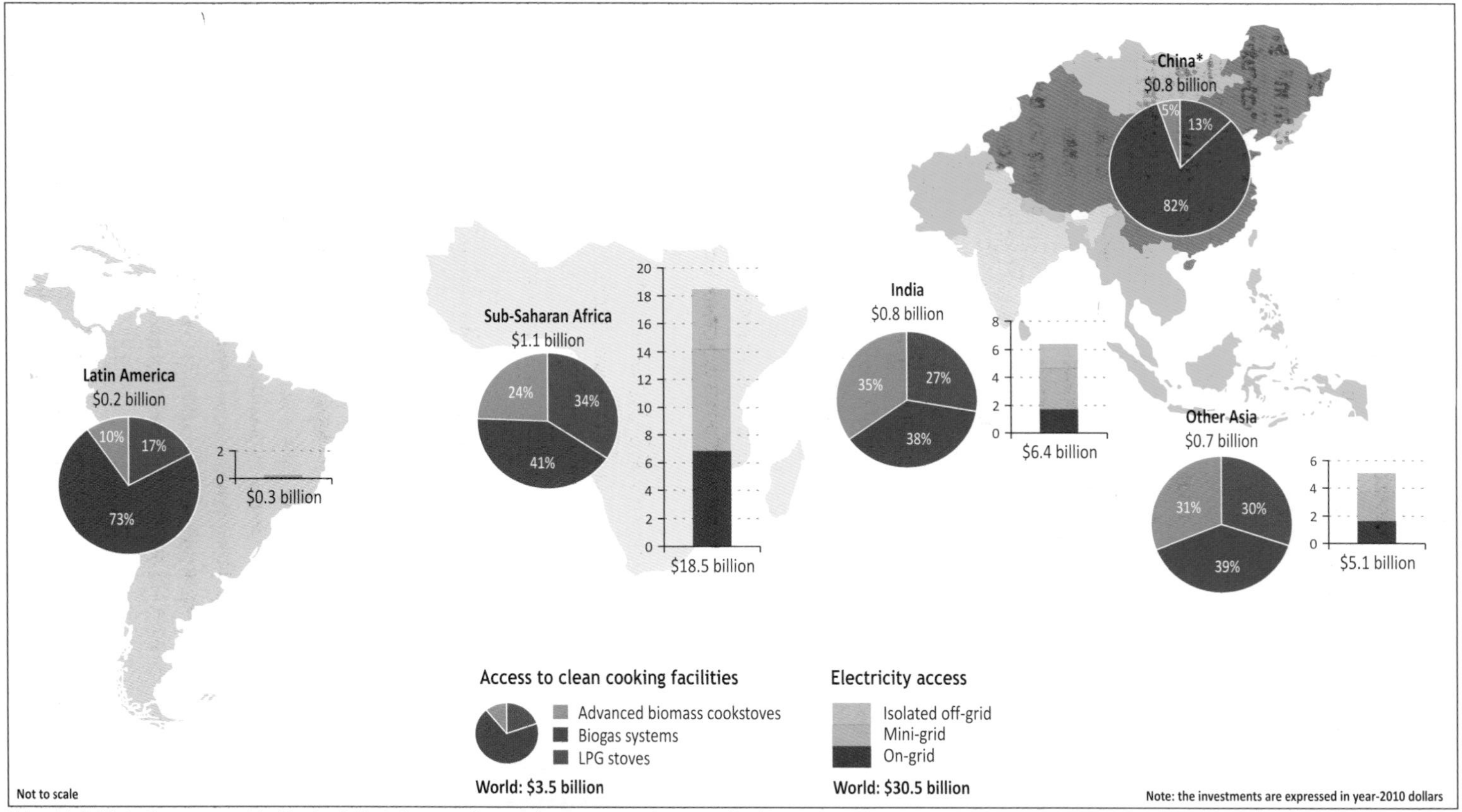

China*
$0.8 billion
5%
13%
82%

Other Asia
$0.7 billion
30%
31%
39%
$5.1 billion

India
$0.8 billion
27%
35%
38%
$6.4 billion

Sub-Saharan Africa
$1.1 billion
34%
24%
41%
$18.5 billion

Latin America
$0.2 billion
17%
10%
73%
$0.3 billion

Access to clean cooking facilities

Advanced biomass cookstoves
Biogas systems
LPG stoves

World: $3.5 billion

Electricity access

Isolated off-grid
Mini-grid
On-grid

World: $30.5 billion

Note: the investments are expressed in year-2010 dollars

Not to scale

This map is for illustrative purposes and is without prejudice to the status of or sovereignty over any territory covered by this map.

*In the Energy for All Case, China's investment in access to electricity is zero and therefore not shown on the map.

additional investment of $20 billion for LPG stoves over the projection period provides clean cooking facilities to nearly 240 million households. Advanced biomass cookstoves and biogas systems represent a relatively greater share of the solution in rural areas, while LPG stoves play a much greater role in urban and peri-urban areas.

Broader implications of achieving modern energy access for all

Achieving the Energy for All Case requires an increase in global electricity generation of 2.5% (around 840 TWh) compared with the New Policies Scenario in 2030, requiring additional electricity generating capacity of around 220 GW. Of the additional electricity needed in 2030, around 45% is expected to be generated and delivered through extensions to national grids, 36% by mini-grid solutions and the remaining 20% by isolated off-grid solutions. More than 60% of the additional on-grid generation comes from fossil fuel sources and coal alone accounts for more than half of the total on-grid additions. In the case of mini-grid and off-grid generation, more than 90% is provided by renewables (Figure 13.8).

Figure 13.8 • *Additional electricity generation by grid solution and fuel in the Energy for All Case compared with the New Policies Scenario, 2030*

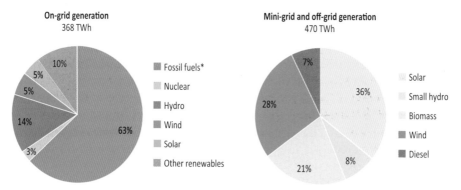

* Coal accounts for more than 80% of the additional on-grid electricity generated from fossil fuels.

Achieving the Energy for All Case is projected to increase global demand for energy by 179 million tonnes of oil equivalent (Mtoe), an increase of 1.1% in 2030, compared with the New Policies Scenario (Table 13.6). Fossil fuels account for around 97 Mtoe, over half of the increase in energy demand in 2030. While an additional 0.88 million barrels per day (mb/d) of LPG is estimated to be required for LPG cookstoves in 2030, this is expected to be available largely as a by-product of increased production of natural gas liquids (NGLs) and refining crude oil. Coal demand increases by almost 60 million tonnes of coal equivalent (Mtce) in 2030, around the current production level of Colombia. Ample coal reserves are available globally to provide this additional fuel to the market (see Chapter 11). Other renewables, mostly solar and wind, enjoy the largest proportional increase in demand in 2030, providing additional deployment opportunities beyond those in the New Policies Scenario.

Table 13.6 ● Additional energy demand in the Energy for All Case compared with the New Policies Scenario, 2020 and 2030

	Additional demand (Mtoe)		Change versus the New Policies Scenario	
	2020	2030	2020	2030
Coal	10	42	0.2%	1.0%
Oil	25	48	0.6%	1.1%
Gas	1	7	0.0%	0.2%
Nuclear	3	3	0.3%	0.2%
Hydro	6	8	1.5%	1.7%
Biomass and waste	8	31	0.5%	1.8%
Other renewables	12	41	4.0%	7.8%
Total	64	179	0.4%	1.1%

In 2030, CO_2 emissions in the Energy for All Case are 239 million tonnes (Mt) higher than in the New Policies Scenario, an increase of only 0.7% (Figure 13.9). Despite this increase, emissions per capita in those countries achieving universal access are still less than one-fifth of the OECD average in 2030. The small size of this increase in emissions is attributable to the low level of energy per-capita consumed by the people provided with modern energy access and to the relatively high proportion of renewable solutions adopted, particularly in rural and peri-urban households. The diversity of factors involved means that the estimate of the total impact on greenhouse-gas emissions of achieving universal access to modern cooking facilities needs to be treated with caution. However, it is widely accepted that advanced stoves and greater conversion efficiency would result in a reduction in emissions and thereby reduce our projection.

Figure 13.9 ● Additional global energy demand and CO_2 emissions in the Energy for All Case compared with the New Policies Scenario, 2030

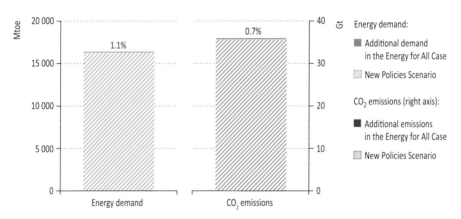

Notes: Percentages are calculated as a share of the total energy demand or CO_2 emissions respectively in 2030. Gt = gigatonnes.

As well as the economic development benefits, the Energy for All Case, if realised, would have a significant impact on the health of those currently cooking with biomass as fuel in basic, inefficient and highly-polluting traditional stoves. Based on World Health Organization (WHO) projections, linked to our projections of the traditional use of biomass in cooking,[24] the number of people who die prematurely each year from the indoor use of biomass could be expected to increase to over 1.5 million in the New Policies Scenario in 2030. The adoption of clean cooking facilities is expected to prevent the majority of deaths attributable to indoor air pollution.[25] The number of premature deaths per year attributable to indoor air pollution is higher than what the WHO projects for deaths from malaria and HIV/AIDS combined in 2030 (Figure 13.10). In addition to avoiding exposure to smoke inhalation, modern energy services can help improve health in other ways, such as refrigeration (improving food quality and storing medicines) and modern forms of communication (supporting health education, training and awareness).[26]

Figure 13.10 ● **Premature annual deaths from household air pollution and selected diseases in the New Policies Scenario**

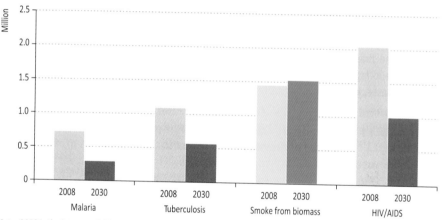

Note: 2008 is the latest available data in WHO database.

Financing to achieve modern energy access for all

The size of the increase in investment that is required in the Energy for All Case is significant. We focus here on how the investment required to achieve the objectives of the Energy for All Case can best be financed. Whatever the possible sources, it is important to recognise

24. Estimates for premature deaths are based on *WEO-2010* projections for biomass use and on Mathers and Loncar (2006); WHO (2008); Smith *et al.*, (2004); and WHO (2004).

25. Evidence of fewer child deaths from Acute Lower Respiratory Infection can be expected soon after reductions in solid fuel air pollution. Evidence of averted deaths from chronic obstructive pulmonary disease would be observed over a period of up to 30 years after adoption of clean cookstoves, due to the long and variable time-scales associated with the disease.

26. See *WEO-2006* and *WEO-2010* for a detailed discussion of the harmful effects of current cooking fuels and technologies on health, the environment and gender equality; and for a broader discussion on the link between energy and the Millennium Developing Goals (IEA, 2006 and IEA, 2010).

loans from international or local banks on the basis of the financial attractiveness of the project, backed by multilateral development bank guarantees. Attracting private investment to such projects depends crucially on investors being able to charge tariffs that generate a reasonable return. In some instances, a state-owned utility mandated by the government to provide universal access may be able to attract private sector loans at competitive rates to supplement internal financing. In other instances, a utility (in private or public ownership) may not be sufficiently creditworthy to raise finance commercially, and may require support, such as through a partial risk guarantee.

Table 13.7 ● **Additional financing for electricity access in the Energy for All Case compared with the New Policies Scenario[31], 2010-2030**

	Additional annual investment ($ billion)	People gaining access annually (million)	Level of household energy expenditure	Main source of financing	Other sources of financing
On-grid	11.0	20	Higher	Private sector	Developing country utilities
			Lower	Government budget	Developing country utilities
Mini-grid	12.2	19	Higher	Government budget, private sector	Multilateral and bilateral guarantees
			Lower	Government budget	Multilateral and bilateral concessional loans
Off-grid	7.4	10	Higher	Multilateral and bilateral guarantees and concessional loans	Private sector, government budget
			Lower	Multilateral and bilateral concessional loans and grants	Government budget

In providing on-grid electrification for lower energy expenditure households, there is a stronger case for explicit public sector funding, such as targeted government subsidies or an equity investment.[32] For example, in the case of Vietnam's successful rural electrification programme, significant cost sharing by local government and the communities being electrified was an important element of the financing model. Cross-subsidisation between higher energy expenditure households or business customers and those with lower energy expenditure may also be pursued (though not desirable on a long-term basis), such as the state-owned utility Eskom has done in South Africa.

---

31. See www.worldenergyoutlook.org/development.asp for more on the methodology related to this table.

32. An equity investment is one in which the investor receives an ownership stake in a project, giving entitlement to a share of the profits (after all associated debts have been paid), but also liability to bear part of any residual losses.

Electricity access – financing mini-grid electrification

In the Energy for All Case, mini-grid electrification requires additional annual investment of $12 billion per year. This area requires the largest increase in financing, relative to the New Policies Scenario, with more than $8 billion per year in additional investment required, on average, to connect an additional 19 million people annually.

Often financed initially by government programmes, mini-grids (diesel and small-hydro) have played an important role in rural electrification in China, Sri Lanka and Mali (World Bank, 2008). Under a Global Environment Fund (GEF) Strategic Energy Programme for West Africa, renewable energy powered mini-grids are being established in eight countries.[33] Hybrid mini-grids, integrating renewable generation with back-up capacity, have expanded rapidly in Thailand (Phuangpornpitak *et al.*, 2005), and are becoming competitive compared with 100% diesel-based generation (ARE, 2011a). In Laos, a successful public-private partnership, has been established to fund a hybrid (hydro, solar PV and diesel) mini-grid, serving more than 100 rural households. In the project, public partners fund the capital assets, while the private local energy provider finances the operating costs (ARE, 2011b).

The most appropriate type of technical and financing solution for mini-grid projects can vary significantly. In some cases, mini-grid projects can be run on a cost-recovery basis with a guaranteed margin, and therefore attract private sector finance on commercial terms (particularly diesel systems). In the case of more marginal projects, output-based subsidies may be used to support private sector activity in the sector. For many high energy expenditure households, an auction for concessions, combined with output-based subsidies, can keep subsidies low while giving concessionaires incentives to complete promised connections. In such cases, electricity providers bid for the value of subsidies that they require (referred to as "viability gap funding") or for the number of electricity connections they will make during a specified period at a pre-determined rate of subsidy per connection. Ideally, such auctions are technology-neutral, as in Senegal's recent programme, allowing providers to determine the most cost-effective solution. Loans or grants to the government from multilateral and bilateral sources could provide financing to support the initial auction and subsidy costs, as the International Development Association (IDA) and GEF grants did in Senegal (GPOBA, 2007). Such sources may also help support end-user financing programmes which offer assistance to cover the connection charges through the concessionaire or local banking system. For example, the IDA and GEF helped Ethiopia's Electric Power Corporation offer credit to rural customers (GPOBA, 2009).

An important form of financing for mini-grid electrification for low energy expenditure households is expected to be government-initiated co-operatives and public-private partnerships.[34] Bangladesh and Nepal provide examples of such co-operatives (Yadoo and

---

33. See www.un-energy.org.

34. Many forms of business co-operatives exist but, in general, the term refers to a company that is owned by a group of individuals who also consume the goods and services it produces and/or are its employees. A utility co-operative is tasked with the generation and/or transmission and distribution of electricity.

Cruickshank, 2010). There is a relatively high assumed capital subsidy from the government in this model, typically with support in the form of concessional loans from multilateral and bilateral donors.

Electricity access – financing off-grid electrification

Isolated off-grid electricity solutions require additional investment of $7 billion per year to 2030. This represents an increase of $5 billion per year, compared to the New Policies Scenario, in order to provide electricity access to an additional 10 million people per year. In general, off-grid connections are less attractive to the private sector and require different technical solutions and related financing. In the Energy for All Case, the main financing model for off-grid electrification of high energy expenditure households involves enhancing the capacity of dealers in solar home systems and lanterns to offer financing to end-users. Examples of this may be found in the Philippines (UNEP, 2007) and Kenya (Yadoo and Cruickshank, 2010). Government and concessional funds could also be used directly to support microfinancing[35] networks or local banks that, in turn, provide loans down the chain to end-users, as has happened, for example, in UNEP's India Solar Loan Programme (UNEP, 2007) and in several African countries under the Rural Energy Foundation, which is supported by the government of the Netherlands (Morris *et al.*, 2007). In some cases, where microfinance is not available, local agricultural co-operatives might be a channel for funds. Government and concessional funds could also be used for output-based subsidies in some countries. Different sources of financing can play complementary roles in different stages of a programme or project to deliver energy access. For example, a programme for small hydro systems in rural areas in Nepal received over 90% of its funding from public sources at the beginning, much of which was dedicated to capacity development. The share of public financing gradually declined to about 50% at a later stage, suggesting that public investments in developing national and local capacities subsequently attracted private financing (UNDP, AEPC, Practical Action, 2010).

Off-grid electrification of low energy expenditure households is the most challenging area in which to raise finance. A potentially attractive solution for many such cases is sustainable solar marketing packages, pioneered by the World Bank and GEF in the Philippines and later introduced in Zambia and Tanzania. They are based on a service contract to install and maintain solar photovoltaic systems to key public service customers, such as schools, clinics and public buildings. Such contracts include an exclusive right to provide such services also to households and commercial customers, and provide a subsidy for each non-public system installed in the concession area. As for many other solutions, the development of end-user financing is also important. The first phase of the "Lighting Africa" programme by the IFC and World Bank saw the most basic needs met through solar home systems (SHS) provided on a fee-for-service basis. While donor-based models remain, and SHS are still an important and growing segment, the lighting market is now entering a new phase that is being led

---

35. The term microfinance typically refers to the provision of financial services to low income people that lack access to such services from mainstream providers, either due to the small sums involved or because they are on terms that are not considered commercially attractive. The stated intention of microfinance organisations is often to provide access to financial services as a means of poverty alleviation.

by entrepreneurs providing solar portable lights. The scale of these operations is currently small, and the cost can still be a barrier, but the technology is improving at a rapid rate and business models are maturing (IFC and World Bank, 2010).

Clean cooking facilities – financing LPG stoves

In the Energy for All Case, of the additional $3.5 billion per year in investment needed to achieve universal access to clean cooking facilities, $0.9 billion is required for LPG stoves to supply an additional 55 million people per year with a first stove and financing for 50% of the first replacement after five years (Table 13.8). Households supplied with LPG stoves are concentrated in urban and peri-urban areas or may be in areas with high levels of deforestation. As in most countries where LPG stoves have been successfully introduced, such as Kenya, Gabon and Senegal, the government has a role to play in market creation, such as developing common standards and the distribution infrastructure. This will require a certain amount of investment on the part of the government, which may be financed in part by concessional loans from multilateral and bilateral institutions. Besides investment in supporting public infrastructure, such as roads, the government may need to ensure that loans are available for entrepreneurs wishing to invest in LPG distribution. This could be done through a guarantee programme for a line of credit made available through participating local banks, possibly ultimately supported by a multilateral development bank.

Table 13.8 • **Additional financing for clean cooking facilities in the Energy for All Case compared with the New Policies Scenario, 2010-2030**

	Additional annual investment ($ billion)	People gaining access annually (million)	Level of household energy expenditure	Main source of financing	Other sources of financing
LPG	0.9	55	Higher	Government budget, private sector	Multilateral and bilateral development banks, microfinance
			Lower	Government budget, multilateral and bilateral development banks	Private sector
Biogas systems	1.8	15	Higher	Private sector	Microfinance, government budget, multilateral and bilateral development banks
			Lower	Government budget, multilateral and bilateral development banks	Private sector, microfinance
Advanced biomass cookstoves	0.8	59	Higher	Private sector	Government budget, multilateral and bilateral development banks
			Lower	Government budget, multilateral and bilateral development banks	Private sector

13

Higher income households are assumed either to purchase their LPG stove and first cylinder directly from their own resources or to obtain credit from banks or microfinance institutions to do so. For example, access to credit through microfinance institutions has helped to promote a relatively rapid uptake of LPG in Kenya (UNDP, 2009). In many countries, urban and peri-urban areas are those where most LPG penetration is expected and also those that are more likely to be served by microfinance institutions. However, some microfinance institutions may initially require a partial credit guarantee provided by the public sector to generate confidence in lending to a new market. Lower income households receiving LPG stoves in the Energy for All Case are expected to benefit from a loan or subsidy that covers the initial cost of the stove and the deposit on the first cylinder. This loan or subsidy is assumed to be funded in part by the government and in part by multilateral and bilateral donors. Experience in Senegal has shown that LPG sometimes requires subsidies to be maintained for a period in order to keep costs below the monthly amounts that households previously spent on competing wood fuel or charcoal. Indonesia has undertaken a programme to distribute free mini-LPG kits to more than 50 million households and small businesses in an attempt to phase out the use of kerosene for cooking (and reduce the fiscal burden of the existing kerosene subsidy). Analysis of the programme indicates that a capital investment of $1.15 billion will result in a subsidy saving of $2.94 billion in the same year (Budya and Yasir Arofat, 2011).

Clean cooking facilities – financing biogas systems

In the Energy for All Case, additional annual investment of $1.8 billion is required in biogas systems over the projection period. This is an increase in investment of $1.2 billion annually, compared with the New Policies Scenario, and provides an additional 15 million people each year with a biogas system for cooking. In the Energy for All Case, an output-based subsidy programme for trained and certified installation companies is assumed to cover about 30% of the cost of a biogas digester. In 2010, a subsidy of 26% of the total cost was available for a home biogas plant of an average size in Bangladesh and Nepal, while in China subsidy levels have been as high as 69% of total costs (SNV, 2011). A subsidy may be provided to the builder via a rural development agency or equivalent after verification of successful installation. In return for receiving the subsidy, the installer can be obliged to guarantee the unit for several years. Assistance from multilateral and bilateral donors or NGOs can help train biogas digester builders, as the Netherlands Development Organisation (SNV) has done in several of countries.

Both higher and lower income households may require a loan to cover part of the cost of a biogas system. For example, the Asian Development Bank (ADB) has worked with the Netherlands' SNV to add a credit component to a biogas programme in Vietnam (ADB, 2009). The Biogas Partnership in Nepal has on-lent donor and government funds to over 80 local banks and microfinance institutions to provide end-user financing (UNDP, 2009 and Ashden Awards, 2006). This programme involved support for the development of local, private sector biogas manufacturing capacity, as well as training and certification facilities to ensure that quality standards were maintained. Between 35% and 50% of the capital costs were subsidised through grants from international donors, such as the German development finance institution, KfW. Loan capital was available for the remaining capital investment.

The government, through a national development bank or rural energy agency, may need to support microfinance institutions or rural agricultural credit co-operatives to expand their coverage and lending to rural areas. This can be done by offering grants, or by temporarily offering partial credit guarantees or loans at below-market interest rates that enable on-lending, until financial institutions are confident to operate in the new market. In some cases, lower income households may also lower the unit price by contributing their labour, which can be around 30% of the cost (Ashden Awards, 2010).

Clean cooking facilities – financing advanced cookstoves

In the Energy for All Case, additional annual financing of $0.8 billion is required in advanced biomass cookstoves. This is an increase of $0.6 billion per year, compared with the New Policies Scenario, and serves to provide a first advanced cookstove to an additional 60 million people per year and financing for 50% of the first replacement after five years. While advanced cookstoves can help cut wood fuel use substantially, the economic arguments alone may not be compelling for many households, especially if wood fuel is considered "free" and the time of the persons collecting it – typically women and girls – is not sufficiently valued. Comprehensive public information and demonstration campaigns to explain the health and other benefits are therefore likely to be required to increase household acceptance. In addition, funding will be required to ensure adequate quality control of cookstoves.[36] Such campaigns are expected to be funded with grants, either from the government or multilateral and bilateral development partners, and will benefit from international support through initiatives such as The Global Alliance for Clean Cookstoves. Public information and demonstration campaigns have successfully led to market transformation in Uganda, Mali and Madagascar among others (AFD, 2011). In Sri Lanka, an estimated 6 million advanced cookstoves have been sold over the last ten years using innovative business models, such as "try before you buy". The programme has been supported by several international donors and the government of Sri Lanka (IEA, 2011).

Of the additional investment in the Energy for All Case, an estimated 70% is directed to lower income households. For these, the provision of credit to help purchase advanced cookstoves may be appropriate in some cases, as successfully implemented by Grameen Shakti in Bangladesh (Ashden Awards, 2008). Unfortunately, use of microcredit may be problematic for advanced stoves, particularly because of the high transaction costs compared with the purchase price and the traditional focus of microcredit on income-generating activities (Marrey and Bellanca, 2010). As an alternative, the government may help develop dealer financing through certified cookstove builders, *e.g.* using a partial credit guarantee with funds provided by the government or by multilateral and bilateral development partners. Experience in some countries has shown that large subsidies (and especially give-aways) can actually undermine the market for advanced stoves and create expectations of a subsidy for replacement stoves (AFD, 2011).

13

---

36. Funding for quality control is not included in our estimates of the required investment costs.

Sources of financing and barriers to scaling up

This section considers the main sources of financing in more detail, and the types of projects and instruments to which they are, or may become, most effectively committed. These sources are summarised under the main categories: multilateral and bilateral development sources, developing country government sources and private sector sources. It is recognised that there are instances in which these categories may overlap or change over time. For example, countries currently focused on investing in energy access domestically may also invest in other countries. Rapidly industrialising countries, such as China and India, may be such cases. Within each broad category, several different types of organisations may offer one or more types of financing instrument to improve energy access. Table 13.9 shows different financing instruments and a summary of the sources that might typically offer them.

Table 13.9 ● **Sources of financing and the financing instruments they provide**

	Grants / credits	Concessionary loans	Market-rate loans	Credit line for on-lending	Partial credit guarantees	Political risk insurance	Equity	Quasi-equity	Carbon financing	Subsidy / cross-subsidy	Feed-in tariff	Technical assistance
Multilateral development banks	✓	✓	✓	✓	✓	✓	✓	✓	✓			✓
Bilateral development agencies	✓	✓	✓	✓					✓			✓
Export-import banks / guarantee agencies			✓			✓						✓
Developing-country governments	✓	✓					✓	✓		✓	✓	
State-owned utilities							✓			✓	✓	
National development banks		✓	✓	✓	✓							✓
Rural energy agencies/ funds	✓									✓		✓
Foundations	✓						✓		✓			
Microfinance			✓									
Local banks			✓									
International banks			✓					✓	✓			
Investment funds							✓		✓			
Private investors							✓		✓			✓

Multilateral and bilateral development sources

Multilateral development sources include the World Bank Group,[37] the regional development banks[38] and major multilateral funds, such as the Organization of the Petroleum Exporting Countries Fund for International Development (OFID) and the Scaling-up Renewable Energy Program for Low-Income Countries (SREP) (Box 13.4 considers the International Energy and Climate Partnership – Energy+). Bilateral sources are primarily official development assistance provided by the 24 OECD countries that are members of the OECD Development Assistance Committee (DAC). These OECD member countries account for the bulk of global development aid (99% of total ODA in 2010), including that provided via multilateral sources, but this situation continues to evolve. The major financing instruments used by these sources for energy access projects are grants, concessional loans and investment guarantees.[39] Carbon financing is another instrument that has begun to be utilised for energy access projects.

Credits from the International Development Association (IDA) have been the main instrument employed by the World Bank Group for energy access projects, followed by grants, including from special funds such as the GEF and the Carbon Funds. However, obtaining grants can require long proposal preparation periods and the need to satisfy multiple criteria. The International Bank for Reconstruction and Development (IBRD) also provides concessional loans to medium-income governments, which are typically applied to large electricity infrastructure projects. The IFC is able to lend to the private sector and organise loan syndications that give international banks greater confidence to invest in projects in developing countries. It also lends to local financial institutions for on-lending to small and medium businesses, and is increasingly creating guarantee products that help develop the capacity of the local banking sector and making equity investments. The UN Development Programme (UNDP) and UN Environment Programme (UNEP) have been particularly active in helping develop schemes for end-user finance.

Political risk insurers, such as the Multilateral Investment Guarantee Agency (MIGA), and bilateral programmes such as Norway's Guarantee Institution for Export Credit (GIEK), have a mission to promote foreign direct investment in developing countries by insuring private investors against risks such as breach of contract, non-fulfilment of government financial obligations and civil disturbances. Obtaining such risk insurance can have leveraging effects, making it easier for projects to obtain commercial finance, or to do so at lower cost. Financing from most multilateral and bilateral development sources is usually accompanied by technical assistance, such as policy and institutional development advice to ensure the efficient use of the provided funds. Such investment in technical assistance can be important in ensuring that an adequate number of private projects enter the financing pipeline.

13

---

37. World Bank Group includes the World Bank, International Development Association, International Bank for Reconstruction and Development, International Finance Corporation and Multilateral Investment Guarantee Agency.

38. Regional development banks include the Asian Development Bank, African Development Bank and Inter-American Development Bank.

39. Bilateral development sources offer many of the same financing products as multilaterals sources.

Potential barriers to scaling up the financing instruments provided by multilateral and bilateral sources for energy access include: the significant amount of regulatory and financial sector reforms that may be necessary to enable some countries to absorb increases in development (and other) financing; the need to satisfy multiple criteria in order to apply much of the available development assistance to energy access projects, particularly those related to renewable sources and climate change; and, the reordering of development priorities that may be required of organisations (and the governments behind them) in order to increase the share of energy-access projects within their portfolios.

Box 13.4 ● International Energy and Climate Partnership – Energy+

The International Energy and Climate Partnership – Energy+, an initiative that aims to increase access to energy and decrease or avoid greenhouse-gas emissions by supporting efforts to scale up investments in renewable energy and energy efficiency, is a pertinent example of the increasing international recognition of the importance of providing modern energy access to the poor. It focuses on the inter-related challenges of access to modern energy services and climate change, recognising that both issues require a serious increase in capital financing. The initiative seeks to engage with developing countries to support large-scale transformative change to energy access and to avoid or reduce energy sector greenhouse-gas emissions. It seeks to apply a results-based sector level approach and to leverage private capital and carbon market financing. The Energy+ Partnership aims to co-operate with governments and to leverage private sector investment, to develop commercially viable renewable energy and energy efficiency business opportunities to meet the challenge of increasing access to energy in a sustainable manner. The intention is to facilitate increased market readiness by creating the necessary technical, policy and institutional frameworks. The government of Norway has initiated dialogue with possible partners to develop the initiative.

Carbon financing

Carbon finance offers a possible source of income for energy access projects that also help reduce greenhouse-gas emissions. The revenue is raised through the sale of carbon credits within the Clean Development Mechanism (CDM) and voluntary mechanisms. The value of carbon credits produced from new CDM projects reached around $7 billion per year prior to the global financial and economic crisis. However, low income regions so far have made little use of carbon finance mechanisms to mobilise capital for investment in energy access. Up to June 2011, only 15 CDM projects, or 0.2% of the total, have been designed to increase or improve energy access for households.[40]

The potential for projects to serve both energy access and climate change purposes in sub-Saharan Africa is estimated to be large, nearly 1 200 TWh (150 GW) of electricity

---

40. Data available at UNEP RISOE CDM Pipeline Analysis and Database at www.uneprisoe.org.

generation at an investment cost of $200 billion. In total, these projects could possibly generate $98 billion in CDM revenue at a carbon offset price of $10 per tonne of CO_2 (World Bank, 2011).

Substantial obstacles must first be overcome. Getting any project approved for CDM is at present often a long, uncertain and expensive process. Upfront costs are incurred to determine the emissions baseline and to get the project assessed, registered, monitored and certified. These high transaction costs mean that CDM is not currently practical for small projects. The CDM Executive Board has taken steps to simplify the requirements for small-scale projects and for projects in the least developed countries and it is hoped that these and other ongoing initiatives, such as standardised project baselines, will facilitate the application of the CDM for energy access projects. The increasing development of programmatic CDM should help reduce transaction costs by consolidating the small carbon savings of individual access projects. National governments in developing countries can act to reap the benefits from such candidates, as recent projects have shown for advanced cookstoves in Togo, Zambia and Rwanda, and for household lighting in Bangladesh and Senegal. Rural electrification agencies or national development banks can act under government direction as co-ordinating and managing entities for bundling energy access projects.

Consensus is building on the importance of using carbon finance to support the development agenda in poor countries. EU legislation provides that carbon credits from new projects registered after 2012 can only be used in the EU Emissions Trading System if the projects are located in the least developed countries. Such steps provide a more bankable basis for raising capital. To get capital flowing into energy access projects backed by carbon finance in low income countries, it remains for national governments fully to empower the relevant national authority to simplify the regulatory requirements and to create the secure commercial environment necessary to win investor confidence. A recent report by the UN Secretary-General's High-level Advisory Group on Climate Financing (AGF)[41] recognised carbon offset development as a stimulant for private sector investment. Building private sector understanding of the process of using carbon finance and confidence in it is an important objective for all parties.

Developing country government sources

Important sources and forms of finance from within developing countries include the balance sheet of state-owned utilities, subsidies provided by the government, grants and loans offered by developing country national development banks, and specialised national institutions and funds, such as rural energy agencies. In many developed countries, grid expansion is financed from the internally generated funds of private or state-owned utilities. This option is not available where state-owned utilities in developing countries often operate at a loss or rely on state subsidies for capital investment and, sometimes, operating

41. The UN Secretary-General established a High-level Advisory Group on Climate Change Financing on 12 February 2010 for a duration of ten months. This Group studied potential sources of revenue that can enable achievement of the level of climate change financing that was promised during the United Nations Climate Change Conference in Copenhagen in December 2009. See www.un.org/climatechnage/agf.

expenditures. Government utilities are nonetheless often a conduit for government funds in practice. While some are able to borrow on the international or local market, based on a government guarantee or good financial track record, many others are not.

Failure by government entities to pay for utility services and politically-imposed limitations on utilities' ability to enforce payment through disconnection are additional important barriers to balance sheet financing by state-owned utilities. Pre-payment meters, which gained widespread use in South Africa's electrification programme over the past decade, have helped power companies in many developing countries address the non-payment issue, although the capital costs of such metering programmes can be a barrier.

Subsidies can be provided from the government budget, sometimes supported by donor funds (Chapter 14 examines developments in energy subsidies). It is important that subsidies are used sparingly and are precisely targeted at those unable to pay and at the item they may have difficulty paying for, usually the connection fee. Unfortunately, many government subsidies in the energy sector are not well targeted (Figure 13.12). A typical example is the provision of consumption subsidies, including "lifeline" tariffs that provide the first 20 to 50 kWh of electricity at below cost to all customers regardless of income. Not only does this waste scarce funds that could be better targeted at poor people, but it foregoes an opportunity to collect cross subsidies from those customers who could afford to pay more. Cross subsidies from customer groups that pay more for their power than the cost to supply them can be an initial source of funds to help provide energy access to the poor, but are not an efficient long-term solution.

Figure 13.12 ● **Fossil-fuel subsidies in selected countries, 2010**

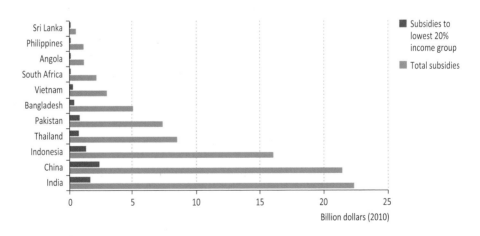

Many developing countries have established development banks to help channel government and donor finance to priority sectors that are not receiving sufficient private investment. National development banks are useful entry points for multilateral and bilateral development institutions seeking to use their lending to the energy-sector, including to end-users, to generate complementary funding from other sources. Some have on-lending and credit guarantee programmes with the local banking sector that might be adapted to lending for energy access projects. National development banks may sometimes be able to serve as an official guarantor for lending programmes supported with donor financing. In recent years, several developing country governments have established agencies specifically to fund and facilitate rural electrification. Some also promote modern cooking facilities, renewables and energy efficiency. Use of rural energy agencies has been particularly prevalent in Africa, including in Mali, Tanzania, Zambia, Senegal and Uganda, though they exist in several Asian countries too, including Cambodia and Nepal.

There are many reasons why domestic governments have difficulty in attracting or repaying financing for energy access. The most notable is poor governance and regulatory frameworks. The absence of good governance increases risk, so discouraging potential investors. The issue must be tackled.

Private sector sources

Private sector financing sources for energy access investments include international banks, local banks and microfinance institutions, as well as international and domestic project developers, concessionaires and contractors. Private finance may also come from specialist risk capital providers, such as venture capital funds, private equity funds and pension funds. The main forms of instruments favoured by private sources include equity, debt and mezzanine finance.[42] An increasingly important instrument, offered through local banks, is the extension of credit to end-users, often with guarantees arranged in partnership with multilateral development banks.

Private investors, enjoying a choice of where to place their money, across countries and across sectors, respond to tradeoffs between risk and reward. Important issues to tackle when seeking to increase private sector investment therefore include the provision of a competitive rate of return that incentivises private sector performance while representing value for money to the public sector, and the clear allocation to the most appropriate party of responsibility for risk. Existing experience reveals that justifying the business case is not always easy, and many private sector participants in energy access projects are doing so on the grounds of broader benefits to the company, such as corporate social responsibility. Despite the challenges, there is significant innovation taking place with several models, products and services in a pilot stage of development. Many potential private sector participants currently view a PPP-type model to be among the most attractive. In instances where the business case for private sector investment is marginal, but there are clear public benefits, government support to enhance or guarantee investment returns may be appropriate.

Countries that are seen to be particularly risky, in terms of macroeconomic, political or regulatory stability, either have to assume more of the risks themselves, by offering credible

13

---

42. Mezzanine finance is a hybrid of debt and equity financing. Mezzanine financing is basically debt capital that gives the lender the right to convert to an ownership or equity interest in the company under certain pre-agreed conditions.

guarantees, or seek to have these risks covered by some form of insurance. A strong track record of introducing and implementing robust and equitable government policies can reduce the need for financing guarantees. Important factors for private investors in the power sector typically include (Lamech and Saeed, 2003):

■ A legal framework that defines the rights and obligations of private investors.

■ Consumer payment discipline and enforcement.

■ Credit enhancement or guarantee from the government or a multilateral agency.

■ Independent regulatory processes, free from arbitrary government interference.

International commercial banks have an established record of financing projects in the energy sector in emerging markets, predominantly in power generation. Pricing finance at market rates according to perceived risk, they offer debt financing, mezzanine finance and, in some cases, equity. They can lend to project developers directly or to a special purpose vehicle set up to conduct the project. Commercial bank financing terms can be less onerous if certain risks are covered by guarantees from a multilateral development bank or the host government.

Local financial institutions in many developing countries are unable or unwilling to provide credit to rural energy projects or to the end-users of such projects. The World Bank, UNDP and UNEP, among others, have been involved in pilot projects to create links between local banks and renewable energy service providers and help them design suitable credit instruments. Understanding households' existing energy expenditures is one important step towards unlocking end-user finance: poor people often are able to afford the full price of modern energy because it costs less than the traditional forms it replaces, *e.g.* kerosene lamps and dry cell batteries, but may be unable to overcome the important hurdle of the initial capital cost.

Microfinance has been used as part of several programmes to tackle the problem of end-user financing for energy access, particularly in India and Bangladesh. It has been found to be particularly useful for grid connection fees and LPG stoves. The scale of the transaction is important. Microfinance has proved more problematic in relation to large solar home systems, where the loan size and consequent payment period can be greater than microfinance institutions are used to handling, or in relation to wood-burning stoves, where the transaction costs can be large relative to the loan amounts involved. Microfinance institutions are often part of networks, which act as wholesale lenders to them: these networks may be able to develop guaranteed lines of credit and related technical assistance from larger organisations. However, microfinance institutions and associated networks are often less prevalent in rural areas, in particular in parts of sub-Saharan Africa.

The main obstacle to obtaining greater private sector financing, apart from uncertain investment and regulatory environments and political risks in many developing countries, is the lack of a strong business case for tackling the worst cases of energy deprivation, because of the inability of users to pay. This issue needs to be squarely faced, through some form of public sector support, if there is to be a breakthrough to universal access to modern energy. In addition, local financial institutions and microfinance institutions find it difficult to be sufficiently expert regarding new technologies and may underestimate the potential credit-worthiness of poor households, based on the large amount they already pay for more traditional sources of energy.

Implications for policy

Modern energy services are crucial to economic and social development; yet escalating global energy prices are pushing this fundamental building block further out of reach of those most in need. Even with the projected level of investment in modern energy access of $14 billion per year in the New Policies Scenario, the absolute numbers of people without access to modern energy in 2030 will be scarcely changed (though the proportion of the global population so deprived will have fallen). In sub-Saharan Africa, the numbers without modern energy access will have actually increased. Neither the policies adopted today nor the plausible new policies allowed for in the New Policies Scenario will do nearly enough to achieve universal access to modern energy services by 2030.

Global energy access is a necessary prerequisite of global energy security. The barriers to achieving modern energy access are surmountable, as many countries have proven. What actions does this analysis suggest that are essential to transform the situation? There are five:

■ Adopt a clear and consistent statement that modern energy access is a political priority and that policies and funding will be reoriented accordingly. National governments need to adopt a specific, staged energy access target, allocate funds to its achievement and define their strategy, implementing measures and the monitoring arrangements to be adopted, with provision for regular public reporting.

■ Mobilise additional investment in universal access, above the $14 billion per year assumed in the New Policies Scenario, of $34 billion per year. The sum is large, but is equivalent to around 3% of global energy infrastructure investment over the period.

■ Draw on all sources and forms of investment finance to reflect the varying risks and returns of the particular solutions adapted to the differing circumstances of those without access to modern energy. To realise the considerable potential for stepping up the proportional involvement of the private sector, national governments need to adopt strong governance and regulatory frameworks and invest in internal capacity building. Multilateral and bilateral institutions need to use their funds, where possible, to leverage greater private sector involvement and encourage the development of replicable business models.

■ Concentrate an important part of multilateral and bilateral direct funding on those difficult areas of access which do not initially offer an adequate commercial return. Provision of end-user finance is required to overcome the barrier of the initial capital cost of gaining access to modern energy services. Operating through local banks and microfinance arrangements, directly or through guarantees, can support the creation of local networks and the necessary capacity in energy sector activity.

13

■ While the *World Energy Outlook* has sought to shed light in this area, it is important that energy access programmes and projects make provision for the collection of robust, regular and comprehensive data to quantify the outstanding challenge and monitor progress towards its elimination. In many ways, providing energy access is an objective well suited to development frameworks such as output-based financing, but accurate data needs to be collected to measure progress.

CHAPTER 14

DEVELOPMENTS IN ENERGY SUBSIDIES
The good, the bad and the ugly?

- Energy subsidies have long been utilised by governments to advance particular political, economic, social and environmental goals, or to address problems in the way that markets operate. When they are well-designed, subsidies to renewables and low-carbon energy technologies can bring long-term economic and environmental benefits. However, the costs of subsidies to fossil fuels generally outweigh the benefits.

- Fossil-fuel consumption subsidies amounted to $409 billion in 2010, with subsidies to oil products representing almost half of the total. Persistently high oil prices have made the cost of subsidies unsustainable in many countries and prompted some governments to act. In a global survey, we have identified 37 economies where subsidies exist, with at least 15 of those having taken steps to phase them out since the start of 2010. Without further reform, the cost of fossil-fuel consumption subsidies is set to reach $660 billion in 2020, or 0.7% of global GDP (MER).

- Relative to a baseline in which rates of subsidisation remain unchanged, if fossil-fuel subsidies were completely phased out by 2020, oil demand savings in 2035 would be equal to 4.4 mb/d. Moreover, global primary energy demand would be cut by nearly 5% and CO_2 emissions by 5.8% (2.6 Gt).

- Only 8% of the $409 billion spent on fossil-fuel subsidies in 2010 was distributed to the poorest 20% of the population, demonstrating that they are an inefficient means of assisting the poor; other direct forms of welfare support would cost much less.

- Renewable energy subsidies grew to $66 billion in 2010, in line with rising production of biofuels and electricity from renewable sources. To meet even existing targets for renewable energy production will involve continuing subsidies. In 2035, subsidies are expected to reach almost $250 billion per year in the New Policies Scenario.

- Unit subsidy costs for renewable energy are expected to decline, due to cost reductions coupled with rising wholesale prices for electricity and transport fuels. Nonetheless, in all three scenarios most renewable energy sources need to be subsidised in order to compete in the market. In the New Policies Scenario, onshore wind becomes competitive around 2020 in the European Union and 2030 in China, but not in the United States by the end of the *Outlook* period.

- By encouraging deployment, renewable-energy subsidies can help cut greenhouse-gas emissions. By 2035, greater use of renewables reduces CO_2 emissions by 3.4 Gt in the New Policies Scenario, compared with the average emissions factor in 2009. Benefits in the 450 Scenario relative to the New Policies Scenario are even greater: additional CO_2 emissions savings of 3.5 Gt and fossil-fuel import-bill savings of $350 billion.

Overview of energy subsidies

Energy subsidies (Box 14.1 provides a definition) have long been used by governments to advance particular political, economic, social and environmental goals, or to address problems in the way markets operate. It is clear that they are costly. Subsidies that artificially reduce end-user prices for fossil fuels amounted to $409 billion in 2010, while subsidies given to renewable energy amounted to $66 billion (Table 14.1).[1] Beyond assessing quantitatively the extent of subsidies to both fossil fuels and renewables, this chapter explores their impact on energy, economic and environmental trends, updating and extending the analyses of previous *World Energy Outlooks (WEOs)*. The chapter does not attempt to deal with all forms of energy subsidies, for example, to nuclear energy; it is most comprehensive in discussing fossil-fuel consumption subsidies (rather than production subsidies, due to data limitations) and subsidies to renewable energy producers.

Table 14.1 ● **Estimated energy subsidies, 2007-2010** ($ billion, nominal)

	2007	2008	2009	2010
Fossil fuels (consumption)	342	554	300	409
Oil	186	285	122	193
Gas	74	135	85	91
Coal	0	4	5	3
Electricity*	81	130	88	122
Renewable energy	39	44	60	66
Biofuels	13	18	21	22
Electricity	26	26	39	44

*Fossil-fuel consumption subsidies designated as "electricity" represent subsidies that result from the under-pricing of electricity generated only by fossil fuels, *i.e.* factoring out the component of electricity price subsidies attributable to nuclear and renewable energy.

Fossil-fuel subsidies are often employed to promote economic development or alleviate energy poverty, but have proven to be an inefficient means of fulfilling these objectives, instead creating market distortions that encourage wasteful consumption and can lead to unintended negative consequences. Moreover, rising international oil prices made their total cost insupportable to many oil-importing countries in 2010. Volatile energy markets and the prospect of higher fossil-fuel prices mean that fossil-fuel subsidies threaten to be a growing liability to state budgets in the future. This prospect has created a strong impetus for reform, strengthened by other associated benefits, including energy savings, lower carbon-dioxide (CO_2) emissions and improved economic efficiency. But fossil-fuel subsidy reform is notoriously difficult as the short-term costs imposed on certain groups of society can be very burdensome and induce fierce political opposition. In the case of fossil-fuel consumption subsidies, rising international fuel prices have frequently outpaced the rate at which domestic fuel prices have risen and presented governments with difficult choices about whether to proceed with reform plans or protect consumers. If removing these subsidies were easy, it would probaly already have happened.

---

1. Although not addressed in this study, nuclear power is also a recipient of subsidies, which are distributed mainly via mechanisms that assist producers. These include loan guarantees, tax incentives, limitation of utilities' financial liability in the event of an accident and grants for research and development.

G-20 and APEC leaders in 2009 and 2010 committed to rationalise and phase out over the medium term inefficient fossil-fuel subsidies (both to producers and consumers) that encourage wasteful consumption. With respect to fossil-fuel consumption subsidies, progress has been mixed. While several countries have successfully brought end-user prices closer to market levels, some of the factors mentioned above have worked against reform. In many countries, the global economic downturn has weakened households' finances and increased reliance on subsidies. Several emerging economies in Asia have been grappling with high rates of inflation and have been wary of superimposing sudden hikes in energy prices. In parts of the Middle East and North Africa – a region where fossil-fuel subsidies are substantial – political and social unrest has delayed, and in some cases reversed, plans to reform energy pricing.

Box 14.1 ● What is an energy subsidy?

The IEA defines an energy subsidy as any government action directed primarily at the energy sector that lowers the cost of energy production, raises the price received by energy producers or lowers the price paid by energy consumers. This broad definition captures many diverse forms of support, direct and indirect. Some are easily recognised, such as arrangements that make fuel available to consumers at prices below international levels. Others may be less so, such as feed-in tariffs that guarantee a premium to power producers for electricity generated from solar panels: the consumer sees no price benefit (and pays a higher average unit price for electricity), but the producer is enabled to draw on a more costly supply. Related types of subsidy arise from support policies that impose mandates on energy supply, such as portfolio standards which oblige utilities to buy a certain volume of renewable generation and implicitly raise the market price for renewables.

Energy subsidies are frequently differentiated according to whether they confer a benefit to consumers or producers. Consumption subsidies benefit consumers by lowering the prices they pay for energy. They are more prevalent in non-OECD countries. Production subsidies typically benefit producers by raising the price they receive, in order to encourage an expansion of domestic energy supply. They remain an important form of subsidisation in developed and developing countries alike.

Subsidies can be further distinguished according to the form of energy they support, for example fossil fuels, renewable energy or nuclear power. Many countries have already or are currently moving to phase out fossil-fuel subsidies on the grounds that they are economically costly to taxpayers and encourage wasteful consumption, which also has the effect of worsening damage to the environment through higher emissions of greenhouse gases and other air pollutants. At the same time, many countries have been introducing subsidies to renewable energy technologies – some of which are still in the early stages of their development – to improve their competitiveness and unlock their potential to reduce greenhouse-gas emissions and to improve energy security in the longer term. Unless well-designed and properly targeted, all types of energy subsidies can lead to an inefficient allocation of resources and market distortions by encouraging excessive production or consumption.

14

The deployment of renewable energy often involves some degree of government intervention (included, here, in our definition of subsidy) because, for many regions and technologies, energy derived from renewable sources remains more costly than energy from fossil fuels. As a result, most renewable electricity and biofuels production capacity built in recent years has received, or is receiving, some form of subsidy, through direct payments or government mandates. Renewable-energy subsidies are often used to encourage the pace and scale of deployment (most are directed toward producers) and, thereby, to accelerate unit-cost reductions through economies of scale and learning-by-doing. The objective is to improve the future competitiveness of renewable energy compared with conventional alternatives. Without such support, many forms of renewable energy are projected to remain uncompetitive throughout the *Outlook* period.

Renewable-energy subsidies have also expanded considerably in recent years as governments have sought to offset the economic distortions present in market pricing, which fails to put an adequate value on the costs of an insecure and insufficiently diverse energy supply, or on the costs of local pollution and emissions of climate-change-inducing greenhouse gases. There are important distinctions that explain moves by governments to increase renewable-energy subsidies, while phasing out fossil-fuel subsidies. Renewable-energy subsidies are intended to aid the deployment of sustainable technologies in order to provide long-lasting energy security and environmental benefits. By contrast, fossil-fuel subsidies, though they may have some near-term benefits, are generally poorly targeted, encourage wasteful consumption and prolong dependence on fuels that are likely to cost more in the future. Most renewable-energy subsidies are intended to make renewable energy ultimately competitive with conventional alternatives, at which point they would be phased out, whereas fossil-fuel subsidies usually have no expiration date.

Fossil-fuel subsidies

Fossil fuels are the recipient of many forms of subsidy, provided through various direct and indirect channels. The most common include tax advantages, direct financial transfers, cheap credit, transfer of risk from the private sector to the government, regulation and trade instruments. The rationale for subsidies to fossil-fuel consumption is typically either social or political, such as alleviating energy poverty, redistributing national resource wealth, or promoting national or regional economic development by conferring an advantage on domestic energy-consuming industries, often to protect jobs. Subsidies to fossil-fuel production are motivated by somewhat different goals that include boosting domestic production, maintaining employment, technology development and social adjustment in declining sectors (Box 14.2).

Both fossil-fuel consumption and production subsidies, by encouraging excessive energy use, lead to an inefficient allocation of resources and market distortions. While they may have well-intentioned objectives, fossil-fuel subsidies have, in practice, usually proved to be an unsuccessful or inefficient means of achieving their stated goals and they invariably have unintended consequences. This is particularly evident in those energy-importing countries that purchase energy at world prices and sell it domestically at lower, regulated prices, where the unsustainable financial burden of fossil-fuel consumption subsidies has become a pressing reason for reform. In energy-exporting nations, high consumption subsidies can erode export availability and foreign currency earnings. In both, consumption subsidies limit financial resources available for investment in the energy sector, discourage efficient energy use, encourage fuel adulteration and smuggling and lead to increased environmental degradation.[2]

Box 14.2 ● Support to fossil-fuel production in OECD countries

The measures that directly or indirectly support the production of fossil fuels in OECD countries vary significantly, depending on the fuel that benefits from such measures. Not all OECD countries produce significant quantities of fossil fuels. Among those that do, some extract crude oil and natural gas, while others mine coal. Some do both, owing to a particular set of geographical and geological conditions. Not surprisingly, this variety of situations is reflected in the equally diverse array of policies in place in OECD countries.

The OECD has recently produced an inventory of those budgetary transfers and tax expenditures that provide support to the production or use of fossil fuels in a selected number of OECD countries (24 countries accounting for about 95% of the OECD's total primary energy supply). This inventory now includes over 250 such measures, which are tracked over time. The OECD estimates that total support to the production and use of fossil fuels has ranged between $45 billion and $75 billion per year over the period 2005 to 2010.

Support to the production of coal is the most visible form of support in OECD countries. It is also the largest. The OECD estimates that coal (including hard coal, lignite and peat) attracted close to 39% of total fossil-fuel producer support in 2010. The importance of coal in total support is largely explained by the need in Europe to use budgetary transfers and price support to allow a gradual restructuring of the coal-mining industry in a socially acceptable manner. Coal also receives support in other countries, such as Australia, Canada, Korea and the United States, but at a lower level. It is provided most notably through tax expenditure and funding for research and development.

The OECD Secretariat has estimated that petroleum and natural gas accounted for about 30% each of total producer support in 2010. Crude oil and natural gas production are supported mainly through tax breaks, typically in the form of advantageous income-tax deductions, such as depletion allowances and the accelerated depreciation of capital expenses. Royalty reductions or credits are also commonly used to encourage extraction at high-cost or marginal wells. These features of countries' tax and royalty regimes are often complex and less transparent than direct expenditures, making country comparisons difficult since the tax expenditure associated with these policies have to be estimated by reference to country-specific baselines. Some of those measures relate to aspects of the tax regime that are specific to the resource sector, making the extent of support obscure.

Source: OECD (2011).

---

2. See *WEO-2010* for a more detailed discussion of the effects of fossil-fuel subsidies and the rationale for reform (IEA, 2010).

Measuring fossil-fuel consumption subsidies

Estimation methodology

The IEA estimates subsidies to fossil fuels that are consumed directly by end-users or consumed as inputs to electricity generation. The price-gap approach, the most commonly applied methodology for quantifying consumption subsidies, is used for this analysis. It compares average end-use prices paid by consumers with reference prices that correspond to the full cost of supply.[3] The price gap is the amount by which an end-use price falls short of the reference price and its existence indicates the presence of a subsidy (Figure 14.1). The methodology is sensitive to the calculation of reference prices. For oil products, natural gas and coal, reference prices are the sum of the international market price, adjusted for quality differences where applicable, the costs of freight and insurance and internal distribution, and any value-added tax.[4] Electricity reference prices are based on the average annual cost of production, which depends on the make up of generating capacity, the unsubsidised cost of fossil-fuel inputs, and transmission and distribution costs. No other costs, such as for investment, are taken into account.

Figure 14.1 ● **Illustration of the price-gap methodology: average reference and retail prices of oil products**

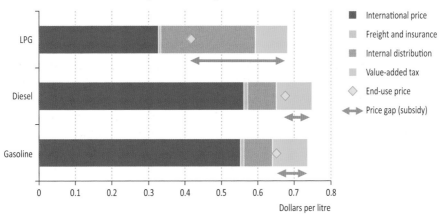

Notes: The sum of international price, freight and insurance, internal distribution and value-added tax bar indicates the total reference price. Gasoline and diesel prices are for the transport sector; liquefied petroleum gas (LPG) prices are for households.

Estimates using the price-gap approach capture only interventions that result in final prices below those that would prevail in a competitive market. While consumer price subsidies account for the vast majority of subsidies to fossil fuels, there are numerous other subsidies which are not captured by the price-gap approach. It does not, for example, capture subsidised research and development. The estimates capture the cost of cross

---

3. Full details of the price-gap methodology used by the IEA in this analysis are available at
www.worldenergyoutlook.org/subsidies.asp.

4. Some countries, such as India, rely on relatively low-quality domestic coal but also import high-quality coal. In such cases, quality differences must be taken into account when determining reference prices, as they affect the market value of a fuel.

subsidies, which may subsidise particular regions, technologies or types of consumers by imposing higher prices on another part of the economy (this offsetting component is not included). Supply-oriented subsidies that do not result in a lower price to end-users or power generators are not picked up by our estimates because of the difficulty in identifying and measuring them. Energy production subsidies have been conservatively estimated to be at least $100 billion per year (GSI, 2010). Our estimates, which cover only consumption subsidies, should be considered a lower bound for the total economic cost of fossil-fuel subsidies and their impact on energy markets.

For countries that export a given product but charge less for it in the domestic markets, the domestic subsidies are implicit; they have no direct budgetary impact so as long as the price covers the cost of production. The subsidy, in this case, is the opportunity cost of pricing domestic energy below international market levels, *i.e.* the rent that could be recovered if consumers paid world prices, adjusting for differences in variables such as transportation costs. For importers, subsidies measured via the price-gap approach may be explicit, representing budget expenditures arising from the domestic sale of imported energy at subsidised prices, or may sometimes be implicit. Many countries, Indonesia for example, rely extensively on domestically produced fuels, but supplement domestic supply by importing the remainder. In such cases, subsidy estimates represent a combination of opportunity costs and direct expenditures.

Estimated costs

Fossil-fuel consumption subsidies worldwide are estimated to have totalled $409 billion in 2010, about $100 billion higher than in 2009 (Figure 14.2). Subsidies were still well below the level of 2008, when they reached more than $550 billion. This estimate is based on a global survey that identified 37 economies as subsidising fossil-fuel consumption; collectively, this group accounted for more than half of global fossil-fuel consumption in 2010 and includes all of the major economies that subsidise end-use prices. Oil products attracted the largest subsidies, totalling $193 billion (or 47% of the total), followed by natural gas at $91 billion. Fossil-fuel subsidies resulting from the under-pricing of electricity were also significant, reaching $122 billion. Subsidies to coal end-use consumption were comparatively small at $3 billion.

Figure 14.2 ● Global economic cost of fossil-fuel consumption subsidies by fuel

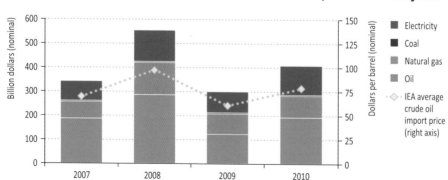

Changes in international fuel prices are chiefly responsible for differences in subsidy costs from year to year. The increase in the global amount of subsidy in 2010 closely tracked the sharp rise in international fuel prices. The average IEA crude-oil import price increased by 28% between 2009 and 2010; natural gas import prices rose modestly. Higher international prices in 2010 negated or reversed the gains that would have occurred in some countries from imposed pricing adjustments, had prices not risen. The series of estimates from 2007 to 2010 demonstrates clearly the risk to which governments are exposed by regulated domestic prices in international energy markets subject to unpredictable price fluctuations. For net-importing countries, fossil-fuel consumption subsidies become particularly difficult to reduce during periods of rising prices, high inflation, uncertain economic growth prospects and fiscal tightening, just when the burden is becoming insupportable. Subsidy estimates can also fluctuate according to changes in exchange rates and demand patterns. The estimate of fossil-fuel consumption subsidies in 2010 would have been higher in the absence of efforts to raise end-user prices towards more competitive levels, thereby limiting the increase in the price gaps for some products.

Fossil-fuel subsidies remain most prevalent in the Middle East, amounting in 2010 to $166 billion, or 41% of the global total. At $81 billion, Iran's subsidies were the highest of any country, although this figure is expected to fall significantly in the coming years if the sweeping energy-pricing reforms that commenced in late 2010 are implemented successfully and prove durable (Figure 14.3). Two leading oil and gas exporters – Saudi Arabia and Russia – had the next-highest subsidies in 2010, at $44 billion and $39 billion. Russian subsidies to gas and electricity remain large, despite continuing reform measures (see Chapter 7). Of the importing countries, the cost of subsidies was highest in India at $22 billion, followed by China at $21 billion. A total of nine member economies of the G-20 were identified as having fossil-fuel consumption subsidies, collectively amounting to $160 billion. In the 21 member economies from the Asia-Pacific region that constitute the Asia-Pacific Economic Co-operation (APEC), ten economies were identified that had fossil-fuel consumption subsidies totalling $105 billion in 2010.

The economic cost of subsidies can be more completely understood when viewed by other measures, such as by percentage of gross domestic product (GDP) or on a per-capita basis (Figure 14.4). In Turkmenistan, for instance, extremely low prices for natural gas mean that subsidies were equivalent to nearly 20% of the country's economic output in 2010. At 10% of GDP or more, subsidies also weigh heavily on the economies of Iran, Iraq and Uzbekistan. In per-capita terms, subsidies tend to be highest in resource-rich countries of the Persian Gulf, ranging from over $350 per person in Iraq to nearly $2 800 in Kuwait. This high level of subsidies is presently paid for in these countries by high revenues from oil and gas exports during the previous years. While the magnitude of fossil-fuel subsidies is large in China and India, they are considerably smaller when viewed as a share of their economic output or relative to their large populations. In China's case, subsidies are comparatively low, at 0.4% of GDP and about $16 per person. The same is true for India, which has subsidies that amount to 1.4% of GDP and $18 per person.

Figure 14.3 ● Economic cost of fossil-fuel consumption subsidies by fuel for top twenty-five economies, 2010

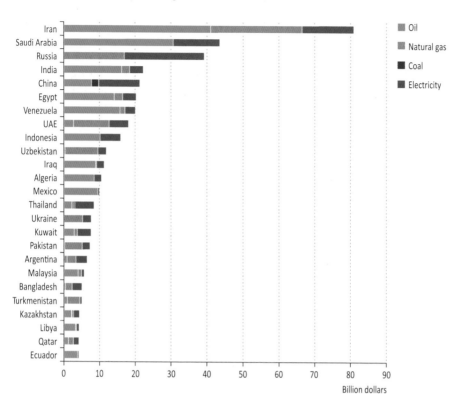

For the 37 economies that were identified as having subsidies, fossil fuels were subsidised at a weighted-average rate of 23%, meaning that their consumers paid roughly 77% of competitive international market reference prices for products (Figure 14.5). The rate of subsidisation was highest among oil and gas exporters in the Middle East, North Africa and parts of Central Asia, many of which set the price of domestic fuels above the cost of indigenous production but well below those that would prevail in the international market. These have fallen in some cases, such as Russia, where pricing reforms are underway to improve the incentives for investment and efficiency. Natural gas and fossil-fuel-based electricity were subsidised at average rates of 53% and 20% in 2010. In addition to the foregone revenues that state-owned companies may face from the under-pricing of energy products to consumers, those in the electricity and natural gas sectors sometimes bear losses from the under-collection of bills, which occurs when consumers cannot afford even subsidised energy prices or there is theft. The subsidisation rate for oil products in the economies studied was about 21% in 2010, although some products, such as kerosene and liquefied petroleum gas (LPG), were more heavily subsidised.

Figure 14.4 • Fossil-fuel consumption subsidies per capita and as a percentage of total GDP in selected economies, 2010

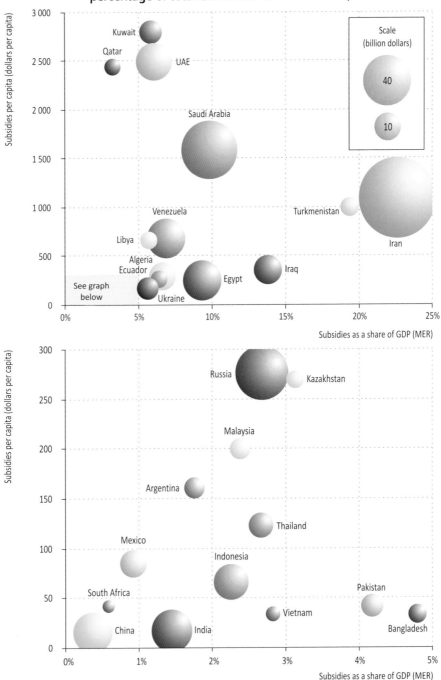

Note: MER = market exchange rate. Circle sizes are proportional to the total value of the subsidy and are comparable across both figures. Uzbekistan is not shown in this figure.

Figure 14.5 ● Rates of subsidisation for fossil-fuel consumption subsidies, 2010

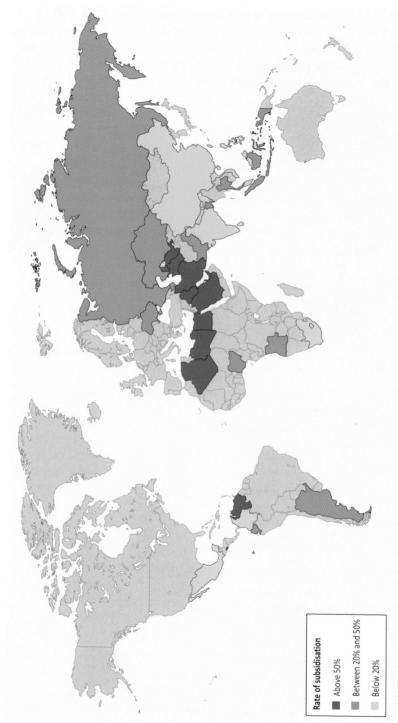

Rate of subsidisation

▮ Above 50%

▮ Between 20% and 50%

▮ Below 20%

This map is for illustrative purposes and is without prejudice to the status of or sovereignty over any territory covered by this map.

14

Fossil-fuel consumption subsidies are more prevalent in net-exporting countries, where they largely represent opportunity costs. For net exporters of oil and gas, subsidies to those fuels totalled $331 billion in 2010, compared with $78 billion in net-importing countries (Figure 14.6). Since 2007, about 80% of the estimated subsidies, on average, have occurred in net exporters of oil and gas. When international fuel prices have dropped, as between 2008 and 2009, subsidies have declined by more in percentage terms in net-importing countries. Many subsidies in net-importing countries are either reduced substantially or eliminated as the gap between reference prices and regulated domestic prices is closed. In some cases, governments seized this opportunity to liberalise prices or raise prices closer to international market levels as lower costs were easier to pass through to consumers. As end-user prices in net-exporting countries are sometimes only enough to cover production costs, falling international prices do not have as large an impact on shrinking the price gap. Rising international fuel prices exert fiscal pressure on net-importing countries that absorb the higher cost of subsidies in their budget, offering strong incentive for reform. In net-exporting countries, however, this incentive is not as strong as they simultaneously have the benefit of higher export revenues (despite the growing opportunity costs).

Figure 14.6 ● **Fossil-fuel consumption subsidies by net oil and gas importer and exporter, 2007-2010**

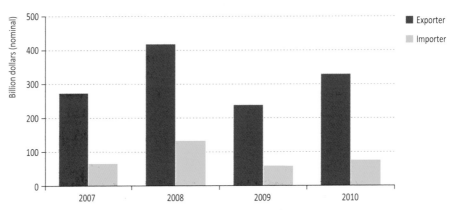

Fossil-fuel subsidies and the poor

One common justification for fossil-fuel subsidies is that they are needed to help the poor gain or maintain access to energy services essential to basic living standards. While making electricity and clean cooking facilities available to the poor is of vital importance (see Chapter 13), studies have found that fossil-fuel subsidies as presently constituted tend to be regressive, disproportionately benefitting higher income groups that can afford higher levels of fuel consumption (Arze del Granado *et al.*, 2010). Poor households may not have access to subsidised energy directly, lacking a connection to electricity or natural gas and owning no vehicle. Low-income households in any case generally spend less in absolute terms on energy than their higher-income counterparts. Without precise targeting, fossil-fuel subsidies are often an inefficient means of assisting the poor.

In practice, the poor capture only a small share of all the subsidies to fossil fuels. We estimate that out of the $409 billion spent on fossil-fuel consumption subsidies in 2010, only $35 billion, or 8% of the total, reached the poorest income group (the bottom 20%).[5] This finding is based on a survey of 11 of the 37 economies identified as having fossil-fuel consumption subsidies and does not take into account subsidies specifically provided to extend access to basic energy services. The eleven economies, which were selected on the basis of data availability for those that have low levels of modern energy access, and have an aggregate population of 3.4 billion, the share of total fossil-fuel consumption subsidies reaching the poorest income group ranged from about 2% to 11% (Figure 14.7). Among the countries surveyed, the share was lowest for South Africa.

Figure 14.7 ● Share of fossil-fuel subsidies received by the lowest 20% income group in selected economies, 2010

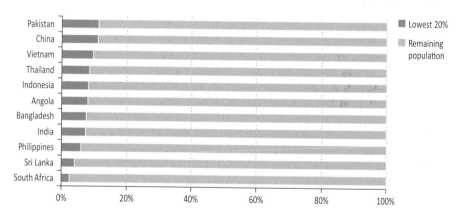

Compared to other fuels, subsidies to kerosene tend to be best targeted on the poor, despite its tendency to be sold in the black market. In 2010, nearly 15% of the kerosene subsidies in the countries analysed reached the lowest income group; subsidies to LPG, gasoline and diesel benefitted the poor least, with only 5% to 6% going to the lowest group (Figure 14.8). Despite the utility of LPG as a clean cooking fuel, the up-front cost for infrastructure connections and the practice of selling LPG in larger quantities than kerosene make this fuel less affordable for the poor, elevating the barrier to their gaining initial access (Shenoy, 2010). Subsidies to electricity and natural gas were in the middle of the range, with shares of 9% and 10% disbursed to the lowest group. These results demonstrate that subsidising fossil fuels is, in practice, an inefficient method of providing assistance to the poor. They also highlight the opportunity for subsidy reform, as the same level of financial support could be distributed more efficiently to low-income households at a lower cost. In general, social welfare programmes are a more effective and less distortionary way of helping the poor than energy subsidies.

14

---

5. More information on the methodology can be found at www.worldenergyoutlook.org/subsidies.asp.

Figure 14.8 ● Share of fossil-fuel subsidies received by the lowest 20% income group by fuel in surveyed economies, 2010

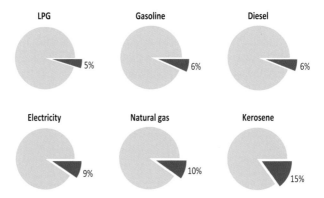

Note: Countries surveyed were Angola, Bangladesh, China, India, Indonesia, Pakistan, Philippines, South Africa, Sri Lanka, Thailand and Vietnam.

Implications of phasing out fossil-fuel consumption subsidies

Fossil-fuel consumption subsidies induce consumers and producers to trade energy products at prices below their true opportunity cost, thereby encouraging additional and often wasteful consumption of finite resources with adverse economic and environmental consequences. Over time, these subsidies can lead to the creation of energy-intensive industries that are unable to compete globally on an unsubsidised basis.

This analysis quantifies the economic and environmental gains that could be obtained from removing fossil-fuel consumption subsidies: energy savings, lower CO_2 emissions and reduced fiscal burdens. It is based on simulations using the IEA's World Energy Model (WEM), assuming the phase-out of all fossil-energy subsidies gradually over the period 2012 to 2020. This timeframe is consistent with the "medium term" as discussed in international forums such as the G-20 and APEC. Because the economic value of subsidies fluctuates from year-to-year, initial subsidisation rates are averaged over the most recent three-year period (2008 to 2010). Savings from eliminating subsidies are presented relative to a baseline case in which average subsidy rates remain unchanged, although the timeframe is extended to 2035 since additional benefits are realised even after the phase-out of subsidies is complete. The analysis is intended to illustrate potential gains and should not be interpreted as a prediction; it is unlikely that all subsidies will, in reality, be removed so quickly.

Our analysis shows that, if fossil-fuel consumption subsidies were completely phased out by 2020, global energy demand would be reduced by 3.9%, or about 600 million tonnes oil equivalent (Mtoe) by that year, and 4.8% or some 900 Mtoe by 2035 (Figure 14.9). In the New Policies Scenario, in which recently announced commitments and plans are assumed to be fully implemented, we project that about three-quarters of this potential would be achieved by 2020.[6] The energy savings from subsidy reform result from the higher prices

---

6. The New Policies Scenario includes other policy measures, in addition to some phase-out of fossil-fuel subsidies, which contribute to achieving these savings.

that would follow from subsidy removal, incentivising energy conservation and efficiency measures. The responsiveness of demand to higher prices varies by country, according to subsidisation levels and the price elasticity of demand.[7]

Figure 14.9 ● Impact of fossil-fuel consumption subsidy phase-out on global fossil-energy demand and CO_2 emissions

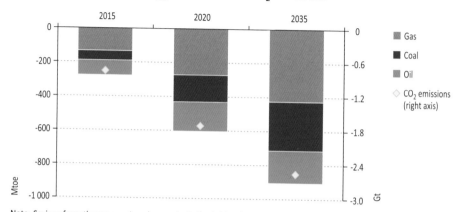

Note: Savings from the progressive phase-out of all subsidies by 2020 are compared with a baseline in which subsidy rates remain unchanged.

Subsidy phase-out would measurably trim global oil demand, by about 3.7 million barrels per day (mb/d) by 2020[8] and 4.4 mb/d, or 4%, by 2035. These cuts stem predominately from the transport sector. Demand for transport fuels is relatively inelastic in the short term, but higher prices (and expectations of such) contribute to greater conservation and the uptake of more efficient vehicles long after the phase-out of subsidies has been completed. Natural gas demand would be cut by 330 billion cubic metres (bcm) in 2020 and 510 bcm, or 9.9% in 2035; the corresponding reduction in coal demand is almost 230 million tonnes of coal equivalent (Mtce) in 2020 and 410 Mtce, or 5.3%, in 2035.

Curbing the growth in energy demand via subsidy reform has several important energy-security implications. In net-importing countries, lower energy demand would reduce import dependence and thereby spending on imports. For net-exporting countries, removing subsidies would boost export availability and earnings. For all countries, it would also improve the competitiveness of renewable energy in relation to conventional fuels and technologies, further diversifying the energy mix. Lower energy demand would also alleviate upward pressure on international energy prices, yet the elimination of subsidies would make

---

7. The price elasticity of demand is the principal determinant of energy and emissions savings from subsidy reform, reflecting the extent to which consumption is responsive to higher prices. Estimated elasticities vary according to the specific region, fuel and uses analysed, but generally increase over time as consumers have opportunities to purchase more energy-efficient equipment and change behaviour.

8. This is less than the oil savings in 2020 (4.7 mb/d) estimated in *WEO-2010* because the projection period is one year shorter and subsidisation rates at the beginning of the phase-out period were reduced in several countries that have undertaken reforms.

14

consumers more responsive to price changes, which should contribute to less volatility in international energy markets.

By encouraging higher levels of consumption and waste, subsidies exacerbate the harmful environmental effects of fossil-fuel use and impede development of cleaner energy technologies. Potential CO_2 savings from subsidy phase-out directly reflect the fossil-energy savings. We estimate that energy-related CO_2 emissions would be cut by 1.7 gigatonnes (Gt), or 4.7%, in 2020 and 2.6 Gt, or 5.8%, in 2035, relative to the prospects if subsidy rates remain unchanged (Figure 14.9). Cumulative CO_2 emissions over the *Outlook* period would be lower by 41 Gt. Although not modelled here, other environmental co-benefits would arise from subsidy phase-out. In particular, slower demand growth for fossil-fuels would reduce emissions of other air pollutants, such as sulphur dioxide (SO_2), nitrogen oxides (NO_x) and particulates, which cause human health and environmental problems.

High and volatile international energy prices in recent years, and the expectation that these conditions will persist, have forced many governments to reconsider the affordability of fossil-fuel subsidies. Without further subsidy reform, we estimate that the total cost of fossil-fuel consumption subsidies would reach $660 billion (in year-2010 dollars) in 2020 (0.7% of projected global GDP [MER]), up about 60% from the average level observed over the period 2007 to 2010. As in the case of the historical estimates, this figure represents a mix of implicit and explicit subsidies to fossil-fuel consumption.

Implementing fossil-fuel subsidy reform

Governments face difficult challenges in reforming inefficient fossil-fuel subsidies. Each country must consider its specific circumstances, the types of subsidies that need reform and their intended purpose and effectiveness pursuing economic, political or social goals. Past reforms have had varying levels of success. Many have achieved positive and lasting change; but some have been planned and implemented poorly, or have succumbed to back-sliding, *i.e.* governments successfully institute reforms only to reintroduce subsidies later. The risk of adverse consequences if subsidy reform is not carefully planned for and executed is real. Previous experiences have illustrated several barriers to implementing fossil-fuels subsidy reform and strategies that can help guide future efforts (Figure 14.10).

Inadequate information about existing subsidies is frequently an impediment to reform. Before taking a decision about reform, governments must precisely identify energy subsidies, including their beneficiaries, and quantify their costs and benefits, in order to determine which subsidies are wasteful or inefficient. If subsidies are failing to serve their defined objectives or are leading to unintended adverse consequences (for example, fuel adulteration and smuggling), the need for remedial action will be clear. Nonetheless, removing subsidies without understanding and providing for the consequences may hurt vulnerable low-income groups that depend on subsidies for access to basic energy services. Making more information on fossil-fuel subsidies available to the general public can help build support for reform. Disclosure may include data on overall costs and benefits as well as on price levels, price composition and price changes (Wagner, 2010).

Are the G-20 and APEC commitments being met?

In September 2009, G-20 leaders, gathered at the Pittsburgh Summit, committed to "rationalize and phase out over the medium term inefficient fossil fuel subsidies that encourage wasteful consumption". In November 2009, APEC leaders meeting in Singapore made a similar pledge, thereby broadening the international commitment to reform. At the request of the G-20, the IEA, the Organisation for Economic Co-operation and Development (OECD), the Organization of the Petroleum Exporting Countries (OPEC) and the World Bank have been collaborating on a series of reports aimed at guiding policy makers through the implementation of fossil-fuel subsidy reform. The latest report in the series (IEA/OECD/OPEC/World Bank, 2011) was submitted to the 2011 G-20 Summit in Cannes and is available at www.worldenergyoutlook.org/subsidies.asp.

Since making these commitments, many G-20 and APEC member economies have publicly identified inefficient fossil-fuel consumption and production subsidies and outlined plans for their removal. These include regulated prices of refined petroleum products that are below international levels, financial support to coal industries and preferential tax provisions related to fossil-fuel production.[9] While this represents an encouraging start, much work remains to be done in order to realise the full extent of benefits from subsidy reform.

Based on our analysis of fossil-fuel consumption subsidies, some members of the G-20 and APEC retain price subsidies that appear to be inefficient, encourage wasteful consumption and are regressive, but are not earmarked for phase-out or better targeting. Of the nine G-20 members identified by the IEA as having price subsidies that benefit consumers, seven have disclosed plans for their phase-out, but only some have been able to raise energy prices and close the price gap. Moreover, members have failed to identify many existing fossil-fuel production subsidies, an equally important part of the group's commitment to reform. APEC member economies (those not also in the G-20) have not as yet collectively submitted lists of inefficient fossil-fuel subsidies. Much remains to be done to fulfil the commitments made in these international forums, both in terms of defining the fossil-fuel subsidies to be phased out and following through with durable and well-designed reform efforts.

The politics of reforming subsidies are challenging. Subsidies create entrenched interests among domestic industries advantaged by cheap energy inputs and those income groups that are accustomed to receiving this form of economic support. Such stakeholders can be expected to resist subsidy phase-out, particularly in the absence of clear plans to compensate losers or make the transition gradual. Resistance to fossil-fuel subsidy reform can be particularly strong in major fossil-fuel-exporting countries, where people may feel entitled to benefit directly

14

---

9. The list of reforms proposed by G-20 members is available at:
www.g20.org/Documents2010/expert/Annexes_of_Report_to_Leaders_G20_Inefficient_Fossil_Fuel_Subsidies.pdf.

from their nation's resource wealth. Although not always necessary, subsidy reform has a better chance of success if supported by key interest groups, sometimes achieved through consultation with stakeholders as reform strategies take shape.

Figure 14.10 ● **Summary of common barriers to fossil-fuel subsidy reform and strategies for successful implementation**

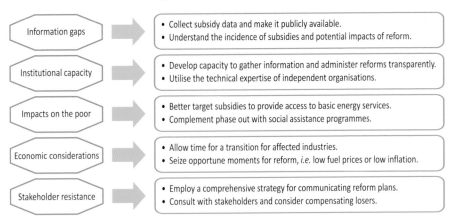

Information gaps	• Collect subsidy data and make it publicly available. • Understand the incidence of subsidies and potential impacts of reform.
Institutional capacity	• Develop capacity to gather information and administer reforms transparently. • Utilise the technical expertise of independent organisations.
Impacts on the poor	• Better target subsidies to provide access to basic energy services. • Complement phase out with social assistance programmes.
Economic considerations	• Allow time for a transition for affected industries. • Seize opportune moments for reform, *i.e.* low fuel prices or low inflation.
Stakeholder resistance	• Employ a comprehensive strategy for communicating reform plans. • Consult with stakeholders and consider compensating losers.

While the removal of fossil-fuel subsidies generally improves long-term economic competitiveness and fiscal balances, fears of negative economic consequences in the short term can be expected. The phase-out of subsidies must be carried out in a way that allows both energy and other industries time to adjust, so as to safeguard the communities affected. Energy-price controls are often an important tool for controlling inflation and their removal may sometimes be accompanied by other macroeconomic policies to reduce inflationary pressures. Once reforms have been implemented, mounting pressure from short-term economic concerns has sometimes led to back-sliding. This risk can be mitigated if governments dissociate themselves from price-setting, such as when energy prices are liberalised or automatic pricing mechanisms are established.

Even where there is interest in pursuing subsidy reform, certain institutional and administrative capacity, and even physical infrastructure, is required for governments to act effectively. It is important to have institutions that are capable of accurate and timely collection of data about the existing subsidies, their distribution and the need for compensation (where necessary) following reform. Governments are well-placed to gather this far-reaching information, although other organisations may have integral technical expertise that can aid in this effort. Targeted compensation is typically more effective where the necessary administrative capacity exists to reach the intended groups and distribute benefits reliably and without fraud. Where subsidies are re-targeted to the poor via direct financial transfers, those eligible to receive support may not be registered with the government or may lack the means for receiving support, such as having a bank account. Creating new infrastructure may sometimes be an important element in subsidy reform. For example, the availability of public transportation reduces the need for personal vehicles and lessens the burden of the rising cost of transport fuels, although its provision is usually only readily practicable in population centres.

Box 14.3 ● Recent experiences implementing subsidy reform

Iran implemented sweeping reforms to energy subsidies in December 2010 in a bid to lessen the burden on its central budget and reverse deep inefficiencies in its energy sector and the larger economy. Prior to the reform, energy prices were often subsidised by over 90%. Under the reform programme, the prices of regular gasoline increased by 300%, premium gasoline by 230%, and diesel and gas oil by 840%. To date, these higher prices have been maintained, despite the population having been previously accustomed to low energy prices and the limited success (with the exception of its use of smart cards for gasoline rationing) of previous reform attempts. To counter the public sense of entitlement to low energy prices and build a case for reform, subsidies were identified and their onerous costs publicised before changes were made. Although details were lacking, the intent to undertake a five-year programme to phase out energy subsidies (that included compensation to consumers) was communicated frequently. To encourage public acceptance and diminish the economic impact of price increases for households, cash payments were made to every citizen prior to the effective date and nearly 90% of the population continue to receive monthly payments. Furthermore, key capabilities were developed to facilitate reform, such as establishing bank accounts for heads of households to receive government compensation payments. Time will tell whether the reforms will be permanent, and certain aspects of the programme can be criticised (for example, the lack of discrimination in compensation payments), but its experience thus far reflects serious intent.

In *El Salvador,* the price of LPG cylinders was previously subsidised, benefitting consumers who could afford to buy larger quantities rather than low-income families. The government raised the price of LPG cylinders to market levels in 2011 and began to provide offsetting payments to the poor through electricity bills. This is an interesting example of targeting assistance to the poor by linking to another household service; the delivery method is not perfect as it fails to offer payments to multiple families on a single connection or those without a connection to the electricity grid.

Inequitable distribution of subsidies and the scale of the associated financial liabilities, particularly as international oil prices have risen, provided a strong impetus for reform in *Indonesia* (Tumiwa *et al.*, 2011). The government accordingly announced in its Medium-Term Plan in 2010 that it would reduce spending on energy subsidies by 40% by 2013 and eliminate subsidies by 2014. Actual implementation measures have been slowed by macroeconomic concerns and lack of public acceptance. Fuel subsidies to private vehicles were to have been phased out, but the programme was postponed in early 2011 because of worries about rising inflation. It will be important for the government to communicate actively the potential benefits to both individuals and the economy (such as improved competitiveness and freeing government revenues for capital investment and social assistance programmes).

14

Recent developments in fossil-fuel subsidies

The burden of fossil-fuel subsidies on public finances in many countries has provided a strong motive for reform, but progress towards phase-out has been uncertain. Of the 37 economies identified in the global survey as having fossil-fuel consumption subsidies, at least 15 have either implemented reforms or announced related plans since the beginning of 2010. Falling prices during the 2008/2009 recession offered a window of opportunity to align regulated domestic prices and international prices. This was seized by several governments, but the obstacles to reform have mounted with the return of higher prices. There are recent examples of progress, but some have struggled to follow-through on subsidy reform (Table 14.2).

Table 14.2 ● **Recent developments in fossil-fuel consumption subsidy policies in selected economies**

	Recent developments
Angola	Increased gasoline and diesel prices by 50% and 38% in September 2010. Plans to reduce fuel subsidies by 20% per year until eliminated.
Argentina	Held gasoline prices steady for part of 2011 to shield citizens from the impacts of rising international prices, despite previous efforts to end fuel subsidies.
Bolivia	Reversed decision to reduce subsidies to gasoline, diesel and jet fuel in January 2011 due to mounting political pressure, strikes and demonstrations.
El Salvador	Reformed LPG subsidies in April 2011, instead using targeted cash payments for households consuming less than 300 kWh of electricity per month.
India	Plans to eliminate cooking gas and kerosene subsidies in a phased manner beginning in April 2012, replacing with direct cash support to the poor. In June 2011, raised domestic prices for gasoil, LPG and kerosene (by 9%, 14% and 20%, respectively).
Indonesia	Plans to reduce spending on energy subsidies by 40% by 2013 and fully eliminate fuel subsidies by 2014, but postponed a restriction of subsidised fuel for private cars in February 2011.
Iran	Significantly reduced energy subsidies in December 2010 as the start of a five-year programme to bring the prices of oil products, natural gas and electricity in line with international market levels. Cash payments are being made to ease the impact of higher fuel prices.
Jordan	Announced an expansion of subsidies in January 2011 by reducing kerosene and gasoline prices.
Malaysia	Cut subsidies to gasoline, diesel and LPG in July 2010 as part of a gradual reform programme. However, total subsidy bill is expected to double in 2011.
Nigeria	Has held gasoline and kerosene prices fixed in 2011 as international prices rose, expanding the subsidy bill.
Pakistan	Raised gasoline, diesel and electricity prices in 2011, but price increases have not kept pace with international prices. Plans are to reduce the electricity subsidy by 23% this year and gradually phase it out.
Qatar	Increased petrol, diesel and kerosene prices by 25% in January 2011.
Syria	Announced an expansion of subsidies in 2011: increasing heating oil allowances for public-sector workers in and lowering diesel prices.
Ukraine	Raised gas prices for households and electricity generation plants by 50% in August 2010 and announced plans to raise them by 30% in 2011.

While the return of higher oil prices in 2010 and 2011 reminded governments of the heavy economic liability associated with subsidised prices, many have not been able even to keep domestic prices increasing in line with the rise in international prices. The cost of subsidies is expected to grow in 2011 in Pakistan, Qatar and the United Arab Emirates, despite their having raised oil product prices to some extent. The increased prices in Qatar and the United Arab Emirates are notable, particularly because both are fossil-energy exporters. Argentina, Malaysia, Morocco, Nigeria, Thailand and Venezuela have frozen prices at times in order to shield citizens from higher prices, thereby increasing subsidies. Political instability has made subsidy reform more challenging across the Middle East and North Africa and, since late 2010, strikes, protests and demonstrations have persuaded several governments to reverse efforts to reduce fuel subsidies. In some cases, such as in Bolivia, public opposition has stemmed from poorly executed reform.

Renewable-energy subsidies

The development and deployment of renewable energy often involves government intervention and subsidies. This occurs because, for many regions and technologies, energy derived from renewable sources is, and is projected to remain throughout the *Outlook* period, more costly than energy from fossil fuels. On the other hand, some forms of renewable energy are already economic in certain locations and need no subsidy to compete, particularly hydropower and geothermal. By facilitating deployment and thereby faster learning, subsidies are intended to lower the cost of renewable energy technologies in general and improve their future competitiveness.

Important distinctions underlie moves by governments to increase renewable-energy subsidies while phasing out fossil-fuel subsidies. Basically, renewable-energy subsidies are intended to help lower to competitive levels the cost of sustainable technologies that can provide durable energy security and environmental benefits, though the extent of these benefits is not universal for the various types of renewables and depend on their uses and the regions in which they are deployed. Although fossil-fuel subsidies may also deliver some near-term benefits (such as energy access), consumer subsidies to fossil fuels are generally poorly targeted, encourage wasteful consumption and prolong dependence on fuels that are likely to cost more in the future. Reflecting this difference, most renewable-energy subsidies are time-limited, the aim being to end the need for the subsidy. Fossil-fuel subsidies, on the other hand, usually have no expiration date.

The majority of subsidies to renewables-based electricity and biofuels are producer subsidies that raise the price received by renewable energy producers for their output. Unlike fossil-fuel subsidies, these typically increase the cost of energy services to consumers. Renewable-energy subsidies take a variety of forms, direct and indirect (Table 14.3). Some mechanisms involve overt financial transfers, including tax credits for investment and production or premiums over market prices to cover higher production costs. Other interventions, such as mandates, quotas and portfolio obligations, involve less visible transfers, but nonetheless support the uptake of renewables at higher cost to society than conventional alternatives. All of these mechanisms fall within the broad definition of energy subsidies adopted by the IEA (Box 14.1).

14

Table 14.3 ● Common mechanisms for subsidising renewable energy

	Description	Examples
Trade instruments	Tariffs and quotas.	Tariffs on imported ethanol.
Regulations	Demand guarantees and mandated deployment rates.	Mandates to blend biofuels with conventional oil-based fuels; renewable electricity standards to reach specified levels of renewables-based electricity supply.
Tax breaks	Direct reductions in tax liability.	Tax credits for renewable energy production (PTCs) and investment (ITCs).
Credit	Low-interest or preferential rates on loans to producers.	Loan guarantees to finance renewable energy projects.
Direct financial transfers	Premiums received by producers on top of the market price.	Feed-in tariffs granted to operators for feeding renewable electricity into the grid.

Note: The list is not exhaustive and is meant to illustrate only the most prevalent mechanisms.

The economic cost of renewable-energy subsidies is borne predominantly by governments and consumers, often in combination. Taxpayers fund support when governments subsidise renewables, for example, using tax credits. Consumers can also fund subsidies, for example, through a levy on electricity prices to fund feed-in tariffs or in cases where the private sector passes through additional costs as a charge on all consumers. Supply mandates, quotas and portfolio obligations that apply proportionately to all suppliers often result in a pass-through of additional costs to consumers. There are occasional instances when these costs cannot be passed through and the burden is absorbed by the private sector, such as when end-use prices are regulated. Renewable energy may involve other associated costs that weigh against potential gains. These include the need for additional flexible electricity-generating capacity and transmission and distribution infrastructure investment to integrate variable resources and the life-cycle impacts of biomass production.[10]

Subsidisation may not be the most cost-effective means of making renewable energy more competitive or to meet broader policy objectives. Internalising the cost of certain externalities, for example by instituting more widespread or higher CO_2 prices, may represent a more economically efficient approach, although there are political hurdles to be overcome. While there have been successes in accelerating the deployment of renewable energy technologies and lowering their costs, constant attention is needed to evaluate the economic efficiency and the effectiveness of the policies being deployed. Choice of the ideal mechanism (or, more often, mechanisms) depends in part on national circumstances. In general, policies should provide a framework of incentives that take account of the maturity of the technology and foster a smooth transition towards mass-market integration, progressively employing market forces (IEA, 2008).

---

10. Chapter 5 of this report includes a more detailed discussion of integration costs for renewable energy in the power sector. *WEO-2010* addressed the issue of sustainability of biomass production.

Measuring renewable-energy subsidies

Estimation methodology

The methodology employed here for calculating the total cost of renewable-energy subsidies is broadly similar to the approach used to estimate fossil-fuel subsidies: price gaps are identified, based on the difference between the price paid per unit of renewable energy and the market value (or reference price) of substitutable fuels or technologies.[11] We have estimated the costs of subsidies to renewables-based electricity generation and biofuels (renewables-based transport fuels) and made projections of the level of subsidies that would be required to support the required growth of renewables in each of the three scenarios.[12] In this manner, our estimates capture most of the subsidies to renewables. The absolute level of renewable-energy subsidies depends on the unit costs of the subsidy needed to make renewables competitive, the quantities subsidised and the reference price. Unit subsidy costs of renewables are expected to fall gradually over time thanks to ongoing research and development and learning-by-doing, coupled with rising fuel and wholesale electricity prices, and higher and more widespread CO_2 prices (Figure 14.11).

Figure 14.11 ● **Illustration of the drivers of unit subsidy costs for renewable energy**

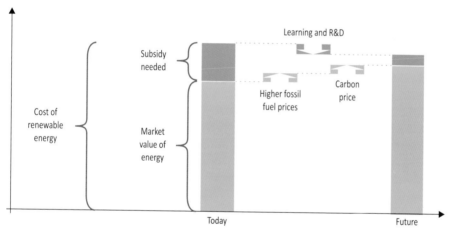

For renewables-based electricity, subsidies are quantified for 27 countries that presently make up about 90% of global electricity generation from biomass, wind, solar photovoltaics (PV) in buildings and geothermal energy. Unit costs derived from these estimates are then applied to the renewables-based generation in the remaining countries to arrive at a global aggregate.[13] For historic years, subsidies to renewable electricity are based on the policies in place in each country. They are measured by calculating the difference between the actual

---

11. A detailed methodology can be found at www.worldenergyoutlook.org/subsidies.asp.

12. Other types of renewable energy – mainly biomass used in buildings – are generally not subsidised.

13. Existing small hydropower is not included, but projected new capacity is included. Large hydropower is not included, as it is assumed that it does not, in most cases, need or receive support.

prices paid to generators of renewables-based electricity and wholesale electricity prices, applying that increment across the total volume of renewables-based electricity supplied under the support policy. To estimate future subsidies, the additional net economic cost of renewables-based output is calculated. The methodology is similar to that used for historic years, whereby the price paid to generators is assumed to match the cost of production for each technology over time, taking account of any cost reductions from learning. In our analysis, the total subsidy for renewables-based electricity technologies falls to zero when the cost of capacity built in previous years – often at higher costs – has been fully repaid, or when wholesale prices have risen to higher levels.

The cost of subsidies to biofuels (mainly ethanol and biodiesel) is assessed for 20 countries, which together account for about 90% of total global biofuels consumption. This figure is then scaled up to account for the remaining countries to arrive at a global estimate. To measure subsidies to biofuels up to 2010, we have multiplied the value of the tax advantage granted to biofuels relative to the oil-based alternatives (gasoline and diesel) by the volume of biofuels supplied. Where blending mandates exist, implicit subsidies are quantified by accounting for tax reductions and differences between the prices of ethanol and biodiesel and oil-based substitutes on an ex-tax energy-equivalent basis. To estimate the cost of subsidies over the *Outlook* period, ex-tax biofuel prices were compared with gasoline and diesel prices before taxes, and the difference was multiplied by the amount of biofuels consumed. Reference prices for biofuels are calculated using biofuels conversion costs and efficiencies and biomass feedstock prices, projected out to 2035. Where biofuels prices reach parity with projected fossil-fuel costs over the *Outlook* period, subsidies are assumed to be eliminated.

The subsidy estimates in this analysis represent the monetary value of most forms of government intervention that support the deployment and application of renewable energy, irrespective of whether the cost is finally carried by the government (and taxpayer) or the consumer. However, some subsidies are not captured, including funding for research and development, grants and loan guarantees. Furthermore, this analysis does not take into account several additional costs associated with renewable energy deployment, such as the costs of integration and additional flexible capacity to support renewable electricity generation (see Chapter 5) or the higher costs for agricultural products.

Estimated costs of renewable-energy subsidies

Total renewable-energy subsidies worldwide are estimated to have been $66 billion in 2010, a 10% increase over 2009. Of the 2010 total, $44 billion went to renewables-based electricity and $22 billion went to biofuels. In the New Policies Scenario, we project total subsidies to renewable energy in 2035 to increase to nearly $250 billion (in year-2010 dollars) (Figure 14.12).

Renewable-energy subsidies in 2010 were highest in the European Union, at $35 billion, almost double the amount in the United States. Together the two regions account for almost 80% of the global total. Subsidies to renewable energy during the *Outlook* period are projected to grow most in the United States, reaching $70 billion in 2035. This rise is driven primarily by the existing biofuels blending mandates and production targets, the latter of which aim to increase volumes of biofuels supply through 2022. Renewable-energy subsidies grow considerably in non-OECD countries towards the end of the *Outlook* period, as they deploy large amounts of renewables-based electricity (along with fossil-fuel generation) to meet burgeoning demand (Figure 14.13).

Figure 14.12 ● **Global subsidies to renewables-based electricity and biofuels by technology and fuel in the New Policies Scenario**

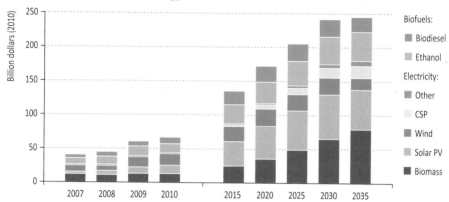

Figure 14.13 ● **Global subsidies to renewables-based electricity and biofuels by region in the New Policies Scenario**

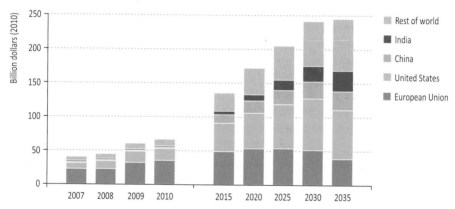

The higher level of renewables deployment in the 450 Scenario would entail $5.8 trillion in total subsidies over the *Outlook* period – $1 trillion more than in the New Policies Scenario. Renewable-energy subsidies grow at comparable rates in the two scenarios until 2020, but accelerate rapidly thereafter in the 450 Scenario, since a substantially higher level of renewables deployment is then needed to achieve deep cuts in CO_2 emissions. Subsidies average some $190 billion per year between 2020 and 2035 in the 450 Scenario, around 28% more than in the New Policies Scenario.

Estimated costs of renewable-electricity subsidies

Subsidies to renewables-based electricity – wind, solar PV in buildings, geothermal and biomass-based technologies – totalled $44 billion in 2010, an increase of 12% over 2009.

This rise was driven by a 22% increase in non-hydro renewables capacity, with wind and solar PV making up more than 90% of the additions. Renewable-electricity subsidies in 2010 were tempered by higher wholesale prices, which effectively narrowed the gap with production costs and lowered unit subsidy costs. Additionally, much of the renewable-electricity capacity deployed throughout the year actually generated little output. Total subsidies to wind were highest in absolute terms, at $18 billion, receiving, on average, $53 per megawatt-hour (MWh) of output. Solar PV, which produces electricity at a higher cost, benefitted from $425/MWh on average. As a result, solar PV received 28% of total renewable-electricity subsidies in 2010, despite accounting for only 4% of subsidised renewable electricity generation.

In the New Policies Scenario, renewable-electricity subsidies increase to nearly $180 billion in 2035, because of increased deployment and more widespread and stronger support policies. These subsidies help boost global renewables-based capacity by around 2 350 GW between today and 2035, allowing generation to nearly triple, increasing by 7 000 terawatt-hours (TWh). The share of subsidies received by each different technology shifts over the next 25 years, reflecting the respective stages of their development and the timing of commercial deployment. For example, unit subsidy costs for wind power fall as it becomes more competitive, causing total subsidies to wind to decline considerably after 2030, even though a significant amount of new capacity comes online. The cumulative extra cost of renewables-based electricity in the New Policies Scenario is estimated to be $3.3 trillion over the *Outlook* period, roughly $750 billion higher than the in *WEO-2010* (Box 14.4).

Subsidy costs per unit of output for renewables-based electricity shrink during the *Outlook* period in all three scenarios, as their production costs decline and wholesale electricity prices increase. Nonetheless, many renewable technologies still rely on government intervention to support their deployment by 2035. Their cost varies widely according to location, as the quality of renewable resources is site specific and investment costs differ from country to country.[14] For wind power, for example, critical factors affecting production costs are the strength and consistency of the wind resource and proximity to existing transmission lines, roads and demand centres. Wholesale electricity prices also vary significantly by region, according to the mix of power generation, the cost of fossil fuels and the policy environment. Wholesale electricity prices are set to increase during the *Outlook* period because of higher fuel prices and more widespread and higher carbon pricing. These effects are most pronounced in the 450 Scenario.

Wind and solar PV technologies experience the strongest growth in output in absolute terms over the *Outlook* period in the New Policies Scenario, accounting for more than 30% of the total global gross additions to generating capacity and one-fifth of incremental generation from 2009 to 2035. Onshore wind and solar PV in buildings in three regions – China, the European Union and the United States – account for two-thirds of the increase in generation. The best sites, *i.e.* with the best resources and lowest costs, are normally

---

14. Comparing average production costs of renewable technologies and wholesale electricity prices does not capture all the characteristics that affect the competitiveness of different generating technologies, including public acceptance and availability at peak load. The generation costs in this section take into account the average costs of the capacity deployed throughout the *Outlook* rather than for the best sites only.

exploited earlier, but there are exceptions. In the European Union, for example, the best sites for solar energy are in the southern countries, such as Spain and Italy, yet most capacity is now being built in Germany.

Box 14.4 • Why is our estimate of renewables-based electricity subsidies higher in this year's *Outlook*?

Our estimate of cumulative renewables-based electricity subsidies in the New Policies Scenario in this year's *Outlook* is 30% higher than we estimated last year. There are two main reasons for this increase:

- Projected wholesale electricity prices are lower, because of the assumption of lower natural-gas prices and postponed implementation of emissions trading schemes in several countries. Both delay the point at which certain renewable energy technologies become competitive. Even when they do, electricity from capacity installed before that date continues to be subsidised until the end of its economic lifetime (typically 20 to 25 years).

- Wind and solar PV capacity additions accelerated in 2010, accompanied by new policies to support further deployment. This acceleration has been included in the latest projections, increasing the amount of subsidised generation throughout the *Outlook* period and, therefore, the overall cost of the subsidy.

We have also revised our methodology for calculating subsidies to solar PV in the buildings sector, which has led to some increase in the estimate. Previously, production costs were compared to end-user prices, but this year subsidies have been calculated with respect to the wholesale electricity price, like other forms of renewable electricity generation. The wholesale price is considered to be a more appropriate reference for quantifying subsidies to solar PV in buildings for two main reasons. First, the difference between the wholesale price and the end-user price is made up of several additional costs that are not affected by the presence of distributed generation, such as metering, billing and the other costs of operating the electricity system. End-user prices also include the cost of transmission and distribution, and although distributed generation, such as solar PV, can reduce the need for some network investment, it does not usually reduce it significantly. Second, end-user prices in many regions include the cost of support to renewable technologies and it would not be appropriate to include this in the calculation of the subsidy required by solar photovoltaic technology.

In the New Policies Scenario, production costs for onshore wind in China decline by about 10% over the *Outlook* period. With wholesale electricity prices expected to rise, we project that onshore wind becomes competitive – and unit subsidy costs are eliminated – by around 2030; in the European Union, onshore wind becomes competitive around 2020 (Figure 14.14a). Onshore wind does not become competitive in the United States by 2035, as wholesale electricity prices are lower than previously expected (due to the increased availability of natural gas) and because of the lack of a price for CO_2. Production costs fall more quickly in the 450 Scenario and wholesale electricity prices are higher, bringing forward the dates at which the unit subsidy cost for onshore wind becomes zero.

Figure 14.14 ● Renewable electricity production cost relative to wholesale prices for selected technologies and regions in the New Policies Scenario

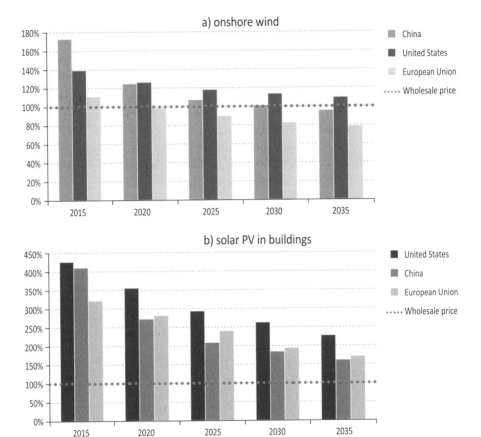

Notes: The cost of production is the average long-run marginal cost at deployed sites over the *Outlook* period; integration costs are excluded. Wholesale electricity prices are taken as the average projected for the respective regions in the New Policies Scenario.

Despite steady declines, high production costs prevent solar PV in buildings from becoming competitive in both the New Policies or 450 Scenario in all three regions. In the European Union, where deployment of solar PV is highest between 2010 and 2035 (and wholesale electricity prices are among the most expensive), average production costs drop by about 60% over the *Outlook* period in the New Policies Scenario (Figure 14.14b). By 2035, unit subsidy costs for solar PV in the European Union are around $75/MWh. In China, unit subsidy costs for solar PV are $40/MWh in 2035: production costs in China are considerably lower than in the European Union, as are wholesale electricity prices. Unit subsidy costs in the United States decline by about 80% in the New Policies Scenario, falling to about $75/MWh in 2035. In the 450 Scenario, unit support costs for solar PV come down moderately for the United States (to about $65/MWh in 2035) and are only marginally lower in the other two regions, as more deployment shifts to lower-quality sites.

Estimated costs of biofuels subsidies

Subsidies to biofuels increased by 6% in 2010 compared with 2009, to $22 billion. This was primarily the result of higher biofuels supply that was brought forward by existing blending mandates and, in some countries, was eligible for tax reductions. While cost reductions over the *Outlook* period improve the competitiveness of conventional biofuels against gasoline and diesel, rising biofuels use (from 1.3 million barrels of oil equivalent [mboe/d] today to 4.4 mboe/d in 2035) results in an increase in the total cost of biofuels subsidies in the New Policies Scenario, reaching $67 billion per year in 2035. Conventional biofuels attract the vast majority of total biofuels subsidies through 2030 while the volumes of advanced biofuels consumed are still small. By 2035, advanced biofuels receive $10 billion in subsidies, or a 15% share of total biofuels subsidies. The cumulative cost of biofuels subsidies between 2011 and 2035 amounts to $1.4 trillion.

With the exception of Brazil, conventional biofuels are generally not cost-competitive with conventional oil-based gasoline, diesel and kerosene-based jet fuel and their deployment is therefore often supported by blending mandates or other policies. The costs of conventional biofuels today vary greatly by region. Sugar cane in Brazil, for example, is generally a cheap feedstock compared with the commonly used corn or sugar beet, even if prices have increased considerably over the past two years as a result of strong El Niño weather phenomena and low cane renewal rates (F.O. Lichts, 2011). Because sugar cane conversion costs are low, sugar cane ethanol can be competitive with gasoline, depending on the prevailing sugar prices. By contrast, biodiesel is often produced using vegetable or palm oil, which are high-quality and high-cost feedstocks.

The future competitiveness of conventional biofuels depends critically on the pace at which technology costs can be reduced, and future feedstock prices. For the former, modest cost reductions are generally possible through technology improvements and economies of scale. As to feedstock costs, the increasing use of conventional biofuels is likely to put upward pressure on prices, due to the competition with other land uses such as food crop production. In addition, rising international oil prices will increase the price of fertilisers and transportation costs. Overall, increasing biomass feedstock prices will offset part of the cost reductions achieved in the case of conventional biofuels, so the cost reductions achieved in the New Policies and 450 Scenarios are modest.

Advanced biofuels potentially offer a way to reduce the problems of environmental sustainability, cost competitiveness and land-use competition that are associated with many conventional biofuels (IEA, 2010). Today, advanced biofuels are not available on a large scale. In the New Policies Scenario, advanced biofuels are assumed to be readily available to the market by 2020, even though developing and deploying advanced biofuels is challenging and will require a large investment in research and development. Like any new technology, advanced biofuels offer potential cost reductions through technological improvements and

14

economies of scale, so that average costs in 2035 are projected to be reduced by 23% in the New Policies Scenario (Figure 14.15). They fall even further in the 450 Scenario, by 29%, where advanced biofuels make up nearly 70% of all biofuels use by 2035 (compared with only 20% in the New Policies Scenario). Consequently, unit subsidy costs in the 450 Scenario drop by 57% between 2020 and 2035.

Figure 14.15 ● **Indicative biofuels production costs and spot oil product prices**

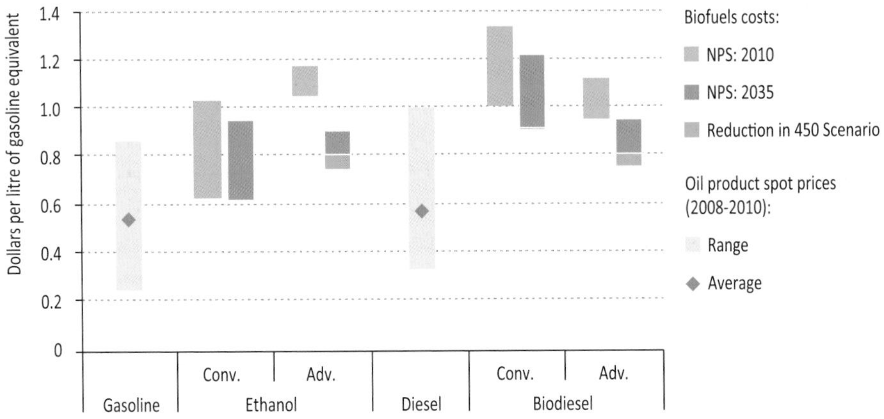

Notes: NPS = New Policies Scenario; "Conv." = conventional; "Adv." = advanced. The range of gasoline and diesel spot prices is taken from the monthly average in the United States from 2008 to 2010. Biofuels costs are not adjusted by subsidies; cost variations can be even larger than depicted here, depending on feedstock and region.

Implications for CO_2 emissions and import bills

By supporting the deployment of renewables and lowering their costs over time, renewable-energy subsidies can support both energy and environmental policy goals, such as lowering energy import dependency, diversifying energy supplies, reducing greenhouse-gas emissions and mitigating local pollution. In the New Policies Scenario, subsidies underpin a rapid expansion of renewable energy, as renewable electricity generation reaches 5 900 TWh (excluding large hydro) and biofuels production 4.4 mb/d in 2035. The share of total renewables in the global energy mix increases from 13% to 18% over the *Outlook* period. This expansion results in considerable CO_2 emissions savings. In 2035, renewables utilisation yield CO_2 emissions savings from the power sector (excluding large hydro) of around 2.9 Gt compared with the emissions that would be generated for the projected level of electricity generation were there no change in the mix of fuels and technologies (see Chapter 5). Biofuels use for transport corresponds to 0.5 Gt of CO_2 emissions savings in 2035, bringing total savings in the two sectors to 3.4 Gt.[15]

---

15. In IEA statistics, CO_2 emissions from biofuels (and biomass more generally) are not included in the data for CO_2 emissions from fuel combustion. This is to avoid the double-counting of biomass CO_2 emissions that are assumed to have been released as soon as the biomass is harvested, in line with the 1996 IPCC Guidelines, and are accounted for under land use, land-use change and forestry (LULUCF). Consequently, emissions savings to offset biomass consumption, such as forest re-plantation, are also reported under LULUCF.

As depicted in the 450 Scenario, more ambitious deployment of renewable energy establishes a more sustainable pathway, which limits the average global temperature increase to 2°C and offers several energy security benefits. In this scenario, government efforts to accelerate renewables deployment are intensified, with cumulative subsidies between 2011 and 2035 amounting to $5.8 trillion, $1 trillion higher than in the New Policies Scenario. In 2035, subsidies in the 450 Scenario reach almost $350 billion per year, about $100 billion higher than in the New Policies Scenario. The additional cost of subsidies in the 450 Scenario would be significantly higher if it were not for the assumption of more widespread and higher pricing of CO_2 (see Chapter 6). Greater displacement of fossil fuels by renewable energy in the power sector (excluding large hydro) and transport sector leads to additional CO_2 emissions reductions of 3.5 Gt in 2035, compared with the New Policies Scenario. Accelerated deployment of renewable energy in the 450 Scenario also reduces imports, lowering import bills by some $350 billion in total (Figure 14.16). About two-thirds of import bill savings result from lower fossil fuel consumption, predominately coal and natural gas, by electricity generators. The remainder of the import bill savings stem from the displacement of oil by biofuels in the transport sector.

Figure 14.16 ● **CO₂ emissions and import bill savings due to renewables subsidies in 2035 in the 450 Scenario relative to the New Policies Scenario**

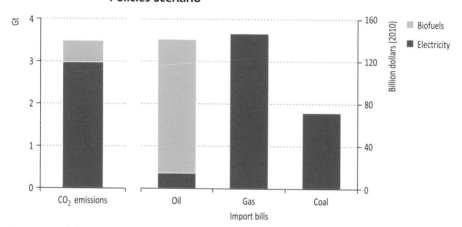

Notes: Import bill savings do not necessarily represent a net cost saving. For example, import bill savings from the displacement of oil by biofuels are roughly matched by the costs paid by consumers for biofuels at market prices similar to their oil-based equivalent. Biofuels subsidies, where they exist, are an additional cost.

For some countries, the immediate environmental gains from mitigating local pollution are a key factor in their decision to subsidise renewables deployment. In place of fossil-fuels, renewable energy can help cut emissions of sulphur dioxide, nitrogen oxides and particulate matter that cause acid rain, soil acidification, ground-level ozone formation and smog. The substitution of fossil fuels by renewable energy can provide other potential benefits that may serve as rationale for government intervention, such as lower adaptation costs from the effects of climate change and benefits for rural development.

14

Impact of renewable-energy subsidies on end-user electricity prices

In many parts of the world, consumers and businesses have been struggling to cope with higher electricity prices in recent years. The impact of renewable-electricity subsidies on residential bills is marginal compared with the impact of rising international coal and natural gas prices and investment in new generating, transmission and distribution capacity. However, in some cases, rising end-user prices can be attributed to mandated increases in the share of renewables-based electricity. Renewable-energy subsidies raise the price that producers receive for their output to cover their higher costs. This additional cost is often passed through to consumers. The impact of renewable-energy subsidies on retail electricity prices varies depending on how support mechanisms are designed, the cost of competing non-renewable fuels, the cost of renewable electricity technologies and their volume deployed over the years. Not all forms of support to renewable energy are reflected in end-user prices, as some are paid for by governments. In a few cases, the cost of subsidies is absorbed by state-owned utilities, for instance when mandates are applied and end-user prices are regulated.

Figure 14.17 ● Cost of renewables-based electricity subsidies as a percentage of average end-user electricity price in the New Policies Scenario

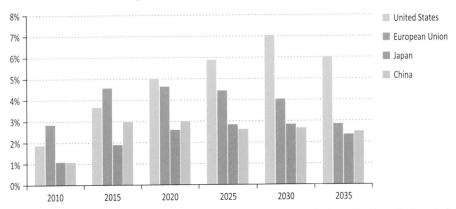

Notes: In some countries, notably in the European Union, most of the subsidy is in the form of feed-in tariffs, the cost of which is recovered in final prices to end-users. Elsewhere, end-users may not pay for these costs directly, such as in the United States where subsidies are in the form of tax credits.

For the United States, the European Union, Japan and China, the cost of renewable-energy subsidies does not exceed (on average) $12/MWh, or 1.2 US cents/kWh, over the projection period in the New Policies Scenario.[16] As a percentage of the electricity price, renewable-energy subsidies are largest in the United States, peaking at over 7% in 2030 (Figure 14.17). In the European Union, renewable-energy subsidies as a percentage of electricity prices lessen after 2020, as an increasing carbon price pushes up the wholesale electricity price, resulting in a reduced subsidy need. The share of subsidies in end-user prices are generally

---

16. These estimates are calculated by dividing the annual total renewable support for a country or region by the average end-user electricity price. They are not a calculation of the impact of renewable-energy subsidies on the end-user electricity price.

lowest in Japan, due partly to electricity prices that are more than double those in the United States and China. Renewable-energy subsidies remain at or below 3% of the average end-user electricity price over time in China, as renewable energy capacity and the electricity price both increase rapidly.

Recent developments in renewable-energy subsidies

Historical figures in this *Outlook* demonstrate that the financial cost of renewable-energy subsidies has grown considerably in recent years. This has been largely due to the expansion of policies that support renewable energy, including in developing countries. By early 2011, at least 118 countries had some type of policy target or renewable-energy support policy at the national level, up from 55 countries in 2005 (REN21, 2011). Policies that support renewable energy and involve subsidies can also be seen at the state, provincial and local levels. Several countries have made notable recent policy changes (Table 14.4).

Table 14.4 • Recent developments in renewable-energy subsidies in selected economies

	Description of recent developments
Australia	Initiated a gradual reduction in tax credits for small-scale solar PV systems, with 20% reductions each year from 2010 to 2013. The state of Queensland has delayed implementation of a 5% biodiesel blending mandate.
Brazil	Plans to triple its capacity of wind, biomass and small hydropower to 27 GW in 2020 by providing $44.5 billion in public investment.
China	Updated guidelines for the "Golden Sun" programme in 2011; eligible solar PV installations are to receive $1.40 or $1.24/watt of polysilicon-based or thin-film modules.
France	Adjusted the feed-in tariff system for solar PV, designating different rates for building-integrated systems and by installation size.
Germany	Reduced feed-in tariffs for solar PV by 13% for installations coming online between July and October 2010, and 16% thereafter. Introduced a premium for renewable energy producers to deliver electricity at peak demand periods. Proposed to dramatically increase geothermal feed-in tariffs in 2012.
India	Established the National Solar Mission targeting 20 GW of solar PV and 2 GW of solar lanterns by 2022. Implemented feed-in tariffs for solar PV and solar thermal to help achieve these goals.
Italy	Announced gradual reductions in feed-in-tariffs for solar plants starting in June 2011 though to 2013 with intent to cap subsidies to solar developers by the end of 2016.
Spain	Lowered feed-in tariffs for wind by 35% and reduced solar PV load-hour payments from January 2011. This applies to new and existing installations.
Turkey	Adopted a new feed-in tariff policy, expanding beyond wind power to solar, geothermal, hydro and biomass plants, with the goal of 600 MW of new capacity per year through 2013, with additional incentives for domestically produced components.
Uganda	Implemented feed-in tariffs for wind, solar, geothermal, biomass, hydro and other renewable-energy plants in 2011 with capacity caps, valid for a period of 20 years.
United Kingdom	Announced that feed-in tariffs for solar plants larger than 50 kW will be cut following higher than expected deployment of large solar projects.
United States	The volumetric ethanol excise tax credit is due to expire at the end of 2011 unless extended again. Production tax credits were extended through the end of 2012 and 2013 for different sources of renewable electricity. Numerous loan guarantees were provided to renewable-energy projects in 2011.

14

While renewable-energy subsidies have been growing strongly, the uncertain outlook for the global economy and the weak state of public finances is beginning to bring pressure to bear on the maintenance of some subsidies, particularly in OECD countries. Whereas many countries in Asia, Europe and North America directed part of their stimulus packages towards renewable energy, in the wake of the 2008/2009 recession, in a bid to boost jobs and economic growth while also addressing climate-change concerns, most of the funds have now been spent. Some renewable-energy subsidies have already been adjusted and scaled back because of cost reductions from technological improvements or reduced manufacturing costs, meaning that unit-subsidy rates (for example, feed-in tariffs for solar power) have declined more quickly than expected. In other cases, renewable-energy subsidy schemes have been adjusted because flaws became apparent in policy design or implementation. This has revealed some subsidies to be overly generous, unnecessarily pushing up electricity prices and having other unintended market consequences.

ANNEXES

TABLES FOR SCENARIO PROJECTIONS

General note to the tables

The tables detail projections for *energy demand*, gross *electricity generation* and *electrical capacity*, and *carbon-dioxide (CO$_2$) emissions* from fuel combustion in the Current Policies, New Policies and 450 Scenarios. The following countries and regions are covered: World, OECD[1], OECD Americas, the United States, OECD Europe, the European Union, OECD Asia Oceania, Japan, non-OECD, Eastern Europe/Eurasia, Russia, non-OECD Asia, China, India, the Middle East, Africa, Latin America and Brazil. The definitions for regions, fuels and sectors can be found in Annex C. By convention, in the table headings *CPS* and *450* refers to Current Policies and 450 Scenarios respectively.

Data for *energy demand*, gross *electricity generation* and *CO$_2$ emissions* from fuel combustion up to 2009 are based on IEA statistics, published in *Energy Balances of OECD Countries*, *Energy Balances of Non-OECD Countries* and *CO$_2$ Emissions from Fuel Combustion*. Historical data for *electrical capacity* is supplemented from the Platts World Electric Power Plants Database (December 2010 version) and the International Atomic Energy Agency PRIS database.

Both in the text of this book and in the tables, rounding may lead to minor differences between totals and the sum of their individual components. Growth rates are calculated on a compound average annual basis and are marked "n.a." when the base year is zero or the value exceeds 200%. Nil values are marked "-".

Definitional note to the tables

Total primary energy demand (TPED) is equivalent to *power generation* plus *other energy sector* excluding electricity and heat, plus *total final consumption (TFC)* excluding electricity and heat. *TPED* does not include ambient heat from heat pumps or electricity trade. Sectors comprising *TFC* include *industry, transport, buildings* (residential and services) and *other (*agriculture and non-energy use). Projected *electrical capacity* is the net result of existing capacity plus additions less retirements. *Total CO$_2$* includes emissions from *other energy sector* in addition to the *power generation* and *TFC* sectors shown in the tables. *CO$_2$ emissions* and *energy demand* from international marine and aviation *bunkers* are included only at the world *transport* level. *CO$_2$ emissions* do not include emissions from industrial waste and non-renewable municipal waste.

---

1. Since the last *Outlook*, the regional groupings have been updated to incorporate the accession of Chile, Estonia, Israel and Slovenia to the OECD; see Annex C for full details of the new groupings.

World: New Policies Scenario

	Energy demand (Mtoe)							Shares (%)		CAAGR (%)
	1990	2009	2015	2020	2025	2030	2035	2009	2035	2009-2035
TPED	**8 785**	**12 132**	**13 913**	**14 769**	**15 469**	**16 206**	**16 961**	**100**	**100**	**1.3**
Coal	2 233	3 294	3 944	4 083	4 104	4 099	4 101	27	24	0.8
Oil	3 226	3 987	4 322	4 384	4 453	4 546	4 645	33	27	0.6
Gas	1 671	2 539	2 945	3 214	3 442	3 698	3 928	21	23	1.7
Nuclear	526	703	796	929	1 036	1 128	1 212	6	7	2.1
Hydro	184	280	334	377	418	450	475	2	3	2.1
Biomass and waste	908	1 230	1 375	1 495	1 622	1 761	1 911	10	11	1.7
Other renewables	36	99	197	287	394	524	690	1	4	7.8
Power generation	**2 988**	**4 572**	**5 342**	**5 850**	**6 276**	**6 712**	**7 171**	**100**	**100**	**1.7**
Coal	1 228	2 144	2 557	2 667	2 705	2 722	2 744	47	38	1.0
Oil	377	267	219	189	161	140	135	6	2	-2.6
Gas	581	1 003	1 139	1 253	1 361	1 486	1 591	22	22	1.8
Nuclear	526	703	796	929	1 036	1 128	1 212	15	17	2.1
Hydro	184	280	334	377	418	450	475	6	7	2.1
Biomass and waste	60	94	130	187	255	335	429	2	6	6.0
Other renewables	32	82	168	248	341	451	586	2	8	7.9
Other energy sector	**897**	**1 263**	**1 423**	**1 490**	**1 534**	**1 584**	**1 629**	**100**	**100**	**1.0**
Electricity	*182*	*284*	*333*	*365*	*393*	*421*	*448*	*22*	*27*	*1.8*
TFC	**6 292**	**8 329**	**9 610**	**10 177**	**10 645**	**11 134**	**11 629**	**100**	**100**	**1.3**
Coal	769	818	990	1 002	976	943	910	10	8	0.4
Oil	2 607	3 461	3 853	3 958	4 073	4 206	4 325	42	37	0.9
Gas	943	1 267	1 490	1 617	1 715	1 816	1 923	15	17	1.6
Electricity	835	1 440	1 790	2 033	2 242	2 453	2 670	17	23	2.4
Heat	333	253	283	292	294	295	295	3	3	0.6
Biomass and waste	801	1 072	1 175	1 234	1 291	1 347	1 402	13	12	1.0
Other renewables	4	17	30	40	53	73	104	0	1	7.1
Industry	**1 809**	**2 279**	**2 858**	**3 089**	**3 205**	**3 298**	**3 388**	**100**	**100**	**1.5**
Coal	473	640	804	820	804	784	765	28	23	0.7
Oil	329	320	352	352	349	341	332	14	10	0.1
Gas	359	442	564	636	673	706	736	19	22	2.0
Electricity	379	580	781	898	979	1 048	1 118	25	33	2.6
Heat	151	110	127	131	127	124	121	5	4	0.4
Biomass and waste	116	186	228	252	272	294	315	8	9	2.0
Other renewables	0	0	1	1	1	1	1	0	0	2.9
Transport	**1 579**	**2 283**	**2 615**	**2 734**	**2 880**	**3 066**	**3 257**	**100**	**100**	**1.4**
Oil	1 483	2 135	2 415	2 499	2 602	2 735	2 863	94	88	1.1
Of which: Bunkers	*199*	*329*	*351*	*376*	*398*	*424*	*454*	*14*	*14*	*1.2*
Electricity	21	23	31	36	41	49	58	1	2	3.6
Biofuels	6	52	84	107	135	167	202	2	6	5.4
Other fuels	69	73	85	92	101	115	134	3	4	2.3
Buildings	**2 254**	**2 844**	**3 108**	**3 282**	**3 449**	**3 622**	**3 804**	**100**	**100**	**1.1**
Coal	241	125	132	124	115	103	90	4	2	-1.2
Oil	329	327	341	338	332	324	313	12	8	-0.2
Gas	431	612	674	716	759	803	850	22	22	1.3
Electricity	404	796	926	1 042	1 157	1 284	1 414	28	37	2.2
Heat	172	140	152	158	164	168	170	5	4	0.8
Biomass and waste	673	828	855	867	873	873	870	29	23	0.2
Other renewables	4	17	28	37	49	68	97	1	3	7.0
Other	**651**	**923**	**1 029**	**1 072**	**1 111**	**1 148**	**1 180**	**100**	**100**	**0.9**

World: Current Policies and 450 Scenarios

	Energy demand (Mtoe)						Shares (%)		CAAGR (%)	
	2020	2030	2035	2020	2030	2035	2035		2009-2035	
	Current Policies Scenario			450 Scenario			CPS	450	CPS	450
TPED	**15 124**	**17 173**	**18 302**	**14 185**	**14 561**	**14 870**	**100**	**100**	**1.6**	**0.8**
Coal	4 416	5 060	5 419	3 716	2 606	2 316	30	16	1.9	-1.3
Oil	4 482	4 808	4 992	4 182	3 905	3 671	27	25	0.9	-0.3
Gas	3 247	3 848	4 206	3 030	3 237	3 208	23	22	2.0	0.9
Nuclear	908	1 024	1 054	973	1 452	1 664	6	11	1.6	3.4
Hydro	366	419	442	391	488	520	2	3	1.8	2.4
Biomass and waste	1 449	1 617	1 707	1 554	2 055	2 329	9	16	1.3	2.5
Other renewables	256	397	481	339	817	1 161	3	8	6.3	9.9
Power generation	**6 070**	**7 297**	**8 002**	**5 553**	**5 815**	**6 149**	**100**	**100**	**2.2**	**1.1**
Coal	2 957	3 547	3 883	2 376	1 400	1 163	49	19	2.3	-2.3
Oil	201	157	152	171	105	95	2	2	-2.1	-3.9
Gas	1 247	1 535	1 729	1 144	1 217	1 130	22	18	2.1	0.5
Nuclear	908	1 024	1 054	973	1 452	1 664	13	27	1.6	3.4
Hydro	366	419	442	391	488	520	6	8	1.8	2.4
Biomass and waste	172	273	332	206	434	576	4	9	5.0	7.2
Other renewables	219	341	410	293	720	1 002	5	16	6.4	10.1
Other energy sector	**1 522**	**1 685**	**1 766**	**1 428**	**1 408**	**1 386**	**100**	**100**	**1.3**	**0.4**
Electricity	*375*	*452*	*493*	*348*	*373*	*386*	*28*	*28*	*2.1*	*1.2*
TFC	**10 348**	**11 619**	**12 303**	**9 841**	**10 259**	**10 400**	**100**	**100**	**1.5**	**0.9**
Coal	1 031	1 031	1 023	944	825	771	8	7	0.9	-0.2
Oil	4 043	4 455	4 663	3 781	3 618	3 421	38	33	1.2	-0.0
Gas	1 651	1 895	2 033	1 560	1 676	1 740	17	17	1.8	1.2
Electricity	2 082	2 598	2 893	1 960	2 241	2 386	24	23	2.7	2.0
Heat	301	318	325	275	259	250	3	2	1.0	-0.0
Biomass and waste	1 204	1 267	1 296	1 275	1 542	1 672	11	16	0.7	1.7
Other renewables	36	56	71	46	98	160	1	2	5.6	8.9
Industry	**3 153**	**3 511**	**3 695**	**2 981**	**3 074**	**3 090**	**100**	**100**	**1.9**	**1.2**
Coal	843	860	865	773	686	650	23	21	1.2	0.1
Oil	361	366	365	344	332	316	10	10	0.5	-0.0
Gas	655	752	802	624	678	695	22	22	2.3	1.8
Electricity	920	1 129	1 243	861	956	989	34	32	3.0	2.1
Heat	135	137	137	125	115	110	4	4	0.9	0.0
Biomass and waste	239	266	282	252	303	326	8	11	1.6	2.2
Other renewables	1	1	1	1	2	3	0	0	2.9	7.7
Transport	**2 776**	**3 218**	**3 466**	**2 628**	**2 746**	**2 744**	**100**	**100**	**1.6**	**0.7**
Oil	2 553	2 924	3 124	2 367	2 248	2 090	90	76	1.5	-0.1
Of which: Bunkers	*375*	*432*	*468*	*364*	*380*	*388*	*14*	*14*	*1.4*	*0.6*
Electricity	35	46	53	39	79	133	2	5	3.2	6.9
Biofuels	97	135	159	127	294	370	5	13	4.4	7.9
Other fuels	92	113	130	94	125	150	4	5	2.2	2.8
Buildings	**3 331**	**3 720**	**3 935**	**3 166**	**3 309**	**3 409**	**100**	**100**	**1.3**	**0.7**
Coal	130	112	100	115	85	69	3	2	-0.8	-2.3
Oil	347	342	337	314	265	239	9	7	0.1	-1.2
Gas	729	835	896	672	690	705	23	21	1.5	0.5
Electricity	1 068	1 346	1 508	1 003	1 137	1 189	38	35	2.5	1.6
Heat	163	177	184	146	140	136	5	4	1.1	-0.1
Biomass and waste	859	854	844	873	901	920	21	27	0.1	0.4
Other renewables	34	53	67	43	91	150	2	4	5.5	8.8
Other	**1 088**	**1 171**	**1 208**	**1 066**	**1 130**	**1 157**	**100**	**100**	**1.0**	**0.9**

World: New Policies Scenario

Electricity generation (TWh)	1990	2009	2015	2020	2025	2030	2035	Shares (%) 2009	Shares (%) 2035	CAAGR (%) 2009-2035
Total generation	11 819	20 043	24 674	27 881	30 640	33 417	36 250	100	100	2.3
Coal	4 425	8 118	10 104	10 860	11 253	11 616	12 035	41	33	1.5
Oil	1 337	1 027	833	713	620	547	533	5	1	-2.5
Gas	1 727	4 299	5 280	6 020	6 676	7 376	7 923	21	22	2.4
Nuclear	2 013	2 697	3 062	3 576	3 984	4 337	4 658	13	13	2.1
Hydro	2 144	3 252	3 887	4 380	4 861	5 231	5 518	16	15	2.1
Biomass and waste	131	288	425	635	879	1 165	1 497	1	4	6.5
Wind	4	273	835	1 282	1 724	2 182	2 703	1	7	9.2
Geothermal	36	67	96	131	174	221	271	0	1	5.5
Solar PV	0	20	126	230	369	551	741	0	2	14.9
CSP	1	1	24	52	92	167	307	0	1	25.5
Marine	1	1	1	2	9	23	63	0	0	20.2

Electrical capacity (GW)	2009	2015	2020	2025	2030	2035	Shares (%) 2009	Shares (%) 2035	CAAGR (%) 2009-2035
Total capacity	4 957	6 196	6 941	7 594	8 293	9 038	100	100	2.3
Coal	1 581	1 975	2 133	2 211	2 289	2 353	32	26	1.5
Oil	431	430	356	303	266	255	9	3	-2.0
Gas	1 298	1 602	1 749	1 868	2 016	2 185	26	24	2.0
Nuclear	393	431	495	546	591	633	8	7	1.8
Hydro	1 007	1 152	1 297	1 439	1 548	1 629	20	18	1.9
Biomass and waste	53	75	109	148	193	244	1	3	6.0
Wind	159	397	582	752	921	1 102	3	12	7.7
Geothermal	11	15	20	27	33	41	0	0	5.0
Solar PV	22	112	184	272	385	499	0	6	12.7
CSP	1	7	14	25	45	81	0	1	20.9
Marine	0	0	1	2	6	17	0	0	17.3

CO_2 emissions (Mt)	1990	2009	2015	2020	2025	2030	2035	Shares (%) 2009	Shares (%) 2035	CAAGR (%) 2009-2035
Total CO_2	20 936	28 844	33 135	34 407	35 045	35 709	36 367	100	100	0.9
Coal	8 314	12 455	14 938	15 423	15 339	15 129	14 949	43	41	0.7
Oil	8 818	10 629	11 524	11 699	11 917	12 226	12 554	37	35	0.6
Gas	3 803	5 760	6 673	7 284	7 789	8 354	8 865	20	24	1.7
Power generation	7 485	11 760	13 570	14 146	14 382	14 556	14 757	100	100	0.9
Coal	4 929	8 562	10 205	10 613	10 689	10 639	10 617	73	72	0.8
Oil	1 199	845	695	598	510	443	426	7	3	-2.6
Gas	1 357	2 353	2 671	2 936	3 184	3 473	3 714	20	25	1.8
TFC	12 446	15 618	17 948	18 559	18 948	19 380	19 791	100	100	0.9
Coal	3 247	3 619	4 398	4 445	4 311	4 150	3 990	23	20	0.4
Oil	7 064	9 142	10 190	10 461	10 761	11 125	11 454	59	58	0.9
Transport	*4 396*	*6 358*	*7 192*	*7 447*	*7 756*	*8 154*	*8 539*	*41*	*43*	*1.1*
Of which: Bunkers	*614*	*1 016*	*1 083*	*1 160*	*1 227*	*1 307*	*1 395*	*7*	*7*	*1.2*
Gas	2 136	2 857	3 360	3 653	3 876	4 104	4 346	18	22	1.6

World: Current Policies and 450 Scenarios

	Electricity generation (TWh)						Shares (%)		CAAGR (%)	
	2020	2030	2035	2020	2030	2035	2035		2009-2035	
	Current Policies Scenario			450 Scenario			CPS	450	CPS	450
Total generation	28 569	35 468	39 368	26 835	30 393	32 224	100	100	2.6	1.8
Coal	12 040	15 110	16 932	9 648	5 943	4 797	43	15	2.9	-2.0
Oil	750	603	591	644	394	360	2	1	-2.1	-4.0
Gas	5 967	7 631	8 653	5 543	6 226	5 608	22	17	2.7	1.0
Nuclear	3 495	3 938	4 053	3 741	5 582	6 396	10	20	1.6	3.4
Hydro	4 254	4 875	5 144	4 547	5 676	6 052	13	19	1.8	2.4
Biomass and waste	579	944	1 150	701	1 515	2 025	3	6	5.5	7.8
Wind	1 130	1 717	2 005	1 486	3 336	4 320	5	13	8.0	11.2
Geothermal	120	174	200	148	311	407	1	1	4.3	7.2
Solar PV	195	360	435	279	903	1 332	1	4	12.5	17.5
CSP	37	101	166	96	471	845	0	3	22.5	30.5
Marine	2	14	39	3	35	82	0	0	18.0	21.4

	Electrical capacity (GW)						Shares (%)		CAAGR (%)	
	2020	2030	2035	2020	2030	2035	2035		2009-2035	
	Current Policies Scenario			450 Scenario			CPS	450	CPS	450
Total capacity	6 955	8 346	9 126	6 923	8 544	9 484	100	100	2.4	2.5
Coal	2 276	2 762	3 030	1 927	1 438	1 268	33	13	2.5	-0.8
Oil	358	273	264	348	236	213	3	2	-1.9	-2.7
Gas	1 767	2 108	2 333	1 727	2 014	2 111	26	22	2.3	1.9
Nuclear	485	539	549	519	758	865	6	9	1.3	3.1
Hydro	1 257	1 434	1 509	1 349	1 690	1 803	17	19	1.6	2.3
Biomass and waste	100	157	189	120	250	329	2	3	5.0	7.3
Wind	521	749	852	661	1 349	1 685	9	18	6.7	9.5
Geothermal	19	27	31	23	47	60	0	1	3.9	6.6
Solar PV	161	268	314	220	625	901	3	10	10.7	15.3
CSP	10	27	44	27	127	226	0	2	18.1	25.7
Marine	1	4	11	1	9	23	0	0	15.2	18.5

	CO_2 emissions (Mt)						Shares (%)		CAAGR (%)	
	2020	2030	2035	2020	2030	2035	2035		2009-2035	
	Current Policies Scenario			450 Scenario			CPS	450	CPS	450
Total CO_2	36 067	40 665	43 320	31 885	24 784	21 574	100	100	1.6	-1.1
Coal	16 725	18 938	20 182	13 917	7 575	5 356	47	25	1.9	-3.2
Oil	11 983	13 036	13 638	11 118	10 289	9 616	31	45	1.0	-0.4
Gas	7 360	8 691	9 500	6 850	6 921	6 603	22	31	1.9	0.5
Power generation	15 331	18 117	19 836	12 582	6 968	4 758	100	100	2.0	-3.4
Coal	11 772	14 026	15 309	9 371	4 054	2 254	77	47	2.3	-5.0
Oil	636	498	481	542	332	302	2	6	-2.1	-3.9
Gas	2 924	3 594	4 046	2 668	2 582	2 203	20	46	2.1	-0.3
TFC	18 993	20 663	21 516	17 686	16 281	15 346	100	100	1.2	-0.1
Coal	4 569	4 529	4 474	4 203	3 245	2 843	21	19	0.8	-0.9
Oil	10 693	11 847	12 441	9 961	9 384	8 769	58	57	1.2	-0.2
Transport	*7 605*	*8 718*	*9 315*	*7 057*	*6 714*	*6 252*	*43*	*41*	*1.5*	*-0.1*
Of which: Bunkers	*1 158*	*1 332*	*1 440*	*1 126*	*1 176*	*1 201*	*7*	*8*	*1.4*	*0.6*
Gas	3 730	4 286	4 602	3 522	3 653	3 733	21	24	1.8	1.0

OECD: New Policies Scenario

	Energy demand (Mtoe)							Shares (%)		CAAGR (%)
	1990	2009	2015	2020	2025	2030	2035	2009	2035	2009-2035
TPED	**4 521**	**5 236**	**5 549**	**5 575**	**5 598**	**5 640**	**5 681**	**100**	**100**	**0.3**
Coal	1 080	1 033	1 097	1 046	984	896	802	20	14	-1.0
Oil	1 870	1 958	1 946	1 856	1 768	1 700	1 640	37	29	-0.7
Gas	843	1 248	1 362	1 404	1 438	1 486	1 516	24	27	0.7
Nuclear	451	585	610	635	662	691	722	11	13	0.8
Hydro	102	114	122	127	131	134	137	2	2	0.7
Biomass and waste	146	242	305	357	415	477	542	5	10	3.1
Other renewables	29	56	107	152	200	255	322	1	6	6.9
Power generation	**1 719**	**2 175**	**2 322**	**2 370**	**2 422**	**2 473**	**2 503**	**100**	**100**	**0.5**
Coal	759	846	870	823	767	686	593	39	24	-1.4
Oil	154	75	51	35	26	24	23	3	1	-4.5
Gas	176	434	479	501	521	556	575	20	23	1.1
Nuclear	451	585	610	635	662	691	722	27	29	0.8
Hydro	102	114	122	127	131	134	137	5	5	0.7
Biomass and waste	53	74	94	115	138	162	185	3	7	3.6
Other renewables	25	48	94	134	176	220	268	2	11	6.8
Other energy sector	**398**	**447**	**453**	**451**	**449**	**446**	**444**	**100**	**100**	**-0.0**
Electricity	*105*	*123*	*129*	*130*	*132*	*133*	*132*	*27*	*30*	*0.3*
TFC	**3 110**	**3 581**	**3 837**	**3 866**	**3 884**	**3 919**	**3 962**	**100**	**100**	**0.4**
Coal	236	115	137	132	125	118	111	3	3	-0.1
Oil	1 593	1 752	1 782	1 719	1 649	1 593	1 545	49	39	-0.5
Gas	589	709	774	787	797	804	810	20	20	0.5
Electricity	552	771	857	902	943	982	1 012	22	26	1.1
Heat	43	59	65	67	70	72	73	2	2	0.8
Biomass and waste	94	167	210	242	276	315	356	5	9	3.0
Other renewables	4	8	12	17	24	36	54	0	1	7.5
Industry	**830**	**773**	**896**	**909**	**908**	**901**	**889**	**100**	**100**	**0.5**
Coal	161	90	111	107	102	97	91	12	10	0.0
Oil	169	114	112	106	101	95	88	15	10	-1.0
Gas	226	236	277	279	277	274	268	31	30	0.5
Electricity	222	239	285	294	297	297	295	31	33	0.8
Heat	15	24	26	26	25	25	24	3	3	0.0
Biomass and waste	37	69	86	96	104	114	123	9	14	2.3
Other renewables	0	0	0	0	1	1	1	0	0	2.8
Transport	**941**	**1 172**	**1 220**	**1 188**	**1 152**	**1 136**	**1 135**	**100**	**100**	**-0.1**
Oil	914	1 106	1 131	1 083	1 027	987	958	94	84	-0.5
Electricity	8	10	12	13	15	17	20	1	2	2.8
Biofuels	0	35	55	68	83	101	120	3	11	4.8
Other fuels	19	21	23	24	27	31	38	2	3	2.3
Buildings	**986**	**1 225**	**1 293**	**1 344**	**1 399**	**1 456**	**1 514**	**100**	**100**	**0.8**
Coal	71	21	23	21	20	18	17	2	1	-0.8
Oil	209	166	160	154	147	139	130	14	9	-1.0
Gas	304	420	441	450	458	465	470	34	31	0.4
Electricity	316	514	552	586	623	660	690	42	46	1.1
Heat	27	35	39	41	44	47	49	3	3	1.3
Biomass and waste	56	61	68	76	85	97	110	5	7	2.3
Other renewables	4	8	11	15	22	32	49	1	3	7.3
Other	**353**	**411**	**428**	**425**	**426**	**425**	**423**	**100**	**100**	**0.1**

OECD: Current Policies and 450 Scenarios

	Energy demand (Mtoe)						Shares (%)		CAAGR (%)	
	2020	2030	2035	2020	2030	2035	2035		2009-2035	
	Current Policies Scenario			450 Scenario			CPS	450	CPS	450
TPED	5 657	5 837	5 946	5 416	5 221	5 196	100	100	0.5	-0.0
Coal	1 126	1 132	1 111	980	498	436	19	8	0.3	-3.3
Oil	1 895	1 793	1 761	1 773	1 411	1 209	30	23	-0.4	-1.8
Gas	1 411	1 516	1 587	1 315	1 322	1 216	27	23	0.9	-0.1
Nuclear	620	636	642	647	828	900	11	17	0.4	1.7
Hydro	126	131	133	129	140	145	2	3	0.6	0.9
Biomass and waste	340	422	468	397	638	760	8	15	2.6	4.5
Other renewables	139	208	245	174	383	530	4	10	5.8	9.0
Power generation	2 409	2 580	2 655	2 300	2 289	2 355	100	100	0.8	0.3
Coal	895	902	876	772	317	263	33	11	0.1	-4.4
Oil	37	26	25	30	17	15	1	1	-4.2	-5.9
Gas	499	558	605	447	472	379	23	16	1.3	-0.5
Nuclear	620	636	642	647	828	900	24	38	0.4	1.7
Hydro	126	131	133	129	140	145	5	6	0.6	0.9
Biomass and waste	109	148	167	119	184	219	6	9	3.2	4.2
Other renewables	123	180	208	154	331	434	8	18	5.8	8.8
Other energy sector	459	466	472	434	400	379	100	100	0.2	-0.6
Electricity	*132*	*139*	*140*	*127*	*123*	*121*	*30*	*32*	*0.5*	*-0.1*
TFC	3 916	4 028	4 107	3 767	3 658	3 608	100	100	0.5	0.0
Coal	138	129	124	122	99	89	3	2	0.3	-1.0
Oil	1 755	1 682	1 662	1 645	1 325	1 143	40	32	-0.2	-1.6
Gas	795	828	847	757	733	725	21	20	0.7	0.1
Electricity	912	1 011	1 059	883	934	958	26	27	1.2	0.8
Heat	69	75	78	63	61	58	2	2	1.1	-0.0
Biomass and waste	231	274	300	277	454	540	7	15	2.3	4.6
Other renewables	16	28	37	20	53	95	1	3	6.0	9.9
Industry	916	927	927	884	859	836	100	100	0.7	0.3
Coal	112	106	102	99	83	76	11	9	0.5	-0.7
Oil	109	100	95	104	94	86	10	10	-0.7	-1.1
Gas	282	283	283	273	260	251	30	30	0.7	0.2
Electricity	298	312	317	286	280	272	34	33	1.1	0.5
Heat	26	26	26	25	23	22	3	3	0.3	-0.3
Biomass and waste	89	99	105	97	119	129	11	15	1.6	2.4
Other renewables	0	1	1	0	1	1	0	0	2.8	3.1
Transport	1 212	1 190	1 203	1 158	1 028	961	100	100	0.1	-0.8
Oil	1 109	1 060	1 055	1 037	783	638	88	66	-0.2	-2.1
Electricity	13	16	17	14	36	68	1	7	2.3	7.9
Biofuels	66	85	97	79	172	207	8	22	4.0	7.0
Other fuels	24	29	34	28	38	48	3	5	1.8	3.2
Buildings	1 357	1 481	1 549	1 298	1 345	1 387	100	100	0.9	0.5
Coal	22	19	19	19	13	10	1	1	-0.4	-2.9
Oil	156	144	136	139	102	83	9	6	-0.8	-2.6
Gas	455	482	497	424	404	395	32	29	0.6	-0.2
Electricity	593	674	716	575	611	611	46	44	1.3	0.7
Heat	43	49	52	37	37	36	3	3	1.5	0.1
Biomass and waste	73	87	95	86	129	162	6	12	1.7	3.8
Other renewables	15	25	34	18	49	90	2	6	5.9	9.9
Other	432	431	428	426	426	424	100	100	0.2	0.1

A

OECD: New Policies Scenario

Electricity generation (TWh)	1990	2009	2015	2020	2025	2030	2035	Shares (%) 2009	Shares (%) 2035	CAAGR (%) 2009-2035
Total generation	7 629	10 394	11 455	11 997	12 504	12 959	13 304	100	100	1.0
Coal	3 093	3 620	3 785	3 673	3 504	3 210	2 882	35	22	-0.9
Oil	697	324	223	149	117	105	101	3	1	-4.4
Gas	770	2 361	2 662	2 827	2 966	3 139	3 182	23	24	1.2
Nuclear	1 729	2 242	2 352	2 445	2 551	2 663	2 779	22	21	0.8
Hydro	1 182	1 321	1 424	1 476	1 522	1 561	1 592	13	12	0.7
Biomass and waste	124	239	315	398	489	582	675	2	5	4.1
Wind	4	223	524	770	994	1 217	1 465	2	11	7.5
Geothermal	29	42	56	72	91	108	124	0	1	4.3
Solar PV	0	20	94	152	211	270	326	0	2	11.4
CSP	1	1	19	33	51	80	117	0	1	20.9
Marine	1	1	1	2	9	23	61	0	0	20.0

Electrical capacity (GW)	2009	2015	2020	2025	2030	2035	Shares (%) 2009	Shares (%) 2035	CAAGR (%) 2009-2035
Total capacity	2 614	3 000	3 149	3 291	3 446	3 595	100	100	1.2
Coal	638	701	688	665	621	548	24	15	-0.6
Oil	218	199	135	101	86	82	8	2	-3.7
Gas	787	894	937	967	1 014	1 067	30	30	1.2
Nuclear	327	332	341	352	366	380	13	11	0.6
Hydro	454	472	489	504	517	527	17	15	0.6
Biomass and waste	41	52	64	78	92	106	2	3	3.8
Wind	120	246	344	424	498	575	5	16	6.2
Geothermal	7	9	11	14	16	19	0	1	3.7
Solar PV	22	89	131	170	207	243	1	7	9.7
CSP	1	5	9	14	21	31	0	1	16.6
Marine	0	0	1	2	6	17	0	0	17.2

CO_2 emissions (Mt)	1990	2009	2015	2020	2025	2030	2035	Shares (%) 2009	Shares (%) 2035	CAAGR (%) 2009-2035
Total CO_2	11 117	11 951	12 399	12 023	11 554	11 064	10 535	100	100	-0.5
Coal	4 155	3 971	4 201	3 980	3 681	3 269	2 824	33	27	-1.3
Oil	5 035	5 097	5 050	4 802	4 555	4 372	4 220	43	40	-0.7
Gas	1 928	2 883	3 149	3 242	3 317	3 424	3 491	24	33	0.7
Power generation	3 964	4 663	4 797	4 590	4 347	4 038	3 660	100	100	-0.9
Coal	3 066	3 407	3 508	3 305	3 044	2 666	2 252	73	62	-1.6
Oil	487	237	164	111	84	75	71	5	2	-4.5
Gas	411	1 018	1 125	1 175	1 219	1 296	1 337	22	37	1.1
TFC	6 559	6 632	6 950	6 777	6 562	6 379	6 224	100	100	-0.2
Coal	1 025	493	595	573	544	512	482	7	8	-0.1
Oil	4 185	4 503	4 569	4 386	4 177	4 011	3 869	68	62	-0.6
Transport	*2 681*	*3 259*	*3 334*	*3 194*	*3 028*	*2 911*	*2 826*	*49*	*45*	*-0.5*
Gas	1 349	1 636	1 786	1 818	1 840	1 857	1 872	25	30	0.5

OECD: Current Policies and 450 Scenarios

	Electricity generation (TWh)						Shares (%)		CAAGR (%)	
	2020	2030	2035	2020	2030	2035	2035		2009-2035	
	Current Policies Scenario			450 Scenario			CPS	450	CPS	450
Total generation	12 143	13 371	13 939	11 743	12 286	12 541	100	100	1.1	0.7
Coal	4 005	4 246	4 262	3 433	1 441	1 161	31	9	0.6	-4.3
Oil	158	112	108	127	72	66	1	1	-4.1	-5.9
Gas	2 830	3 138	3 343	2 572	2 746	2 057	24	16	1.3	-0.5
Nuclear	2 389	2 450	2 472	2 495	3 189	3 463	18	28	0.4	1.7
Hydro	1 461	1 525	1 547	1 505	1 632	1 683	11	13	0.6	0.9
Biomass and waste	371	523	600	412	664	801	4	6	3.6	4.8
Wind	696	999	1 140	891	1 745	2 165	8	17	6.5	9.1
Geothermal	70	93	100	77	139	174	1	1	3.4	5.6
Solar PV	135	212	243	176	397	533	2	4	10.2	13.6
CSP	26	60	88	51	228	362	1	3	19.6	26.3
Marine	2	13	37	3	34	76	0	1	17.7	21.1

	Electrical capacity (GW)						Shares (%)		CAAGR (%)	
	2020	2030	2035	2020	2030	2035	2035		2009-2035	
	Current Policies Scenario			450 Scenario			CPS	450	CPS	450
Total capacity	3 127	3 383	3 512	3 184	3 628	3 912	100	100	1.1	1.6
Coal	717	707	676	637	364	282	19	7	0.2	-3.1
Oil	135	87	84	131	73	64	2	2	-3.6	-4.6
Gas	943	1 033	1 099	934	1 040	1 072	31	27	1.3	1.2
Nuclear	334	339	337	349	435	468	10	12	0.1	1.4
Hydro	484	504	511	499	542	561	15	14	0.5	0.8
Biomass and waste	60	83	94	67	105	126	3	3	3.3	4.4
Wind	316	426	472	391	681	807	13	21	5.4	7.6
Geothermal	11	14	15	12	21	26	0	1	2.9	5.0
Solar PV	119	169	189	149	297	391	5	10	8.7	11.8
CSP	7	16	24	14	60	95	1	2	15.3	21.6
Marine	1	3	10	1	9	21	0	1	15.0	18.2

	CO_2 emissions (Mt)						Shares (%)		CAAGR (%)	
	2020	2030	2035	2020	2030	2035	2035		2009-2035	
	Current Policies Scenario			450 Scenario			CPS	450	CPS	450
Total CO_2	12 467	12 381	12 336	11 284	7 386	5 958	100	100	0.1	-2.6
Coal	4 297	4 234	4 085	3 679	1 034	552	33	9	0.1	-7.3
Oil	4 911	4 651	4 589	4 576	3 532	2 965	37	50	-0.4	-2.1
Gas	3 258	3 497	3 662	3 029	2 819	2 441	30	41	0.9	-0.6
Power generation	4 884	4 969	4 945	4 190	1 589	853	100	100	0.2	-6.3
Coal	3 594	3 582	3 452	3 052	600	194	70	23	0.1	-10.4
Oil	118	81	78	96	55	50	2	6	-4.2	-5.8
Gas	1 171	1 306	1 415	1 041	934	610	29	71	1.3	-2.0
TFC	6 915	6 738	6 705	6 468	5 226	4 573	100	100	0.0	-1.4
Coal	597	556	536	531	352	279	8	6	0.3	-2.2
Oil	4 481	4 268	4 211	4 189	3 244	2 713	63	59	-0.3	-1.9
Transport	*3 269*	*3 127*	*3 113*	*3 059*	*2 309*	*1 883*	*46*	*41*	*-0.2*	*-2.1*
Gas	1 836	1 914	1 958	1 748	1 629	1 581	29	35	0.7	-0.1

OECD Americas: New Policies Scenario

	Energy demand (Mtoe)							Shares (%)		CAAGR (%)
	1990	2009	2015	2020	2025	2030	2035	2009	2035	2009-2035
TPED	**2 260**	**2 620**	**2 780**	**2 787**	**2 803**	**2 835**	**2 864**	**100**	**100**	**0.3**
Coal	490	520	558	534	517	483	445	20	16	-0.6
Oil	920	1 003	1 021	974	923	894	866	38	30	-0.6
Gas	517	664	698	719	738	760	779	25	27	0.6
Nuclear	180	243	260	272	283	293	297	9	10	0.8
Hydro	52	59	63	64	66	68	69	2	2	0.6
Biomass and waste	82	109	138	161	192	228	265	4	9	3.5
Other renewables	19	23	42	62	84	110	143	1	5	7.3
Power generation	**853**	**1 047**	**1 123**	**1 151**	**1 188**	**1 216**	**1 229**	**100**	**100**	**0.6**
Coal	421	469	500	479	464	433	392	45	32	-0.7
Oil	47	27	17	13	9	8	7	3	1	-5.0
Gas	95	205	215	228	240	255	266	20	22	1.0
Nuclear	180	243	260	272	283	293	297	23	24	0.8
Hydro	52	59	63	64	66	68	69	6	6	0.6
Biomass and waste	41	23	29	37	48	62	76	2	6	4.8
Other renewables	18	21	39	57	77	98	121	2	10	7.0
Other energy sector	**191**	**235**	**233**	**233**	**234**	**236**	**239**	**100**	**100**	**0.1**
Electricity	*56*	*62*	*65*	*67*	*68*	*69*	*70*	*26*	*29*	*0.5*
TFC	**1 547**	**1 788**	**1 917**	**1 920**	**1 922**	**1 946**	**1 975**	**100**	**100**	**0.4**
Coal	61	28	33	31	29	27	25	2	1	-0.4
Oil	809	910	948	911	870	846	825	51	42	-0.4
Gas	361	380	402	403	405	407	409	21	21	0.3
Electricity	272	375	415	438	461	483	500	21	25	1.1
Heat	3	7	8	7	7	6	5	0	0	-1.2
Biomass and waste	41	86	110	124	143	166	188	5	10	3.1
Other renewables	0	2	3	5	7	12	22	0	1	10.1
Industry	**361**	**345**	**400**	**405**	**403**	**398**	**390**	**100**	**100**	**0.5**
Coal	51	26	31	29	28	26	24	7	6	-0.3
Oil	60	41	43	41	39	37	35	12	9	-0.6
Gas	138	138	157	156	154	150	145	40	37	0.2
Electricity	94	95	113	117	117	117	115	27	29	0.7
Heat	1	6	6	6	6	5	5	2	1	-0.8
Biomass and waste	17	39	50	55	59	63	66	11	17	2.0
Other renewables	0	0	0	0	0	0	0	0	0	0.9
Transport	**562**	**690**	**731**	**706**	**685**	**687**	**697**	**100**	**100**	**0.0**
Oil	543	649	676	644	608	591	580	94	83	-0.4
Electricity	1	1	2	2	2	3	3	0	0	4.4
Biofuels	-	23	34	42	54	69	85	3	12	5.2
Other fuels	18	17	18	19	21	24	29	3	4	2.0
Buildings	**460**	**572**	**591**	**611**	**635**	**661**	**689**	**100**	**100**	**0.7**
Coal	10	2	2	1	1	1	1	0	0	-2.7
Oil	64	57	52	48	43	39	33	10	5	-2.0
Gas	184	209	210	211	213	216	218	37	32	0.2
Electricity	176	278	298	318	340	361	380	49	55	1.2
Heat	2	1	1	1	1	1	1	0	0	-3.7
Biomass and waste	24	24	25	27	30	33	37	4	5	1.7
Other renewables	0	2	3	4	6	11	20	0	3	10.0
Other	**164**	**181**	**196**	**198**	**200**	**200**	**199**	**100**	**100**	**0.4**

OECD Americas: Current Policies and 450 Scenarios

	Energy demand (Mtoe)						Shares (%)		CAAGR (%)	
	2020	2030	2035	2020	2030	2035	2035		2009-2035	
	Current Policies Scenario			450 Scenario			CPS	450	CPS	450
TPED	2 833	2 919	2 980	2 724	2 613	2 608	100	100	0.5	-0.0
Coal	568	577	583	527	242	239	20	9	0.4	-3.0
Oil	994	937	919	932	732	621	31	24	-0.3	-1.8
Gas	724	770	795	667	722	660	27	25	0.7	-0.0
Nuclear	270	279	283	277	350	368	10	14	0.6	1.6
Hydro	64	67	68	65	68	70	2	3	0.5	0.6
Biomass and waste	156	201	226	182	317	383	8	15	2.9	5.0
Other renewables	57	88	105	73	181	266	4	10	6.1	9.9
Power generation	1 174	1 261	1 300	1 131	1 122	1 164	100	100	0.8	0.4
Coal	510	520	518	476	201	197	40	17	0.4	-3.3
Oil	14	8	7	11	6	6	1	0	-5.0	-5.8
Gas	227	252	267	196	266	214	21	18	1.0	0.2
Nuclear	270	279	283	277	350	368	22	32	0.6	1.6
Hydro	64	67	68	65	68	70	5	6	0.5	0.6
Biomass and waste	37	57	68	38	68	86	5	7	4.3	5.3
Other renewables	53	78	89	67	162	223	7	19	5.7	9.5
Other energy sector	237	245	252	224	211	202	100	100	0.3	-0.6
Electricity	*68*	*72*	*73*	*65*	*64*	*63*	*29*	*31*	*0.7*	*0.1*
TFC	1 944	1 987	2 025	1 876	1 813	1 786	100	100	0.5	-0.0
Coal	33	31	30	28	20	18	1	1	0.3	-1.7
Oil	931	888	879	874	693	593	43	33	-0.1	-1.6
Gas	408	415	421	388	365	357	21	20	0.4	-0.2
Electricity	442	492	515	430	460	474	25	27	1.2	0.9
Heat	8	7	6	7	6	5	0	0	-0.7	-1.4
Biomass and waste	119	144	158	144	249	297	8	17	2.4	4.9
Other renewables	5	10	16	6	20	42	1	2	8.9	13.0
Industry	407	408	405	392	372	358	100	100	0.6	0.1
Coal	31	29	28	26	20	17	7	5	0.3	-1.5
Oil	43	40	39	40	37	34	10	9	-0.2	-0.7
Gas	158	156	154	152	138	131	38	37	0.4	-0.2
Electricity	118	122	123	113	109	104	30	29	1.0	0.4
Heat	6	6	6	6	5	5	1	1	-0.2	-1.0
Biomass and waste	51	55	57	55	64	66	14	19	1.4	2.1
Other renewables	0	0	0	0	0	0	0	0	0.9	0.9
Transport	723	710	722	692	621	586	100	100	0.2	-0.6
Oil	661	629	627	619	464	381	87	65	-0.1	-2.0
Electricity	2	2	3	2	13	32	0	6	3.3	13.8
Biofuels	41	57	67	50	117	141	9	24	4.2	7.2
Other fuels	19	22	25	21	27	32	3	5	1.4	2.4
Buildings	616	669	700	594	619	643	100	100	0.8	0.4
Coal	2	2	2	1	1	0	0	0	-0.3	-8.7
Oil	48	41	36	43	26	18	5	3	-1.7	-4.4
Gas	214	221	226	198	184	179	32	28	0.3	-0.6
Electricity	320	365	387	312	337	335	55	52	1.3	0.7
Heat	1	1	1	1	1	1	0	0	-3.6	-4.0
Biomass and waste	27	31	34	33	52	70	5	11	1.4	4.3
Other renewables	4	9	15	5	18	40	2	6	9.0	13.1
Other	198	199	198	199	200	200	100	100	0.4	0.4

A

OECD Americas: New Policies Scenario

Electricity generation (TWh)	1990	2009	2015	2020	2025	2030	2035	Shares (%) 2009	Shares (%) 2035	CAAGR (%) 2009-2035
Total generation	3 819	5 090	5 584	5 881	6 169	6 432	6 636	100	100	1.0
Coal	1 796	2 029	2 193	2 172	2 141	2 057	1 931	40	29	-0.2
Oil	211	117	76	58	44	39	34	2	1	-4.7
Gas	406	1 130	1 226	1 323	1 411	1 506	1 575	22	24	1.3
Nuclear	687	931	998	1 047	1 089	1 125	1 142	18	17	0.8
Hydro	602	691	730	749	769	788	802	14	12	0.6
Biomass and waste	91	86	111	149	197	256	317	2	5	5.1
Wind	3	79	191	284	372	464	576	2	9	7.9
Geothermal	21	24	33	43	55	67	77	0	1	4.6
Solar PV	0	2	19	39	62	88	118	0	2	17.4
CSP	1	1	8	16	25	37	55	0	1	17.6
Marine	0	0	0	0	2	5	8	0	0	23.8

Electrical capacity (GW)	2009	2015	2020	2025	2030	2035	Shares (%) 2009	Shares (%) 2035	CAAGR (%) 2009-2035
Total capacity	1 289	1 396	1 466	1 531	1 598	1 660	100	100	1.0
Coal	354	372	390	390	369	330	27	20	-0.3
Oil	93	84	54	37	31	27	7	2	-4.7
Gas	467	486	499	514	538	569	36	34	0.8
Nuclear	121	128	134	140	144	146	9	9	0.7
Hydro	193	199	205	211	216	220	15	13	0.5
Biomass and waste	16	20	26	33	42	51	1	3	4.6
Wind	39	85	121	152	182	216	3	13	6.8
Geothermal	4	5	7	8	10	12	0	1	4.0
Solar PV	2	13	26	40	56	73	0	4	15.4
CSP	0	2	4	6	9	14	0	1	14.0
Marine	0	0	0	1	1	2	0	0	19.3

CO_2 emissions (Mt)	1990	2009	2015	2020	2025	2030	2035	Shares (%) 2009	Shares (%) 2035	CAAGR (%) 2009-2035
Total CO_2	5 578	6 152	6 418	6 233	6 050	5 865	5 650	100	100	-0.3
Coal	1 920	1 967	2 116	2 021	1 941	1 786	1 596	32	28	-0.8
Oil	2 469	2 655	2 691	2 558	2 416	2 337	2 272	43	40	-0.6
Gas	1 189	1 530	1 611	1 653	1 693	1 742	1 782	25	32	0.6
Power generation	2 019	2 415	2 526	2 449	2 395	2 277	2 113	100	100	-0.5
Coal	1 647	1 845	1 965	1 874	1 806	1 659	1 476	76	70	-0.9
Oil	150	90	57	43	32	27	23	4	1	-5.0
Gas	222	480	503	532	558	590	613	20	29	0.9
TFC	3 213	3 361	3 521	3 406	3 277	3 202	3 141	100	100	-0.3
Coal	269	115	138	131	123	114	105	3	3	-0.3
Oil	2 115	2 369	2 456	2 344	2 220	2 149	2 091	70	67	-0.5
Transport	*1 585*	*1 899*	*1 980*	*1 885*	*1 781*	*1 732*	*1 698*	*56*	*54*	*-0.4*
Gas	829	877	928	931	935	939	945	26	30	0.3

OECD Americas: Current Policies and 450 Scenarios

	Electricity generation (TWh)						Shares (%)		CAAGR (%)	
	2020	2030	2035	2020	2030	2035	2035		2009-2035	
	Current Policies Scenario			450 Scenario			CPS	450	CPS	450
Total generation	5 931	6 556	6 847	5 763	6 102	6 252	100	100	1.1	0.8
Coal	2 301	2 458	2 513	2 138	939	882	37	14	0.8	-3.2
Oil	61	38	33	50	30	27	0	0	-4.7	-5.5
Gas	1 305	1 466	1 547	1 148	1 605	1 236	23	20	1.2	0.3
Nuclear	1 037	1 074	1 089	1 066	1 346	1 414	16	23	0.6	1.6
Hydro	745	775	785	752	795	814	11	13	0.5	0.6
Biomass and waste	143	231	279	153	283	360	4	6	4.6	5.7
Wind	251	365	421	338	717	938	6	15	6.6	10.0
Geothermal	42	57	60	46	87	111	1	2	3.6	6.1
Solar PV	33	60	74	46	144	213	1	3	15.4	20.1
CSP	12	27	39	27	152	249	1	4	16.1	24.6
Marine	0	3	6	0	5	8	0	0	22.2	23.5

	Electrical capacity (GW)						Shares (%)		CAAGR (%)	
	2020	2030	2035	2020	2030	2035	2035		2009-2035	
	Current Policies Scenario			450 Scenario			CPS	450	CPS	450
Total capacity	1 457	1 569	1 622	1 460	1 627	1 771	100	100	0.9	1.2
Coal	403	409	393	352	180	150	24	8	0.4	-3.2
Oil	54	31	28	52	28	23	2	1	-4.5	-5.2
Gas	497	536	567	504	579	594	35	34	0.7	0.9
Nuclear	133	138	139	137	172	180	9	10	0.5	1.5
Hydro	204	212	215	206	218	223	13	13	0.4	0.6
Biomass and waste	25	38	45	26	46	58	3	3	4.1	5.2
Wind	109	149	166	139	263	331	10	19	5.7	8.6
Geothermal	7	9	9	7	13	16	1	1	3.1	5.4
Solar PV	22	39	47	29	88	130	3	7	13.5	18.0
CSP	3	7	10	7	39	64	1	4	12.6	20.8
Marine	0	1	1	0	1	2	0	0	17.7	19.1

	CO_2 emissions (Mt)						Shares (%)		CAAGR (%)	
	2020	2030	2035	2020	2030	2035	2035		2009-2035	
	Current Policies Scenario			450 Scenario			CPS	450	CPS	450
Total CO_2	6 438	6 405	6 414	5 950	3 759	3 017	100	100	0.2	-2.7
Coal	2 153	2 169	2 146	1 975	361	164	33	5	0.3	-9.1
Oil	2 622	2 472	2 443	2 445	1 870	1 562	38	52	-0.3	-2.0
Gas	1 663	1 765	1 824	1 530	1 529	1 291	28	43	0.7	-0.7
Power generation	2 572	2 639	2 650	2 334	818	430	100	100	0.4	-6.4
Coal	1 997	2 025	2 005	1 843	277	95	76	22	0.3	-10.8
Oil	45	27	23	37	21	19	1	4	-5.1	-5.8
Gas	530	587	622	454	520	316	23	74	1.0	-1.6
TFC	3 481	3 363	3 347	3 257	2 601	2 267	100	100	-0.0	-1.5
Coal	139	129	124	117	72	57	4	2	0.3	-2.7
Oil	2 401	2 275	2 251	2 245	1 718	1 431	67	63	-0.2	-1.9
Transport	_1 936_	_1 842_	_1 837_	_1 811_	_1 360_	_1 116_	_55_	_49_	_-0.1_	_-2.0_
Gas	941	959	972	895	810	780	29	34	0.4	-0.5

A

United States: New Policies Scenario

	Energy demand (Mtoe)							Shares (%)		CAAGR (%)
	1990	2009	2015	2020	2025	2030	2035	2009	2035	2009-2035
TPED	**1 915**	**2 160**	**2 285**	**2 264**	**2 257**	**2 262**	**2 265**	**100**	**100**	**0.2**
Coal	460	485	517	493	479	451	419	22	19	-0.6
Oil	757	801	817	768	717	684	653	37	29	-0.8
Gas	438	534	558	562	568	577	583	25	26	0.3
Nuclear	159	216	230	241	249	255	257	10	11	0.7
Hydro	23	24	24	25	25	26	26	1	1	0.4
Biomass and waste	62	84	108	128	155	186	218	4	10	3.7
Other renewables	14	16	31	47	64	84	109	1	5	7.6
Power generation	**750**	**898**	**957**	**972**	**997**	**1 012**	**1 013**	**100**	**100**	**0.5**
Coal	396	440	464	444	432	407	372	49	37	-0.6
Oil	27	11	8	5	3	3	3	1	0	-4.9
Gas	90	173	178	183	188	197	203	19	20	0.6
Nuclear	159	216	230	241	249	255	257	24	25	0.7
Hydro	23	24	24	25	25	26	26	3	3	0.4
Biomass and waste	40	19	24	32	41	51	62	2	6	4.6
Other renewables	14	15	29	43	58	73	90	2	9	7.2
Other energy sector	**150**	**173**	**170**	**163**	**157**	**151**	**148**	**100**	**100**	**-0.6**
Electricity	*49*	*48*	*49*	*50*	*51*	*51*	*51*	*28*	*35*	*0.3*
TFC	**1 294**	**1 463**	**1 562**	**1 551**	**1 542**	**1 553**	**1 569**	**100**	**100**	**0.3**
Coal	56	24	28	27	25	23	20	2	1	-0.5
Oil	683	741	766	727	684	656	633	51	40	-0.6
Gas	303	312	331	330	330	329	330	21	21	0.2
Electricity	226	313	343	360	378	394	405	21	26	1.0
Heat	2	7	7	7	6	5	5	0	0	-1.4
Biomass and waste	23	65	84	97	114	135	156	4	10	3.4
Other renewables	0	2	3	4	6	11	19	0	1	10.0
Industry	**284**	**259**	**301**	**301**	**296**	**288**	**279**	**100**	**100**	**0.3**
Coal	46	22	27	25	23	22	20	8	7	-0.4
Oil	44	27	28	26	24	22	20	11	7	-1.2
Gas	110	106	123	121	118	114	108	41	39	0.1
Electricity	75	69	81	81	80	77	74	27	27	0.3
Heat	-	5	6	5	5	5	4	2	2	-0.9
Biomass and waste	9	30	37	42	46	49	52	11	19	2.2
Other renewables	-	0	0	0	0	0	0	0	0	0.7
Transport	**488**	**578**	**610**	**583**	**560**	**559**	**564**	**100**	**100**	**-0.1**
Oil	472	540	560	526	489	470	455	94	81	-0.7
Electricity	0	1	1	1	1	2	2	0	0	5.2
Biofuels	-	22	33	40	51	66	82	4	15	5.2
Other fuels	15	15	16	16	18	21	25	3	4	2.1
Buildings	**389**	**482**	**497**	**511**	**529**	**549**	**570**	**100**	**100**	**0.7**
Coal	10	2	2	1	1	1	1	0	0	-2.7
Oil	48	38	33	28	23	18	13	8	2	-4.0
Gas	164	183	184	183	184	186	187	38	33	0.1
Electricity	152	244	262	278	296	314	329	51	58	1.2
Heat	2	1	1	1	1	1	0	0	0	-4.4
Biomass and waste	14	13	13	15	17	19	22	3	4	2.1
Other renewables	0	1	2	4	6	10	18	0	3	10.2
Other	**133**	**144**	**155**	**157**	**157**	**157**	**155**	**100**	**100**	**0.3**

United States: Current Policies and 450 Scenarios

	Energy demand (Mtoe)						Shares (%)		CAAGR (%)	
	2020	2030	2035	2020	2030	2035	2035		2009-2035	
	Current Policies Scenario			450 Scenario			CPS	450	CPS	450
TPED	**2 304**	**2 327**	**2 354**	**2 215**	**2 076**	**2 066**	**100**	**100**	**0.3**	**-0.2**
Coal	526	534	541	488	219	228	23	11	0.4	-2.9
Oil	786	719	695	740	563	469	30	23	-0.5	-2.0
Gas	562	573	578	520	560	495	25	24	0.3	-0.3
Nuclear	238	243	244	244	306	324	10	16	0.5	1.6
Hydro	25	26	26	25	26	27	1	1	0.4	0.5
Biomass and waste	124	165	188	143	256	308	8	15	3.2	5.1
Other renewables	44	68	82	56	146	216	3	10	6.4	10.5
Power generation	**991**	**1 048**	**1 070**	**956**	**924**	**957**	**100**	**100**	**0.7**	**0.2**
Coal	474	483	483	443	183	191	45	20	0.4	-3.2
Oil	5	4	3	4	3	3	0	0	-4.7	-5.2
Gas	178	185	188	157	223	169	18	18	0.3	-0.1
Nuclear	238	243	244	244	306	324	23	34	0.5	1.6
Hydro	25	26	26	25	26	27	2	3	0.4	0.5
Biomass and waste	32	49	59	32	55	67	6	7	4.4	4.9
Other renewables	40	59	67	51	129	177	6	19	6.0	10.1
Other energy sector	**166**	**158**	**158**	**157**	**135**	**125**	**100**	**100**	**-0.4**	**-1.3**
Electricity	*51*	*53*	*54*	*49*	*47*	*46*	*34*	*37*	*0.4*	*-0.2*
TFC	**1 573**	**1 583**	**1 604**	**1 515**	**1 446**	**1 419**	**100**	**100**	**0.4**	**-0.1**
Coal	28	26	25	23	17	14	2	1	0.2	-2.0
Oil	744	692	677	700	539	455	42	32	-0.3	-1.9
Gas	334	336	338	316	292	284	21	20	0.3	-0.4
Electricity	363	399	415	354	375	382	26	27	1.1	0.8
Heat	7	6	5	6	5	4	0	0	-0.9	-1.6
Biomass and waste	93	116	129	111	202	241	8	17	2.7	5.2
Other renewables	4	9	15	5	17	38	1	3	9.0	13.0
Industry	**302**	**295**	**289**	**289**	**267**	**252**	**100**	**100**	**0.4**	**-0.1**
Coal	27	25	24	22	16	14	8	6	0.3	-1.7
Oil	27	24	23	25	22	20	8	8	-0.7	-1.3
Gas	122	118	115	117	103	96	40	38	0.3	-0.4
Electricity	82	81	79	78	71	66	27	26	0.5	-0.2
Heat	6	5	5	5	5	4	2	2	-0.4	-1.1
Biomass and waste	38	42	43	41	50	53	15	21	1.5	2.2
Other renewables	0	0	0	0	0	0	0	0	0.7	0.7
Transport	**599**	**578**	**582**	**572**	**506**	**476**	**100**	**100**	**0.0**	**-0.7**
Oil	542	502	495	508	371	300	85	63	-0.3	-2.2
Electricity	1	1	2	1	10	26	0	5	3.6	15.1
Biofuels	40	55	65	44	102	121	11	25	4.2	6.7
Other fuels	16	19	21	18	24	28	4	6	1.4	2.6
Buildings	**515**	**555**	**578**	**497**	**516**	**535**	**100**	**100**	**0.7**	**0.4**
Coal	2	2	1	1	0	0	0	0	-0.3	-9.2
Oil	28	19	15	25	11	4	3	1	-3.6	-8.7
Gas	186	190	193	171	157	151	33	28	0.2	-0.7
Electricity	280	316	334	274	294	290	58	54	1.2	0.7
Heat	1	1	0	1	1	0	0	0	-4.3	-4.7
Biomass and waste	14	18	20	20	37	52	3	10	1.8	5.6
Other renewables	4	9	15	4	17	38	3	7	9.2	13.2
Other	**156**	**156**	**155**	**157**	**157**	**156**	**100**	**100**	**0.3**	**0.3**

A

United States: New Policies Scenario

Electricity generation (TWh)								Shares (%)		CAAGR (%)
	1990	2009	2015	2020	2025	2030	2035	2009	2035	2009-2035
Total generation	3 203	4 165	4 544	4 756	4 967	5 157	5 292	100	100	0.9
Coal	1 700	1 893	2 026	2 009	1 992	1 936	1 837	45	35	-0.1
Oil	131	50	38	23	17	16	15	1	0	-4.6
Gas	382	950	1 011	1 059	1 102	1 156	1 195	23	23	0.9
Nuclear	612	830	883	927	957	980	990	20	19	0.7
Hydro	273	276	278	287	295	301	306	7	6	0.4
Biomass and waste	86	72	94	128	170	217	266	2	5	5.1
Wind	3	74	165	242	315	388	472	2	9	7.4
Geothermal	16	17	23	31	40	48	55	0	1	4.6
Solar PV	0	2	18	38	58	81	106	0	2	17.2
CSP	1	1	8	14	20	30	45	0	1	16.7
Marine	-	-	0	0	1	3	5	-	0	n.a.

Electrical capacity (GW)							Shares (%)		CAAGR (%)
	2009	2015	2020	2025	2030	2035	2009	2035	2009-2035
Total capacity	1 083	1 162	1 213	1 260	1 307	1 342	100	100	0.8
Coal	333	347	366	367	349	314	31	23	-0.2
Oil	69	61	33	22	19	17	6	1	-5.2
Gas	422	431	434	437	450	468	39	35	0.4
Nuclear	106	112	117	121	123	124	10	9	0.6
Hydro	101	104	107	110	112	114	9	8	0.5
Biomass and waste	12	15	21	27	34	42	1	3	5.0
Wind	35	73	102	127	151	176	3	13	6.4
Geothermal	3	4	5	6	7	8	0	1	3.9
Solar PV	2	13	25	37	50	64	0	5	15.1
CSP	0	2	4	5	8	11	0	1	13.1
Marine	-	0	0	0	1	1	-	0	n.a.

CO$_2$ emissions (Mt)								Shares (%)		CAAGR (%)
	1990	2009	2015	2020	2025	2030	2035	2009	2035	2009-2035
Total CO$_2$	4 850	5 168	5 379	5 159	4 962	4 765	4 539	100	100	-0.5
Coal	1 797	1 832	1 953	1 864	1 795	1 664	1 498	35	33	-0.8
Oil	2 042	2 101	2 133	1 995	1 855	1 771	1 700	41	37	-0.8
Gas	1 011	1 235	1 292	1 300	1 312	1 329	1 341	24	30	0.3
Power generation	1 848	2 171	2 268	2 181	2 132	2 025	1 875	100	100	-0.6
Coal	1 550	1 729	1 824	1 738	1 681	1 559	1 399	80	75	-0.8
Oil	88	39	29	16	12	11	11	2	1	-4.9
Gas	210	403	415	426	439	455	465	19	25	0.6
TFC	2 730	2 739	2 862	2 734	2 596	2 507	2 433	100	100	-0.5
Coal	245	97	118	111	102	94	85	4	3	-0.5
Oil	1 788	1 920	1 976	1 858	1 729	1 650	1 584	70	65	-0.7
Transport	1 376	1 580	1 639	1 539	1 432	1 375	1 331	58	55	-0.7
Gas	697	723	768	765	764	763	765	26	31	0.2

United States: Current Policies and 450 Scenarios

	Electricity generation (TWh)						Shares (%)		CAAGR (%)	
	2020	2030	2035	2020	2030	2035	2035		2009-2035	
	Current Policies Scenario			450 Scenario			CPS	450	CPS	450
Total generation	4 794	5 233	5 427	4 667	4 878	4 956	100	100	1.0	0.7
Coal	2 133	2 281	2 344	1 985	853	853	43	17	0.8	-3.0
Oil	24	17	16	20	15	14	0	0	-4.4	-4.9
Gas	1 024	1 072	1 089	912	1 347	975	20	20	0.5	0.1
Nuclear	917	934	937	938	1 176	1 244	17	25	0.5	1.6
Hydro	285	298	303	288	304	310	6	6	0.4	0.5
Biomass and waste	123	204	246	131	234	292	5	6	4.8	5.5
Wind	215	302	339	295	612	763	6	15	6.0	9.4
Geothermal	31	43	44	32	64	82	1	2	3.8	6.2
Solar PV	31	57	70	43	129	187	1	4	15.4	19.8
CSP	12	25	35	23	141	232	1	5	15.6	24.3
Marine	0	2	4	0	3	5	0	0	n.a.	n.a.

	Electrical capacity (GW)						Shares (%)		CAAGR (%)	
	2020	2030	2035	2020	2030	2035	2035		2009-2035	
	Current Policies Scenario			450 Scenario			CPS	450	CPS	450
Total capacity	1 207	1 283	1 314	1 205	1 317	1 415	100	100	0.7	1.0
Coal	379	384	369	328	162	137	28	10	0.4	-3.4
Oil	33	20	18	31	16	14	1	1	-5.0	-6.0
Gas	431	445	461	441	495	497	35	35	0.3	0.6
Nuclear	116	118	118	118	148	156	9	11	0.4	1.5
Hydro	107	111	113	108	113	115	9	8	0.4	0.5
Biomass and waste	20	32	39	21	37	46	3	3	4.7	5.4
Wind	92	123	135	120	223	267	10	19	5.3	8.1
Geothermal	5	7	7	5	10	12	1	1	3.2	5.4
Solar PV	21	37	44	27	77	111	3	8	13.5	17.6
CSP	3	6	9	6	36	59	1	4	12.1	20.5
Marine	0	0	1	0	1	1	0	0	n.a.	n.a.

	CO_2 emissions (Mt)						Shares (%)		CAAGR (%)	
	2020	2030	2035	2020	2030	2035	2035		2009-2035	
	Current Policies Scenario			450 Scenario			CPS	450	CPS	450
Total CO_2	5 339	5 206	5 162	4 946	2 896	2 257	100	100	-0.0	-3.1
Coal	1 990	2 000	1 986	1 829	292	147	38	7	0.3	-9.2
Oil	2 050	1 883	1 839	1 918	1 422	1 167	36	52	-0.5	-2.2
Gas	1 300	1 322	1 336	1 199	1 182	944	26	42	0.3	-1.0
Power generation	2 290	2 321	2 316	2 095	663	327	100	100	0.2	-7.0
Coal	1 856	1 879	1 868	1 718	223	91	81	28	0.3	-10.7
Oil	17	12	11	14	10	10	0	3	-4.7	-5.2
Gas	416	430	437	363	429	226	19	69	0.3	-2.2
TFC	2 799	2 642	2 601	2 618	2 030	1 746	100	100	-0.2	-1.7
Coal	118	108	103	98	59	45	4	3	0.2	-2.9
Oil	1 908	1 755	1 715	1 788	1 321	1 080	66	62	-0.4	-2.2
Transport	*1 586*	*1 469*	*1 448*	*1 488*	*1 085*	*880*	*56*	*50*	*-0.3*	*-2.2*
Gas	774	778	783	732	650	621	30	36	0.3	-0.6

A

OECD Europe: New Policies Scenario

	Energy demand (Mtoe)							Shares (%)		CAAGR (%)
	1990	2009	2015	2020	2025	2030	2035	2009	2035	2009-2035
TPED	1 630	1 766	1 863	1 876	1 882	1 890	1 904	100	100	0.3
Coal	452	290	294	268	242	209	185	16	10	-1.7
Oil	615	617	596	572	548	523	503	35	26	-0.8
Gas	260	441	495	515	528	547	551	25	29	0.9
Nuclear	205	230	230	226	222	224	228	13	12	-0.0
Hydro	38	44	48	50	52	54	55	3	3	0.8
Biomass and waste	54	117	147	172	195	217	240	7	13	2.8
Other renewables	5	26	53	73	94	117	143	1	8	6.7
Power generation	626	739	775	785	794	809	823	100	100	0.4
Coal	278	219	204	181	159	130	110	30	13	-2.6
Oil	51	25	17	12	9	7	7	3	1	-5.0
Gas	41	154	173	185	194	210	213	21	26	1.3
Nuclear	205	230	230	226	222	224	228	31	28	-0.0
Hydro	38	44	48	50	52	54	55	6	7	0.8
Biomass and waste	9	45	58	68	78	85	92	6	11	2.8
Other renewables	3	21	45	63	81	99	119	3	14	6.9
Other energy sector	151	143	148	144	140	136	131	100	100	-0.4
Electricity	*39*	*44*	*44*	*43*	*44*	*44*	*43*	*30*	*33*	*-0.1*
TFC	1 131	1 241	1 330	1 354	1 373	1 386	1 403	100	100	0.5
Coal	126	49	57	55	52	48	46	4	3	-0.2
Oil	523	547	540	522	504	482	464	44	33	-0.6
Gas	201	263	298	307	313	316	318	21	23	0.7
Electricity	193	259	287	302	317	330	341	21	24	1.1
Heat	40	47	52	55	58	60	62	4	4	1.1
Biomass and waste	45	71	89	104	117	131	147	6	11	2.8
Other renewables	2	5	8	10	13	18	24	0	2	6.0
Industry	324	275	321	326	328	328	326	100	100	0.7
Coal	71	28	36	34	33	31	29	10	9	0.1
Oil	59	38	36	34	32	29	26	14	8	-1.4
Gas	78	76	95	96	96	95	93	28	29	0.8
Electricity	88	95	111	115	117	118	118	34	36	0.9
Heat	14	15	17	17	17	17	17	6	5	0.3
Biomass and waste	14	22	27	30	34	37	42	8	13	2.5
Other renewables	0	0	0	0	0	0	0	0	0	7.8
Transport	268	343	351	350	341	330	325	100	100	-0.2
Oil	262	322	321	313	300	284	274	94	84	-0.6
Electricity	5	6	8	8	9	10	12	2	4	2.4
Biofuels	0	12	20	25	29	31	33	3	10	4.1
Other fuels	1	2	3	3	4	5	6	1	2	3.7
Buildings	406	484	518	544	570	595	619	100	100	1.0
Coal	51	18	19	18	17	15	15	4	2	-0.7
Oil	97	71	69	67	64	62	58	15	9	-0.8
Gas	105	170	185	192	198	201	203	35	33	0.7
Electricity	96	153	163	174	186	197	207	32	33	1.2
Heat	24	31	35	38	40	43	45	6	7	1.4
Biomass and waste	30	35	40	46	52	60	68	7	11	2.6
Other renewables	2	5	7	10	13	17	23	1	4	5.9
Other	133	140	139	134	133	133	132	100	100	-0.2

OECD Europe: Current Policies and 450 Scenarios

	Energy demand (Mtoe)						Shares (%)		CAAGR (%)	
	2020	2030	2035	2020	2030	2035	2035		2009-2035	
	Current Policies Scenario			450 Scenario			CPS	450	CPS	450
TPED	1 897	1 960	1 995	1 804	1 759	1 746	100	100	0.5	-0.0
Coal	302	299	280	233	133	106	14	6	-0.1	-3.8
Oil	587	564	558	545	440	380	28	22	-0.4	-1.9
Gas	514	563	600	479	430	395	30	23	1.2	-0.4
Nuclear	215	189	178	228	274	296	9	17	-1.0	1.0
Hydro	50	52	53	52	57	59	3	3	0.7	1.1
Biomass and waste	164	196	213	186	270	313	11	18	2.3	3.9
Other renewables	67	97	114	81	155	198	6	11	5.8	8.1
Power generation	793	843	866	746	750	768	100	100	0.6	0.1
Coal	213	218	202	151	64	44	23	6	-0.3	-6.0
Oil	13	8	8	11	6	6	1	1	-4.2	-5.6
Gas	182	215	243	164	119	89	28	12	1.8	-2.1
Nuclear	215	189	178	228	274	296	21	39	-1.0	1.0
Hydro	50	52	53	52	57	59	6	8	0.7	1.1
Biomass and waste	64	78	85	70	98	111	10	15	2.5	3.5
Other renewables	57	83	97	70	131	162	11	21	6.1	8.2
Other energy sector	146	141	138	139	122	113	100	100	-0.1	-0.9
Electricity	*44*	*46*	*46*	*42*	*40*	*40*	*33*	*35*	*0.2*	*-0.4*
TFC	1 372	1 434	1 470	1 313	1 295	1 280	100	100	0.7	0.1
Coal	56	51	49	50	41	36	3	3	0.0	-1.1
Oil	536	519	515	498	405	350	35	27	-0.2	-1.7
Gas	310	328	338	293	291	287	23	22	1.0	0.3
Electricity	306	342	359	294	312	322	24	25	1.3	0.8
Heat	56	63	66	51	50	48	4	4	1.3	0.1
Biomass and waste	100	117	128	115	172	201	9	16	2.3	4.1
Other renewables	9	14	17	11	24	36	1	3	4.5	7.6
Industry	328	334	336	319	317	312	100	100	0.8	0.5
Coal	35	32	31	33	28	26	9	8	0.3	-0.4
Oil	34	30	27	33	28	25	8	8	-1.3	-1.6
Gas	96	97	97	93	93	90	29	29	0.9	0.7
Electricity	116	123	125	112	111	109	37	35	1.1	0.5
Heat	17	17	18	16	16	15	5	5	0.5	0.0
Biomass and waste	29	34	38	31	40	46	11	15	2.1	2.9
Other renewables	0	0	0	0	0	0	0	0	7.8	8.5
Transport	354	354	360	339	302	280	100	100	0.2	-0.8
Oil	319	313	315	298	231	191	87	68	-0.1	-2.0
Electricity	8	10	11	9	15	23	3	8	2.0	5.1
Biofuels	24	27	29	28	48	55	8	19	3.5	6.0
Other fuels	3	4	5	4	7	11	1	4	3.4	6.6
Buildings	549	605	635	522	544	557	100	100	1.1	0.5
Coal	18	16	16	16	11	9	2	2	-0.5	-2.7
Oil	67	63	60	61	46	37	9	7	-0.6	-2.4
Gas	194	211	220	182	177	172	35	31	1.0	0.0
Electricity	177	204	218	169	182	186	34	33	1.4	0.8
Heat	39	45	48	34	34	33	8	6	1.7	0.2
Biomass and waste	44	53	58	50	72	86	9	15	1.9	3.5
Other renewables	9	13	16	11	23	34	3	6	4.4	7.5
Other	141	140	139	134	132	132	100	100	-0.0	-0.2

A

OECD Europe: New Policies Scenario

Electricity generation (TWh)	1990	2009	2015	2020	2025	2030	2035	Shares (%) 2009	Shares (%) 2035	CAAGR (%) 2009-2035
Total generation	2 683	3 508	3 835	4 003	4 178	4 335	4 450	100	100	0.9
Coal	1 040	897	853	764	697	575	498	26	11	-2.2
Oil	216	94	62	43	32	27	26	3	1	-4.8
Gas	168	831	911	966	1 023	1 094	1 057	24	24	0.9
Nuclear	787	884	883	866	850	859	874	25	20	-0.0
Hydro	446	515	556	584	606	622	636	15	14	0.8
Biomass and waste	21	127	171	207	241	267	291	4	7	3.2
Wind	1	135	310	449	567	675	784	4	18	7.0
Geothermal	4	11	14	17	21	25	29	0	1	4.0
Solar PV	0	14	64	90	116	139	156	0	4	9.7
CSP	-	0	9	14	22	36	51	0	1	34.7
Marine	1	0	1	2	5	14	47	0	1	19.1

Electrical capacity (GW)	2009	2015	2020	2025	2030	2035	Shares (%) 2009	Shares (%) 2035	CAAGR (%) 2009-2035
Total capacity	900	1 130	1 180	1 233	1 294	1 356	100	100	1.6
Coal	179	220	189	169	150	123	20	9	-1.4
Oil	67	64	48	37	31	30	7	2	-3.0
Gas	206	264	272	286	303	320	23	24	1.7
Nuclear	138	131	126	121	122	124	15	9	-0.4
Hydro	193	204	213	220	226	231	21	17	0.7
Biomass and waste	21	26	32	37	41	44	2	3	3.0
Wind	76	151	209	252	289	324	8	24	5.7
Geothermal	2	2	3	3	4	4	0	0	3.2
Solar PV	17	64	84	101	115	127	2	9	8.2
CSP	0	3	4	6	10	15	0	1	20.4
Marine	0	0	1	1	4	13	0	1	16.5

CO_2 emissions (Mt)	1990	2009	2015	2020	2025	2030	2035	Shares (%) 2009	Shares (%) 2035	CAAGR (%) 2009-2035
Total CO_2	3 964	3 778	3 841	3 712	3 555	3 374	3 229	100	100	-0.6
Coal	1 713	1 132	1 130	1 019	899	749	651	30	20	-2.1
Oil	1 674	1 638	1 579	1 515	1 445	1 371	1 314	43	41	-0.8
Gas	578	1 008	1 132	1 179	1 211	1 254	1 264	27	39	0.9
Power generation	1 398	1 337	1 301	1 210	1 113	1 011	931	100	100	-1.4
Coal	1 139	896	841	740	634	500	415	67	45	-2.9
Oil	164	80	55	38	27	23	21	6	2	-5.0
Gas	95	361	405	432	451	488	495	27	53	1.2
TFC	2 389	2 258	2 362	2 329	2 275	2 201	2 141	100	100	-0.2
Coal	535	208	249	239	226	212	201	9	9	-0.1
Oil	1 394	1 443	1 426	1 382	1 326	1 260	1 207	64	56	-0.7
Transport	775	965	961	936	897	850	820	43	38	-0.6
Gas	460	607	687	708	722	730	734	27	34	0.7

OECD Europe: Current Policies and 450 Scenarios

	Electricity generation (TWh)						Shares (%)		CAAGR (%)	
	2020	2030	2035	2020	2030	2035	2035		2009-2035	
	Current Policies Scenario			450 Scenario			CPS	450	CPS	450
Total generation	4 056	4 499	4 700	3 899	4 082	4 193	100	100	1.1	0.7
Coal	925	1 010	982	647	262	179	21	4	0.4	-6.0
Oil	45	31	32	37	23	20	1	0	-4.1	-5.7
Gas	974	1 138	1 259	887	612	379	27	9	1.6	-3.0
Nuclear	824	727	681	877	1 050	1 137	14	27	-1.0	1.0
Hydro	576	606	616	600	658	683	13	16	0.7	1.1
Biomass and waste	192	242	264	217	311	358	6	9	2.8	4.1
Wind	409	568	642	495	862	1 021	14	24	6.2	8.1
Geothermal	16	21	24	18	31	38	1	1	3.2	5.0
Solar PV	83	118	130	99	182	231	3	6	8.9	11.4
CSP	12	29	42	21	64	87	1	2	33.7	37.5
Marine	1	8	28	2	25	61	1	1	16.8	20.3

	Electrical capacity (GW)						Shares (%)		CAAGR (%)	
	2020	2030	2035	2020	2030	2035	2035		2009-2035	
	Current Policies Scenario			450 Scenario			CPS	450	CPS	450
Total capacity	1 163	1 257	1 313	1 215	1 429	1 534	100	100	1.5	2.1
Coal	199	180	166	180	120	81	13	5	-0.3	-3.0
Oil	48	30	30	48	27	24	2	2	-3.1	-3.9
Gas	277	317	347	275	303	319	26	21	2.0	1.7
Nuclear	121	105	96	129	150	160	7	10	-1.4	0.6
Hydro	210	220	223	218	239	248	17	16	0.6	1.0
Biomass and waste	29	37	40	33	48	55	3	4	2.6	3.8
Wind	194	255	279	231	363	410	21	27	5.1	6.7
Geothermal	3	3	4	3	5	6	0	0	2.6	4.2
Solar PV	78	100	108	91	150	189	8	12	7.5	9.8
CSP	3	8	12	6	18	25	1	2	19.4	22.8
Marine	0	2	8	1	7	17	1	1	14.3	17.7

	CO_2 emissions (Mt)						Shares (%)		CAAGR (%)	
	2020	2030	2035	2020	2030	2035	2035		2009-2035	
	Current Policies Scenario			450 Scenario			CPS	450	CPS	450
Total CO_2	3 881	3 905	3 899	3 407	2 398	1 996	100	100	0.1	-2.4
Coal	1 156	1 127	1 049	871	339	221	27	11	-0.3	-6.1
Oil	1 548	1 484	1 468	1 441	1 128	950	38	48	-0.4	-2.1
Gas	1 177	1 294	1 382	1 096	930	824	35	41	1.2	-0.8
Power generation	1 337	1 398	1 395	1 026	423	251	100	100	0.2	-6.2
Coal	872	871	803	611	154	69	58	28	-0.4	-9.4
Oil	41	27	26	35	20	18	2	7	-4.2	-5.6
Gas	425	500	566	380	249	164	41	65	1.7	-3.0
TFC	2 369	2 340	2 338	2 215	1 832	1 616	100	100	0.1	-1.3
Coal	244	220	211	222	151	120	9	7	0.1	-2.1
Oil	1 411	1 362	1 347	1 316	1 034	869	58	54	-0.3	-1.9
Transport	*955*	*937*	*941*	*893*	*692*	*570*	*40*	*35*	*-0.1*	*-2.0*
Gas	714	758	780	677	646	627	33	39	1.0	0.1

A

European Union: New Policies Scenario

	Energy demand (Mtoe)							Shares (%)		CAAGR (%)
	1990	2009	2015	2020	2025	2030	2035	2009	2035	2009-2035
TPED	1 633	1 654	1 731	1 734	1 727	1 724	1 731	100	100	0.2
Coal	455	267	260	228	198	163	140	16	8	-2.5
Oil	601	575	551	524	497	469	446	35	26	-1.0
Gas	295	416	468	485	498	513	515	25	30	0.8
Nuclear	207	233	230	231	227	230	237	14	14	0.1
Hydro	25	28	30	31	32	33	34	2	2	0.7
Biomass and waste	46	115	146	171	193	214	236	7	14	2.8
Other renewables	3	20	45	64	83	103	125	1	7	7.4
Power generation	644	698	725	730	731	741	754	100	100	0.3
Coal	286	210	191	162	135	105	85	30	11	-3.4
Oil	61	26	18	12	9	7	7	4	1	-5.1
Gas	54	139	158	169	177	192	194	20	26	1.3
Nuclear	207	233	230	231	227	230	237	33	31	0.1
Hydro	25	28	30	31	32	33	34	4	4	0.7
Biomass and waste	8	44	56	66	75	82	88	6	12	2.7
Other renewables	3	18	41	59	76	92	110	3	15	7.3
Other energy sector	150	133	134	129	125	120	115	100	100	-0.6
Electricity	*39*	*41*	*40*	*39*	*39*	*38*	*37*	*31*	*33*	*-0.3*
TFC	1 124	1 155	1 230	1 247	1 258	1 262	1 271	100	100	0.4
Coal	123	36	40	38	34	31	29	3	2	-0.9
Oil	500	505	496	477	455	430	408	44	32	-0.8
Gas	224	258	291	299	305	307	308	22	24	0.7
Electricity	185	234	256	268	279	290	298	20	23	0.9
Heat	54	49	54	57	59	62	64	4	5	1.0
Biomass and waste	38	71	90	105	117	131	147	6	12	2.9
Other renewables	1	2	3	5	7	10	15	0	1	7.8
Industry	341	255	295	298	297	295	292	100	100	0.5
Coal	69	23	27	26	24	22	20	9	7	-0.5
Oil	57	36	34	32	30	27	24	14	8	-1.5
Gas	97	74	92	93	93	92	90	29	31	0.7
Electricity	85	85	98	101	102	101	101	33	34	0.7
Heat	19	15	16	16	16	16	16	6	5	0.2
Biomass and waste	14	22	27	30	33	37	42	9	14	2.5
Other renewables	-	0	0	0	0	0	0	0	0	2.6
Transport	258	322	331	329	317	301	292	100	100	-0.4
Oil	252	302	301	291	275	256	241	94	83	-0.9
Electricity	5	6	7	8	9	10	11	2	4	2.3
Biofuels	0	12	20	26	29	31	33	4	11	4.0
Other fuels	1	2	3	3	4	5	6	1	2	3.9
Buildings	394	451	478	501	525	547	569	100	100	0.9
Coal	50	11	11	10	8	7	7	3	1	-1.9
Oil	89	64	61	60	58	55	52	14	9	-0.8
Gas	107	166	179	186	191	193	195	37	34	0.6
Electricity	90	139	147	155	165	176	183	31	32	1.1
Heat	33	34	37	40	43	46	48	7	8	1.4
Biomass and waste	23	35	40	46	53	60	69	8	12	2.6
Other renewables	1	2	3	5	7	10	15	0	3	7.8
Other	132	127	125	119	119	118	118	100	100	-0.3

European Union: Current Policies and 450 Scenarios

	Energy demand (Mtoe)						Shares (%)		CAAGR (%)	
	2020	2030	2035	2020	2030	2035	2035		2009-2035	
	Current Policies Scenario			450 Scenario			CPS	450	CPS	450
TPED	1 754	1 791	1 818	1 670	1 618	1 606	100	100	0.4	-0.1
Coal	257	243	221	200	114	92	12	6	-0.7	-4.0
Oil	541	512	503	500	395	336	28	21	-0.5	-2.0
Gas	486	530	566	450	402	367	31	23	1.2	-0.5
Nuclear	219	197	188	233	273	296	10	18	-0.8	0.9
Hydro	31	32	33	32	35	36	2	2	0.6	1.0
Biomass and waste	162	193	209	184	265	306	12	19	2.3	3.8
Other renewables	58	84	99	71	135	172	5	11	6.4	8.7
Power generation	734	770	788	695	697	717	100	100	0.5	0.1
Coal	190	183	165	138	63	47	21	7	-0.9	-5.6
Oil	13	8	8	12	7	6	1	1	-4.5	-5.5
Gas	166	197	225	147	108	80	29	11	1.9	-2.1
Nuclear	219	197	188	233	273	296	24	41	-0.8	0.9
Hydro	31	32	33	32	35	36	4	5	0.6	1.0
Biomass and waste	62	75	81	69	94	107	10	15	2.4	3.5
Other renewables	53	76	89	65	118	145	11	20	6.4	8.5
Other energy sector	131	125	122	124	108	99	100	100	-0.3	-1.1
Electricity	*40*	*41*	*40*	*38*	*36*	*35*	*33*	*35*	*-0.0*	*-0.5*
TFC	1 266	1 311	1 339	1 210	1 182	1 165	100	100	0.6	0.0
Coal	39	32	30	35	26	22	2	2	-0.7	-1.9
Oil	491	469	461	455	361	307	34	26	-0.3	-1.9
Gas	302	320	328	286	281	276	25	24	0.9	0.3
Electricity	271	300	314	261	275	284	23	24	1.1	0.7
Heat	58	65	68	52	51	50	5	4	1.3	0.1
Biomass and waste	100	117	128	115	171	199	10	17	2.3	4.1
Other renewables	5	8	10	6	16	27	1	2	6.0	10.1
Industry	299	300	301	291	285	280	100	100	0.6	0.4
Coal	26	22	20	24	20	18	7	6	-0.5	-0.9
Oil	32	28	25	31	26	23	8	8	-1.4	-1.8
Gas	94	94	94	90	89	87	31	31	0.9	0.6
Electricity	101	106	108	97	95	93	36	33	0.9	0.4
Heat	16	16	17	16	15	14	5	5	0.4	-0.1
Biomass and waste	29	34	38	31	40	46	13	16	2.1	2.8
Other renewables	0	0	0	0	0	0	0	0	2.6	5.3
Transport	333	326	328	319	278	254	100	100	0.1	-0.9
Oil	298	287	284	278	208	168	87	66	-0.2	-2.2
Electricity	8	9	10	8	14	21	3	8	1.9	4.9
Biofuels	24	26	28	28	48	54	8	21	3.3	6.0
Other fuels	3	4	5	4	7	11	2	4	3.6	6.2
Buildings	507	559	587	481	502	514	100	100	1.0	0.5
Coal	10	8	8	8	4	3	1	0	-1.5	-5.6
Oil	61	57	55	55	42	34	9	7	-0.6	-2.4
Gas	188	203	211	175	169	163	36	32	0.9	-0.1
Electricity	158	181	192	151	163	167	33	33	1.3	0.7
Heat	41	48	51	36	36	35	9	7	1.7	0.2
Biomass and waste	45	54	60	50	72	86	10	17	2.1	3.5
Other renewables	5	8	10	6	16	26	2	5	6.0	10.1
Other	127	125	124	119	118	117	100	100	-0.1	-0.3

A

European Union: New Policies Scenario

	Electricity generation (TWh)							Shares (%)		CAAGR (%)
	1990	2009	2015	2020	2025	2030	2035	2009	2035	2009-2035
Total generation	2 568	3 170	3 439	3 566	3 700	3 820	3 904	100	100	0.8
Coal	1 050	849	783	667	581	454	373	27	10	-3.1
Oil	221	96	63	42	30	26	24	3	1	-5.2
Gas	191	726	800	848	903	965	920	23	24	0.9
Nuclear	795	894	883	885	872	882	907	28	23	0.1
Hydro	286	328	353	365	375	384	393	10	10	0.7
Biomass and waste	20	124	167	201	233	257	279	4	7	3.2
Wind	1	133	306	440	551	646	737	4	19	6.8
Geothermal	3	6	9	12	15	18	21	0	1	5.2
Solar PV	0	14	64	89	114	137	153	0	4	9.6
CSP	-	0	9	14	22	36	51	0	1	34.7
Marine	1	0	1	2	5	14	47	0	1	19.1

	Electrical capacity (GW)						Shares (%)		CAAGR (%)
	2009	2015	2020	2025	2030	2035	2009	2035	2009-2035
Total capacity	834	1 049	1 090	1 132	1 180	1 230	100	100	1.5
Coal	177	215	180	158	137	108	21	9	-1.9
Oil	66	61	45	33	27	26	8	2	-3.6
Gas	193	246	252	263	277	292	23	24	1.6
Nuclear	139	131	129	124	125	129	17	10	-0.3
Hydro	146	151	156	161	165	169	17	14	0.6
Biomass and waste	20	26	31	36	39	43	2	3	2.9
Wind	75	150	207	248	280	307	9	25	5.6
Geothermal	1	1	2	2	3	3	0	0	4.3
Solar PV	17	64	83	100	114	126	2	10	8.1
CSP	0	3	4	6	10	15	0	1	20.4
Marine	0	0	1	1	4	13	0	1	16.5

	CO_2 emissions (Mt)							Shares (%)		CAAGR (%)
	1990	2009	2015	2020	2025	2030	2035	2009	2035	2009-2035
Total CO_2	4 035	3 529	3 542	3 377	3 186	2 981	2 827	100	100	-0.8
Coal	1 735	1 045	1 005	868	728	571	480	30	17	-2.9
Oil	1 643	1 533	1 469	1 399	1 320	1 236	1 167	43	41	-1.0
Gas	658	950	1 068	1 109	1 138	1 174	1 180	27	42	0.8
Power generation	1 491	1 270	1 215	1 096	978	864	789	100	100	-1.8
Coal	1 170	861	788	663	538	397	318	68	40	-3.8
Oil	195	83	58	39	27	23	21	7	3	-5.1
Gas	127	326	369	394	412	444	449	26	57	1.2
TFC	2 372	2 093	2 172	2 133	2 066	1 981	1 909	100	100	-0.4
Coal	527	159	182	171	156	141	131	8	7	-0.7
Oil	1 332	1 340	1 320	1 272	1 208	1 132	1 068	64	56	-0.9
Transport	*744*	*905*	*902*	*873*	*824*	*766*	*723*	*43*	*38*	*-0.9*
Gas	513	594	670	689	702	708	710	28	37	0.7

European Union: Current Policies and 450 Scenarios

	Electricity generation (TWh)						Shares (%)		CAAGR (%)	
	2020	2030	2035	2020	2030	2035	2035		2009-2035	
	Current Policies Scenario			450 Scenario			CPS	450	CPS	450
Total generation	3 610	3 960	4 119	3 471	3 613	3 707	100	100	1.0	0.6
Coal	808	839	787	579	249	183	19	5	-0.3	-5.7
Oil	45	30	29	39	23	20	1	1	-4.5	-5.8
Gas	862	1 016	1 133	764	529	307	28	8	1.7	-3.3
Nuclear	840	757	722	893	1 045	1 136	18	31	-0.8	0.9
Hydro	362	375	379	372	404	422	9	11	0.6	1.0
Biomass and waste	187	233	254	211	297	339	6	9	2.8	3.9
Wind	400	541	601	481	781	905	15	24	6.0	7.7
Geothermal	10	15	17	12	24	29	0	1	4.3	6.6
Solar PV	82	117	128	97	175	223	3	6	8.9	11.2
CSP	12	29	42	20	61	82	1	2	33.7	37.2
Marine	1	9	28	2	25	61	1	2	16.8	20.3

	Electrical capacity (GW)						Shares (%)		CAAGR (%)	
	2020	2030	2035	2020	2030	2035	2035		2009-2035	
	Current Policies Scenario			450 Scenario			CPS	450	CPS	450
Total capacity	1 072	1 143	1 188	1 123	1 291	1 380	100	100	1.4	2.0
Coal	188	161	142	174	118	83	12	6	-0.8	-2.9
Oil	45	26	26	45	23	20	2	1	-3.6	-4.5
Gas	257	293	323	256	274	286	27	21	2.0	1.5
Nuclear	124	109	101	131	150	161	9	12	-1.2	0.5
Hydro	155	161	162	160	174	182	14	13	0.4	0.9
Biomass and waste	29	36	39	32	45	52	3	4	2.5	3.7
Wind	192	246	265	226	333	369	22	27	5.0	6.3
Geothermal	2	2	3	2	4	4	0	0	3.5	5.5
Solar PV	77	99	107	90	146	183	9	13	7.5	9.7
CSP	3	8	12	6	17	24	1	2	19.4	22.6
Marine	0	2	8	1	7	17	1	1	14.3	17.7

	CO_2 emissions (Mt)						Shares (%)		CAAGR (%)	
	2020	2030	2035	2020	2030	2035	2035		2009-2035	
	Current Policies Scenario			450 Scenario			CPS	450	CPS	450
Total CO_2	3 533	3 469	3 434	3 107	2 162	1 796	100	100	-0.1	-2.6
Coal	987	901	808	749	269	175	24	10	-1.0	-6.6
Oil	1 435	1 351	1 325	1 333	1 017	842	39	47	-0.6	-2.3
Gas	1 110	1 217	1 301	1 025	877	778	38	43	1.2	-0.8
Power generation	1 209	1 208	1 189	936	400	259	100	100	-0.3	-5.9
Coal	778	723	641	558	145	81	54	31	-1.1	-8.7
Oil	42	27	25	37	22	19	2	7	-4.5	-5.5
Gas	389	458	523	342	233	159	44	61	1.8	-2.7
TFC	2 173	2 120	2 107	2 028	1 644	1 432	100	100	0.0	-1.4
Coal	175	146	137	159	94	67	6	5	-0.6	-3.2
Oil	1 302	1 238	1 213	1 212	928	765	58	53	-0.4	-2.1
Transport	*894*	*858*	*852*	*833*	*624*	*503*	*40*	*35*	*-0.2*	*-2.2*
Gas	696	736	756	658	623	600	36	42	0.9	0.0

A

OECD Asia Oceania: New Policies Scenario

	Energy demand (Mtoe)							Shares (%)		CAAGR (%)
	1990	2009	2015	2020	2025	2030	2035	2009	2035	2009-2035
TPED	**631**	**850**	**906**	**912**	**913**	**914**	**912**	**100**	**100**	**0.3**
Coal	138	223	245	243	225	204	172	26	19	-1.0
Oil	335	338	329	310	297	284	271	40	30	-0.8
Gas	66	144	168	170	172	179	186	17	20	1.0
Nuclear	66	111	120	136	157	174	197	13	22	2.2
Hydro	11	10	12	12	13	13	13	1	1	1.2
Biomass and waste	10	16	19	24	28	33	37	2	4	3.2
Other renewables	4	7	12	16	22	28	36	1	4	6.2
Power generation	**241**	**389**	**424**	**434**	**440**	**448**	**451**	**100**	**100**	**0.6**
Coal	60	159	166	164	144	123	91	41	20	-2.1
Oil	56	22	17	10	8	8	9	6	2	-3.6
Gas	40	74	91	88	87	91	95	19	21	1.0
Nuclear	66	111	120	136	157	174	197	29	44	2.2
Hydro	11	10	12	12	13	13	13	3	3	1.2
Biomass and waste	3	7	8	10	12	15	17	2	4	3.7
Other renewables	3	6	10	14	18	23	28	2	6	5.9
Other energy sector	**57**	**68**	**73**	**74**	**75**	**74**	**74**	**100**	**100**	**0.3**
Electricity	*11*	*18*	*20*	*20*	*20*	*20*	*19*	*26*	*26*	*0.4*
TFC	**431**	**552**	**590**	**591**	**589**	**587**	**584**	**100**	**100**	**0.2**
Coal	49	39	47	46	44	42	40	7	7	0.1
Oil	261	295	295	286	276	266	255	53	44	-0.6
Gas	26	65	74	77	79	81	83	12	14	1.0
Electricity	86	137	155	161	165	169	171	25	29	0.9
Heat	0	5	5	5	5	6	6	1	1	0.6
Biomass and waste	7	10	11	14	16	18	20	2	3	2.8
Other renewables	2	1	2	3	4	5	8	0	1	7.6
Industry	**145**	**152**	**174**	**177**	**177**	**176**	**173**	**100**	**100**	**0.5**
Coal	39	36	44	43	42	40	38	24	22	0.2
Oil	51	34	33	31	30	28	26	22	15	-1.0
Gas	11	22	25	27	28	29	30	14	17	1.2
Electricity	40	50	60	62	63	63	62	33	36	0.9
Heat	-	3	3	3	3	3	3	2	1	-0.2
Biomass and waste	5	7	9	10	12	13	15	5	9	2.7
Other renewables	0	0	0	0	0	0	0	0	0	1.8
Transport	**110**	**139**	**139**	**132**	**125**	**119**	**113**	**100**	**100**	**-0.8**
Oil	109	135	134	126	119	112	104	97	92	-1.0
Electricity	2	2	3	3	3	4	4	2	4	2.8
Biofuels	-	0	1	1	1	1	1	0	1	2.4
Other fuels	0	1	2	2	2	3	3	1	3	3.5
Buildings	**120**	**170**	**184**	**189**	**194**	**199**	**206**	**100**	**100**	**0.7**
Coal	10	2	2	2	2	1	1	1	1	-0.9
Oil	47	38	39	39	39	39	39	23	19	0.0
Gas	15	40	45	46	47	48	49	24	24	0.7
Electricity	44	84	91	95	98	101	104	50	50	0.8
Heat	0	2	2	3	3	3	3	1	2	1.3
Biomass and waste	2	2	2	3	3	4	5	1	2	3.1
Other renewables	1	1	1	2	3	4	6	1	3	7.4
Other	**56**	**91**	**93**	**93**	**93**	**92**	**92**	**100**	**100**	**0.0**

OECD Asia Oceania: Current Policies and 450 Scenarios

	Energy demand (Mtoe)						Shares (%)		CAAGR (%)	
	2020	2030	2035	2020	2030	2035	2035		2009-2035	
	Current Policies Scenario			450 Scenario			CPS	450	CPS	450
TPED	**927**	**959**	**971**	**888**	**849**	**843**	**100**	**100**	**0.5**	**-0.0**
Coal	256	256	249	220	123	92	26	11	0.4	-3.3
Oil	314	292	283	296	239	209	29	25	-0.7	-1.8
Gas	174	183	191	168	170	161	20	19	1.1	0.4
Nuclear	135	167	181	142	204	236	19	28	1.9	2.9
Hydro	12	12	13	13	15	16	1	2	0.9	1.9
Biomass and waste	21	26	29	29	50	63	3	8	2.1	5.3
Other renewables	15	23	26	20	47	66	3	8	4.9	8.7
Power generation	**442**	**475**	**489**	**423**	**417**	**423**	**100**	**100**	**0.9**	**0.3**
Coal	172	164	155	145	51	22	32	5	-0.1	-7.3
Oil	11	9	9	8	4	4	2	1	-3.3	-6.3
Gas	90	91	95	87	87	75	19	18	1.0	0.1
Nuclear	135	167	181	142	204	236	37	56	1.9	2.9
Hydro	12	12	13	13	15	16	3	4	0.9	1.9
Biomass and waste	9	12	14	11	17	21	3	5	3.0	4.6
Other renewables	13	19	22	17	38	48	4	11	4.9	8.2
Other energy sector	**77**	**80**	**81**	**72**	**67**	**64**	**100**	**100**	**0.7**	**-0.2**
Electricity	*21*	*21*	*21*	*20*	*19*	*18*	*26*	*28*	*0.7*	*0.1*
TFC	**599**	**608**	**612**	**577**	**551**	**541**	**100**	**100**	**0.4**	**-0.1**
Coal	49	47	46	44	38	35	8	6	0.7	-0.4
Oil	289	275	268	273	227	200	44	37	-0.4	-1.5
Gas	78	85	89	76	78	80	14	15	1.2	0.8
Electricity	165	178	185	159	162	162	30	30	1.2	0.7
Heat	5	6	6	5	5	5	1	1	0.6	0.2
Biomass and waste	12	13	14	18	33	42	2	8	1.4	5.7
Other renewables	2	4	4	3	9	17	1	3	5.1	10.8
Industry	**181**	**185**	**185**	**174**	**170**	**167**	**100**	**100**	**0.8**	**0.3**
Coal	46	44	43	41	35	33	23	20	0.7	-0.4
Oil	32	30	28	31	29	27	15	16	-0.7	-0.9
Gas	27	30	32	27	29	30	17	18	1.4	1.2
Electricity	64	68	69	61	60	58	37	35	1.3	0.6
Heat	3	3	3	3	2	2	1	1	-0.2	-0.6
Biomass and waste	9	10	11	11	14	16	6	10	1.3	3.0
Other renewables	0	0	0	0	0	0	0	0	1.8	1.8
Transport	**134**	**125**	**122**	**128**	**106**	**94**	**100**	**100**	**-0.5**	**-1.5**
Oil	129	118	114	121	87	67	93	71	-0.6	-2.7
Electricity	3	3	4	3	8	13	3	13	2.3	7.1
Biofuels	1	1	1	1	7	11	1	11	1.5	12.9
Other fuels	2	3	4	3	4	4	3	5	3.6	4.4
Buildings	**192**	**206**	**214**	**182**	**182**	**187**	**100**	**100**	**0.9**	**0.4**
Coal	2	2	2	2	1	1	1	1	0.3	-1.3
Oil	40	40	40	35	30	28	19	15	0.2	-1.2
Gas	47	50	52	44	43	44	24	24	1.0	0.3
Electricity	96	105	110	93	93	90	52	48	1.1	0.3
Heat	3	3	3	3	3	3	2	2	1.5	0.9
Biomass and waste	2	3	3	3	5	6	2	3	1.8	4.5
Other renewables	2	3	3	2	8	15	2	8	5.0	11.4
Other	**93**	**92**	**91**	**93**	**93**	**92**	**100**	**100**	**0.0**	**0.1**

A

OECD Asia Oceania: New Policies Scenario

Electricity generation (TWh)	1990	2009	2015	2020	2025	2030	2035	Shares (%) 2009	Shares (%) 2035	CAAGR (%) 2009-2035
Total generation	1 127	1 797	2 036	2 113	2 157	2 192	2 219	100	100	0.8
Coal	257	695	738	737	666	577	453	39	20	-1.6
Oil	270	114	85	48	41	40	41	6	2	-3.9
Gas	197	400	526	537	532	540	550	22	25	1.2
Nuclear	255	428	471	532	612	678	763	24	34	2.3
Hydro	133	114	139	143	147	151	154	6	7	1.2
Biomass and waste	12	25	33	42	50	59	67	1	3	3.8
Wind	-	9	22	37	55	79	105	0	5	9.9
Geothermal	4	8	10	12	14	16	18	0	1	3.3
Solar PV	0	4	11	22	33	43	52	0	2	10.8
CSP	-	0	1	3	4	7	10	0	0	35.3
Marine	-	-	0	1	2	3	5	-	0	n.a.

Electrical capacity (GW)	2009	2015	2020	2025	2030	2035	Shares (%) 2009	Shares (%) 2035	CAAGR (%) 2009-2035
Total capacity	426	474	503	526	553	579	100	100	1.2
Coal	104	110	109	106	102	94	25	16	-0.4
Oil	57	51	32	27	25	25	13	4	-3.1
Gas	115	143	166	168	174	178	27	31	1.7
Nuclear	69	73	80	91	99	110	16	19	1.8
Hydro	67	69	71	73	75	77	16	13	0.5
Biomass and waste	4	6	7	8	10	11	1	2	3.5
Wind	5	9	14	20	27	35	1	6	8.1
Geothermal	1	2	2	2	2	3	0	0	3.2
Solar PV	3	11	21	29	36	43	1	7	10.3
CSP	0	0	1	1	2	3	0	0	32.5
Marine	-	0	0	0	1	1	-	0	n.a.

CO_2 emissions (Mt)	1990	2009	2015	2020	2025	2030	2035	Shares (%) 2009	Shares (%) 2035	CAAGR (%) 2009-2035
Total CO_2	1 574	2 021	2 140	2 078	1 949	1 825	1 655	100	100	-0.8
Coal	521	872	955	940	841	733	577	43	35	-1.6
Oil	892	805	780	728	695	664	633	40	38	-0.9
Gas	161	344	406	410	413	429	445	17	27	1.0
Power generation	548	911	970	931	839	750	617	100	100	-1.5
Coal	280	666	702	691	604	507	362	73	59	-2.3
Oil	174	67	51	30	26	25	27	7	4	-3.5
Gas	94	178	217	211	209	218	228	20	37	1.0
TFC	958	1 013	1 068	1 041	1 010	977	942	100	100	-0.3
Coal	221	171	209	203	195	186	176	17	19	0.1
Oil	676	691	687	660	632	602	572	68	61	-0.7
Transport	*321*	*396*	*394*	*372*	*351*	*329*	*307*	*39*	*33*	*-1.0*
Gas	61	151	172	179	183	188	194	15	21	1.0

OECD Asia Oceania: Current Policies and 450 Scenarios

	Electricity generation (TWh)						Shares (%)		CAAGR (%)	
	2020	2030	2035	2020	2030	2035	2035		2009-2035	
	Current Policies Scenario			450 Scenario			CPS	450	CPS	450
Total generation	2 156	2 317	2 391	2 081	2 102	2 096	100	100	1.1	0.6
Coal	779	778	767	648	239	101	32	5	0.4	-7.2
Oil	52	42	43	40	20	19	2	1	-3.7	-6.6
Gas	552	534	537	537	528	443	22	21	1.1	0.4
Nuclear	528	650	702	552	793	912	29	44	1.9	3.0
Hydro	140	144	146	154	179	186	6	9	0.9	1.9
Biomass and waste	36	50	56	43	69	83	2	4	3.1	4.7
Wind	36	66	77	59	166	206	3	10	8.6	12.8
Geothermal	11	14	16	13	21	24	1	1	2.7	4.5
Solar PV	19	34	39	31	71	89	2	4	9.6	13.1
CSP	2	5	7	3	13	26	0	1	33.3	40.1
Marine	0	1	3	1	4	7	0	0	n.a.	n.a.

	Electrical capacity (GW)						Shares (%)		CAAGR (%)	
	2020	2030	2035	2020	2030	2035	2035		2009-2035	
	Current Policies Scenario			450 Scenario			CPS	450	CPS	450
Total capacity	506	556	577	510	572	607	100	100	1.2	1.4
Coal	114	118	117	105	64	51	20	8	0.4	-2.7
Oil	32	26	26	31	19	17	5	3	-3.0	-4.5
Gas	169	181	186	154	157	160	32	26	1.9	1.3
Nuclear	80	96	102	83	114	128	18	21	1.5	2.4
Hydro	70	72	73	75	86	89	13	15	0.3	1.1
Biomass and waste	6	8	9	7	11	13	2	2	2.8	4.3
Wind	13	23	26	22	55	66	5	11	6.9	10.8
Geothermal	2	2	2	2	3	4	0	1	2.6	4.5
Solar PV	19	30	33	29	59	71	6	12	9.3	12.5
CSP	1	1	2	1	3	6	0	1	30.5	37.2
Marine	0	0	1	0	1	2	0	0	n.a.	n.a.

	CO_2 emissions (Mt)						Shares (%)		CAAGR (%)	
	2020	2030	2035	2020	2030	2035	2035		2009-2035	
	Current Policies Scenario			450 Scenario			CPS	450	CPS	450
Total CO_2	2 148	2 071	2 024	1 927	1 229	945	100	100	0.0	-2.9
Coal	989	938	889	834	334	167	44	18	0.1	-6.2
Oil	741	696	678	690	535	453	33	48	-0.7	-2.2
Gas	418	438	457	403	360	325	23	34	1.1	-0.2
Power generation	974	932	900	829	347	172	100	100	-0.0	-6.2
Coal	726	686	644	598	169	30	72	17	-0.1	-11.3
Oil	32	27	28	25	13	13	3	7	-3.3	-6.2
Gas	217	218	227	207	165	129	25	75	1.0	-1.2
TFC	1 065	1 035	1 020	996	793	690	100	100	0.0	-1.5
Coal	215	207	201	192	129	103	20	15	0.6	-1.9
Oil	670	631	613	628	491	413	60	60	-0.5	-2.0
Transport	378	348	335	355	257	197	33	29	-0.6	-2.6
Gas	181	197	206	176	173	173	20	25	1.2	0.5

A

Japan: New Policies Scenario

	1990	2009	2015	2020	2025	2030	2035	Shares (%) 2009	Shares (%) 2035	CAAGR (%) 2009-2035
				Energy demand (Mtoe)						
TPED	439	472	498	490	485	481	478	100	100	0.0
Coal	77	101	116	111	101	91	81	21	17	-0.9
Oil	250	200	190	174	165	157	149	42	31	-1.1
Gas	44	81	98	101	101	103	104	17	22	1.0
Nuclear	53	73	73	78	88	95	104	15	22	1.4
Hydro	8	6	8	8	8	8	8	1	2	1.0
Biomass and waste	5	7	8	10	12	13	14	1	3	3.1
Other renewables	3	4	5	7	10	13	17	1	3	6.0
Power generation	174	214	230	229	233	238	243	100	100	0.5
Coal	25	58	63	59	51	44	36	27	15	-1.8
Oil	51	17	13	7	6	6	7	8	3	-3.5
Gas	33	52	63	65	63	64	64	24	26	0.8
Nuclear	53	73	73	78	88	95	104	34	43	1.4
Hydro	8	6	8	8	8	8	8	3	3	1.0
Biomass and waste	2	4	6	7	8	9	10	2	4	3.6
Other renewables	1	3	4	6	8	11	13	1	5	5.9
Other energy sector	38	35	37	36	34	32	30	100	100	-0.6
Electricity	*7*	*9*	*10*	*10*	*10*	*10*	*10*	*26*	*33*	*0.3*
TFC	300	314	332	329	324	318	314	100	100	0.0
Coal	32	26	33	32	31	29	28	8	9	0.3
Oil	184	172	167	159	152	144	136	55	43	-0.9
Gas	15	32	38	40	41	42	43	10	14	1.1
Electricity	64	80	90	93	95	96	98	26	31	0.8
Heat	0	1	1	1	1	1	1	0	0	1.8
Biomass and waste	3	2	2	3	3	4	4	1	1	2.1
Other renewables	1	1	1	1	2	2	4	0	1	6.5
Industry	103	82	96	96	94	92	89	100	100	0.3
Coal	31	25	31	31	30	28	27	30	30	0.3
Oil	37	24	23	22	20	19	18	29	20	-1.2
Gas	4	7	10	11	11	12	13	9	14	2.1
Electricity	29	23	30	30	30	29	28	28	31	0.7
Heat	-	-	-	-	-	-	-	-	-	n.a.
Biomass and waste	3	2	2	3	3	3	4	3	4	1.8
Other renewables	-	-	-	-	-	-	-	-	-	n.a.
Transport	72	76	73	68	63	57	52	100	100	-1.4
Oil	70	74	71	65	60	54	48	98	92	-1.6
Electricity	1	2	2	2	2	3	3	2	6	2.6
Biofuels	-	-	-	-	0	0	0	-	0	n.a.
Other fuels	-	-	0	0	0	0	1	-	1	n.a.
Buildings	84	112	120	123	126	129	133	100	100	0.7
Coal	1	1	1	1	1	1	1	0	1	1.7
Oil	36	30	31	31	31	32	32	27	24	0.2
Gas	11	25	28	29	29	29	29	22	22	0.7
Electricity	34	55	59	61	63	65	67	50	50	0.7
Heat	0	1	1	1	1	1	1	0	1	1.8
Biomass and waste	0	0	0	0	0	0	0	0	0	7.1
Other renewables	1	1	1	1	1	2	3	1	2	6.7
Other	41	44	43	42	41	40	39	100	100	-0.4

Japan: Current Policies and 450 Scenarios

	Energy demand (Mtoe)						Shares (%)		CAAGR (%)	
	2020	2030	2035	2020	2030	2035	2035		2009-2035	
	Current Policies Scenario			450 Scenario			CPS	450	CPS	450
TPED	496	502	505	474	448	442	100	100	0.3	-0.2
Coal	116	113	109	98	55	42	22	10	0.3	-3.3
Oil	177	163	157	165	132	117	31	26	-0.9	-2.1
Gas	102	103	106	99	88	77	21	17	1.1	-0.2
Nuclear	78	94	100	81	119	137	20	31	1.2	2.5
Hydro	8	8	8	8	9	10	2	2	0.9	1.6
Biomass and waste	8	11	12	12	19	22	2	5	2.3	4.7
Other renewables	7	10	12	10	26	38	2	8	4.6	9.4
Power generation	233	251	259	223	226	229	100	100	0.7	0.3
Coal	62	62	60	50	13	4	23	2	0.1	-9.7
Oil	7	7	7	5	2	2	3	1	-3.2	-7.7
Gas	65	64	65	63	51	39	25	17	0.8	-1.1
Nuclear	78	94	100	81	119	137	39	60	1.2	2.5
Hydro	8	8	8	8	9	10	3	4	0.9	1.6
Biomass and waste	6	8	9	7	10	11	4	5	3.1	3.8
Other renewables	6	9	10	9	21	25	4	11	4.7	8.6
Other energy sector	37	34	32	34	28	25	100	100	-0.3	-1.3
Electricity	11	11	10	10	10	9	32	36	0.5	-0.0
TFC	332	329	328	319	297	289	100	100	0.2	-0.3
Coal	33	32	31	30	26	24	9	8	0.7	-0.3
Oil	161	149	144	152	124	110	44	38	-0.7	-1.7
Gas	40	43	44	39	40	41	14	14	1.2	0.9
Electricity	94	101	104	92	92	91	32	31	1.0	0.5
Heat	1	1	1	1	1	1	0	0	1.9	1.3
Biomass and waste	2	2	3	4	9	11	1	4	0.3	6.0
Other renewables	1	2	2	2	6	12	1	4	3.8	11.7
Industry	97	96	95	93	88	85	100	100	0.5	0.1
Coal	32	31	30	29	25	23	31	27	0.7	-0.3
Oil	22	20	19	21	20	18	20	22	-1.0	-1.1
Gas	10	12	13	11	13	14	14	16	2.2	2.3
Electricity	31	31	31	29	27	26	32	31	1.1	0.5
Heat	-	-	-	-	-	-	-	-	n.a.	n.a.
Biomass and waste	2	2	2	3	4	4	3	5	0.3	2.0
Other renewables	-	-	-	-	-	-	-	-	n.a.	n.a.
Transport	69	61	57	66	52	45	100	100	-1.1	-2.0
Oil	66	58	54	63	44	34	94	76	-1.2	-3.0
Electricity	2	3	3	2	5	8	5	17	2.0	6.1
Biofuels	-	-	-	-	2	2	-	5	n.a.	n.a.
Other fuels	0	0	1	0	1	1	1	2	n.a.	n.a.
Buildings	125	133	137	118	117	120	100	100	0.8	0.3
Coal	1	1	1	1	1	1	1	1	1.9	1.3
Oil	32	33	33	27	24	23	24	19	0.4	-1.1
Gas	29	30	31	27	26	26	22	22	0.8	0.3
Electricity	62	67	70	60	60	57	51	47	0.9	0.1
Heat	1	1	1	1	1	1	1	1	1.9	1.3
Biomass and waste	0	0	0	0	0	1	0	1	2.2	14.0
Other renewables	1	1	2	2	5	12	1	10	4.1	12.2
Other	42	40	39	42	40	40	100	100	-0.4	-0.4

A

Japan: New Policies Scenario

Electricity generation (TWh)	1990	2009	2015	2020	2025	2030	2035	Shares (%) 2009	Shares (%) 2035	CAAGR (%) 2009-2035
Total generation	836	1 041	1 174	1 207	1 225	1 241	1 255	100	100	0.7
Coal	117	279	308	290	256	223	185	27	15	-1.6
Oil	248	92	71	37	32	31	33	9	3	-3.8
Gas	167	285	369	407	401	396	382	27	30	1.1
Nuclear	202	280	287	308	346	373	408	27	33	1.5
Hydro	89	75	89	91	94	96	99	7	8	1.0
Biomass and waste	11	21	28	34	39	43	47	2	4	3.1
Wind	-	3	10	18	28	41	56	0	4	12.0
Geothermal	2	3	3	4	4	5	6	0	0	2.7
Solar PV	0	3	9	18	26	32	37	0	3	10.5
CSP	-	-	-	-	-	-	-	-	-	n.a.
Marine	-	-	-	-	0	1	2	-	0	n.a.

Electrical capacity (GW)	2009	2015	2020	2025	2030	2035	Shares (%) 2009	Shares (%) 2035	CAAGR (%) 2009-2035
Total capacity	274	300	316	327	340	351	100	100	1.0
Coal	46	48	46	43	42	39	17	11	-0.7
Oil	50	45	26	23	21	21	18	6	-3.2
Gas	72	91	112	112	114	114	26	32	1.8
Nuclear	50	49	51	57	60	65	18	18	1.0
Hydro	47	48	49	50	52	53	17	15	0.4
Biomass and waste	4	5	6	6	7	8	1	2	2.8
Wind	2	4	7	11	15	20	1	6	9.1
Geothermal	1	1	1	1	1	1	0	0	2.7
Solar PV	3	9	17	24	28	31	1	9	10.0
CSP	-	-	-	-	-	-	-	-	n.a.
Marine	-	-	-	0	0	1	-	0	n.a.

CO$_2$ emissions (Mt)	1990	2009	2015	2020	2025	2030	2035	Shares (%) 2009	Shares (%) 2035	CAAGR (%) 2009-2035
Total CO$_2$	1 063	1 088	1 160	1 101	1 037	981	918	100	100	-0.7
Coal	293	392	450	426	386	348	305	36	33	-1.0
Oil	655	492	464	422	399	376	355	45	39	-1.2
Gas	115	204	246	253	251	257	258	19	28	0.9
Power generation	363	431	470	435	396	368	332	100	100	-1.0
Coal	128	255	277	258	224	193	157	59	47	-1.8
Oil	157	49	39	20	18	18	20	11	6	-3.4
Gas	78	127	153	157	153	156	155	29	47	0.8
TFC	655	616	648	627	604	578	553	100	100	-0.4
Coal	150	121	153	149	144	138	131	20	24	0.3
Oil	470	420	406	385	365	343	321	68	58	-1.0
Transport	*208*	*219*	*209*	*192*	*176*	*159*	*142*	*36*	*26*	*-1.6*
Gas	35	75	90	94	95	98	100	12	18	1.1

Japan: Current Policies and 450 Scenarios

	Electricity generation (TWh)						Shares (%)		CAAGR (%)	
	2020	2030	2035	2020	2030	2035	2035		2009-2035	
	Current Policies Scenario			450 Scenario			CPS	450	CPS	450
Total generation	1 223	1 295	1 329	1 187	1 184	1 164	100	100	0.9	0.4
Coal	309	315	313	244	66	18	24	2	0.4	-10.1
Oil	41	34	36	29	11	12	3	1	-3.6	-7.6
Gas	409	382	376	400	323	234	28	20	1.1	-0.8
Nuclear	308	368	392	319	466	535	29	46	1.3	2.5
Hydro	91	94	96	96	109	114	7	10	0.9	1.6
Biomass and waste	29	37	42	35	46	51	3	4	2.6	3.4
Wind	17	33	39	35	100	123	3	11	10.4	15.4
Geothermal	4	4	5	4	9	11	0	1	1.8	5.2
Solar PV	16	27	30	26	53	61	2	5	9.7	12.7
CSP	-	-	-	-	-	-	-	-	n.a.	n.a.
Marine	-	0	1	-	1	4	0	0	n.a.	n.a.

	Electrical capacity (GW)						Shares (%)		CAAGR (%)	
	2020	2030	2035	2020	2030	2035	2035		2009-2035	
	Current Policies Scenario			450 Scenario			CPS	450	CPS	450
Total capacity	317	343	353	320	348	364	100	100	1.0	1.1
Coal	48	49	48	44	20	17	14	5	0.2	-3.8
Oil	27	22	22	25	15	14	6	4	-3.0	-4.9
Gas	114	118	119	102	94	90	34	25	2.0	0.9
Nuclear	51	59	62	53	72	81	18	22	0.8	1.9
Hydro	49	51	52	51	57	59	15	16	0.3	0.8
Biomass and waste	5	6	7	6	7	8	2	2	2.2	3.1
Wind	7	13	14	14	34	41	4	11	7.8	12.2
Geothermal	1	1	1	1	2	2	0	1	1.9	5.0
Solar PV	16	25	27	25	46	52	8	14	9.4	12.1
CSP	-	-	-	-	-	-	-	-	n.a.	n.a.
Marine	-	0	0	-	0	1	0	0	n.a.	n.a.

	CO$_2$ emissions (Mt)						Shares (%)		CAAGR (%)	
	2020	2030	2035	2020	2030	2035	2035		2009-2035	
	Current Policies Scenario			450 Scenario			CPS	450	CPS	450
Total CO$_2$	1 132	1 088	1 069	1 019	660	521	100	100	-0.1	-2.8
Coal	448	437	425	375	155	93	40	18	0.3	-5.4
Oil	431	395	381	398	303	260	36	50	-1.0	-2.4
Gas	254	256	263	246	202	169	25	32	1.0	-0.7
Power generation	454	444	441	385	162	86	100	100	0.1	-6.0
Coal	274	270	263	217	47	4	60	5	0.1	-14.8
Oil	22	20	22	15	6	6	5	7	-3.1	-7.6
Gas	158	154	157	153	110	75	36	88	0.8	-2.0
TFC	638	607	592	596	467	408	100	100	-0.2	-1.6
Coal	154	148	144	139	93	74	24	18	0.7	-1.9
Oil	390	360	344	366	285	243	58	60	-0.8	-2.1
Transport	196	170	158	186	131	100	27	25	-1.2	-3.0
Gas	93	100	104	91	89	91	17	22	1.2	0.7

A

Non-OECD: New Policies Scenario

	\multicolumn{7}{c}{Energy demand (Mtoe)}	Shares (%)		CAAGR (%)
	1990	2009	2015	2020	2025	2030	2035	2009	2035	2009-2035
TPED	4 065	6 567	8 013	8 818	9 472	10 141	10 826	100	100	1.9
Coal	1 153	2 260	2 847	3 038	3 120	3 203	3 299	34	30	1.5
Oil	1 157	1 700	2 025	2 153	2 287	2 421	2 551	26	24	1.6
Gas	828	1 291	1 583	1 810	2 004	2 213	2 412	20	22	2.4
Nuclear	74	119	185	295	374	437	490	2	5	5.6
Hydro	83	166	212	250	287	316	338	3	3	2.8
Biomass and waste	762	988	1 070	1 137	1 207	1 283	1 368	15	13	1.3
Other renewables	8	43	91	136	194	268	368	1	3	8.7
Power generation	1 269	2 397	3 020	3 480	3 854	4 239	4 667	100	100	2.6
Coal	469	1 298	1 686	1 844	1 938	2 036	2 150	54	46	2.0
Oil	223	192	168	154	135	116	112	8	2	-2.1
Gas	405	569	660	752	839	930	1 016	24	22	2.3
Nuclear	74	119	185	295	374	437	490	5	10	5.6
Hydro	83	166	212	250	287	316	338	7	7	2.8
Biomass and waste	7	19	36	72	116	174	244	1	5	10.2
Other renewables	8	33	73	113	165	231	318	1	7	9.0
Other energy sector	498	816	970	1 039	1 086	1 139	1 185	100	100	1.4
Electricity	*76*	*161*	*204*	*234*	*261*	*288*	*316*	*20*	*27*	*2.6*
TFC	2 984	4 419	5 422	5 935	6 362	6 790	7 213	100	100	1.9
Coal	533	703	854	871	851	825	799	16	11	0.5
Oil	815	1 381	1 720	1 863	2 025	2 189	2 327	31	32	2.0
Gas	354	558	716	830	919	1 012	1 112	13	15	2.7
Electricity	283	669	933	1 132	1 298	1 471	1 658	15	23	3.6
Heat	291	194	218	225	225	223	222	4	3	0.5
Biomass and waste	707	905	965	993	1 015	1 032	1 044	20	14	0.6
Other renewables	0	9	17	22	29	37	50	0	1	6.8
Industry	979	1 506	1 962	2 181	2 297	2 397	2 498	100	100	2.0
Coal	313	550	694	714	702	687	674	37	27	0.8
Oil	160	207	240	246	248	247	244	14	10	0.6
Gas	134	206	287	356	396	432	467	14	19	3.2
Electricity	157	340	497	604	681	751	823	23	33	3.5
Heat	136	86	101	105	102	99	97	6	4	0.5
Biomass and waste	79	117	142	156	168	180	192	8	8	1.9
Other renewables	-	0	0	0	0	0	0	0	0	3.3
Transport	440	783	1 043	1 170	1 329	1 505	1 667	100	100	2.9
Oil	370	701	933	1 040	1 177	1 323	1 451	89	87	2.8
Electricity	13	14	19	23	27	32	39	2	2	4.1
Biofuels	6	16	29	39	52	66	81	2	5	6.4
Other fuels	50	53	62	67	74	84	96	7	6	2.3
Buildings	1 267	1 619	1 815	1 938	2 050	2 166	2 290	100	100	1.3
Coal	170	104	109	103	96	85	73	6	3	-1.3
Oil	121	161	181	183	185	185	184	10	8	0.5
Gas	126	192	234	266	301	338	380	12	17	2.7
Electricity	88	281	374	456	534	624	724	17	32	3.7
Heat	145	105	113	117	119	121	121	6	5	0.6
Biomass and waste	617	767	788	791	788	777	760	47	33	-0.0
Other renewables	0	9	17	22	28	36	48	1	2	6.8
Other	298	512	602	647	686	722	757	100	100	1.5

Non-OECD: Current Policies and 450 Scenarios

	Energy demand (Mtoe)						Shares (%)		CAAGR (%)	
	2020	2030	2035	2020	2030	2035	2035		2009-2035	
	Current Policies Scenario			450 Scenario			CPS	450	CPS	450
TPED	9 091	10 903	11 887	8 404	8 941	9 254	100	100	2.3	1.3
Coal	3 289	3 929	4 308	2 736	2 108	1 880	36	20	2.5	-0.7
Oil	2 212	2 582	2 763	2 045	2 114	2 074	23	22	1.9	0.8
Gas	1 836	2 332	2 619	1 715	1 916	1 991	22	22	2.8	1.7
Nuclear	288	388	413	325	624	764	3	8	4.9	7.4
Hydro	240	288	309	262	348	376	3	4	2.4	3.2
Biomass and waste	1 109	1 194	1 239	1 157	1 398	1 537	10	17	0.9	1.7
Other renewables	117	190	236	164	434	632	2	7	6.8	10.9
Power generation	3 662	4 717	5 348	3 253	3 526	3 795	100	100	3.1	1.8
Coal	2 062	2 645	3 007	1 604	1 083	900	56	24	3.3	-1.4
Oil	164	132	128	141	88	80	2	2	-1.6	-3.3
Gas	748	977	1 124	696	745	751	21	20	2.7	1.1
Nuclear	288	388	413	325	624	764	8	20	4.9	7.4
Hydro	240	288	309	262	348	376	6	10	2.4	3.2
Biomass and waste	63	126	165	87	250	356	3	9	8.6	11.9
Other renewables	97	161	202	139	389	568	4	15	7.2	11.5
Other energy sector	1 063	1 219	1 294	994	1 007	1 007	100	100	1.8	0.8
Electricity	*243*	*314*	*353*	*221*	*250*	*265*	*27*	*26*	*3.1*	*1.9*
TFC	6 057	7 158	7 727	5 710	6 202	6 372	100	100	2.2	1.4
Coal	894	902	898	823	726	682	12	11	0.9	-0.1
Oil	1 913	2 340	2 533	1 772	1 913	1 891	33	30	2.4	1.2
Gas	856	1 067	1 186	803	943	1 016	15	16	2.9	2.3
Electricity	1 170	1 587	1 834	1 077	1 307	1 428	24	22	4.0	3.0
Heat	232	243	247	212	198	191	3	3	0.9	-0.0
Biomass and waste	974	992	996	998	1 069	1 100	13	17	0.4	0.8
Other renewables	20	28	34	25	45	64	0	1	5.2	7.8
Industry	2 237	2 584	2 768	2 096	2 214	2 253	100	100	2.4	1.6
Coal	731	755	763	674	603	574	28	25	1.3	0.2
Oil	253	266	271	240	238	230	10	10	1.0	0.4
Gas	373	469	519	351	418	444	19	20	3.6	3.0
Electricity	622	816	927	575	677	718	33	32	3.9	2.9
Heat	108	111	111	100	92	88	4	4	1.0	0.1
Biomass and waste	150	167	177	155	184	198	6	9	1.6	2.0
Other renewables	0	0	0	0	1	2	0	0	3.3	13.4
Transport	1 189	1 595	1 794	1 106	1 319	1 363	100	100	3.2	2.2
Oil	1 068	1 432	1 600	966	1 085	1 064	89	78	3.2	1.6
Electricity	22	31	36	25	43	65	2	5	3.8	6.2
Biofuels	31	49	61	48	103	131	3	10	5.2	8.4
Other fuels	67	84	96	67	87	103	5	8	2.3	2.6
Buildings	1 974	2 239	2 385	1 867	1 964	2 022	100	100	1.5	0.9
Coal	108	93	82	96	72	59	3	3	-0.9	-2.1
Oil	192	199	200	175	163	156	8	8	0.8	-0.1
Gas	274	353	399	248	286	309	17	15	2.9	1.9
Electricity	475	672	793	428	526	578	33	29	4.1	2.8
Heat	120	129	132	109	103	100	6	5	0.9	-0.2
Biomass and waste	786	767	749	788	772	759	31	38	-0.1	-0.0
Other renewables	20	27	32	24	42	60	1	3	5.1	7.7
Other	656	740	780	641	704	734	100	100	1.6	1.4

A

Non-OECD: New Policies Scenario

	Electricity generation (TWh)							Shares (%)		CAAGR (%)
	1990	2009	2015	2020	2025	2030	2035	2009	2035	2009-2035
Total generation	4 190	9 649	13 219	15 884	18 136	20 457	22 946	100	100	3.4
Coal	1 332	4 498	6 320	7 187	7 749	8 406	9 153	47	40	2.8
Oil	640	703	610	564	503	442	433	7	2	-1.8
Gas	957	1 938	2 618	3 194	3 710	4 237	4 741	20	21	3.5
Nuclear	283	454	710	1 130	1 433	1 674	1 879	5	8	5.6
Hydro	962	1 931	2 463	2 904	3 339	3 670	3 926	20	17	2.8
Biomass and waste	7	49	110	238	390	583	822	1	4	11.4
Wind	0	50	311	513	730	965	1 237	1	5	13.1
Geothermal	8	25	40	58	83	112	146	0	1	7.1
Solar PV	0	1	32	78	157	281	415	0	2	28.4
CSP	-	-	5	19	40	87	191	-	1	n.a.
Marine	-	-	-	-	0	1	2	-	0	n.a.

	Electrical capacity (GW)						Shares (%)		CAAGR (%)
	2009	2015	2020	2025	2030	2035	2009	2035	2009-2035
Total capacity	2 342	3 195	3 791	4 303	4 848	5 443	100	100	3.3
Coal	943	1 274	1 445	1 546	1 668	1 805	40	33	2.5
Oil	213	230	221	201	180	173	9	3	-0.8
Gas	511	708	812	901	1 002	1 118	22	21	3.1
Nuclear	66	99	154	194	226	252	3	5	5.3
Hydro	553	680	808	936	1 031	1 102	24	20	2.7
Biomass and waste	12	23	45	70	101	138	1	3	9.7
Wind	39	151	238	328	422	527	2	10	10.5
Geothermal	4	6	9	13	17	22	0	0	6.8
Solar PV	1	23	53	103	177	257	0	5	26.9
CSP	-	1	5	11	23	49	-	1	n.a.
Marine	-	-	-	0	0	1	-	0	n.a.

	CO_2 emissions (Mt)							Shares (%)		CAAGR (%)
	1990	2009	2015	2020	2025	2030	2035	2009	2035	2009-2035
Total CO_2	9 205	15 876	19 652	21 224	22 264	23 338	24 438	100	100	1.7
Coal	4 160	8 483	10 737	11 443	11 658	11 860	12 125	53	50	1.4
Oil	3 170	4 515	5 391	5 738	6 134	6 548	6 939	28	28	1.7
Gas	1 876	2 878	3 524	4 043	4 472	4 930	5 374	18	22	2.4
Power generation	3 521	7 097	8 773	9 556	10 035	10 518	11 097	100	100	1.7
Coal	1 863	5 155	6 697	7 308	7 645	7 973	8 364	73	75	1.9
Oil	712	608	531	487	425	368	355	9	3	-2.0
Gas	946	1 334	1 545	1 761	1 965	2 177	2 377	19	21	2.2
TFC	5 273	7 970	9 914	10 622	11 159	11 694	12 172	100	100	1.6
Coal	2 222	3 126	3 802	3 872	3 767	3 638	3 508	39	29	0.4
Oil	2 265	3 623	4 538	4 916	5 356	5 808	6 190	45	51	2.1
Transport	*1 101*	*2 083*	*2 774*	*3 094*	*3 500*	*3 936*	*4 318*	*26*	*35*	*2.8*
Gas	786	1 221	1 574	1 835	2 036	2 248	2 474	15	20	2.8

Non-OECD: Current Policies and 450 Scenarios

	Electricity generation (TWh)						Shares (%)		CAAGR (%)	
	2020	2030	2035	2020	2030	2035	2035		2009-2035	
	Current Policies Scenario			450 Scenario			CPS	450	CPS	450
Total generation	16 426	22 097	25 429	15 092	18 107	19 683	100	100	3.8	2.8
Coal	8 035	10 864	12 670	6 215	4 503	3 636	50	18	4.1	-0.8
Oil	593	491	483	517	322	294	2	1	-1.4	-3.3
Gas	3 136	4 493	5 310	2 970	3 480	3 551	21	18	4.0	2.4
Nuclear	1 105	1 488	1 582	1 246	2 393	2 932	6	15	4.9	7.4
Hydro	2 793	3 350	3 597	3 042	4 044	4 369	14	22	2.4	3.2
Biomass and waste	208	422	550	289	851	1 224	2	6	9.7	13.1
Wind	435	718	865	594	1 591	2 155	3	11	11.6	15.6
Geothermal	50	81	100	71	172	234	0	1	5.6	9.0
Solar PV	60	148	192	103	506	798	1	4	24.7	31.7
CSP	11	41	77	45	243	484	0	2	n.a.	n.a.
Marine	-	1	2	0	2	6	0	0	n.a.	n.a.

	Electrical capacity (GW)						Shares (%)		CAAGR (%)	
	2020	2030	2035	2020	2030	2035	2035		2009-2035	
	Current Policies Scenario			450 Scenario			CPS	450	CPS	450
Total capacity	3 829	4 963	5 614	3 739	4 916	5 571	100	100	3.4	3.4
Coal	1 560	2 054	2 354	1 289	1 075	986	42	18	3.6	0.2
Oil	223	186	181	217	163	149	3	3	-0.6	-1.4
Gas	824	1 075	1 234	793	974	1 039	22	19	3.5	2.8
Nuclear	151	200	212	170	322	396	4	7	4.6	7.1
Hydro	773	929	997	850	1 148	1 242	18	22	2.3	3.2
Biomass and waste	40	74	94	54	145	203	2	4	8.1	11.3
Wind	205	322	380	270	668	878	7	16	9.1	12.7
Geothermal	8	12	15	11	26	34	0	1	5.3	8.6
Solar PV	42	99	125	70	328	510	2	9	23.4	30.3
CSP	3	11	20	13	67	131	0	2	n.a.	n.a.
Marine	-	0	1	0	0	2	0	0	n.a.	n.a.

	CO_2 emissions (Mt)						Shares (%)		CAAGR (%)	
	2020	2030	2035	2020	2030	2035	2035		2009-2035	
	Current Policies Scenario			450 Scenario			CPS	450	CPS	450
Total CO_2	22 442	26 951	29 543	19 476	16 222	14 416	100	100	2.4	-0.4
Coal	12 428	14 704	16 097	10 237	6 541	4 804	54	33	2.5	-2.2
Oil	5 913	7 053	7 609	5 417	5 580	5 450	26	38	2.0	0.7
Gas	4 101	5 195	5 838	3 822	4 101	4 162	20	29	2.8	1.4
Power generation	10 447	13 148	14 890	8 392	5 379	3 905	100	100	2.9	-2.3
Coal	8 177	10 444	11 857	6 319	3 454	2 060	80	53	3.3	-3.5
Oil	517	417	403	446	277	252	3	6	-1.6	-3.3
Gas	1 753	2 288	2 630	1 627	1 647	1 593	18	41	2.6	0.7
TFC	10 920	12 593	13 371	10 093	9 880	9 572	100	100	2.0	0.7
Coal	3 972	3 973	3 938	3 673	2 892	2 564	29	27	0.9	-0.8
Oil	5 054	6 247	6 789	4 646	4 964	4 856	51	51	2.4	1.1
Transport	*3 177*	*4 259*	*4 761*	*2 872*	*3 229*	*3 169*	*36*	*33*	*3.2*	*1.6*
Gas	1 894	2 373	2 644	1 774	2 024	2 152	20	22	3.0	2.2

A

E. Europe/Eurasia: New Policies Scenario

	\multicolumn{7}{c}{Energy demand (Mtoe)}	Shares (%)		CAAGR (%)
	1990	2009	2015	2020	2025	2030	2035	2009	2035	2009-2035
TPED	1 543	1 051	1 163	1 211	1 263	1 314	1 371	100	100	1.0
Coal	367	193	217	213	212	211	209	18	15	0.3
Oil	474	223	241	247	251	253	258	21	19	0.6
Gas	602	516	574	595	627	655	683	49	50	1.1
Nuclear	59	75	81	99	105	112	118	7	9	1.8
Hydro	23	25	26	27	29	30	32	2	2	0.9
Biomass and waste	17	18	21	24	28	34	45	2	3	3.7
Other renewables	0	1	3	7	11	17	25	0	2	15.2
Power generation	743	519	562	582	603	623	646	100	100	0.8
Coal	197	128	142	135	133	130	126	25	19	-0.1
Oil	127	21	17	15	12	10	9	4	1	-2.9
Gas	333	264	287	292	304	310	315	51	49	0.7
Nuclear	59	75	81	99	105	112	118	15	18	1.8
Hydro	23	25	26	27	29	30	32	5	5	0.9
Biomass and waste	4	5	6	7	9	14	22	1	3	5.5
Other renewables	0	0	3	6	11	17	24	0	4	16.5
Other energy sector	200	165	189	192	196	199	204	100	100	0.8
Electricity	*35*	*37*	*41*	*42*	*43*	*45*	*46*	*22*	*23*	*0.8*
TFC	1 073	675	747	787	826	865	905	100	100	1.1
Coal	114	39	40	42	42	43	43	6	5	0.4
Oil	281	169	187	196	205	214	223	25	25	1.1
Gas	261	216	246	259	274	291	310	32	34	1.4
Electricity	127	99	117	126	135	144	151	15	17	1.6
Heat	277	138	143	147	150	152	154	21	17	0.4
Biomass and waste	13	12	14	17	19	21	23	2	3	2.4
Other renewables	-	0	0	0	0	1	1	0	0	7.1
Industry	394	195	220	233	243	252	262	100	100	1.2
Coal	56	28	29	31	31	32	32	14	12	0.5
Oil	51	21	22	23	23	24	26	11	10	0.8
Gas	85	49	61	65	69	72	76	25	29	1.7
Electricity	75	44	56	61	64	67	69	23	26	1.7
Heat	125	50	50	52	53	54	55	26	21	0.4
Biomass and waste	0	1	2	2	3	3	4	1	1	4.2
Other renewables	-	0	0	0	0	0	0	0	0	0.0
Transport	171	134	160	169	179	189	200	100	100	1.6
Oil	122	92	105	111	119	125	131	68	66	1.4
Electricity	12	9	11	12	13	14	16	7	8	2.2
Biofuels	0	1	1	2	2	3	3	1	2	5.9
Other fuels	37	33	43	44	45	48	50	24	25	1.7
Buildings	387	259	278	290	305	317	329	100	100	0.9
Coal	56	9	9	9	9	9	8	3	3	-0.1
Oil	39	15	17	17	16	15	15	6	4	-0.0
Gas	111	98	106	113	121	129	137	38	42	1.3
Electricity	26	42	45	48	51	55	57	16	17	1.2
Heat	142	85	90	92	94	95	95	33	29	0.4
Biomass and waste	12	10	11	12	13	14	15	4	5	1.7
Other renewables	-	0	0	0	0	1	1	0	0	7.0
Other	121	87	90	94	100	106	114	100	100	1.0

E. Europe/Eurasia: Current Policies and 450 Scenarios

	Energy demand (Mtoe)						Shares (%)		CAAGR (%)	
	2020	2030	2035	2020	2030	2035	2035		2009-2035	
	Current Policies Scenario			450 Scenario			CPS	450	CPS	450
TPED	**1 240**	**1 388**	**1 478**	**1 156**	**1 185**	**1 209**	**100**	**100**	**1.3**	**0.5**
Coal	223	239	255	195	160	142	17	12	1.1	-1.2
Oil	252	262	270	238	230	223	18	18	0.7	0.0
Gas	611	704	760	551	531	516	51	43	1.5	0.0
Nuclear	98	111	114	109	139	159	8	13	1.6	2.9
Hydro	27	30	32	30	39	41	2	3	0.9	1.9
Biomass and waste	22	29	34	26	57	81	2	7	2.5	6.1
Other renewables	6	11	14	8	27	46	1	4	12.5	17.9
Power generation	**599**	**672**	**720**	**554**	**560**	**575**	**100**	**100**	**1.3**	**0.4**
Coal	144	155	166	120	87	67	23	12	1.0	-2.4
Oil	17	12	11	14	9	8	2	1	-2.3	-3.4
Gas	300	342	370	265	228	203	51	35	1.3	-1.0
Nuclear	98	111	114	109	139	159	16	28	1.6	2.9
Hydro	27	30	32	30	39	41	4	7	0.9	1.9
Biomass and waste	7	11	14	8	32	51	2	9	3.8	9.1
Other renewables	6	11	13	7	27	45	2	8	13.8	19.4
Other energy sector	**195**	**210**	**219**	**185**	**178**	**175**	**100**	**100**	**1.1**	**0.2**
Electricity	*44*	*49*	*53*	*40*	*40*	*40*	*24*	*23*	*1.4*	*0.3*
TFC	**805**	**908**	**968**	**751**	**783**	**797**	**100**	**100**	**1.4**	**0.6**
Coal	43	46	47	40	39	39	5	5	0.8	0.0
Oil	199	221	234	188	192	190	24	24	1.3	0.4
Gas	267	304	326	244	259	268	34	34	1.6	0.8
Electricity	132	161	176	122	133	138	18	17	2.2	1.3
Heat	149	159	164	139	134	132	17	17	0.7	-0.2
Biomass and waste	15	18	19	18	25	30	2	4	1.7	3.4
Other renewables	0	0	1	0	1	1	0	0	4.9	7.9
Industry	**239**	**269**	**287**	**222**	**231**	**234**	**100**	**100**	**1.5**	**0.7**
Coal	31	33	33	30	29	29	12	12	0.7	0.1
Oil	23	25	27	22	23	24	9	10	0.9	0.4
Gas	70	79	84	61	66	67	29	29	2.1	1.2
Electricity	63	76	83	58	60	60	29	26	2.4	1.2
Heat	51	54	56	49	50	50	20	21	0.4	-0.0
Biomass and waste	2	3	3	3	4	5	1	2	3.2	5.1
Other renewables	0	0	0	0	0	0	0	0	0.0	0.0
Transport	**169**	**192**	**205**	**162**	**168**	**169**	**100**	**100**	**1.6**	**0.9**
Oil	113	129	139	106	107	103	68	61	1.6	0.4
Electricity	12	14	15	13	16	19	7	11	2.1	3.0
Biofuels	1	1	1	2	4	5	1	3	3.0	8.2
Other fuels	44	47	49	41	41	41	24	24	1.6	0.9
Buildings	**300**	**338**	**359**	**275**	**280**	**283**	**100**	**100**	**1.3**	**0.3**
Coal	10	10	11	8	8	7	3	3	0.8	-0.6
Oil	17	17	16	16	14	13	5	4	0.4	-0.6
Gas	116	134	145	106	111	114	41	40	1.5	0.6
Electricity	51	62	68	46	49	50	19	18	1.9	0.7
Heat	95	101	104	86	81	79	29	28	0.8	-0.3
Biomass and waste	11	13	14	13	16	19	4	7	1.3	2.4
Other renewables	0	0	1	0	1	1	0	0	5.0	7.8
Other	**96**	**110**	**118**	**92**	**104**	**111**	**100**	**100**	**1.2**	**1.0**

E. Europe/Eurasia: New Policies Scenario

	Electricity generation (TWh)							Shares (%)		CAAGR (%)
	1990	2009	2015	2020	2025	2030	2035	2009	2035	2009-2035
Total generation	1 894	1 608	1 857	1 982	2 102	2 214	2 323	100	100	1.4
Coal	429	375	440	420	409	398	383	23	16	0.1
Oil	271	39	26	21	14	10	9	2	0	-5.7
Gas	702	611	759	810	885	929	952	38	41	1.7
Nuclear	226	287	309	377	402	430	453	18	19	1.8
Hydro	266	292	304	317	332	350	371	18	16	0.9
Biomass and waste	0	3	7	12	20	36	65	0	3	12.6
Wind	-	1	10	18	27	43	64	0	3	20.0
Geothermal	0	0	3	5	10	15	21	0	1	15.8
Solar PV	-	0	1	2	3	4	6	0	0	29.6
CSP	-	-	-	-	-	-	-	-	-	n.a.
Marine	-	-	-	-	0	0	0	-	0	n.a.

	Electrical capacity (GW)						Shares (%)		CAAGR (%)
	2009	2015	2020	2025	2030	2035	2009	2035	2009-2035
Total capacity	410	438	450	462	476	497	100	100	0.7
Coal	103	105	98	94	87	81	25	16	-0.9
Oil	25	23	18	12	8	7	6	1	-4.7
Gas	147	164	170	179	186	189	36	38	1.0
Nuclear	42	45	54	57	60	62	10	12	1.5
Hydro	90	94	98	102	107	112	22	23	0.9
Biomass and waste	1	2	2	4	6	11	0	2	9.3
Wind	0	4	7	11	17	25	0	5	16.8
Geothermal	0	0	1	2	2	3	0	1	14.6
Solar PV	0	1	2	3	4	5	0	1	23.9
CSP	-	-	-	-	-	-	-	-	n.a.
Marine	-	-	-	0	0	0	-	0	n.a.

	CO_2 emissions (Mt)							Shares (%)		CAAGR (%)
	1990	2009	2015	2020	2025	2030	2035	2009	2035	2009-2035
Total CO_2	3 997	2 476	2 705	2 749	2 827	2 883	2 945	100	100	0.7
Coal	1 336	769	821	802	796	783	769	31	26	0.0
Oil	1 257	550	593	610	624	638	656	22	22	0.7
Gas	1 404	1 158	1 291	1 337	1 407	1 462	1 519	47	52	1.1
Power generation	1 982	1 222	1 322	1 299	1 309	1 299	1 290	100	100	0.2
Coal	799	535	592	563	554	539	521	44	40	-0.1
Oil	405	66	57	50	40	34	31	5	2	-2.9
Gas	778	620	673	686	715	727	738	51	57	0.7
TFC	1 901	1 124	1 241	1 302	1 362	1 421	1 485	100	100	1.1
Coal	526	217	214	223	226	228	231	19	16	0.3
Oil	784	427	477	500	523	541	561	38	38	1.1
Transport	*361*	*270*	*308*	*328*	*350*	*368*	*386*	*24*	*26*	*1.4*
Gas	591	480	549	579	614	651	692	43	47	1.4

E. Europe/Eurasia: Current Policies and 450 Scenarios

	Electricity generation (TWh)						Shares (%)		CAAGR (%)	
	2020	2030	2035	2020	2030	2035	2035		2009-2035	
	Current Policies Scenario			450 Scenario			CPS	450	CPS	450
Total generation	2 069	2 465	2 695	1 913	2 038	2 095	100	100	2.0	1.0
Coal	462	509	558	368	254	165	21	8	1.5	-3.1
Oil	22	11	9	20	8	6	0	0	-5.3	-6.9
Gas	860	1 095	1 223	716	577	456	45	22	2.7	-1.1
Nuclear	376	426	435	416	532	608	16	29	1.6	2.9
Hydro	316	352	373	348	452	475	14	23	0.9	1.9
Biomass and waste	12	29	39	17	98	166	1	8	10.4	16.7
Wind	15	30	43	20	88	171	2	8	18.1	24.6
Geothermal	5	9	10	6	21	34	0	2	12.7	18.0
Solar PV	2	3	4	3	7	11	0	1	27.4	32.8
CSP	-	-	-	-	-	-	-	-	n.a.	n.a.
Marine	-	0	0	0	0	1	0	0	n.a.	n.a.

	Electrical capacity (GW)						Shares (%)		CAAGR (%)	
	2020	2030	2035	2020	2030	2035	2035		2009-2035	
	Current Policies Scenario			450 Scenario			CPS	450	CPS	450
Total capacity	467	514	543	448	503	560	100	100	1.1	1.2
Coal	103	101	103	90	60	55	19	10	-0.0	-2.4
Oil	18	8	7	18	8	7	1	1	-4.9	-4.8
Gas	185	218	232	159	164	161	43	29	1.8	0.4
Nuclear	53	59	59	59	74	84	11	15	1.3	2.6
Hydro	97	107	112	106	135	141	21	25	0.9	1.8
Biomass and waste	2	5	7	3	17	27	1	5	7.3	13.2
Wind	7	13	17	8	35	69	3	12	15.1	21.4
Geothermal	1	2	2	1	3	5	0	1	12.0	16.6
Solar PV	1	3	3	3	6	10	1	2	21.7	27.2
CSP	-	-	-	-	-	-	-	-	n.a.	n.a.
Marine	-	0	0	0	0	0	0	0	n.a.	n.a.

	CO$_2$ emissions (Mt)						Shares (%)		CAAGR (%)	
	2020	2030	2035	2020	2030	2035	2035		2009-2035	
	Current Policies Scenario			450 Scenario			CPS	450	CPS	450
Total CO$_2$	2 841	3 134	3 337	2 549	2 184	1 995	100	100	1.2	-0.8
Coal	843	895	946	730	504	397	28	20	0.8	-2.5
Oil	624	669	698	584	562	541	21	27	0.9	-0.1
Gas	1 374	1 570	1 692	1 235	1 119	1 056	51	53	1.5	-0.4
Power generation	1 360	1 479	1 585	1 167	831	665	100	100	1.0	-2.3
Coal	601	639	682	499	311	219	43	33	0.9	-3.4
Oil	56	40	36	47	30	28	2	4	-2.3	-3.3
Gas	703	801	867	621	489	418	55	63	1.3	-1.5
TFC	1 332	1 484	1 571	1 239	1 209	1 186	100	100	1.3	0.2
Coal	227	240	247	216	178	163	16	14	0.5	-1.1
Oil	508	564	595	478	474	457	38	39	1.3	0.3
Transport	*332*	*381*	*408*	*312*	*316*	*304*	*26*	*26*	*1.6*	*0.4*
Gas	597	680	730	545	557	565	46	48	1.6	0.6

A

Russia: New Policies Scenario

	Energy demand (Mtoe)							Shares (%)		CAAGR (%)
	1990	2009	2015	2020	2025	2030	2035	2009	2035	2009-2035
TPED	**880**	**648**	**719**	**744**	**771**	**799**	**833**	**100**	**100**	**1.0**
Coal	191	95	115	116	119	119	117	15	14	0.8
Oil	264	138	148	150	151	150	153	21	18	0.4
Gas	367	350	384	393	407	422	435	54	52	0.8
Nuclear	31	43	49	57	61	67	71	7	9	2.0
Hydro	14	15	15	15	16	18	19	2	2	0.9
Biomass and waste	12	6	6	7	9	12	19	1	2	4.2
Other renewables	0	0	2	4	8	12	18	0	2	15.7
Power generation	**444**	**344**	**376**	**390**	**404**	**418**	**435**	**100**	**100**	**0.9**
Coal	105	64	78	79	82	82	81	19	19	0.9
Oil	62	13	12	11	9	7	7	4	2	-2.4
Gas	228	204	217	219	222	223	224	59	52	0.4
Nuclear	31	43	49	57	61	67	71	13	16	2.0
Hydro	14	15	15	15	16	18	19	4	4	0.9
Biomass and waste	4	4	4	4	6	9	15	1	4	5.4
Other renewables	0	0	2	4	8	12	18	0	4	15.7
Other energy sector	**127**	**101**	**120**	**121**	**123**	**125**	**128**	**100**	**100**	**0.9**
Electricity	*21*	*25*	*28*	*28*	*30*	*31*	*32*	*25*	*25*	*1.0*
TFC	**625**	**423**	**460**	**478**	**498**	**517**	**538**	**100**	**100**	**0.9**
Coal	55	18	18	18	18	17	17	4	3	-0.4
Oil	145	106	112	116	120	123	128	25	24	0.7
Gas	143	129	146	153	161	170	181	30	34	1.3
Electricity	71	59	70	75	80	85	90	14	17	1.6
Heat	203	109	112	114	116	118	118	26	22	0.3
Biomass and waste	8	2	2	3	3	3	3	1	1	1.4
Other renewables	-	-	0	0	0	0	0	-	0	n.a.
Industry	**209**	**124**	**134**	**140**	**145**	**151**	**158**	**100**	**100**	**0.9**
Coal	15	14	13	13	13	13	13	11	8	-0.2
Oil	25	16	14	14	15	15	16	12	10	0.1
Gas	30	29	35	36	38	41	43	24	27	1.5
Electricity	41	27	34	37	39	41	43	22	27	1.9
Heat	98	39	38	38	39	41	42	31	27	0.3
Biomass and waste	-	0	1	1	1	1	1	0	1	4.0
Other renewables	-	-	-	-	-	-	-	-	-	n.a.
Transport	**116**	**90**	**107**	**111**	**116**	**120**	**127**	**100**	**100**	**1.3**
Oil	73	55	61	64	67	69	72	61	57	1.0
Electricity	9	7	9	9	10	11	13	8	10	2.4
Biofuels	-	-	-	-	-	-	-	-	-	n.a.
Other fuels	34	28	37	38	39	40	42	31	33	1.7
Buildings	**228**	**147**	**160**	**166**	**172**	**175**	**178**	**100**	**100**	**0.7**
Coal	40	4	5	4	4	4	3	3	2	-1.3
Oil	12	6	8	7	7	6	5	4	3	-0.7
Gas	57	44	49	53	57	60	64	30	36	1.4
Electricity	15	24	25	27	29	30	31	16	17	1.0
Heat	98	67	71	73	74	74	73	45	41	0.3
Biomass and waste	7	2	2	2	2	2	2	1	1	0.3
Other renewables	-	-	0	0	0	0	0	-	0	n.a.
Other	**72**	**61**	**60**	**62**	**65**	**70**	**75**	**100**	**100**	**0.8**

Russia: Current Policies and 450 Scenarios

	Energy demand (Mtoe)						Shares (%)		CAAGR (%)	
	2020	2030	2035	2020	2030	2035	2035		2009-2035	
	Current Policies Scenario			450 Scenario			CPS	450	CPS	450
TPED	759	849	908	706	708	722	100	100	1.3	0.4
Coal	121	134	142	105	77	67	16	9	1.5	-1.3
Oil	151	153	158	145	138	136	17	19	0.5	-0.1
Gas	403	460	497	361	334	318	55	44	1.4	-0.4
Nuclear	57	66	70	65	84	96	8	13	1.9	3.1
Hydro	16	18	19	17	23	24	2	3	1.0	1.8
Biomass and waste	7	11	13	9	31	48	1	7	2.8	8.1
Other renewables	4	7	9	5	20	33	1	5	12.6	18.5
Power generation	399	451	486	368	370	381	100	100	1.3	0.4
Coal	83	95	103	68	46	38	21	10	1.8	-2.0
Oil	11	7	7	11	7	7	1	2	-2.4	-2.5
Gas	224	249	268	198	164	144	55	38	1.1	-1.3
Nuclear	57	66	70	65	84	96	14	25	1.9	3.1
Hydro	16	18	19	17	23	24	4	6	1.0	1.8
Biomass and waste	5	8	10	5	25	40	2	11	3.5	9.3
Other renewables	4	7	9	5	20	33	2	9	12.6	18.5
Other energy sector	123	132	138	116	109	106	100	100	1.2	0.2
Electricity	29	34	37	27	27	27	27	26	1.6	0.4
TFC	489	547	583	454	460	464	100	100	1.2	0.4
Coal	18	19	19	17	14	13	3	3	0.1	-1.2
Oil	117	126	133	111	112	111	23	24	0.9	0.2
Gas	157	180	193	143	148	153	33	33	1.6	0.7
Electricity	78	96	107	72	78	80	18	17	2.3	1.2
Heat	115	123	127	107	102	99	22	21	0.6	-0.4
Biomass and waste	3	3	4	3	6	7	1	2	1.6	4.6
Other renewables	0	0	0	0	0	0	0	0	n.a.	n.a.
Industry	144	164	176	134	138	140	100	100	1.4	0.4
Coal	13	13	13	13	11	10	8	7	-0.1	-1.0
Oil	14	16	17	14	14	15	10	10	0.4	-0.2
Gas	40	47	50	35	37	39	28	28	2.1	1.1
Electricity	38	47	52	35	37	37	29	27	2.6	1.3
Heat	38	41	43	37	37	37	24	27	0.4	-0.2
Biomass and waste	1	1	1	1	1	2	1	1	3.8	6.1
Other renewables	-	-	-	-	-	-	-	-	n.a.	n.a.
Transport	111	121	128	106	107	108	100	100	1.4	0.7
Oil	64	70	75	61	60	58	58	53	1.2	0.2
Electricity	9	11	12	10	13	15	10	14	2.3	3.1
Biofuels	-	-	-	-	1	2	-	1	n.a.	n.a.
Other fuels	38	40	41	35	33	33	32	31	1.5	0.7
Buildings	171	191	201	154	147	143	100	100	1.2	-0.1
Coal	5	5	5	4	3	3	2	2	0.4	-2.1
Oil	8	7	6	7	5	4	3	3	0.0	-1.5
Gas	54	63	69	48	49	49	34	34	1.7	0.4
Electricity	29	35	38	25	25	25	19	17	1.8	0.1
Heat	75	79	81	67	62	59	40	41	0.7	-0.5
Biomass and waste	2	2	2	2	3	4	1	3	1.1	3.2
Other renewables	0	0	0	0	0	0	0	0	n.a.	n.a.
Other	63	71	77	60	68	74	100	100	0.9	0.7

Russia: New Policies Scenario

Electricity generation (TWh)	1990	2009	2015	2020	2025	2030	2035	Shares (%) 2009	Shares (%) 2035	CAAGR (%) 2009-2035
Total generation	1 082	990	1 148	1 219	1 295	1 368	1 443	100	100	1.5
Coal	157	164	218	221	230	226	225	17	16	1.2
Oil	129	16	13	11	7	3	2	2	0	-8.4
Gas	512	469	551	572	601	624	631	47	44	1.1
Nuclear	118	164	186	217	234	255	273	17	19	2.0
Hydro	166	174	169	179	191	205	218	18	15	0.9
Biomass and waste	0	3	4	7	12	23	46	0	3	11.6
Wind	-	0	4	8	12	20	30	0	2	41.0
Geothermal	0	0	2	4	8	12	17	0	1	14.9
Solar PV	-	-	0	0	1	1	1	-	0	n.a.
CSP	-	-	-	-	-	-	-	-	-	n.a.
Marine	-	-	-	-	-	-	-	-	-	n.a.

Electrical capacity (GW)	2009	2015	2020	2025	2030	2035	Shares (%) 2009	Shares (%) 2035	CAAGR (%) 2009-2035
Total capacity	225	243	249	257	265	277	100	100	0.8
Coal	47	50	48	47	43	41	21	15	-0.5
Oil	6	6	5	3	2	1	3	0	-7.3
Gas	101	107	108	110	113	113	45	41	0.5
Nuclear	23	27	31	33	35	37	10	13	1.8
Hydro	48	49	51	54	58	61	21	22	1.0
Biomass and waste	1	1	2	2	4	8	0	3	8.4
Wind	0	1	3	4	7	11	0	4	28.4
Geothermal	0	0	1	1	2	3	0	1	13.7
Solar PV	-	0	1	1	1	2	-	1	n.a.
CSP	-	-	-	-	-	-	-	-	n.a.
Marine	-	-	-	-	-	-	-	-	n.a.

CO_2 emissions (Mt)	1990	2009	2015	2020	2025	2030	2035	Shares (%) 2009	Shares (%) 2035	CAAGR (%) 2009-2035
Total CO_2	2 179	1 517	1 655	1 687	1 730	1 756	1 787	100	100	0.6
Coal	687	405	444	451	464	461	457	27	26	0.5
Oil	625	327	348	352	353	354	362	22	20	0.4
Gas	866	785	863	884	914	941	968	52	54	0.8
Power generation	1 162	799	879	885	900	896	894	100	100	0.4
Coal	432	278	331	336	350	349	346	35	39	0.8
Oil	198	41	39	36	30	24	22	5	2	-2.3
Gas	532	480	509	513	520	524	525	60	59	0.3
TFC	960	652	701	726	750	774	804	100	100	0.8
Coal	253	121	108	109	108	107	105	19	13	-0.5
Oil	389	251	272	278	285	291	300	39	37	0.7
Transport	*217*	*162*	*180*	*187*	*196*	*203*	*212*	*25*	*26*	*1.0*
Gas	318	280	322	338	357	377	399	43	50	1.4

Russia: Current Policies and 450 Scenarios

	Electricity generation (TWh)						Shares (%)		CAAGR (%)	
	2020	2030	2035	2020	2030	2035	2035		2009-2035	
	Current Policies Scenario			450 Scenario			CPS	450	CPS	450
Total generation	1 269	1 532	1 692	1 171	1 237	1 272	100	100	2.1	1.0
Coal	240	282	315	187	111	83	19	6	2.5	-2.6
Oil	9	2	1	10	2	1	0	0	-9.1	-8.8
Gas	604	748	834	507	383	283	49	22	2.2	-1.9
Nuclear	217	254	266	246	323	366	16	29	1.9	3.1
Hydro	181	210	226	196	270	279	13	22	1.0	1.8
Biomass and waste	7	18	25	10	77	132	1	10	9.1	16.2
Wind	6	10	15	9	52	96	1	8	37.2	47.4
Geothermal	4	8	9	5	17	28	1	2	11.9	17.1
Solar PV	0	0	1	1	2	3	0	0	n.a.	n.a.
CSP	-	-	-	-	-	-	-	-	n.a.	n.a.
Marine	-	-	-	-	0	0	-	0	n.a.	n.a.

	Electrical capacity (GW)						Shares (%)		CAAGR (%)	
	2020	2030	2035	2020	2030	2035	2035		2009-2035	
	Current Policies Scenario			450 Scenario			CPS	450	CPS	450
Total capacity	261	290	308	246	279	306	100	100	1.2	1.2
Coal	51	51	55	42	23	20	18	7	0.6	-3.2
Oil	5	2	1	5	2	1	0	0	-7.6	-7.4
Gas	117	134	140	101	96	91	46	30	1.3	-0.4
Nuclear	31	35	36	35	45	50	12	16	1.7	3.0
Hydro	52	59	63	56	75	78	21	25	1.1	1.9
Biomass and waste	2	3	4	2	13	22	1	7	6.1	12.7
Wind	2	4	6	3	20	37	2	12	25.2	34.4
Geothermal	1	1	1	1	3	4	0	1	11.1	15.8
Solar PV	0	1	1	1	2	4	0	1	n.a.	n.a.
CSP	-	-	-	-	-	-	-	-	n.a.	n.a.
Marine	-	-	-	-	0	0	-	0	n.a.	n.a.

	CO_2 emissions (Mt)						Shares (%)		CAAGR (%)	
	2020	2030	2035	2020	2030	2035	2035		2009-2035	
	Current Policies Scenario			450 Scenario			CPS	450	CPS	450
Total CO_2	1 732	1 915	2 046	1 551	1 232	1 102	100	100	1.2	-1.2
Coal	470	525	561	401	230	172	27	16	1.3	-3.2
Oil	355	364	379	338	314	302	19	27	0.6	-0.3
Gas	907	1 026	1 106	811	688	627	54	57	1.3	-0.9
Power generation	914	1 015	1 093	788	520	412	100	100	1.2	-2.5
Coal	354	407	441	290	154	109	40	26	1.8	-3.5
Oil	36	24	23	35	23	22	2	5	-2.3	-2.5
Gas	525	584	629	464	343	282	58	68	1.0	-2.0
TFC	740	811	857	689	638	617	100	100	1.1	-0.2
Coal	111	113	114	106	72	59	13	10	-0.2	-2.7
Oil	281	300	315	267	255	245	37	40	0.9	-0.1
Transport	*188*	*207*	*220*	*179*	*175*	*170*	*26*	*27*	*1.2*	*0.2*
Gas	349	398	428	316	312	313	50	51	1.6	0.4

A

Non-OECD Asia: New Policies Scenario

	\multicolumn{7}{c}{Energy demand (Mtoe)}	Shares (%)		CAAGR (%)
	1990	2009	2015	2020	2025	2030	2035	2009	2035	2009-2035
TPED	**1 591**	**3 724**	**4 761**	**5 341**	**5 775**	**6 226**	**6 711**	**100**	**100**	**2.3**
Coal	697	1 942	2 484	2 669	2 744	2 826	2 929	52	44	1.6
Oil	311	794	991	1 083	1 188	1 307	1 416	21	21	2.2
Gas	71	293	437	565	657	761	877	8	13	4.3
Nuclear	10	35	92	177	239	283	325	1	5	9.0
Hydro	24	74	108	135	161	178	188	2	3	3.7
Biomass and waste	471	549	575	603	634	670	712	15	11	1.0
Other renewables	7	37	76	110	151	201	264	1	4	7.8
Power generation	**331**	**1 413**	**1 937**	**2 311**	**2 599**	**2 897**	**3 236**	**100**	**100**	**3.2**
Coal	229	1 101	1 462	1 621	1 714	1 816	1 939	78	60	2.2
Oil	45	47	33	27	22	20	19	3	1	-3.5
Gas	16	122	163	213	253	303	365	9	11	4.3
Nuclear	10	35	92	177	239	283	325	2	10	9.0
Hydro	24	74	108	135	161	178	188	5	6	3.7
Biomass and waste	0	6	18	47	83	126	177	0	5	14.2
Other renewables	7	29	62	92	128	171	224	2	7	8.2
Other energy sector	**162**	**412**	**500**	**537**	**559**	**584**	**610**	**100**	**100**	**1.5**
Electricity	*24*	*83*	*115*	*138*	*159*	*180*	*201*	*20*	*33*	*3.5*
TFC	**1 223**	**2 451**	**3 137**	**3 492**	**3 761**	**4 037**	**4 319**	**100**	**100**	**2.2**
Coal	393	636	778	793	773	747	721	26	17	0.5
Oil	240	686	893	992	1 105	1 229	1 340	28	31	2.6
Gas	32	129	218	294	345	397	452	5	10	4.9
Electricity	85	404	613	771	898	1 030	1 174	17	27	4.2
Heat	14	55	74	78	75	71	68	2	2	0.8
Biomass and waste	460	532	546	545	541	533	524	22	12	-0.1
Other renewables	0	8	14	19	24	30	40	0	1	6.2
Industry	**397**	**978**	**1 338**	**1 512**	**1 594**	**1 661**	**1 731**	**100**	**100**	**2.2**
Coal	236	502	640	657	645	630	617	51	36	0.8
Oil	54	100	119	121	122	121	118	10	7	0.7
Gas	9	55	102	155	181	202	222	6	13	5.6
Electricity	51	237	369	461	527	587	649	24	37	4.0
Heat	11	36	51	54	50	46	42	4	2	0.6
Biomass and waste	36	48	58	64	70	76	82	5	5	2.0
Other renewables	-	0	0	0	0	0	0	0	0	3.3
Transport	**114**	**331**	**480**	**566**	**676**	**806**	**933**	**100**	**100**	**4.1**
Oil	100	315	458	536	635	750	858	95	92	3.9
Electricity	1	4	7	10	13	17	22	1	2	6.6
Biofuels	-	2	7	11	16	22	31	1	3	10.7
Other fuels	12	10	8	10	12	17	23	3	2	3.3
Buildings	**600**	**863**	**983**	**1 051**	**1 108**	**1 169**	**1 239**	**100**	**100**	**1.4**
Coal	111	87	92	86	79	69	57	10	5	-1.6
Oil	33	91	101	101	102	101	99	11	8	0.3
Gas	5	36	63	83	103	126	152	4	12	5.8
Electricity	24	140	208	267	321	384	454	16	37	4.6
Heat	3	19	23	24	25	26	26	2	2	1.1
Biomass and waste	423	482	482	471	455	435	412	56	33	-0.6
Other renewables	0	8	14	18	23	29	39	1	3	6.3
Other	**112**	**279**	**336**	**363**	**383**	**401**	**416**	**100**	**100**	**1.5**

Non-OECD Asia: Current Policies and 450 Scenarios

	Energy demand (Mtoe)						Shares (%)		CAAGR (%)	
	2020	2030	2035	2020	2030	2035	2035		2009-2035	
	Current Policies Scenario			450 Scenario			CPS	450	CPS	450
TPED	**5 543**	**6 785**	**7 478**	**5 081**	**5 411**	**5 653**	**100**	**100**	**2.7**	**1.6**
Coal	2 901	3 498	3 855	2 392	1 815	1 621	52	29	2.7	-0.7
Oil	1 111	1 375	1 510	1 040	1 158	1 180	20	21	2.5	1.5
Gas	554	765	896	560	747	849	12	15	4.4	4.2
Nuclear	171	245	266	197	430	532	4	9	8.1	11.1
Hydro	127	153	164	143	198	214	2	4	3.1	4.2
Biomass and waste	585	607	620	617	743	819	8	14	0.5	1.6
Other renewables	94	141	168	132	319	439	2	8	6.0	9.9
Power generation	**2 456**	**3 289**	**3 782**	**2 143**	**2 333**	**2 531**	**100**	**100**	**3.9**	**2.3**
Coal	1 822	2 381	2 727	1 401	933	784	72	31	3.5	-1.3
Oil	32	26	24	25	17	15	1	1	-2.6	-4.2
Gas	187	278	346	207	296	350	9	14	4.1	4.2
Nuclear	171	245	266	197	430	532	7	21	8.1	11.1
Hydro	127	153	164	143	198	214	4	8	3.1	4.2
Biomass and waste	40	87	115	59	178	250	3	10	12.3	15.7
Other renewables	77	118	141	111	281	386	4	15	6.3	10.5
Other energy sector	**553**	**633**	**675**	**513**	**514**	**516**	**100**	**100**	**1.9**	**0.9**
Electricity	*144*	*196*	*225*	*129*	*152*	*163*	*33*	*32*	*3.9*	*2.6*
TFC	**3 571**	**4 268**	**4 641**	**3 366**	**3 686**	**3 822**	**100**	**100**	**2.5**	**1.7**
Coal	814	820	814	748	655	613	18	16	1.0	-0.1
Oil	1 015	1 295	1 435	951	1 085	1 111	31	29	2.9	1.9
Gas	309	426	491	297	393	442	11	12	5.3	4.8
Electricity	798	1 111	1 298	728	898	986	28	26	4.6	3.5
Heat	83	84	83	74	64	60	2	2	1.6	0.3
Biomass and waste	534	509	495	547	554	558	11	15	-0.3	0.2
Other renewables	17	23	27	21	37	53	1	1	4.6	7.3
Industry	**1 554**	**1 803**	**1 934**	**1 452**	**1 530**	**1 555**	**100**	**100**	**2.7**	**1.8**
Coal	674	695	702	619	552	524	36	34	1.3	0.2
Oil	123	125	126	117	116	112	7	7	0.9	0.4
Gas	165	223	252	164	216	234	13	15	6.1	5.8
Electricity	475	636	728	437	524	560	38	36	4.4	3.4
Heat	58	57	55	51	42	38	3	2	1.7	0.2
Biomass and waste	59	66	70	64	77	85	4	5	1.4	2.2
Other renewables	0	0	0	0	1	2	0	0	3.3	13.4
Transport	**580**	**852**	**1 001**	**543**	**720**	**792**	**100**	**100**	**4.3**	**3.4**
Oil	550	799	930	505	626	655	93	83	4.3	2.9
Electricity	10	16	20	11	23	38	2	5	6.3	8.9
Biofuels	10	19	26	15	45	62	3	8	10.0	13.8
Other fuels	10	19	25	11	25	37	3	5	3.6	5.2
Buildings	**1 071**	**1 203**	**1 279**	**1 010**	**1 042**	**1 067**	**100**	**100**	**1.5**	**0.8**
Coal	90	74	63	80	57	44	5	4	-1.3	-2.6
Oil	107	111	110	97	87	82	9	8	0.7	-0.4
Gas	87	131	156	75	100	116	12	11	5.9	4.6
Electricity	280	414	498	247	310	344	39	32	5.0	3.5
Heat	25	27	27	22	22	21	2	2	1.3	0.3
Biomass and waste	465	424	398	468	431	411	31	38	-0.7	-0.6
Other renewables	16	22	26	21	35	49	2	5	4.6	7.2
Other	**367**	**409**	**428**	**362**	**394**	**408**	**100**	**100**	**1.7**	**1.5**

A

Non-OECD Asia: New Policies Scenario

Electricity generation (TWh)	1990	2009	2015	2020	2025	2030	2035	Shares (%) 2009	Shares (%) 2035	CAAGR (%) 2009-2035
Total generation	1 271	5 660	8 459	10 565	12 286	14 055	15 977	100	100	4.1
Coal	729	3 852	5 547	6 398	6 953	7 616	8 386	68	52	3.0
Oil	163	158	103	86	72	65	64	3	0	-3.4
Gas	58	578	808	1 103	1 350	1 641	1 989	10	12	4.9
Nuclear	39	133	352	678	916	1 086	1 248	2	8	9.0
Hydro	274	859	1 250	1 568	1 871	2 065	2 181	15	14	3.7
Biomass and waste	0	14	55	160	282	428	601	0	4	15.6
Wind	0	46	284	463	650	835	1 021	1	6	12.7
Geothermal	7	20	30	42	58	76	96	0	1	6.3
Solar PV	0	1	25	58	116	207	305	0	2	27.2
CSP	-	-	4	8	16	35	84	-	1	n.a.
Marine	-	-	-	-	0	1	2	-	0	n.a.

Electrical capacity (GW)	2009	2015	2020	2025	2030	2035	Shares (%) 2009	Shares (%) 2035	CAAGR (%) 2009-2035
Total capacity	1 360	2 004	2 493	2 913	3 350	3 797	100	100	4.0
Coal	794	1 112	1 283	1 383	1 508	1 646	58	43	2.8
Oil	66	67	62	57	54	50	5	1	-1.0
Gas	151	219	280	336	404	486	11	13	4.6
Nuclear	19	48	90	122	145	166	1	4	8.8
Hydro	284	382	481	577	640	677	21	18	3.4
Biomass and waste	6	14	32	52	75	103	0	3	11.7
Wind	37	139	218	296	370	442	3	12	10.0
Geothermal	3	5	6	9	11	14	0	0	6.0
Solar PV	0	18	40	77	133	192	0	5	25.8
CSP	-	1	2	4	9	21	-	1	n.a.
Marine	-	-	-	0	0	1	-	0	n.a.

CO_2 emissions (Mt)	1990	2009	2015	2020	2025	2030	2035	Shares (%) 2009	Shares (%) 2035	CAAGR (%) 2009-2035
Total CO_2	3 527	9 990	12 950	14 195	14 938	15 771	16 687	100	100	2.0
Coal	2 540	7 321	9 438	10 131	10 353	10 586	10 900	73	65	1.5
Oil	858	2 025	2 544	2 799	3 108	3 466	3 799	20	23	2.4
Gas	130	644	968	1 265	1 478	1 718	1 989	6	12	4.4
Power generation	1 078	4 781	6 263	6 976	7 398	7 869	8 452	100	100	2.2
Coal	897	4 347	5 776	6 390	6 735	7 097	7 539	91	89	2.1
Oil	144	150	105	87	70	63	60	3	1	-3.5
Gas	37	284	382	498	593	709	853	6	10	4.3
TFC	2 298	4 818	6 217	6 714	7 036	7 376	7 689	100	100	1.8
Coal	1 583	2 793	3 445	3 502	3 394	3 264	3 133	58	41	0.4
Oil	656	1 748	2 297	2 559	2 873	3 224	3 541	36	46	2.8
Transport	*300*	*939*	*1 365*	*1 598*	*1 893*	*2 237*	*2 558*	*19*	*33*	*3.9*
Gas	60	277	475	653	768	888	1 015	6	13	5.1

Non-OECD Asia: Current Policies and 450 Scenarios

	Electricity generation (TWh)						Shares (%)		CAAGR (%)	
	2020	2030	2035	2020	2030	2035	2035		2009-2035	
	Current Policies Scenario			450 Scenario			CPS	450	CPS	450
Total generation	10 948	15 193	17 704	9 959	12 198	13 351	100	100	4.5	3.4
Coal	7 172	9 873	11 586	5 498	3 971	3 264	65	24	4.3	-0.6
Oil	99	80	75	82	58	55	0	0	-2.8	-4.0
Gas	940	1 436	1 796	1 079	1 664	1 999	10	15	4.5	4.9
Nuclear	656	942	1 020	756	1 651	2 040	6	15	8.1	11.1
Hydro	1 475	1 785	1 903	1 659	2 307	2 491	11	19	3.1	4.2
Biomass and waste	138	296	386	202	613	867	2	6	13.7	17.3
Wind	387	622	732	537	1 313	1 638	4	12	11.3	14.8
Geothermal	35	53	64	53	119	155	0	1	4.7	8.2
Solar PV	43	97	121	74	390	598	1	4	22.8	30.5
CSP	3	10	18	18	110	243	0	2	n.a.	n.a.
Marine	-	1	2	-	1	3	0	0	n.a.	n.a.

	Electrical capacity (GW)						Shares (%)		CAAGR (%)	
	2020	2030	2035	2020	2030	2035	2035		2009-2035	
	Current Policies Scenario			450 Scenario			CPS	450	CPS	450
Total capacity	2 509	3 396	3 886	2 444	3 327	3 726	100	100	4.1	4.0
Coal	1 388	1 866	2 153	1 138	952	863	55	23	3.9	0.3
Oil	63	56	52	62	53	48	1	1	-0.9	-1.2
Gas	270	388	469	282	414	466	12	13	4.5	4.4
Nuclear	87	125	135	101	220	274	3	7	7.9	10.9
Hydro	451	548	585	510	721	781	15	21	2.8	4.0
Biomass and waste	28	53	68	39	106	146	2	4	9.9	13.2
Wind	186	284	328	247	556	668	8	18	8.7	11.7
Geothermal	5	8	10	8	18	23	0	1	4.4	7.9
Solar PV	31	67	81	52	259	394	2	11	21.7	29.3
CSP	1	2	4	5	29	62	0	2	n.a.	n.a.
Marine	-	0	1	-	0	1	0	0	n.a.	n.a.

	CO_2 emissions (Mt)						Shares (%)		CAAGR (%)	
	2020	2030	2035	2020	2030	2035	2035		2009-2035	
	Current Policies Scenario			450 Scenario			CPS	450	CPS	450
Total CO_2	15 165	18 617	20 651	12 945	10 388	9 111	100	100	2.8	-0.4
Coal	11 041	13 209	14 525	9 025	5 756	4 238	70	47	2.7	-2.1
Oil	2 885	3 682	4 097	2 669	3 000	3 064	20	34	2.7	1.6
Gas	1 239	1 726	2 029	1 251	1 633	1 809	10	20	4.5	4.1
Power generation	7 733	10 101	11 600	6 051	3 702	2 579	100	100	3.5	-2.3
Coal	7 192	9 369	10 713	5 487	2 988	1 787	92	69	3.5	-3.4
Oil	103	83	77	81	54	50	1	2	-2.6	-4.1
Gas	437	649	810	483	660	742	7	29	4.1	3.8
TFC	6 909	7 951	8 457	6 411	6 229	6 082	100	100	2.2	0.9
Coal	3 595	3 580	3 537	3 315	2 595	2 292	42	38	0.9	-0.8
Oil	2 628	3 415	3 817	2 438	2 777	2 836	45	47	3.0	1.9
Transport	*1 641*	*2 383*	*2 774*	*1 505*	*1 868*	*1 956*	*33*	*32*	*4.3*	*2.9*
Gas	687	956	1 103	658	856	955	13	16	5.5	4.9

A

China: New Policies Scenario

	Energy demand (Mtoe)							Shares (%)		CAAGR (%)
	1990	2009	2015	2020	2025	2030	2035	2009	2035	2009-2035
TPED	872	2 271	3 002	3 345	3 522	3 687	3 835	100	100	2.0
Coal	534	1 525	1 925	2 004	1 988	1 976	1 974	67	51	1.0
Oil	113	383	533	586	638	686	703	17	18	2.4
Gas	13	78	165	251	306	363	420	3	11	6.7
Nuclear	-	18	63	142	188	221	249	1	6	10.6
Hydro	11	53	78	96	109	115	118	2	3	3.1
Biomass and waste	200	204	202	210	216	224	238	9	6	0.6
Other renewables	0	10	36	56	78	102	134	0	3	10.3
Power generation	181	920	1 319	1 568	1 719	1 867	2 022	100	100	3.1
Coal	153	822	1 093	1 182	1 213	1 249	1 294	89	64	1.8
Oil	16	7	8	7	6	5	5	1	0	-1.5
Gas	1	16	44	72	97	127	157	2	8	9.3
Nuclear	-	18	63	142	188	221	249	2	12	10.6
Hydro	11	53	78	96	109	115	118	6	6	3.1
Biomass and waste	-	1	9	30	50	72	98	0	5	18.2
Other renewables	0	2	23	39	57	77	101	0	5	15.3
Other energy sector	94	295	352	367	365	362	360	100	100	0.8
Electricity	*12*	*54*	*74*	*85*	*93*	*101*	*108*	*18*	*30*	*2.7*
TFC	668	1 441	1 914	2 122	2 233	2 328	2 400	100	100	2.0
Coal	316	519	621	610	573	532	491	36	20	-0.2
Oil	86	340	484	539	595	646	669	24	28	2.6
Gas	9	50	104	160	189	215	241	3	10	6.2
Electricity	43	267	425	539	616	687	761	19	32	4.1
Heat	13	55	73	77	74	70	67	4	3	0.8
Biomass and waste	200	202	193	180	166	152	140	14	6	-1.4
Other renewables	0	8	13	17	20	25	32	1	1	5.6
Industry	242	682	939	1 042	1 066	1 073	1 081	100	100	1.8
Coal	178	406	505	497	469	440	411	59	38	0.0
Oil	21	49	60	59	59	57	54	7	5	0.4
Gas	3	16	42	80	91	97	101	2	9	7.4
Electricity	30	175	282	352	397	435	473	26	44	3.9
Heat	11	36	51	53	49	45	42	5	4	0.6
Biomass and waste	-	-	-	-	-	-	-	-	-	n.a.
Other renewables	-	0	0	0	0	0	0	0	0	3.3
Transport	38	163	276	331	393	457	496	100	100	4.4
Oil	28	155	268	318	374	430	458	95	92	4.3
Electricity	1	3	5	7	10	14	18	2	4	7.4
Biofuels	-	1	2	5	8	12	16	1	3	10.5
Other fuels	10	4	0	1	1	2	3	2	1	-0.2
Buildings	316	443	521	559	579	599	621	100	100	1.3
Coal	96	68	69	63	55	45	33	15	5	-2.7
Oil	7	46	52	51	48	44	39	10	6	-0.6
Gas	2	25	49	65	81	98	116	6	19	6.0
Electricity	9	77	125	165	193	223	253	17	41	4.7
Heat	2	19	22	24	24	24	25	4	4	1.0
Biomass and waste	200	201	191	175	158	140	123	45	20	-1.9
Other renewables	0	8	13	16	20	25	32	2	5	5.6
Other	71	153	178	190	195	199	201	100	100	1.0

China: Current Policies and 450 Scenarios

	Energy demand (Mtoe)						Shares (%)		CAAGR (%)	
	2020	2030	2035	2020	2030	2035	2035		2009-2035	
	Current Policies Scenario			450 Scenario			CPS	450	CPS	450
TPED	**3 465**	**4 068**	**4 361**	**3 186**	**3 148**	**3 152**	**100**	**100**	**2.5**	**1.3**
Coal	2 149	2 431	2 596	1 817	1 283	1 074	60	34	2.1	-1.3
Oil	595	717	751	563	607	588	17	19	2.6	1.7
Gas	248	365	431	260	377	434	10	14	6.8	6.8
Nuclear	135	188	201	159	323	396	5	13	9.7	12.6
Hydro	93	107	112	99	117	121	3	4	2.9	3.2
Biomass and waste	199	188	186	223	278	310	4	10	-0.4	1.6
Other renewables	47	71	84	65	162	229	2	7	8.4	12.6
Power generation	**1 658**	**2 140**	**2 403**	**1 463**	**1 485**	**1 543**	**100**	**100**	**3.8**	**2.0**
Coal	1 305	1 627	1 816	1 039	667	527	76	34	3.1	-1.7
Oil	8	6	6	6	4	4	0	0	-0.9	-2.5
Gas	60	108	139	75	137	169	6	11	8.8	9.6
Nuclear	135	188	201	159	323	396	8	26	9.7	12.6
Hydro	93	107	112	99	117	121	5	8	2.9	3.2
Biomass and waste	26	51	66	39	105	140	3	9	16.4	19.8
Other renewables	32	52	62	46	132	186	3	12	13.2	18.1
Other energy sector	**379**	**401**	**413**	**352**	**310**	**291**	**100**	**100**	**1.3**	**-0.1**
Electricity	*89*	*112*	*124*	*80*	*83*	*82*	*30*	*28*	*3.3*	*1.7*
TFC	**2 169**	**2 485**	**2 621**	**2 043**	**2 099**	**2 090**	**100**	**100**	**2.3**	**1.4**
Coal	623	581	552	575	453	397	21	19	0.2	-1.0
Oil	548	680	721	517	568	554	28	27	2.9	1.9
Gas	169	235	269	167	220	244	10	12	6.7	6.3
Electricity	559	750	857	510	591	624	33	30	4.6	3.3
Heat	82	83	81	73	63	58	3	3	1.5	0.2
Biomass and waste	173	136	120	184	173	170	5	8	-2.0	-0.7
Other renewables	15	19	22	19	31	43	1	2	3.9	6.7
Industry	**1 076**	**1 189**	**1 245**	**1 003**	**977**	**949**	**100**	**100**	**2.3**	**1.3**
Coal	507	483	466	468	371	329	37	35	0.5	-0.8
Oil	60	59	57	58	57	52	5	6	0.6	0.2
Gas	86	113	125	92	119	125	10	13	8.3	8.2
Electricity	364	478	541	334	386	403	43	42	4.4	3.3
Heat	57	57	55	51	42	38	4	4	1.7	0.2
Biomass and waste	-	-	-	-	-	-	-	-	n.a.	n.a.
Other renewables	0	0	0	0	1	2	0	0	3.3	13.4
Transport	**335**	**480**	**534**	**318**	**418**	**442**	**100**	**100**	**4.7**	**3.9**
Oil	324	456	501	300	361	357	94	81	4.6	3.3
Electricity	7	13	17	9	18	29	3	7	7.1	9.5
Biofuels	4	9	12	8	30	40	2	9	9.2	14.3
Other fuels	1	2	4	1	8	15	1	3	0.3	5.6
Buildings	**566**	**612**	**634**	**534**	**512**	**507**	**100**	**100**	**1.4**	**0.5**
Coal	66	49	37	58	37	25	6	5	-2.3	-3.8
Oil	52	46	42	49	37	32	7	6	-0.3	-1.4
Gas	68	102	119	59	74	84	19	17	6.1	4.7
Electricity	173	243	282	153	172	177	44	35	5.1	3.3
Heat	24	26	26	22	21	20	4	4	1.2	0.2
Biomass and waste	169	127	107	176	143	129	17	25	-2.4	-1.7
Other renewables	15	19	21	18	29	41	3	8	4.0	6.6
Other	**191**	**204**	**208**	**188**	**192**	**192**	**100**	**100**	**1.2**	**0.9**

A

China: New Policies Scenario

Electricity generation (TWh)								Shares (%)		CAAGR (%)
	1990	2009	2015	2020	2025	2030	2035	2009	2035	2009-2035
Total generation	650	3 735	5 812	7 264	8 249	9 169	10 100	100	100	3.9
Coal	471	2 941	4 156	4 704	4 976	5 281	5 631	79	56	2.5
Oil	49	17	18	15	12	12	11	0	0	-1.5
Gas	3	62	204	355	499	673	843	2	8	10.6
Nuclear	-	70	241	544	721	850	956	2	9	10.6
Hydro	127	616	909	1 112	1 266	1 337	1 375	16	14	3.1
Biomass and waste	-	2	30	109	178	250	339	0	3	21.1
Wind	0	27	237	388	526	642	735	1	7	13.6
Geothermal	-	0	1	3	6	10	14	0	0	19.1
Solar PV	0	0	14	29	54	89	127	0	1	25.9
CSP	-	-	3	6	11	26	67	-	1	n.a.
Marine	-	-	-	-	0	1	1	-	0	n.a.

Electrical capacity (GW)							Shares (%)		CAAGR (%)
	2009	2015	2020	2025	2030	2035	2009	2035	2009-2035
Total capacity	931	1 379	1 728	1 970	2 179	2 378	100	100	3.7
Coal	650	859	984	1 044	1 099	1 159	70	49	2.3
Oil	15	15	14	12	12	12	2	1	-0.9
Gas	33	73	107	135	169	204	4	9	7.2
Nuclear	9	32	71	94	111	125	1	5	10.7
Hydro	197	270	330	376	397	408	21	17	2.8
Biomass and waste	1	6	20	32	44	58	0	2	17.1
Wind	26	114	180	236	280	312	3	13	10.1
Geothermal	0	0	0	1	1	2	0	0	18.1
Solar PV	0	10	20	36	58	81	0	3	24.0
CSP	-	1	2	3	6	16	-	1	n.a.
Marine	-	-	-	0	0	0	-	0	n.a.

CO_2 emissions (Mt)								Shares (%)		CAAGR (%)
	1990	2009	2015	2020	2025	2030	2035	2009	2035	2009-2035
Total CO_2	2 244	6 877	9 065	9 727	9 920	10 113	10 253	100	100	1.5
Coal	1 914	5 751	7 332	7 638	7 547	7 456	7 398	84	72	1.0
Oil	305	958	1 366	1 522	1 681	1 833	1 900	14	19	2.7
Gas	26	169	367	566	693	824	955	2	9	6.9
Power generation	652	3 324	4 467	4 872	5 027	5 200	5 404	100	100	1.9
Coal	598	3 262	4 336	4 679	4 779	4 884	5 021	98	93	1.7
Oil	52	25	27	24	20	19	17	1	0	-1.6
Gas	2	36	103	168	227	298	367	1	7	9.3
TFC	1 507	3 289	4 280	4 516	4 562	4 578	4 510	100	100	1.2
Coal	1 265	2 315	2 790	2 741	2 562	2 369	2 174	70	48	-0.2
Oil	225	863	1 257	1 411	1 571	1 722	1 789	26	40	2.8
Transport	*83*	*463*	*798*	*949*	*1 115*	*1 279*	*1 365*	*14*	*30*	*4.2*
Gas	17	110	233	364	429	487	548	3	12	6.4

China: Current Policies and 450 Scenarios

Electricity generation (TWh)							Shares (%)		CAAGR (%)	
	2020	2030	2035	2020	2030	2035	2035		2009-2035	
	Current Policies Scenario			450 Scenario			CPS	450	CPS	450
Total generation	7 537	10 023	11 407	6 858	7 840	8 216	100	100	4.4	3.1
Coal	5 194	6 766	7 749	4 079	2 808	2 127	68	26	3.8	-1.2
Oil	16	13	11	14	10	9	0	0	-1.4	-2.4
Gas	286	541	704	379	763	962	6	12	9.8	11.1
Nuclear	520	723	772	611	1 238	1 519	7	18	9.7	12.6
Hydro	1 079	1 249	1 302	1 146	1 364	1 402	11	17	2.9	3.2
Biomass and waste	95	183	231	138	366	491	2	6	19.3	22.8
Wind	318	492	560	441	977	1 168	5	14	12.4	15.6
Geothermal	2	5	7	3	14	22	0	0	15.8	21.0
Solar PV	23	43	53	35	222	340	0	4	21.7	30.7
CSP	3	8	16	13	77	175	0	2	n.a.	n.a.
Marine	-	1	2	-	0	1	0	0	n.a.	n.a.

Electrical capacity (GW)							Shares (%)		CAAGR (%)	
	2020	2030	2035	2020	2030	2035	2035		2009-2035	
	Current Policies Scenario			450 Scenario			CPS	450	CPS	450
Total capacity	1 764	2 294	2 563	1 687	2 120	2 280	100	100	4.0	3.5
Coal	1 072	1 364	1 532	886	711	616	60	27	3.4	-0.2
Oil	15	13	12	14	12	11	0	0	-0.8	-1.3
Gas	103	164	201	113	184	206	8	9	7.2	7.3
Nuclear	68	94	101	80	161	198	4	9	9.8	12.6
Hydro	320	370	386	340	405	416	15	18	2.6	2.9
Biomass and waste	18	32	40	25	63	83	2	4	15.5	18.8
Wind	151	223	249	201	411	476	10	21	9.1	11.9
Geothermal	0	1	1	0	2	3	0	0	15.2	19.8
Solar PV	17	30	36	25	151	230	1	10	20.2	29.0
CSP	1	2	4	3	19	41	0	2	n.a.	n.a.
Marine	-	0	0	-	0	0	0	0	n.a.	n.a.

CO$_2$ emissions (Mt)							Shares (%)		CAAGR (%)	
	2020	2030	2035	2020	2030	2035	2035		2009-2035	
	Current Policies Scenario			450 Scenario			CPS	450	CPS	450
Total CO$_2$	10 314	12 014	12 897	8 921	6 398	4 979	100	100	2.4	-1.2
Coal	8 207	9 254	9 862	6 882	3 994	2 539	76	51	2.1	-3.1
Oil	1 549	1 931	2 053	1 453	1 590	1 548	16	31	3.0	1.9
Gas	558	829	981	585	814	893	8	18	7.0	6.6
Power generation	5 339	6 699	7 507	4 278	2 329	1 243	100	100	3.2	-3.7
Coal	5 172	6 423	7 161	4 082	2 019	899	95	72	3.1	-4.8
Oil	26	22	20	22	15	13	0	1	-0.9	-2.6
Gas	140	253	327	174	295	331	4	27	8.8	8.9
TFC	4 622	4 946	5 004	4 319	3 788	3 469	100	100	1.6	0.2
Coal	2 803	2 594	2 453	2 596	1 823	1 504	49	43	0.2	-1.6
Oil	1 436	1 816	1 938	1 345	1 482	1 440	39	42	3.2	2.0
Transport	964	1 358	1 493	894	1 076	1 065	30	31	4.6	3.3
Gas	384	536	613	378	482	524	12	15	6.8	6.2

A

India: New Policies Scenario

	Energy demand (Mtoe)							Shares (%)		CAAGR (%)
	1990	2009	2015	2020	2025	2030	2035	2009	2035	2009-2035
TPED	319	669	810	945	1 092	1 256	1 464	100	100	3.1
Coal	106	280	363	434	490	544	618	42	42	3.1
Oil	61	159	178	202	237	288	356	24	24	3.1
Gas	11	49	63	82	100	124	154	7	11	4.5
Nuclear	2	5	12	17	29	38	48	1	3	9.2
Hydro	6	9	13	18	25	29	30	1	2	4.7
Biomass and waste	133	165	176	184	195	208	220	25	15	1.1
Other renewables	0	2	5	9	16	25	36	0	2	12.2
Power generation	73	253	317	392	471	553	657	100	100	3.7
Coal	58	203	248	292	327	360	412	80	63	2.8
Oil	4	11	9	7	5	5	4	4	1	-3.6
Gas	3	23	28	44	56	72	91	9	14	5.4
Nuclear	2	5	12	17	29	38	48	2	7	9.2
Hydro	6	9	13	18	25	29	30	4	5	4.7
Biomass and waste	-	1	3	7	15	26	38	0	6	14.4
Other renewables	0	2	5	8	14	23	33	1	5	12.5
Other energy sector	20	59	79	98	118	140	166	100	100	4.1
Electricity	*7*	*18*	*27*	*36*	*46*	*57*	*69*	*30*	*42*	*5.4*
TFC	252	435	528	604	690	794	923	100	100	2.9
Coal	42	60	89	107	119	131	143	14	16	3.4
Oil	53	129	149	176	214	264	331	30	36	3.7
Gas	6	21	29	31	35	42	51	5	6	3.5
Electricity	18	60	88	113	141	173	212	14	23	4.9
Heat	-	-	-	-	-	-	-	-	-	n.a.
Biomass and waste	133	164	173	177	180	181	182	38	20	0.4
Other renewables	0	0	1	1	1	2	4	0	0	10.6
Industry	70	132	183	219	251	285	321	100	100	3.5
Coal	29	45	72	91	103	116	130	34	40	4.1
Oil	10	23	28	30	32	33	34	18	11	1.4
Gas	0	7	9	9	12	15	18	5	6	4.0
Electricity	9	28	42	56	69	85	102	21	32	5.1
Heat	-	-	-	-	-	-	-	-	-	n.a.
Biomass and waste	23	29	31	33	35	37	38	22	12	1.1
Other renewables	-	-	-	-	-	-	-	-	-	n.a.
Transport	27	51	57	74	104	147	211	100	100	5.6
Oil	24	48	52	68	95	134	192	94	91	5.4
Electricity	0	1	2	2	2	2	3	2	1	3.5
Biofuels	-	0	1	2	3	5	8	0	4	16.1
Other fuels	2	2	2	3	4	6	9	4	4	6.1
Buildings	137	195	214	227	242	257	276	100	100	1.3
Coal	11	15	16	16	16	15	14	7	5	-0.2
Oil	11	24	27	29	32	36	40	12	14	2.0
Gas	0	0	1	1	2	4	5	0	2	23.3
Electricity	4	21	29	38	48	62	78	11	28	5.2
Heat	-	-	-	-	-	-	-	-	-	n.a.
Biomass and waste	111	136	141	142	142	140	136	69	49	0.0
Other renewables	0	0	1	1	1	2	3	0	1	9.9
Other	17	56	74	85	94	104	114	100	100	2.8

India: Current Policies and 450 Scenarios

	Energy demand (Mtoe)						Shares (%)		CAAGR (%)	
	2020	2030	2035	2020	2030	2035	2035		2009-2035	
	Current Policies Scenario			450 Scenario			CPS	450	CPS	450
TPED	**1 000**	**1 372**	**1 622**	**877**	**1 078**	**1 223**	**100**	**100**	**3.5**	**2.3**
Coal	489	680	804	372	349	365	50	30	4.1	1.0
Oil	214	312	388	196	254	289	24	24	3.5	2.3
Gas	77	118	146	75	122	145	9	12	4.3	4.3
Nuclear	17	33	41	20	66	90	3	7	8.5	11.9
Hydro	14	20	23	20	40	48	1	4	3.6	6.6
Biomass and waste	181	196	205	183	211	230	13	19	0.8	1.3
Other renewables	7	13	16	12	36	57	1	5	8.7	14.2
Power generation	**432**	**633**	**767**	**342**	**420**	**495**	**100**	**100**	**4.4**	**2.6**
Coal	343	481	578	238	174	169	75	34	4.1	-0.7
Oil	10	9	8	6	3	3	1	1	-1.2	-4.8
Gas	35	61	78	38	70	83	10	17	4.8	5.0
Nuclear	17	33	41	20	66	90	5	18	8.5	11.9
Hydro	14	20	23	20	40	48	3	10	3.6	6.6
Biomass and waste	5	18	26	9	32	50	3	10	12.8	15.6
Other renewables	7	11	14	11	33	52	2	11	8.7	14.5
Other energy sector	**100**	**147**	**175**	**92**	**127**	**147**	**100**	**100**	**4.3**	**3.6**
Electricity	*38*	*61*	*75*	*33*	*49*	*59*	*43*	*40*	*5.7*	*4.7*
TFC	**621**	**833**	**977**	**581**	**733**	**822**	**100**	**100**	**3.2**	**2.5**
Coal	110	143	160	100	124	136	16	17	3.8	3.2
Oil	185	284	358	171	234	269	37	33	4.0	2.9
Gas	34	46	56	30	41	50	6	6	3.8	3.3
Electricity	115	180	223	104	152	182	23	22	5.1	4.3
Heat	-	-	-	-	-	-	-	-	n.a.	n.a.
Biomass and waste	176	178	178	174	179	180	18	22	0.3	0.4
Other renewables	1	2	2	1	3	4	0	1	8.9	11.3
Industry	**222**	**295**	**335**	**206**	**268**	**301**	**100**	**100**	**3.6**	**3.2**
Coal	93	127	146	86	112	126	43	42	4.6	4.0
Oil	30	34	35	28	31	32	10	11	1.5	1.2
Gas	11	16	19	9	13	16	6	5	4.2	3.6
Electricity	56	85	103	52	75	88	31	29	5.1	4.5
Heat	-	-	-	-	-	-	-	-	n.a.	n.a.
Biomass and waste	31	32	33	32	37	39	10	13	0.5	1.2
Other renewables	-	-	-	-	-	-	-	-	n.a.	n.a.
Transport	**81**	**164**	**234**	**74**	**129**	**166**	**100**	**100**	**6.0**	**4.6**
Oil	74	150	213	67	111	138	91	84	5.9	4.1
Electricity	2	2	2	2	4	6	1	3	3.3	6.7
Biofuels	2	5	8	2	7	11	3	7	15.9	17.6
Other fuels	3	7	10	3	7	10	4	6	6.6	6.5
Buildings	**233**	**269**	**292**	**217**	**233**	**242**	**100**	**100**	**1.6**	**0.8**
Coal	17	16	15	15	12	10	5	4	-0.0	-1.3
Oil	31	40	46	27	31	34	16	14	2.5	1.3
Gas	1	4	6	1	3	4	2	2	23.9	22.5
Electricity	40	67	86	34	50	60	29	25	5.6	4.1
Heat	-	-	-	-	-	-	-	-	n.a.	n.a.
Biomass and waste	143	141	138	140	134	130	47	54	0.1	-0.2
Other renewables	1	1	2	1	2	4	1	1	8.6	10.5
Other	**85**	**106**	**117**	**84**	**103**	**113**	**100**	**100**	**2.9**	**2.7**

A

India: New Policies Scenario

	Electricity generation (TWh)							Shares (%)		CAAGR (%)
	1990	2009	2015	2020	2025	2030	2035	2009	2035	2009-2035
Total generation	**289**	**899**	**1 319**	**1 723**	**2 162**	**2 671**	**3 264**	**100**	**100**	**5.1**
Coal	192	617	898	1 080	1 231	1 431	1 716	69	53	4.0
Oil	10	26	24	19	14	13	12	3	0	-2.9
Gas	10	111	148	243	319	418	536	12	16	6.2
Nuclear	6	19	44	65	110	145	184	2	6	9.2
Hydro	72	107	147	208	290	338	352	12	11	4.7
Biomass and waste	-	2	7	19	47	83	122	0	4	17.2
Wind	0	18	41	63	98	139	183	2	6	9.3
Geothermal	-	-	0	0	1	1	2	-	0	n.a.
Solar PV	-	0	9	23	49	94	140	0	4	39.0
CSP	-	-	1	2	5	9	15	-	0	n.a.
Marine	-	-	-	-	0	0	1	-	0	n.a.

	Electrical capacity (GW)						Shares (%)		CAAGR (%)
	2009	2015	2020	2025	2030	2035	2009	2035	2009-2035
Total capacity	**176**	**303**	**390**	**498**	**636**	**779**	**100**	**100**	**5.9**
Coal	92	174	201	221	267	325	52	42	5.0
Oil	7	8	8	8	8	7	4	1	-0.0
Gas	20	34	50	67	85	108	11	14	6.7
Nuclear	4	7	10	17	22	28	2	4	7.6
Hydro	39	49	68	95	111	115	22	15	4.2
Biomass and waste	2	3	4	9	14	20	1	3	9.6
Wind	11	23	33	49	67	85	6	11	8.2
Geothermal	-	0	0	0	0	0	-	0	n.a.
Solar PV	0	7	15	31	58	85	0	11	27.9
CSP	-	0	1	1	3	4	-	1	n.a.
Marine	-	-	-	0	0	0	-	0	n.a.

	CO_2 emissions (Mt)							Shares (%)		CAAGR (%)
	1990	2009	2015	2020	2025	2030	2035	2009	2035	2009-2035
Total CO_2	**593**	**1 548**	**1 924**	**2 286**	**2 618**	**3 004**	**3 535**	**100**	**100**	**3.2**
Coal	406	1 043	1 347	1 599	1 789	1 969	2 227	67	63	3.0
Oil	166	401	443	508	610	758	961	26	27	3.4
Gas	21	105	134	179	219	277	347	7	10	4.7
Power generation	**245**	**873**	**1 056**	**1 256**	**1 414**	**1 578**	**1 824**	**100**	**100**	**2.9**
Coal	226	785	961	1 131	1 268	1 396	1 598	90	88	2.8
Oil	11	33	29	23	15	14	13	4	1	-3.6
Gas	8	54	66	102	130	168	212	6	12	5.4
TFC	**330**	**625**	**808**	**958**	**1 122**	**1 327**	**1 594**	**100**	**100**	**3.7**
Coal	175	255	382	459	512	562	617	41	39	3.5
Oil	146	330	373	440	542	681	871	53	55	3.8
Transport	*74*	*145*	*156*	*204*	*285*	*404*	*577*	*23*	*36*	*5.4*
Gas	9	40	54	59	68	85	106	6	7	3.8

India: Current Policies and 450 Scenarios

	Electricity generation (TWh)						Shares (%)		CAAGR (%)	
	2020	2030	2035	2020	2030	2035	2035		2009-2035	
	Current Policies Scenario			450 Scenario			CPS	450	CPS	450
Total generation	1 769	2 796	3 449	1 586	2 335	2 790	100	100	5.3	4.4
Coal	1 229	1 878	2 312	910	754	731	67	26	5.2	0.7
Oil	25	23	22	16	10	9	1	0	-0.7	-4.0
Gas	192	351	460	212	429	512	13	18	5.6	6.1
Nuclear	67	126	156	75	253	346	5	12	8.5	11.9
Hydro	168	235	267	234	466	558	8	20	3.6	6.6
Biomass and waste	15	56	82	27	104	163	2	6	15.3	18.5
Wind	58	87	99	78	178	242	3	9	6.8	10.5
Geothermal	0	1	1	1	2	5	0	0	n.a.	n.a.
Solar PV	15	40	50	28	108	159	1	6	33.5	39.6
CSP	0	0	1	5	31	65	0	2	n.a.	n.a.
Marine	-	0	1	-	0	1	0	0	n.a.	n.a.

	Electrical capacity (GW)						Shares (%)		CAAGR (%)	
	2020	2030	2035	2020	2030	2035	2035		2009-2035	
	Current Policies Scenario			450 Scenario			CPS	450	CPS	450
Total capacity	362	574	699	380	629	763	100	100	5.5	5.8
Coal	200	313	386	168	164	166	55	22	5.7	2.3
Oil	8	8	7	8	7	7	1	1	-0.0	-0.4
Gas	45	80	102	50	97	115	15	15	6.5	6.9
Nuclear	10	19	23	12	38	51	3	7	6.8	10.2
Hydro	55	77	88	77	153	184	13	24	3.1	6.1
Biomass and waste	4	10	14	6	18	27	2	3	7.9	10.7
Wind	30	42	47	39	77	96	7	13	5.7	8.7
Geothermal	0	0	0	0	0	1	0	0	n.a.	n.a.
Solar PV	10	26	32	19	66	96	5	13	23.2	28.6
CSP	0	0	0	1	9	21	0	3	n.a.	n.a.
Marine	-	0	0	-	0	0	0	0	n.a.	n.a.

	CO_2 emissions (Mt)						Shares (%)		CAAGR (%)	
	2020	2030	2035	2020	2030	2035	2035		2009-2035	
	Current Policies Scenario			450 Scenario			CPS	450	CPS	450
Total CO_2	2 523	3 581	4 320	2 012	2 040	2 159	100	100	4.0	1.3
Coal	1 812	2 485	2 933	1 362	1 121	1 093	68	51	4.1	0.2
Oil	545	836	1 061	487	652	750	25	35	3.8	2.4
Gas	166	260	327	163	267	317	8	15	4.5	4.4
Power generation	1 444	2 034	2 446	1 028	798	771	100	100	4.0	-0.5
Coal	1 331	1 865	2 239	921	627	575	92	75	4.1	-1.2
Oil	31	27	24	19	10	9	1	1	-1.2	-4.8
Gas	82	142	183	88	161	187	7	24	4.8	4.9
TFC	1 004	1 443	1 750	914	1 153	1 287	100	100	4.0	2.8
Coal	471	608	681	432	484	506	39	39	3.9	2.7
Oil	467	742	954	425	586	678	55	53	4.2	2.8
Transport	*222*	*451*	*641*	*201*	*335*	*417*	*37*	*32*	*5.9*	*4.1*
Gas	66	93	115	57	82	102	7	8	4.1	3.7

A

Middle East: New Policies Scenario

	Energy demand (Mtoe)							Shares (%)		CAAGR (%)
	1990	2009	2015	2020	2025	2030	2035	2009	2035	2009-2035
TPED	208	589	705	775	856	936	1 000	100	100	2.1
Coal	1	1	2	2	2	2	2	0	0	1.9
Oil	133	299	360	383	407	418	428	51	43	1.4
Gas	72	287	336	376	425	483	520	49	52	2.3
Nuclear	-	-	2	5	8	11	13	-	1	n.a.
Hydro	1	1	3	3	4	4	5	0	0	5.6
Biomass and waste	0	1	1	2	4	6	10	0	1	10.7
Other renewables	0	0	2	4	6	12	23	0	2	21.0
Power generation	60	194	207	233	264	298	328	100	100	2.0
Coal	0	0	0	0	1	1	1	0	0	5.8
Oil	27	79	79	78	73	65	65	41	20	-0.8
Gas	32	113	122	144	172	204	218	58	66	2.5
Nuclear	-	-	2	5	8	11	13	-	4	n.a.
Hydro	1	1	3	3	4	4	5	1	1	5.6
Biomass and waste	-	0	1	1	2	5	8	0	2	37.3
Other renewables	0	0	0	2	4	9	19	0	6	30.2
Other energy sector	21	67	86	99	110	126	139	100	100	2.9
Electricity	*4*	*13*	*16*	*19*	*22*	*24*	*27*	*19*	*19*	*2.8*
TFC	147	393	494	540	594	641	678	100	100	2.1
Coal	0	0	0	1	1	1	1	0	0	2.2
Oil	102	214	276	296	323	342	350	55	52	1.9
Gas	29	126	149	163	176	189	204	32	30	1.9
Electricity	15	52	66	78	90	104	118	13	17	3.2
Heat	-	-	-	-	-	-	-	-	-	n.a.
Biomass and waste	0	1	1	1	1	1	2	0	0	4.2
Other renewables	0	0	1	2	2	3	4	0	1	13.9
Industry	43	110	133	143	150	157	163	100	100	1.5
Coal	0	0	0	1	1	1	1	0	0	2.3
Oil	20	41	48	50	50	48	47	37	29	0.5
Gas	20	58	70	76	82	87	92	53	56	1.8
Electricity	3	10	13	16	19	21	24	9	15	3.2
Heat	-	-	-	-	-	-	-	-	-	n.a.
Biomass and waste	0	0	0	0	0	0	0	0	0	1.4
Other renewables	-	0	0	0	0	0	0	0	0	3.3
Transport	47	111	153	168	193	210	219	100	100	2.7
Oil	47	107	149	164	187	204	211	97	96	2.6
Electricity	-	0	0	0	0	0	0	0	0	0.9
Biofuels	-	-	-	-	-	-	-	-	-	n.a.
Other fuels	-	3	4	5	5	7	8	3	4	3.6
Buildings	33	100	118	132	146	162	178	100	100	2.2
Coal	-	-	-	-	-	-	-	-	-	n.a.
Oil	17	21	23	23	23	23	22	21	12	0.1
Gas	3	40	44	48	53	57	63	40	35	1.7
Electricity	12	39	49	58	67	78	88	38	50	3.2
Heat	-	-	-	-	-	-	-	-	-	n.a.
Biomass and waste	0	0	0	1	1	1	1	0	1	5.3
Other renewables	0	0	1	2	2	3	4	0	2	14.0
Other	23	72	90	97	104	111	118	100	100	2.0

Middle East: Current Policies and 450 Scenarios

	Energy demand (Mtoe)						Shares (%)		CAAGR (%)	
	2020	2030	2035	2020	2030	2035	2035		2009-2035	
	Current Policies Scenario			450 Scenario			CPS	450	CPS	450
TPED	804	1 014	1 105	729	813	822	100	100	2.5	1.3
Coal	2	3	3	2	2	2	0	0	3.1	1.7
Oil	396	470	490	360	360	330	44	40	1.9	0.4
Gas	393	515	580	347	390	394	52	48	2.7	1.2
Nuclear	5	11	11	5	16	23	1	3	n.a.	n.a.
Hydro	3	4	4	3	5	5	0	1	5.5	5.7
Biomass and waste	2	4	6	5	15	22	1	3	8.4	14.2
Other renewables	3	8	11	5	25	46	1	6	17.8	24.3
Power generation	245	313	351	215	249	278	100	100	2.3	1.4
Coal	0	1	1	1	0	0	0	0	8.9	4.1
Oil	77	70	71	70	46	41	20	15	-0.4	-2.5
Gas	156	219	251	131	155	157	72	56	3.1	1.3
Nuclear	5	11	11	5	16	23	3	8	n.a.	n.a.
Hydro	3	4	4	3	5	5	1	2	5.5	5.7
Biomass and waste	1	3	4	1	6	10	1	3	33.7	38.6
Other renewables	2	5	9	3	21	42	3	15	26.5	34.3
Other energy sector	101	136	155	91	105	108	100	100	3.3	1.9
Electricity	*19*	*26*	*29*	*18*	*22*	*24*	*19*	*22*	*3.2*	*2.4*
TFC	557	702	757	513	574	568	100	100	2.6	1.4
Coal	1	1	1	1	1	1	0	0	2.9	2.8
Oil	309	389	407	282	303	276	54	49	2.5	1.0
Gas	166	198	217	152	164	168	29	30	2.1	1.1
Electricity	80	111	129	73	94	107	17	19	3.6	2.9
Heat	-	-	-	-	-	-	-	-	n.a.	n.a.
Biomass and waste	1	1	2	4	9	12	0	2	4.1	12.2
Other renewables	2	2	3	2	3	4	0	1	11.9	14.2
Industry	149	174	187	138	144	144	100	100	2.1	1.0
Coal	1	1	1	1	1	1	0	0	3.0	2.9
Oil	53	57	58	52	51	49	31	34	1.3	0.6
Gas	78	94	102	70	72	72	54	50	2.2	0.8
Electricity	16	23	27	15	20	22	14	15	3.7	2.9
Heat	-	-	-	-	-	-	-	-	n.a.	n.a.
Biomass and waste	0	0	0	0	0	0	0	0	1.4	2.4
Other renewables	0	0	0	0	0	0	0	0	3.3	3.3
Transport	174	242	256	158	185	163	100	100	3.3	1.5
Oil	169	236	249	150	168	141	97	86	3.3	1.0
Electricity	0	0	0	0	1	4	0	3	0.6	21.9
Biofuels	-	-	-	3	8	10	-	6	n.a.	n.a.
Other fuels	4	6	7	5	8	8	3	5	2.9	3.5
Buildings	135	170	190	123	142	153	100	100	2.5	1.6
Coal	-	-	-	-	-	-	-	-	n.a.	n.a.
Oil	24	24	23	22	20	19	12	12	0.4	-0.5
Gas	50	60	67	45	50	53	35	35	2.0	1.1
Electricity	59	82	96	54	68	75	50	49	3.6	2.6
Heat	-	-	-	-	-	-	-	-	n.a.	n.a.
Biomass and waste	1	1	1	1	1	2	1	1	5.1	5.7
Other renewables	2	2	3	2	3	4	1	3	12.0	14.2
Other	100	116	124	94	104	108	100	100	2.1	1.6

A

Middle East: New Policies Scenario

Electricity generation (TWh)								Shares (%)		CAAGR (%)
	1990	2009	2015	2020	2025	2030	2035	2009	2035	2009-2035
Total generation	219	743	947	1 115	1 293	1 483	1 669	100	100	3.2
Coal	-	0	1	2	3	2	2	0	0	7.3
Oil	101	300	302	302	290	265	267	40	16	-0.5
Gas	106	430	598	732	881	1 036	1 126	58	67	3.8
Nuclear	-	-	7	18	31	41	52	-	3	n.a.
Hydro	12	13	32	40	47	51	54	2	3	5.6
Biomass and waste	-	0	2	4	8	16	26	0	2	37.1
Wind	0	0	3	6	13	27	56	0	3	23.5
Geothermal	-	-	-	-	-	-	-	-	-	n.a.
Solar PV	-	-	2	6	12	24	38	-	2	n.a.
CSP	-	-	-	4	9	20	50	-	3	n.a.
Marine	-	-	-	-	-	-	-	-	-	n.a.

Electrical capacity (GW)							Shares (%)		CAAGR (%)
	2009	2015	2020	2025	2030	2035	2009	2035	2009-2035
Total capacity	205	295	327	347	369	414	100	100	2.7
Coal	0	0	0	1	0	0	0	0	2.2
Oil	65	79	81	78	70	68	32	17	0.2
Gas	127	197	216	226	237	251	62	61	2.7
Nuclear	-	1	2	4	6	7	-	2	n.a.
Hydro	12	15	20	23	25	26	6	6	2.8
Biomass and waste	0	0	1	1	2	4	0	1	21.5
Wind	0	1	2	5	11	22	0	5	23.4
Geothermal	-	-	-	-	-	-	-	-	n.a.
Solar PV	-	1	3	6	13	20	-	5	n.a.
CSP	-	-	1	3	6	15	-	4	n.a.
Marine	-	-	-	-	-	-	-	-	n.a.

CO_2 emissions (Mt)								Shares (%)		CAAGR (%)
	1990	2009	2015	2020	2025	2030	2035	2009	2035	2009-2035
Total CO_2	557	1 510	1 778	1 925	2 089	2 238	2 333	100	100	1.7
Coal	1	4	5	8	9	9	9	0	0	3.5
Oil	392	856	1 026	1 081	1 144	1 168	1 186	57	51	1.3
Gas	163	650	746	836	936	1 061	1 139	43	49	2.2
Power generation	159	513	534	584	637	685	717	100	100	1.3
Coal	0	1	2	4	6	5	5	0	1	5.6
Oil	85	247	246	243	230	204	203	48	28	-0.8
Gas	74	265	286	336	401	476	509	52	71	2.5
TFC	346	843	1 058	1 142	1 244	1 321	1 371	100	100	1.9
Coal	1	1	2	2	2	2	2	0	0	2.1
Oil	281	566	731	785	858	906	924	67	67	1.9
Transport	*141*	*318*	*441*	*485*	*555*	*604*	*624*	*38*	*46*	*2.6*
Gas	64	275	325	355	384	413	444	33	32	1.9

Middle East: Current Policies and 450 Scenarios

Electricity generation (TWh)	2020	2030	2035	2020	2030	2035	Shares (%) 2035		CAAGR (%) 2009-2035	
	Current Policies Scenario			450 Scenario			CPS	450	CPS	450
Total generation	1 142	1 584	1 829	1 047	1 337	1 514	100	100	3.5	2.8
Coal	2	4	5	2	1	1	0	0	10.5	5.1
Oil	302	288	294	272	182	165	16	11	-0.1	-2.3
Gas	766	1 150	1 357	684	853	853	74	56	4.5	2.7
Nuclear	18	41	41	18	62	88	2	6	n.a.	n.a.
Hydro	38	50	52	40	53	55	3	4	5.5	5.7
Biomass and waste	4	9	13	5	19	32	1	2	33.5	38.3
Wind	6	15	23	8	76	139	1	9	19.3	27.9
Geothermal	-	-	-	-	-	-	-	-	n.a.	n.a.
Solar PV	4	14	19	8	37	69	1	5	n.a.	n.a.
CSP	3	14	24	10	54	112	1	7	n.a.	n.a.
Marine	-	-	-	-	-	0	-	0	n.a.	n.a.

Electrical capacity (GW)	2020	2030	2035	2020	2030	2035	Shares (%) 2035		CAAGR (%) 2009-2035	
	Current Policies Scenario			450 Scenario			CPS	450	CPS	450
Total capacity	326	383	427	327	408	485	100	100	2.9	3.4
Coal	0	1	1	0	0	0	0	0	4.8	-0.1
Oil	82	74	74	79	58	51	17	11	0.5	-0.9
Gas	217	260	294	215	247	262	69	54	3.3	2.8
Nuclear	2	6	6	2	8	12	1	2	n.a.	n.a.
Hydro	18	24	25	19	25	26	6	5	2.8	2.9
Biomass and waste	1	1	2	1	3	5	0	1	18.4	22.6
Wind	2	6	9	3	32	58	2	12	19.0	28.0
Geothermal	-	-	-	-	-	-	-	-	n.a.	n.a.
Solar PV	2	8	10	4	19	36	2	7	n.a.	n.a.
CSP	1	4	7	3	16	34	2	7	n.a.	n.a.
Marine	-	-	-	-	-	0	-	0	n.a.	n.a.

CO_2 emissions (Mt)	2020	2030	2035	2020	2030	2035	Shares (%) 2035		CAAGR (%) 2009-2035	
	Current Policies Scenario			450 Scenario			CPS	450	CPS	450
Total CO_2	1 996	2 464	2 659	1 795	1 822	1 714	100	100	2.2	0.5
Coal	8	12	15	10	8	8	1	0	5.6	2.8
Oil	1 112	1 320	1 372	1 013	987	884	52	52	1.8	0.1
Gas	875	1 131	1 271	771	828	822	48	48	2.6	0.9
Power generation	612	740	819	533	495	478	100	100	1.8	-0.3
Coal	4	8	11	7	4	3	1	1	8.7	3.9
Oil	242	220	222	219	143	129	27	27	-0.4	-2.5
Gas	365	512	587	307	348	346	72	72	3.1	1.0
TFC	1 181	1 474	1 563	1 078	1 139	1 055	100	100	2.4	0.9
Coal	2	3	3	2	3	3	0	0	2.9	2.5
Oil	817	1 038	1 085	744	794	708	69	67	2.5	0.9
Transport	501	699	738	444	496	417	47	39	3.3	1.0
Gas	362	434	474	331	343	344	30	33	2.1	0.9

A

Africa: New Policies Scenario

	Energy demand (Mtoe)							Shares (%)		CAAGR (%)
	1990	2009	2015	2020	2025	2030	2035	2009	2035	2009-2035
TPED	391	665	739	790	835	878	915	100	100	1.2
Coal	74	106	119	125	129	130	126	16	14	0.7
Oil	89	151	167	171	173	173	178	23	19	0.6
Gas	30	83	94	109	119	129	136	13	15	1.9
Nuclear	2	3	3	3	8	14	15	1	2	6.1
Hydro	5	8	10	13	16	19	23	1	3	3.9
Biomass and waste	190	312	342	362	379	394	407	47	45	1.0
Other renewables	0	1	3	6	12	19	30	0	3	12.5
Power generation	69	138	157	178	198	216	233	100	100	2.0
Coal	39	63	72	77	78	76	73	46	31	0.5
Oil	11	20	18	16	13	10	9	14	4	-2.8
Gas	11	41	48	58	65	68	70	30	30	2.0
Nuclear	2	3	3	3	8	14	15	2	7	6.1
Hydro	5	8	10	13	16	19	23	6	10	3.9
Biomass and waste	0	1	3	5	7	11	15	0	7	13.1
Other renewables	0	1	3	6	11	18	28	1	12	12.5
Other energy sector	60	91	98	102	105	109	111	100	100	0.8
Electricity	*6*	*10*	*12*	*13*	*15*	*16*	*17*	*11*	*15*	*1.9*
TFC	289	491	550	587	619	650	680	100	100	1.3
Coal	20	19	23	23	23	23	22	4	3	0.6
Oil	70	122	142	150	158	166	174	25	26	1.4
Gas	9	28	29	32	34	37	41	6	6	1.5
Electricity	21	45	54	63	72	82	92	9	14	2.8
Heat	-	-	-	-	-	-	-	-	-	n.a.
Biomass and waste	169	277	302	318	330	341	349	56	51	0.9
Other renewables	0	0	0	0	1	1	2	0	0	13.4
Industry	60	84	99	105	109	115	121	100	100	1.4
Coal	14	10	13	13	13	13	13	12	11	0.9
Oil	14	12	13	14	14	14	14	15	12	0.5
Gas	5	14	15	17	18	20	22	16	18	1.7
Electricity	12	20	23	26	27	29	30	23	25	1.7
Heat	-	-	-	-	-	-	-	-	-	n.a.
Biomass and waste	16	28	34	36	37	39	41	33	34	1.5
Other renewables	-	-	-	-	-	-	-	-	-	n.a.
Transport	36	79	93	99	106	112	117	100	100	1.5
Oil	36	77	91	96	103	108	113	98	96	1.5
Electricity	0	0	1	1	1	1	1	1	1	2.0
Biofuels	-	-	1	1	1	1	1	-	1	n.a.
Other fuels	0	1	1	1	1	2	2	2	2	2.2
Buildings	178	302	331	352	372	390	407	100	100	1.1
Coal	3	8	8	8	8	8	8	3	2	-0.0
Oil	13	17	21	23	24	25	27	6	7	1.7
Gas	1	6	6	7	7	8	9	2	2	1.5
Electricity	8	23	28	34	41	49	57	8	14	3.6
Heat	-	-	-	-	-	-	-	-	-	n.a.
Biomass and waste	152	248	267	280	291	300	305	82	75	0.8
Other renewables	0	0	0	0	0	1	1	0	0	11.1
Other	15	26	28	30	32	33	35	100	100	1.2

Africa: Current Policies and 450 Scenarios

	Energy demand (Mtoe)						Shares (%)		CAAGR (%)	
	2020	2030	2035	2020	2030	2035	2035		2009-2035	
	Current Policies Scenario			450 Scenario			CPS	450	CPS	450
TPED	799	897	945	765	815	838	100	100	1.4	0.9
Coal	132	148	154	120	105	91	16	11	1.4	-0.6
Oil	173	177	184	157	138	131	19	16	0.8	-0.5
Gas	112	137	147	100	99	90	16	11	2.2	0.3
Nuclear	3	8	9	3	21	33	1	4	3.8	9.2
Hydro	12	18	21	13	20	24	2	3	3.6	4.1
Biomass and waste	361	394	408	363	397	412	43	49	1.0	1.1
Other renewables	6	14	22	8	35	58	2	7	11.2	15.5
Power generation	185	225	250	171	197	214	100	100	2.3	1.7
Coal	83	91	97	73	56	43	39	20	1.7	-1.5
Oil	17	11	10	16	10	9	4	4	-2.5	-2.9
Gas	61	75	79	51	43	29	32	14	2.5	-1.3
Nuclear	3	8	9	3	21	33	4	15	3.8	9.2
Hydro	12	18	21	13	20	24	8	11	3.6	4.1
Biomass and waste	3	9	14	6	14	20	5	9	12.7	14.3
Other renewables	5	13	21	8	34	56	8	26	11.2	15.5
Other energy sector	103	112	115	100	103	103	100	100	0.9	0.5
Electricity	*14*	*17*	*18*	*13*	*15*	*15*	*16*	*15*	*2.2*	*1.5*
TFC	591	661	696	570	606	622	100	100	1.4	0.9
Coal	23	23	23	22	20	19	3	3	0.7	-0.1
Oil	152	171	181	137	131	127	26	20	1.5	0.1
Gas	32	38	42	31	36	40	6	6	1.6	1.4
Electricity	65	86	98	62	77	85	14	14	3.1	2.5
Heat	-	-	-	-	-	-	-	-	n.a.	n.a.
Biomass and waste	318	343	351	318	341	349	50	56	0.9	0.9
Other renewables	0	1	1	0	1	2	0	0	11.7	14.8
Industry	106	118	125	102	107	111	100	100	1.5	1.1
Coal	13	13	13	13	11	10	11	9	1.0	-0.2
Oil	14	15	16	12	11	11	13	10	0.9	-0.5
Gas	17	19	21	16	18	20	17	18	1.7	1.5
Electricity	27	31	33	25	27	28	27	26	2.1	1.4
Heat	-	-	-	-	-	-	-	-	n.a.	n.a.
Biomass and waste	36	39	41	36	39	41	33	37	1.5	1.6
Other renewables	-	-	-	-	-	-	-	-	n.a.	n.a.
Transport	99	113	120	90	86	82	100	100	1.6	0.2
Oil	97	110	117	85	78	71	97	87	1.6	-0.3
Electricity	1	1	1	1	1	2	1	2	1.4	5.3
Biofuels	0	0	0	2	5	6	0	7	n.a.	n.a.
Other fuels	1	2	2	2	2	3	2	4	2.4	3.8
Buildings	355	396	414	348	380	394	100	100	1.2	1.0
Coal	8	8	8	8	7	7	2	2	0.1	-0.3
Oil	23	26	28	22	24	25	7	6	1.9	1.5
Gas	7	9	10	6	8	9	2	2	1.8	1.3
Electricity	35	51	60	33	45	51	14	13	3.8	3.2
Heat	-	-	-	-	-	-	-	-	n.a.	n.a.
Biomass and waste	281	302	308	278	296	300	74	76	0.8	0.7
Other renewables	0	0	1	0	1	1	0	0	9.4	13.0
Other	30	34	36	30	33	35	100	100	1.3	1.2

A

Africa: New Policies Scenario

	Electricity generation (TWh)							Shares (%)		CAAGR (%)
	1990	2009	2015	2020	2025	2030	2035	2009	2035	2009-2035
Total generation	316	630	762	885	999	1 124	1 259	100	100	2.7
Coal	165	250	291	321	328	330	327	40	26	1.0
Oil	41	79	73	67	55	45	41	13	3	-2.5
Gas	45	186	245	294	329	355	380	29	30	2.8
Nuclear	8	13	13	13	32	53	59	2	5	6.1
Hydro	56	98	120	148	183	222	266	16	21	3.9
Biomass and waste	0	1	7	15	24	36	50	0	4	17.4
Wind	-	2	6	10	16	25	41	0	3	13.1
Geothermal	0	1	2	4	7	10	14	0	1	9.3
Solar PV	-	0	2	6	13	24	38	0	3	32.3
CSP	-	-	1	6	13	24	45	-	4	n.a.
Marine	-	-	-	-	-	-	-	-	-	n.a.

	Electrical capacity (GW)						Shares (%)		CAAGR (%)
	2009	2015	2020	2025	2030	2035	2009	2035	2009-2035
Total capacity	141	185	210	238	272	313	100	100	3.1
Coal	41	48	55	59	62	68	29	22	1.9
Oil	24	28	27	23	20	18	17	6	-1.1
Gas	47	69	74	80	86	91	33	29	2.6
Nuclear	2	2	2	4	7	8	1	3	5.8
Hydro	25	31	39	47	57	68	18	22	4.0
Biomass and waste	1	2	3	5	6	8	1	3	9.0
Wind	1	3	5	7	11	17	1	6	12.7
Geothermal	0	0	1	1	2	2	0	1	10.0
Solar PV	0	2	4	8	14	21	0	7	32.7
CSP	-	0	2	3	6	11	-	4	n.a.
Marine	-	-	-	-	-	-	-	-	n.a.

	CO_2 emissions (Mt)							Shares (%)		CAAGR (%)
	1990	2009	2015	2020	2025	2030	2035	2009	2035	2009-2035
Total CO_2	545	928	1 054	1 127	1 154	1 165	1 170	100	100	0.9
Coal	235	322	369	392	384	365	337	35	29	0.2
Oil	248	426	481	500	513	527	545	46	47	1.0
Gas	62	179	203	234	257	274	288	19	25	1.8
Power generation	212	405	449	487	485	464	438	100	100	0.3
Coal	152	246	279	299	292	272	246	61	56	-0.0
Oil	35	62	57	51	41	33	30	15	7	-2.8
Gas	25	97	113	136	151	159	163	24	37	2.0
TFC	302	485	558	589	616	644	672	100	100	1.3
Coal	83	76	90	91	91	90	89	16	13	0.6
Oil	201	349	405	428	451	473	493	72	73	1.3
Transport	*105*	*230*	*270*	*287*	*305*	*322*	*337*	*47*	*50*	*1.5*
Gas	18	60	63	69	74	81	90	12	13	1.6

Africa: Current Policies and 450 Scenarios

Electricity generation (TWh)							Shares (%)		CAAGR (%)	
	2020	2030	2035	2020	2030	2035	2035		2009-2035	
	Current Policies Scenario			450 Scenario			CPS	450	CPS	450
Total generation	912	1 181	1 341	863	1 051	1 157	100	100	3.0	2.4
Coal	345	402	440	307	246	181	33	16	2.2	-1.2
Oil	70	48	44	67	44	40	3	3	-2.2	-2.6
Gas	307	402	439	265	236	177	33	15	3.4	-0.2
Nuclear	13	32	34	13	81	126	3	11	3.8	9.2
Hydro	142	206	245	154	232	277	18	24	3.6	4.1
Biomass and waste	10	30	45	18	45	66	3	6	16.9	18.7
Wind	10	21	28	12	48	97	2	8	11.5	16.9
Geothermal	4	8	11	4	14	22	1	2	8.5	11.2
Solar PV	6	18	27	10	39	61	2	5	30.6	34.8
CSP	4	15	28	14	66	109	2	9	n.a.	n.a.
Marine	-	-	-	-	0	0	-	0	n.a.	n.a.

Electrical capacity (GW)							Shares (%)		CAAGR (%)	
	2020	2030	2035	2020	2030	2035	2035		2009-2035	
	Current Policies Scenario			450 Scenario			CPS	450	CPS	450
Total capacity	214	282	323	216	298	369	100	100	3.3	3.8
Coal	59	74	85	53	55	61	26	17	2.8	1.5
Oil	27	20	18	27	20	18	6	5	-1.0	-1.0
Gas	78	100	109	74	81	81	34	22	3.3	2.1
Nuclear	2	4	5	2	11	17	1	5	3.5	8.8
Hydro	37	53	63	40	60	72	19	19	3.6	4.2
Biomass and waste	2	5	8	4	8	11	2	3	8.6	10.1
Wind	5	9	12	5	21	42	4	11	11.1	16.6
Geothermal	1	1	2	1	2	3	1	1	9.7	11.8
Solar PV	4	11	16	6	22	34	5	9	31.2	35.2
CSP	1	4	7	4	19	30	2	8	n.a.	n.a.
Marine	-	-	-	-	0	0	-	0	n.a.	n.a.

CO_2 emissions (Mt)							Shares (%)		CAAGR (%)	
	2020	2030	2035	2020	2030	2035	2035		2009-2035	
	Current Policies Scenario			450 Scenario			CPS	450	CPS	450
Total CO_2	1 165	1 284	1 352	1 045	818	675	100	100	1.5	-1.2
Coal	415	448	468	372	198	98	35	15	1.4	-4.5
Oil	508	545	570	458	418	400	42	59	1.1	-0.2
Gas	241	292	313	215	202	177	23	26	2.2	-0.0
Power generation	517	563	591	453	249	118	100	100	1.5	-4.6
Coal	321	353	374	283	123	29	63	24	1.6	-7.9
Oil	54	35	32	51	32	29	5	25	-2.5	-2.9
Gas	142	175	185	119	94	61	31	51	2.5	-1.8
TFC	595	661	697	543	520	508	100	100	1.4	0.2
Coal	92	92	91	88	74	68	13	13	0.7	-0.4
Oil	434	487	515	388	368	353	74	69	1.5	0.0
Transport	*288*	*328*	*347*	*253*	*230*	*212*	*50*	*42*	*1.6*	*-0.3*
Gas	69	82	91	67	78	87	13	17	1.6	1.4

A

Latin America: New Policies Scenario

	Energy demand (Mtoe)							Shares (%)		CAAGR (%)
	1990	2009	2015	2020	2025	2030	2035	2009	2035	2009-2035
TPED	333	538	644	700	743	787	829	100	100	1.7
Coal	15	18	26	28	32	34	32	3	4	2.3
Oil	150	233	266	268	268	269	271	43	33	0.6
Gas	52	112	142	166	176	186	197	21	24	2.2
Nuclear	2	6	7	11	14	17	18	1	2	4.6
Hydro	30	58	65	71	78	84	91	11	11	1.8
Biomass and waste	82	109	131	146	162	178	194	20	23	2.2
Other renewables	1	3	6	9	13	19	26	1	3	8.6
Power generation	66	134	157	175	190	206	225	100	100	2.0
Coal	3	5	10	11	13	13	12	4	5	3.0
Oil	14	26	22	18	14	11	10	20	5	-3.6
Gas	14	29	38	45	45	46	49	22	22	2.0
Nuclear	2	6	7	11	14	17	18	4	8	4.6
Hydro	30	58	65	71	78	84	91	43	40	1.8
Biomass and waste	2	8	9	12	15	18	23	6	10	4.2
Other renewables	1	3	5	8	11	17	23	2	10	8.6
Other energy sector	56	81	97	110	116	120	122	100	100	1.6
Electricity	*8*	*18*	*20*	*22*	*23*	*24*	*25*	*22*	*21*	*1.4*
TFC	253	410	493	530	562	598	630	100	100	1.7
Coal	7	9	12	12	12	12	12	2	2	1.1
Oil	122	189	223	229	233	238	240	46	38	0.9
Gas	24	59	74	82	89	98	106	14	17	2.2
Electricity	35	69	83	93	102	112	123	17	19	2.2
Heat	-	-	-	-	-	-	-	-	-	n.a.
Biomass and waste	65	83	101	112	124	136	146	20	23	2.2
Other renewables	-	0	1	1	2	2	3	0	1	8.4
Industry	85	139	172	188	201	212	221	100	100	1.8
Coal	6	9	11	12	12	12	11	6	5	1.0
Oil	20	32	37	39	39	40	39	23	18	0.8
Gas	15	30	39	43	47	51	55	22	25	2.4
Electricity	16	29	36	41	44	47	50	21	23	2.1
Heat	-	-	-	-	-	-	-	-	-	n.a.
Biomass and waste	27	40	49	54	58	62	65	28	29	1.9
Other renewables	-	-	-	-	-	-	-	-	-	n.a.
Transport	71	128	158	166	175	187	198	100	100	1.7
Oil	65	109	131	133	133	136	138	85	70	0.9
Electricity	0	0	0	0	0	0	1	0	0	3.3
Biofuels	6	13	21	26	33	40	46	10	23	4.9
Other fuels	0	5	6	7	9	11	12	4	6	3.3
Buildings	70	94	105	113	120	128	137	100	100	1.4
Coal	0	0	0	0	0	0	0	0	0	2.1
Oil	17	17	19	20	20	21	21	18	16	0.9
Gas	6	12	14	15	16	18	19	12	14	2.0
Electricity	17	38	44	49	54	60	66	41	48	2.1
Heat	-	-	-	-	-	-	-	-	-	n.a.
Biomass and waste	29	27	28	28	28	27	27	28	19	-0.0
Other renewables	-	0	1	1	2	2	3	0	2	8.4
Other	27	49	58	63	67	71	74	100	100	1.6

Latin America: Current Policies and 450 Scenarios

	Energy demand (Mtoe)						Shares (%)		CAAGR (%)	
	2020	2030	2035	2020	2030	2035	2035		2009-2035	
	Current Policies Scenario			450 Scenario			CPS	450	CPS	450
TPED	**705**	**820**	**881**	**672**	**718**	**731**	**100**	**100**	**1.9**	**1.2**
Coal	31	41	41	26	26	24	5	3	3.3	1.2
Oil	279	297	309	250	226	210	35	29	1.1	-0.4
Gas	166	212	237	156	149	141	27	19	2.9	0.9
Nuclear	11	12	13	11	17	18	2	2	3.5	4.7
Hydro	71	82	88	72	86	92	10	13	1.6	1.8
Biomass and waste	138	160	171	146	185	203	19	28	1.8	2.4
Other renewables	8	16	21	11	29	43	2	6	7.7	10.6
Power generation	**177**	**217**	**245**	**171**	**187**	**197**	**100**	**100**	**2.3**	**1.5**
Coal	12	16	17	9	7	6	7	3	4.5	0.1
Oil	20	13	12	15	6	5	5	3	-3.0	-6.0
Gas	45	64	78	42	24	11	32	6	3.9	-3.5
Nuclear	11	12	13	11	17	18	5	9	3.5	4.7
Hydro	71	82	88	72	86	92	36	47	1.6	1.8
Biomass and waste	11	15	18	12	21	26	7	13	3.4	4.7
Other renewables	7	14	19	9	26	39	8	20	7.8	10.8
Other energy sector	**112**	**127**	**131**	**105**	**107**	**106**	**100**	**100**	**1.9**	**1.0**
Electricity	_22_	_25_	_27_	_21_	_22_	_23_	_21_	_22_	_1.7_	_1.0_
TFC	**533**	**619**	**665**	**509**	**551**	**563**	**100**	**100**	**1.9**	**1.2**
Coal	13	13	14	12	11	11	2	2	1.6	0.8
Oil	238	264	276	214	202	187	41	33	1.5	-0.0
Gas	82	100	111	79	91	98	17	17	2.4	2.0
Electricity	94	119	133	91	105	112	20	20	2.5	1.9
Heat	-	-	-	-	-	-	-	-	n.a.	n.a.
Biomass and waste	105	121	129	112	139	151	19	27	1.7	2.3
Other renewables	1	2	3	1	2	4	0	1	7.3	9.0
Industry	**189**	**220**	**235**	**182**	**203**	**209**	**100**	**100**	**2.0**	**1.6**
Coal	12	13	13	11	11	11	6	5	1.6	0.8
Oil	40	43	45	37	37	36	19	17	1.3	0.4
Gas	43	54	60	41	46	50	25	24	2.7	2.0
Electricity	41	50	55	39	45	47	23	23	2.5	1.9
Heat	-	-	-	-	-	-	-	-	n.a.	n.a.
Biomass and waste	53	60	63	53	63	66	27	32	1.8	2.0
Other renewables	-	-	-	-	-	-	-	-	n.a.	n.a.
Transport	**167**	**196**	**213**	**153**	**160**	**157**	**100**	**100**	**2.0**	**0.8**
Oil	139	157	166	120	107	94	78	60	1.6	-0.6
Electricity	0	0	1	1	1	2	0	1	3.1	7.9
Biofuels	20	28	34	26	41	48	16	31	3.6	5.0
Other fuels	7	11	12	8	11	13	6	8	3.3	3.5
Buildings	**113**	**132**	**143**	**111**	**120**	**125**	**100**	**100**	**1.6**	**1.1**
Coal	0	0	0	0	0	0	0	0	2.3	1.9
Oil	20	22	23	19	18	18	16	14	1.1	0.1
Gas	15	18	20	15	17	18	14	14	2.2	1.7
Electricity	50	63	71	48	54	57	49	46	2.4	1.6
Heat	-	-	-	-	-	-	-	-	n.a.	n.a.
Biomass and waste	28	27	27	28	28	28	19	22	-0.0	0.2
Other renewables	1	2	3	1	2	4	2	3	7.3	9.0
Other	**63**	**71**	**75**	**62**	**69**	**72**	**100**	**100**	**1.7**	**1.5**

Latin America: New Policies Scenario

	Electricity generation (TWh)							Shares (%)		CAAGR (%)
	1990	2009	2015	2020	2025	2030	2035	2009	2035	2009-2035
Total generation	489	1 009	1 194	1 337	1 456	1 581	1 717	100	100	2.1
Coal	9	20	41	46	57	60	55	2	3	3.9
Oil	64	127	106	87	71	57	53	13	3	-3.3
Gas	45	135	208	254	265	276	294	13	17	3.1
Nuclear	10	21	28	44	54	64	68	2	4	4.6
Hydro	354	669	757	830	905	980	1 054	66	61	1.8
Biomass and waste	7	32	38	46	56	68	81	3	5	3.7
Wind	-	2	9	16	24	37	55	0	3	14.0
Geothermal	1	3	5	7	9	12	16	0	1	6.5
Solar PV	-	0	2	7	13	21	28	0	2	40.6
CSP	-	-	-	-	2	7	12	-	1	n.a.
Marine	-	-	-	-	-	-	-	-	-	n.a.

	Electrical capacity (GW)						Shares (%)		CAAGR (%)
	2009	2015	2020	2025	2030	2035	2009	2035	2009-2035
Total capacity	227	273	311	344	380	422	100	100	2.4
Coal	4	7	8	10	11	10	2	2	3.2
Oil	33	34	34	32	28	29	14	7	-0.5
Gas	40	60	72	79	89	100	17	24	3.6
Nuclear	3	4	6	7	9	9	1	2	4.5
Hydro	142	157	172	187	203	218	62	52	1.7
Biomass and waste	5	6	7	8	10	12	2	3	3.8
Wind	1	4	6	9	14	20	0	5	12.6
Geothermal	1	1	1	1	2	2	0	1	5.7
Solar PV	0	2	5	9	14	18	0	4	74.2
CSP	-	-	-	1	2	3	-	1	n.a.
Marine	-	-	-	-	-	-	-	-	n.a.

	CO_2 emissions (Mt)							Shares (%)		CAAGR (%)
	1990	2009	2015	2020	2025	2030	2035	2009	2035	2009-2035
Total CO_2	578	972	1 165	1 229	1 255	1 281	1 303	100	100	1.1
Coal	47	67	104	110	116	117	110	7	8	1.9
Oil	415	658	746	749	745	749	753	68	58	0.5
Gas	116	247	316	370	394	415	440	25	34	2.2
Power generation	90	176	205	211	207	200	199	100	100	0.5
Coal	15	26	48	51	58	59	53	15	27	2.8
Oil	44	82	67	55	44	35	32	47	16	-3.6
Gas	32	68	90	105	105	107	114	39	57	2.0
TFC	426	700	841	876	900	932	956	100	100	1.2
Coal	29	39	52	53	53	53	52	6	5	1.1
Oil	342	532	629	644	651	664	670	76	70	0.9
Transport	193	326	390	396	397	406	412	47	43	0.9
Gas	54	129	161	179	196	215	233	18	24	2.3

Latin America: Current Policies and 450 Scenarios

	Electricity generation (TWh)						Shares (%)		CAAGR (%)	
	2020	2030	2035	2020	2030	2035	2035		2009-2035	
	Current Policies Scenario			450 Scenario			CPS	450	CPS	450
Total generation	1 354	1 674	1 859	1 310	1 483	1 567	100	100	2.4	1.7
Coal	54	77	81	39	31	25	4	2	5.5	0.9
Oil	99	64	60	77	31	27	3	2	-2.9	-5.8
Gas	264	410	495	226	150	65	27	4	5.1	-2.7
Nuclear	42	47	51	43	67	70	3	4	3.5	4.7
Hydro	823	957	1 023	841	1 000	1 072	55	68	1.6	1.8
Biomass and waste	44	58	67	47	76	92	4	6	2.9	4.2
Wind	16	31	40	17	65	110	2	7	12.5	17.0
Geothermal	7	11	14	7	17	23	1	1	6.1	8.1
Solar PV	5	16	21	8	34	60	1	4	39.1	44.7
CSP	-	3	7	4	13	20	0	1	n.a.	n.a.
Marine	-	-	-	-	0	2	-	0	n.a.	n.a.

	Electrical capacity (GW)						Shares (%)		CAAGR (%)	
	2020	2030	2035	2020	2030	2035	2035		2009-2035	
	Current Policies Scenario			450 Scenario			CPS	450	CPS	450
Total capacity	312	389	434	304	379	431	100	100	2.5	2.5
Coal	10	13	13	8	7	7	3	2	4.5	1.7
Oil	34	29	29	31	25	25	7	6	-0.5	-1.1
Gas	75	109	130	63	68	69	30	16	4.7	2.2
Nuclear	6	7	7	6	9	9	2	2	3.5	4.7
Hydro	170	198	212	174	207	222	49	52	1.6	1.7
Biomass and waste	7	9	10	7	11	14	2	3	3.1	4.3
Wind	6	11	14	7	24	41	3	9	11.1	15.7
Geothermal	1	2	2	1	2	3	0	1	5.3	7.2
Solar PV	4	11	15	6	21	36	3	8	72.6	78.8
CSP	-	1	2	1	3	5	0	1	n.a.	n.a.
Marine	-	-	-	-	0	0	-	0	n.a.	n.a.

	CO_2 emissions (Mt)						Shares (%)		CAAGR (%)	
	2020	2030	2035	2020	2030	2035	2035		2009-2035	
	Current Policies Scenario			450 Scenario			CPS	450	CPS	450
Total CO_2	1 275	1 451	1 545	1 143	1 010	921	100	100	1.8	-0.2
Coal	121	140	143	101	75	63	9	7	2.9	-0.3
Oil	783	837	871	693	614	560	56	61	1.1	-0.6
Gas	372	475	532	349	320	298	34	32	3.0	0.7
Power generation	226	265	296	189	102	64	100	100	2.0	-3.8
Coal	59	75	76	44	28	21	26	33	4.3	-0.8
Oil	63	39	37	48	18	16	12	26	-3.0	-6.0
Gas	104	150	183	97	56	26	62	42	3.9	-3.6
TFC	903	1 023	1 083	822	784	741	100	100	1.7	0.2
Coal	56	59	60	52	43	38	6	5	1.7	-0.1
Oil	668	742	777	598	552	503	72	68	1.5	-0.2
Transport	*415*	*468*	*494*	*357*	*319*	*280*	*46*	*38*	*1.6*	*-0.6*
Gas	179	221	246	173	189	201	23	27	2.5	1.7

A

Brazil: New Policies Scenario

	Energy demand (Mtoe)							Shares (%)		CAAGR (%)
	1990	2009	2015	2020	2025	2030	2035	2009	2035	2009-2035
TPED	**138**	**237**	**300**	**336**	**364**	**393**	**421**	**100**	**100**	**2.2**
Coal	10	11	16	16	16	15	14	5	3	1.0
Oil	59	95	113	116	115	116	116	40	28	0.8
Gas	3	17	34	50	59	66	76	7	18	5.9
Nuclear	1	3	4	6	9	11	12	1	3	5.1
Hydro	18	34	37	40	42	45	47	14	11	1.3
Biomass and waste	48	76	95	106	121	135	147	32	35	2.6
Other renewables	-	1	1	2	3	5	7	0	2	10.8
Power generation	**22**	**50**	**65**	**75**	**83**	**93**	**104**	**100**	**100**	**2.9**
Coal	1	3	6	5	5	4	4	5	4	1.5
Oil	1	3	3	2	2	2	2	6	2	-2.1
Gas	0	2	9	13	14	16	21	5	20	8.5
Nuclear	1	3	4	6	9	11	12	7	12	5.1
Hydro	18	34	37	40	42	45	47	68	46	1.3
Biomass and waste	1	4	6	7	9	11	12	9	12	4.2
Other renewables	-	0	1	1	2	3	5	0	5	16.3
Other energy sector	**26**	**40**	**51**	**62**	**67**	**68**	**69**	**100**	**100**	**2.1**
Electricity	*3*	*9*	*10*	*11*	*12*	*12*	*13*	*21*	*19*	*1.8*
TFC	**112**	**191**	**237**	**259**	**279**	**303**	**326**	**100**	**100**	**2.1**
Coal	4	5	6	6	7	7	7	3	2	1.3
Oil	53	86	103	106	106	107	108	45	33	0.9
Gas	2	10	15	18	23	28	34	5	10	5.0
Electricity	18	35	43	48	53	58	65	18	20	2.4
Heat	-	-	-	-	-	-	-	-	-	n.a.
Biomass and waste	34	55	70	79	90	102	112	29	34	2.7
Other renewables	-	0	1	1	1	1	2	0	1	5.9
Industry	**40**	**71**	**91**	**102**	**113**	**123**	**133**	**100**	**100**	**2.4**
Coal	4	5	6	6	6	6	6	7	5	1.3
Oil	8	12	15	15	15	16	16	17	12	1.0
Gas	1	7	11	14	17	21	25	10	19	5.2
Electricity	10	16	20	23	25	27	30	23	22	2.4
Heat	-	-	-	-	-	-	-	-	-	n.a.
Biomass and waste	17	31	39	44	49	52	56	44	42	2.2
Other renewables	-	-	-	-	-	-	-	-	-	n.a.
Transport	**33**	**63**	**80**	**85**	**91**	**98**	**106**	**100**	**100**	**2.0**
Oil	27	48	59	60	59	59	59	76	55	0.8
Electricity	0	0	0	0	0	0	0	0	0	4.4
Biofuels	6	13	19	22	29	36	42	21	40	4.6
Other fuels	0	2	2	2	3	4	5	3	5	3.9
Buildings	**23**	**34**	**37**	**40**	**43**	**47**	**50**	**100**	**100**	**1.6**
Coal	0	-	-	-	-	-	-	-	-	n.a.
Oil	6	7	7	7	8	8	8	21	17	0.7
Gas	0	0	1	1	1	1	1	1	3	5.1
Electricity	8	17	21	23	25	28	32	52	63	2.3
Heat	-	-	-	-	-	-	-	-	-	n.a.
Biomass and waste	9	8	8	8	8	8	7	25	15	-0.5
Other renewables	-	0	1	1	1	1	2	1	4	5.9
Other	**16**	**24**	**29**	**31**	**33**	**35**	**37**	**100**	**100**	**1.8**

Brazil: Current Policies and 450 Scenarios

	Energy demand (Mtoe)						Shares (%)		CAAGR (%)	
	2020	2030	2035	2020	2030	2035	2035		2009-2035	
	Current Policies Scenario			450 Scenario			CPS	450	CPS	450
TPED	337	403	434	323	354	365	100	100	2.4	1.7
Coal	18	18	17	15	11	10	4	3	1.7	-0.5
Oil	119	130	134	108	93	84	31	23	1.3	-0.5
Gas	52	81	95	45	48	49	22	13	6.8	4.2
Nuclear	6	8	9	6	12	13	2	3	3.9	5.2
Hydro	39	44	46	40	46	48	11	13	1.2	1.4
Biomass and waste	101	119	127	106	139	153	29	42	2.0	2.7
Other renewables	2	4	5	2	6	9	1	3	9.6	11.9
Power generation	75	97	111	73	83	88	100	100	3.1	2.2
Coal	6	6	5	5	1	1	5	1	2.8	-4.7
Oil	2	2	2	2	1	1	2	1	-1.4	-4.2
Gas	14	26	35	11	7	5	31	5	10.6	2.3
Nuclear	6	8	9	6	12	13	8	14	3.9	5.2
Hydro	39	44	46	40	46	48	42	55	1.2	1.4
Biomass and waste	7	8	9	8	12	14	8	15	3.1	4.6
Other renewables	1	3	4	1	5	7	4	9	14.9	17.8
Other energy sector	63	72	72	58	60	59	100	100	2.3	1.5
Electricity	11	13	14	11	11	12	20	20	2.0	1.2
TFC	259	308	334	250	277	289	100	100	2.2	1.6
Coal	7	7	8	6	6	6	2	2	1.8	0.6
Oil	109	120	125	99	86	77	37	27	1.5	-0.4
Gas	19	29	36	18	25	30	11	10	5.2	4.5
Electricity	48	61	68	47	54	58	20	20	2.6	2.0
Heat	-	-	-	-	-	-	-	-	n.a.	n.a.
Biomass and waste	75	90	96	79	104	116	29	40	2.1	2.9
Other renewables	1	1	2	1	1	2	0	1	5.4	6.1
Industry	103	125	137	100	119	126	100	100	2.5	2.2
Coal	6	7	7	6	6	5	5	4	1.8	0.6
Oil	16	17	18	15	15	14	13	11	1.5	0.6
Gas	14	22	27	14	19	22	20	18	5.4	4.7
Electricity	23	28	31	22	26	28	23	22	2.6	2.1
Heat	-	-	-	-	-	-	-	-	n.a.	n.a.
Biomass and waste	44	51	54	44	53	56	39	45	2.1	2.3
Other renewables	-	-	-	-	-	-	-	-	n.a.	n.a.
Transport	84	99	107	78	81	82	100	100	2.1	1.1
Oil	62	69	72	54	41	33	68	41	1.6	-1.3
Electricity	0	0	0	0	0	1	0	1	4.2	7.7
Biofuels	19	26	29	22	37	44	28	53	3.1	4.7
Other fuels	2	4	5	2	3	4	5	5	4.0	3.3
Buildings	41	48	53	40	42	44	100	100	1.8	1.0
Coal	-	-	-	-	-	-	-	-	n.a.	n.a.
Oil	7	8	9	7	6	6	16	13	0.9	-0.8
Gas	1	1	2	1	1	1	3	3	5.3	4.4
Electricity	23	30	34	23	25	27	64	61	2.6	1.6
Heat	-	-	-	-	-	-	-	-	n.a.	n.a.
Biomass and waste	8	8	7	8	8	8	14	19	-0.4	0.1
Other renewables	1	1	2	1	1	2	3	4	5.4	6.1
Other	31	36	37	31	35	37	100	100	1.8	1.7

A

Brazil: New Policies Scenario

	Electricity generation (TWh)							Shares (%)		CAAGR (%)
	1990	2009	2015	2020	2025	2030	2035	2009	2035	2009-2035
Total generation	223	466	573	647	709	781	866	100	100	2.4
Coal	5	10	25	23	22	19	17	2	2	2.2
Oil	5	15	12	10	9	8	9	3	1	-1.9
Gas	1	13	59	85	94	107	136	3	16	9.4
Nuclear	2	13	15	25	33	43	47	3	5	5.1
Hydro	207	391	427	460	492	523	552	84	64	1.3
Biomass and waste	4	23	29	34	40	45	50	5	6	3.0
Wind	-	1	5	8	13	20	31	0	4	13.1
Geothermal	-	-	-	-	-	-	-	-	-	n.a.
Solar PV	-	-	1	4	8	13	18	-	2	n.a.
CSP	-	-	-	-	0	3	6	-	1	n.a.
Marine	-	-	-	-	-	-	-	-	-	n.a.

	Electrical capacity (GW)						Shares (%)		CAAGR (%)
	2009	2015	2020	2025	2030	2035	2009	2035	2009-2035
Total capacity	108	130	150	169	191	216	100	100	2.7
Coal	2	4	4	4	4	4	2	2	2.0
Oil	6	7	8	8	8	8	5	4	1.4
Gas	9	18	25	31	37	48	8	22	6.7
Nuclear	2	2	3	4	6	6	2	3	4.7
Hydro	86	93	100	107	113	119	80	55	1.3
Biomass and waste	3	4	4	5	6	7	3	3	3.4
Wind	1	2	3	5	7	11	1	5	11.9
Geothermal	-	-	-	-	-	-	-	-	n.a.
Solar PV	-	1	3	5	9	12	-	6	n.a.
CSP	-	-	-	0	1	1	-	1	n.a.
Marine	-	-	-	-	-	-	-	-	n.a.

	CO$_2$ emissions (Mt)							Shares (%)		CAAGR (%)
	1990	2009	2015	2020	2025	2030	2035	2009	2035	2009-2035
Total CO$_2$	194	338	452	493	510	525	545	100	100	1.9
Coal	29	38	63	61	60	56	53	11	10	1.3
Oil	159	261	310	316	315	317	318	77	58	0.8
Gas	6	39	79	115	135	152	174	12	32	5.9
Power generation	12	30	62	65	65	66	74	100	100	3.5
Coal	8	14	32	29	26	23	20	47	27	1.4
Oil	4	10	8	6	5	5	6	33	8	-2.1
Gas	0	6	22	30	33	38	48	19	65	8.5
TFC	167	280	346	363	372	387	401	100	100	1.4
Coal	18	21	27	29	30	30	30	8	7	1.3
Oil	144	237	284	292	291	293	294	85	73	0.8
Transport	81	143	176	180	176	176	176	51	44	0.8
Gas	4	22	34	42	52	64	78	8	19	5.0

Brazil: Current Policies and 450 Scenarios

	Electricity generation (TWh)						Shares (%)		CAAGR (%)	
	2020	2030	2035	2020	2030	2035	2035		2009-2035	
	Current Policies Scenario			450 Scenario			CPS	450	CPS	450
Total generation	651	817	917	637	727	772	100	100	2.6	2.0
Coal	28	26	25	20	5	3	3	0	3.6	-4.0
Oil	10	10	11	8	5	5	1	1	-1.2	-3.9
Gas	92	174	230	70	46	28	25	4	11.6	2.9
Nuclear	24	31	35	25	45	48	4	6	3.9	5.2
Hydro	455	510	536	464	530	559	59	72	1.2	1.4
Biomass and waste	32	38	40	35	49	55	4	7	2.2	3.4
Wind	8	16	20	9	24	41	2	5	11.3	14.4
Geothermal	-	-	-	-	-	-	-	-	n.a.	n.a.
Solar PV	3	10	14	5	16	24	2	3	n.a.	n.a.
CSP	-	2	4	1	6	9	0	1	n.a.	n.a.
Marine	-	-	-	-	0	1	-	0	n.a.	n.a.

	Electrical capacity (GW)						Shares (%)		CAAGR (%)	
	2020	2030	2035	2020	2030	2035	2035		2009-2035	
	Current Policies Scenario			450 Scenario			CPS	450	CPS	450
Total capacity	150	192	218	145	176	194	100	100	2.7	2.3
Coal	5	4	4	4	3	3	2	1	2.6	0.8
Oil	8	8	8	7	6	6	4	3	1.5	0.1
Gas	26	47	61	19	19	19	28	10	7.7	3.0
Nuclear	3	4	5	3	6	6	2	3	3.7	4.9
Hydro	99	111	116	101	115	121	53	62	1.2	1.3
Biomass and waste	4	5	5	4	6	7	2	4	2.4	3.8
Wind	3	6	7	3	9	15	3	8	10.0	13.0
Geothermal	-	-	-	-	-	-	-	-	n.a.	n.a.
Solar PV	2	7	10	3	11	15	5	8	n.a.	n.a.
CSP	-	0	1	0	1	2	0	1	n.a.	n.a.
Marine	-	-	-	-	0	0	-	0	n.a.	n.a.

	CO_2 emissions (Mt)						Shares (%)		CAAGR (%)	
	2020	2030	2035	2020	2030	2035	2035		2009-2035	
	Current Policies Scenario			450 Scenario			CPS	450	CPS	450
Total CO_2	516	611	657	453	380	343	100	100	2.6	0.1
Coal	68	68	66	57	31	26	10	8	2.1	-1.5
Oil	327	358	373	293	243	212	57	62	1.4	-0.8
Gas	121	184	217	103	106	105	33	31	6.8	3.9
Power generation	74	98	117	56	26	18	100	100	5.4	-2.0
Coal	34	31	28	26	6	4	24	22	2.7	-4.8
Oil	6	6	7	5	3	3	6	18	-1.4	-4.2
Gas	33	61	81	25	17	11	70	60	10.6	2.3
TFC	375	431	461	339	300	277	100	100	1.9	-0.0
Coal	30	33	34	28	22	19	7	7	1.8	-0.4
Oil	302	332	346	271	225	195	75	70	1.5	-0.7
Transport	*188*	*207*	*216*	*162*	*123*	*100*	*47*	*36*	*1.6*	*-1.3*
Gas	43	66	82	41	54	63	18	23	5.2	4.1

A

POLICIES AND MEASURES BY SCENARIO

The *World Energy Outlook-2011* presents projections for three scenarios. The **Current Policies Scenario** includes all policies in place and supported through enacted measures as of mid-2011. A number of the policy commitments and plans that were included in the New Policies Scenario in *WEO-2010* have since been enacted, so are now included in the Current Policies Scenario in this *Outlook*. Some of the policies modelled under the Current Policies Scenario include:

- China's 12th Five-Year Plan for the period 2011 to 2015;

- a new scheme that enables trading of renewable energy certificates and a new programme of support for alternative fuel vehicles in India;

- new European Union (EU) directives covering the energy performance of buildings and emissions standards for light-commercial vehicles;

- new appliance standards in the United States;

- early retirement of all nuclear plants in Germany by the end of 2022;

- no lifetime extension for existing nuclear plants and no new ones in Switzerland;

- and an emissions trading scheme (ETS) in New Zealand from 2010.

The **New Policies Scenario** is based on broad policy commitments and plans that have been announced by countries around the world to address energy security, climate change and local pollution, and other pressing energy-related challenges, even where the specific measures to implement these commitments have yet to be announced. More specifically, the New Policies Scenario takes into account all policies and measures included in the Current Policies Scenario, as well as the following:

- Cautious implementation of recently announced commitments and plans, including the Cancun Agreements.

- Continuation of an ETS in Europe and New Zealand; introduction of CO_2 pricing through taxes/ETS in Australia from mid-2012, in Korea from 2015 and in China from 2020. Shadow price of carbon adopted from 2015 in Canada, Japan and the United States, influencing solely new investment decisions in power generation. Access to international offset credits is assumed for all countries participating in an ETS.

- Phase-out of fossil-fuel consumption subsidies in all net-importing regions by 2020 (and, as in the Current Policies Scenario, in net-exporting regions where specific policies have already been introduced).

- Extension by five years of the nuclear plant lifetimes assumed in the Current Policies Scenario, in those countries that have confirmed their intention to continue or expand the use of nuclear energy, accounting for about 50% of the plants currently in operation.

- Extended support for renewables-based electricity-generation technologies and biofuels for transport.

- Heavy-duty vehicle fuel efficiency standards in the United States; light-commercial vehicle and passenger light-duty vehicle (PLDV) emissions per kilometre targets in the European Union; PLDV fuel efficiency standards in India and China.

- For 2020 to 2035, additional measures that maintain the pace of the global decline in carbon intensity – measured as emissions per dollar of gross domestic product, in purchasing power parity terms – established in the period 2009 to 2020.

The **450 Scenario** sets out an energy pathway that is consistent with a 50% chance of meeting the goal of limiting the increase in average global temperature to 2°C compared with pre-industrial levels. For the period to 2020, the 450 Scenario assumes more vigorous policy action to implement fully the Cancun Agreements than is assumed in the New Policies Scenario (which assumes cautious implementation). After 2020, OECD countries and other major economies are assumed to set economy-wide emissions targets for 2035 and beyond that collectively ensure an emissions trajectory consistent with stabilisation of the greenhouse-gas concentration at 450 parts per million. In addition, the 450 Scenario also includes the following specific policies:

- Implementation of the high-end of the range of the commitments arising from the Cancun Agreements, where they are expressed as ranges.

- Strengthening of the ETS in Europe, and staggered introduction of CO_2 pricing through taxes or ETS, at latest by 2025 in all OECD countries and, from 2020, in China (with higher CO_2 pricing than the New Policies Scenario), Russia, Brazil and South Africa. Access to international offset credits is assumed for all countries participating in an ETS.

- International sectoral agreements for the iron and steel, and the cement industries.

- International agreements on fuel-economy standards for PLDVs, aviation and shipping.

- National policies and measures, such as efficiency standards for buildings and labelling of appliances.

- The complete phase-out of fossil-fuel consumption subsidies in all net-importing regions by 2020 (at the latest) and in all net-exporting regions by 2035 (at the latest), except for the Middle East where it is assumed that the average subsidisation rate declines to 20% by 2035.

- Extension by five years of the nuclear plant lifetimes assumed in the New Policies Scenario, in those countries that have confirmed their intention to continue or expand the use of nuclear energy, accounting for about 75% of the plants currently in operation.

- Further strengthening (compared to the New Policies Scenario) of support for renewables-based electricity-generation technologies and biofuels for transport.

The specific policies adopted for different sectors by selected countries and regions for the Current Policies, New Policies and 450 Scenarios are outlined below. The policies are cumulative. That is, measures listed under the New Policies Scenario supplement those under the Current Policies Scenario and measures listed under the 450 Scenario supplement those under the New Policies Scenario.

B

Table B.1 ● Power sector policies and measures as modelled by scenario in selected regions

	Current Policies Scenario	New Policies Scenario	450 Scenario
OECD			
United States	– State-level support for renewables. – American Recovery and Reinvestment Act (2009): tax credits for renewable and other clean energy sources, prolonged over the entire projection period. – Lifetimes of US nuclear plants extended beyond 60 years.	– Shadow price of carbon adopted for investment decisions from 2015. – EPA regulations including Maximum Achievable Control Technology (MACT) for mercury and other pollutants. – Extension of support for nuclear, including loan guarantees. – Funding for CCS (demonstration-scale).	– Shadow price for investment decisions from 2015 to 2019; CO_2 pricing implemented from 2020. – Extended support to renewables, nuclear and CCS.
Japan	– Support for renewable generation. – Decommissioning of units 1-4 and no construction of new units at the Fukushima Daiichi site. – Strategic Energy Plan[1]: – Increasing share of renewable energy to 10% by 2020; – Shift to more advanced coal power generation technologies.	– Shadow price of carbon adopted for investment decisions from 2015.	– Shadow price for investment decisions from 2015 to 2019; CO_2 pricing implemented from 2020. – Strategic Energy Plan[1]: – Share of low-carbon electricity generation to increase by 2020 and expand further by 2030; – Expansion of renewables support; – Introduction of CCS to coal-fired power generation.
European Union	– Climate and Energy Package: – Emissions Trading System; – Support for renewables sufficient to reach 20% share of energy demand in 2020; – Financial support for CCS, including use of credits from the ETS New Entrants' Reserve; – Early retirement of all nuclear plants in Germany by the end of 2022.	– Extended support to renewable-based electricity-generation technologies.	– Emissions Trading System strengthened in line with the 2050 roadmap. – Reinforcement of government support in favour of renewables. – Expanded support measures for CCS.

Notes: CCS = carbon capture and storage; FIT = feed-in tariffs.

1. Following the Great East Japan Earthquake, Japan is undertaking a full review of its Strategic Energy Plan with results expected in 2012.

Table B.1 ● **Power sector policies and measures as modelled by scenario in selected regions** (continued)

	Current Policies Scenario	New Policies Scenario	450 Scenario
Non-OECD			
Russia	– Competitive wholesale electricity market.	– State support to the nuclear and hydropower sectors; a support mechanism for non-hydro renewables introduced from 2014.	– CO_2 pricing implemented from 2020. – Stronger support for nuclear power and renewables.
China	– Implementation of measures in 12th Five-Year Plan; solar additions of 5 GW by 2015; wind additions of 70 GW by 2015 and start construction of 120 GW of hydropower by 2015. Delays in nuclear capacity additions resulting from the temporary suspension of approval for new projects. – A 15% share of non-fossil energy in total energy supply by 2020.	– CO_2 pricing implemented from 2020. – Nuclear capacity target of 70 to 80 GW by 2020. – 100 GW installed wind capacity by 2015, 150 GW installed wind capacity by 2020. – Subsidies for building-integrated photovoltaic (PV) projects. – Solar FiT for utility-scale solar PV plants. – Local pollution reduction goals.	– Higher CO_2 pricing. – Enhanced support for renewables. – Continued support to nuclear capacity additions post 2020. – Deployment of CCS from around 2020.
India	– Renewable Energy Certificate trade for all eligible grid-connected renewable-based electricity-generation technologies. – Achievement of the national solar mission phase one and phase two targets (an additional 4 GW by 2017). – Increased use of supercritical coal technology.	– Various renewable energy support policies and targets, including small hydro. – Achievement of the national solar mission target of 20 GW of solar PV capacity by 2022.	– Renewables (excluding large hydro) to reach 15% of installed capacity by 2020. – Support to renewables, nuclear and efficient coal. – Deployment of CCS from around 2020.
Brazil	– Increase of wind, biomass, solar and hydro (small and large) capacity.	– Enhanced deployment of renewables technologies.	– CO_2 pricing implemented from 2020. – Further increases of generation from renewable sources.

B

Table B.2 ● Transport sector policies and measures as modelled by scenario in selected regions

	Current Policies Scenario	New Policies Scenario	450 Scenario
OECD			**OECD**
United States	– CAFE standards: 35.5 miles-per-gallon for PLDVs by 2016, and further strengthening thereafter. – Renewable Fuel Standard.	– New heavy-duty vehicle standards for each model year from 2014 to 2018. – Renewable Fuel Standard. – Support to natural gas in road freight. – Increase of ethanol blending mandates.	On-road emission targets for PLDVs in 2035 — 65 g CO_2/km Light-commercial vehicles — Full technology spill-over from PLDVs.
Japan	– Top Runner Program: improvement in fuel efficiency of PLDVs by 23.5% by 2015 compared to 2004. – Promotion of demand-side measures such as intelligent transport systems and modal shifts. – Fiscal incentives for hybrid and electric vehicles.	– Promotion of modal shift. – Target shares of new car sales according to Next Generation Vehicle Strategy 2010: |	2020	2030
---	---	---
Conventional ICE vehicles	50%-80%	30%-50%
Hybrid vehicles	20%-30%	30%-40%
Electric vehicles and plug-in hybrids	15%-20%	20%-30%
Fuel cell vehicles	<1%	<3%
Clean diesel vehicles	<5%	5%-10%		Medium- and heavy-freight traffic — 20% more efficient by 2035 than in NPS. Aviation — 45% efficiency improvements by 2035 (compared to 2010) and support for the use of biofuels. Other sectors such as maritime and rail — National policies and measures.
European Union	– Climate and Energy Package: renewable energy to reach 10% share of transport energy demand in 2020. – Emissions Trading System to include aviation from 2012. – Climate and Energy Package: CO_2 emission standards for PLDVs by 2020 (120 to 130 g CO_2/km). – Support to biofuels.	– Emissions Trading System. – More stringent emission target for PLDVs (95 g CO_2/km by 2020), and further strengthening post 2020. – Emission target for light-commercial vehicles (135 g CO_2/km by 2020), and further strengthening post 2020. – Road map on transport: CO_2-free city logistics in major urban centres by 2030. – Enhanced support to alternative fuels.	Fuels — Retail fuel prices kept at a level similar to Current Policies Scenario. Alternative clean fuels — Enhanced support to alternative fuels.

Table B.2 • Transport sector policies and measures as modelled by scenario in selected regions (continued)

	Current Policies Scenario	New Policies Scenario	450 Scenario
Non-OECD			**Non-OECD**
China	– Subsidies for hybrid and electric vehicles. – Promotion for fuel efficient cars. – Ethanol blending mandates 10% in selected provinces.	– Fuel economy standard for PLDVs (7 litres/100 km) by 2015, and further strengthening post 2015. – Extended subsidies on the purchase of alternative vehicles. – Increased biofuels blending.	On-road emission targets for PLDVs in 2035 — 100 g CO_2/km Light-commercial vehicles — Full technology spill-over from PLDVs. Medium- and heavy-freight vehicles — 20% more efficient by 2035 than in NPS. Aviation — 45% efficiency improvements by 2035 (compared to 2010) and support for the use of biofuels. Other sectors such as maritime and rail — National policies and measures. Fuels — Retail fuel prices kept at a level similar to Current Policies Scenario. Alternative clean fuels — Enhanced support for alternative fuels.
India	– Support for alternative fuel vehicles.	– Extended support for alternative fuel vehicles. – Proposed auto fuel efficiency standards. – Increased utilisation of natural gas in road transport.	
Brazil	– Ethanol targets in road transport 20% to 25%.	– Increase of ethanol blending mandates.	

Notes: CAFE = corporate average fuel economy; PLDVs = passenger light-duty vehicles; ICE = internal combustion engines.

Table B.3 ● Industry sector policies and measures as modelled by scenario in selected regions

	Current Policies Scenario	New Policies Scenario	450 Scenario
OECD			**OECD**
United States	– Support for high-energy efficiency technologies.	– Tax reduction and funding for efficient technologies. – R&D in low-carbon technologies.	– CO$_2$ pricing introduced from 2025, at the latest, in all countries. – International sectoral agreements with targets for iron, steel and cement industries. – Enhanced energy efficiency standards. – Policies to support the introduction of CCS in industry.
Japan	– Long-Term Outlook on Energy Supply and Demand (2009), including reforms in steel manufacturing and chemical industry technology.	– Maintenance and strengthening of top-end/low-carbon efficiency standards by: – Higher efficiency combined heat and power systems; – Promotion of state-of-the-art technology and faster replacement of aging equipments; – Fuel switching to gas.	
European Union	– Emissions Trading System.	– Directive on energy end-use efficiency and energy efficiency, including the development of: – Inverters for electric motors; – High-efficiency co-generation; – Mechanical vapour compression; – Innovations in industrial processes.	

Table B.3 ● Industry sector policies and measures as modelled by scenario in selected regions (continued)

	Current Policies Scenario	New Policies Scenario	450 Scenario
Non-OECD			**Non-OECD**
Russia	– Competitive wholesale electricity market price for industry.	– Improvements in energy efficiency. – Industrial gas prices reach the equivalent of export prices (minus taxes and transportation) in 2020. – Elaboration of comprehensive federal and regional legislation on energy savings.	– CO_2 pricing introduced as of 2020 in Russia, China, Brazil and South Africa. – Wider hosting of international offset projects. – International sectoral agreements with targets for iron, steel and cement industries. – Enhanced energy-efficiency standards. – Policies to support the introduction of CCS in industry.
China	– Priority given to gas use to 2015 (12th Five-Year plan). – Scrapping of small, energy-inefficient plants.	– CO_2 pricing implemented from 2020. – Contain the expansion of energy-intensive industries. – Enhanced use of Energy Service Companies and energy performance contracting.	
India	– National Mission for Enhanced Energy Efficiency, targeting a 5% reduction in energy use by 2015 (compared to 2010).	– Further implementation of National Mission for Enhanced Energy Efficiency recommendations including: – Enhancement of cost-effective improvements in energy efficiency in energy-intensive large industries and facilities, through certification of energy savings that could be traded; – Creation of mechanisms that would help finance demand-side management programmes by capturing future energy savings; – Development of fiscal instruments to promote energy efficiency.	
Brazil	– Encourage investment and R&D in energy efficiency.	– More use of charcoal in iron production to substitute for coal. – Implementation of measures included in the 2010 energy efficiency state programme.	

Note: R&D = research and develpment.

B

Table B.4 ● **Buildings sector policies and measures as modelled by scenario in selected regions**

	Current Policies Scenario	New Policies Scenario	450 Scenario
OECD			
United States	– AHAM-ACEEE Multi-Product Standards Agreement. – American Recovery and Reinvestment Act (2009): Funding energy efficiency and renewables. – Energy Star: Federal Tax Credits for Consumer Energy Efficiency; new appliance efficiency standards. – Energy Improvement and Extension Act of 2008.	– Extensions to 2025 of tax credit for energy-efficient equipment (including furnaces, boilers, air conditioners, air and ground source heat pumps, water heaters and windows), and for solar PV and solar thermal water heaters. – Budget proposals 2011: institute programmes to make commercial buildings 20% more efficient by 2020; tax credit for renewable energy deployment.	– More stringent mandatory building codes by 2020. – Extension of energy-efficiency grants to end of projection period. – Zero-energy buildings initiative.
Japan	– Long-Term Outlook on Energy Supply and Demand (2009): Energy savings using demand-side management. – Basic Energy Plan 2010: Promotion of energy efficient appliances and equipment.	– Strategic Energy Plan:[2] – Environmental Efficiency (CASBEE) for all buildings by 2030; – High-efficiency lighting: 100% of purchases by 2020; 100% in use by 2030; – Deployment of high-efficiency heating, cooling and water heating systems; – Net zero-energy buildings by 2030 for new construction; – Increased introduction of gas and renewable energy;	– Net zero-energy buildings by 2025 for new construction. – Mandatory standards for high-efficiency heating, cooling and water heating systems.
European Union	– Energy Performance of Buildings Directive. – New EU-US Energy Star Agreement: energy labelling of appliances.	– Nearly zero-energy buildings standards mandatory for new public buildings in the EU after 2018 and all new homes and offices from 2020.	– Zero-carbon footprint for all new buildings as of 2018. – Enhanced energy efficiency in all existing buildings.

Notes: AHAM = Association of Home Appliance Manufacturers; ACEEE = American Council for an Energy-Efficient Economy; CASBEE = Comprehensive Assessment System for Built Environment Efficiency.

2. Following the Great East Japan Earthquake, Japan is undertaking a full review of its Strategic Energy Plan with results expected in 2012.

Table B.4 ● Buildings sector policies and measures as modelled by scenario in selected regions (continued)

	Current Policies Scenario	New Policies Scenario	450 Scenario
Non-OECD			
Russia	– Implementation of 2009 energy efficiency legislation.	– Gradual above-inflation increase in residential electricity and gas prices. – New building codes, meter installations and refurbishment programmes, leading to efficiency gains in space heating (relative to Current Policies Scenario). – Efficiency standards for appliances.	– Faster liberalisation of gas and electricity prices. – Extension and reinforcement of all measures included in the 2010 energy efficiency state programme; mandatory building codes by 2030 and phase out inefficient lighting equipment and appliances by 2030.
China	– Civil Construction Energy Conservation Design Standard. – Minimum Energy Performance Standards for selected devices.	– Civil Construction Energy Conservation Design Standard: heating energy consumption per unit area of existing buildings to be reduced by 65% in cold and very cold regions; 50% in hot-in-summer and cold-in-winter regions compared to 1980/1981 level; new buildings to have 65% improvement in all regions. – Energy Price Policy (reform heating price to be based on actual consumption, rather than on living area supplied). – Mandatory energy efficiency labels for refrigerators and air conditioners.	– More stringent implementation of Civil Construction Energy Conservation Design Standard. – Mandatory energy efficiency labels for all appliances and also for building shell.
India	– Measures under national solar mission.	– Mandatory minimum efficiency requirements and labelling requirements for all equipment and appliances by 2035. – Phase out of incandescent light bulbs by 2025.	– Mandatory energy conservation standards and labelling requirements for all equipment and appliances by 2030. – Increased penetration of energy-efficient lighting.

B

UNITS, DEFINITIONS, REGIONAL AND COUNTRY GROUPINGS, ABBREVIATIONS AND ACRONYMS

This annex provides general information on terminology used throughout *WEO-2011* including: units and general conversion factors; definitions on fuels, processes and sectors; regional and country groupings; and, abbreviations and acronyms.

Units

Area	Ha	hectare
	GHa	giga-hectare (1 hectare x 10^9)
	km²	square kilometre
Coal	Mtce	million tonnes of coal equivalent
Emissions	ppm	parts per million (by volume)
	Gt CO_2-eq	gigatonne of carbon-dioxide equivalent (using 100-year global warming potentials for different greenhouse gases)
	kg CO_2-eq	kilogramme of carbon-dioxide equivalent
	g CO_2/km	grammes of carbon dioxide per kilometre
	g CO_2/kWh	grammes of carbon dioxide per kilowatt-hour
Energy	Mtce	million tonnes of coal equivalent (equals 0.7 Mtoe)
	boe	barrel of oil equivalent
	Mboe	million barrels of oil equivalent
	toe	tonne of oil equivalent
	Mtoe	million tonnes of oil equivalent
	MBtu	million British thermal units
	kcal	kilocalorie (1 calorie x 10^3)
	Gcal	gigacalorie (1 calorie x 10^9)
	MJ	megajoule (1 joule x 10^6)
	GJ	gigajoule (1 joule x 10^9)
	TJ	terajoule (1 joule x 10^{12})
	PJ	petajoule (1 joule x 10^{15})
	kWh	kilowatt-hour
	MWh	megawatt-hour

	GWh	gigawatt-hour
	TWh	terawatt-hour
Gas	mcm	million cubic metres
	bcm	billion cubic metres
	tcm	trillion cubic metres
	scf	standard cubic foot
Mass	kg	kilogramme (1 000 kg = 1 tonne)
	kt	kilotonne (1 tonne x 10^3)
	Mt	million tonnes (1 tonne x 10^6)
	Gt	gigatonne (1 tonne x 10^9)
Monetary	$ million	1 US dollar x 10^6
	$ billion	1 US dollar x 10^9
	$ trillion	1 US dollar x 10^{12}
Oil	b/d	barrels per day
	kb/d	thousand barrels per day
	mb/d	million barrels per day
	mpg	miles per gallon
Power	W	watt (1 joule per second)
	kW	kilowatt (1 Watt x 10^3)
	MW	megawatt (1 Watt x 10^6)
	GW	gigawatt (1 Watt x 10^9)
	GW_{th}	gigawatt thermal (1 Watt x 10^9)
	TW	terawatt (1 Watt x 10^{12})

General conversion factors for energy

Convert to:	TJ	Gcal	Mtoe	MBtu	GWh
From:	*multiply by:*				
TJ	1	238.8	2.388×10^{-5}	947.8	0.2778
Gcal	4.1868×10^{-3}	1	10^{-7}	3.968	1.163×10^{-3}
Mtoe	4.1868×10^{4}	10^7	1	3.968×10^7	11 630
MBtu	1.0551×10^{-3}	0.252	2.52×10^{-8}	1	2.931×10^{-4}
GWh	3.6	860	8.6×10^{-5}	3 412	1

Definitions

Advanced biofuels

Advanced biofuels comprise different emerging and novel conversion technologies that are currently in the research and development, pilot or demonstration phase. This definition differs from the one used for "Advanced Biofuels" in the US legislation, which is based on a minimum 50% lifecycle greenhouse-gas reduction and which, therefore, includes sugar cane ethanol.

Advanced biomass cookstoves

Advanced biomass cookstoves are biomass gasifier-operated cooking stoves that run on solid biomass, such as wood chips and briquettes. These cooking devices have significantly lower emissions and higher efficiencies than the traditional biomass cookstoves (three-stone fires) currently used largely in developing countries.

Agriculture

Agriculture includes all energy used on farms, in forestry and for fishing.

Biodiesel

Biodiesel is a diesel-equivalent, processed fuel made from the transesterification (a chemical process which removes the glycerine from the oil) of both vegetable oils and animal fats.

Biofuels

Biofuels are fuels derived from biomass or waste feedstocks and include ethanol and biodiesel. They can be classified as conventional and advanced biofuels according to the technologies used to produce them and their respective maturity.

Biogas

Biogas is a mixture of methane and CO_2 produced by bacterial degradation of organic matter and used as a fuel.

Biomass and waste

Biomass and waste includes solid biomass, gas and liquids derived from biomass, industrial waste and the renewable part of municipal waste. Includes both traditional and modern biomass.

Biomass-to-liquids

Biomass-to-liquids (BTL) refers to a process featuring biomass gasification into syngas (a mixture of hydrogen and carbon monoxide) followed by synthesis of liquid products (such as diesel, naphtha or gasoline) from the syngas using Fischer-Tropsch catalytic synthesis or a methanol-to-gasoline reaction path. The process is similar to those used in coal-to-liquids or gas-to-liquids.

C

Brown coal

Brown coal includes lignite and sub-bituminous coal where lignite is defined as non-agglomerating coal with a gross calorific value less than 4 165 kilocalories per kilogramme (kcal/kg) and sub-bituminous coal is defined as non-agglomerating coal with a gross calorific value between 4 165 kcal/kg and 5 700 kcal/kg.

Buildings

Buildings includes energy used in residential, commercial and institutional buildings. Building energy use includes space heating and cooling, water heating, lighting, appliances and cooking equipment.

Bunkers

Bunkers includes both international marine bunkers and international aviation bunkers.

Capacity credit

Capacity credit refers to the proportion of capacity that can be reliably expected to generate electricity during times of peak demand in the grid to which it is connected.

Clean coal technologies

Clean coal technologies (CCTs) are designed to enhance the efficiency and the environmental acceptability of coal extraction, preparation and use.

Coal

Coal includes both primary coal (including hard coal and brown coal) and derived fuels (including patent fuel, brown-coal briquettes, coke-oven coke, gas coke, gas-works gas, coke-oven gas, blast-furnace gas and oxygen steel furnace gas). Peat is also included.

Coalbed methane

Coalbed methane (CBM), found in coal seams, is a source of unconventional natural gas.

Coal-to-liquids

Coal-to-liquids (CTL) refers to the transformation of coal into liquid hydrocarbons. It can be achieved through either coal gasification into syngas (a mixture of hydrogen and carbon monoxide), combined with Fischer-Tropsch or methanol-to-gasoline synthesis to produce liquid fuels, or through the less developed direct-coal liquefaction technologies in which coal is directly reacted with hydrogen.

Coking coal

Coking coal is a type of hard coal that can be used in the production of coke, which is capable of supporting a blast furnace charge.

Condensates

Condensates are liquid hydrocarbon mixtures recovered from associated or non-associated gas reservoirs. They are composed of C5 and higher carbon number hydrocarbons and normally have an API gravity between 50° and 85°.

Conventional biofuels

Conventional biofuels include well-established technologies that are producing biofuels on a commercial scale today. These biofuels are commonly referred to as first-generation and include sugar cane ethanol, starch-based ethanol, biodiesel, Fatty Acid Methyl Esther (FAME) and Straight Vegetable Oil (SVO). Typical feedstocks used in these mature processes include sugar cane and sugar beet, starch bearing grains, like corn and wheat, and oil crops, like canola and palm, and in some cases animal fats.

Electricity generation

Electricity generation is defined as the total amount of electricity generated by power only or combined heat and power plants including generation required for own use. This is also referred to as *gross* generation.

Ethanol

Although ethanol can be produced from a variety of fuels, in this book, ethanol refers to bio-ethanol only. Ethanol is produced from fermenting any biomass high in carbohydrates. Today, ethanol is made from starches and sugars, but second generation technologies will allow it to be made from cellulose and hemicellulose, the fibrous material that makes up the bulk of most plant matter.

Gas

Gas includes natural gas, both associated and non-associated with petroleum deposits, but excludes natural gas liquids.

Gas-to-liquids

Gas-to-liquids (GTL) refers to a process featuring reaction of methane with oxygen or steam to produce syngas (a mixture of hydrogen and carbon monoxide) followed by synthesis of liquid products (such as diesel and naphtha) from the syngas using Fischer-Tropsch catalytic synthesis. The process is similar to those used in coal-to-liquids or biomass-to-liquids.

C

Hard coal

Hard coal is coal of gross calorific value greater than 5 700 kilocalories per kilogramme on an ash-free but moist basis. Hard coal can be further disaggregated into anthracite, coking coal and other bituminous coal.

Heat energy

Heat energy is obtained from the combustion of fuels, nuclear reactors, geothermal reservoirs, capture of sunlight, exothermic chemical processes and heat pumps which can extract it from ambient air and liquids. It may be used for heating or cooling, or converted into mechanical energy for transport vehicles or electricity generation. Commercial heat sold is reported under total final consumption with the fuel inputs allocated under power generation.

Heavy petroleum products

Heavy petroleum products include heavy fuel oil.

Hydropower

Hydropower refers to the energy content of the electricity produced in hydropower plants, assuming 100% efficiency. It excludes output from pumped storage and marine (tide and wave) plants.

Industry

Industry includes fuel used within the manufacturing and construction industries. Key industry sectors include iron and steel, chemical and petrochemical, non-metallic minerals, and pulp and paper. Use by industries for the transformation of energy into another form or for the production of fuels is excluded and reported separately under other energy sector. Consumption of fuels for the transport of goods is reported as part of the transport sector.

International aviation bunkers

International aviation bunkers includes the deliveries of aviation fuels to aircraft for international aviation. Fuels used by airlines for their road vehicles are excluded. The domestic/international split is determined on the basis of departure and landing locations and not by the nationality of the airline. For many countries this incorrectly excludes fuels used by domestically owned carriers for their international departures.

International marine bunkers

International marine bunkers covers those quantities delivered to ships of all flags that are engaged in international navigation. The international navigation may take place at sea, on inland lakes and waterways, and in coastal waters. Consumption by ships engaged in domestic navigation is excluded. The domestic/international split is determined on the basis of port of departure and port of arrival, and not by the flag or nationality of the ship. Consumption by fishing vessels and by military forces is also excluded and included in residential, services and agriculture.

Light petroleum products

Light petroleum products include liquefied petroleum gas (LPG), naphtha and gasoline.

Lower heating value

Lower heating value is the heat liberated by the complete combustion of a unit of fuel when the water produced is assumed to remain as a vapour and the heat is not recovered.

Middle distillates

Middle distillates include jet fuel, diesel and heating oil.

Modern biomass

Modern biomass includes all biomass with the exception of traditional biomass.

Modern renewables

Modern renewables includes all types of renewables with the exception of traditional biomass.

Natural decline rate

Natural decline rate is the base production decline rate of an oil or gas field without intervention to enhance production.

Natural gas liquids

Natural gas liquids (NGLs) are the liquid or liquefied hydrocarbons produced in the manufacture, purification and stabilisation of natural gas. These are those portions of natural gas which are recovered as liquids in separators, field facilities, or gas processing plants. NGLs include but are not limited to ethane, propane, butane, pentane, natural gasoline and condensates.

Non-energy use

Non-energy use refers to fuels used for chemical feedstocks and non-energy products. Examples of non-energy products include lubricants, paraffin waxes, coal tars and oils as timber preservatives.

C

Nuclear

Nuclear refers to the primary heat equivalent of the electricity produced by a nuclear plant with an average thermal efficiency of 33%.

Observed decline rate

Observed decline rate is the production decline rate of an oil or gas field after all measures have been taken to maximise production. It is the aggregation of all the production increases and declines of new and mature oil or gas fields in a particular region.

Oil

Oil includes crude oil, condensates, natural gas liquids, refinery feedstocks and additives, other hydrocarbons (including emulsified oils, synthetic crude oil, mineral oils extracted from bituminous minerals such as oil shale, bituminous sand and oils from coal liquefaction) and petroleum products (refinery gas, ethane, LPG, aviation gasoline, motor gasoline, jet fuels, kerosene, gas/diesel oil, heavy fuel oil, naphtha, white spirit, lubricants, bitumen, paraffin waxes and petroleum coke).

Other energy sector

Other energy sector covers the use of energy by transformation industries and the energy losses in converting primary energy into a form that can be used in the final consuming sectors. It includes losses by gas works, petroleum refineries, coal and gas transformation and liquefaction. It also includes energy used in coal mines, in oil and gas extraction and in electricity and heat production. Transfers and statistical differences are also included in this category

Power generation

Power generation refers to fuel use in electricity plants, heat plants and combined heat and power (CHP) plants. Both main activity producer plants and small plants that produce fuel for their own use (autoproducers) are included.

Renewables

Renewable includes biomass and waste, geothermal, hydropower, solar PV, concentrating solar power (CSP), wind and marine (tide and wave) energy for electricity and heat generation.

Total final consumption

Total final consumption (TFC) is the sum of consumption by the different end-use sectors. TFC is broken down into energy demand in the following sectors: industry (including manufacturing and mining), transport, buildings (including residential and services) and other (including agriculture and non-energy use). It excludes international marine and aviation bunkers, except at world level where it is included in the transport sector.

Total primary energy demand

Total primary energy demand (TPED) represents domestic demand only and is broken down into power generation, other energy sector and total final consumption.

Traditional biomass

Traditional biomass refers to the use of fuelwood, charcoal, animal dung and agricultural residues in stoves with very low efficiencies.

Transport

Transport refers to fuels and electricity used in the transport of goods or persons within the national territory irrespective of the economic sector within which the activity occurs. This includes fuel and electricity delivered to vehicles using public roads or for use in rail vehicles; fuel delivered to vessels for domestic navigation; fuel delivered to aircraft for domestic aviation; and energy consumed in the delivery of fuels through pipelines. Fuel delivered to international marine and aviation bunkers is presented only at the world level and is excluded from the transport sector at the domestic level.

Regional and country groupings

Africa

Algeria, Angola, Benin, Botswana, Cameroon, Congo, Democratic Republic of Congo, Côte d'Ivoire, Egypt, Eritrea, Ethiopia, Gabon, Ghana, Kenya, Libya, Morocco, Mozambique, Namibia, Nigeria, Senegal, South Africa, Sudan, United Republic of Tanzania, Togo, Tunisia, Zambia, Zimbabwe and other African countries (Burkina Faso, Burundi, Cape Verde, Central African Republic, Chad, Comoros, Djibouti, Equatorial Guinea, Gambia, Guinea, Guinea-Bissau, Lesotho, Liberia, Madagascar, Malawi, Mali, Mauritania, Mauritius, Niger, Reunion, Rwanda, Sao Tome and Principe, Seychelles, Sierra Leone, Somalia, Swaziland and Uganda).

Annex I Parties to the United Nations Framework Convention on Climate Change

Australia, Austria, Belarus, Belgium, Bulgaria, Canada, Croatia, Czech Republic, Denmark, Estonia, Finland, France, Germany, Greece, Hungary, Iceland, Ireland, Italy, Japan, Latvia, Liechtenstein, Lithuania, Luxembourg, Monaco, Netherlands, New Zealand, Norway, Poland, Portugal, Romania, Russian Federation, Slovak Republic, Slovenia, Spain, Sweden, Switzerland, Turkey, Ukraine, United Kingdom and United States.

C

ASEAN

Brunei Darussalam, Cambodia, Indonesia, Laos, Malaysia, Myanmar, Philippines, Singapore, Thailand and Vietnam.

Caspian

Armenia, Azerbaijan, Georgia, Kazakhstan, Kyrgyz Republic, Tajikistan, Turkmenistan and Uzbekistan.

China

Refers to the People's Republic of China, including Hong Kong.

Developing countries

Non-OECD Asia, Middle East, Africa and Latin America regional groupings.

Eastern Europe/Eurasia

Albania, Armenia, Azerbaijan, Belarus, Bosnia and Herzegovina, Bulgaria, Croatia, Georgia, Kazakhstan, Kyrgyz Republic, Latvia, Lithuania, Former Yugoslav Republic of Macedonia, Republic of Moldova, Romania, Russian Federation, Serbia[1], Tajikistan, Turkmenistan, Ukraine and Uzbekistan. For statistical reasons, this region also includes Cyprus, Gibraltar and Malta.

European Union

Austria, Belgium, Bulgaria, Cyprus, Czech Republic, Denmark, Estonia, Finland, France, Germany, Greece, Hungary, Ireland, Italy, Latvia, Lithuania, Luxembourg, Malta, Netherlands, Poland, Portugal, Romania, Slovak Republic, Slovenia, Spain, Sweden and United Kingdom.

G-8

Canada, France, Germany, Italy, Japan, Russian Federation, United Kingdom and United States.

G-20

G-8 countries and Argentina, Australia, Brazil, China, India, Indonesia, Mexico, Saudi Arabia, South Africa, Korea, Turkey and the European Union.

---

1. Serbia includes Montenegro until 2004 and Kosovo until 1999.

Latin America

Argentina, Bolivia, Brazil, Chile, Colombia, Costa Rica, Cuba, Dominican Republic, Ecuador, El Salvador, Guatemala, Haiti, Honduras, Jamaica, Netherlands Antilles, Nicaragua, Panama, Paraguay, Peru, Trinidad and Tobago, Uruguay, Venezuela and other Latin American countries (Antigua and Barbuda, Aruba, Bahamas, Barbados, Belize, Bermuda, British Virgin Islands, Cayman Islands,. Dominica, Falkland Islands, French Guyana, Grenada, Guadeloupe, Guyana, Martinique, Montserrat, St. Kitts and Nevis, Saint Lucia, Saint Pierre et Miquelon, St. Vincent and the Grenadines, Suriname and Turks and Caicos Islands).

Middle East

Bahrain, Islamic Republic of Iran, Iraq, Jordan, Kuwait, Lebanon, Oman, Qatar, Saudi Arabia, Syrian Arab Republic, United Arab Emirates and Yemen. It includes the neutral zone between Saudi Arabia and Iraq.

Non-OECD Asia

Bangladesh, Brunei Darussalam, Cambodia, China, Chinese Taipei, India, Indonesia, Democratic People's Republic of Korea, Malaysia, Mongolia, Myanmar, Nepal, Pakistan, Philippines, Singapore, Sri Lanka, Thailand, Vietnam and other non-OECD Asian countries (Afghanistan, Bhutan, Cook Islands, East Timor, Fiji, French Polynesia, Kiribati, Laos, Macau, Maldives, New Caledonia, Papua New Guinea, Samoa, Solomon Islands, Tonga and Vanuatu).

North Africa

Algeria, Egypt, Libya, Morocco and Tunisia.

OECD[2]

Includes OECD Europe, OECD Americas and OECD Asia Oceania regional groupings.

OECD Americas

Canada, Chile, Mexico and United States.

OECD Asia Oceania

Includes OECD Asia, comprising Japan and Korea, and OECD Oceania, comprising Australia and New Zealand.

---

2. Chile, Estonia, Israel and Slovenia joined the OECD in 2010, and in the *WEO-2011*, unlike previous editions, these countries are included in the OECD.

OECD Europe

Austria, Belgium, Czech Republic, Denmark, Estonia, Finland, France, Germany, Greece, Hungary, Iceland, Ireland, Italy, Luxembourg, Netherlands, Norway, Poland, Portugal, Slovak Republic, Slovenia, Spain, Sweden, Switzerland, Turkey and United Kingdom. For statistical reasons, this region also includes Israel.

OPEC

Algeria, Angola, Ecuador, Islamic Republic of Iran, Iraq, Kuwait, Libya, Nigeria, Qatar, Saudi Arabia, United Arab Emirates and Venezuela.

Other Asia

Non-OECD Asia regional grouping excluding China and India.

Sub-Saharan Africa

Africa regional grouping excluding the North African regional grouping and South Africa.

Abbreviations and Acronyms

APEC	Asia-Pacific Economic Cooperation
API	American Petroleum Institute
ASEAN	Association of Southeast Asian Nations
BTL	biomass-to-liquids
CAAGR	compound average annual growth rate
CAFE	corporate average fuel economy (standards in the United States)
CBM	coalbed methane
CER	Certified Emission Reduction
CCGT	combined-cycle gas turbine
CCS	carbon capture and storage
CDM	Clean Development Mechanism (under the Kyoto Protocol)
CFL	compact fluorescent lamp
CH_4	methane
CHP	combined heat and power; the term co-generation is sometimes used
CMM	coal mine methane

CNG	compressed natural gas
CO	carbon monoxide
CO$_2$	carbon dioxide
CO$_2$-eq	carbon-dioxide equivalent
COP	Conference of Parties (UNFCCC)
CPC	Caspian Pipeline Consortium
CSP	concentrating solar power
CSS	cyclic steam stimulation
CTL	coal-to-liquids
CV	calorific value
E&P	exploration and production
EDI	Energy Development Index
EOR	enhanced oil recovery
EPA	Environmental Protection Agency (United States)
ESCO	energy service company
EU	European Union
EUA	European Union allowances
EU ETS	European Union Emissions Trading System
EV	electric vehicle
FAO	Food and Agriculture Organization (United Nations)
FDI	foreign direct investment
FFV	flex-fuel vehicle
FOB	free on board
GCV	gross calorific value
GDP	gross domestic product
GHG	greenhouse gases
GTL	gas-to-liquids
HDI	Human Development Index
HDV	heavy-duty vehicles
HIV/AIDS	human immunodeficiency virus/acquired immune deficiency syndrome
IAEA	International Atomic Energy Agency

C

ICE	internal combustion engine
IGCC	integrated gasification combined-cycle
IMF	International Monetary Fund
IOC	international oil company
IPCC	Intergovernmental Panel on Climate Change
IPP	independent power producer
LCV	light-commercial vehicle
LDV	light-duty vehicle
LHV	lower heating value
LNG	liquefied natural gas
LPG	liquefied petroleum gas
LRMC	long-run marginal cost
LULUCF	land use, land-use change and forestry
MER	market exchange rate
MDGs	Millennium Development Goals
MEPS	minimum energy performance standards
N_2O	nitrous oxide
NCV	net calorific value
NEA	Nuclear Energy Agency (an agency within the OECD)
NGL	natural gas liquids
NGV	natural gas vehicle
NOC	national oil company
NO_x	nitrogen oxides
OCGT	open-cycle gas turbine
OECD	Organisation for Economic Co-operation and Development
OPEC	Organization of the Petroleum Exporting Countries
PHEV	plug-in hybrid electric vehicle
PLDV	passenger light-duty vehicle
PM	particulate matter
$PM_{2.5}$	particulate matter with a diameter of 2.5 micrometres or less
PPP	purchasing power parity

PSA	production-sharing agreement
PV	photovoltaic
RD&D	research, development and demonstration
RDD&D	research, development, demonstration and deployment
SAGD	steam-assisted gravity drainage
SCO	synthetic crude oil
SO$_2$	sulphur dioxide
SRMC	short-run marginal cost
T&D	transmission and distribution
TFC	total final consumption
TPED	total primary energy demand
TPES	total primary energy supply
UAE	United Arab Emirates
UCG	underground coal gasification
UN	United Nations
UNDP	United Nations Development Programme
UNEP	United Nations Environment Programme
UNFCCC	United Nations Framework Convention on Climate Change
UNIDO	United Nations Industrial Development Organization
US	United States
USC	ultra-supercritical
USGS	United States Geological Survey
WEO	World Energy Outlook
WEM	World Energy Model
WHO	World Health Organization
WTI	West Texas Intermediate
WTO	World Trade Organization
WTW	well-to-wheel

C

REFERENCES

PART A: Global energy trends

Chapter 1: Context and analytical framework

IEA (International Energy Agency) (2010a), *World Energy Outlook 2010*, OECD/IEA, Paris.

— (2010b), *Energy Technology Perspectives 2010*, OECD/IEA, Paris.

— (2011), *Medium-Term Oil and Gas Markets 2011*, OECD/IEA, Paris.

IMF (International Monetary Fund) (2011), *World Economic Outlook: Slowing Growth, Risks Ahead*, IMF, Washington, DC, September.

OECD (Organisation for Economic Co-operation and Development) (2011), *OECD Economic Outlook No. 89*, OECD, Paris, May.

UNPD (United Nations Population Division) (2011), *World Population Prospects: The 2010 Revision*, United Nations, New York.

Chapter 2: Energy projections to 2035

BGR (*Bundesanstalt für Geowissenschaften und Rohstoffe* – German Federal Institute for Geosciences and Natural Resources) (2010), *Energierohstoffe 2010, Reserven, Ressourcen, Verfügbarkeit, Tabellen [Energy Resources 2010, Reserves, Resources, Availability, Tables]*, BGR, Hannover, Germany.

Cedigaz (2010), *Natural Gas in the World*, Institut Français du Pétrole, Rueil-Malmaison, France.

IEA (International Energy Agency) (2007), *World Energy Outlook 2007*, OECD/IEA, Paris.

IIASA (International Institute for Applied Systems Analysis) (2011), *Emissions of Air Pollutants for the World Energy Outlook 2011 Energy Scenarios,* report prepared for the IEA using the GAINS model, IIASA, Laxenberg, Austria, *www.worldenergyoutlook.org*.

NEA (Nuclear Energy Agency, OECD) and IAEA (International Atomic Energy Agency) (2010), *Uranium 2009: Resources, Production and Demand*, OECD, Paris.

O&GJ (*Oil and Gas Journal*) (2010), "Worldwide Look at Reserves and Production", *Oil and Gas Journal*, Pennwell Corporation, Oklahoma City, United States, December.

Chapter 3: Oil market outlook

API (American Petroleum Institute) (2011), *The Economic Impacts of the Oil and Natural Gas Industry on the U.S. Economy in 2009*, Discussion Paper No. 119, API, Washington, DC.

APICORP (Arab Petroleum Investment Corporation) (2011), *Economic Commentary*, APICORP, Vol. 6, No. 3, Dammam, Saudi Arabia.

BGR (*Bundesanstalt für Geowissenschaften und Rohstoffe* – German Federal Institute for Geosciences and Natural Resources) (2010), *Energierohstoffe 2010, Reserven, Ressourcen, Verfügbarkeit, Tabellen [Energy Resources 2010, Reserves, Resources, Availability, Tables]*, BGR, Hannover, Germany.

BH (Baker Hughes) (2011), *Rig Count Database, www.gis.bakerhughesdirect.com/Reports/ StandardReport.aspx*, accessed June 2011.

BP (2011), *BP Statistical Review of World Energy 2011*, BP, London.

CGES (Centre for Global Energy Studies) (2011), *Commentary: Saudi Arabia's Target Oil Price in 2011, www.cges.co.uk*, accessed March 2011.

Credit Suisse (2011), *Commodity Prices: How Much of a Threat?*, Equity Research, 1 March, Credit Suisse, London.

Deutsche Bank (2011), *European Oils Weekly*, 20 June, Deutsche Bank, London.

GGFRP (Global Gas Flaring Reduction Partnership) (2010), "Estimated Flared Volumes from Satellite Data, 2006-2010", US National Oceanographic and Atmospheric Administration, Washington, DC, *http://www.go.worldbank.org/D03ET1BVD0*, accessed June 2011.

Greene, D. (2010), *Why the Market for New Passenger Cars Generally Undervalues Fuel Economy*, Discussion Paper No. 2010-6, International Transport Forum, OECD, Paris.

ILO (International Labour Organisation) (2010), *Saudi Arabia Labour Statistics,* G-20 Meeting of Labour and Employment Ministers, 20-21 April 2010, Washington, DC.

IMF (International Monetary Fund) (2011), *Kuwait – 2011 Article IV Consultation Concluding Statement, www.imf.org/external/np/ms/2011/050911.htm*, accessed July 2011.

IMOO (Iraq Ministry of Oil) (2011), *Iraq Crude Oil Export Statistics, www.oil.gov.iq/ EXPORT%20CAPACITIES.php*, accessed July 2011.

IEA (International Energy Agency) (2008), *World Energy Outlook 2008*, OECD/IEA, Paris.

— (2010), *World Energy Outlook 2010*, OECD/IEA, Paris.

— (2011a), "Are We Entering a Golden Age of Gas?", *World Energy Outlook 2011 Special Report*, OECD/IEA, Paris.

— (2011b), *Technology Roadmap – Electric and Plug-in Hybrid Electric Vehicles*, OECD/IEA, Paris.

— (2011c), *Medium Term Oil & Gas Markets Report*, OECD/IEA, Paris.

— (2011d), *Oil Market Report*, OECD/IEA, Paris, August.

NDSG (North Dakota State Government) (2011), *2011 Monthly Statistical Update*, Industrial Commission of North Dakota Oil & Gas Division, *www.dmr.nd.gov/oilgas/stats/ 2011monthlystats.pdf*, accessed September 2011.

O&GJ (*Oil and Gas Journal*) (2010), *Worldwide Look at Reserves and Production*, Pennwell Corporation, Oklahoma City, United States, December.

OICA (International Organization of Motor Vehicle Manufacturers), (2011), *World Motor Vehicle Production*, OICA Correspondents' Survey, OICA, Paris.

Otkritie (2011), *Oil and Gas Yearbook 2011*, Otkritie Capital, Moscow, Russia.

OPEC (Organization of Petroleum Exporting Countries) (2011), *Annual Statistical Bulletin 2010/2011 Edition*, OPEC, Vienna.

PFC (PFC Energy) (2011), *OPEC Members' Breakeven Price Continues to Rise*, Press Release, 27 April, PFC Energy, Washington, DC.

Rosstat (Russian Federation Federal State Statistics Service) (2010), *Russian Employment Statistics*, www.gks.ru, accessed July 2011.

RRC (Railroad Commission of Texas) (2011), Eagle Ford Information on Production, RRC, *www.rrc.state.tx.us/eagleford/index.php*, accessed July 2011.

SA (Saudi Aramco) (2011), *Saudi Aramco Annual Review*, Dhahran, Saudi Arabia.

US DOE/EIA (US Department of Energy/Energy Information Agency) (2010), *Summary: US Crude Oil Natural Gas and Natural Gas Liquids Proved Reserves 2009*, US DOE, Washington, DC.

— (2011a), *Review of Emerging Resources: US Shale Gas and Shale Oil Plays*, US DOE, Washington, DC.

— (2011b), *Oil Well Drilling Count*, *www.eia.gov/dnav/ng/ng_enr_wellend_s1_a.htm*, accessed June 2011.

— (2011c), *Producing Oil Well Count*, *www.eia.gov/pub/oil_gas/petrosystem/us_table.html*, accessed June 2011.

USGS (United States Geological Survey) (2000), *World Petroleum Assessment*, USGS, Boulder, Colorado.

— (2008a), "Circum-Arctic Resource Appraisal: Estimates of Undiscovered Oil and Gas North of the Arctic Circle", *Fact Sheet 2008-3049*, USGS, Boulder, Colorado.

— (2008b), "Assessment of Undiscovered Oil Resources in the Devonian-Mississippian Bakken Formation, Williston Basin Province, Montana and North Dakota", *Fact Sheet 2008–3021*, USGS, Boulder, Colorado.

— (2009a), "An Estimate of Recoverable Heavy Oil Resources of the Orinoco Oil Belt, Venezuela", *Fact Sheet 2009-3028 - October*, USGS, Boulder, Colorado.

— (2009b), "Assessment of In-place Oil Shale Resources of the Green River Formation, Piceance Basin, Western Colorado", *Fact Sheet 2009-3012 - March*, USGS, Boulder, Colorado.

— (2010), "Assessment of In-place Oil Shale Resources of the Green River Formation, Uinta Basin, Utah and Colorado", *Fact Sheet 2010-3010 - May*, USGS, Boulder, Colorado.

Chapter 4: Natural gas market outlook

BGR (*Bundesanstalt für Geowissenschaften und Rohstoffe* – German Federal Institute for Geosciences and Natural Resources) (2010), *Energierohstoffe 2010, Reserven, Ressourcen, Verfügbarkeit, Tabellen [Energy Resources 2010, Reserves, Resources, Availability, Tables]*, BGR, Hannover, Germany.

Cedigaz (2010), *Natural Gas in the World*, Institut Français du Pétrole, Rueil-Malmaison, France.

IEA (International Energy Agency) (2011), "Are We Entering a Golden Age of Gas?", *World Energy Outlook 2011 Special Report*, OECD/IEA, Paris.

US DOE/EIA (US Department of Energy/Energy Information Administration) (2011), *World Shale Gas Resources: An Initial Assessment of 14 Regions Outside the United States*, US DOE, Washington, DC.

USGS (United States Geological Survey) (2000), *World Petroleum Assessment*, USGS, Boulder, Colorado.

— (2008), "Circum-Arctic Resource Appraisal: Estimates of Undiscovered Oil and Gas North of the Arctic Circle", *Fact Sheet 2008-3049*, USGS, Boulder, Colorado.

Chapter 5: Power and renewables outlook

ABS Energy Research (2010), *Global Transmission & Distribution Report Ed. 9-2010*, London.

Dena (Deutsche Energie Agentur) (2010), *dena Grid Study II – Integration of Renewable Energy Sources in the German Power Supply System from 2015-2020 with an Outlook to 2025*, Dena, Berlin.

ECF (European Climate Foundation) (2010), *Roadmap 2050: A Practical Guide to a Prosperous Low-Carbon Europe. Volume 1 – Technical and Economic Analysis*, ECF, The Hague.

ENTSO-E (European Network of Transmission System Operators for Electricity) (2011), *European Network of Transmission and System Operators for Electricity: Factsheet 2011*, Brussels.

Heide, D., *et al.* (2010), "Seasonal Optimal Mix of Wind and Solar Power in a Future, Highly Renewable Europe", *Renewable Energy*, Vol. 35, pp. 2483-2589, data provided to IEA by Siemens.

IEA (International Energy Agency) (forthcoming), *System Effects of Variable Renewable Energies to Thermal Power Plants*, IEA Working Paper, OECD/IEA, Paris.

— (2011a), *Harnessing Renewables: A Guide to the Balancing Challenge*, OECD/IEA, Paris.

— (2011b), *Technology Roadmap – Smart Grids*, OECD/IEA, Paris.

IPCC (Intergovernmental Panel on Climate Change) (2011). *Special Report on Renewable Energy Sources and Climate Change Mitigation*, IPCC, Geneva.

NREL (National Renewable Energy Laboratory) (2010), *Western Wind and Solar Integration Study, prepared by GE Energy*, NREL, Golden, United States.

— (2011a), *Eastern Wind Integration and Transmission Study, prepared by EnerNex Corporation*, NREL, Golden, United States.

— (2011b), *National Solar Radiation Base, 1991-2005 Update, http://rredc.nrel.gov/solar*, accessed July 2011.

World Wind Atlas (2011), Sander + Partner GmbH: Global wind speed database, *www.sander-partner.ch/en/atlas.html*, accessed October 2011.

Chapter 6: Climate Change and the 450 Scenario

Anderson, K. and A. Bows (2011), "Beyond 'Dangerous' Climate Change: Emission Scenarios for a New World", *Philosophical Transactions of the Royal Society*, Vol. 369, No. 1934, pp. 20-44.

Davis, S., K. Caldeira and H. Matthews (2010), "Future CO_2 Emissions and Climate Change from Existing Energy Infrastructure", *Science*, Vol. 329, No. 5997, Washington, DC, pp. 1330-1333.

Hansen, J., *et al.* (2008), "Target Atmospheric CO_2: Where Should Humanity Aim?", *The Open Atmospheric Science Journal*, Vol. 2, pp. 217-231.

IEA (International Energy Agency) (2010a), *World Energy Outlook 2010,* OECD/IEA, Paris.

— (2010b), *Energy Technology Perspectives 2010,* OECD/IEA, Paris.

— (2011), *IEA Scoreboard 2011: Implementing Energy Efficiency Policy: Progress and challenges in IEA Member Countries,* OECD/IEA, Paris.

IIASA (International Institute for Applied Systems Analysis) (2011), *Emissions of Air Pollutants for the World Energy Outlook 2011 Energy Scenarios,* report prepared for the IEA using the GAINS model, IIASA, Laxenberg, Austria, *www.worldenergyoutlook.org*.

IPCC (Intergovernmental Panel on Climate Change) (2007a), "Climate Change 2007: Synthesis Report", contribution of Working Groups I, II, and III to the *Fourth Assessment Report of the IPCC*, R. Pachauri and A. Reisinger (eds.), IPCC, Geneva.

— (2007b), "Climate Change 2007: Impacts, Adaptation and Vulnerability", contribution of Working Group II to the *Fourth Assessment Report of the IPCC*, M. Parry *et al.* (eds.), Cambridge University Press, Cambridge, United Kingdom.

Lenton, T., *et al.* (2008), "Tipping Elements in the Earth's Climate System", *Proceedings of the National Academy of Sciences*, Vol. 105, No. 6, Washington, DC.

Lewis, S., *et al.* (2011), "The 2010 Amazon Drought", *Science*, Vol. 331, No. 6017, Washington, DC, pp. 554.

Meinshausen, M., *et al.* (2009), "Greenhouse Gas Emissions Targets for Limiting Global Warming to 2°C, *Nature*, Vol. 458, No. 7242, pp. 1158-1162.

Parry, M., *et al.* (2009), *Assessing the Costs of Adaptation to Climate Change: A Review of the UNFCCC and Other Recent Estimates,* International Institute for Environment and Development, London.

Rockström, J., *et al.* (2009), "A Safe Operating Space for Humanity", *Nature*, Vol. 461, pp. 472-475.

Schaefer, K., *et al.* (2011), *Amount and Timing of Permafrost Carbon Release in Response to Climate Warming*, Tellus B, Vol. 63, No. 2, International Meteorological Institute, Stockholm, pp. 165-180.

Smith, J., *et al.* (2009), "Assessing Dangerous Climate Change through an Update of the Intergovernmental Panel on Climate Change 'Reasons for Concern'", *Proceedings of the National Academy of Sciences*, Vol. 106, No.11, Washington, DC.

Stern, N. (2006), *Stern Review: Economics of Climate Change*, Cambridge University Press, Cambridge, United Kingdom.

UNEP (United Nations Environment Programme) (2010), *The Emissions Gap Report*, UNEP, Paris.

PART B: Outlook for Russian energy

Chapter 7: Russian domestic energy prospects

Bashmakov, I. (2011), *Energy Efficiency Policies and Developments in Russia*, Centre for Energy Efficiency, Moscow.

CENEF (Centre for Energy Efficiency) (2008), *Resource of Energy Efficiency in Russia: Scale, Costs and Benefits*, CENEF, Moscow.

Government of Russia (2008), *Transport Strategy of Russia for the Period to 2030*, Government of Russia, Moscow (in Russian).

— (2009), *Energy Strategy of Russia for the Period to 2030*, Government of Russia, Moscow.

— (2010), *State Programme on Energy Savings and Increased Energy Efficiency in the Period to 2020*, Government of Russia, Moscow (in Russian).

— (2011), *Climate Doctrine Action Plan*, Government of Russia, Moscow (in Russian).

IAEA (International Atomic Energy Agency) (2011), *PRIS Database*, ww.iaea.org/ programmes/a2/, accessed August 2011.

IMF (International Monetary Fund) (2011a), *Concluding Statement for the 2011 Article IV Consultation Mission*, Moscow, June, www.imf.org/external/np/ms/2011/ 061411b.htm.

— (2011b), *World Economic Outlook*, IMF, Washington, DC, April.

Institute of Energy Strategy (2010), *Russian Energy; A View to the Future: Background Materials to the Energy Strategy of Russia to 2030*, Institute of Energy Strategy, Moscow (in Russian).

IEA (International Energy Agency) (2006), *Optimising Russian Natural Gas*, OECD/IEA, Paris.

— (2009), *Implementing Energy Efficiency Policies – Are IEA Member Countries on Track?*, OECD/IEA, Paris.

— (2010), *Energy Technology Perspectives 2010*, OECD/IEA, Paris.

— (2011), *Development of Energy Efficiency Indicators in Russia*, IEA Working Paper, OECD/IEA, Paris.

RosHydroMet (2011), *Report on Climatic Conditions in Russia in 2010*, Federal Service for Hydrometeorology and Environmental Monitoring (RosHydroMet), Moscow (in Russian).

Russian Academy of Sciences (2009), *Development of Economic Mechanisms for Stimulating Investment in Energy Efficient Technologies*, Institute of Energy Research of the Russian Academy of Sciences, Moscow (in Russian).

Solanko L. (2011), *How to Succeed with a Thousand TWh Reform?*, FIIA Working Paper, The Finnish Institute of International Affairs, Helsinki.

Transparency International (2010), *Corruption Perceptions Index*, Berlin, *www.transparency.org/policy_research/surveys_indices/cpi/2010,* accessed July 2011.

UNIDO (United Nations Industrial Development Organisation) (2010), *Global Industrial Energy Efficiency Benchmarking, An Energy Policy Tool*, UNIDO Working Paper, Vienna.

World Bank (2008), *Energy Efficiency in Russia: Untapped Reserves*, World Bank, Washington, DC.

— (2010), *Lights Out: the Outlook for Energy in Eastern Europe and the Former Soviet Union*, World Bank, Washington, DC.

— (2011), *Ease of Doing Business Index*, World Bank, *www.doingbusiness.org/rankings*.

World Steel Association (2000), *Steel Statistical Yearbook 2000*, World Steel Association, Brussels.

— (2010), *Steel Statistical Yearbook 2010*, World Steel Association, Brussels.

Chapter 8: Russian resources and supply potential

BGR (*Bundesanstalt für Geowissenschaften und Rohstoffe* – German Federal Institute for Geosciences and Natural Resources) (2010), *Energierohstoffe 2010, Reserven, Ressourcen, Verfügbarkeit, Tabellen [Energy Resources 2010, Reserves, Resources, Availability, Tables]*, BGR, Hannover, Germany.

BP (2011), *BP Statistical Review of World Energy 2011*, BP, London.

Cedigaz (2011), *Natural Gas in the World, 2010 Edition*, Cedigaz, Rueil-Malmaison, France.

Efimov, A., *et al.* (2009), *Accelerating the Development of the Hydrocarbon Resource Base in Eastern Siberia and the Republic of Sakha*, First Break 27, European Association of Geoscientists and Engineers, Houten, The Netherlands, pp. 69.

Everett, M. (2010), *Characterizing the Pre-Cambrian Petroleum Systems of Eastern Siberia: Evidence from Geochemistry and Basin Modelling*, SPE 136334, SPE (Society of Petroleum Engineers), Richardson, United States.

Gerasimov, Y. and T. Karjalainen (2011), "Energy Wood Resources in Northwest Russia", *Biomass and Bioenergy*, Vol. 35, Elsevier, Amsterdam, pp. 1655-1662.

GGFRP (Global Gas Flaring Reduction Partnership) (2010), "Estimated Flared Volumes from Satellite Data, 2006-2010", US National Oceanographic and Atmospheric Administration, Washington, DC, *http://go.worldbank.org/D03ET1BVD0*, accessed June 2011.

Government of Russia (2008), *Concept of Long-term Social and Economic Development of the Russian Federation for the Period to 2020*, Government of Russia, Moscow (in Russian).

— (2009), *Energy Strategy of Russia for the Period to 2030*, Government of Russia, Moscow.

— (2010), *State Programme on Energy Saving and Increased Energy Efficiency in the Period to 2020*, Government of Russia, Moscow (in Russian).

— (2011), *Climate Doctrine Action Plan*, Government of Russia, Moscow (in Russian).

Henderson, J. (2010), *Non-Gazprom Gas Producers in Russia*, Oxford Institute for Energy Studies, Oxford, United Kingdom.

IEA (International Energy Agency) (2003), *Renewables in Russia*, OECD/IEA, Paris.

— (2008), *World Energy Outlook 2008*, OECD/IEA, Paris.

— (2009), *World Energy Outlook 2009*, OECD/IEA, Paris.

— (2010a), *World Energy Outlook 2010*, OECD/IEA, Paris.

— (2010b), *Natural Gas Information*, OECD/IEA, Paris.

Kontorovich, A., *et al.* (2010), "Geology and Hydrocarbon Resources of the Continental Shelf in Russian Arctic Seas and the Prospects of their Development", *Russian Geology and Geophysics*, Vol. 51, Elsevier, Amsterdam, pp. 3-11.

Ministry of Energy of the Russian Federation (2011), Approval of the General Scheme for Development of the Oil Industry to 2020, *www.minenergo.gov.ru/ press/min_news/7473. html*, accessed September 2011 (in Russian).

Ministry of Natural Resources and Environment of the Russian Federation (2010), *5th National Communication to the UNFCCC, www.unfccc.int/national_reports/annex_i_ natcom/submitted_natcom/items/4903.php*, accessed June 2011 (in Russian).

— (2011), *On the State and Use of Mineral Resources in the Russian Federation in 2009*, *www.mnr.gov.ru/part/?act=more&id=6555&pid=153*, accessed May 2011 (in Russian).

O&GJ (*Oil and Gas Journal*) (2010), "Worldwide Look, at Reserves and Production" *Oil and Gas Journal*, Pennwell Corporation, Oklahoma City, United States, December.

PFC Energy (2007), *Using Russia's Associated Gas*, report for the Global Gas Flaring Reduction Partnership and the World Bank, PFC, Washington, DC.

Piskarev, A. and M. Shkatov (2009), *Probable Reserves and Prospects for Exploration and Development of Oil and Gas Deposits in the Russian Arctic Seas*, IPTC-13290, SPE (Society of Petroleum Engineers), Richardson, United States.

Popel, O., *et al.,* (2010), *An Atlas of the Solar Energy Resource in Russia*, United Institute of High Temperatures of the Russian Academy of Sciences, Moscow (in Russian).

Ragner, C. (2008), *The Northern Sea Route*, in T. Hallberg (ed.), Barents – ett gränsland i Norden, Arena Norden, Stockholm, pp. 114-127.

Rogner, H. (1997), "An Assessment of World Hydrocarbon Resources", *Annual Reviews of Energy and Environment*, Vol. 22, Palo Alto, United States, pp. 217-262.

Shakhova, N. and I. Semiletov (2010), "Methane Release from the East Siberian Arctic Shelf and the Potential for Abrupt Climate Change", presented at the US DoD Partners in Environmental Technology Symposium, Washington, DC, 30 November, *http://symposium2010.serdp-estcp.org/Technical-Sessions/1A*, accessed July 2011.

South Stream (2011), "South Stream" project presentation, Brussels, 25 May, *www.south-stream.info/index.php?id=28&L=1*, accessed July 2011.

USGS (United States Geological Survey) (2000), *World Petroleum Assessment*, USGS, Boulder, Colorado.

WEC (World Energy Council) (2010), *2010 Survey of Energy Resources*, World Energy Council, London.

Chapter 9: Implications of Russia's energy development

Government of Russia (2008), *Concept of Long-term Social and Economic Development of the Russian Federation for the Period to 2020*, Government of Russia, Moscow (in Russian).

Gurvich, E. (2010), "Oil and Gas Rent in the Russian Economy", *Questions of Economics*, Vol. 11, Russian Academy of Sciences, Moscow (in Russian).

IIASA (International Institute for Applied Systems Analysis) (2011), *Emissions of Air Pollutants for the World Energy Outlook 2011 Energy Scenarios*, report prepared for the IEA using the GAINS model, IIASA, Laxenburg, Austria,*www.worldenergyoutlook.org*.

IEA (International Energy Agency) (2010), *World Energy Outlook 2010*, OECD/IEA, Paris.

— (2011a), *Medium-Term Oil and Gas Market Report,* OECD/IEA, Paris.

— (2011b), *Are We Entering a Golden Age of Gas?*, World Energy Outlook 2011 Special Report, OECD/IEA, Paris.

Kuboniwa M., S. Tabata and N. Ustinova (2005), "How Large is the Oil and Gas Sector of Russia?" *Eurasian Geography and Economics*, Vol. 46, No. 1, pp. 68-76.

OECD (Organisation for Economic Co-operation and Development) (2011), *Economic Survey of Russia*, OECD, Paris, forthcoming.

World Bank (2005), *From Transition to Development, A Country Economic Memorandum for the Russian Federation*, World Bank, Moscow.

PART C: Outlook for coal markets

Chapter 10: Coal demand prospects

GCCSI (Global CCS Institute) (2011), *The Global Status of CCS: 2011,* GCCSI, Canberra, Australia.

IEA (International Energy Agency) (2009), *Technology Roadmap – Carbon Capture and Storage*, OECD/IEA, Paris.

— (2010a), *Power Generation from Coal*, OECD/IEA, Paris.

— (2010b), *Energy Technology Perspectives 2010*, OECD/IEA, Paris.

— (2010c), *World Energy Outlook 2010*, OECD/IEA, Paris.

Chapter 11: Coal supply and investment prospects

ABARES (2011), *Minerals and Energy: Major Development Projects – April 2011 Listing*, Australian Bureau of Agricultural and Resource Economics and Sciences, Canberra.

Barry Rogliano Salles (2011), "Outlook for the Bulk Carrier Market", presented at the IEA Workshop on Outlook for Coal Industry and Markets, Beijing, 14 April, *www.iea.org/work/2011/ WEO_Coal/04_02_FIKKERS_XING.pdf.*

Baruya, P. (2009), *Prospects for Coal and Clean Coal Technologies in Indonesia,* IEA Clean Coal Centre, London.

BGR (*Bundesanstalt für Geowissenschaften und Rohstoffe* – German Federal Institute for Geosciences and Natural Resources) (2010), *Energierohstoffe 2010, Reserven, Ressourcen, Verfügbarkeit, Tabellen [Energy Resources 2010, Reserves, Resources, Availability, Tables],* BGR, Hannover, Germany.

BP (2011), *BP Statistical Review of World Energy 2011*, BP, London.

DnB NOR (2010), *Dry Bulk Outlook: Iron Ore and Coal*, DnB NOR, Norway.

Eberhard, A. (2011), *The Future of South African Coal: Market, Investment and Policy Challenges*, Freeman Spogli Institute for International Studies at Stanford University, Working Paper No. 100, Stanford, United States.

IEA (International Energy Agency) (2011), *Medium-Term Coal Market Report 2011*, OECD/IEA, Paris, forthcoming.

McCloskey (2007-2011), *McCloskey Coal Reports 2007 to 2011,* McCloskey Group, *http://cr.mccloskeycoal.com/.*

Meister, W. (2008), "Cost Trends in Mining", presented at McCloskey European Coal Outlook Conference, Nice, France, 18-19 May.

Mills, J. (2007), *Prospects for Coal and Clean Coal Technologies in India,* IEA Clean Coal Centre, London.

Minchener, A. (2007), *Coal Supply Challenges for China*, IEA Clean Coal Centre, London.

Morse, R. and G. He (2010), *The World's Greatest Coal Arbitrage: China's Coal Import Behaviour and Implication for the Global Coal Market,* Freeman Spogli Institute for International Studies at Stanford University, Working Paper No. 94, Stanford, United States.

NBS (National Bureau of Statistics of China) (2010), *China Energy Statistical Yearbook 2010*, NBS, Beijing.

NSW (New South Wales) Department of Primary Industries (2011), *Summary of NSW Coal Statistics*, Government of New South Wales, *www.dpi.nsw.gov.au/minerals/resources/coal/summary-of-nsw-coal-statistics*, accessed July 2011.

Paulus, M. and J. Trüby (2011), "Coal Lumps vs. Electrons: How do Chinese Bulk Energy Transport Decisions Affect the Global Steam Coal Market?" *Energy Economics*, forthcoming.

Queensland Department of Mines and Energy (2011), *Coal Industry Review 2009-2010 Statistical Tables,* Government of Queensland, *http://mines.industry.qld.gov.au/mining/coal-statistics.htm*, accessed July 2011.

Rademacher, M. and R. Braun (2011), "The Impact of the Financial Crisis on the Global Seaborne Hard Coal Market: Are there Implications for the Future?", *Zeitschrift für Energiewirtschaft,* Vol. 35, No. 2, Wiesbaden, Germany, pp. 89-104.

Rui, H., R. Morse and G. He (2010), *Remaking the World's Largest Coal Market: The Quest to Develop Large Coal-Power Bases in China*, Freeman Spogli Institute for International Studies at Stanford University Working Paper No. 98, Stanford, United States.

Schiffer, H. (2011), "Rolle der Kohle im weltweiten Energiemix" [The Role of Coal in the Global Energy Mix], *Zeitschrift für Energiewirtschaft,* Vol. 35, No. 1, Wiesbaden, Germany, pp. 1-13 (in German).

Trüby, J. and M. Paulus (2011), *Market Structure Scenarios in International Steam Coal Trade*, Institute of Energy Economics, Working Paper Series 11/02, University of Cologne, Cologne, Germany.

Tu, J. (2010), *Industrial Organization of the Chinese Coal Industry*, Freeman Spogli Institute for International Studies at Stanford University, Working Paper No. 103, Stanford, United States.

US DOE/EIA (United States Department of Energy/Energy Information Administration) (2011), *Annual Energy Review 2009*, US DOE, Washington, DC.

VDKI (*Verein der Kohlenimporteure* – German Hard Coal Importer's Association) (2010), *Annual Report 2010 – German Hard Coal Importer's Association*, VDKI, Hamburg, Germany.

Wang, H. and N. Horii (2008), *Chinese Energy Market & Reform on Pricing Mechanisms – Case Study and Analysis of Shanxi's Experiment*, Institute of Developing Economies, Chiba, Japan.

World Coal Institute (2005), *The Coal Resource: A Comprehensive Overview of Coal*, World Coal Institute, London.

D

PART D: SPECIAL TOPICS

Chapter 12: The implications of less nuclear power

IAEA (International Atomic Energy Agency) (2011), *PRIS Database*, *www.iaea.org/ programmes/a2/*, accessed August 2011.

IEA (International Energy Agency) (2010), *Technology Roadmap – Nuclear Energy*, OECD/ IEA, Paris.

NEI (Nuclear Energy Institute) (2010), *Factsheet: Nuclear Industry's Comprehensive Approach Develops Skilled Work Force for the Future*, NEI, Washington, DC, September.

ORISE (Oak Ridge Institute for Science and Education) (2010), *Nuclear Engineering Enrolments and Degrees Survey, 2009 Data*, ORISE, Knoxville, United States.

WNA (World Nuclear Association) (2011), *Nuclear Power in the USA*, WNA, London, June.

Chapter 13: Energy for all

ADB (Asia Development Bank) (2009), *Energy for All: Vietnam: Boosting Biogas*, ADB, Manila.

AFD (Agence Française de Développement) (2011), "The Paris-Nairobi Climate Initiative: Access to Clean Energy for All in Africa and Countries Vulnerable to Climate Change", background document to Clean Energy for All Conference, Paris, 21 April.

AGECC (Secretary-General's Advisory Group on Energy and Climate Change) (2010), *Energy for a Sustainable Future Summary Report and Recommendations*, United Nations, New York.

ARE (Alliance for Rural Electrification) (2011a), *Hybrid Mini-Grid for Rural Electrification: Lessons Learned*, ARE, Brussels.

— (2011b) *Rural Electrification with Renewable Technologies, Quality Standards and Business Models*, ARE, Brussels.

Ashden Awards (2006), *Case Study: Biogas Sector Partnership*, Ashden Awards, London.

— (2008), *Grameen Shakti, Bangladesh: Rapidly Growing Solar Installer Provides Clean Cooking as Well*, Ashden Awards, London.

— (2010), *Case Study Summary: Ministry of Agriculture and Rural Development (MARD), Vietnam and Netherlands Development Organisation (SNV)*, Ashden Awards, London.

Budya, H. and M. Yasir Arofat (2011), "Providing Cleaner Energy Access in Indonesia through the Mega-project of Kerosene Conversion to LPG", *Energy Policy*, Elsevier, Amsterdam, pp. 1-12.

Government of India (2008), *Household Consumer Expenditure in India, 2006-07*, Ministry of Statistics and Programme Implementation, Government of India, New Delhi.

— (2011), *Key Indicators of Household Consumer Expenditure in India, 2009-10*, Ministry of Statistics and Programme Implementation, Government of India, New Delhi.

GPOBA (Global Partnership on Output-Based Aid) (2007), "Output-Based Aid in Senegal – Designing Technology-Neutral Concessions for Rural Electrification", *OBApproaches* Note No. 14, GPOBA, Washington, DC.

— (2009), "Output-Based Aid in Ethiopia: Dealing with the 'Last Mile' Paradox in Rural Electrification", *OBApproaches* Note No. 27, GPOBA, Washington, DC.

IEA (International Energy Agency) (2006), *World Energy Outlook 2006*, OECD/IEA, Paris.

— (2010), *World Energy Outlook 2010*, OECD/IEA, Paris.

— (2011), *Advantage Energy, Emerging Economies, Developing Countries and the Private-Public Sector Interface,* OECD/IEA, Paris.

IFC (International Finance Corporation) (forthcoming), *Energy Access Business Models: Leveraging the Private Sector to Serve the Poor*, IFC, Washington, DC, forthcoming.

IFC and World Bank (2010), "Solar Lighting for the Base of the Pyramid – Overview of an Emerging Market", *Lighting Africa Report*, IFC and World Bank, Washington, DC.

IJHD (International Journal of Hydropower and Dams) (2010), *International Journal of Hydropower and Dams World Atlas and Industry Guide*, Aqua Media International Ltd, Wallington, United Kingdom.

IPCC (International Panel on Climate Change) (2011), "Hydropower", *Special Report on Renewable Energy Sources and Climate Change Mitigation*, IPCC, Geneva.

Lamech, R. and K. Saeed (2003), "What International Investors Look for When Investing in Developing Countries: Results from a Survey of International Investors in the Power Sector", *Energy and Mining Sector Board Discussion Paper*, No. 6, World Bank, Washington, DC.

Liu, H., and X. Hu (2010), "The Development and Practice of Small Hydropower Clean Development Mechanism Project in China", *China Water Power and Electrification*, Vol. 69, No. 9, pp. 8-14.

Marrey, C. and R. Bellanca (2010), *GTZ-HERA Cooking Energy Compendium*, GTZ-HERA, Eschborn, Germany.

Mathers, C. and D. Loncar (2006), "Projections of Global Mortality and Burden of Disease from 2002 to 2030", *PLoS Medicine*, Vol. 3, No. 11, 28 November, Public Library of Science, Cambridge, United Kingdom.

Morris, E., *et al.* (2007), *Using Microfinance to Expand Access to Energy Services: Summary of Findings*, Small Enterprise Education and Promotion Network (SEEP) Network, Washington, DC.

Phuangpornpitak, N. and S. Kumar (2007), "PV Hybrid Systems for Rural Electrification in Thailand", *Renewable and Sustainable Energy Reviews*, Vol. 11, Asian Institute of Technology, Klong Luang, Thailand, pp. 1530-1543.

Sanchez, T. (2010), *The Hidden Energy Crisis: How Policies Are Failing the World's Poor,* Practical Action Publishing, London.

D

SNV Netherlands Development Organisation (2011), *"Domestic Biogas Newsletter"*, Issue 5, September, SNV, The Hague, Netherlands.

Smith, K., S. Mehta and M. Feuz (2004), *"Indoor Air Pollution from Household Use of Solid Fuels"* in M. Ezzati *et al.* (eds.), *Comparative Quantification of Health Risks: Global and Regional Burden of Disease Attributable to Selected Major Risk Factors,* World Health Organization, Geneva.

UNEP (United Nations Environment Programme) (2007), *Financing Mechanisms and Public/Private Risk Sharing Instruments for Financing Small-Scale Renewable Energy Equipment and Projects*, UNEP, Paris.

UNDP (United Nations Development Programme) (2009), *Bringing Small-Scale Finance to the Poor for Modern Energy Services: What is the Role of Government?,* UNDP, New York.

UNDP and World Health Organization (2009), *The Energy Access Situation in Developing Countries – A Review on the Least Developed Countries and Sub-Saharan Africa*, UNDP, New York.

UNDP, AEPC (Alternative Energy Promotion Centre) and Practical Action (2010), *Capacity Development for Scaling Up Decentralised Energy Access Programmes: Lessons from Nepal on its Role, Costs, and Financing,* Practical Action Publishing, Rugby, United Kingdom.

WHO (World Health Organization) (2004), "Indoor Smoke from Solid Fuel Use: Assessing the Environmental Burden of Disease", *Environmental Burden of Disease Series*, No. 4, WHO, Geneva.

— (2008), *The Global Burden of Disease: 2004 Update*, WHO, Geneva.

World Bank (2008), *Designing Sustainable Off-Grid Rural Electrification Projects: Principles and Practices*, World Bank, Washington, DC.

— (2009a), *Africa Electrification Initiative Workshop*, proceedings of conference held 9-12 June, Maputo, Mozambique, World Bank, Washington, DC.

— (2009b) "Africa's Infrastructure, a Time for Transformation", *World Bank Africa Infrastructure Country Diagnostic*, World Bank, Washington, DC.

— (2011), *State and Trends of the Carbon Market 2011*, World Bank, Washington, DC.

Yadoo, A. and H. Cruickshank (2010), "The Value of Co-operatives in Rural Electrification", *Energy Policy,* Vol. 38, No. 6, Elsevier, Amsterdam, pp. 2941-2947.

Chapter 14: Developments in energy subsidies

Arze del Granado, J., D. Coady and R. Gillingham (2010), "The Unequal Benefits of Fuel Subsidies: A Review of Evidence for Developing Countries", *IMF Working Paper* No. 10/202, International Monetary Fund, Washington, DC, pp. 1-23.

IEA (International Energy Agency) (2008), "Deploying Renewables, Principles for Effective Policies", OECD/IEA, Paris.

— (2010), *World Energy Outlook 2010*, OECD/IEA, Paris.

IEA, OECD (Organisation for Economic Co-operation and Development), World Bank and OPEC (Organization of the Petroleum Exporting Countries) (2011), *Joint Report by IEA, OPEC, OECD and World Bank on Fossil-Fuel and Other Energy Subsidies: An Update of the G-20 Pittsburgh and Toronto Commitments*, prepared for the G-20 Meeting of Finance Ministers and Central Bank Governors, Cannes, France.

F.O. Lichts (2011), "Revamping Brazil's Sugar and Ethanol Industry Will Require Patience", World Ethanol and Biofuels Report, Vol. 9, No. 21.

GSI (Global Subsidies Initiative) (2010), *Defining Fossil-Fuel Subsidies for the G-20: Which Approach is Best,* Global Subsidies Initiative of the International Institute for Sustainable Development, Geneva.

OECD (2011), *Inventory of Estimated Budgetary Support and Tax Expenditures Relating to Fossil Fuels in Selected OECD Countries*, OECD, Paris.

REN21 (2011), *Renewables 2011 Global Status Report*, REN21, Paris.

Shenoy, B. (2010), *Lessons Learned from Attempts to Reform India's Kerosene Subsidy*, International Institute for Sustainable Development, Winnipeg, Canada.

Tumiwa, F., *et al.* (2011), "A Citizen's Guide to Energy Subsidies in Indonesia", Global Subsidies Initiative of the International Institute for Sustainable Development, Geneva, and the Institute for Essential Services Reform, Jakarta.

Wagner, A. (2010), "Three Dimensions of Fuel Pricing: Political Steps and Principles of Setting Effective Fuel Pricing Mechanisms", presented at the GSI-UNEP Conference on Increasing the Momentum of Fossil-Fuel Subsidy Reform, Geneva, 14-15 October, *www.globalsubsidies. org/files/assets/ffs_gsiunepconf_sess3_awagner.pdf*.

D

International
Energy Agency

Online
bookshop

**Buy IEA publications
online:**

www.iea.org/books

**PDF versions available
at 20% discount**

Books published before January 2010
- except statistics publications -
are freely available in pdf

International Energy Agency • 9 rue de la Fédération • 75739 Paris Cedex 15, France

iea

Tel: +33 (0)1 40 57 66 90

E-mail:
books@iea.org

FSC
www.fsc.org
MIXTE
Issu de sources
responsables
FSC® C031289

The paper used for this document and the forest from which it comes have received
FSC certification for meeting a set of strict environmental and social standards.
The FSC is an international, membership-based, non-profit
organisation that supports environmentally appropriate, socially beneficial,
and economically viable management of the world's forests.

IEA PUBLICATIONS, 9 rue de la Fédération, 75739 PARIS CEDEX 15
PRINTED IN FRANCE BY SOREGRAPH, November 2011
(61 2011 24 1P1) ISBN: 978 92 64 12413 4

Cover design: IEA. Photo credits: © Image100/GraphicObsession